*Chemical and Botanical Guide
to Lichen Products*

Chemical and Botanical Guide to Lichen Products

Chicita F. Culberson

The University of North Carolina Press · Chapel Hill

The preparation and publication of this book were made possible by a subsidy under U.S. Public Health Service Research Grant GM-08345 to Duke University from the Division of General Medical Sciences, National Institutes of Health.

Preface

The known lichen products include hundreds of compounds many of which are restricted to the lichens. This guide to lichen products has been completed at a moment of renewed research activity in all aspects of the chemistry, taxonomy, and physiology of these symbiotic organisms. Its aim is to summarize the vast and widely scattered literature on the nature and occurrence of lichen substances and to organize the information for the use of (*1*) botanists interested in the application of chemistry to problems of biology and taxonomy, (*2*) chemists looking for sources of substances peculiar to the lichens, and (*3*) medical researchers seeking biologically active products as potential medicinals.

Until now it has been enormously time-consuming — often prohibitively so — to determine the natural products reported for a given lichen species because no complete listing of the constituents of lichens was available. The findings of many important studies made by botanists and based on the microchemical examination of herbarium materials are scattered through a literature that was never thoroughly indexed by the chemical abstracting services. Consequently botanists are rarely apt to search the chemical literature before initiating research on lichen chemistry and chemists have no means of knowing which lichens produce specific compounds. It is hoped that the present guide to the chemical products of the lichens will help to avoid repetitious research and to direct future work toward the many real problems that exist.

I am indebted to Mrs. Elaine Kauvar, who did bibliographic work, and to the following undergraduates, who assisted in checking the manuscript: Elaine Conner, Ann Coulter, William Ishmael, Rees Shearer, Carol Watkins, and Ann Wilder. I am especially indebted to Paul Blake for careful proofreading for two years.

I am also indebted to a number of friends and colleagues who read portions of the manuscript. Dr. Peter Jeffs reviewed Chapters II and III. Dr. Siegfried Huneck checked the descriptions of higher aliphatic acids, xanthones, and triterpenes. Dr. Wolfgang Maass reviewed the pulvinic acid derivatives, Dr. Bengt Lindberg provided helpful suggestions concerning the carbohydrate derivatives, and Dr. Johan Santesson contributed information on his most recent studies. Dr. Mason E. Hale, Jr. reviewed the nomenclature used in the genus *Parmelia*, and his comments on the botanical index helped to clarify certain points.

I thank Mrs. Patricia James whose meticulous typing of the manuscript could be appreciated only by the proofreaders and editors who worked with it.

This book would never have been undertaken, continued, or completed without the suggestions, encouragement, and assistance of my husband, William Louis Culberson. His bibliography of lichen papers abstracted by biological and chemical abstracting services yielded the core of the references used. He generously gave of his own research time over a period of ten years to check every lichen name cited in the botanical index so that reports on synonyms could be cross-referenced to the current botanical name. His critical comments and editorial advice improved the presentation and clarity of every section of this book.

Department of Botany
Duke University
Durham, N.C.
August, 1968

C.F.C.

Table of Contents

Chemical and Botanical Guide
to Lichen Products

Chapter I

Introduction:
The Organization of Information in This Book

This guide contains two major reference sections. Chapter III lists some 300 lichen constituents with physical data, chemical structures, and annotated references. Chapter V lists the occurrence of the substances in approximately 2,000 species, varieties, and forms of lichens. Chapter II describes the nature of lichen products, their biogenesis, and their relationships to the products of nonlichen-forming fungi. A summary of the occurrence of the compounds in the genera of lichens is given in Chapter IV.

A. Explanation of the Chemical Guide to Lichen Products

The organic and inorganic constituents reported from lichens prior to 1968 are treated in Chapter III. It was originally intended to include only the secondary products of the lichens, and the present list is thought to cover completely the known constituents of this category. During the years that this catalogue was being compiled, however, it became obvious that interest in lichen physiology and metabolism had developed to the point that it would be desirable to extend a major coverage to materials such as elements and isotopes, amino acids, vitamins, and acids of the tricarboxylic acid cycle.

The organic compounds are arranged in groups of structurally and biosynthetically related substances. The biosynthetic classification of certain lichen substances proposed by Shibata (1965) was used in part in establishing the present classification. Empirical and structural formulas, physical properties, and the occurrence of the product in organisms other than lichens are indicated for most of the substances. In addition, there is a selected and annotated bibliography dealing with (*1*) historical information, (*2*) structural studies, (*3*) chemotaxonomic investigations, (*4*) antibiotic

3

properties, (5) biosynthesis, and (6) isolation from particular species. Publications that have appeared during the past 10 years are particularly emphasized. Although the references are indeed complete for many compounds, the aim has been to include a diversity of material rather than to provide an exhaustive bibliography.

The extensive references to microchemical methods are intended to help botanists identify chemical constituents for taxonomic purposes. Such microchemical studies greatly elucidate lichen systematics and also serve as a valuable tool in the search for new natural products.

References to the occurrence of compounds in particular lichens should help workers select standard sources of substances and give an appreciation of how widespread individual compounds may be among the lichens. Frequently additional information for a compound will be found under the pertinent species name in the botanical guide (Chapter V). It should be pointed out, however, that the species names used in the chemical guide are the ones in current botanical use. If the species name in the original report is no longer in botanical use, it is cross-referenced in the botanical guide.

Some compounds of unknown structure but of known chemical class are included in the appropriate sections. A few unknown substances that cannot be classified by structural type are reported in Part E of Chapter III (pages 218–222). A list of 363 species containing unidentified substances is given at the end of Chapter III (pages 222–227). Since some of the compounds occur in several species, this list probably represents about 250 additional compounds, many of which will doubtless prove to be new natural products.

There are about 113 compounds discovered before 1923 but not studied since that time. These substances are mentioned in the botanical guide but not in the chemical guide because no new information on them has appeared since the reviews by Zopf (1907) or Brieger (1923). Some of these compounds must surely be identical to substances of known structure, others are probably mixtures of substances, and some may be degradation products formed during extraction; but a number of these compounds are doubtless new lichen products. The unknown compounds are listed below with one or more species in which they have been reported. References can be obtained from review articles cited under the corresponding species names in Chapter V (pages 247–552).

B. List of Unknown Compounds Described in the Pre-1923 Literature

Acanthellin. *Cornicularia aculeata.*
Acolic acid. *Cyphelium tigillare.*
Acromelidin. *Teloschistes flavicans* var. *acromelus.*

Akromelin. *Teloschistes flavicans.*
Armoric acid. *Ramalina armorica.*
Armoricaic acid. *Ramalina armorica.*
Apoolivetoric acid. *Pseudevernia olivetorina.*
Apoolivoric acid. *Pseudevernia furfuracea.*
Areolatin. *Pertusaria rupestris.*
Areolin. *Pertusaria rupestris.*
Articulatic acid. *Usnea articulata* var. *asperula.*
Aspicilic acid. *Lecanora calcarea, L. gibbosula.*
Atraic acid. *Lecanora atra* "var. *panormitata.*"
Atranorinic acid. *Cladonia rangiformis.*

Barbatin. *Usnea ceratina.*
Blastenin. *Caloplaca arenaria, C. percrocata.*
Bryopogonic acid. *Alectoria cana, A. implexa.*

Calyciarin. *Lepraria candelaris.*
Caninin. *Peltigera canina.*
Caperidin. *Parmelia caperata.*
Caperin. *Cetraria endocrocea, Parmelia caperata.*
Cervicornic acid = cervicornin. *Cladonia subcervicornis.*
Cetrarialic acid. *Cetraria alvarensis.*
Cetrataic acid. *Parmelia reticulata.*
Chiodectin. *Herpothallon sanguineum.*
Cladestin. *Cladonia bacillaris, C. destricta.*
Cladestinic acid. *Cladonia destricta.*
Cladonin. *Cladonia crispata, Pycnothelia papillaria.*
Coccinic acid. *Haematomma coccineum.*
Coniocybic acid. *Coniocybe furfuracea.*
Conspersaic acid. *Parmelia conspersa.*
Cornicularin. *Cladonia impexa, Cornicularia aculeata, C. muricata.*
Cuspidatic acid. *Ramalina siliquosa.*

Destrictaic acid. *Cladonia destricta.*
Diffusin. *Parmelia sorediosa.*
Diffusinic acid. *Lecidea mollis, Parmelia sorediosa, Parmeliopsis hyperopta.*

β-Erythrin. *Roccella fuciformis.*
Everninic acid. *Evernia prunastri.*
Everniin. *Evernia prunastri.*

Furevernic acid. *Pseudevernia furfuracea.*
Fureverninic acid. *Pseudevernia furfuracea.*
Furevernuric acid. *Pseudevernia furfuracea.*
Furfuracinic acid. *Pseudevernia ceratea, P. furfuracea.*

Glomellic acid. *Parmelia glomellifera.*
Gyrophorin. *Umbilicaria vellea.*

Haematommidin. *Haematomma coccineum.*
Haematommin. *Haematomma coccineum.*
Hirtaic acid. *Usnea hirta.*
Hirtinic acid. *Usnea hirta.*
Hydrohaematommin. *Haematomma coccineum.*
Hydrosolorinol. *Solorina crocea.*
Hydroxyroccellic acid. *Lepraria farinosa, L. latebrarum, Pannaria pityrea, Roccella babingtonii, R. fuciformis, R. montagnei, R. phycopsis, R. tinctoria.*
Hymenorhodin. *Haematomma coccineum, H. porphyrium.*
Hypogymnol I. *Hypogymnia physodes.*

Icmadophilaic acid. *Icmadophila ericetorum.*
Isidic acid. *Pseudevernia isidiophora.*

Kamtschadalic acid. *Parmelia camtschadalis.*

Landroensin. *Ramalina landroensis.*
Latebrid. *Lepraria latebrarum.*
Lecasterid. *Lecanora rupicola.*
Lecasterinic acid. *Lecanora rupicola.*
Lecidic acid. *Lecidea cinereoatra.*
Lecidol. *Lecidea cinereoatra.*
Leiphaemic acid. *Haematomma leiphaemum.*
Leiphaemin. *Haematomma coccineum, H. leiphaemum, H. porphyrium.*
Lepranthaic acid. *Arthonia impolita.*
Lepranthin. *Arthonia impolita.*
Leprariaic acid. *Lepraria latebrarum.*
Lepraric acid. *Calicium chlorinum, Lepraria chlorina.*
Leprarin. *Calicium chlorinum, Lepraria latebrarum.*

Menegazziaic acid. *Menegazzia terebrata.*

Nivalic acid. *Cetraria nivalis.*

Olivaceaic acid. *Parmelia olivacea, P. prolixa.*
Olivacein. *Parmelia olivacea, P. prolixa.*
Olivetorin. *Cetrelia olivetorum.*
Olivoric acid. *Pseudevernia furfuracea, P. olivetorina.*
Orbiculatic acid. *Pertusaria globulifera.*

Paralichesterinic acid. *Cetraria islandica.*
Peltidactylin. *Peltigera polydactyla.*
Pertusaric acid. *Pertusaria globulifera.*
Pertusaridin. *Pertusaria globulifera.*
Pertusarin. *Pertusaria globulifera.*
Physodin. *Hypogymnia physodes.*
Physodylic acid. *Hypogymnia physodes, Pseudevernia furfuracea.*
Physol. *Hypogymnia physodes.*

Picropertusarin. *Pertusaria globulifera.*
Picrusunidic acid. *Usnea hirta.*
Pilosellic acid. *Parmelia crinita.*
Placodin. *Lecanora melanaspis.*
Placodiolic acid. *Lecanora melanophthalma, L. rubina.*
Plicatic acid. *Usnea plicata.*
Porin. *Pertusaria tuckermanii.*
Porinic acid. *Pertusaria tuckermanii.*
Proto-α-lichesterinic acid. *Cetraria islandica, Cornicularia muricata, Pycnothelia papillaria* var. *molariformis.*
Pulveraric acid. *Lepraria farinosa.*
Pulverin. *Lepraria latebrarum.*

Rhizoplacic acid. *Lecanora melanophthalma.*
Roccellinin. *Reinkella lirellina, Roccella tinctoria.*

Santhomic acid. *Usnea hirta.*
Saxatic acid. *Parmelia omphalodes, P. saxatilis* var. *retiruga.*
Soloric acid. *Solorina crocea.*
Solorinin. *Solorina crocea.*
Sordidin. *Lecanora sordida, L. sulphurea.*
Stictalbin. *Sticta glaucolurida.*
Stictinin. *Sticta gilva.*
Stigmatidin. *Enterographa crassa.*
Stuppeaic acid. *Cornicularia muricata.*
Subauriferin. *Parmelia subaurifera.*

Talebraric acid. *Lepraria latebrarum.*
Terrestrin. *Cetraria tilesii.*

Usnarinic acid. *Usnea hirta.*
Usnidic acid. *Usnea hirta.*

Zeoric acid. *Lecanora rupicola.*

C. Explanation of the Chemical Summary of the Genera

The chemical summary of the genera is intended to give some evaluation of chemical variation through the lichens. This list includes all of the genera for which there is any information about the production of secondary products as well as a few for which only primary substances are known. The genera are by families and the families by a tentative phylogenetic system (Hale and Culberson, 1966). The number of species or taxa for which information is reported in the botanical guide (Chapter V) is indicated in parentheses after the name of each genus. Data from studies made since the completion of the botanical guide are included in the chemical summary of the genera. Dr. Siegfried Huneck and Dr. Johan Santesson kindly provided information on work recently completed and currently in progress.

The substances are listed in groups of related compounds beginning with those secondary products most characteristic of lichens and most used in lichen taxonomy. Orcinol and β-orcinol compounds are listed separately, and the typical aromatic substances are separated by extra space from groups of compound types which are not specific to lichens. Amino acids and elemental constituents are not included, but a few miscellaneous compounds and some compounds of unknown structure are included in a single category at the end of the listing for each genus.

D. Explanation of the Botanical Guide to Lichen Products

The substances in the species treated in Chapter V were described in papers appearing between 1923, the year of Brieger's excellent review, and 1965, with the inclusion of a number of reports from 1966 and 1967. The species are listed alphabetically and the authority citations are abbreviated. The full names of the authorities for the epithets are given in Appendix 2 (pages 609–616).

1. CRITICAL SUMMARY OF THE CONSTITUENTS

A critical summary of the constituents reported for each species immediately follows the species name. The chemicals are listed in related groups and alphabetically within each group. The groups of compounds, separated by semicolons (;), are arranged so that the most characteristic lichen substances (and consequently those most frequently referred to in the taxonomic literature) are at the beginning of the list. The summary excludes reports known or suspected to be based upon (*1*) incorrectly identified plant specimens, (*2*) mixtures of species, (*3*) constituents inferred from color tests alone, and (*4*) compounds known to have been artifacts of the chemical procedure. Specific but less complete reports supporting the critical summary are given in separate paragraphs. The compounds not included in the main summary list are cited under OTHER REPORTS.

Certainly not all species have been uniformly studied chemically. Some lichens have been worked on in detail by many researchers using a variety of chemical methods. Ideally microchemical results obtained by lichenologists testing material from single plants of accurately determined collections, perhaps even authentic or type specimens, substantiate the results of chemical studies in which the compounds have been purified and precisely identified.

If no summary list appears after the name of the species, I consider the reports either too unreliable or too conflicting to permit any conclusion about the plant's constituents. Reports of amorphous pigmented products such as "parmelia-brown" are usually not included in the critical summary nor are a number of compounds isolated before 1923 and recorded in the early reviews but for which chemical information is scant or uncertain.

In the botanical guide, the compounds are listed by the names used in the chemical guide (Chapter III). If synonymous names of compounds were used in any of the original reports, the synonyms are indicated in parentheses in the summary list — a contingency in the event that the synonymy is disputed in the future. A list of synonyms encountered in the botanical guide is given in Part E of this section (pages 12–15).

2. SPECIFIC REPORTS

The paragraphs following the critical summary list substances in the order in which they occur in the summary list and cite the documenting references. Some of the reports are not original chemical determinations but refer to chemical constituents in discussions of other subjects — taxonomy or biosynthesis for example. The nature of the reference is usually obvious from the title of the article as given in the bibliography to the botanical guide (pages 552–597) and from annotations in the text. The annotations give (*1*) the geographic origin of the samples studied, (*2*) the chemical method used to identify the substances, (*3*) the yield of substance by extraction or the results of quantitative spectrophotometric analysis, (*4*) miscellaneous chemical or taxonomic conclusions of the article, and (*5*) the current scientific name of the plant extracted if the epithet used in the original report differs from presently accepted botanical nomenclature. A more detailed description of these annotations follows:

1. Geographic origins of the samples. — Many otherwise uniform lichens show variations in their major products and most chemical variants exhibit distinct geographic ranges. Consequently, reports on lichen constituents must include the origin of the samples studied. In the present listing, the provenance of foreign specimens is indicated by country or by major region (e.g. Himalayas, Gobi Desert, etc.) and that of American specimens by state, whenever this information is given in the article cited.

2. Chemical method. — The methods used to study lichen constituents include thallus color tests, microcrystal tests, paper and thin-layer chromatography, and macroextraction. Early studies of lichen chemistry were primarily based upon the extraction of large quantities of plant material and the isolation of the pure constituents. This is still the definitive method to confirm the identity of substances and to establish the structures of new constituents. But macroextraction is not suited to studies of chemical variation in morphologically similar plants and is obviously impractical for surveying the constituents of many collections for taxonomic purposes. Although the identity of the compounds isolated by macroextractions is certain, many macrochemical studies inadvertently report contaminants from small amounts of foreign species accidentally included in the bulk collections used.

Hale (1967) gives directions for the preparation and use of the routine spot-test reagents and microcrystal solutions. The standard abbreviations used in Chapters III and V refer to the following:

K	A color test with aqueous KOH.
C	A color test with aqueous $Ca(OCl)_2$.
KC	A color test with aqueous KOH followed by aqueous $Ca(OCl)_2$.
PD	A color test with alcoholic *p*-phenylenediamine or with Steiner's stable *p*-phenylenediamine reagent.
$FeCl_3$	A color test with aqueous or alcoholic $FeCl_3$.
GE	A microcrystal test solution for recrystallizations from acetic acid-glycerol.
GA*o*-T	A microcrystal test solution for preparing *o*-toluidine condensation products of aldehydes.
GAQ	A microchemical test solution for recrystallizing or preparing quinoline salts

3. Yields and quantitative data. — For many of the more recent studies involving macroextractions, the yields of constituents isolated are indicated. These yields represent the minimum content of the substance and do not necessarily measure real concentrations in the plant. Spectrophotometric methods have been used in lichen studies for quantitative measurements of pigments. The yellow usnic acid can be measured easily, and anthraquinones have been analyzed. Although a number of elements and isotopes have been determined quantitatively, only the qualitative results are included here.

4. Chemical or taxonomic comments. — This information notes, for example, analyses upon (nomenclatural) type specimens, the use of the species as an authentic source for a given compound, and studies on the structure or biosynthesis of a constituent. Corrections or reinterpretations of earlier studies and special taxonomic treatments suggested on the basis of the chemical data are also mentioned. If the constituents of a species were identified in taxonomic forms or varieties, these minor taxa are listed.

5. Botanical nomenclature. — Reports of constituents originally made under species names now considered to be synonymous are united under the currently accepted name. Cross-referencing is complete and every species name originally used in the literature summarized can be associated with the name currently accepted as correct and the one, consequently, used in this catalogue. Since botanical nomenclature is constantly reinterpreted as taxonomic progress is made, some taxa listed in this catalogue as distinct may be combined or divided further in the future.

Some morphologically very similar or indistinguishable lichens accumulate different metabolites. For constituents of sporadic occurrence — the "accessory substances" in lichenology — the sign "($\pm$)" is used. This symbol immediately follows the name of the compound and is not to be confused with the same symbol preceding a chemical name and indicating a racemic mixture. There has been a tendency here to arrange the strains involving "accessory substances" as variations under single species and to arrange the strains involving outright replacements or substitutions as distinct species in cases in which conflicting botanical views force a decision for the present list. This procedure should not be interpreted as advocating a particular taxonomy but is followed only for the sake of simplicity in presenting data. The abundant cross references should alert the reader to such situations.

3. OTHER REPORTS

Information listed under OTHER REPORTS includes conflicting data, physiological or cultural studies related to metabolism, and reports of poorly known compounds mentioned in the early literature. Some conflicting reports result from errors in the botanical or chemical identifications or from artifacts caused by decomposition of a constituent during analysis. But future studies will surely show that many of the apparent discrepancies are true chemical variations. For this reason it should not be assumed that relegation of a reference to the category of OTHER REPORTS necessarily represents a value judgment.

4. REVIEWS

Four reviews on lichen substances deal with the pre-1923 literature. The best known is Zopf's (1907) excellent book on lichen substances. A review by Hesse (1912) contains information and references not in Zopf.

Later Brieger (1923) reorganized these earlier treatments and added references for the intervening years. In 1932, Thies drew up a list of lichen substances unique among the reviews in being a botanical index arranged by species. He also included some poorly understood amorphous pigments not considered in the other papers. Unfortunately, his work is without bibliographic references. Citation of all these reviews throughout the present index substitutes for a coverage of the literature prior to 1923, a date nearly coinciding with the beginning of Asahina's remarkable contributions and the opening of a new era in the field of lichen chemistry.

Four other reviews are cited. The book by Asahina and Shibata, first published in Japanese in 1948 and translated into English with revisions six years later (Asahina and Shibata, 1954), is a general description including data from numerous studies originally published in obscure Japanese journals. Two reviews by Shibata (1958, 1963) are general accounts of the lichen products, and one by Neelakantan (1965) concerns studies between 1954 and 1964 on the elucidation of the chemical structures of lichen compounds.

E. Synonyms of Names of Compounds

For maximum clarity and usability of the botanical guide (Chapter V), compounds are always indicated by a single name even if they were reported under synonymous names in the original literature. The synonymous names of lichen substances are listed below. Since some of this synonymy may be changed by future studies, the synonyms as well are indicated in the botanical guide in the entries for the particular species involved. The names given in parentheses below refer to substances that are probably degradation products formed during extraction rather than natural constituents of lichens.

alectoric acid = barbatolic acid
atranoric acid = atranorin
atranorinic acid = atrinic acid

betaorcinol = β-orcinol

(callopisminic acid = ethylpulvinic acid from the reaction of ethanol with
    pulvinic dilactone)
capraric acid = protocetraric acid
carbousnic acid = usnic acid
catolechin = diploicin
cenomycin = barbatic acid

ceruchdiol = α-hydroxykaurane
ceruchinol = α-hydroxykaurane
(cetraric acid = cetrarin = α-ethyletherprotocetraric acid)
chlorophaeic acid = grayanic acid
chrysocetraric acid = pinastric acid
chrysophanic acid = parietin; chrysophanic acid is a name currently in
 use for 3-methyl-1,8-dihydroxyanthraquinone.
cinchol = β-sitosterol
coccellic acid = barbatic acid
coenomycin = cenomycin = barbatic acid
γ-collatolic acid = α-collatolic acid + β-collatolic acid (the enol lactone of
 α-collatolic acid)
confluentin = confluentinic acid = confluentic acid
coquimboic acid = hypoprotocetraric acid

deschlorothiophanic acid = arthothelin
dirhizonic acid = diffractaic acid

emodin methyl ether = parietin
γ-ergostenol = fungisterol
ergot sugar = trehalose
erythrinic acid = erythrin
meso-erythritol = erythritol
i-erythritol = erythritol
erythroglucinol = erythritol
Evans' substance A = usnic acid
Evans' substance B = usnic acid
Evans' substance C = pseudonorrangiformic acid
Evans' substance D = rangiformic acid
Evans' substance E = ursolic acid
evernuric acid = physodic acid

fallacin = fallacinal + teloschistin
fallacinol = teloschistin
farinacic acid = physodic acid
farinacinic acid = physodic acid
fimbriatic acid = fumarprotocetraric acid
fiscinic acid = parietin
flavicanone = thiophaninic acid (Zopf)
fuciformic acid = lepraric acid

3-(β-D-galactofuranosido)-D-arabinose = umbilicin
glabratic acid = lecanoric acid
1-(α-D-glucosido)-α-D-glucose = trehalose

hirtellic acid = thamnolic acid
hypogymnol II = ventosic acid (perhaps)

incrassatic acid = didymic acid
iso-squamatic acid = squamatic acid

kullensisic acid = protocetraric acid

lauropurpone = xanthorin
lecanorin = lecanoric acid
lecanorol = α-collatolic acid
lecanorolic acid = α-collatolic acid
lepranthic acid = lepranthaic acid
(lichenol = ethyl everninate from the reaction of ethanol with evernic
   acid)
lichen starch = lichenin
lutein = xanthophyll

malol = ursolic acid
D-β-mannoheptitol = volemitol
6-methoxy-1,8-dihydroxy-3-methylanthraquinone = parietin
methylcrysophanic acid = parietin
methylemodin = parietin
"α-methylethersalazinic acid" = impure stictic acid
"β-methylethersalazinic acid" = impure stictic acid
"γ-methylethersalazinic acid" = a questionable report
monoacetylprotocetraric acid = physodalic acid
moss starch = lichenin
mycose = trehalose

nemoxynic acid = homosekikaic acid
nephrosteraic acid = nephrosteranic acid
nephrosteric acid = nephrosterinic acid

ocellatic acid = thamnolic acid
ochrolechiaic acid = variolaric acid
α-orcinol = orcinol
α-orsellinic acid = lecanoric acid
orygmaeaic acid = polyporic acid

parellic acid (Hesse) = psoromic acid
parellic acid (Schunck) = variolaric acid
parietic acid = parietinic acid
parmatic acid = salazinic acid
parmelialic acid = lecanoric acid
parmelin = atranorin
peltigerin = tenuiorin
phycitol = erythritol
physciaic acid = parietin
physcianic acid = parietin
physcion = parietin

physodalin = physodic acid
picroerythrin = montagnetol
pleopsidic acid = acaranoic acid + acarenoic acid
protovitamin D_2 = ergosterol
prunol = ursolic acid
pseudopsoromic acid = stictic acid
pulvic anhydride = pulvinic dilactone
pulvinic (acid) anhydride = pulvinic dilactone
pulvinic lactone = pulvinic dilactone
pustulin = pustulan

ramalic acid = obtusatic acid
ramalinic acid = protocetraric acid
rheochrysidin = parietin
(rhizocarpinic acid = ethyl norrhizocarpate)
rhizoic acid = barbatic acid
rhizonic acid = barbatic acid
rupicolin = rupicolone = sordidone

saxatilic acid = salazinic acid
scopuloric acid = stictic acid
α-sedoheptitol = D-volemitol
silvatic acid = rangiformic acid
sphaerophoric acid = squamatic acid
squamaric acid = psoromic acid
stereocaulic acid = lobaric acid
stereocaulic acid = psoromic acid
stereocaulonic acid = stictic acid
stictaic acid = stictic acid
stictaurin = calycin + pulvinic dilactone (1:1)
sulcatic acid = psoromic acid

thiophaninic acid (Hesse) = sordidone ≠ thiophaninic acid (Zopf)
thuringion = thuringione
α,α-trehalose = trehalose
tumidulin = methyl 3,5-dichlorolecanorate

uncinatic acid = squamatic acid
urson = ursolic acid
usnaric acid = salazinic acid
usnarinic acid = thamnolic acid
usnein = usnic acid
usnetic acid = usnetinic acid = lobaric acid
usninic acid = usnic acid

ventosaric acid = thamnolic acid
verrucosic acid = scrobiculin
vitamin C = ascorbic acid

Literature Cited in This Chapter

ASAHINA, Y. AND S. SHIBATA. 1954. Chemistry of Lichen Substances. Japan Society for the Promotion of Science, Tokyo.

BRIEGER, W. 1923. Synthetische Versuche auf dem Gebiete der Flechtenstoffe und ihrer Bausteine. Pp. 205–438 *in* E. ABERHALDEN [ed.], Handbuch der biochemischen Arbeitsmethoden, Abteilung I: Chemische Methoden, Teil 10. Urban & Schwarzenberg, Berlin.

HALE, M. E., Jr. 1967. The Biology of Lichens. Edward Arnold, Ltd., London.

HALE, M. E., Jr. AND W. L. CULBERSON. 1966. A third checklist of the lichens of the continental United States and Canada. *Bryologist* **69**, 141–182.

HESSE, O. 1912. Die Flechtenstoffe. Pp. 32–144 *in* E. ABERHALDEN [ed.], Biochemisches Handlexikon. VII. Band. Julius Springer, Berlin.

NEELAKANTAN, S. 1965. Recent developments in the chemistry of lichen substances. Pp. 35–84 *in* Advancing Frontiers in the Chemistry of Natural Products. Hindustan Publishing Corp., Delhi.

SHIBATA, S. 1958. Especial compounds of lichens. Pp. 560–623 *in* W. RUHLAND [ed.], Handbuch der Pflanzenphysiologie. Vol. X. Springer-Verlag, Berlin.

SHIBATA, S. 1963. Lichen substances. Pp. 155–193 *in* H. F. LINSKENS AND M. V. TRACEY [eds.], Modern Methods of Plant Analysis, Vol. VI. Springer-Verlag, Berlin.

SHIBATA, S. 1965. Biogenetical and chemotaxonomical aspects of lichen substances. Pp. 451–465 *in* Beitrage zur Biochemie und Physiologie von Naturstoffen, Festschrift Kurt Mothes zum 65. Geburtstag. Gustav Fischer Verlag, Jena.

THIES, W. 1932. Systematische Verbreitung und Vorkommen der Flechtenstoffe (Flechtensäuren). Pp. 429–452 *in* G. KLEIN [ed.], Handbuch der Pflanzenanalyse. 3. Band. Spezielle Analyse. 2. Teil. Organische Stoffe II. Julius Springer, Wein.

ZOPF, W. 1907. Die Flechtenstoffe in chemischer, botanischer, pharmakologischer und technischer Beziehung. Gustav Fischer, Jena.

Chapter II

The Lichen Substances, Their Biogenesis, and Their Relationship to the Products of Free-Living Fungi

Biosynthetic studies on some of the compounds treated in this book and on structurally similar products from other natural sources allow some suggestions concerning the biogenesis of many lichen products. And a compiled list of fungal products (Shibata *et al.*, 1964) permits a comparison of the substances produced by the symbiotic lichen fungi and by the nonlichen-forming ones. The purely chemical aspects of lichen substances and the proofs of structural formulae are the subject of a review by Siegfried Huneck (1968) and will not be described in the present chapter. In addition, the chemistry and biological significance of the lichen polyols have recently been discussed by Lewis and Smith (1967*a*, 1967*b*). The present chapter deals primarily with structural and possible biosynthetic relationships of secondary lichen products and some other natural products, particularly those from free-living fungi.

A. Background

1. HISTORICAL BACKGROUND

Lichens produce many unusual secondary products not found in other plants. Chemical interest in lichen substances was early generated by the uniqueness of many of the aromatic products. In more recent years this interest has been increased by the discovery that some of the lichen products are biologically active and that a great many of them can contribute to the solution of the enormous task of identifying and classifying the some 18,000 known species of lichens. Today the interest in lichen chemistry is heightened even further by a large series of biologically intriguing studies on the biosynthesis of the products, the distribution of chemical races in nature, and the physiology of both intact lichens and the isolated fungal and algal components in culture.

Although serious scientific study of lichen products started about the middle of the nineteenth century, man had known the lichens as a source of useful chemical products for centuries. Perhaps even in classical times, and certainly in the Middle Ages, lichens were used as a source of dye-stuffs for textiles, an application that persists to the present day in the rural handweaving industry of Europe. The chemist's familiar litmus paper was obtained by fermenting species of *Ochrolechia* and *Roccella* with sodium or potassium carbonate in urine saturated with ammonium carbonate, although synthetic litmus was prepared from orcinol as early as 1864. Certain lichen products are still an important commodity in the perfume and soap industries. Lichen compounds have not been widely exploited as medicinals, but the sodium salt of usnic acid is apparently finding a relatively wide application as an antibiotic in the Soviet Union (Lazarev and Savicz, 1957). For an exhaustive review of the historical applications of lichen substances, the reader is referred to an excellent account by Llano (Perez-Llano, 1944).

In early scientific studies on lichen substances, large quantities of plant material were extracted to obtain the pure compounds. As a result of these studies well over 200 compounds were isolated and named. The high point of this period came in 1907 with the publication of Zopf's some 450-page book, *Die Flechtenstoffe in chemischer, botanischer, pharmakologischer und technischer Beziehung*, summarizing the findings from the entire period and describing the chemical properties of the then-known lichen products.

Accelerated progress toward the elucidation of the structural formulae of the compounds came in a long series of papers by Yasuhiko Asahina and his students in Japan. The chemistry of lichen products with special emphasis on the problems of determining the chemical structures was summarized in Asahina and Shibata's book *Chemistry of Lichen Substances* of which the English translation appeared in 1954.

Our knowledge of the products of the lichens, however, would be much smaller if it came exclusively from macroextractions of plant material in bulk. In reality most of the information on the occurrence of specific substances in the lichens has resulted from microchemical identifications. Although in recent years the natural products of many plant groups have been identified microchemically, methods for such determinations in the lichens were devised much earlier and as a result a great body of information has accumulated.

2. MICROCHEMICAL METHODS

In the 1930's Asahina expanded and developed a technique to identify the major lichen metabolites by microcrystallizations. A crude extract of a minute lichen fragment is recrystallized or converted to a crystalline derivative by treating it with a drop of reagent on a microscope slide. The compound is identified by the color and microscopic morphology of the resultant crystals (for a recent review, see Hale, 1967). The technique of microcrystallization was used with remarkable success by Asahina and others to identify substances from herbarium specimens, a rich source of lichen materials found to contain the characteristic compounds even in the oldest museum samples. The simplicity of the tests and the rapidity with which they can be performed subsequently allowed botanists to determine the major constituents of hundreds of species.

An interest in chemical variations among closely related, sometimes morphologically indistinguishable lichens spurred on the determination of the constituents of many species. Long before the advent of paper chromatography, a large background of information already existed on chemical variation in lichens. Paper chromatography (Wachmeister, 1959) confirmed many of these findings, allowed more rapid studies of large numbers of specimens, and facilitated the mapping of the geographic distributions of many chemical strains. Still, for many workers the microcrystal method remained the quickest and surest one for preliminary identifications of the common lichen products.

Within the past five years, thin-layer chromatography (Santesson, 1967*b*) has replaced paper chromatography as the major method for the identification of lichen constituents. The sensitivity of this method also allows the recognition of certain categories of previously unsuspected joint occurrences of substances and permits the identification of some compounds resistant to earlier microchemical methods.

Gas chromatography (Shibata *et al.*, 1965) has had limited use, but will certainly be an important tool in some areas of lichen chemistry in the future. Unlike thin-layer chromatography and microcrystal tests, however, gas chromatography cannot be applied by the herbarium taxonomist lacking special laboratory facilities.

3. COMPOUNDS CONSIDERED LICHEN PRODUCTS

In the past, the terms "lichen substances," "lichen acids," and "lichenic acids" have been used by some authors to refer to the unique compounds in lichens and by others to indicate all compounds in lichens. In this book the designations "lichen substances" and "lichen products" are used for any substance found in lichens, just as the designation "fungal products" is generally used to mean all compounds known from nonlichen-forming fungi. The familiar but ambiguous designations "lichen acids," "lichenic acids," and "so-called lichen acids" will not be used at all. The vast majority of known compounds unique to lichens are indeed acidic for reasons to be considered in the discussion that follows, but there is no evidence to indicate that compounds peculiar to lichens must necessarily contain acidic functional groups.

Most compounds known from lichens are doubtless synthesized primarily by the fungus from carbohydrates supplied by the alga. Although recent studies suggest that the alga may participate in terminal stages of the synthesis of some of the characteristic lichen products (Mosbach and Jakobsson, 1968), it would seem that the major steps in the synthesis of most of the complex organic products are related to the metabolism of the fungus. Outside of the lichens, natural products most similar to the complex lichen products occur in free-living fungi but not in free-living algae (Shibata *et al.*, 1964). Purely algal constituents are probably rarely isolated from lichens because the proportion of the alga to the fungus is generally quite small. Bednar (1963) calculated that the alga of one species represented only 3.1% of the dry weight of the whole lichen. Wilhelmsen (1959) found that the chlorophyll content of the lichens that he studied was only 10–25% of that in the leaves of flowering plants. Many compounds expected to occur in the algal partner — such as chlorophyll, carotenes, vitamins, and saccharides — have been identified in pure cultures of the isolated algal constituent. It is also possible that at least some of these compounds are elaborated by the fungus. The present book, therefore, treats all the lichen constituents without separation of algal and fungal products.

4. The Characteristic Lichen Products

Lichen-forming fungi produce compounds many of which are identical or closely related to those produced by nonlichen-forming fungi. The most distinguishing feature of the compounds produced only by lichens is the high proportion of acetate-polymalonate-derived aromatic phenols composed of two or three phenolic units joined by esterification or by oxidative coupling or by both. The depsides, depsidones, dibenzofurans, usnic acids, and the depsone picrolichenic acid all appear to be produced by such mechanisms and all are peculiar to lichens. Other aromatic compounds of acetate-polymalonate origin — such as chromones, xanthones, and anthraquinones, which probably form by internal cyclizations of a folded polyketide chain — are often identical to or only minor derivatives of products of nonlichen-forming fungi or higher plants. Whether or not the coupled nature of the products most characteristic of lichen-forming fungi has a biological significance related to the symbiotic nature of the lichen itself is unknown, but this question will be considered briefly below and in the next section (pages 22–24).

Information on the synthesis of lichen products by normally lichen-forming fungi cultured in the absence of the algal partner is limited. The substances detected from cultures have been few and include pulvinic acid derivatives (Mosbach, 1967; E. A. Thomas, 1939), phenolic acid units (Hess, 1959), a chromone (Cecil Fox, private communication), and anthraquinones (E. A. Thomas, 1936, 1939). It will be noted that none of these compound types are peculiar to lichen-forming fungi. The only report of coupled products of the sort characteristically associated with lichens (Castle and Kubsch, 1949) has not been verified in spite of many attempts to do so (e.g., Ahmadjian, 1964).

Although the characteristic lichen substances have not been found in cultures of the isolated lichen fungus, the compounds form in the intact lichen under even bizarre experimental conditions. For example, recent studies on the biosynthesis of lichen substances have repeatedly demonstrated that intact lichens can synthesize depsides, tridepsides, and usnic acid under conditions such as total submersion in a liquid (Mosbach 1964a), an environmental situation completely alien to the normal ecology of the plants. Furthermore, Mosbach and Jakobsson (1968) recently prepared a cell-free system from *Evernia prunastri* that synthesized evernic acid from acetyl coenzyme A and malonyl coenzyme A. It would seem that if the mechanism for synthesis exists in the plant, purely physical factors like periodic desiccation may not be critical to at least the short-term production of these compounds.

In an earlier study, a depside-hydrolyzing enzyme was isolated from both the algal and the fungal components of *Lasallia pustulata* (Mosbach and Ehrensvärd, 1966). It was suggested that depsides may be formed by the enzymatic action of the algal partner on aromatic units synthesized by the fungus. If this theory proves to be correct, it would become obvious why depsides, depsidones, and perhaps some of the other diaryl derivatives formed by coupling of phenolic units have not been found in cultures of the isolated fungal components. But it would still be difficult to understand why the phenolic units are not commonly observed in these cultures. Possibly the units are not formed under the environmental conditions of the cultures, or perhaps they are produced only in small amounts or are not liberated from their enzyme-bound precursors unless they are converted to coupled products. Perhaps the phenolic units are destroyed in cultures by conversion to other products such as the highly pigmented water-soluble compounds observed in cultures of lichen fungi but apparently not produced in lichens under natural conditions.

5. The Biological Role of the Characteristic Lichen Substances

The biological role of the characteristic lichen substances is not known, but it seems reasonable that either their biosynthesis or their function must pertain to the symbiotic state since (*1*) they do not appear to form in cultures of the isolated symbionts, (*2*) they have not been identified in nonlichen-forming fungi, and (*3*) they are produced by many lichens distantly related taxonomically (see Chapter IV). The production of the characteristic lichen compounds is not required to maintain the symbiotic state, at least not in all lichens, since many genera and families produce none of these compounds. None of the basidiolichens are known to produce characteristic lichen substances. In some genera of ascolichens, certain individuals produce the typical lichen compounds while others do not, even though such races may be closely related morphologically.

The characteristic aromatic lichen substances appear to be formed in the most vigorous phase of hyphal growth since color tests for the compounds are usually most intense in the regions of active growth, such as the youngest lobes of Parmelias and the podetial tips of Cladonias. Spectrophotometric analysis of the concentration of usnic acid in successive segments of a *Cladonia* species showed the highest percentage of this substance in the tips of the branchlets (Fedoseev and Yakimov, 1960; Laakso and Gustafsson, 1952). The decreasing concentration of usnic acid in the older parts of the lichen may be due to an increase in the weight of the tissues in those parts (by accumulation of other products or by cell enlargement) or to the sloughing off of the outer layers of tissue richest in the substance. Paper chromatographic analyses of individuals of various ages of the foliose *Lasallia papulosa* revealed no differences with age in the concentration of the medullary tridepside gyrophoric acid (C. F. Culberson and Culberson, 1958). This result agrees with the notion that gyrophoric acid is formed at a relatively constant rate during active growth, even in the youngest individuals tested, and that once precipitated it remains unchanged throughout the life of the plant. The concentration of gyrophoric acid in the related species *Lasallia pustulata* is not affected by the death of the lichen (Měrka, 1951), and there is no evidence that the characteristic aromatic products that precipitate on the outer surfaces of the hyphae are ever reabsorbed and utilized. The only indication that certain lichen products may be degraded in the oldest parts of the thallus is the production of pigmented compounds noticeable in some herbarium specimens. It has been suggested that this pigmentation might arise from the action of an alkaline atmosphere on oxidizable aromatic products or by the formation with alkalis of bright red complex salts of salazinic acid and norstictic acid.

The general resistance of herbarium specimens of lichens to attack by insects and microorganisms normally encountered as pests in mycological collections has led to the suggestion that the characteristic lichen products may have a protective function in nature. Most lichens grow very slowly, persist in their habitat for years, and are resistant to the microorganisms that attack and destroy many nonlichen-forming fungi.

Many phenolic compounds are either toxic or at least inhibit growth (for a review, see Cruickshank and Perrin, 1964). Their toxicity may be reduced when the phenolic hydroxyl groups are removed or chemically bound. Synthesis of acetate-polymalonate-derived phenols containing fewer than the number of phenolic hydroxyls predicted to result from cyclization of a polyacetyl chain occurs in free-living fungi but rarely in lichens. The formation of glycosides, in which the phenolic hydroxyl is bound to a sugar, occurs in higher plants but is unknown in lichens. If large quantities of phenolic acids were produced by the lichen, then economical ways to bind phenolic and acid hydroxyl groups would be (*1*) intermolecular esterification of the units to form depsides, (*2*) oxidative cyclization of these esters to form depsidones, or (*3*) dehydrative coupling of the units to form dibenzofurans. In higher plants, glycosides of phenols are thought to be hydrolyzed to the toxic phenolic compounds when the plant is invaded by a pathogen. In lichens, the depsides might be hydrolyzed to release the free phenolic acids, but it is less likely that the ether linkages of depsidones, dibenzofurans, and the usnic acids would be easily cleaved. Crude enzymes, isolated from both the fungal and the algal components of *Lasallia pustulata*, were capable of hydrolyzing the depside evernic acid and two tridepsides (Mosbach and Ehrensvärd, 1966). Possibly such enzymes may also catalyze the synthesis of depsides. If the alga is responsible for the synthesis of depsides from phenolic units produced by the fungus, it could be a protective response to the toxicity of these compounds. When the characteristic coupled products are absent from a lichen, the aromatic units have never been observed to accumulate in their place.

Regardless of where coupling of the phenolic acid units occurs, the production of these units by the fungus remains unexplained. Possible functional roles for phenolic compounds currently perplex biochemists concerned with animal and plant metabolism (Finkle and Runeckles, 1967). The difficulties encountered in trying to ascribe a particular function to these compounds and to most other secondary products have led to the idea that they may be synthesized by the same enzymes normally involved in basic physiological processes.

Perhaps the phenolic compounds in lichens are actually metabolic by-products produced in abundance as a result of the unusual nutrient balance controlled by the alga. Algae are noted for their production of copious and often unusual carbohydrates, and the C/N ratio of the food supply to the fungus may be very high. Except in lichens containing nitrogen-fixing blue-green algae, the nitrogen available to most lichen fungi is doubtless very low. Metabolic pathways through the amino acids and those leading to nitrogen-containing end-products would not be expected to be favored.

When *Penicillium patulum* is grown under conditions involving a decrease in protein synthesis, it produces 6-methylsalicylic acid which is closely related to the orsellinic acid unit of many lichen depsides. A recent study showed that 6-methylsalicylic acid is synthesized in the fungus by a multi-enzyme complex, and it was suggested that this complex may also be involved in some other biosynthetic or physiological processes (Light, 1968). The concept of secondary substances in plants as the products of enzymes primarily involved in fundamental physiological reactions may well prove to be correct.

It has been suggested that lichen products may increase the permeability of the cell wall of the alga (Follmann, 1960; Follmann and Villagrán, 1965) and thereby allow a flow of nutrients from the alga to the fungus. When the *Nostoc* algal component of *Peltigera polydactyla* was removed from the lichen, it stopped releasing glucose into the medium within 48 hours (Drew and Smith, 1967a). Although aromatic lichen products could be involved here, the cause of this observed change in permeability has not been demonstrated and may well be due to other factors, especially since *Peltigera polydactyla* is one of the many species of lichens associated with blue-green algae and characterized by a relatively low concentration of aromatic lichen products. Huneck and Tümmler (1965) found 1% of the tridepside tenuiorin in this species.

The accumulated lichen substances could alter the intensity and quality of light reaching the algal cells. Reduced light intensity may protect the alga in some environments (Hale, 1967) and the fluorescence maximum of the depside atranorin corresponds to an absorption maximum of chlorophyll (Rao and LeBlanc, 1965). But these factors, which may be significant side effects of the production of the typical lichen substances in some species, probably cannot be called upon to explain the known production of characteristic lichen compounds in hundreds of species from a wide variety of habitats.

B. Secondary Products of the Acetate-polymalonate Pathway

1. AROMATIC PRODUCTS ARISING FROM INTERMOLECULAR ESTERIFICATION OR OXIDATIVE COUPLING

While some aromatic lichen products are unique to lichens, others are found in nonlichen-forming fungi, and many are common in other plant groups as well. For example, although some anthraquinones are peculiar to lichen-forming fungi, this class of compounds is common in nature. The synthesis of the particular anthraquinones found in lichens does not reflect a notable diversion from the synthetic routes in free-living fungi or even in higher plants. Similarly, for example, lichens produce chromones, xanthones, fatty acids, terpenes, terphenylquinones, and pulvinic acid derivatives, and all these classes of compounds are represented in non-lichen-forming fungi. The aim of the following discussion is to show how the characteristic lichen products and the superficially similar compounds found in free-living fungi may in reality have different biosynthetic origins.

The compounds most characteristic of the lichen-forming fungi are (*1*) the aromatic esters represented by depsides, [1] and [2], and tridepsides [3], (*2*) the depsidones [4], which are derived from *para*-depsides, (*3*) the dibenzofuran derivatives [5], including the usnic acids [6], and (*4*) the depsone derivative picrolichenic acid [7].

In lichens, all of these compound types are formed by joining two or occasionally three phenolic units, which are always derived by the acetate-polymalonate pathway. That intermolecular esterification or oxidative coupling reactions are required for the synthesis of the depsides, tridepsides, depsidones, dibenzofurans, and the depsone seems to be a significant feature that all of these compounds share and that distinguishes them from many of the structurally similar substances that are not unique to lichens. In the acetate-polymalonate-derived compounds, like anthraquinones [8] and xanthones [9] for example, the two aromatic rings which are joined together arise by intramolecular cyclizations of a folded polyketide chain. Although alternariol [10], a well known fungal product from *Alternaria tenuis*, has an ester linkage between two aromatic rings, it is not closely related biosynthetically to lichen depsides. Its aromatic rings are joined by a carbon-carbon bond remaining from a long-chain acetate-polymalonate precursor, and the ester linkage is derived by an intramolecular rather than intermolecular reaction (R. Thomas, 1961). Another fungal product, geodoxin [17] from *Aspergillus terreus*, has a phenolic ring joined to a quinoid ring by ester and ether bondings. But even in this case where no carbon-carbon linkage between the rings suggests a single-chain precursor,

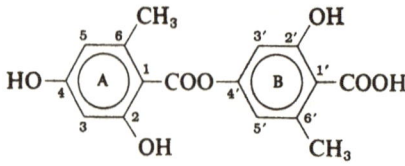

para-depside: lecanoric acid

[1]

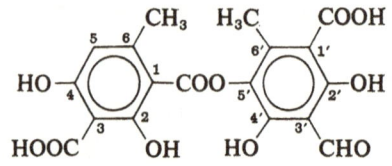

meta-depside: thamnolic acid

[2]

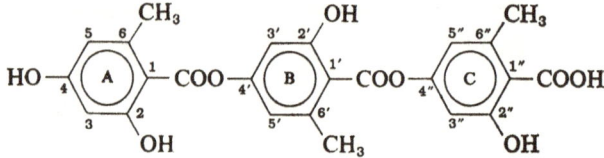

tridepside: gyrophoric acid

[3]

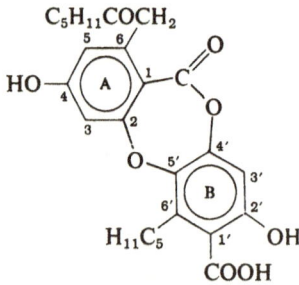

depsidone: physodic acid

[4]

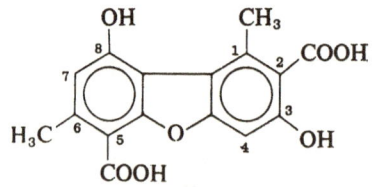

dibenzofuran: pannaric acid

[5]

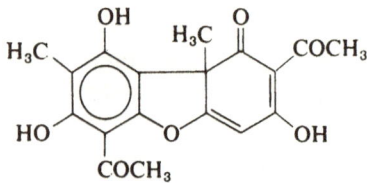

usnic acid

[6]

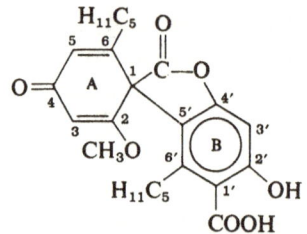

depsone: picrolichenic acid

[7]

the compound is, nevertheless, believed to arise from such a single poly-ketide chain. The biosynthesis begins with the formation of the anthra-quinone emodin [11], the decarboxylation product of endocrocin [8]. The 1-O-methyl derivative of emodin, questin [12], is converted to sulochrin

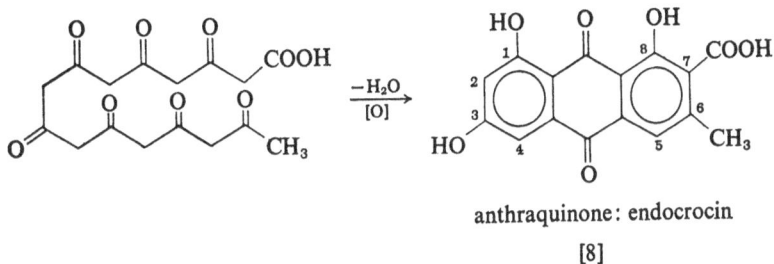

anthraquinone: endocrocin

[8]

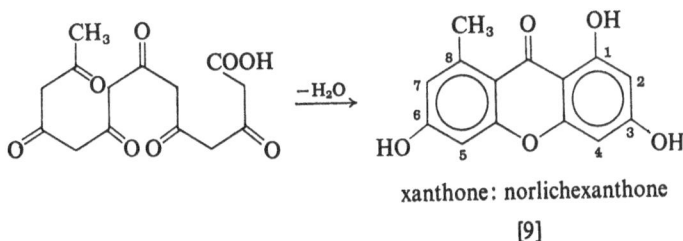

xanthone: norlichexanthone

[9]

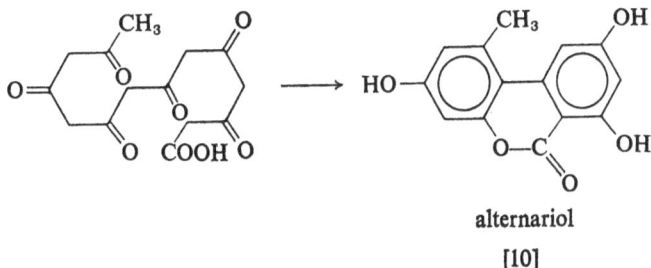

alternariol

[10]

[13] (Gatenbeck, 1968) and then by intramolecular oxidative coupling to bisdechlorogeodin [14] (Misconi and Stickings, 1968) in *Penicillium frequentans*. Emodin and sulochrin are converted to geodin [15] in *Aspergillus terreus* (Gatenbeck, 1968) — although (curiously) one study interpreted the biosynthesis of sulochrin in this fungus as a coupling of separate units (Curtis, Hassall, and Pike, 1968). The ketone carbonyl joining the rings in bisdechlorogeodin [14] may then be converted directly to the ester linkage in geodoxin [17] by a Baeyer-Villiger peracid rearrangement or through a carboxylic acid like the fungal product asterric acid [16] and then to geodoxin [17] by oxidative coupling. Whatever the details of this biosynthetic sequence may be, it is most probable that geodoxin and the related diaryl fungal products are formed by intramolecular conversions and are consequently distinct from the characteristic diaryl and triaryl lichen products that are formed by intermolecular esterification and coupling reactions. For a more complete discussion of the biosynthesis of these fungal metabolites, the reader is referred to reviews by Richards and Hendrickson (1964) and by Whalley (1967).

endocrocin [8] $\xrightarrow{-CO_2}$

emodin (R = −H) [11]
questin (R = −CH₃) [12]

bisdechlorogeodin (X = −H) [14]
geodin   (X = −Cl) [15]

sulochrin

[13]

asterric acid

[16]

geodoxin

[17]

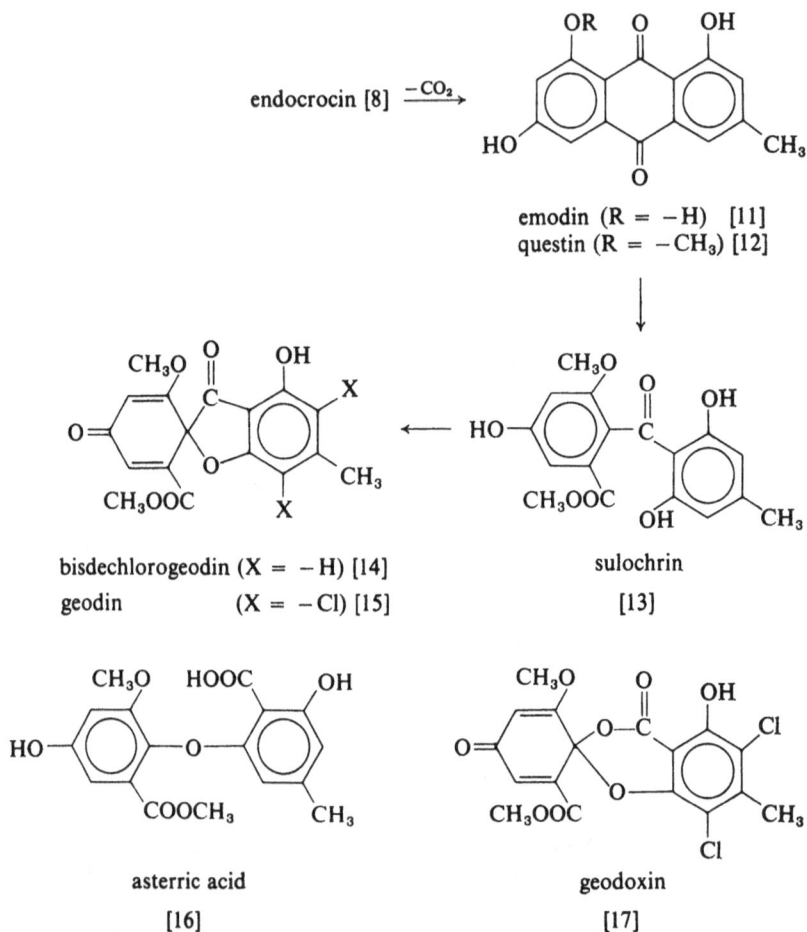

Depsidones in lichens are believed to arise by oxidative cyclization of depsides. This type of cyclization reaction is an intramolecular dehydrogenation and is not unique to lichens. Intramolecular oxidative coupling of phenols is encountered again and again among fungal products of many classes, the conversion of sulochrin to bisdechlorogeodin for example, and in many important natural products such as alkaloids (Scott, 1965). The fact that depsidones are characteristic of lichens is probably the result of the uniqueness of lichen depsides from which they form. Although depsidones are chemically a distinct class of compounds, biosynthetically they are simply depside derivatives.

No depsides have been found in free-living fungi. The only depsidones known in the nonlichen-forming fungi are nidulin, nornidulin, and dechloronornidulin [18–20], constituents of a mutant strain of *Aspergillus*

nidulin   (X = −Cl; R = −CH₃) [18]

nornidulin   (X = −Cl; R = −H) [19]

dechloronornidulin   (X = −H; R = −H) [20]

3,4-dihydro-6,8-dihydroxy-
3,4,5-trimethylisocoumarin-
7-carboxylic acid   [21]

[22]

[20]

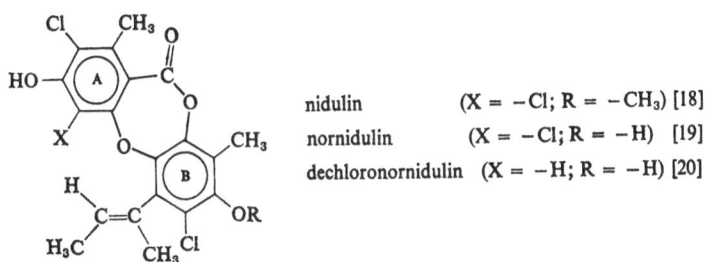

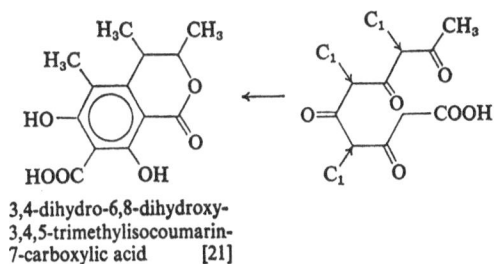

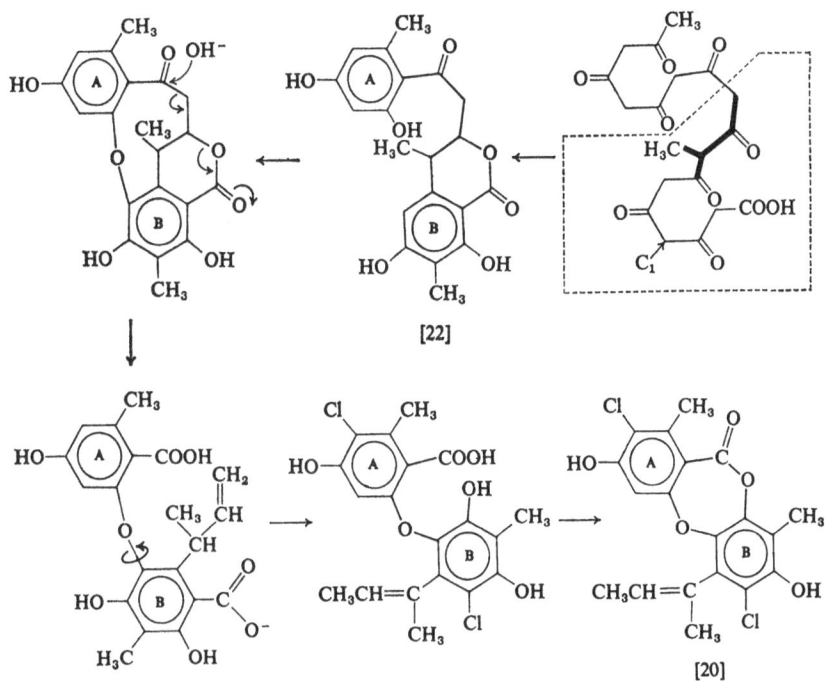

nidulans grown in liquid culture (Dean *et al.*, 1953, 1954, 1956; Hogeboom and Craig, 1946). The A ring of nidulin and nornidulin is identical to the A ring of the lichen depside methyl 3,5-dichlorolecanorate and the lichen depsidones gangaleoidin and diploicin. But the B ring of the nidulins differs from any phenolic acid unit known in lichens by having a branched, unsaturated, four-carbon substituent *para* to a methyl group. An isocoumarin [21] produced by *Aspergillus terreus* closely resembles the B ring of the nidulans.

The simplest mechanism for the production of the fungal depsidones would be through depside precursors by the route proposed for the biosynthesis of lichen depsidones. But, since no compounds resembling lichen depsides are known in free-living fungi, one might consider the possibility that fungal depsidones form from a single polyketide chain and not from a depside precursor. At least one such mechanism can be imagined. The nonlinear side chain of a substance resembling the isocoumarin derivative [21] from *Aspergillus terreus* could be preceded by eight carbons that can cyclize to a second aromatic ring (ring A in [22]). The five carbon atoms joined by bold lines in the hypothetical polyketide precursor illustrated are derived from isoleucine, possibly by way of tigyl coenzyme A or methyl acetoacetyl coenzyme A (Beach and Richards, 1963). The involvement of fragments derived from amino acid metabolism is completely unlike the exclusively acetate-malonate pathways which seem to account for the synthesis of the characteristic lichen substances. Cyclodehydrogenation, between rings A and B in [22], could yield the ether linkage of the nidulins. Cleavage at the α-keto carbon of ring A, decarboxylation, and chlorination, could yield a diaryl ether which requires only a rearrangement of the exomethylenic double bond and an internal esterification to be converted to dechloronornidulin [20]. The possibility of such an alternative to a depside precursor for the nidulin compounds could be eliminated at once if incorporation of labeled monochloro or dichloroorsellinic acid could be demonstrated. But, until depsides are found in nonlichen-forming fungi or experimental evidence shows an intermolecular joining of the aromatic rings of the nidulin depsidones, a depside precursor for these compounds and a close biosynthetic relationship to lichen depsidones should be considered with caution.

The polyacetyl chains envisioned in the biosynthesis of aromatic compounds by the acetate-polymalonate pathway can cyclize to produce phenolic products either by an orsellinic acid-type cyclization or by a phloroglucinol-type cyclization (Chart I). Except in the case of usnic acid, the aromatic rings of the characteristic lichen compounds are formed by an orsellinic acid-type cyclization which accounts for the large proportion of carboxylic acids among these compounds and the large number of strongly acidic aromatic lichen substances. The phloroglucinol-type cyclization, which does not yield carboxylic acids, occurs widely in fungal products, in the lichen pigment usnic acid, and in a number of xanthones and chromones found in both lichens and nonlichen-forming fungi. A

depsides
depsidones
dibenzofurans
depsone

orsellinic acid-type cyclization

usnic acids

phloroglucinol-type cyclization
(illustrated with C-methylation to yield methylphloroacetophenone)

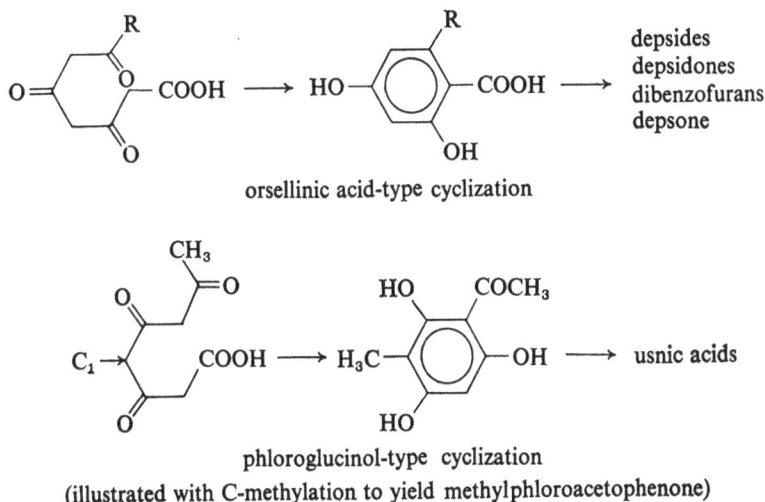

Chart I. Comparison of the orsellinic acid-type cyclization with the phloroglucinol-type cyclization.

significant consequence of this process is that aromatic rings formed by the phloroglucinol-type cyclization and lacking the carboxylic acid substituent cannot be esterified to depsides and tridepsides, depsidones, or the depsone picrolichenic acid. Thus, the single choice of the phloroglucinol-type cyclization pattern eliminates the possibility for the production of four types of characteristic lichen compounds, leaving only the dibenzofurans and usnic acid which do not have an ester linkage joining the two phenolic units.

a. DIARYL AND TRIARYL ESTERS — ORCINOL SERIES

The commonest phenolic acid units derived by the acetate-polymalonate pathway and combined to form the characteristic lichen substances are of two types: The orcinol-type units [23] are represented by orsellinic acid and a number of related units; the β-orcinol-type units [24] have an extra C_1 substituent at the 3-position. While compounds formed from these two types of units are similar in many ways, differences in their structure and especially in their distribution among the lichens suggest that the usual tendency to consider the orcinol and β-orcinol compounds separately probably has a biosynthetic justification.

The most common fate of acetate-polymalonate-derived phenolic acids in lichens is intermolecular esterification of two or three similar or identical units. In the simplest case, the carboxylic acid of one unit is joined to the hydroxyl *para* to the carboxylic acid of the second unit. Such esterifications

orcinol-type unit
[23]

β-orcinol-type unit
[24]

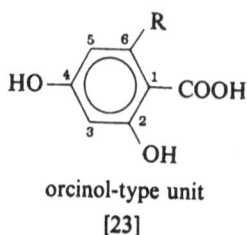

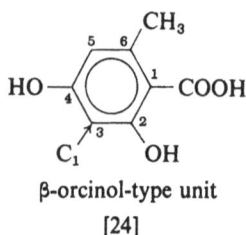

lead to the *para*-depsides. A second esterification reaction leads to tridepsides. If an ester linkage joins the first unit to a position *meta* to the carboxylic acid of the second ring, a *meta*-depside results.

i. para-*Depsides*

In the orcinol series, phenolic acid units have unbranched side chains of 1, 3, 5, or 7 carbons. The C_5 and C_7 side chains may retain from their acetate-polymalonate origin an oxygen function as a β-keto group [23, $R = -CH_2COC_3H_7$ or $-CH_2COC_5H_{11}$] or, very rarely, a C_5 side chain may have an α-keto group [23, $R = -COC_4H_9$]. Partial reduction to alcohols and dehydration to unsaturated side chains, variations relatively common in fungal products, are now unknown among the lichen substances, although the formation of the α-keto C_5 side chain may require such transformations at the α- and β-carbons.

The C_3, C_5, and C_7 side chains may arise by acetate-malonate condensations accompanied at each stage by enol reduction, dehydration, and double bond reduction as in the synthesis of fatty acids. But instead of continuing this process to give a fatty acid, four more malonate units condense, without reduction of the carbonyl groups, and cyclize to a phenolic acid. Perhaps preformed coenzyme A derivatives of C_4, C_6, and C_8 acids take the place of acetyl coenzyme A to initiate the condensations with four malonate units which then cyclize to phenolic acids having three-, five-, and seven-carbon side chains respectively. An α-keto C_6 acid could initiate the formation of the phenolic acid with the oxidized side chain found in lobaric acid and norlobaridone, and β-keto C_6 and C_8 acids could yield units like the A rings of glomelliferic acid and olivetoric acid respectively. In any case, variations leading to different lengths of side chains must arise early in the synthesis of the aromatic rings.

A comparison of the occurrence of orcinol-type phenolic acid units with side chains of various lengths in depsides, depsidones, and dibenzofurans, shows that the C_1-substituted units, derived from orsellinic acid, are by far the most common. Compounds containing the orsellinic acid unit occur widely in lichens of genera believed to be evolutionarily primitive as well as those believed to be more advanced. The orsellinic acid unit is the simplest possible aromatic phenolic acid derivable from the condensation of acetate and malonate units. The biosynthesis of this unit does not require reduction or dehydration steps that must be involved in the synthesis of higher homologues with saturated side chains. It seems reasonable to suppose that the orsellinic acid unit is more primitive than the units with longer side chains. Furthermore one might expect that after the ability to produce units with longer side chains arose, these units would be more similar to each other biosynthetically than they would be to the orsellinic acid unit. Compounds derived from the C_3, C_5, and C_7 units are more common in somewhat advanced genera than in the most primitive ones. Many genera of lichens produce orsellinic acid-derived products, but no compounds containing longer side chains, and a number of genera produce compounds with C_3, C_5, and often C_7 side chains, but few if any C_1 side chains. The orsellinic acid unit is rarely combined with a unit having a longer side chain, the two known examples, both rare, being the depside sphaerophorin and the depsidone grayanic acid. On the other hand, units with C_3 and C_5 side chains or C_5 and C_7 side chains are frequently combined in depsides and depsidones. Compounds composed of orsellinic acid units are rarely found in lichens with ones composed of units having longer side chains, one example being the joint occurrence of imbricaric acid and gyrophoric acid in *Parmelia locarnensis* Zopf. On the other hand, homologous compounds — or minor variants of them with C_3, C_5, and C_7 side chains — quite frequently occur in the same species or in closely related species. But there are some notable and interesting exceptions. In *Pseudevernia*, the European species produce compounds with five- and seven-carbon side chains while the three North American species produce lecanoric acid (Hale, 1968). This relatively large chemical change is accompanied by only minor morphological differences. In the genus *Evernia*, the commonest and most widespread species produces evernic acid, which has only C_1 substituents, while the other species all produce divaricatic acid, which has C_3 substituents. Such cases could arise if a species retains a primitive synthetic mechanism or retrogressively evolves the mechanism leading to the same products.

ii. Tridepsides

The known tridepsides in lichens are all derivatives of orsellinic acid. No tridepside with β-orcinol-type units or orcinol-type units having C_3, C_5, or C_7 side chains, the common substituents of the diaryl esters, are known. Erythrin is a depside which is further esterified with the sugar alcohol erythritol rather than with a third phenolic acid. Erythrin occurs in genera producing its depside precursor lecanoric acid and the tridepside gyrophoric acid.

iii. meta-*Depsides*

No hydroxyl exists at a position *meta* to the carboxylic acid in units derived by the acetate-polymalonate pathway. Esterification of two units to form a *meta*-depside must be preceded by a nuclear hydroxylation (Chart II). If nuclear hydroxylation does not occur, the mechanism joining the units of *meta*-depsides differs fundamentally from the one involved in the synthesis of *para*-depsides. The *meta*-depsides could also form, for example, by oxidative coupling of a carboxylate radical of ring A or by rearrangement from a *para*-depside with hydroxylation at the carbonyl carbon (Chart II). The structure of one *meta*-depside, sekikaic acid, does not

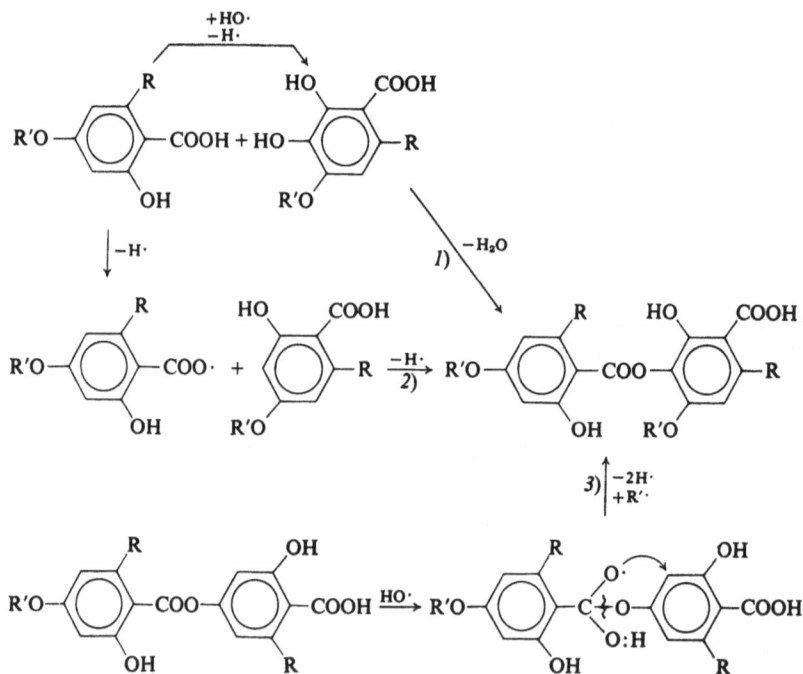

Chart II. Three alternative biogenetic mechanisms for *meta*-depside synthesis by (*1*) prior hydroxylation, (*2*) direct attack of a carboxylate radical, and (*3*) rearrangement of a *para*-depside.

support the latter alternative. Until there is evidence on this point, it is simplest to assume that the second phenolic acid unit of *meta*-depsides has undergone a nuclear hydroxylation, but it must be kept in mind that alternate mechanisms are conceivable. The nuclear hydroxylation mechanism would be the most attractive were it not relatively rare for known lichen substances derived by the acetate-polymalonate pathway to show any other evidence of such substitution.

Nuclear hydroxylation at the 3- and 5-positions of the orsellinic acid unit is observed in diploschistesic acid and hiascic acid, but no *meta*-depside involving an orsellinic acid unit is known. If orcinol-type phenolic acid units with longer side chains are ever hydroxylated, the products are known only as the B rings of *meta*-depsides. Assuming that hydroxylation occurs, there are two nonequivalent *meta*-positions available for reaction in the orcinol-type units. In all known orcinol-type *meta*-depsides, the ester linkage joins at the 3-position between the oxygen-substituted aryl carbons and *para* to the alkyl side chain. In the β-orcinol *meta*-depsides, this site is blocked by a C_1 substituent and the ester linkage joins at the 5-position, *para* to a phenolic hydroxyl.

The orcinol-type *meta*-depsides are all formed from units with reduced side chains of three or five carbons, and they differ from each other simply by the combination of units esterified and the methylation patterns of the phenolic hydroxyls. Nevertheless, *meta*-depsides seem to be basically more complex than *para*-depsides and tend to be more common in relatively advanced lichen genera. While orsellinic acid and rings with oxidized side chains are still unknown in *meta*-depsides, it is precisely these units that occur as the A ring in nearly all of the orcinol-type depsidones. The depsidones are derived from *para*-depsides by oxidative coupling and cannot be formed by the same mechanism from *meta*-depsides. In the genera *Cladonia* and *Ramalina*, the best known sources of *meta*-depsides, orcinol derivatives with long partially oxidized side chains are completely unknown. *Cladonia* has yielded a single orcinol-type depsidone — a rather unusual one — and *Ramalina* seems to contain no orcinol-type depsidones. No orcinol-type *meta*-depside has ever been found in a species containing an orcinol-type depsidone. Related *meta*- and *para*-depsides with reduced side chains do occur in the same species and in closely related species. For example, the *para*-depside divaricatic acid and the *meta*-depside sekikaic acid are two of the principal constituents of the genus *Anzia*. In the genus *Physcia* these same two compounds are also found in related species.

iv. Depsidones

Most known orcinol-type depsidones have an α- or a β-keto group in the side chain of the first ring. It is well known that this functional group has a strong effect upon the ester linkage between the two rings, since enol lactones form readily. A similar interaction may aid depsidone formation by stabilizing a conformation favorable to cyclization. But in diploicin, variolaric acid, and grayanic acid, the first ring is an orsellinic acid unit lacking any oxygen function that could conceivably affect the cyclization. In this respect, these compounds resemble depsidones in the β-orcinol series.

Oxidative cyclization of depsides to depsidones usually joins the 2-hydroxyl of ring A and the 5'-position of ring B, in spite of the fact that most orcinol depsides also have a free position at 3'. The mechanism of the cyclization reaction may involve intramolecular coupling of a phenoxyl radical and an aryl radical leading to the formation of the ether linkage of the depsidone structure. Olivetoric and physodic acids, sphaerophorin and grayanic acid, and 4-O-demethylbarbatic and hypoprotocetraric acids are depside-depsidone pairs with exactly corresponding structures. Of these three pairs of compounds only olivetoric acid and physodic acid have been found together in individual specimens. 4-O-Demethylbarbatic acid and hypoprotocetraric acid both occur in the genus *Ramalina*, and sphaerophorin and grayanic acid are known only in the widely separated genera *Sphaerophorus* and *Cladonia*.

The orcinol depsidone variolaric acid has an ether linkage to the 3'-position and is a rare exception to the rule of depsidone synthesis in lichens. Variolaric acid is known only in *Ochrolechia*. No other depsidones have been found in this genus and all of the orcinol-type products from it — lecanoric acid, erythrin, and gyrophoric acid — are derived from orsellinic acid. It is probable that the biosynthesis of the depsidone variolaric acid has evolved separately from that of other orcinol-type depsidones. *Ochrolechia* also contains a number of unidentified products, and the structures of these compounds may give some clues to the origin of variolaric acid. The phenolic units in variolaric acid also occur in two dibenzofurans, strepsilin from *Cladonia* and porphyrilic acid from *Haematomma*. Neither of these dibenzofurans shows evidence of a coupling reaction *para* to the methylene of the lactone. Since a *para*-quinoid structure in ring B could favor depsidone formation, the methylene of the benzyl lactone might be involved in an oxidative coupling reaction at the 3'-position in variolaric acid. Although the occurrence of an orcinol-type depsidone in the genus *Ochrolechia* would be exceptional, it would be less surprising if depsidone synthesis in variolaric acid had an evolutionary origin independent from that of other orcinol-type depsidones.

Of the orcinol-type compounds, derivatives of orsellinic acid frequently show variations of structure that are also observed among the β-orcinol-type compounds, but are not yet known for the orcinol-type units with longer side chains. For example, the only chlorinated orcinol-type compounds are methyl 3,5-dichlorolecanorate and diploicin, both derivatives of orsellinic acid. Chlorinated compounds are rather common among the β-orcinol derivatives.

v. Depsone

The depsone, picrolichenic acid, has a methylated 2-hydroxyl, and carbon-carbon coupling links the 1-position of the A ring and the 5-position of the B ring. Picrolichenic acid has been regarded as closely related to depsidones, being formed by a cyclodehydrogenation of 2-O-methylanziaic acid. But depsidones with long and completely reduced side chains, such as the one to which picrolichenic acid would be related, are unknown in lichens. Picrolichenic acid may be more closely related to the dibenzofuran didymic acid. The carbonyl group in ring A of didymic acid is lost during its biosynthesis, but the corresponding carbonyl group can be retained in a quinoid structure in picrolichenic acid since the *para*-hydroxyl is unmethylated. By this analogy, the ester bond in picrolichenic acid would be roughly equivalent to the ether bond in didymic acid, and it would arise by displacement of hydroxyl or coenzyme A from the carbonyl of the A ring. Although picrolichenic acid may exist in entiomeric forms, only the racemic mixture has been isolated. A model of this compound shows that the plane of the benzene ring is perpendicular to the quinoid ring. In the joining of these rings, apparently there is no difference in the energy requirement for reaction from one side of the quinoid ring as opposed to the other. This suggests that during synthesis the ring is not held flat against an enzyme surface that could distinguish a C_5 side chain from a methoxyl substituent.

b. DIARYL ESTERS — β-ORCINOL SERIES

The orcinol-type compounds discussed in the previous section form a closely related series of substances in which changes in the length and oxidation state of the 6-alkyl substituents are major sources of variation among the phenolic units. The compounds synthesized by various combinations of these units show secondary modifications attributable to O-methylation, chlorination, decarboxylation, and lactonization. The β-orcinol compounds may undergo all of the same secondary reactions, but the commonest variation is in the oxidation state of the C_1 substituents at the 3- and 6-positions of the phenolic acid units. Examples of all four possible oxidation states — $-CH_3$, $-CH_2OH$, $-CHO$, and $-COOH$ — occur in the known β-orcinol derivatives, although the frequencies of occurrence of particular oxidation states at the two 3-, 3'-, 6-, and 6'-positions differ.

The C_1 substituents at the 6- and 6'-positions are derived from the methyl of the original acetylcoenzyme A which initiated the formation of the aromatic ring. It is analogous to the 6-methyl substituent of orsellinic acid. The C_1 substituents at the 3- and 3'-positions are "extra" carbons, which are probably added before cyclization to the aromatic ring. Studies on the origin of the "extra" carbons in atranorin and chloroatranorin have shown that formate is incorporated (Yamazaki, Matsuo, and Shibata, 1965), but degradation experiments to determine the position of the radioactive carbons in these compounds have not yet been reported.

Presumably the C_1 substituent at the 3-position of a β-orcinol unit is always added in the fully reduced state, as a methyl group, which may then be oxidized in subsequent steps. This point, which remains to be proven, bears upon the biosynthetic relationships among the various β-orcinol derivatives. At least one unit with either a methyl or an aldehyde at the 3-position and a methyl at the 6-position occurs in all but one of the known β-orcinol derivatives, gangaleoidin being the exception, and both are present in the commonest β-orcinol compound known in lichens, the *para*-depside atranorin.

A simple biosynthetic connection between the 3-methyl and the 3-aldehydic derivatives is suggested by the extremely common occurrence of these two units. It is possible that the 3-methyl derivative is oxidized directly to the aldehyde without a stable hydroxymethyl intermediate by the action of a peroxidase enzyme. Benzaldehyde has been isolated by peroxide oxidation of toluene, and a methyl substituent in mesitol is oxidized to an aldehyde by the action of peroxide and peroxidase. In the latter case, the hydroxymethylene derivative is also converted immediately to the aldehyde (Saunders *et al.*, 1964).

No tridepsides are known in the β-orcinol series. Formation of tridepsides in lichens seems to be entirely restricted to orsellinic acid units. Since the β-orcinol ring is substituted at the 3-position, reactions at this position in orcinol-type compounds have no direct counterparts in the β-orcinol series, which may explain in part the absence of pannaric acid-type dibenzofurans with a β-orcinol substitution pattern. But dibenzofurans like didymic acid and strepsilin do not directly involve the 3-position of either aromatic ring, and still no β-orcinol analog is known. Certainly, if such compounds could be formed in lichens, they would be expected in the Cladoniaceae, which produces didymic acid and strepsilin as well as numerous β-orcinol derivatives.

None of the β-orcinol units represented in the lichen depsides and depsidones has been found in free-living fungi. The most similar compounds in fungi are cyclopaldic acid [25] and cyclopolic acid [26] in *Penicillium cyclopium* var. *album* and quadrilineatin [27] in *Aspergillus quadrilineatus*.

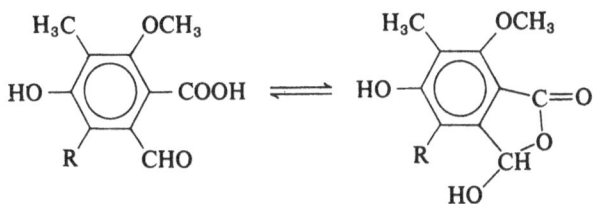

cyclopaldic acid (R = −CHO)   [25]
cyclopolic acid (R = −CH$_2$OH)   [26]

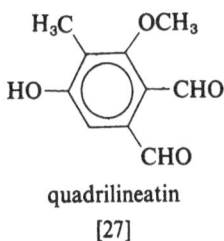

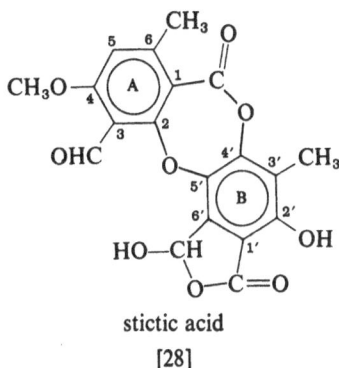

quadrilineatin
[27]

stictic acid
[28]

In the *Aspergillus* compound the 5-position is unsubstituted as in lichen compounds, but the 1-position has an aldehyde rather than a carboxylic acid substituent. Otherwise the structure of quadrilineatin is identical to that of cyclopaldic and cyclopolic acids. Cyclopaldic acid has 3-methyl and 6-aldehydic substituents and resembles the B ring of stictic acid [28] and norstictic acid except in having a C$_1$ substituent at the 5-position and a methylated 2-hydroxyl. Cyclopaldic and cyclopolic acids may not be good starting materials for the A ring of depsides because the carboxylic acid is bound in a cyclic structure. The similar compounds gladiolic acid and dihydrogladiolic acid, known in *Penicillium gladioli*, lack the 4-hydroxyl as well.

i. para-Depsides

Of the 16 units theoretically derivable by changing the oxidation states of the C$_1$ substituents, only eight are represented among the known β-orcinol compounds (Chart III). Six *para-* and three *meta*-depsides are derived from only three β-orcinol units, all of which have the original fully reduced methyl group at the 6-position and either a methyl, an aldehydic, or more rarely a carboxylic acid group at the 3-position. All the *para*-depsides have the same B ring substituted with methyls at both the 3- and the 6-positions. The C_1 substituents of the B rings of *meta*-depsides and depsidones are usually more highly oxidized than those of *para*-depsides. These units may be more susceptible to further oxidations, such as oxidative cyclization to depsidones or attack by hydroxyl or carboxylate radicals which are probably required for *meta*-depside synthesis.

Charts III and IV summarize the chemical relationships among the β-orcinol derivatives. The A and B units are numbered solely upon the basis of the oxidation state of the C_1 substituents at the 3- and 6-positions. Decarboxylated units are indicated by "$(-CO_2)$" after the number of the unit. Oxidation of the 6-methyl substituent is not observed in the A ring of any of the β-orcinol esters, possibly because such oxidation would influence the reactivity of the adjacent carbonyl group which must be esterified. For example, the cyclic structure of the B rings [6 and 7] of stictic acid, norstictic acid, and salazinic acid might prevent that unit from participating as the A ring in esterification.

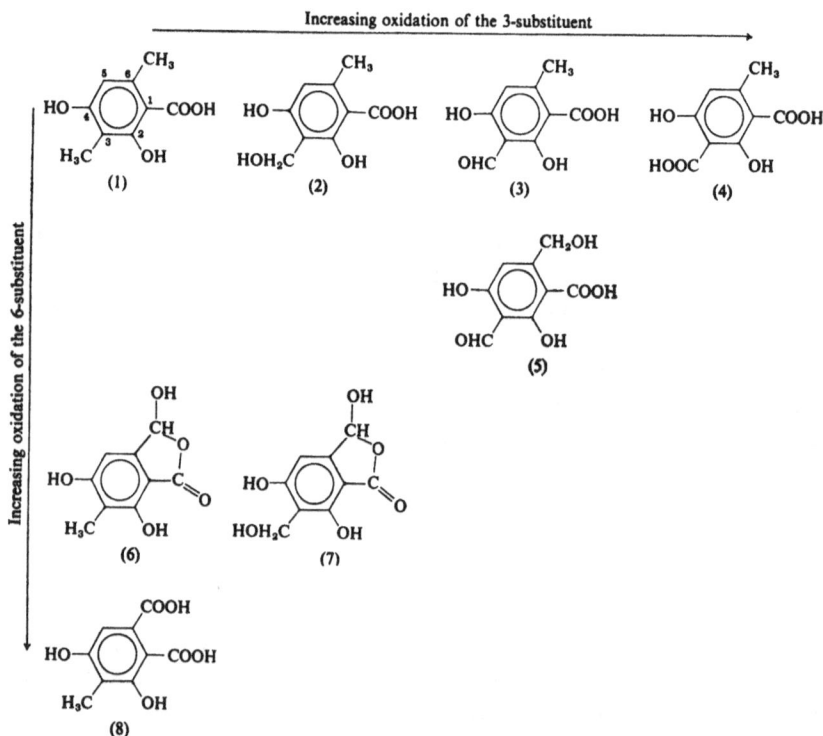

Chart III. β-Orcinol-type phenolic units involved in diaryl esters arranged in an order by oxidation state of the C_1 substituents. Variations arising by O-methylation and chlorination are not considered.

Atranorin may be a very primitive compound. Nuclear chlorination relates it to chloroatranorin and a change in O-methylation pattern could give baeomycesic acid. The latter compound seems to be closely related to squamatic acid with which it occurs in a number of lichen species. These compounds differ only by one oxidation step in the A ring. Baeomycesic acid and squamatic acid frequently occur together in species closely related to ones producing thamnolic acid, a situation possibly to be explained by a loss of the ability to produce the phenolic unit (number 1) with methyls in the 3- and 6-positions. Thamnolic acid is the most common *meta*-depside presently known in lichens. As will be described later, *meta*-depsides may form more readily than *para*-depsides when the B ring is more highly oxidized.

Loss of the ability to oxidize the C_1 substituent at the 3-position of the β-orcinol unit could yield *para*-depsides like 4-O-demethylbarbatic acid. These compounds are all related by O-methylation pattern, the most common being the 4-O-methylated compound, barbatic acid. These compounds have not been found in the most primitive families and probably reach their maximum development in relatively advanced genera like *Parmelia* and *Usnea*.

Phenolic Units of the A and B rings (*Chart III*) A–B	para-*Depsides*	*Depsidones*	meta-*Depsides*
1–1	barbatic acid, diffractaic acid, 4-O-demethylbarbatic acid	hypoprotocetraric acid	
1–1 (− CO_2)		vicanicin	
3–1	atranorin, chloroatranorin, baeomycesic acid	virensic acid	
4–1	squamatic acid		hypothamnolic acid
3–2		fumarprotocetraric acid, protocetraric acid, physodalic acid	
3–3			haemathamnolic acid
3–6		norstictic acid, stictic acid	
4–3			thamnolic acid
3–7		salazinic acid	
3–8 (− CO_2)		psoromic acid	

Chart IV. The distribution of phenolic acid units among the β-orcinol-type diaryl esters.

Obtusatic acid combines a β-orcinol and an orcinol unit. It is classified here with the orcinol compounds because it appears to be related to evernic acid, a compound with which it often occurs in the relatively advanced genus *Ramalina*. Obtusatic acid may not represent a biosynthetic link between the orcinol and β-orcinol compounds but rather a modification of a pre-existing pathway to esterified orsellinic acid compounds such as evernic acid and lecanoric acid.

ii. meta-*Depsides*

Only three *meta*-depsides of the β-orcinol series are known. There have been no biosynthetic studies on any of these compounds. As with the *meta*-depsides of the orcinol series, it is not known if these compounds form by hydroxylation of the B ring followed by esterification or by some other mechanism such as direct attack by a carboxylate radical of the A ring. No hydroxylated rings are known among the β-orcinol derivatives. If the 6-methyl substituent is oxidized to an aldehyde or to a carboxylic acid, then hydrogen-bonding between the new carbonyl and the 4-hydroxyl group would be favored. Hydrogen-bonding of the 4-hydroxyl might account for the fact that such units are not known as the B ring in any of the *para*-depsides or depsidones. In the *meta*-depsides, this hydrogen-bonding could increase the stability of an *ortho*-quinoid structure in addition to the *para*-quinoid structure already possible through the 2-hydroxyl, and electrophilic or radical attack at position 5 would be even more favored. Although these factors may assist the formation of thamnolic acid and haemathamnolic acid, they could not affect the synthesis of hypothamnolic acid which has the same oxidation pattern as the *para*-depside squamatic acid.

All of the known *meta*-depsides of the β-orcinol series are methylated at the 4-hydroxyl. None of these compounds is chlorinated or further modified in any other way.

iii. *Depsidones*

Except in gangaleoidin, the A ring of all β-orcinol depsidones has either a 3-aldehydic substituent like the A ring of the *para*-depside atranorin or a 3-methyl substituent like the B ring of atranorin (Chart IV). The B rings of the depsidones are much more variable, however, and some show oxidation of the 6-methyl substituent, a condition never observed among the phenolic acid units of the known *para*- and *meta*-depsides. If the C_1 substituents of the B ring of a *para*-depside were oxidized, the tendency to form depsidones could increase, either because the ring becomes more susceptible to the oxidative cyclization or because internal assisting mechanisms become possible.

A molecular model of protocetraric acid shows that the hydroxyl of the 3'-hydroxymethyl group can be in close proximity to the carbonyl oxygen of the ester linkage. An interaction here may assist hydrolysis of the ester linkage and explain the red coloration observed with protocetraric acid by treatment with potassium hydroxide followed by saturated aqueous calcium hypochlorite (the taxonomist's "KC" test). This reaction requires the presence of free *meta*-hydroxyl groups, which are present in protocetraric acid after hydrolysis of the depsidone ester linkage. Physodalic acid and fumarprotocetraric acid, benzyl esters of protocetraric acid, do not show this color reaction.

A few depsidones seem very closely related to known depsides. Hypoprotocetraric acid exactly corresponds to the *para*-depside 4-O-demethylbarbatic acid, but these two compounds have not been found together in the same lichen. In the genus *Ramalina* both 4-O-demethylbarbatic and hypoprotocetraric acid are known. *Cladonia*, *Parmelia*, and *Usnea*, which have many species producing barbatic acid and its close derivatives, are not known to synthesize hypoprotocetraric acid. If 4-O-methylation of *para*-depsides can block depsidone formation (W. L. Culberson and Culberson, 1968), then hypoprotocetraric acid could form from 4-O-demethylbarbatic acid, but barbatic acid would not be expected to yield a 4-O-methylated hypoprotocetraric acid.

Virensic acid is a depsidone with an oxidation pattern identical to that of atranorin and baeomycesic acid but lacking O-methyl substituents. Virensic acid is known only in one relatively advanced genus (*Alectoria*) in spite of the fact that the related depsides are extremely common throughout the genera of lichens.

Vicanicin is known only in *Teloschistes*. The C_1 substituents in this compound are all methyl groups as in hypoprotocetraric acid, but vicanicin is decarboxylated, fully chlorinated, and has an O-methyl substituent at the 2'-position. This compound is analogous to the orcinol derivative diploicin which occurs in *Buellia*. Gangaleoidin has the same A ring as diploicin, a fully chlorinated orsellinic acid unit with no O-methyl substituent, but the B ring is anomalous. If the structure of this compound is correct, then it must be derived from a β-orcinol ring in which a C_1 substituent is lost, possibly by oxidation and decarboxylation. Gangaleoidin is known only in *Lecanora gangaleoides*.

Pannarin is another chlorinated depsidone of unusual structure. The A ring is chlorinated at the 5-position like the A ring in vicanicin, but the 4-hydroxyl is methylated. The B ring is most unusual among lichen substances; it appears to have been derived from an orsellinic acid unit by reduction of the carboxylic acid function to an aldehyde. The depsidone linkage occurs at the 3'-position instead of the normal 5'-position. Pannarin may actually have been derived from a β-orcinol ring system by loss of the C_1 substituent at the 3'-position, a possibility that could explain its anomalous depsidone linkage. The structures of pannarin and gangaleoidin should be confirmed, since the postulated biosynthetic pathways to these compounds seem unusual.

iv. Benzyl Ester

One benzyl ester, barbatolic acid, is known in three usneaceous genera producing a number of the usual β-orcinol esters previously described. The A ring of barbatolic acid is like the A ring of atranorin but the B ring is substituted at the 6-position with the hydroxymethyl group involved in the ester linkage. Several lactones formed from a similarly oxidized orsellinic acid are units in orcinol-type compounds. Barbatolic acid is the only 6-hydroxymethyl-substituted unit known among the compounds derived from β-orcinol-type phenolic acids. Its formation could represent a variation, as suggested for *meta*-depside synthesis, related to a reduced activity of the 4-hydroxyl of the B ring that can hydrogen-bond with the 3-aldehyde. Perhaps the ester linkage forms by direct attack upon an unhydroxylated 6-methyl of a unit exactly like the A ring. In that case barbatolic acid would be a biosynthetic alternative to the *meta*-depside haemathamnolic acid.

C. DIBENZOFURANS

Unlike usnic acid, a compound to be discussed later, the phenolic units involved in the production of the true lichen dibenzofurans are derived by the orsellinic acid-type cyclization. These units are potential starting materials for either depsides or dibenzofurans, and in fact are found in both groups of compounds. Didymic acid and a number of *meta*-depsides are formed from the same basic units as the *para*-depside imbricaric acid. The dibenzofurans strepsilin and porphyrilic acid have the same phenolic acid units as the depsidone variolaric acid. Schizopeltic and pannaric acids are derived from orsellinic acid, a common unit in many *para*-depsides and tridepsides. The dibenzofurans appear to form by carbon-carbon coupling and cyclodehydration of two such acetate-polymalonate-derived phenolic acid units. Similar compounds are unknown in nonlichen-forming fungi and appear to be extremely rare in all living systems. Dibenzofuran has been identified in shale oils, petroleum fractions, and coal (Eisen *et al.*, 1964; Hartung and Jewell, 1962; Ōuchi and Imuta, 1963). A tetrahydroxy dimethyl di-*iso*-valeryl dibenzofuran was tentatively identified in extracts of the fruits of the Australian finger cherry, *Rhodomyrtus macro-carpa* (Trippett, 1957). It has been suggested that this toxic constituent may actually be the product of a fungus associated with the fruits.

In the dibenzofuran derivatives, the ether linkage forms by a cyclode-hydration rather than by a cyclodehydrogenation or oxidative coupling, and it is biosynthetically distinct from the ether linkage in depsidones. In fact, the reaction involved in the synthesis of the ether linkage of depsi-dones is more allied to the carbon-carbon coupling reaction in the synthesis of dibenzofurans. Both reactions are oxidative couplings, possibly free-radical in nature, and usually involve the 5-position of one of the phenolic units [29 and 30].

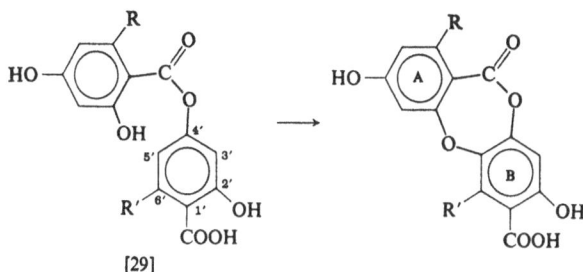

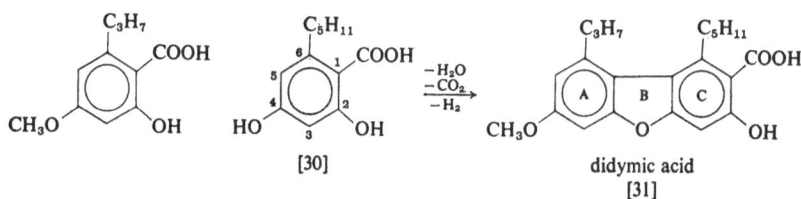

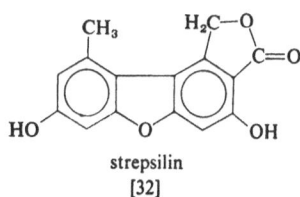

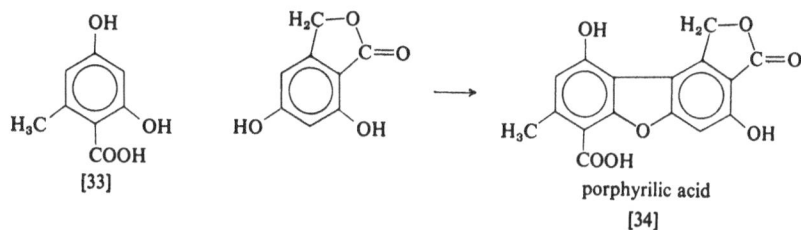

Two types of lichen dibenzofurans can be distinguished by the orientation of the aromatic rings. The first type is presently known only in the Cladoniaceae; the second type is known in the Lecideaceae, Lecanoraceae, Roccellaceae, and the genus *Lepraria*. In all cases, the 5- and 4-positions of one phenolic unit, which becomes ring C in the dibenzofuran structure, are involved in the furan ring. In didymic acid [31] and strepsilin [32] the other phenolic acid unit is decarboxylated so that the positions on the ring cannot be correlated with certainty to those of a phenolic acid precursor. But if decarboxylation occurs at a carbon which also undergoes oxidative coupling, the 1- and 2-positions of the latter unit are involved. The methylation pattern of didymic acid slightly favors this assignment since the 4-hydroxyl is most frequently the one methylated in lichen phenolic acids.

In the second type of dibenzofuran, the carbon-carbon coupling occurs at the 3-position of the first ring [33] instead of the 1-position, and the unit is not decarboxylated. This second orientation corresponds to a 180° rotation about an axis passing through the 2-hydroxyl and the 5-position of the A ring of dibenzofurans of the first type. Strepsilin [32] and porphyrilic acid [34] are related exactly in the sense just described. Pannaric acid differs from porphyrilic acid by lacking the lactone of the second phenolic unit and schizopeltic acid is a methylated pannaric acid. Except in didymic acid, the only phenolic acid units found in the known dibenzofurans are orsellinic acid and a lactone, as in ring C of strepsilin and porphyrilic acid, probably derived from orsellinic acid.

d. USNIC ACIDS

The best-known lichen products are the usnic acids [38 and 39], yellow pigments produced in the upper cortex of many species from phylogenetically widely separated families. These pigments are known only in lichens. A laboratory culture of the fungal component of *Cladonia cristatella* was claimed to synthesize an usnic acid (Castle and Kubsch, 1949), but repeated attempts (e.g., Ahmadjian, 1964) to confirm this report have failed. The usnic acids appear, therefore, to be among the compounds which (*1*) are most characteristic of lichens, (*2*) are apparently not synthesized in cultures of the isolated fungal components, and (*3*) are formed by reactions uniting two or three phenolic units.

Cyclization to the aromatic ring of the methylphloroacetophenone precursor [36] of the usnic acids is basically different from the orsellinic acid cyclizations leading to precursors of dibenzofuran derivatives, but the oxidative coupling and cyclodehydration reactions of pairs of units are similar in the two cases. With the usnic acids, a C-methyl substituent prevents rearomatization of one ring. The structurally isomeric usnic

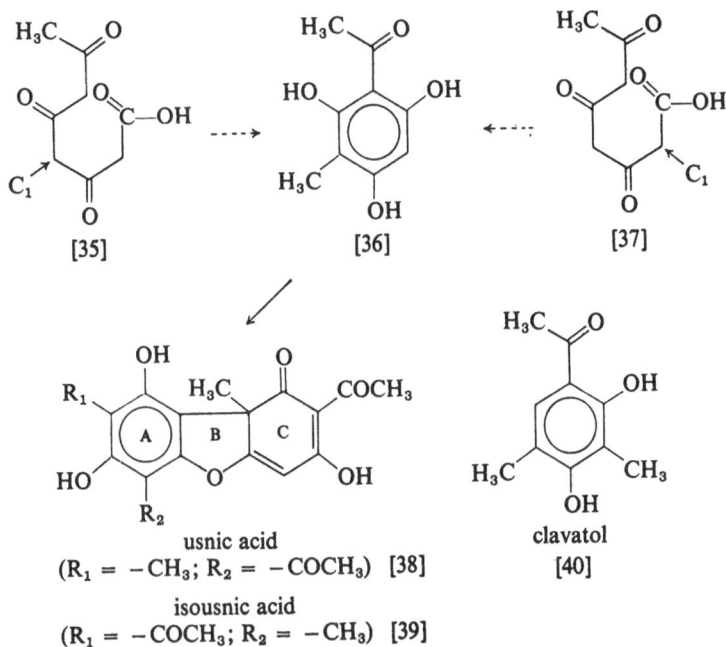

[35]

[36]

[37]

usnic acid
(R_1 = $-CH_3$; R_2 = $-COCH_3$) [38]

isousnic acid
(R_1 = $-COCH_3$; R_2 = $-CH_3$) [39]

clavatol
[40]

acids [38 and 39] are related by the orientation of the A ring about the carbon-carbon bond. In the two types of dibenzofuran derivatives [32 and 34] known in lichens, the second unit also remains fixed but the A ring is reoriented with respect to the ether linkage.

The orientation of the second phenolic unit in the usnic acids resembles that of all the known dibenzofuran derivatives in that the position involved in the carbon-carbon coupling is *meta* to a carbonyl substituent and *ortho* and *para* to O-substituted positions. It differs from the dibenzofurans in having a second *ortho*-hydroxyl substituent instead of an alkyl group. Dibenzofuran derivatives could form from the methylphloroacetaphenone precursor of usnic acid by a change in the orientation of the C ring with respect to the ether linkage, and such an orientation would still allow a *para*-quinoid radical.

The methylphloroacetophenone precursor of the usnic acids is formed by condensation of one acetate and three malonate units with C-methylation prior to cyclization (Taguchi, Sankawa, and Shibata, 1966). Clavatol [40], a product from *Aspergillus clavatus*, is related to the phenolic unit of the usnic acids. But it lacks a hydroxyl at one position *ortho* to the methyl ketone function and it is C-methylated at both *meta* positions. Clavatol could not couple to form an aryl-quinoid ring system resembling the usnic acids because it has no free aryl position *ortho* to a ring hydroxyl.

Until recently, it was thought that two usnic acid pigments were produced in lichens, (+)-usnic acid and its optical isomer. Now structural isomers have also been found in which the carbon-carbon bond remains the same, but the ether linkage forms by dehydrative cyclization at a different hydroxyl group of the A ring (Shibata and Taguchi, 1967). Both optical isomers of these isousnic acids have been found. Dextrorotary isousnic acid accompanies dextrorotatory usnic acid and levorotatory usnic and isousnic acids occur together. Since the structures of usnic acid [38] and isousnic acid [39] are very similar, it is probable that the pairs of compounds rotating polarized light in the same direction also have the same configuration about the asymmetric carbon.

Although the phenolic units of usnic and isousnic acids appear to be identical, there is no proof and they are biosynthetically equivalent. The C-methyl substituent could be attached to the last (as in [35]) or the next to last (as in [37]) malonate unit. In one case the C-methyl is adjacent to the carbon involved in ring closure to the phloroacetophenone unit [36], and in the other it is *para* to this carbon. The position of the hydroxyl attached to the carbon involved in ring closure might determine some aspects of the orientation of this ring on an enzyme surface and the isomer of usnic acid formed by the subsequent coupling reaction.

The biosynthesis of the usnic acids is pictured as an initial carbon-carbon coupling of an *ortho*-quinoid radical of the first unit and a *para*-quinoid radical of the second unit (Chart V). Presumably, the *trans* configuration would be preferred unless some special property of the enzyme favors *cis* coupling. The four *trans* configurations are determined by the relative orientations of substituents on the two rings and by whether ring A joins ring C from the top or from the bottom. In the same way, there are four possible *cis* configurations. For any given orientation of the substituents on ring A, the sign of rotation of the product is determined by whether this ring joins from above or below the plane of ring C. When ring A rearomatizes, the difference between *cis* and *trans* structures vanishes and only four possibilities exist, one pair leading by cyclization to the optical antipodes of usnic acid and the other pair leading to optical antipodes of isousnic acid. The second of these pairs may be derived from the first by rotation of the aromatic ring (ring A) about the bond joining the two rings. Cyclization following at 180° rotation would yield an isomeric usnic acid with the same sign of rotation.

The isousnic acids are known only in *Cladonia*, but extensive surveys for the distribution of these compounds have not yet been reported. In *Cladonia* both (+)- and (−)-usnic acids occur, but in this genus as in others both optical isomers are never found in a single plant. Some genera, such as *Ramalina* and *Evernia*, produce only one optical isomer. It is probable that

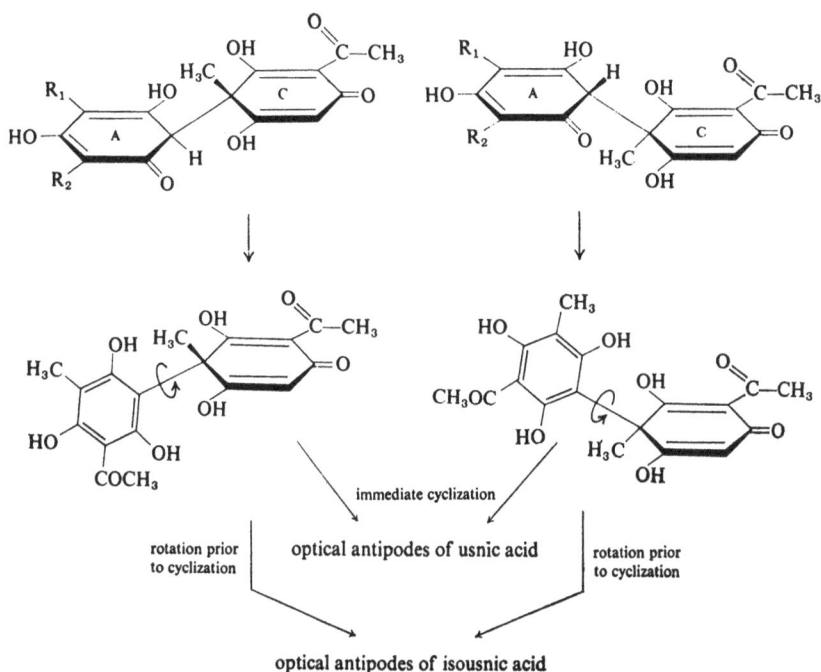

Chart V.

the methylphloroacetophenone unit of usnic acid is a common aromatic lichen product for which mechanisms of coupling and cyclodehydration have arisen evolutionarily many times. The ubiquitous usnic acids have had little use in taxonomy. But when the details of the mechanism for their production become known, these compounds, like so many other phenolic lichen constituents, may be found to have meaningful distributions among the lichens.

e. DIQUINONE

Quinone dimers and polymers are well known in nonlichen-forming fungi. Dibenzoquinones form by a carbon-carbon coupling of two quinone units. The phenolic precursors of the mono-quinones should have *para*-oriented hydroxyls and a position on the ring available for oxidative coupling. Although *para*- and *ortho*-dihydroxy phenols are frequently derived by the shikimic acid pathway, the structures of the fungal dibenzo-quinones oosporein [41] and phoenicin [42] and the related monoquinones and hydroquinones suggest that the monomers form by the acetate-poly-malonate pathway and are subsequently modified by hydroxylation, decarboxylation, and oxidation.

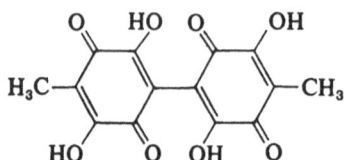

oosporein

[41]

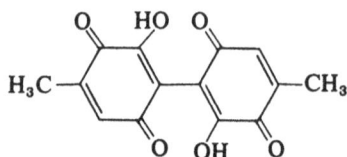

phoenicin

[42]

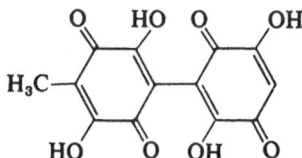

pyxiferin

[43]

Although diquinones derived by intermolecular coupling of two acetate-polymalonate-derived phenolic units might be expected to occur in lichens, only one lichen compound, pyxiferin [43], has been tentatively identified as a diquinone. The apparent rarity of such compounds in lichens may be related to the fact that a phenolic unit correctly substituted for diquinone formation and derived by the acetate-polymalonate pathway requires biosynthetic modifications rarely encountered in lichens.

If pyxiferin were formed by oxidative coupling of acetate-polymalonate-derived units, the first ring would arise by decarboxylation, hydroxylation, and oxidation of orsellinic acids. The second ring would also require oxidative removal of the C_1 substituent. The tentative structure of this compound requires a considerable biosynthetic departure from pathways leading to most of the orsellinic acid derivatives in lichens.

2. Aromatic Products Derived from a Single Polyacetyl Chain

a. Monoaryl Products Related to Phenolic Acids

Although simple monocyclic derivatives are quite common in free-living fungi, they are relatively rare in lichens. The occurrence of these substances always requires careful verification since they could be derived by accidental decomposition of known constituents in the lichens in which they are reported. Methyl β-orsellinate could form from atranorin, orcinol from any orsellinic acid-derived depside or tridepside, and montagnetol from erythrin. But the discovery of an orsellinic acid decarboxylase (Mosbach and Ehrensvärd, 1966) in lichens supports the possibility that orcinol is produced in the plant, and the occurrence of montagnetol seems to be well established (see the numerous references under this compound in Chapter III).

All of the lichen monoaryl derivatives have lost their free carboxylic acid groups either by esterification or by decarboxylation. These compounds might be synthesized in the lichen from the coenzyme A derivatives of the initially formed phenolic acid units or by biological decomposition of depsides. There have been no experiments to show that depsides formed in lichens are later decomposed, but hydrolase enzymes which can attack the ester linkage of some depsides *in vitro* have been reported (Mosbach and Ehrensvärd, 1966).

In fungi the simple monoaryl phenolic acids occur free or modified by having (*1*) variously oxidized C_1 substituents, (*2*) nuclear hydroxyl or chloro substituents, (*3*) C- or O-methyl substituents, (*4*) internal ester groups, and (*5*) fewer than the predicted number of nuclear hydroxyl substituents. These variations, occurring singly or in combination, lead to a large number of important monoaryl fungal products unknown in lichens.

Orsellinic acid and sparassol (= everninic acid) are produced by free-living fungi. These same units must also be synthesized by lichen fungi which produce (*1*) depsides like lecanoric acid, evernic acid, erythrin, obtusatic acid, diploschistesic acid, and methyl 3,5-dichlorolecanorate; (*2*) tridepsides like gyrophoric acid and its several known O-methylated and hydroxylated derivatives; (*3*) the depsidones variolaric acid and grayanic acid; and (*4*) dibenzofurans like porphyrilic acid, strepsilin, and pannaric acid. Monocyclic derivatives of the orcinol type with C_3, C_5, or C_7 side chains occur in lichens only as esterified or coupled products — depsides, depsidones, dibenzofurans, and a depsone. In fungi, related monocyclic compounds occur that are not intermolecularly esterified or coupled. For example, ustic acid [44] (Raistrick and Stickings, 1951; Vora, 1954) and a series of compounds [45] with partially oxidized C_3 side chains from *Penicillium brevicompactum* (Clutterbuck *et al.*, 1932; Oxford and Raistrick, 1933a, 1933b) are similar to the orcinol-type compound divaric acid [46]. Divaric acid is involved in the synthesis of numerous lichen depsides.

Since the simple aromatic units of lichen substances are so rarely found in lichens, these compounds may be synthesized and esterified or coupled without being liberated from the surface of the enzyme or multienzyme complex catalyzing the reactions leading to the phenolic units. The first product of the synthesis of the aromatic units must be an acetyl

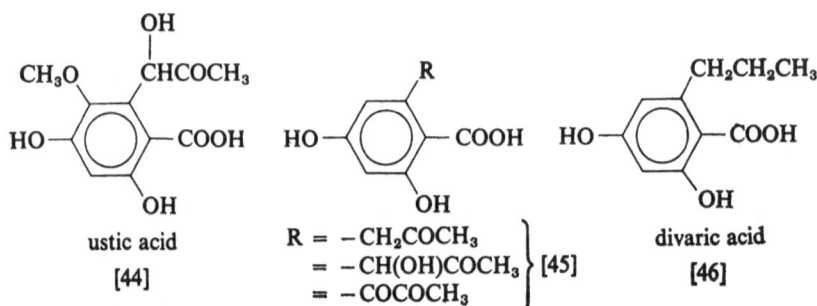

ustic acid

[44]

R = –CH$_2$COCH$_3$
= –CH(OH)COCH$_3$ } [45]
= –COCOCH$_3$

divaric acid

[46]

coenzyme A derivative of the acid (Richards and Hendrickson, 1964, pp. 132–133) and this derivative rather than the free acids might be involved in subsequent transformations leading to either monoaryl or diaryl products. The coupling or esterifying reactions may serve to regulate the concentration of phenolic acid units (see page 23) and simultaneously free the bound enzymes for repetition of the synthesis.

b. CHROMONES

Three chromones are known in lichens. Siphulin has two aromatic rings following the orsellinic acid cyclization pattern. The entire molecule conforms to a structure explainable by linear condensations of acetate and malonate units. Twelve such two-carbon units would be involved, and if the compound is synthesized by this mechanism it requires the longest carbon chain of any known aromatic substance in lichens. The anthraquinone pigment solorinic acid would require ten units for its synthesis.

Siphulin has been found only in the genus *Siphula*, which is best known for its production of β-orcinol *para*- and *meta*-depsides. But siphulin is probably more similar to the anthraquinones, which are also formed by internal condensations of long acetate-polymalonate chains, than it is to the typical lichen esters and dibenzofurans.

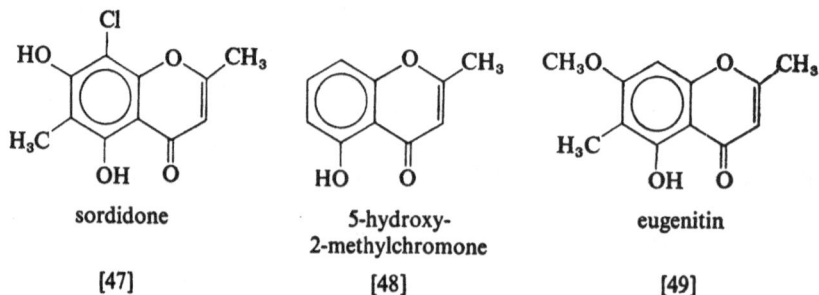

sordidone

[47]

5-hydroxy-
2-methylchromone

[48]

eugenitin

[49]

The chromone sordidone [47] follows the phloroglucinol cyclization pattern rather than the orsellinic acid cyclization pattern indicated for siphulin. In sordidone only five acetate and malonate units are required, a C_1 substituent is added to a ring position, and the product is chlorinated. The same ring system modified by the loss of a hydroxyl occurs as 5-hydroxy-2-methylchromone [48] in the fungus *Daldinia concentrica* (Allport and Bu'Lock, 1960). But an even more closely related chromone is eugenitin [49] from the flowering plant *Eugenia caryophyllata* (Schmid, 1949). Sordidone is a chlorinated O-demethylated derivative of eugenitin. Lepraric acid is a second lichen chromone resembling eugenitin, but it is esterified, at the C_1 substituent of the aromatic ring, to a β-methylglutaconic acid.

C. XANTHONES

Xanthones are known in free-living fungi, and recent studies indicate that they are rather common in lichens too. Although no xanthone derivative is known to occur in both free-living and lichen-forming fungi, the lichen xanthones are so closely related to griseoxanthone C [50] from *Penicillium urticae* (McMaster *et al.*, 1960), that it is probable that the latter compound will also be found in lichens. Xanthone derivatives should be expected in pure cultures of lichen fungi. Unlike the fungal xanthones, many lichen xanthones have one or more nuclear chlorine substituents. Highly chlorinated derivatives may occur in the same species with less chlorinated ones, and the availability of chloride in the environment probably affects the production of these compounds.

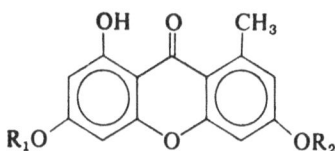

griseoxanthone C (R_1 = $-CH_3$; R_2 = $-H$)   [50]
norlichexanthone (R_1 = $-H$; R_2 = $-H$)   [51]
lichexanthone (R_1 = R_2 = $-CH_3$)   [52]

The fundamental structure of the known lichen xanthones could be derived directly by linear condensation of seven acetate and malonate units with one orsellinic acid-type cyclization and one phloroglucinol-type cyclization. The two rings are joined by a ketonic carbon and by an ether-oxygen arising from cyclodehydration. The first product would be nor-lichexanthone [51], a compound recently discovered in lichens, which would be converted to the commonest lichen xanthone, lichexanthone

[52], and to other known derivatives by O-methylation and chlorination. The hydroxyl which can hydrogen-bond to the ketone oxygen is not methylated in the known xanthones, but both of the remaining hydroxyls are susceptible to O-methylation. All four unsubstituted nuclear positions are available for chlorination. Although there have been no biosynthetic studies on the lichen xanthones, a possible product-precursor relationship for xanthones and anthraquinones (Whalley, 1967) would not seem to apply to the lichen xanthones and anthraquinones.

d. ANTHRAQUINONES

The most widespread anthraquinone in lichens is parietin (often called physcion), which is also known in fungi and higher plants. Parietin probably forms by cyclization of a polyketide chain of eight two-carbon units. The cyclization, involving three carbonyl-methylene condensations, must proceed by an orsellinic acid-type ring closure, and the first product would be an anthrone of endocrocin. This anthrone could be oxidized to endocrocin or oxidized with decarboxylation to emodin. Emodin, recently found in lichens and a common anthraquinone in nature, would yield parietin by O-methylation. Chrysophanol, resembling emodin but lacking one hydroxyl group, is known in both fungi and lichens (Shoji Shibata, personal communication). Skyrin and rugulosin, two anthraquinone dimers common in free-living fungi and derived from emodin, have recently been found in lichens (Shoji Shibata, personal communication). Three additional anthraquinone derivatives may form by stepwise oxidation of the C-methyl substituent of parietin — to a hydroxymethylene substituent in teloschistin, an aldehydic substituent in fallacinal, and a carboxylic acid group in parietinic acid. Several anthraquinone pigments directly related to emodin by nuclear hydroxylation, O-methylation, and chlorination have also recently been found in lichens.

Solorinic acid [53] and norsolorinic acid [54] are known only in the genus *Solorina*. In these compounds, cyclization seems to proceed by the phloroglucinol mechanism and ten two-carbon units are required for the polyacyl chain. A pigment, averantin [55], related to norsolorinic acid but with the ketonic group in the side chain reduced to an alcohol occurs in *Aspergillus versicolor* (Birkinshaw *et al.*, 1966). It was isolated from a culture directly descended from one that several years earlier had yielded the corresponding *trans*-olefinic dehydration product, averythrin [56]. A lower homologue of solorinic acid, rhodocomatulin, is known along with a monomethyl ether derivative in the crinoid *Comatula pectinata* (Sutherland and Wells, 1959).

Rhodocladonic acid [57] is the familiar red pigment in the apothecia of *Cladoniae* of the subsection *Cocciferae*. The substitution pattern of the

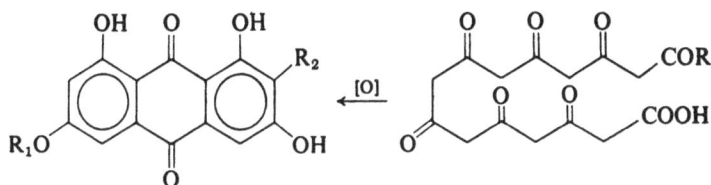

solorinic acid (R_1 = $-CH_3$; R_2 = $-COC_5H_{11}$)   [53]
norsolorinic acid (R_1 = $-H$; R_2 = $-COC_5H_{11}$)   [54]
averantin (R_1 = $-H$; R_2 = $-CHOHC_5H_{11}$)   [55]
averythrin (R_1 = $-H$; R_2 = $-CH{=}CHC_4H_9$)   [56]

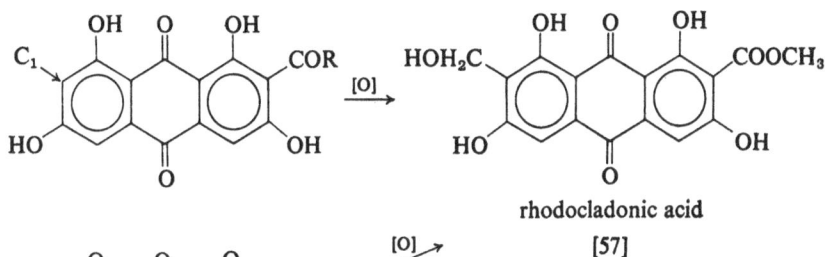

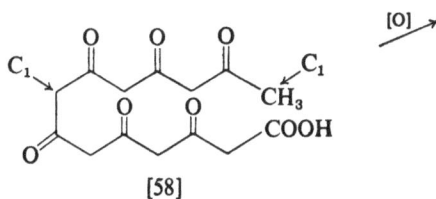

rhodocladonic acid

[57]

[58]

tentative structure of rhodocladonic acid suggests a variation from the previously described biosynthetic routes. The compound could form by a cyclization like that in solorinic acid, but with C-methylation and oxidation to a hydroxymethylene substituent, oxidation of an α-keto side chain to a carboxylic acid, and esterification. Richards and Hendrickson (1964) proposed that rhodocladonic acid could form by a phloroglucinol-type ring closure of a polyacyl chain composed of only seven acetate units [58]. Two C-methylations with oxidation and esterification would then yield rhodocladonic acid.

3. ALIPHATIC PRODUCTS DERIVED FROM A SINGLE POLYACETYL CHAIN — HIGHER ALIPHATIC ACIDS

The fatty acids of lichens show some resemblance to those in nonlichen-forming fungi, but none are identical. Two fungal products are closely related to caperatic acid, a relatively common lichen product. Agaricic acid

[59] is a demethylated higher homologue in the fungus *Fomes officinalis*. Fatty acids like agaricic acid and caperatic acid, which closely resemble citric acid, have been thought to be produced by the condensation of a long-chain aliphatic acid with oxalacetic acid by the same mechanism by

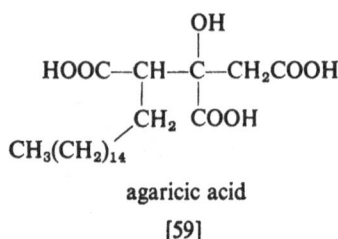

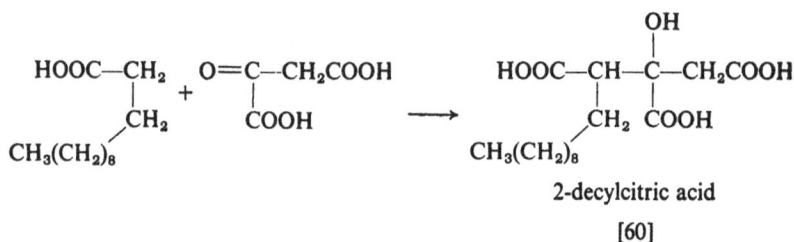

$$
\begin{array}{c}
\text{OH} \\
| \\
\text{HOOC}-\text{CH}-\text{C}-\text{CH}_2\text{COOH} \\
| \quad | \\
\text{CH}_2 \;\; \text{COOH} \\
\diagup \\
\text{CH}_3(\text{CH}_2)_{14}
\end{array}
$$

agaricic acid

[59]

$$
\begin{array}{c}
\text{HOOC}-\text{CH}_2 \quad \text{O}=\text{C}-\text{CH}_2\text{COOH} \\
| \qquad\qquad + \qquad | \\
\text{CH}_2 \qquad\qquad \text{COOH} \\
\diagup \\
\text{CH}_3(\text{CH}_2)_8
\end{array}
\quad\longrightarrow\quad
\begin{array}{c}
\text{OH} \\
| \\
\text{HOOC}-\text{CH}-\text{C}-\text{CH}_2\text{COOH} \\
| \quad | \\
\text{CH}_2 \;\; \text{COOH} \\
\diagup \\
\text{CH}_3(\text{CH}_2)_8
\end{array}
$$

2-decylcitric acid

[60]

which acetic acid enters the tricarboxylic acid cycle. A demethylated lower homologue of caperatic acid, 2-decylcitric acid [60], was recently isolated from a culture of *Penicillium spiculisporum* that had lost the ability to synthesize spiculisporic acid (Måhlén and Gatenbeck, 1968). Måhlén and Gatenbeck (1968) were able to isolate an enzyme that catalyzed the reaction of oxalacetic acid with lauryl coenzyme A, lower homologues of lauryl coenzyme A, and even phenylacetyl coenzyme A. A recent study on the biosynthesis of a seemingly related lichen lactonic acid, (+)-protolichesterinic acid (Bloomer *et al.*, 1968), showed incorporation of [1-^{14}C] acetate in agreement with a synthesis by condensation of a straight-chain fatty acid with a smaller fragment. But attempts to observe incorporation of [1,4-[14]C$_2$] succinic acid failed, even though this compound would be expected to yield labeled oxalacetic acid via the tricarboxylic acid cycle. Severe difficulties were encountered in this study in achieving incorporation of ^{14}C, and further experiments now in progress should clarify this biosynthesis (James Bloomer, personal communication).

Neither the absolute configuration of caperatic acid nor the particular carboxylic acid group which is methylated is yet known. The methylated carboxyl group may be the one thought to be derived from an enzyme-bound palmitic acid precursor. Mass spectral evidence indicates that the methyl ester is located at the corresponding position in a closely related substance, rangiformic acid. Dehydration and *cis*-hydrogenation could, perhaps, convert caperatic acid or a stereoisomer of caperatic acid to rangiformic acid. The synthesis of roccellic acid might proceed in a similar way from a precursor derived from myristic acid rather than palmitic acid. Roccellic acid, rangiformic acid, and norrangiformic acid all have the same absolute configurations.

The lactonic acids known in lichens may form by condensation of enzyme-bound β-hydroxyl or β-keto derivatives of palmitic and myristic acids rather than the fully reduced compounds. For example, (−)-*allo*-protolichesterinic acid could form by condensation of β-keto palmitic acid with a C_3 fragment like oxalacetic acid, followed by (*1*) enol lactonization, (*2*) loss of carbon dioxide and water to form the *exo*-methylene substituent, and (*3*) hydrogenation of the enolic double bond to give the 3,4-*cis* configuration. The steric repulsion of the *cis*-oriented groups might favor a rearrangement of the *exo*-methylenic double bond into the ring to produce (−)-lichesterinic acid, a compound with which it sometimes occurs in *Cetraria ericetorum*.

Although (−)-protolichesterinic acid might be formed by another rearrangement of the double bond in (−)-lichesterinic acid, returning it to the *exo*-methylenic position with inversion at carbon 3, there are no obvious steric or electronic factors that would assist such a reaction. Both the (+)- and (−)-isomers of protolichesterinic acid are known. The (+)-isomer is more common, although (+)-*allo*-protolichesterinic and (+)-lichesterinic acids have not been found. This difference of occurrence of the compounds of the (+)- and (−)-series may reflect a basic divergence in the mechanisms leading to the optical isomers of protolichesterinic acid.

A *cis*-addition of hydrogen to the double bonds in (−)-*allo*-protolichesterinic acid and (+)-protolichesterinic acid could give rise to (−)-nephromopsinic acid and (+)-roccellaric acid respectively. The configurations of (+)-nephrosteranic acid and (+)-nephrosterinic acid are unknown, but these compounds are lower homologues differing by two carbons in the side chains from (+)-protolichesterinic acid and (+)-roccellaric acid respectively. (+)-Nephrosteranic acid and (+)-nephrosterinic acid, which occur together in *Cetraria endocrocea*, may have the same absolute configurations as (+)-protolichesterinic acid and (+)-roccellaric acid since the optical rotations of the homologous pairs of compounds are very similar.

Acarenoic acid and acaranoic acid in *Acarospora oxytona* might form by loss of carbon dioxide and water from the 2- and 3-carbons rather than from the methylene carboxylic acid side chain as suggested for compounds like protolichesterinic acid. Reduction of the endocyclic bond of acarenoic acid would yield acaranoic acid. This mechanism requires that these two acids have the same configuration at the 4-carbon, but their absolute configurations are unknown.

Since both fatty acids and typical aromatic lichen substances can be biosynthesized by acetate-polymalonate pathways, it is not too surprising that some chemical races of lichens are known in which one produces aliphatic products and the other makes aromatic products. Relatively few lichen species are known to accumulate large amounts of both fatty acids and aromatic products. Studies on a strain of *Aspergillus nidulans* capable of producing the depsidones nidulin, nornidulin, and dechloronornidulin showed that at higher temperatures more fatty acids accumulated and at lower temperatures the depsidones formed. This may reflect different temperature optima for some enzymatic reactions, but there is no experimental evidence relating to similar variations in lichens. We cannot say whether *Cetraria islandica* accumulates more fatty acids in the summer and more fumarprotocetraric acid in the winter, for example, or whether individuals in colder regions produce more aromatic products than individuals of the same species in warmer regions.

Long straight-chain aliphatic compounds are poorly known from the lichens. The fatty acids palmitic acid and myristic acid involved in the synthesis of the complex fatty acids described above have not themselves yet been reported from lichens. Some unsaturated acids were found by Wagner and Friedrich (1965) by thin-layer chromatography, but these substances were not identified. Tetrahydroxy fatty acids were examined by Solberg (1960) and have been obtained in extracts of numerous species. However, long-chain alcohols and saturated hydrocarbons, known from nonlichen-forming fungi and thought to be derived from straight-chain fatty acids, are not known from lichens. Similarly acetylenic compounds and straight-chain fungal products with conjugated double and triple bonds and with high antibiotic activity are unknown in lichens. In the nonlichen-forming fungi these constituents frequently occur in small concentrations and in mixtures that are difficult to separate. Many allenes in higher fungi probably form by dehydrogenations and oxidative chain shortening reactions of fatty acids such as crepenynic acid derived from linoleic acid (Bu'Lock, 1966). But even the common fatty acids linoleic acid and oleic acid have been reported only once in lichens (Klima, 1933).

C. Secondary Products of the Mevalonic Acid Pathway — Terpenes, Steroids, Carotenes

All of the acetate-derived aliphatic products described to this point have arisen by head-to-tail condensations leading to unbranched products.

Compounds derived by the mevalonic acid pathway involve condensation of an acetyl coenzyme A unit at the methylenic carbon of an acetoacetyl coenzyme A [61] unit to produce a branched precursor, β-hydroxy-β-methylglutaryl coenzyme A [62], which is then reduced to mevalonic acid [63]. Compounds derived by the mevalonic acid pathway resemble those formed by the acetate-polymalonate pathway by requiring acetate and coenzyme A, but malonate units formed by carboxylation of acetate units are not involved and the mode of condensation differs.

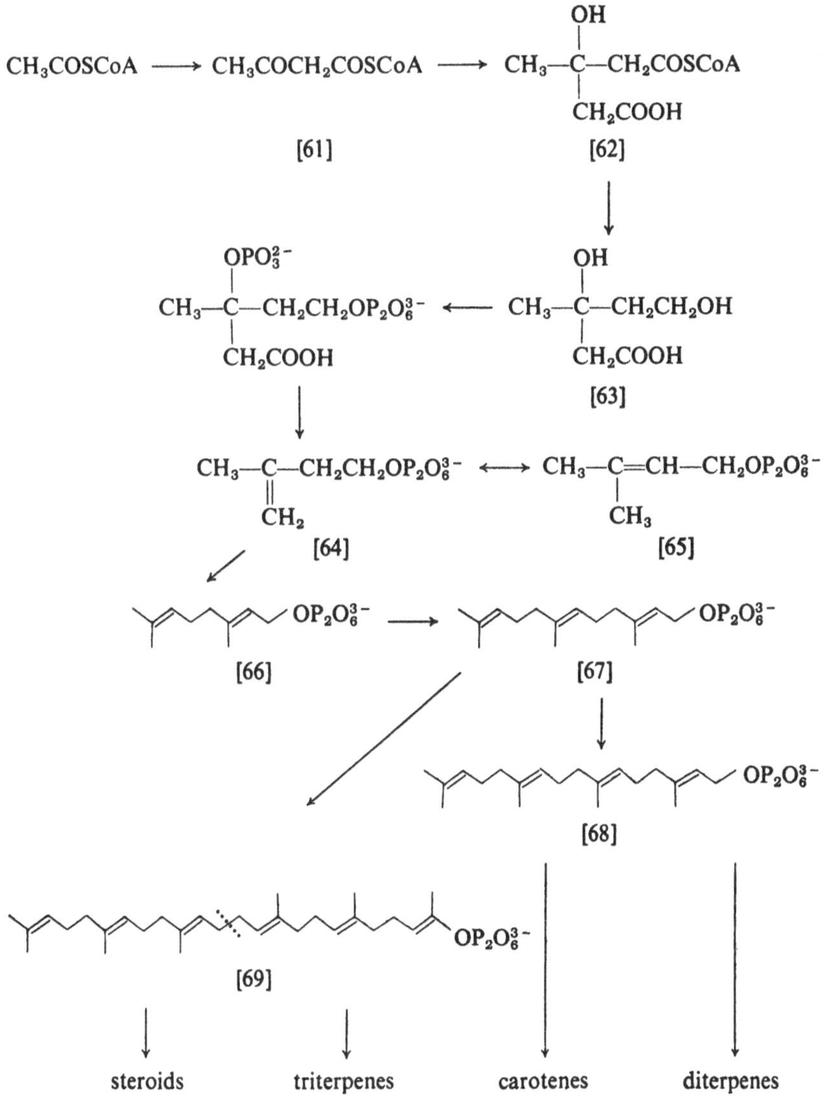

$$CH_3COSCoA \longrightarrow CH_3COCH_2COSCoA \longrightarrow \overset{\displaystyle OH}{\underset{\displaystyle CH_2COOH}{CH_3-\overset{|}{\underset{|}{C}}-CH_2COSCoA}}$$

[61]   [62]

$$\underset{\displaystyle CH_2COOH}{CH_3-\overset{\displaystyle OPO_3^{2-}}{\underset{|}{\overset{|}{C}}}-CH_2CH_2OP_2O_6^{3-}} \longleftarrow \underset{\displaystyle CH_2COOH}{CH_3-\overset{\displaystyle OH}{\underset{|}{\overset{|}{C}}}-CH_2CH_2OH}$$

[63]

$$CH_3-\overset{\displaystyle \parallel}{\underset{\displaystyle CH_2}{C}}-CH_2CH_2OP_2O_6^{3-} \longleftrightarrow CH_3-\overset{|}{\underset{\displaystyle CH_3}{C}}=CH-CH_2OP_2O_6^{3-}$$

[64]   [65]

[66]   [67]

[68]

[69]

steroids   triterpenes   carotenes   diterpenes

Once formed, the primary and tertiary alcohol groups of mevalonic acid are converted to pyrophosphate and phosphate esters. Decarboxylation, assisted by elimination of phosphate anion, gives isopentenyl pyrophosphate [64] which self-condenses with its dimethylallyl pyrophosphate form [65]. The condensation, assisted by elimination of pyrophosphate anion, produces geranyl pyrophosphate [66] which condenses with a third unit of dimethylallyl pyrophosphate producing the key intermediate, farnesyl pyrophosphate [67]. The latter either condenses again with dimethylallyl pyrophosphate to give a 20-carbon intermediate, called geranylgeranyl pyrophosphate [68], or it dimerizes to the 30-carbon intermediate, squalene [69]. The diterpenes are derived from geranyl-geranyl pyrophosphate, an intermediate that also leads to the carotene pigments by dimerization. Squalene is the precursor of triterpenes and, via lanosterol, of the steroids. The key intermediate in the cyclization of squalene to lanosterol is squalene 2,3-oxide [70] and it appears that oxides may be quite generally involved in the biosynthetic reactions leading to cyclic terpenes (Tamelen, 1968).

One diterpene, (−)-16α-hydroxykaurane, is known in lichens, namely in *Ramalina* and *Roccella*, but not in the genera best known for triterpene synthesis. This diterpene is also found in nonlichen-forming fungi and the dehydrated derivative (−)-kaurene is a precursor of gibberellic acid in *Gibberella fujikuroi*.

Internal cyclizations of folded forms of a long squalene-type precursor, usually accompanied by rearrangements within the carbon chain and of hydrogen and methyl groups along the chain, are believed to account for the production of all the natural triterpenes. The all-chair configuration of hydroxyhopanone [71] could form without complex rearrangements. The lichen triterpene zeorin [72] is directly related to hydroxyhopanone [71] since it has the same configuration of the rings as well as of the side chain at C-21. The lichen compound leucotylin [72] also has the all-chair configuration, but an α-orientation of the side chain. All of the triterpenes known in lichens that are not also found in other plants have the basic structure of either zeorin with a β-side chain or leucotylin with an α-side chain.

A long misunderstanding of the configuration of the C-21 side chain in zeorin was recently corrected by X-ray analysis of 16β-O-*p*-bromobenzoyl-6-ketoleucotylin (Yosioka, Nakanishi, and Kitagawa, 1968; Nakanishi, Fujiwara, and Tomita, 1968). The latter compound was proven to have an α-quasi-equatorial orientation of the side chain at C-21. Since the basic skeleton of zeorin was known to differ from that of leucotylin only in the orientation of this side chain, it followed that zeorin must have the β-oriented side chain. This correction revises the stereochemistry of all sub-

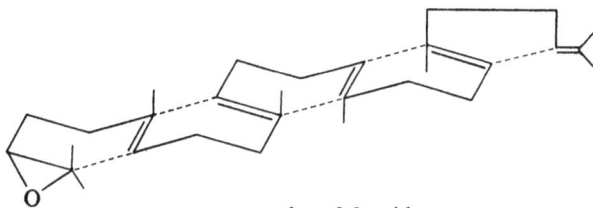

squalene 2,3-oxide
[70]

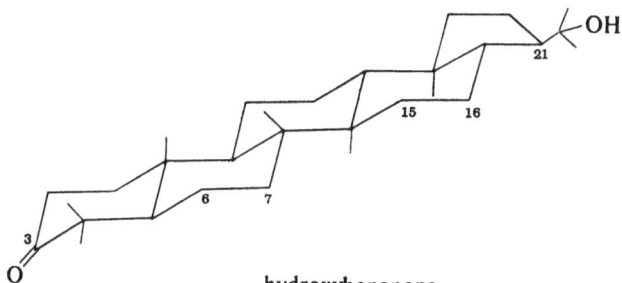

hydroxyhopanone
[71]

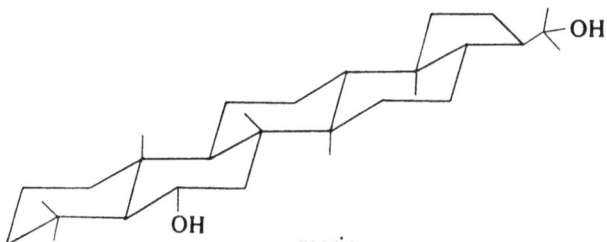

zeorin
[72]

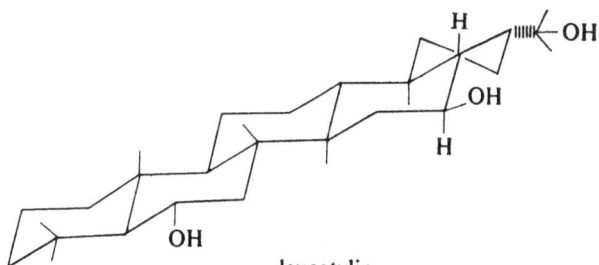

leucotylin
[73]

stances for which the structural proof relates back to the hydroxyhopane system. All of the leucotylin derivatives with an α-oriented side chain at C-21 carry a β-hydroxyl or a β-acetoxyl function at the 16-position, a coincidence undoubtedly related to the biosynthetic mechanism leading to these products.

Zeorin is the best known and apparently the most widespread triterpene in lichens although the simple microchemical test used for its identification may be responsible for this impression. A number of natural derivatives of zeorin and leucotylin show various patterns of hydroxylation, acetoxylation, and oxidation of C-methyl groups. Derivatives that might form by hydrogen or methyl rearrangements of these compounds have not been reported. All of these closely related triterpenes appear to be specific to lichens, but they probably form by relatively direct mechanisms closely resembling or identical to those leading to the triterpenes of other plants.

The triterpenes known from both lichens and other plants are ursolic acid, taraxerene, and friedelin. Ursolic acid is common in flowering plants, especially in the cuticle of fruits and leaves. It is frequently associated with β-sitosterol, a compound also known in lichens, and with oleanolic acid, which may be a precursor to ursolic acid. Taraxerene and friedelin may be related biosynthetically, since both could form from lupeol. These triterpenes, especially friedelin, occur in lichens in small concentration and it is difficult to say whether they are elaborated by the fungus or the alga. They are probably not detected by the usual microchemical procedures and may be more widespread than is presently thought.

Three related sterols are known in lichens. Like triterpenes, these compounds are thought to be derived by cyclizations of a folded squalene-type precursor. Ergosterol and fungisterol require methylation of C-24. β-Sitosterol, the most widely distributed phytosterol, requires an ethyl group at the C-24 position which is apparently acquired by two successive C-methylations. Identification of fungisterol in lichens has not been repeated since Zellner's discovery of this product in *Pseudevernia furfuracea* in 1935.

D. Secondary Products of the Shikimic Acid Pathway — Terphenylquinones and Pulvinic Acid Derivatives

The terphenylquinones and pulvinic acid pigments are well documented examples of secondary lichen products derived by the shikimic acid pathway. Terphenylquinones are also well known in nonlichen-forming Basidiomycetes. Polyporic acid occurs in *Polyporus* and *Peniophora* and thelephoric acid in *Thelephora, Hydnum, Cantharellus,* and *Polystictus.* Pulvinic acid pigments, which were only recently discovered in the nonlichen-forming fungi (Santesson, 1967*b*), are very widespread in lichens and are also produced in cultures of some isolated fungal components of lichens (Mosbach, 1967; E. A. Thomas, 1939).

The genera of lichens best known for their production of pulvinic acid pigments belong to the Stictaceae in which the algal symbionts are frequently blue-green rather than green. Most families of lichens containing only green algae are best known for producing typical acetate-polymalonate-derived aromatic esters and coupled products.

Lichens with green algae generally show a very low nitrogen content unless they grow on high-nitrogen substrates (Massé, 1966*b*). Although nitrogen metabolism in lichens is poorly understood, the small number of secondary substances containing nitrogen and the complete absence of alkaloids in lichens suggest that in many species the supply of available nitrogen to the fungus is severely limited. Some lichens show a strong habitat preference for barnyards, rookeries, or other places high in organic nitrogen. In contrast to the green algae, many blue-green algae can fix nitrogen and many lichens containing blue-green algae show a much enhanced nitrogen content compared to that of their substrates (Massé, 1966*a*).

Although nitrogen fixation by a blue-green algal symbiont must profoundly affect the metabolism of its fungal partner, it now appears that green and blue-green algae may also supply entirely different carbohydrates as the carbon source for the fungal metabolism. *Peltigera polydactyla* contains a blue-green alga (*Nostoc*) that supplies glucose to the fungus (Drew and Smith, 1967*a*, 1967*b*), but the fungus of the lichen *Xanthoria parietina* receives adonitol rather than glucose from its green algal partner (*Trebouxia*) (Bednar and Smith, 1966; Richardson *et al.*, 1967). Thus the lichen fungi associated with blue-green algae may have a nutrition that is high in nitrogen and glucose, both of which are required for the synthesis of the shikimic acid-derived amino acid phenylalanine (Chart VI). It has been demonstrated, moreover, that phenylalanine can serve as a precursor to polyporic acid and to several pulvinic acid derivatives.

The biosynthesis of vulpinic acid was studied by Mosbach (1964*b*), who found that [1-^{14}C]-phenylalanine is efficiently incorporated into vulpinic acid by *Letharia vulpina* and that the labelling pattern in the product is consistent with a biosynthesis by oxidative ring opening of a polyporic acid precursor. An alternative mechanism involving rearrangement of the side chain of the phenylalanine precursor prior to condensation could still not be ruled out. Studies on *Pseudocyphellaria crocata* (Maass *et al.*, 1964) also support the biosynthesis of pulvinic acid from polyporic acid derived

GLUCOSE

pentose phosphate pathway

glycolysis

$CH_2OPO_3^{2-}$

$O=C$ H

CHOH

CH

OH

D-erythrose-
4-phosphate

OPO_3^{2-}

$CH_2=C-COOH$

phosphoenol
pyruvate

HO COOH

O

OH

OH

COOH

HO OH

OH

shikimic acid

COOH

$O=C$

H_2C COOH

OH

OPO_3^{2-}

$CH_2=C-COOH$ +

COOH

$^{2-}O_3PO$ OH

OH

$CH_2COCOOH$

phenylpyruvic
acid

NH_2

$CH_2CHCOOH$

phenylalanine

O OH

HO O

polyporic acid

COOH

OH

O

O

pulvinic acid

O C=O

OH

O

O

calycin

O

O

O

O

pulvinic dilactone

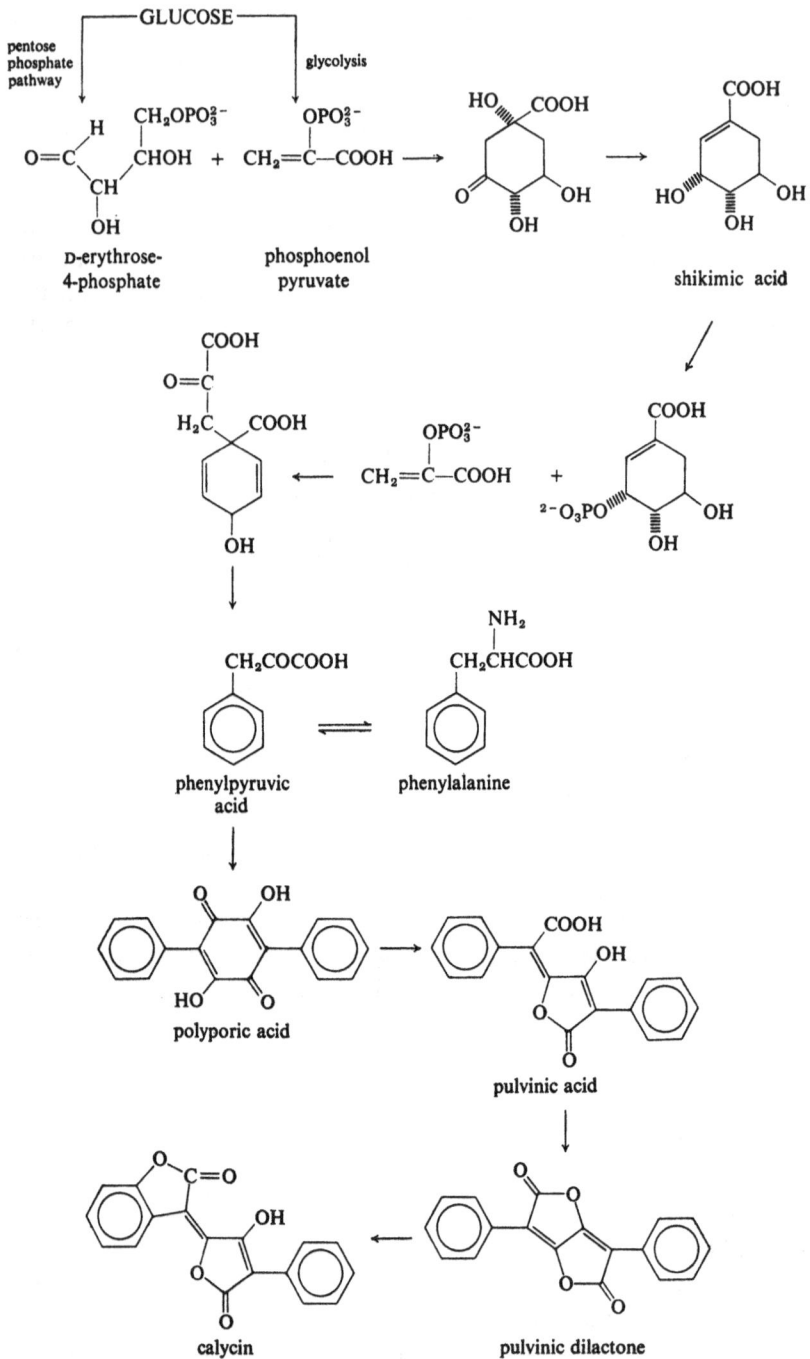

Chart VI. The shikimic acid pathway and the biosynthesis of some pulvinic acid
derivatives.

from phenylalanine. Furthermore, phenylalanine and polyporic acid were incorporated with about the same efficiency in both calycin and pulvinic dilactone while *o*-hydroxyphenylalanine was not incorporated (Maass and Neish, 1967). These results and a pulse-labelling experiment described in the same study suggested that pulvinic dilactone or some immediate product of pulvinic dilactone may be hydroxylated and form calycin by lactonization. A sequence of reactions involved in the conversion of phenyl-alanine to calycin was postulated as follows: phenylalanine → phenyl-pyruvic acid → polyporic acid → pulvinic acid → pulvinic dilactone → calycin (Chart VI).

A terphenylquinone, thelephoric acid, is known in *Lobaria* but polyporic acid and pulvinic acid pigments are not. Thelephoric acid might arise by tetrahydroxylation of polyporic acid and oxidative ring closure to produce two furan rings; but the symmetry of the product suggests that, unlike the case of calycin previously described, hydroxylation may occur before con-densation to the terphenylquinone. The production of thelephoric acid is associated with the rhizines in *Lobaria* while the pulvinic acid pigments are generally produced in the medulla in *Sticta* and *Pseudocyphellaria*.

Not all lichens with blue-green algae contain pulvinic acid derivatives, and many lichens with pulvinic acid pigments or terphenylquinones do not contain nitrogen-fixing algae. Although phenylalanine can serve as a pre-cursor to pulvinic acid derivatives in the species studied, the compounds other than rhizocarpic acid and epanorin do not contain nitrogen in the molecule. They might arise from non-nitrogenous, prephenic acid de-rivatives or by some mechanism allowing a small concentration of nitrogen to be constantly recycled. Perhaps some lichens which produce pulvinic acid compounds or terphenylquinones, but which contain green algae, once had blue-green algal components and subsequently evolved the capacity for symbiosis with a green algal constituent instead. Such an evolutionary modification would not necessarily involve the loss of the more primitive physiological capacity for the production of these pigments in spite of the change in the organic nutrients supplied by the new algal partner.

The diketopiperizine picroroccellin (classified as a primary product in Chapter III) is almost identical in structure to L-phenylalanine anhydride known from the fungus *Penicillium nigricans*. While the same or similar precursors may be involved in the synthesis of picroroccellin and the pulvinic acid derivatives, in the former the nitrogen is maintained in the molecule as peptide linkages while in the latter it is expelled from the ring system. Picroroccellin is known only in the Roccellaceae and no pulvinic acid derivatives are known in that family. Many Roccellaceae inhabit sea-shore boulders visited by birds, but they contain the non-nitrogen-fixing green alga *Trentepohlia*.

E. Elemental Constituents

The elemental constituents of some lichens have been analyzed qualitatively and also quantitatively by spectrophotometric methods. The essential elements in the nutrition are poorly known because it is difficult to grow lichens under controlled laboratory conditions. Probably the functional approach, which deduces the mineral requirements from the elements present in obligatory enzyme systems, will provide the information as a by-product of current studies on lichen enzymes.

Morphological effects induced by different calcium concentrations in the natural substrates have been observed (Schade, 1966a, 1966b), and in one study high calcium oxalate concentrations in 14 species of *Cladonia* and three species of other genera were related to the operation of a carbide factory in the area 50 years earlier (Schade, 1967). Many botanists have noted high concentrations of calcium as precipitates of calcium oxalate on fungal hyphae. Recent studies using the technique of differential thermal analysis in a nitrogen atmosphere showed that the highest concentrations of calcium oxalate appear to be associated with species of lichens confined to limestone and that ubiquitous species contain far less (Mitchell and Birnie, 1966; Syers *et al.*, 1967).

Strontium is chemically very similar to calcium, which it can replace in the fungus *Allomyces arbuscula* if its concentration in the culture medium is greatly increased relative to the concentration of calcium. Accumulation by lichens of ^{90}Sr from radioactive fallout causes a serious problem in northern latitudes where lichens serve as fodder for caribou and reindeer, which in turn are eaten by man. A nonmetabolic mechanism for uptake of ^{90}Sr in lichens was demonstrated recently (Tuominen, 1967).

The chelating ability of *ortho*-hydroxy phenolic acids and esters in lichens as a possible explanation of the etching of glass and the deterioration of rock substrates has received attention. It is not clear whether in nature the ions are chelated by relatively insoluble depsides and depsidones or by low concentrations of simpler phenolic acids which would have a greater solubility in water. Nor is it known whether the particularly high concentrations of minerals accumulated by some lichens have any function or whether they are the unavoidable consequence of repeated wetting and drying in the presence of chelate-forming constituents.

A relatively large number of organic lichen products contain chlorine. The chlorine substituents on aromatic rings of several depsides, depsidones, xanthones, and anthraquinones appear to be analogous to products from certain free-living fungi cultured on media containing varying concentrations of chloride ion. Under these conditions, chlorine-substituted aryl constituents are produced with varying degrees of chlorine substitution

(Birkinshaw, 1965). In lichens, chloroatranorin and atranorin frequently occur together, and recent studies have shown the joint occurrence of xanthone pigments with various chlorination patterns (Johan Santesson, personal communication) and of anthraquinone pigments likewise with various chlorination patterns (Bendz *et al.*, 1967). It seems probable that at least some cases of variation in these chlorination patterns are strongly dependent upon environmental factors.

Nitrogen as a constituent of lichen products has been previously discussed (page 63).

Literature Cited in This Chapter

AHMADJIAN, V. 1964. Further studies on lichenized fungi. *Bryologist* **67**, 87–98.

ALLPORT, D. C. AND J. D. BU'LOCK. 1960. Biosynthetic pathways in *Daldinia concentrica*. *J. Chem. Soc.* **1960**, 654–662.

ASAHINA, Y. 1936–1940. Mikrochemischer Nachweis der Flechtenstoffe I–XI. *J. Japan. Botany* **12**, 516–525, 859–872; **13**, 529–536, 855–861; **14**, 39–44, 244–250, 318–323, 650–659, 767–773; **15**, 467–472; **16**, 185–193.

ASAHINA, Y. AND S. SHIBATA. 1954. Chemistry of Lichen Substances. Japan Society for the Promotion of Science, Tokyo.

BEACH, W. F. AND J. H. RICHARDS. 1963. The structure and biosynthesis of nidulin. *J. Org. Chem.* **28**, 2746–2751.

BEDNAR, T. W. 1963. Physiological Studies on the Isolated Components of the Lichen *Peltigera aphthosa*. Ph.D. Dissertation, University of Wisconsin. (Reference from V. AHMADJIAN, 1967. The Lichen Symbiosis. Blaisdell Publishing Co., Waltham, Massachusetts.)

BEDNAR, T. W. AND D. C. SMITH. 1966. Studies in the physiology of lichens. VI. Preliminary studies of photosynthesis and carbohydrate metabolism of the lichen *Xanthoria aureola*. *New Phytol.* **65**, 211–220.

BENDZ, GERD, GERD BOHMAN, AND J. SANTESSON. 1967. Chemical studies on lichens. 9. Chlorinated anthraquinones from *Nephroma laevigatum*. *Acta Chem. Scand.* **21**, 2889–2890.

BIRKINSHAW, J. H. 1965. Chemical constituents of the fungal cell. Pp. 179–228 *in* G. C. AINSWORTH AND A. S. SUSSMAN [eds.], The Fungi, An Advanced Treatise, Vol. I. The Fungal Cell. Academic Press, New York.

BIRKINSHAW, J. H., J. C. ROBERTS, AND P. ROFFEY. 1966. Studies in mycological chemistry. Part XIX. "Product B" (averantin) [1,3,6,8-tetrahydroxy-2-(1-hydroxyhexyl)anthraquinone], a pigment from *Aspergillus versicolor* (Vuillemin) Tiraboschi. *J. Chem. Soc.* **1966C**, 855–857.

BLOOMER, J. L., W. R. EDER, AND W. F. HOFFMAN. 1968. The biosynthesis of (+)-protolichesterinic acid. *Chem. Commun.* **1968**, 354–355.

BU'LOCK, J. D. 1966. The biogenesis of natural acetylenes. Pp. 79–95 *in* T. SWAIN [ed.], Comparative Phytochemistry. Academic Press, New York.

CASTLE, H. and FLORA KUBSCH. 1949. The production of usnic, didymic, and rhodocladonic acids by the fungal component in the lichen *Cladonia cristatella. Arch. Biochem.* **23**, 158–159.

CLUTTERBUCK, P. W., A. E. OXFORD, H. RAISTRICK, AND G. SMITH. 1932. CLXXI. Studies in the biochemistry of micro-organisms. XXIV. The metabolic products of the *Penicillium brevi-compactum* series. *Biochem. J.* **26**, 1441–1458.

CRUICKSHANK, I. A. M. AND DAWN R. PERRIN. 1964. Pathological function of phenolic compounds in plants. Pp. 511–544 *in* J. B. HARBORNE [ed.], Biochemistry of Phenolic Compounds. Academic Press, New York.

CULBERSON, CHICITA F. AND W. L. CULBERSON. 1958. Age and chemical constituents of individuals of the lichen *Lasallia papulosa. Lloydia* **21**, 189–192.

CULBERSON, W. L. AND CHICITA F. CULBERSON. 1968. The lichen genera *Cetrelia* and *Platismatia* (Parmeliaceae). *Contrib. U.S. Natl. Herb.* **34**, i–iv + 449–558.

CURTIS, R. F., C. H. HASSALL, AND R. K. PIKE. 1968. The biosynthesis of phenols. Part XVI. Further carbon-14 labelling studies relating to the biosynthesis of sulochrin. *J. Chem. Soc.* **1968C**, 1807–1810.

DEAN, F. M., A. D. T. ERNI, AND A. ROBERTSON. 1956. The chemistry of fungi. Part XXVI. Dechloronornidulin. *J. Chem. Soc.* **1956**, 3545–3548.

DEAN, F. M., J. C. ROBERTS, AND A. ROBERTSON. 1954. The chemistry of fungi. Part XXII. Nidulin and nornidulin ("ustin"): chlorine-containing metabolic products of *Aspergillus nidulans. J. Chem. Soc.* **1954**, 1432–1439.

DEAN, F. M., A. ROBERTSON, J. C. ROBERTS, AND K. B. RAPER. 1953. Nidulin and "ustin": two chlorine-containing metabolic products of *Aspergillus nidulans. Nature* **172**, 344.

DREW, E. A. AND D. C. SMITH. 1967a. Studies in the physiology of lichens. VII. The physiology of the *Nostoc* symbiont of *Peltigera polydactyla* compared with cultured and free-living forms. *New Phytol.* **66**, 379–388.

DREW, E. A. AND D. C. SMITH. 1967b. Studies in the physiology of lichens. VIII. Movement of glucose from alga to fungus during photosynthesis in the thallus of *Peltigera polydactyla. New Phytol.* **66**, 389–400.

EISEN, O., E. ARUMEEL, J. EISEN, H. RAUDE, I. PÕDER, O. KIRRET, L. LAHE, AND P. M. VÄNIKVER. 1964. [Determination of individual composition of middle fractions of shale tar by gas chromatography and spectral analysis.] *Eesti NSV Teaduste Akad. Toimetised, Fuusikalis-Mat. ja Tehn. Teaduste Seer.* **13**, 135–142; *Chem. Abstr.* **62**, 2644h (1965).

FEDOSEEV, K. G. AND P. A. YAKIMOV. 1960. [Preparation of usnic acid from lichens. I. A study of conditions of chemical extraction of usnic acid from *Cladonia* lichens.] *Tr. Leningr. Khim. Farmatsevt. Inst.* **9**, 139–149.

FINKLE, B. J. AND V. C. RUNECKLES [eds.]. 1967. Phenolic Compounds and Metabolic Regulation. Appleton-Century-Crofts, New York.

FOLLMANN, G. 1960. Die Durchlässigkeitseigenschaften der Protoplasten von Phycobionten aus *Cladonia furcata* (Huds.) Schrad. *Naturwissenschaften* **47**, 405–406.

FOLLMANN, G. AND VILMA VILLAGRÁN. 1965. Flechtenstoffe und Zellpermeabilität. *Z. Naturforsch.* **20b**, 723.

GATENBECK, S. 1968. On the biosynthesis of sulochrin and geodin. P. 114 *in* Abstracts of Papers Presented at the 5th International Symposium on the Chemistry of Natural Products, London, 8–13 July, 1968.

HALE, M. E., JR. 1967. The Biology of Lichens. Edward Arnold Ltd., London.

HALE, M. E., JR. 1968. A synopsis of the lichen genus *Pseudevernia*. *Bryologist* **71**, 1–11.

HARTUNG, G. K. AND D. M. JEWELL. 1962. Carbazoles, phenazines and dibenzofuran in petroleum products; methods of isolation, separation and determination. *Anal. Chim. Acta* **26**, 514–527.

HESS, D. 1959. Untersuchungen über die Bildung von Phenolkörpern durch isolierte Flechtenpilze. *Z. Naturforsch*, **14b**, 345–347.

HOGEBOOM, G. H. AND L. C. CRAIG. 1946. Identification by distribution studies. VI. Isolation of antibiotic principles from *Aspergillus ustus*. *J. Biol. Chem.* **162**, 363–368.

HUNECK, S. 1968. Lichen substances. Pp. 223–346 *in* L. REINHOLD AND Y. LIWSCHITZ [eds.], Progress in Phytochemistry, Volume 1. John Wiley & Sons, London.

HUNECK, S. AND R. TÜMMLER. 1965. Flechteninhaltsstoffe, XII. Die Struktur von Peltigerin. *Ann. Chem.* **685**, 128–133.

KLIMA, J. 1933. Zur Chemie der Flechten II. *Alectoria ochroleuca* Ehrh. *Monatsh. Chem.* **62**, 209–213.

LAAKSO, P. V. AND MARGIT GUSTAFSSON. 1952. The colorimetric determination of usnic acid in the lichen *Cladonia alpestris*. *Suomen Kemistilehti* **25B** (2), 7–10.

LAZAREV, N. V. AND V. P. SAVICZ. 1957. [The New Antibiotic Binan, or the Sodium Salt of Usnic Acid (Botanical and Medical Investigations).] Akademiia Nauk SSSR, Botanicheskii Institut im. V. L. Komarova, Moscow.

LEWIS. D. H. AND D. C. SMITH. 1967a. Sugar alcohols (polyols) in fungi and green plants. I. Distribution, physiology and metabolism. *New Phytol.* **66**, 143–184.

LEWIS, D. H. AND D. C. SMITH. 1967b. Sugar alcohols (polyols) in fungi and green plants. II. Methods of detection and quantitative estimation in plant extracts. *New Phytol.* **66**, 185–204.

LIGHT, R. 1968. Biosynthesis of 6-methylsalicylic acid. P. 108 *in* Abstracts of Papers Published at the 5th International Symposium on the Chemistry of Natural Products, London, 8–13 July, 1968.

MAASS, W. S. G. AND A. C. NEISH. 1967. Lichen substances. II. Biosynthesis of calycin and pulvinic dilactone by the lichen, *Pseudocyphellaria crocata*. *Can. J. Botany* **45**, 59–72.

MAASS, W. S. G., G. H. N. TOWERS, AND A. C. NEISH. 1964. Flechten-stoffe: I. Untersuchungen zur Biogenese des Pulvinsäureanhydrids. *Ber. Deut. Botan. Ges.* **77**, 157–161.

MÅHLÉN, A. AND S. GATENBECK. 1968. (+)- And (−)-decylcitric acids from *Penicillium spiculisporum*. Identification and enzymic synthesis. Pp. 115–116 *in* Abstracts of Papers Presented at the 5ᵗʰ International Symposium on the Chemistry of Natural Products, London, 8–13 July, 1968.

MASSÉ, L. 1966a. Étude comparée des teneurs en azote total des lichens et de leur substrat: les espèces à gonidies Cyanophycées. *Compt. Rend.,* Sér. D, **263**, 781–784.

MASSÉ, L. 1966b. Étude comparée des teneurs en azote total des lichens et de leur substrat: les espèces "ornithocoprophiles." *Compt. Rend.,* Sér. D, **262**, 1721–1724.

McMASTER, W. J., A. I. SCOTT, AND S. TRIPPETT. 1960. Metabolic products of *Penicillium patulum*. *J. Chem. Soc.* **1960**, 4628–4631.

MĚRKA, V. 1951. [Changes in buffering action and gyrophoric acid content in the lichen *Umbilicaria pustulata* by dying off.] *Spisy Vydávané Přírodovědeckou Fak. Masary. Univ.* **327**, 97–119.

MISCONI, L. Y. AND C. E. STICKINGS. 1968. The biosynthesis of questin and (+)-bisdechlorogeodin in *Penicillium frequentans*. Pp. 184–185 *in* Abstracts of Papers Presented at the 5ᵗʰ International Symposium on the Chemistry of Natural Products, London, 8–13 July, 1968.

MITCHELL, B. D. AND A. C. BIRNIE. 1966. The thermal analysis of lichens growing on limestone. *Analyst* **91**, 783–789.

MOSBACH, K. 1964a. On the biosynthesis of lichen substances. Part 1. The depside gyrophoric acid. *Acta Chem. Scand.* **18**, 329–334.

MOSBACH, K. 1964b. On the biosynthesis of lichen substances. Part 2. The pulvic acid derivative vulpinic acid. *Biochem. Biophys. Res. Commun.* **17**, 363–367.

MOSBACH, K. 1967. On the biosynthesis of lichen substances. Part 4. The formation of pulvic acid derivatives by isolated lichen fungi. *Acta Chem. Scand.* **21**, 2331–2334.

MOSBACH, K. AND URSULA EHRENSVÄRD. 1966. Studies on lichen enzymes. Part I. Preparation and properties of a depside hydrolysing esterase and of orsellinic acid decarboxylase. *Biochem. Biophys. Res. Commun.* **22**, 145–150.

MOSBACH, K. AND GUNILLA JACKOBSSON. 1968. The biosynthesis of aromatic lichen compounds in cell-free systems. P. 109 *in* Abstracts of Papers Presented at the 5ᵗʰ International Symposium on the Chemistry of Natural Products, London, 8–13 July, 1968.

NAKANISHI, T., T. FUJIWARA, AND K. TOMITA. 1968. The crystal structure of 16β-O-p-bromobenzoate of 6-keto-leucotylin. *Tetrahedron Letters* **1968**, 1491–1495.

ŌUCHI, K. AND K. IMUTA. 1963. The analysis of benzene extracts of Yūbari coal. II — Analysis by gas chromatography. *Fuel* **42**, 445–456.

OXFORD, A. E. AND H. RAISTRICK. 1933a. LXXXV. Studies in the biochemistry of micro-organisms. XXX. The molecular constitution of the metabolic products of *Penicillium brevi-compactum* Dierckx and related species. *Biochem. J.* **27**, 634–653.

OXFORD, A. E. AND H. RAISTRICK. 1933*b*. CXCIX. Studies in the biochemistry of micro-organisms. XXXIV. A note on the mechanism of the production of phenolic acids from glucose by *Penicillium brevi-compactum* Dierckx. *Biochem. J.* **27**, 1473–1478.

PEREZ-LLANO, G. A. 1944. Lichens: their biological and economical significance. *Botan. Rev.* **10**, 1–65.

RAISTRICK, H. AND C. E. STICKINGS. 1951. Studies in the biochemistry of micro-organisms. 82. Ustic acid, a metabolic product of *Aspergillus ustus* (Bainier) Thom & Church. *Biochem. J.* **48**, 53–66.

RAO, D. N. AND F. LeBLANC. 1965. A possible role of atranorin in the lichen thallus. *Bryologist* **68**, 284–289.

RICHARDS, J. H. AND J. B. HENDRICKSON. 1964. The biosynthesis of steroids, terpenes, and acetogenins. W. A. Benjamin, Inc., New York.

RICHARDSON, D. H. S., D. C. SMITH, AND D. H. LEWIS. 1967. Carbohydrate movement between the symbionts of lichens. *Nature* **214**, 879–882.

SANTESSON, J. 1967*a*. Chemical studies on lichens — III. The pigments of *Thelocarpon epibolum*, *T. laureri* and *Ahlesia lichenicola*. *Phytochemistry* **6**, 685–686.

SANTESSON, J. 1967*b*. Chemical studies on lichens. 4. Thin layer chromatography of lichen substances. *Acta Chem. Scand.* **21**, 1162–1172.

SAUNDERS, B. C., A. G. HOLMES-SIEDLE, AND B. P. STARK. 1964. Peroxidase. Butterworth Inc., Washington.

SCHADE, A. 1966*a*. Über kalkanzeigende Flechten von Spitzbergen. *Ber. Deut. Botan. Ges.* **79**, 463–473.

SCHADE, A. 1966*b*. Über die Artberechtigung der *Cladonia subrangiformis* Sandst. sowie das Auftreten von Calciumoxalat-Exkreten bei ihr und einigen anderen Flechten. *Nova Hedwigia* **11**, 285–308.

SCHADE, A. 1967. Über das Vorkommen von Calciumoxalat-Exkreten in Bodenflechten der Kiefern-Heidewälder um Schwarze Pumpe (NL) und seine Ursache. *Abhandl. Ber. Naturkundemus. Görlitz* **42** (8), 1–20.

SCHMID, H. 1949. Über die Inhaltstoffe von *Eugenia caryphyllata* (L.) Thunbg. III. Isolierung und Konstitution des Eugenitins. *Helv. Chim. Acta* **32**, 813–820.

SCOTT, A. I. 1965. Oxidative coupling of phenolic compounds. *Quart. Rev.* (London) **1965**, 1–35.

SHIBATA, S., T. FURUYA, AND H. IIZUKA. 1965. Gas-liquid chromatography of lichen substances. I. Studies on zeorin. *Chem. Pharm. Bull.* (Tokyo) **13**, 1254–1257.

SHIBATA, S., S. NATORI, AND S. UDAGAWA. 1964. List of Fungal Products. Charles C Thomas, Springfield.

SHIBATA, S. AND H. TAGUCHI. 1967. Occurrence of isousnic acid in lichens with reference to isodihydrousnic acid derived from dihydrousnic acid. *Tetrahedron Letters* **1967**, 4867–4871.

SOLBERG, Y. J. 1960. Studies on the chemistry of lichens. III. Long-chain tetrahydroxy fatty acids from some Norwegian lichens. *Acta Chem. Scand.* **14**, 2152–2160.

SUTHERLAND, M. D. AND J. W. WELLS. 1959. Anthraquinone pigments from the crinoid *Comatula pectinata. Chem. Ind.* (London) **1959**, 291–292.

SYERS, J. K., A. C. BIRNIE, AND B. D. MITCHELL. 1967. The calcium oxalate content of some lichens growing on limestone. *Lichenologist* **3**, 409–414.

TAGUCHI, H., U. SANKAWA, AND S. SHIBATA. 1966. Biosynthesis of usnic acid in lichens. *Tetrahedron Letters* **1966**, 5211–5214.

TAMELEN, E. E. VAN. 1968. Bioorganic chemistry: sterols and acyclic terpene terminal epoxides. *Accounts Chem. Res.* **1**, 111–120.

THOMAS, E. A. 1936. Die Spezifizität des Parietin als Flechtenstoff. *Ber. Schweiz. Botan. Ges.* **45**, 191–197.

THOMAS, E. A. 1939. Über die Biologie von Flechtenbildnern. *Beitr. Kryptogamenflora Schweiz* **9** (1), 1–208.

THOMAS, R. 1961. Studies in the biosynthesis of fungal metabolites. 2. The biosynthesis of alternariol and its relation to other fungal phenols. *Biochem. J.* **78**, 748–758.

TRIPPETT, S. 1957. Toxic constituents of the Australian finger cherry, *Rhodomyrtus macrocarpa* Benth. *J. Chem. Soc.* **1957**, 414–419.

TUOMINEN, Y. 1967. Studies on the strontium uptake of the *Cladonia alpestris* thallus. *Ann. Botan. Fennica* **4**, 1–28.

VORA, V. C. 1954. Metabolic products of *Paecilomyces victoriae* V. Szilvinyi. *J. Sci. Ind. Res.* (India) **13B**, 842–844.

WACHTMEISTER, C. A. 1959. Flechtensäuren. Pp. 135–141 *in* H. F. LINSKENS [ed.], Papierchromatographie in der Botanik. Zweite, erweiterte Auflage. Springer-Verlag, Berlin.

WAGNER, H. AND H. FRIEDRICH. 1965. Über die ungesattigten Fettsäuren von Moosen, Bärlappgewächsen und Flechten. *Naturwissenschaften* **52**, 303.

WHALLEY, B. W. 1967. The biosynthesis of fungal metabolites. Pp. 1025–1063 *in* P. BERNFELD [ed.], Biogenesis of Natural Compounds. Pergamon Press, Oxford.

WILHELMSEN, J. B. 1959. Chlorophylls in the lichens *Peltigera, Parmelia,* and *Xanthoria. Botan. Tidsskr.* **55**, 30–36.

YAMAZAKI, M., M. MATSUO, AND S. SHIBATA. 1965. Biosynthesis of lichen depsides, lecanoric acid and atranorin. *Chem. Pharm. Bull.* (Tokyo) **13**, 1015–1017.

YOSIOKA, I., T. NAKANISHI, AND I. KITAGAWA. 1968. On the stereostructures of zeorin and leucotylin. *Tetrahedron Letters* **1968**, 1485–1490.

ZELLNER, J. 1935. Zur Chemie der Flechten IV. *Gyrophora Dillenii* (Tuck.) Müll. Arg. und *Parmelia furfuracea* L. *Monatsh. Chem.* **66**, 81–86.

ZOPF, W. 1907. Die Flechtenstoffe in chemischer, botanischer, pharmakologischer und technischer Beziehung. Gustav Fischer, Jena.

Chapter III

Chemical Guide to Lichen Products

Primary Metabolic Substances, Polysaccharides, Elemental Constituents, etc.

Polyols, Saccharides, and Related Substances

POLYOLS

adonitol (= ribitol)
$C_5H_{12}O_5$
m.p. 102° (water), prisms; [α] 0°.
Also known from algae and higher plants.

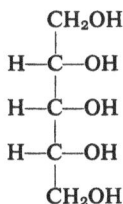

CH₂OH
|
H—C—OH
|
H—C—OH
|
H—C—OH
|
CH₂OH

BEDNAR, T. W. AND D. C. SMITH, *New Phytol.* **65**, 211–220 (1966). A pentitol, provisionally identified as adonitol, found to be the first product accumulating during photosynthesis in the *Trebouxia*-containing lichen *Xanthoria aureola*.

LEWIS, D. H. AND D. C. SMITH, *New Phytol.* **66**, 143–184 (1967). Review on occurrence, physiological role, and metabolism.

LEWIS, D. H. AND D. C. SMITH, *New Phytol.* **66**, 185–204 (1967). Extensive review and discussion of qualitative and quantitative analytical methods.

RICHARDSON, D. H. S., D. C. SMITH, AND D. H. LEWIS, *Nature* **214**, 879–882 (1967). Adonitol produced by the algal partner and converted to mannitol and arabitol by the fungus; in *Lobaria amplissima*, *L. laetevirens*, and *L. pulmonaria*.

73

D-arabitol

$C_5H_{12}O_5$

m.p. 102–103° (ethanol-acetone), prisms; $[\alpha]_D^{20}$ +7.8° (satd. $Na_2B_4O_7$ solution); $[\alpha]_D^{29}$ +11.3° (c = 1.13, satd. $Na_2B_4O_7$ solution); $[\alpha]_D^{20}$ +12° (c = 2.0, satd. $Na_2B_4O_7$ solution).

Also known from *Boletus bovinus* and *Fistulina hepatica*.

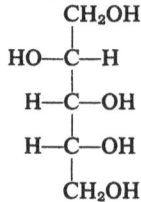

$$
\begin{array}{c}
CH_2OH \\
| \\
HO-C-H \\
| \\
H-C-OH \\
| \\
H-C-OH \\
| \\
CH_2OH
\end{array}
$$

AGARWAL, S. C., K. AGHORAMURTHY, K. G. SARMA, AND T. R. SESHADRI, *J. Sci. Ind. Res.* (India) **20B**, 613–615 (1961). From *Lobaria isidiosa*.

AGHORAMURTHY, K., K. G. SARMA, AND T. R. SESHADRI, *J. Sci. Ind. Res.* (India) **20B**, 166–168 (1961). From *Stereocaulon foliolosum*, *S. tomentosum*, and *Thamnolia vermicularis*.

ASAHINA, Y. AND M. YANAGITA, *Chem. Ber.* **67**, 799–803 (1934). From *Lobaria pulmonaria* and *Ramalina crassa*.

BREADEN, T. W., J. KEANE, AND T. J. NOLAN, *Sci. Proc. Roy. Dublin Soc.* **23**, 6–9 (1942). From *Cladonia impexa*.

BRINER, G. P., G. E. GREAM, AND N. V. RIGGS, *Australian J. Chem.* **13**, 277–284 (1960). From *Ramalina fraxinea*.

CULBERSON, CHICITA F., *Phytochemistry* **4**, 951–961 (1965). From *Ramalina druidarum*.

CURD, F. H. AND A. ROBERTSON, *J. Chem. Soc.* **1935**, 1379–1381 (1935). From *Ramalina curnowii*.

DHAR, M. L., S. NEELAKANTAN, S. RAMANUJAM, AND T. R. SESHADRI, *J. Sci. Ind. Res.* (India) **18B**, 111–113 (1959). From *Usnea longissima*.

HARPER, S. H. AND R. M. LETCHER, *Proc. Trans. Rhodesian Sci. Assoc.* **51**, 156–184 (1966). In *Dermatiscum thunbergii*, *Parmelia dilatata*, *P. gossweileri*, *Pertusaria* sp., and *Usnea implicita*, paper chromatography.

LEWIS, D. H. AND D. C. SMITH, *New Phytol.* **66**, 143–184 (1967). Review on occurrence, physiological role, and metabolism.

LEWIS, D. H. and D. C. SMITH, *New Phytol.* **66**, 185–204 (1967). Extensive review and discussion of qualitative and quantitative analytical methods.

LINDBERG, B., *Acta Chem. Scand.* **9**, 917–919 (1955). From *Roccella fuciformis*, *R. hypomecha*, and *R. linearis*.

LINDBERG, B. AND H. MEIER, *Acta Chem. Scand.* **16**, 543–547 (1962). From *Siphula ceratites*.

LINDBERG, B., A. MISIORNY, AND C. A. WACHTMEISTER, *Acta Chem. Scand.* **7**, 591–595 (1953). Found in a number of Discomycete lichens, paper chromatography.

LINDBERG, B., B. G. SILVANDER, AND C. A. WACHTMEISTER, *Acta Chem. Scand.* **17**, 1348–1350 (1963). From *Peltigera aphthosa*.

LINDBERG, B. AND B. WICKBERG, *Acta Chem. Scand.* **7**, 140–142 (1953). From *Lasallia pustulata*, column chromatography.

MITTAL, O. P., S. NEELAKANTAN, AND T. R. SESHADRI, *J. Sci. Ind. Res.* (India) **11B**, 386–387 (1952). From *Ramalina calicaris* and *R. sinensis*.

NOLAN, T. J. AND J. KEANE, *Nature* **132**, 281 (1933). From *Lobaria pulmonaria*.

NOLAN, T. J. AND J. KEANE, *Sci. Proc. Roy. Dublin Soc.* **22**, 199–209 (1940). From *Lecanora gangaleoides*.

PUEYO, G., *Rev. Bryol. Lichénol.* **29**, 124–129 (1960). From *Parmelia caperata*.

PUEYO, G., *Bull. Centre Études Rech. Sci. Biarritz* **5**, 97–101 (1964). In *Xanthoria parietina*, paper chromatography.

PUEYO, G., *Bull. Centre Études Rech. Sci. Biarritz* **5**, 103–107 (1964). From *Xanthoria parietina*, cellulose-powder column chromatography.

PUEYO, G., *Rev. Bryol. Lichénol.* **33**, 592–594 (1964–1965). From *Evernia prunastri*, column chromatography.

PUEYO, G., *Bryologist* **68**, 334–336 (1965). From *Cladonia rangiferina*, column chromatography.

RAO, P. S., K. G. SARMA, AND T. R. SESHADRI, *Current Sci.* (India) **34**, 9–11 (1965). From *Lobaria retigera*.

RAO, P. S., K. G. SARMA, AND T. R. SESHADRI, *Current Sci.* (India) **35**, 147–148 (1966). From *Lobaria isidiosa* var. *subisidiosa* and *L. subretigera*.

SARMA, K. G. AND S. HUNECK, *Pharmazie*, in press. From *Anaptychia neoleucomelaena*.

SOLBERG, Y. J., *Acta Chem. Scand.* **9**, 1234–1235 (1955). From *Alectoria chalybeiformis*.

STARK, J. B., E. D. WALTER, AND H. S. OWENS, *J. Am. Chem. Soc.* **72**, 1819–1820 (1950). From *Ramalina reticulata*.

meso-erythritol
$C_4H_{10}O_4$
m.p. 121.5° (methanol), prisms; $[\alpha]_D$ 0°.
Also known in algae, nonlichen-forming fungi, and higher plants.

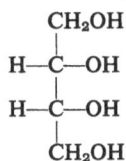

$$
\begin{array}{c}
CH_2OH \\
| \\
H\!-\!C\!-\!OH \\
| \\
H\!-\!C\!-\!OH \\
| \\
CH_2OH
\end{array}
$$

HUNECK, S. AND G. FOLLMANN, *Z. Naturforsch.* **22b**, 362–363 (1967). From *Ingaderia pulcherrima*.

HUNECK, S. AND G. FOLLMANN, *Z. Naturforsch.* **22b**, 666–670 (1967). From *Roccellaria mollis*.

HUNECK, S. AND G. FOLLMANN, Z. *Naturforsch.* **22b**, 1369–1370 (1967). From *Roccella fucoides* (1.1%) and *R. vicentina* (1.5%).

HUNECK, S., G. FOLLMANN, AND H. ULLRICH, Z. *Naturforsch.* **23b**, 292–293 (1968). From *Roccella boergesenii* (2.0%), *R. canariensis* (1.1%), and *R. teneriffensis* (0.25%).

HUNECK, S., G. FOLLMANN, W. A. WEBER, AND G. TROTET, Z. *Naturforsch.* **22b**, 671–673 (1967). From *Roccella babingtonii, R. fimbriata, R. gayana,* and a sample of *R. portentosa* from Chile.

HUNECK, S., ANNICK MATHEY, AND G. TROTET, Z. *Naturforsch.* **22b**, 1367–1368 (1967). From *Roccella fuciformis* from the English Channel (0.9%) and from Morocco (3%).

HUNECK, S. AND G. TROTET, Z. *Naturforsch.* **22b**, 363 (1967). From *Dirina repanda.*

LEWIS, D. H. AND D. C. SMITH, *New Phytol.* **66**, 143–184 (1967). Review on occurrence, physiological role, and metabolism, found in *Roccella* spp. and the algal component *Trentepohlia.*

LEWIS, D. H. AND D. C. SMITH, *New Phytol.* **66**, 185–204 (1967). Extensive review and discussion of qualitative and quantitative analytical methods.

LINDBERG, B., *Acta Chem. Scand.* **9**, 917–919 (1955). From *Roccella fuciformis, R. hypomecha,* and *R. linearis.*

PUEYO, G., *Bull. École Natl. Supér. Agronom. Nancy* **5**, 195–198 (1963). From *Roccella fuciformis.*

RAO, V. S. AND T. R. SESHADRI, *Proc. Indian Acad. Sci.* **13A**, 199–202 (1941). From *Roccella montagnei.*

ZELLNER, J., *Monatsh. Chem.* **66**, 81–86 (1935). From *Pseudevernia furfuracea.*

glycerol

$C_3H_8O_3$

Syrup.

Component of oils and fats and found free in some algae.

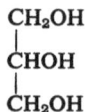

$$CH_2OH$$
$$|$$
$$CHOH$$
$$|$$
$$CH_2OH$$

LESTANG LAISNÉ, GENEVIÈVE DE, *Rev. Bryol. Lichénol.* **34**, 346–369 (1966). From *Lichina pygmaea.*

myo-inositol

$C_6H_{12}O_6$

m.p. 225° for the anhydrous substance; m.p. 218° for the dihydrate (water).

Widely distributed in plants and animals.

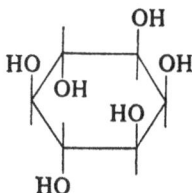

LINDBERG, B., B. G. SILVANDER, AND C. A. WACHTMEISTER, *Acta Chem. Scand.* **17**, 1348–1350 (1963). From *Peltigera aphthosa.*

D-mannitol

$C_6H_{14}O_6$

m.p. 166–167° (water), prisms; $[\alpha]_D$ +28.3° (satd. $Na_2B_4O_7$ solution). Well known in higher plants, algae, and nonlichen-forming fungi.

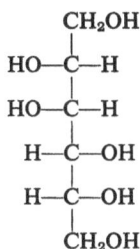

ASAHINA, Y. AND Y. TANASE, *Chem. Ber.* **67**, 766–773 (1934). From *Parmelia zollingeri.*

BEDNAR, T. W. AND D. C. SMITH, *New Phytol.* **65**, 211–220 (1966). In *Xanthoria aureola*, ^{14}C accumulation in a pentitol (probably adonitol) and then in mannitol during photosynthesis.

BREEN, J., J. KEANE, AND T. J. NOLAN, *Sci. Proc. Roy. Dublin Soc.* **21**, 587–592 (1937). From *Pertusaria pseudocorallina.*

BRINER, G. P., G. E. GREAM, AND N. V. RIGGS, *Australian J. Chem.* **13**, 277–284 (1960). From *Parmelia scabrosa* Tayl.

DHAR, M. L., S. NEELAKANTAN, S. RAMANUJAM, AND T. R. SESHADRI, *J. Sci. Ind. Res.* (India) **18B**, 111–113 (1959). From *Parmelia taractica* and *Xanthoria parietina.*

DREW, E. A. AND D. C. SMITH, *New Phytol.* **66**, 389–400 (1967). Conversion of glucose, formed by the alga during photosynthesis, to mannitol by the fungus in *Peltigera polydactyla* and the effects of added glucose in the medium.

HARPER, S. H. AND R. M. LETCHER, *Proc. Trans. Rhodesian Sci. Assoc.* **51**, 156–184 (1966). In *Dermatiscum thunbergii, Parmelia dilatata, P. gossweileri, Pertusaria* sp., and *Usnea implicita*, paper chromatography.

KENNEDY, G., J. BREEN, J. KEANE, AND T. J. NOLAN, *Sci. Proc. Roy. Dublin Soc.* **21**, 557–566 (1937). From *Lecanora rupicola.*

LESTANG LAISNÉ, GENEVIÈVE DE, *Rev. Bryol. Lichénol.* **34**, 346–369 (1966). In *Lichina pygmaea*.

LEWIS, D. H. AND D. C. SMITH, *New Phytol.* **66**, 143–184 (1967). Review on occurrence, physiological role, and metabolism.

LEWIS, D. H. AND D. C. SMITH, *New Phytol.* **66**, 185–204 (1967). Extensive review and discussion of qualitative and quantitative analytical methods.

LINDBERG, B., *Acta Chem. Scand.* **9**, 917–919 (1955). From *Dermatocarpon miniatum, Roccella fuciformis, R. hypomecha*, and *R. linearis*.

LINDBERG, B. AND H. MEIER, *Acta Chem. Scand.* **16**, 543–547 (1962). From *Siphula ceratites*.

LINDBERG, B., A. MISIORNY, AND C. A. WACHTMEISTER, *Acta Chem. Scand.* **7**, 591–595 (1953). Identified in a large number of species, paper chromatography.

LINDBERG, B., B. G. SILVANDER, AND C. A. WACHTMEISTER, *Acta Chem. Scand.* **17**, 1348–1350 (1963). From *Peltigera aphthosa*.

LINDBERG, B., C. A. WACHTMEISTER, AND B. WICKBERG, *Acta Chem. Scand.* **6**, 1052–1055 (1952). From *Lasallia pustulata*.

LINDBERG, B. AND B. WICKBERG, *Acta Chem. Scand.* **7**, 140–142 (1953). From *Lasallia pustulata*, column chromatography.

MURPHY, D., J. KEANE, AND T. J. NOLAN, *Sci. Proc. Roy. Dublin Soc.* **23**, 71–82 (1943). From *Ochrolechia parella*.

NOLAN, T. J., *Sci. Proc. Roy. Dublin Soc.* **21**, 67–71 (1934). From *Buellia canescens*.

PUEYO, G., *Bryologist* **66**, 74–76 (1963). From *Lichina pygmaea*.

PUEYO, G., *Bull. Centre Études Rech. Sci. Biarritz* **5**, 97–101 (1964). In *Xanthoria parietina*, paper chromatography.

PUEYO, G., *Bull. Centre Études Rech. Sci. Biarritz* **5**, 103–107 (1964). From *Xanthoria parietina*, cellulose-powder column chromatography.

PUEYO, G., *Bull. École Natl. Supér. Agronom. Nancy* **6**, 153–156 (1964). From *Usnea comosa*.

PUEYO, G., *Rev. Bryol. Lichénol.* **33**, 592–594 (1964–1965). From *Evernia prunastri*, column chromatography.

PUEYO, G., *Rev. Bryol. Lichénol.* **33**, 595–596 (1964–1965). From *Cladonia endiviaefolia*.

PUEYO, G., *Bryologist* **68**, 334–336 (1965). From *Cladonia rangiferina*, column chromatography.

RAO, P. S., K. G. SARMA, AND T. R. SESHADRI, *Current Sci.* (India) **34**, 9–11 (1965). From *Lobaria retigera*.

RAO, P. S., K. G. SARMA, AND T. R. SESHADRI, *Current Sci.* (India) **35**, 147–148 (1966). From *Lobaria subretigera*.

SMITH, D. C., *Lichenologist* **1**, 209–226 (1961). Synthesis in *Peltigera polydactyla* from labelled bicarbonate or glucose.

SMITH, D. C. AND E. A. DREW, *New Phytol.* **64**, 195–200 (1965). Translocation of ^{14}C from alga to fungus in *Peltigera polydactyla* and accumulation of mannitol.

ZELLNER, J., *Monatsh. Chem.* **59**, 300–304 (1932). From *Peltigera canina*.
ZELLNER, J., *Monatsh. Chem.* **66**, 81–86 (1935). From *Umbilicaria mammulata*.
ZOPF, W., *Ann. Chem.* **364**, 273–313 (1909). From *Nephroma laevigatum*, *N. parile*, and *N. resupinatum*.

siphulitol (=1-deoxy-D-*glycero*-D-*talo*-heptitol)
$C_7H_{16}O_6$
m.p. 122–123° (methanol); $[\alpha]_D^{20}$ −8° (c = 1.5, H_2O).

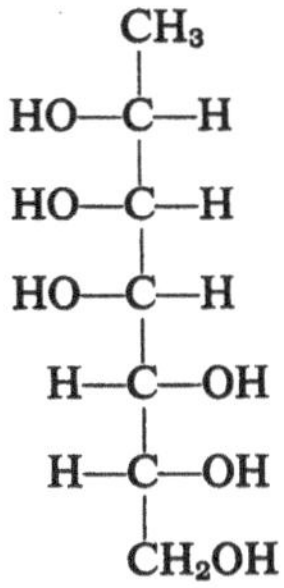

$$
\begin{array}{c}
CH_3 \\
HO-C-H \\
HO-C-H \\
HO-C-H \\
H-C-OH \\
H-C-OH \\
CH_2OH
\end{array}
$$

LINDBERG, B. AND H. MEIER, *Acta Chem. Scand.* **16**, 543–547 (1962). From *Siphula ceratites*, first isolation, structure proved, paper chromatography.

volemitol (=D-*glycero*-D-*talo*-heptitol)
$C_7H_{16}O_7$
m.p. 151–153° (ethanol), needles; $[\alpha]_D^{20}$ +2.0 to +2.5° (H_2O).
Also known in nonlichen-forming fungi, in algae, and in species of the genus *Primula*.

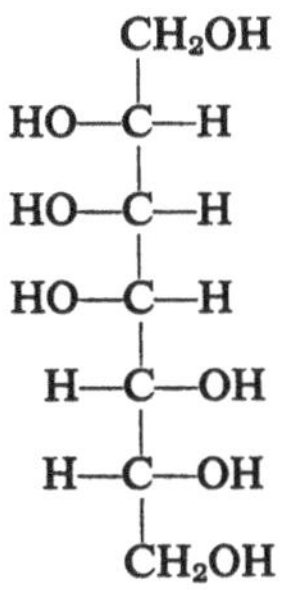

$$
\begin{array}{c}
CH_2OH \\
HO-C-H \\
HO-C-H \\
HO-C-H \\
H-C-OH \\
H-C-OH \\
CH_2OH
\end{array}
$$

ASAHINA, Y. AND M. KAGITANI, *Chem. Ber.* **67**, 804–805 (1934). From *Dermatocarpon miniatum*.
DHAR, M. L., S. NEELAKANTAN, S. RAMANUJAM, AND T. R. SESHADRI, *J. Sci. Ind. Res.* (India) **18B**, 111–113 (1959). From *Physcia setosa*.
LEWIS, D. H. AND D. C. SMITH, *New Phytol.* **66**, 143–184 (1967). Review on occurrence, physiological role, and metabolism.

LEWIS, D. H. AND D. C. SMITH, *New Phytol.* **66**, 185–204 (1967). Extensive review and discussion of qualitative and quantitative analytical methods.

LINDBERG, B., *Acta Chem. Scand.* **9**, 917–919 (1955). From *Dermatocarpon miniatum*.

LINDBERG, B., A. MISIORNY, AND C. A. WACHTMEISTER, *Acta Chem. Scand.* **7**, 591–595 (1953). In many species of Pyrenomycete lichens, paper chromatography.

SARMA, K. G. AND S. HUNECK, *Pharmazie*, in press. From *Dermatocarpon vellereum*.

MONOSACCHARIDES

arabinose

$C_5H_{10}O_5$

m.p. 157–160°.

Common in plants, often in polysaccharides.

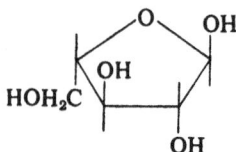

LESTANG LAISNÉ, GENEVIÈVE DE, *Rev. Bryol. Lichénol.* **34**, 346–369 (1966). In *Lichina pygmaea*.

D-fructose

$C_6H_{12}O_6$

m.p. 103–105° d; $[\alpha]_D^{20}$ −132° → −92° (c = 2, H_2O), mutarotation.

Common in plants.

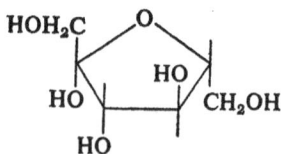

LESTANG LAISNÉ, GENEVIÈVE DE, *Rev. Bryol. Lichénol.* **34**, 346–369 (1966). In *Lichina pygmaea*.

LEWIS, D. H. AND D. C. SMITH, *New Phytol.* **66**, 185–204 (1967). Mentioned in an extensive review and discussion of qualitative and quantitative analytical methods for polyols.

PUEYO, G., *Rev. Bryol. Lichénol.* **32**, 279–284 (1963). In *Lichina pygmaea* and *Parmelia saxatilis*.
PUEYO, G., *Bull. Centre Études Rech. Sci. Biarritz* **5**, 97–101 (1964). In *Xanthoria parietina*, paper chromatography.

D-galactose
$C_6H_{12}O_6$
m.p. 166° (ethanol), prisms or plates; $[\alpha]_D^{20}$ $+144° \rightarrow +79.3°$ (24 hrs., c = 2.0, H_2O), mutarotation.
Widely distributed in natural polysaccharides.

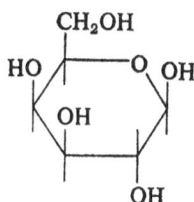

LESTANG LAISNÉ, GENEVIÈVE DE, *Rev. Bryol. Lichénol.* **34**, 346–369 (1966). In *Lichina pygmaea*.
LEWIS, D. H. AND D. C. SMITH, *New Phytol.* **66**, 185–204 (1967). Extensive review and discussion of qualitative and quantitative analytical methods.
PUEYO, G., *Rev. Bryol. Lichénol.* **32**, 279–284 (1963). In *Parmelia saxatilis* and *Umbilicaria murina*.
PUEYO, G., *Bull. Centre Études Rech. Sci. Biarritz* **5**, 97–101 (1964). In *Xanthoria parietina*, paper chromatography.

D-glucose
$C_6H_{12}O_6$
m.p. 147° d for anhydrous α-D-glucose (methanol); $[\alpha]_D^{20}$ (final value) $+52.5°$ (c = 10, H_2O).
Common in nature.

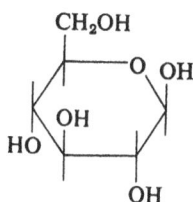

DREW, E. A. AND D. C. SMITH, *New Phytol.* **66**, 379–388 (1967). Physiology of the *Nostoc* alga isolated from *Peltigera polydactyla* compared with a free-living strain.

DREW, E. A. AND D. C. SMITH, *New Phytol.* **66**, 389–400 (1967). Movement from the *Nostoc* algal partner to the fungus during photosynthesis in *Peltigera polydactyla*.

LESTANG LAISNÉ, GENEVIÈVE DE, *Rev. Bryol. Lichénol.* **34**, 346–369 (1966). In *Lichina pygmaea*.

LEWIS, D. H. AND D. C. SMITH, *New Phytol.* **66**, 185–204 (1967). Mentioned in an extensive review and discussion of qualitative and quantitative analytical methods for polyols.

PUEYO, G., *Rev. Bryol. Lichénol.* **32**, 279–284 (1963). In *Lichina pygmaea* and *Parmelia saxatilis*.

PUEYO, G., *Bull. Centre Études Rech. Sci. Biarritz* **5**, 97–101 (1964). In *Xanthoria parietina*, paper chromatography.

RICHARDSON, D. H. S., D. C. SMITH, AND D. H. LEWIS, *Nature* **214**, 879–882 (1967). Glucose produced by the *Nostoc* algal partner and converted to mannitol by the fungus, in *Lobaria scrobiculata* and *Sticta fuliginosa*.

ZELLNER, J., *Monatsh. Chem.* **66**, 81–86 (1935). From *Umbilicaria mammulata*.

D-tagatose
$C_6H_{12}O_6$
m.p. 134–135° (ethanol-water); $[\alpha]_D^{20}$ −2.3° (c = 2.19, H_2O).
Also obtained by hydrolysis of *Sterulia setigera* gum.

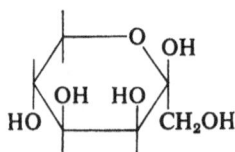

LINDBERG, B., *Acta Chem. Scand.* **9**, 917–919 (1955). From *Roccella linearis* and *R. fuciformis*.

D-xylose
$C_5H_{10}O_5$
m.p. 153–154°.
Common in plants as glycosides and xylan.

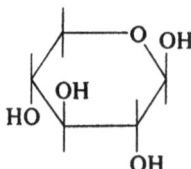

LESTANG LAISNÉ, GENEVIÈVE DE, *Rev. Bryol. Lichénol.* **34**, 346–369 (1966). Xylose possibly present in *Lichina pygmaea*.

OLIGOSACCHARIDES

3-O-β-D-glucopyranosyl-D-mannitol
$C_{12}H_{24}O_{11}$
m.p. 97–100° (ethanol-water), as a dihydrate; $[\alpha]_D^{20}$ −6° (c = 2.0, H_2O).

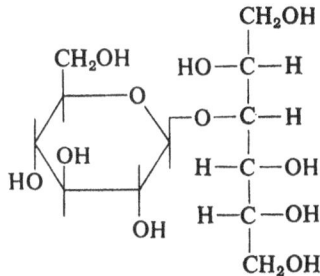

LINDBERG, B., B. G. SILVANDER, AND C. A. WACHTMEISTER, *Acta Chem. Scand.* **17**, 1348–1350 (1963). From *Peltigera aphthosa*, first extraction, structure proved, synthesis.

LINDBERG, B., B. G. SILVANDER, AND C. A. WACHTMEISTER, *Acta Chem. Scand.* **18**, 213–216 (1964). From *Peltigera horizontalis* and several other *Peltigera* species.

peltigeroside (= 3-O-β-D-galactofuranosyl-D-mannitol)
$C_{12}H_{24}O_{11}$
m.p. 161–163° (96% ethanol), as a monohydrate; $[\alpha]_D^{20}$ −61° (c = 2.0, H_2O), for the monohydrate.

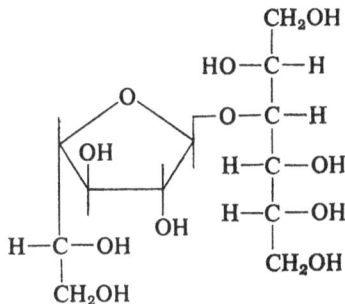

LINDBERG, B., B. G. SILVANDER, AND C. A. WACHTMEISTER, *Acta Chem. Scand.* **18**, 213–216 (1964). From *Peltigera horizontalis*, structure proved.

PUEYO, G., *Rev. Bryol. Lichénol.* **29**, 124–129 (1960). From *Peltigera horizontalis*, first isolation.

sucrose (= 2-(α-D-glucopyranosyl)-β-D-fructofuranose)
$C_{12}H_{22}O_{11}$
m.p. 185–187° (ethanol or water); m.p. 169–170° (methanol); $[\alpha]_D^{20}$ +66.5° (H_2O).
Widely distributed in plants.

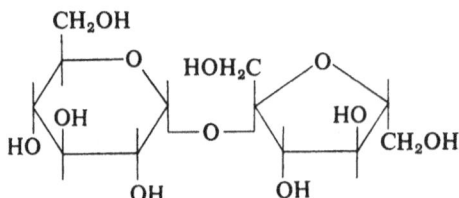

LEWIS, D. H. AND D. C. SMITH, *New Phytol.* **66**, 185–204 (1967). Extensive review and discussion of qualitative and quantitative analytical methods.

LINDBERG, B., *Acta Chem. Scand.* **9**, 917–919 (1955). From *Dermatocarpon miniatum*.

LINDBERG, B., A. MISIORNY, AND C. A. WACHTMEISTER, *Acta Chem. Scand.* **7**, 591–595 (1953). From *Agyrophora rigida, Cetraria islandica, Cladonia rangiferina, Lasallia pustulata*, and *Haematomma ventosum* and in *Lecanora atra*.

LINDBERG, B., B. G. SILVANDER, AND C. A. WACHTMEISTER, *Acta Chem. Scand.* **17**, 1348–1350 (1963). From *Peltigera aphthosa*.

LINDBERG, B. AND B. WICKBERG, *Acta Chem. Scand.* **7**, 140–142 (1953). From *Lasallia pustulata*, column chromatography.

PUEYO, G., *Rev. Bryol. Lichénol.* **32**, 279–284 (1963). In *Parmelia saxatilis*.

PUEYO, G., *Rev. Bryol. Lichénol.* **32**, 285–289 (1963). From *Lichina pygmaea* and *Umbilicaria murina*.

PUEYO, G., *Bull. Centre Études Rech. Sci. Biarritz* **5**, 97–101 (1964). In *Xanthoria parietina*, paper chromatography.

trehalose (= 1-(α-D-glucopyranosyl)-α-D-glucopyranose)
$C_{12}H_{22}O_{11}$
m.p. 203° for anhydrous crystals; $[\alpha]_D^{20}$ +197° (H_2O).
Also known in nonlichen-forming fungi, bacteria, algae, and some higher plants.

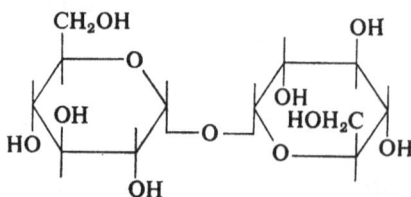

LEWIS, D. H. AND D. C. SMITH, *New Phytol.* **66**, 185–204 (1967). Extensive review and discussion of qualitative and quantitative analytical methods.

LINDBERG, B., *Acta Chem. Scand.* **9**, 917–919 (1955). From *Dermatocarpon miniatum.*

LINDBERG, B., A. MISIORNY, AND C. A. WACHTMEISTER, *Acta Chem. Scand.* **7**, 591–595 (1953). From *Agyrophora rigida, Cetraria islandica, Cladonia rangiferina, Lasallia pustulata, Lecanora atra,* and *Haematomma ventosum.*

LINDBERG, B. AND B. WICKBERG, *Acta Chem. Scand.* **7**, 140–142 (1953). From *Lasallia pustulata,* column chromatography.

umbilicin (= 2-O-β-D-galactofuranosyl-D-arabitol)

$C_{11}H_{22}O_{10}$

m.p. 138–139° (ethanol); $[\alpha]_D^{20}$ −81° (c = 2.0, H_2O).

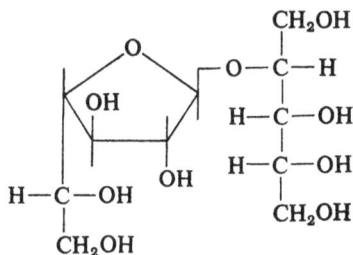

BEVING, H. F. G., H. B. BOREN, AND P. J. GAREGG, *Acta Chem. Scand.* **22**, 193–196 (1968). Synthesis.

LINDBERG, B., A. MISIORNY, AND C. A. WACHTMEISTER, *Acta Chem. Scand.* **7**, 591–595 (1953). From *Agyrophora rigida, Cetraria islandica, Lasallia pustulata,* and *Haematomma ventosum.*

LINDBERG, B., C. A. WACHTMEISTER, AND B. WICKBERG, *Acta Chem. Scand.* **6**, 1052–1055 (1952). From *Lasallia pustulata,* first isolation, structure studied.

LINDBERG, B. AND B. WICKBERG, *Acta Chem. Scand.* **7**, 140–142 (1953). From *Lasallia pustulata,* column chromatography, structure revised.

LINDBERG, B. AND B. WICKBERG, *Acta Chem. Scand.* **8**, 821–824 (1954). Structure studied.

LINDBERG, B. AND B. WICKBERG, *Acta Chem. Scand.* **16**, 2240–2244 (1962). Structure proved.

POLYSACCHARIDES

isolichenin

$(C_6H_{12}O_6)_n$

Soluble in cold water; I_2KI+ blue but less intense than starch.

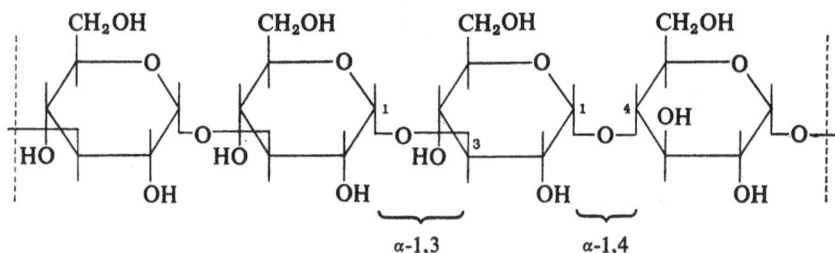

α-1,3   α-1,4

CHANDA, N. B., E. L. HIRST, AND D. J. MANNERS, *J. Chem. Soc.* **1957**, 1951–1958 (1957). Methylation and periodate oxidation, linear α-1,3- and α-1,4-glucosidic linkages in an approximate proportion of 3:2.

DHAR, M. L., S. NEELAKANTAN, S. RAMANUJAM, AND T. R. SESHADRI, *J. Sci. Ind. Res.* (India) **18B**, 111–113 (1959). From *Xanthoria parietina*.

FLEMING, M. AND D. J. MANNERS, *Biochem. J.* **100** (2), 24P (1966). Smith degradation procedure revealing sequences of single or paired α-1,3-linked glucose residues flanked by α-1,4 linkages.

LAL, B. M. AND K. R. RAO, *J. Sci. Ind. Res.* (India) **15C**, 71–73 (1956). In several species of *Parmelia* and in *Peltigera canina*.

MEYER, K. H. AND P. GÜRTLER, *Helv. Chim. Acta* **30**, 761–765 (1947). From *Cetraria islandica*.

MITTAL, O. P. AND T. R. SESHADRI, *J. Sci. Ind. Res.* (India) **13A**, 174–177 (1954). A review of the chemistry.

PEAT, S., W. J. WHELAN, J. R. TURVEY, AND K. MORGAN, *J. Chem. Soc.* **1961**, 623–629 (1961). Partial acid hydrolysis of isolichenin showing an α-glucan structure with 1,4 and 1,3 linkages.

RAO, V. S. AND T. R. SESHADRI, *Proc. Indian Acad. Sci.* **13A**, 199–202 (1941). From *Roccella montagnei*.

lichenin

$(C_6H_{12}O_6)_n$

MW $\sim$ 20,000–40,000; insoluble in cold water, soluble in hot water; $[\alpha]_D^{20}$ −8.3° (2N NaOH).

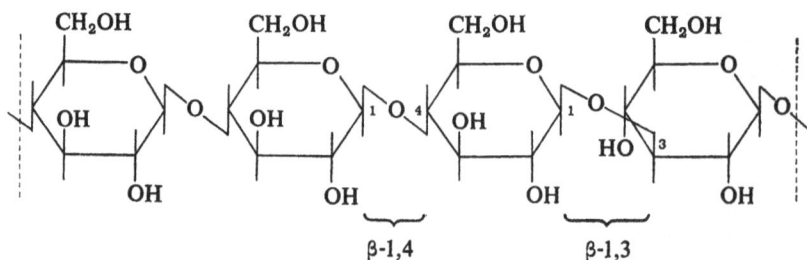

β-1,4   β-1,3

BOISSONNAS, R. A., *Helv. Chim. Acta* **30**, 1703–1704 (1947). The linkages found to be 68 ± 4% 1,4 and 32 ± 4% 1,3.

BUSTON, H. W. AND V. H. CHAMBERS, *Biochem. J.* **27**, 1691–1702 (1933). From *Cetraria islandica*.

DHAR, M. L., S. NEELAKANTAN, S. RAMANUJAM, AND T. R. SESHADRI, *J. Sci. Ind. Res.* (India) **18B**, 111–113 (1959). From *Xanthoria parietina*.

FLEMING, M. AND D. J. MANNERS, *Biochem. J.* **100** (1), 4P–5P (1966). Barley glucan found to have blocks of adjacent 1,3-linkages, similar to oat glucan but differing from lichenin where 3-substituted glucose units are interspersed between 4-substituted units.

HESS, K. AND H. FRIES, *Ann. Chem.* **455**, 180–205 (1927). From *Cetraria islandica*.

HESS, K. AND L. W. LAURIDSEN, *Chem. Ber.* **73**, 115–126 (1940). From *Cetraria islandica*, structure studied.

HULTIN, E., *Acta Chem. Scand.* **10**, 157 (1956). A quantitative viscosimetric method, detection of lichenase activity.

HUNECK, S., *Chem. Ber.* **94**, 614–622 (1961). From *Lecanora muralis*.

KARRER, P., B. JOOS, AND M. STAUB, *Helv. Chim. Acta* **6**, 800–816 (1923). Cellobiose obtained from lichenin.

KARRER, P., M. STAUB, AND J. STAUB, *Helv. Chim. Acta* **7**, 159–162 (1924). Occurrence of lichenin.

KLASON, P., *Chem. Ber.* **19**, 2541 (1886). Lichenin composed entirely of glucose units.

LAL, B. M. AND K. R. RAO, *J. Sci. Ind. Res.* (India) **15C**, 71–73 (1956). Food value of some lichens.

MEYER, K. H. AND P. GÜRTLER, *Helv. Chim. Acta* **30**, 751–761 (1947). From *Cetraria islandica*, structure studied, 1,4 (74 ± 4%) and 1,3 linkages between glucose residues proposed.

MEYER, K. H. AND P. GÜRTLER, *Helv. Chim. Acta* **30**, 761–765 (1947). From *Cetraria islandica*.

MITTAL, O. P., S. NEELAKANTAN, AND T. R. SESHADRI, *J. Sci. Ind. Res.* (India) **11B**, 386–387 (1952). From *Ramalina calicaris* and *R. sinensis*.

MITTAL, O. P. AND T. R. SESHADRI, *J. Sci. Ind. Res.* (India) **13A**, 174–177 (1954). A review of the chemistry.

MITTAL, O. P. AND T. R. SESHADRI, *J. Sci. Ind. Res.* (India) **13B**, 244–245 (1954). From *Usnea longissima*, improved method of isolation.

PEAT, S., W. J. WHELAN, AND J. G. ROBERTS, *J. Chem. Soc.* **1957**, 3916–3924 (1957). Partial acid hydrolysis of lichenin showing a β-glucan structure with 1,4 and 1,3 linkages, thought to be identical to a glucan from oats.

PERLIN, A. S. AND S. SUZUKI, *Can. J. Chem.* **40**, 50–56 (1962). Enzymatic degradation, differs from barley glucan.

VARRY, G., *Ann. Chem.* **13**, 71–75 (1835). Suggests the name lichenin.

YANOVSKY, E. AND R. M. KINGSBURY, *J. Assoc. Offic. Agr. Chemists* **21**, 648–665 (1938). From *Alectoria jubata*.

ZELLNER, J., *Monatsh. Chem.* **64**, 6–11 (1934). From *Hypogymnia physodes.*

pustulan

$(C_6H_{12}O_6)_n$
$[\alpha]_D -46°$ (c = 2.0, H_2O).

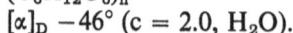

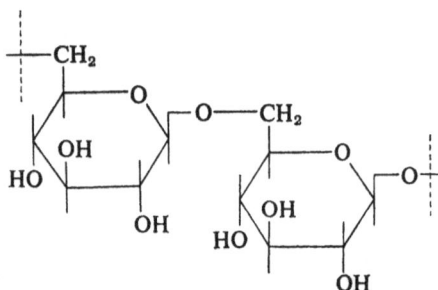

DRAKE, B., *Biochem. Z.* **313**, 388–399 (1943). From *Lasallia pustulata* and *Umbilicaria hirsuta*, first isolation, as pustulin.
LINDBERG, B. AND J. MCPHERSON, *Acta Chem. Scand.* **8**, 985–988 (1954). From *Lasallia pustulata*, isolation, a linear glucan with only β-1,6-linkages.

RELATED SUBSTANCES

ASPINALL, G. O., E. L. HIRST, AND MARGARET WARBURTON, *J. Chem. Soc.* **1955**, 651–655 (1955). From *Cladonia alpestris*, highly branched polysaccharides of D-galactose, D-glucose, and D-mannose.
BEDNAR, T. W. AND D. C. SMITH, *New Phytol.* **65**, 211–220 (1966). Accumulation of ^{14}C from $NaH^{14}CO_3$ in an unknown β-glycoside and in hexose monophosphate in *Xanthoria aureola*.
DREW, E. A. AND D. C. SMITH, *New Phytol.* **66**, 389–400 (1967). Glucose polysaccharides as a minor product of photosynthesis in *Peltigera polydactyla*.
LESTANG LAISNÉ, GENEVIÈVE DE, *Rev. Bryol. Lichénol.* **34**, 346–369 (1966). In *Lichina pygmaea*, a mannitol mannoside, a glycogen, polysaccharides of mannose and galactose, of xylose and galactose, of mannose, xylose, galactose, and glucuronic acid, possibly of arabinose and galactose, and a complex polysaccharide of mannose, xylose, arabinose, glucose, galactose, and glucuronic acid.
PUEYO, G., *Bull. Centre Études Rech. Sci. Biarritz* **5**, 97–101 (1964). In *Xanthoria parietina*, paper chromatography.
VOTOČEK, E. AND J. BURDA, *Bull. Soc. Chim. France* **39**, 248–254 (1926). From *Hypogymnia physodes, Lasallia pustulata, Peltigera canina, Peltigera polydactyla,* and *Pseudevernia furfuracea,* and a polysaccharide (named bryopogon) of galactose, glucose, and mannose from *Alectoria jubata*.

TRICARBOXYLIC ACID CYCLE COMPOUNDS AND RELATED SUBSTANCES

citric acid
 $C_6H_8O_7$
 m.p. 153°.
 Common in nature.

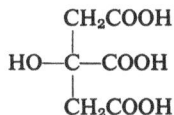

$$
\begin{array}{c}
CH_2COOH \\
| \\
HO-C-COOH \\
| \\
CH_2COOH
\end{array}
$$

BEDNAR, T. W. AND D. C. SMITH, *New Phytol.* **65**, 211–220 (1966). Accumulation of ^{14}C from $NaH^{14}CO_3$ as citric acid by *Xanthoria aureola*.

glyceric acid
 $C_3H_6O_4$
 Syrup.
 Known from higher plants.

$$
\begin{array}{c}
COOH \\
| \\
CHOH \\
| \\
CH_2OH
\end{array}
$$

BEDNAR, T. W. AND D. C. SMITH, *New Phytol.* **65**, 211–220 (1966). Accumulation of ^{14}C from $NaH^{14}CO_3$ as glyceric acid by *Xanthoria aureola*.

malic acid
 $C_4H_6O_5$
 m.p. 99–100°.
 Common in plants.

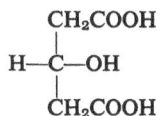

$$
\begin{array}{c}
CH_2COOH \\
| \\
H-C-OH \\
| \\
CH_2COOH
\end{array}
$$

BEDNAR, T. W. AND D. C. SMITH, *New Phytol.* **65**, 211–220 (1966). Accumulation of ^{14}C from $NaH^{14}CO_3$ as malic acid by *Xanthoria aureola*.

oxalic acid
 $C_2H_2O_4$
 m.p. 189.5° anhydrous (water), prisms.

In lichens as the calcium or potassium salt. Widely distributed in higher plants, algae, and nonlichen-forming fungi. (See also calcium.)

$$\begin{array}{c} \text{COOH} \\ | \\ \text{COOH} \end{array}$$

ASAHINA, Y., *J. Japan. Botany* **19**, 47–56 (1943). In *Cladonia capitata* and *C. symphicarpia*, as calcium oxalate.
SCHADE, A., *Nova Hedwigia* **11**, 285–308 (1966). Calcium oxalate excretion by numerous species.
SCHULTE, F., *Beih. Botan. Zentralblatt* **18**, 1–22 (1905). In numerous species of the genus *Usnea*, as the potassium salt.

phosphoglyceric acid
$C_3H_7O_7P$

BEDNAR, T. W. AND D. C. SMITH, *New Phytol.* **65**, 211–220 (1966). Accumulation of ^{14}C from $NaH^{14}CO_3$ as PGA by *Xanthoria aureola*.

succinic acid
$C_4H_6O_4$
m.p. 185–187° (water), prisms
Widely distributed in plants.

$$HOOCCH_2CH_2COOH$$

ASAHINA, Y., *J. Japan. Botany* **18**, 663–683 (1942); *ibid.* **18**, 549–552 (1942). From *Cladonia hondoensis*.
ASAHINA, Y., Lichens of Japan. Vol. I. Genus *Cladonia*. Tokyo. 1950. In *Cladonia hondoensis*.

NITROGEN COMPOUNDS (EXCEPT VITAMINS AND PULVINIC ACID DERIVATIVES)

AMMONIA AND AMINES

ammonia
$$NH_3$$

choline
$$((CH_3)_3\overset{+}{N}CH_2CH_2OH)OH^-$$

choline sulfate ester
$C_5H_{13}NO_4S$
m.p. ~ 300° d; ir (KBr), ν cm^{-1}: 4500 (w), 3450 (w), 2995 (w), 1650 (w), 1515 (s), 1490 (s), 1420 (w), 1265 (s), 1235 (s), 1102 (m), 1065 (m),

1042 (s), 965 (s), 912 (m), 890 (m), 768 (s), and 720 (w); uv: no absorption above 205 mμ; nmr (D$_2$O), τ ppm: 6.80 (N$^+$-methyls), 6.28 (N$^+$-methylene), and 5.54 (O-methylene).
Also known from fungi and from red algae.

$$(CH_3)_3\overset{+}{N}CH_2CH_2OSO_3^-$$

ethanolamine

$$HOCH_2CH_2NH_2$$

methylamine

$$CH_3NH_2$$

trimethylamine

$$(CH_3)_3N$$

HARPER, S. H. AND R. M. LETCHER, *Chem. Ind.* (London) **1966**, 419–420 (1966). Choline sulfate ester from *Dermatiscum thunbergii*.

KLIMA, J., *Monatsh. Chem.* **62**, 209–213 (1933). Choline from *Alectoria ochroleuca*.

LINDBERG, B., *Acta Chem. Scand.* **9**, 917–919 (1955). Choline sulfate ester from *Roccella fuciformis*, *R. hypomecha*, and *R. linearis*.

LINKO, P., M. ALFTHAN, J. K. MIETTINEN, AND A. I. VIRTANEN, *Acta Chem. Scand.* **7**, 1310–1311 (1953). Ethanolamine in *Cladonia arbuscula*.

STEIN VON KAMIENSKI, E., *Planta* **50**, 331–352 (1958). Ammonia, methylamine, and trimethylamine in *Sticta fuliginosa* and *S. sylvatica*.

AMINO ACIDS

(See also Vitamins and Growth Factors)

alanine

$$CH_3CH(NH_2)COOH$$

α-aminobutyric acid

$$CH_3CH_2CH(NH_2)COOH$$

γ-aminobutyric acid

$$NH_2CH_2CH_2CH_2COOH$$

arginine

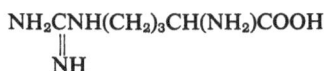

asparagine

$$NH_2COCH_2CH(NH_2)COOH$$

aspartic acid

$$HOOCCH_2CH(NH_2)COOH$$

betaine

$$(CH_3)_3\overset{+}{N}CH_2COO^-$$

cystine

$$HOOCCH(NH_2)CH_2SSCH_2CH(NH_2)COOH$$

glutamic acid

$$HOOCCH_2CH_2CH(NH_2)COOH$$

glutamine

$$NH_2COCH_2CH_2CH(NH_2)COOH$$

glycine

$$NH_2CH_2COOH$$

isoleucine

$$CH_3CH_2CH(CH_3)CH(NH_2)COOH$$

leucine

$$(CH_3)_2CHCH_2CH(NH_2)COOH$$

lysine

$$NH_2(CH_2)_4CH(NH_2)COOH$$

methionine

$$CH_3SCH_2CH_2CH(NH_2)COOH$$

phenylalanine

proline

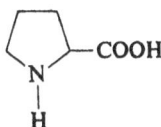

sarcosine

$$CH_3NHCH_2COOH$$

serine

$$HOCH_2CH(NH_2)COOH$$

threonine

$$CH_3CH(OH)CH(NH_2)COOH$$

tryptophan

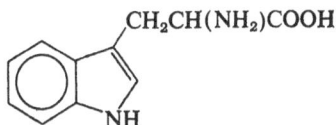

tyrosine

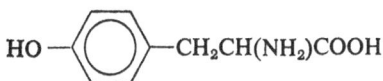

valine

$$(CH_3)_2CHCH(NH_2)COOH$$

BEDNAR, T. W. AND D. C. SMITH, *New Phytol.* **65**, 211–220 (1966). Incorporation of ^{14}C from $NaH^{14}CO_3$ into several amino acids by *Xanthoria aureola.*

DELLA TORRE, B. AND G. POMA, *Atti Ist. Botan. Lab. Crittogam. Univ. Pavia*, Ser. 6, **2**, 167–179 (1966). Production of amino acids by the isolated fungal partner of *Sarcogyne similis* being fed single amino acids or KNO_3.

LINKO, P., M. ALFTHAN, J. K. MIETTINEN, AND A. I. VIRTANEN, *Acta Chem. Scand.* **7**, 1310–1311 (1953). Amino acids in *Cladonia arbuscula.*

RAMAKRISHNAN, S. AND S. S. SUBRAMANIAN, *Indian J. Chem.* **2**, 467 (1964). Amino acids in *Parmelia tinctorum, Roccella montagnei, Usnea flexilis, U. orientalis,* and *U. venosa.*

RAMAKRISHNAN, S. AND S. S. SUBRAMANIAN, *Current Sci.* (India) **34**, 345–347 (1965). Amino acids in *Cladonia gracilis* var. *chordalis, C. rangiferina,* and *Lobaria isidiosa.*

RAMAKRISHNAN, S. AND S. S. SUBRAMANIAN, *Current Sci.* (India) **35**, 124–125 (1966). Amino acids in *Lasallia pustulata, Lobaria isidiosa* var. *subisidiosa, Parmelia cirrhata,* and *Ramalina sinensis,* paper chromatography.

RAMAKRISHNAN, S. AND S. S. SUBRAMANIAN, *Current Sci.* (India) **35**, 284–285 (1966). Amino acids in *Dermatocarpon moulinsii,* paper chromatography.

SUBRAMANIAN, S. S. AND S. RAMAKRISHNAN, *Current Sci.* (India) **33**, 522 (1964). Amino acids in *Peltigera canina.*

OLIGOPEPTIDE

picroroccellin
$C_{20}H_{22}N_2O_4$
m.p. 190–220°, heating-rate dependent (ethanol); $[\alpha]_D$ +12.5° ($CHCl_3$).

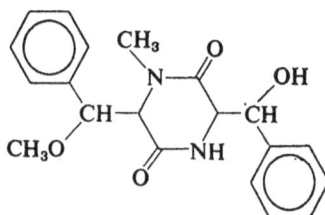

Substitution on nitrogens may be reversed.

FORSTER, M. O. AND W. B. SAVILLE, *J. Chem. Soc.* **121**, 816–827 (1922). Structure studied.

STENHOUSE, J. AND C. E. GROVES, *Ann. Chem.* **185**, 14–25 (1877). From *Roccella fuciformis,* first extraction.

ENZYMES

amylase	invertase
asparaginase	lichenase
catalase	lipase
cellulase	orsellinic acid decarboxylase
depside hydrolyzing esterase	oxidase

peroxidase	tannase
phenolase	tyrosinase
protease	urease
ribonuclease	zymase

KOLLER, G. AND G. PFEIFFER, *Monatsh. Chem.* **62**, 359–372 (1933). Depside-hydrolyzing activity of powdered lichen demonstrated.
MOISSEJEVA, E. N., [Biochemical Properties of Lichens and Their Practical Importance.] Moscow-Leningrad. 1961. Activities of various enzymes in numerous species of lichens.
MOSBACH, K. AND URSULA EHRENSVÄRD, *Biochem. Biophys. Res. Commun.* **22**, 145–150 (1966). Depside-hydrolyzing esterase and orsellinic acid decarboxylase activity isolated from *Lasallia pustulata* and from a *Trebouxia* culture from *L. papulosa*.
RENNERT, ALDONA AND M. GUBANSKI, *Naturwissenschaften* **47**, 18–19 (1960). Ribonuclease in *Cetraria islandica*.

CHLOROPHYLL

chlorophyll
GIUDICI DE NICOLA, MARINA AND R. TOMASELLI, *Boll. Ist. Botan. Univ. Catania*, Ser. 3, **2**, 29–34 (1961). Total chlorophyll and chlorophyll *a* in the lichen alga *Trebouxia decolorans*.
GIUDICI DE NICOLA, MARINA AND G. DI BENEDETTO, *Boll. Ist. Botan. Univ. Catania*, Ser. 3, **3**, 22–33 (1962). Chlorophylls *a* and *b* in the algal partner of *Xanthoria parietina*, chlorophyll to carotene ratio measured and discussed.
GODNEV, T. N., É. V. KHODASEVICH, AND A. I. ARNAUTOVA, *Dokl., Biochem. Sect., Proc. Acad. Sci. USSR* (Engl. Transl.) **167**, 90–91 (1966). ^{14}C incorporation at low temperatures, chlorophyll *a* and chlorophyll *b*, in *Hypogymnia physodes*.
WILHELMSEN, J. B., *Botan. Tidsskr.* **55**, 30–36 (1959). In *Hypogymnia physodes*, *Peltigera canina*, and *Xanthoria parietina*, column chromatography.

SULFUR COMPOUND (EXCEPT WITH NITROGEN)

dimethyl sulfone
$C_2H_6O_2S$
m.p. 107–109°.
Also found in species of *Equisetum*.

$(CH_3)_2SO_2$

BRUUN, T. AND N. A. SÖRENSEN, *Acta Chem. Scand.* **8**, 703 (1954). From *Cladonia deformis*.

VITAMINS AND GROWTH FACTORS

ascorbic acid (= vitamin C)
$C_6H_8O_6$

m.p. 190–192° d, crystals; $[\alpha]_D^{23}$ +48° (c = 1.0, MeOH); uv (acid solution), λ_{max} mμ: 245; uv (neutral solution), λ_{max} mμ: 265.

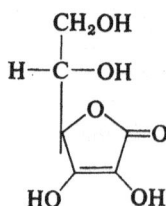

biotin
$C_{10}H_{16}N_2O_3S$

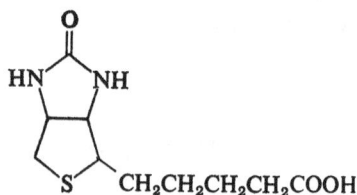

folic acid
$C_{19}H_{19}N_7O_6$

Decomposes from 250°, yellow-orange crystals.

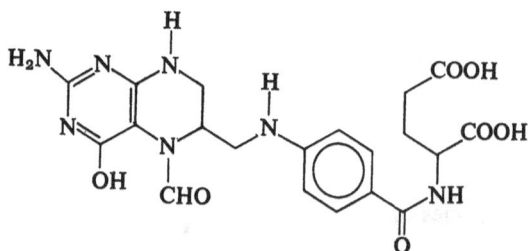

folinic acid

$C_{20}H_{19}N_7O_7$

m.p. 248–250° d, yellowish crystals.

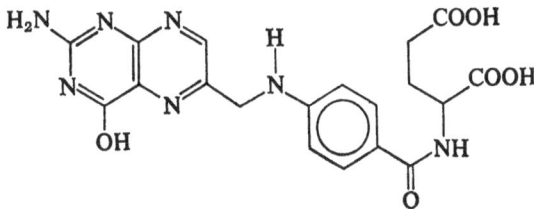

nicotinic acid (=niacin)

$C_6H_5NO_2$

m.p. 225–227° (ethanol or water), needles.

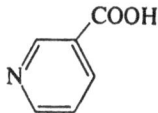

pantothenic acid

$C_9H_{17}NO_5$

An oil; $[\alpha]_D^{25}$ +37.5°.

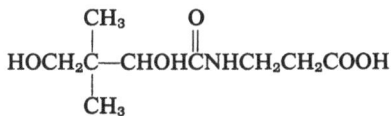

riboflavin (=vitamin B$_2$)

$C_{17}H_{20}N_4O_6$

m.p. 282–292° d (water), orange-yellow needles; $[\alpha]_D^{21}$ −118° (c = 0.50, 0.1N NaOH).

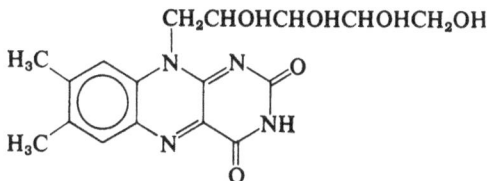

vitamin B$_1$ (= thiamine)
$C_{12}H_{17}N_4OS^+$

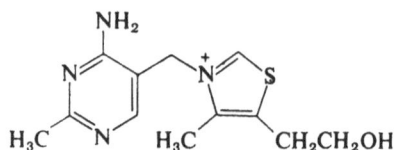

vitamin B$_{12}$
$C_{63}H_{90}N_{14}O_{14}PCo$
m.p. >300°, darkens 210–220°, dark red crystals; $[\alpha]_{656}^{23}$ −59 ± 9° (H_2O); visible and uv (H_2O), λ_{max} ($A_{1\,cm}^{1\%}$) mμ: 550 (64), 361 (204), and 278 (115).

BEDNAR, T. W. AND O. HOLM-HANSEN, *Plant Cell Physiol.* (Tokyo) **5**, 297–303 (1964). Biotin liberated by the alga (*Coccomyxa* sp.) of *Peltigera aphthosa*.

BOURNE, G. AND R. ALLEN, *Nature* **136**, 185–186 (1935). Ascorbic acid in algae, bacteria, fungi, lichens, and protozoa.

BOURNE, G. AND R. ALLEN, *Australian J. Exptl. Biol. Med. Sci.* **13**, 165–174 (1935). Ascorbic acid in lower organisms including lichens.

HENRIKSSON, ELISABET, *Physiol. Plant.* **14**, 813–817 (1961). Biotin, nicotinic acid, pantothenic acid, riboflavin, and vitamin B$_1$ produced by the isolated alga of *Collema tenax*.

GUSTAFSON, F. G., *Bull. Torrey Botan. Club* **81**, 313–322 (1954). Ascorbic acid, nicotinic acid, riboflavin, and vitamin B$_1$ in *Cetraria delisei* and *Stereocaulon* sp.

KAREV, G. I. AND V. P. KOCHEVYKH, *Botan. Zh.* **47**, 1686–1688 (1962). Ascorbic acid in tundra fodder lichens, i.e. *Alectoria chalybeiformis*, *Cetraria cucullata*, and *C. delisei*.

LAL, B. M. AND K. R. RAO, *J. Sci. Ind. Res.* (India) **15C**, 71–73 (1956). Food value, including ascorbic acid, and riboflavin content of some lichens.

SJÖSTRÖM, A. G. M. AND L.-E. ERICSON, *Acta Chem. Scand.* **7**, 870–872 (1953). Folic acid-, folinic acid-, and vitamin B$_{12}$-group factors in numerous species.

ELEMENTS, ANIONS, AND ISOTOPES

antimony-125	calcium	cobalt
beryllium	cesium-137	copper
boron	chloride	chromium

germanium	nickel	sodium
iron	nitrogen	strontium-90
lead	phosphorus	sulfate
lead-210	polonium-210	sulfur
magnesium	potassium	tin
manganese	potassium-40	titanium
manganese-54	radium-226	zinc
molybdenum	ruthenium-106	

BEASLEY, T. M. AND H. E. PALMER, *Science* **152**, 1062–1064 (1966). ^{210}Pb and ^{210}Po in a composite sample of lichen from Alaska.

BURLEY, J. W. A., G. E. GILBERT, AND L. C. CLUM, *Neotoma Ecol. Bioclim. Lab., Ohio State Univ. Ohio Agr. Expt. Sta., Spec. Rept. No. 10*, 1962. ^{40}K and ^{137}Cs in *Umbilicaria mammulata*.

HÄSÄNEN, E. AND J. K. MIETTINEN, *Nature* **212**, 379–382 (1966). ^{137}Cs uptake by lichens.

HOLTZMAN, R. B., *Nature* **210**, 1094–1097 (1966). ^{210}Pb, ^{210}Po, and ^{226}Ra uptake.

HUNECK, S. AND G. FOLLMANN, *Z. Naturforsch.* **20b**, 496 (1965). K^+, Mg^{++}, Na^+, Cl^-, and $SO_4^=$, in *Dolichocarpus chilensis*.

JENKINS, D. A. AND R. I. DAVIES, *Nature* **210**, 1296–1297 (1966). Trace elements in *Parmelia omphalodes*, atmospheric origin proposed.

LAL, B. M. AND K. R. RAO, *J. Sci. Ind. Res.* (India), **15C**, 71–73 (1956). Ca, Fe, P, and N from various species.

LAMBINON, J., A. MAQUINAY, AND J. L. RAMAUT, *Bull. Jardin Botan. État Bruxelles* **34**, 273–282 (1964). Zn in various species.

LANGE, O. L. AND H. ZIEGLER, *Mitteil. Florist-soziolog. Arbeitsgemeinschaft*, N. F. Heft 10 (Festschr. für Prof. Dr. Otto Stocker), 156–183 (1963). Fe and Cu in numerous species.

LOUNAMAA, K. J., *Ann. Botan. Fennici* **2**, 127–137 (1965). Fe, Mn, and Zn in numerous species.

MAQUINAY, A., I. M. LAMB, J. LAMBINON, AND J. L. RAMAUT, *Physiol. Plant.* **14**, 284–289 (1961). Zn in *Stereocaulon nanodes* f. *tyroliense*.

SCOTTER, G. W., *Can. J. Plant Sci.* **45**, 246–250 (1965). Ca and P in numerous species.

SOLBERG, Y. J., *Ann. Botan. Fennici* **4**, 29–34 (1967). B, Ca, K, Mg, Mn, Na, P, and S in 45 species of lichens from Norway.

TUOMINEN, Y., *Ann. Botan. Fennici* **4**, 1–28 (1967). ^{90}Sr uptake by *Cladonia alpestris* found to be nonmetabolic, relationship to uronic acid content, pH and temperature dependence, geographic variability, cation selectivity, and other factors studied.

Secondary Products Originating by the Acetate-Polymalonate Pathway

HIGHER ALIPHATIC ACIDS AND RELATED SUBSTANCES

acaranoic acid
$C_{17}H_{30}O_4$
m.p. 154–155° (ether); $[\alpha]_D^{25}$ −30° (c=0.29, $CHCl_3$); ir (KBr), ν cm^{-1}: 1755 (C=O) and 1695 (C=O).

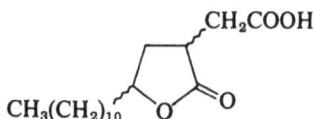

BENDZ, GERD, J. SANTESSON, AND L. TIBELL, *Acta Chem. Scand.* **20**, 1181 (1966). Thin-layer chromatography, "pleopsidic acid" proved to be a mixture, from *Acarospora chlorophana*.

SANTESSON, J., *Acta Chem. Scand.* **21**, 1162–1172 (1967). Thin-layer chromatography.

SANTESSON, J., *Acta Chem. Scand.* **21**, 1993–1996 (1967). From *Acarospora chlorophana*, structure proved.

SARMA, K. G. AND S. HUNECK, *Pharmazie*, in press. From *Acarospora oxytona*.

acarenoic acid
$C_{17}H_{28}O_4$
m.p. 144–146.5° (benzene); $[\alpha]_D^{25}$ −39° (c = 0.25, $CHCl_3$); ir (KBr), ν cm^{-1}: 1740 (C=O), 1692 (C=O), and 1630 (C=C).

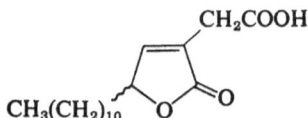

BENDZ, GERD, J. SANTESSON, AND L. TIBELL, *Acta Chem. Scand.* **20**, 1181 (1966). Thin-layer chromatography, "pleopsidic acid" proved to be a mixture, from *Acarospora chlorophana*.

SANTESSON, J., *Acta Chem. Scand.* **21**, 1162–1172 (1967). Thin-layer chromatography.

SANTESSON, J., *Acta Chem. Scand.* **21**, 1993–1996 (1967). From *Acarospora chlorophana*, structure proved.

SARMA, K. G. AND S. HUNECK, *Pharmazie*, in press. From *Acarospora oxytona*.

(+)-aspicilin

$C_{18}H_{30}O_4$

m.p. 153–154° (methanol), colorless prismatic needles; $[\alpha]_D^{20} + 32°$ (c = 2.31, CHCl$_3$); ir, ν cm^{-1}: 3400 and 3250 (OH), 1710 and 1655 (C=O), and 885 (>C=CH$_2$); nmr, τ ppm: 3.78 (>C=CH$_2$); neutral compound.

Structure not known.

HESSE, O., *J. Prakt. Chem.* **70**, 449–502 (1904). From *Lecanora gibbosula*, first extraction.

HUNECK, S., *Z. Naturforsch.* **21b**, 888–890 (1966). From *Lecanora viridula*, an aliphatic lactone.

HUNECK, S. AND G. FOLLMANN, *Z. Naturforsch.* **22b**, 110–111 (1967). From *Ramalina ecklonii* var. *ambigua*.

HUNECK, S. AND G. FOLLMANN, unpublished. From *Lecanora caesiocinerea* f. *plumbea*, *L. cenisea* var. *atrynea*, *L. coarctata* var. *ornata*, and *L. contorta*.

HUNECK, S., G. FOLLMANN, W. A. WEBER, AND G. TROTET, *Z. Naturforsch.* **22b**, 671–673 (1967). From *Roccella fucoides*.

(−)-caperatic acid

$C_{21}H_{38}O_7$

m.p. 132–133.5° (methanol-water); $[\alpha]_D^{10} - 3.85°$ (CHCl$_3$); ir (nujol), ν cm^{-1}: 3600 (OH), 1750 (COOCH$_3$), and 1720 and 1690 (COOH).

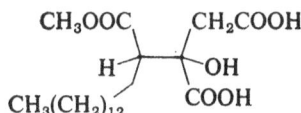

ASAHINA, Y., Lichens of Japan. Vol. II. Genus *Parmelia*. Tokyo. 1952. Microcrystal test.

ASAHINA, Y., *J. Japan. Botany* **34**, 225–230 (1959). Microcrystal test and paper chromatography.

ASANO, M. AND T. AZUMI, *Chem. Ber.* **68**, 995–997 (1935). From *Cetraria stracheyi* f. *ectocarpisma*.

ASANO, M., Y. KAMEDA, AND O. TAMEMASA, *Yakugaku Zasshi* **64**, 203–206 (1944); reference from *Chem. Abstr.* **45**, 2871g (1951). Structure studied.

ASANO, M. AND Z. OHTA, *Chem. Ber.* **66**, 1020–1023 (1933). From *Parmelia caperata*, structure studied.

ASANO, M. AND Z. OHTA, *Chem. Ber.* **67**, 1842–1845 (1934). Structure studied.

BENDZ, GERD, J. SANTESSON, AND L. TIBELL, *Acta Chem. Scand.* **20**, 1181 (1966). Thin-layer chromatography.

CULBERSON, CHICITA F., *J. Pharm. Sci.* **54**, 1815–1816 (1965). From *Parmelia cryptochlorophaea.*

CULBERSON, W. L. AND CHICITA F. CULBERSON, *Contrib. U.S. Natl. Herb.* **34**, 449–558 (1968). Distribution in the genus *Platismatia*, microcrystal tests.

DHAR, M. L., S. NEELAKANTAN, S. RAMANUJAM, AND T. R. SESHADRI, *J. Sci. Ind. Res.* (India) **18B**, 111–113 (1959). From *Usnea orientalis.*

HALE, M. E., JR., *Contrib. U.S. Natl. Herb.* **36**, 193–358 (1965). Distribution in the genus *Parmelia.*

HESSE, O., *J. Prakt. Chem.* **57**, 409–447 (1898). From *Parmelia caperata*, first isolation.

KAMEDA, Y., *Yakugaku Zasshi* **61**, 266–270 (1941); reference from *Chem. Abstr.* **44**, 9356c (1950). Structure studied.

KROG, HILDUR, *Nytt Mag. Naturvidenskapene* **88**, 57–85 (1951). Microcrystal tests.

SANTESSON, J., *Acta Chem. Scand.* **21**, 1162–1172 (1967). Thin-layer chromatography.

(−)-lichesterinic acid
$C_{19}H_{32}O_4$

m.p. 123–124° (acetic acid or ethanol), plates; $[\alpha]_D^{15}$ − 30° (CHCl₃).

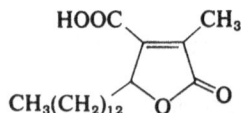

ASAHINA, Y. AND M. YASUE, *Chem. Ber.* **70**, 1053–1059 (1937). From *Cetraria ericetorum.*

ASANO, M. AND T. AZUMI, *Chem. Ber.* **68**, 991–994 (1935). Structure studied.

ASANO, M. AND T. AZUMI, *Chem. Ber.* **68**, 995–997 (1935). From *Nephromopsis stracheyi* f. *ectocarpisma.*

ASANO, M. AND T. KANEMATSU, *Chem. Ber.* **65**, 1175–1178 (1932). Structure proved.

ASANO, M. AND Z. OHTA, *Yakugaku Zasshi* **51**, 395–401 (1931); German summary, pp. 36–37 (1931); reference from *Chem. Abstr.* **25**, 4267 (1931). Synthesis.

BENDZ, GERD, J. SANTESSON, AND L. TIBELL, *Acta Chem. Scand.* **20**, 1181 (1966). Thin-layer chromatography.

CAVALTITO, C. J., DOROTHY M. FRUEHAUF, AND J. H. BAILEY, *J. Am. Chem. Soc.* **70**, 3724–3726 (1948). Synthesis from (−)-protolichesterinic acid, antibacterial activity.

SANTESSON, J., *Acta Chem. Scand.* **21**, 1162–1172 (1967). Thin-layer chromatography.

SHIBATA, S., Y. MIURA, H. SUGIMURA, AND Y. TOYOIZUMI, *Yakugaku Zasshi* **68**, 300–303 (1948). Antibacterial activity.

TAMELEN, E. E. VAN, C. E. OSBORNE, JR., AND SHIRLEY R. BACH, *J. Am. Chem. Soc.* **77**, 4625–4629 (1955). Synthesis of methyl ($\pm$)-lichesterinate.

TUNMANN, O., *Apotheker Ztg.* **88**, 892–893 (1913). Microcrystal tests.

linoleic acid

$C_{18}H_{32}O_2$

m.p. $-5°$; b.p. $202°/1.4$ mm; oxidized by air.

Widely distributed in plant fats.

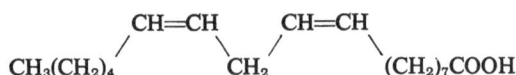

KLIMA, J., *Monatsh. Chem.* **62**, 209–213 (1933). From *Alectoria ochroleuca*, isolated as tetrahydroxystearic acid after oxidation.

(−)-nephromopsinic acid

$C_{19}H_{34}O_4$

m.p. $137°$ (ethanol), leaflets; $[\alpha]_D^{12}$ $-85.1°$ (CHCl$_3$).

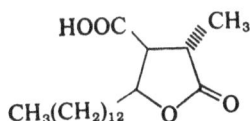

ASANO, M. AND T. AZUMI, *Chem. Ber.* **68**, 995–997 (1935). From *Cetraria stracheyi* f. *ectocarpisma*, structure studied.

ASANO, M. AND T. AZUMI, *Chem. Ber.* **72**, 35–39 (1939). Stereochemistry and structure studied.

ASANO, M. AND T. HASUSUMI, *Yakugaku Zasshi* **59**, 377–383 (1939). Structure studied.

BENDZ, GERD, J. SANTESSON, AND L. TIBELL, *Acta Chem. Scand.* **20**, 1181 (1966). Thin-layer chromatography.

HUNECK, S. AND G. FOLLMANN, *Z. Naturforsch.* **22b**, 666–670 (1967). Absolute configuration proved.

SANTESSON, J., *Acta Chem. Scand.* **21**, 1162–1172 (1967). Thin-layer chromatography.

(+)-nephrosteranic acid

$C_{17}H_{30}O_4$

m.p. 95° (petroleum ether), plates; $[\alpha]_D^{21}$ +38.4° $(CHCl_3)$.

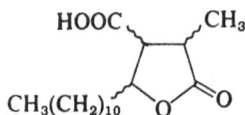

$$HOOC \diagdown \diagup CH_3$$
$$CH_3(CH_2)_{10} \diagdown O \diagup O$$

ASAHINA, Y. AND M. YANAGITA, *Chem. Ber.* **70**, 227–235 (1937). From *Cetraria endocrocea*, column chromatographic separation, structure proved.

BENDZ, GERD, J. SANTESSON, AND L. TIBELL, *Acta Chem. Scand.* **20**, 1181 (1966). Thin-layer chromatography.

(+)-nephrosterinic acid
$C_{17}H_{28}O_4$
m.p. 96° (dilute acetic acid), leaflets; $[\alpha]_D^{10}$ +10.8° $(CHCl_3)$.

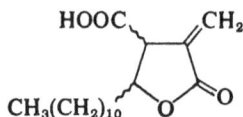

$$HOOC \diagdown \diagup CH_2$$
$$CH_3(CH_2)_{10} \diagdown O \diagup O$$

ASAHINA, Y. AND M. YANAGITA, *Chem. Ber.* **70**, 227–235 (1937). From *Cetraria endocrocea*, column chromatographic separation, structure proved.

BENDZ, GERD, J. SANTESSON, AND L. TIBELL, *Acta Chem. Scand.* **20**, 1181 (1966). Thin-layer chromatography.

(+)-norrangiformic acid
$C_{20}H_{36}O_6$
m.p. 114–122° (from the hydrolysis of rangiformic acid); $[\alpha]_D^{18}$ +12.9°.

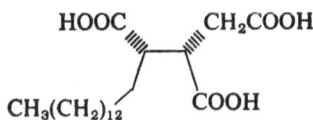

$$HOOC \diagdown \diagup CH_2COOH$$
$$CH_3(CH_2)_{12} \diagdown COOH$$

ÅKERMARK, B., *Arkiv Kemi* **27**(2), 11–17 (1967). Stereospecific synthesis of (±)-norrangiformic acid, ir data.

AOKI, M. *Yakugaku Zasshi* **66**, 52–55 (1946); reference from *Chem. Abstr.* **45**, 6584e (1951). Structure of norrangiformic acid proved.

ASAHINA, Y., *J. Japan. Botany* **17**, 620–630 (1941). In *Cladonia mitis*, microcrystal test.

BENDZ, GERD, J. SANTESSON, AND L. TIBELL, *Acta Chem. Scand.* **20**, 1181 (1966). Thin-layer chromatography.

SANTESSON, J., *Acta Chem. Scand.* **21**, 1162–1172 (1967). Thin-layer chromatography.

oleic acid

$C_{18}H_{34}O_2$

b.p. 286°/100 mm, colorless liquid; oxidized by air.
Widely distributed in plant and animal fats.

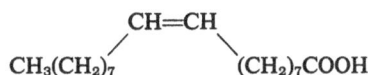

$$CH_3(CH_2)_7 \diagup \overset{CH=CH}{} \diagdown (CH_2)_7COOH$$

KLIMA, J., *Monatsh. Chem.* **62**, 209–213 (1933). From *Alectoria ochroleuca*, isolated as dihydroxystearic acid after oxidation.

(−)-protolichesterinic acid

$C_{19}H_{32}O_4$

m.p. 106° (acetic acid); $[\alpha]_D^{27}$ −12.7° (CHCl$_3$); $[\alpha]_D^{22}$ −15° (CHCl$_3$).

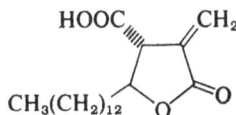

For references see (+)-protolichesterinic acid.

(+)-protolichesterinic acid

$C_{19}H_{32}O_4$

m.p. 107.5° (benzene or acetic acid); $[\alpha]_D^{19.5}$ +12.1° (CHCl$_3$).

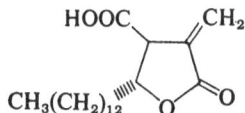

AGHORAMURTHY, K., S. NEELAKANTAN, AND T. R. SESHADRI, *J. Sci. Ind. Res.* (India) **13B**, 326–328 (1954). From *Parmelia cirrhata*, (+)-protolichesterinic acid.

ASAHINA, Y., *J. Japan. Botany* **17**, 71–76 (1941). Microcrystal test.

ASAHINA, Y., *J. Japan. Botany* **18**, 489–502 (1942). In *Pycnothelia papillaria*.

ASAHINA, Y., Lichens of Japan. Vol. I. Genus *Cladonia*. Tokyo. 1950. Microcrystal test.

ASAHINA, Y., Lichens of Japan. Vol. II. Genus *Parmelia*. Tokyo. 1952. Microcrystal test.

ASAHINA, Y., *J. Japan. Botany* **34**, 225–230 (1959). Microcrystal test and paper chromatography.

ASANO, M., *Yakugaku Zasshi* **539**, 1–6 (1927). From *Cetraria ericetorum*, structure studied.

ASANO, M. AND T. AZUMI, *Chem. Ber.* **68**, 995–997 (1935). From *Cetraria stracheyi* f. *ectocarpisma*, as unknown substance B, (−)-protolichesterinic acid.

ASANO, M. AND T. KANEMATSU, *Yakugaku Zasshi* **51**, 390–395 (1931). From *Cetraria ericetorum*, (−)-protolichesterinic acid.

ASANO, M. AND T. KANEMATSU, *Chem. Ber.* **65**, 1175–1178 (1932). From *Cetraria ericetorum*, structure proved, (−)-protolichesterinic acid.

ASANO, M. AND M. TANIGUTI, *Yakugaku Zasshi* **59**, 607–609 (1939); German summary, pp. 216 (1939). From *Cetraria stracheyi* f. *ectocarpisma*, (−)-protolichesterinic acid.

BENDZ, GERD, J. SANTESSON, AND L. TIBELL, *Acta Chem. Scand.* **20**, 1181 (1966). Thin-layer chromatography.

CAVALTITO, C. J., DOROTHY M. FRUEHAUF, AND J. H. BAILEY, *J. Am. Chem. Soc.* **70**, 3724–3726 (1948). Antibacterial activity, (−)-protolichesterinic acid.

GERTIG, H., *Dissertationes Pharm.* **15**, 235–240 (1963). From *Cetraria islandica*, (+)-protolichesterinic acid.

HALE, M. E., JR., *Contrib. U.S. Natl. Herb.* **36**, 193–358 (1965). Distribution in the genus *Parmelia*.

HUNECK, S. AND G. FOLLMANN, *Z. Naturforsch.* **22b**, 666–670 (1967). Absolute configuration proved.

KROG, HILDUR, *Blyttia* **8**, 91–98 (1951). In *Parmelia fraudans*, microcrystal test.

RANGASWAMI, S. AND V. S. RAO, *Indian J. Pharm.* **17**, 50–53 (1955). From *Parmelia cirrhata*, (+)-protolichesterinic acid.

SANTESSON, J., *Acta Chem. Scand.* **21**, 1162–1172 (1967). Thin-layer chromatography.

TAMELEN, E. E. VAN AND SHIRLEY R. BACH, *Chem. Ind.* **1956**, 1308 (1956); *J. Am. Chem. Soc.* **80**, 3079–3086 (1958). Synthesis of (±)-protolichesterinic acid.

ZOPF, W., *Ann. Chem.* **306**, 282–321 (1899). From *Cetraria cucullata*.

ZOPF, W., *Ann. Chem.* **327**, 317–354 (1903). Structure studied.

(−)-*allo*-protolichesterinic acid

$C_{19}H_{32}O_4$

m.p. 107° (acetic acid); $[\alpha]_D^{18}$ −102° ($CHCl_3$).

$$HOOC \quad CH_2$$

$$CH_3(CH_2)_{12} \quad O \quad O$$

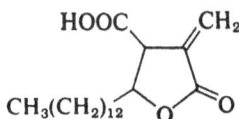

ASAHINA, Y. AND M. YANAGITA, *Chem. Ber.* **69**, 120–125 (1936). From *Cetraria islandica*, first extraction.
ASAHINA, Y. AND M. YASUE, *Chem. Ber.* **70**, 1053–1059 (1937). From *Cetraria ericetorum*, structure studied.
HUNECK, S. AND G. FOLLMANN, *Z. Naturforsch.* **22b**, 666–670 (1967). Absolute configuration proved.

(+)-pseudonorrangiformic acid
$C_{20}H_{36}O_6$
m.p. 189° d, sintering at 184–185° (ethanol-water or ethyl acetate); $[\alpha]_D^{21.8}$ +1.5° (c = 0.2001, H_2O).
Structure not known.

ASAHINA, Y., *J. Japan. Botany* **33**, 1–5 (1958). In *Cladonia submitis*, microcrystal test.
ASAHINA, Y. AND Y. SAKURAI, *Yakugaku Zasshi* **71**, 1166 (1951). From *Cladonia submitis*, named pseudonorrangiformic acid, a tricarboxylic acid isomeric with norrangiformic acid.
EVANS, A. W., *Rhodora* **45**, 417–438 (1943). In *Cladonia alpestris* and *C. submitis*, as unidentified substance C, microcrystal test.
EVANS, A. W., *Trans. Conn. Acad. Arts Sci.* **35**, 519–626 (1944). In *Cladonia alpestris* (as an accessory) and *C. submitis* from Connecticut.
EVANS, A. W., *Bryologist* **58**, 93–112 (1955). In *Cladonia submitis*.

(+)-rangiformic acid
$C_{21}H_{38}O_6$
m.p. 104–105° (acetic acid), colorless needles; $[\alpha]_D^{20}$ +18° (c = 1.92, EtOH).

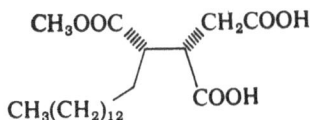

$$CH_3OOC \quad CH_2COOH$$

$$CH_3(CH_2)_{12} \quad COOH$$

ÅKERMARK, B., *Acta Chem. Scand.* **21**, 589–590 (1967). Absolute configuration proved, stereospecific synthesis of (−)-norrangiformic acid.
ÅKERMARK, B., *Svensk Kem. Tidskr.* **79**, 3–19 (1967). Review.
AOKI, M., *Yakugaku Zasshi* **66**, 52–55 (1946); reference from *Chem. Abstr.* **45**, 6584e (1951). Structure of norrangiformic acid proved.

ASAHINA, Y., *J. Japan. Botany* 17, 620–630 (1941); *ibid.* 33, 1–5 (1958). Microcrystal test.

ASAHINA, Y., Lichens of Japan. Vol. I. Genus *Cladonia*. Tokyo. 1950. Microcrystal test.

ASAHINA, Y. AND T. SASAKI, *Bull. Chem. Soc. Japan* 17, 495–498 (1942). From *Cladonia mitis*, structure studied.

BENDZ, GERD, J. SANTESSON, AND L. TIBELL, *Acta Chem. Scand.* 20, 1181 (1966). Thin-layer chromatography.

EVANS, A. W., *Rhodora* 45, 417–438 (1943). Microcrystal tests, and as compound D, in *Cladonia mitis*.

HUNECK, S., *Z. Naturforsch.* 21b, 888–890 (1966). From *Lecanora polytropa*, mass spectral evidence for the position of the methyl ester.

NUNO, MARIKO, *J. Japan. Botany* 37, 77–80 (1962). In *Cladonia retipora*.

PATERNÒ, E., *Gazz. Chim. Ital.* 12, 231–261 (1882). From *Cladonia rangiformis*, first isolation.

SANTESSON, J., *Acta Chem. Scand.* 21, 1162–1172 (1967). Thin-layer chromatography.

(+)-roccellaric acid

$C_{19}H_{34}O_4$

m.p. 110–111° (methanol), colorless prisms; $[\alpha]_D^{20}$ +35° (c = 1.73, $CHCl_3$); ir (KBr), ν cm^{-1}: 3440 (OH), 3360 (OH), 2910, 2845, 1740 (lactone C=O), 1710 (C=O), 1465, 1258, 1205, 1170, 970, and 710.

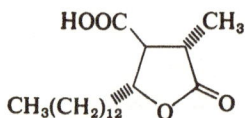

HUNECK, S. AND G. FOLLMANN, *Z. Naturforsch.* 22b, 666–670 (1967). From *Roccellaria mollis*, structure and absolute configuration proved, mass spectrum, nmr spectrum of methyl ester.

(+)-roccellic acid

$C_{17}H_{32}O_4$

m.p. 129–130° (ethyl acetate); $[\alpha]_D^{26}$ +16.8°; ir, ν cm^{-1}: 3000 (very broad, OH), 1695 (broad, C=O), 1275, and 1200 (C—O).

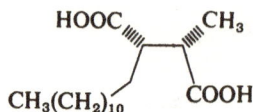

ÅKERMARK, B., *Acta Chem. Scand.* 16, 599–606 (1962). Absolute configuration proved.

Chemical Guide   109

ÅKERMARK, B., *Svensk Kem. Tidskr.* **79**, 3–19 (1967). Review.

ÅKERMARK, B., H. ERDTMAN, AND C. A. WACHTMEISTER, *Acta Chem. Scand.* **13**, 1855–1862 (1959). From *Lepraria membranacea*, structure proved, paper chromatography.

ÅKERMARK, B. AND N.-G. JOHANSSON, *Arkiv Kemi* **27** (1), 1–10 (1967). Stereospecific synthesis of (±)-roccellic acid, absolute configuration confirmed.

BARRY, V. C. AND P. A. MCNALLY, *Nature* **156**, 48–49 (1945). From *Lecanora rupicola*, inhibition of acid-fast bacteria.

BENDZ, GERD, J. SANTESSON, AND L. TIBELL, *Acta Chem. Scand.* **20**, 1181 (1966). Thin-layer chromatography.

HESSE, O., *J. Prakt. Chem.* **57**, 232–318 (1898); *ibid.* **57**, 409–447 (1898). Formula determined.

HUNECK, S., *Tetrahedron Letters* **1966**, 3547–3549 (1966). From *Lecanora rupicola*.

HUNECK, S. AND G. FOLLMANN, *Z. Naturforsch.* **19b**, 658–659 (1964). From *Dirina lutosa*.

HUNECK, S. AND G. FOLLMANN, *Z. Naturforsch.* **22b**, 1185–1188 (1967). From *Roccellina condensata*.

HUNECK, S. AND G. FOLLMANN, *Z. Naturforsch.* **22b**, 1369–1370 (1967). *Roccella fucoides* and *R. gayana*.

HUNECK, S. AND G. FOLLMANN, *Ber. Deut. Botan. Ges.*, in press. From *Roccella hypomecha*.

HUNECK, S., G. FOLLMANN, W. A. WEBER, AND G. TROTET, *Z. Naturforsch.* **22b**, 671–673 (1967). From *Roccella fucoides* and *R. portentosa*.

HUNECK, S., G. FOLLMANN, AND H. ULLRICH, *Z. Naturforsch.* **23b**, 292–293 (1968). From *Roccella boergesenii*.

HUNECK, S., ANNICK MATHEY, AND G. TROTET, *Z. Naturforsch.* **22b**, 1367–1368 (1967). From *Roccella fuciformis*.

KENNEDY, G., J. BREEN, J. KEANE, AND T. J. NOLAN, *Sci. Proc. Roy. Dublin Soc.* **21**, 557–566 (1937). From *Lecanora rupicola*, structure proved.

RAO, V. S. AND T. R. SESHADRI, *Proc. Indian Acad. Sci.* **13A**, 199–202 (1941). From *Roccella montagnei*.

SANTESSON, J., *Acta Chem. Scand.* **21**, 1162–1172 (1967). Thin-layer chromatography.

SUBRAMANIAN, S. S. AND M. N. SWAMY, *J. Sci. Ind. Res.* (India) **20C**, 275–276 (1961). From *Roccella montagnei*.

VARTIA, K. O., *Ann. Med. Exptl. Biol. Fenniae* (Helsinki) **28** (Suppl. 7), 1–82 (1950). From *Lecanora cenisea*.

9,10,12,13-tetrahydroxyheneicosanoic acid
$C_{21}H_{42}O_6$

$$CH_3(CH_2)_7\overset{OH}{CH}\overset{OH}{CH}CH_2\overset{OH}{CH}\overset{OH}{CH}(CH_2)_7COOH$$

SOLBERG, Y. J., *Acta Chem. Scand.* **14**, 2152–2160 (1960). Mixed with ventosic acid from *Haematomma ventosum*, presence indicated by HIO_4 oxidation experiments.

tetrahydroxytricosanoic acid
$C_{23}H_{46}O_6$
Structure incompletely known.

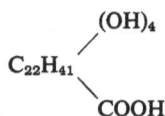

$$C_{22}H_{41} \overset{\diagup (OH)_4}{\underset{\diagdown COOH}{}}$$

SOLBERG, Y. J., *Acta Chem. Scand.* **14**, 2152–2160 (1960). From *Parmelia centrifuga*, product (m.p. 191°) probably containing a small amount of a related substance, structure studied, infrared spectrum.

ventosic acid (=9,10,12,13-tetrahydroxydocosanoic acid)
$C_{22}H_{44}O_6$
m.p. 183–185° (dioxane; ethanol; glacial acetic acid), white amorphous powder; nearly insoluble in NaOH solution; ir (KBr), ν cm^{-1}: 3280 (OH), 1710 (dimeric aliphatic acid), 1382 (C—CH_3), and 2920 and 1445–1470 (aliphatic C—H and CH_2).

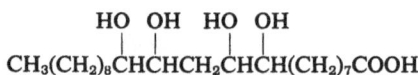

$$\overset{HO \quad OH \quad HO \quad OH}{CH_3(CH_2)_8CHCHCH_2CHCH(CH_2)_7COOH}$$

AGHORAMURTHY, K., K. G. SARMA, AND T. R. SESHADRI, *J. Sci. Ind. Res.* (India) **20B**, 166–168 (1961). From *Stereocaulon tomentosum* ssp. *myriocarpum* var. *orizabae* and *Usnea pectinata*.

SOLBERG, Y. J., *Acta Chem. Scand.* **11**, 1477–1484 (1957). From *Haematomma ventosum*, structure studied, X-ray diffraction diagram, similar substances found in *Cladonia alpestris*, *Cetraria nivalis*, and *Parmelia centrifuga*, infrared spectra.

SOLBERG, Y. J., *Acta Chem. Scand.* **14**, 2152–2160 (1960). Structure proposed, HIO_4 oxidation.

unidentified aliphatic acids
ASAHINA, Y., *J. Japan. Botany* **42**, 1–9 (1967). In *Usnea bayleyi* and subspecies, microcrystal test.

CULBERSON, CHICITA F., *Phytochemistry* **4**, 951–961 (1965). Unidentified tetrahydroxy aliphatic acid, from *Ramalina druidarum* (as *R. siliquosa*).

HUNECK, S., G. FOLLMANN, AND H. ULLRICH, *Z. Naturforsch.* **23b**, 292–293 (1968). Tetrahydroxy fatty acid mixtures from *Roccella boergesenii* (0.5%) and *R. immutata* (0.06%).

HUNECK, S., G. FOLLMANN, W. A. WEBER, AND G. TROTET, *Z. Natur-forsch.* **22b**, 671–673 (1967). Unidentified tetrahydroxy fatty acids from *Roccella fimbriata, R. gayana,* and *R. portentosa.*

HUNECK, S. AND G. TROTET, *Z. Naturforsch.* **21b**, 904 (1966). Tetrahydroxy aliphatic acid mixture (m.p. 185–186°) from *Ramalina boulhautiana.*

SOLBERG, Y. J., *Acta Chem. Scand.* **11**, 1477–1484 (1957). Tetrahydroxy aliphatic acids from *Cladonia alpestris, Cetraria nivalis,* and *Parmelia centrifuga,* infrared spectra.

SOLBERG, Y. J., *Acta Chem. Scand.* **14**, 2152–2160 (1960). Tetrahydroxy aliphatic acids from *Alectoria ochroleuca, Cetraria delisei, Cetraria nivalis, Cladonia alpestris, Hypogymnia encausta, Pseudevernia furfuracea,* and *Siphula ceratites,* infrared spectra.

WAGNER, H. AND H. FRIEDRICH, *Naturwissenschaften* **52**, 305 (1965). C_{16}, C_{18}, C_{20}, and C_{22} unsaturated aliphatic acids in *Hypogymnia physodes,* thin-layer chromatography.

ZELLNER, J., *Monatsh. Chem.* **64**, 6–11 (1934). From *Hypogymnia physodes,* as hypogymnol I (m.p. 190°) and hypogymnol II (m.p. 218°).

NONAROMATIC CARBOCYCLIC PRODUCTS

acetylportentol
$C_{19}H_{28}O_6$
m.p. 215–216°; $[\alpha]_D$ −35°.

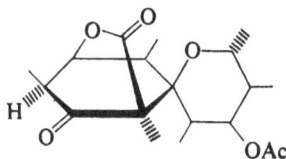

ABERHART, D. J., S. HUNECK, AND K. H. OVERTON, Abstracts of papers presented at the 5th International Symposium on the Chemistry of Natural Products, London, 8–13 July, 1968, p. 110. Structure proved.

HUNECK, S., ANNICK MATHEY, AND G. TROTET, *Z. Naturforsch.* **22b**, 1367–1368 (1967). From *Roccella fuciformis* from the English Channel (1.3%).

portentol
$C_{17}H_{26}O_5$
m.p. 240–241° d (ethanol); m.p. 244–245° (chloroform-ethanol); $[\alpha]_D$

$+21°$; ir (KBr), ν cm^{-1}: 3450 (OH), 1754 (C=O), 1720 (C=O); uv (MeOH), λ_{max} (log ε) mμ: 295 (2.47).

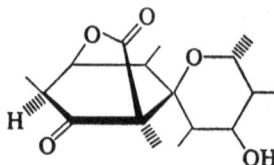

Aberhart, D. J., S. Huneck, and K. H. Overton, Abstracts of papers presented at the 5th International Symposium on the Chemistry of Natural Products, London, 8–13 July, 1968, p. 110. Structure proved.

Huneck, S. and G. Follmann, *Z. Naturforsch.* **22b**, 1185–1188 (1967). From *Roccellina condensata* (0.3%).

Huneck, S. and G. Trotet, *Z. Naturforsch.* **22b**, 363 (1967). From *Dirina repanda*, first extraction.

Huneck, S., G. Follmann, W. A. Weber, and G. Trotet, *Z. Naturforsch.* **22b**, 671–673 (1967). From *Roccella portentosa*.

Huneck, S., Annick Mathey, and G. Trotet, *Z. Naturforsch.* **22b**, 1367–1368 (1967). From a sample of *Roccella fuciformis* (0.1%).

Phenolic Carboxylic Acid Derivatives: Monocyclic Derivatives

methyl β-orcinolcarboxylate
$C_{10}H_{12}O_4$
m.p. 143–144° (benzene), prismatic rods.

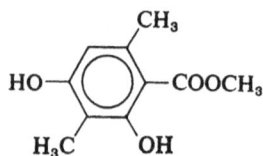

Murty, T. K., *J. Sci. Ind. Res.* (India) **19B**, 508–509 (1960). From *Parmelia tinctorum* (0.05%).

(±)-montagnetol
$C_{12}H_{16}O_7$
m.p. 153–154° (ethyl acetate-light petroleum ether, 2:1 v/v); ir (KBr), ν cm^{-1}: 3448, 1639, 1449, 1307, 1242, 1190, 1163, and 1099; uv (MeOH), λ_{max} (log ε) mμ: 262 (3.9) and 298 (4.3); λ_{min} (log ε) mμ: 237 (3.7) and 282 (3.7); soluble in water.
For structure and references see (+)-montagnetol.

(+)-montagnetol

$C_{12}H_{16}O_7$

m.p. 135–136°; $[\alpha]_D$ +16.0° (H_2O), +12.6° (acetone), racemized by boiling in water 3 hrs.

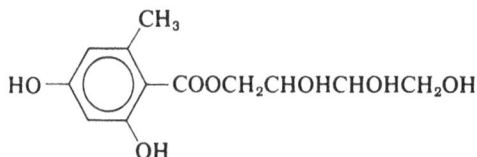

MANAKTALA, S. K., S. NEELAKANTAN, AND T. R. SESHADRI, *Tetrahedron*
22, 2373–2376 (1966). Synthesis of (±)-montagnetol.

RAO, V. S. AND T. R. SESHADRI, *Proc. Indian Acad. Sci.* **13A**, 199–202
(1941). From *Roccella montagnei*, first isolation of the racemate.

RAO, V. S. AND T. R. SESHADRI, *Proc. Indian Acad. Sci.* **15A**, 18–23
(1942). Structure proved.

RAO, V. S. AND T. R. SESHADRI, *Proc. Indian Acad. Sci.* **15A**, 429–431
(1942). Discovery of optically active montagnetol in *Roccella montagnei*.

SUBRAMANIAN, S. S. AND M. N. SWAMY, *J. Sci. Ind. Res.* (India) **20C**,
275–276 (1961). (+)-Montagnetol from *Roccella montagnei*.

orcinol

$C_7H_8O_2$

m.p. 106.5–108.0° for anhydrous crystals (forms a hydrate from water);
ir ($CHCl_3$), ν cm⁻¹: 3597, 3311, 1603, 1497, 1340, 1148, 1033, and
831 (for the monohydrate); nmr (acetone), τ ppm: 7.87 (aryl CH_3),
5.62 (H_2O), 3.78 (aryl protons), and 0.73 (phenolic protons).

Sometimes formed during extraction by decomposition of depsides of
orsellinic acid.

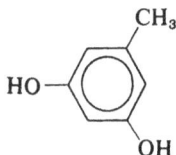

MILLER, E. V., C. E. GRIFFIN, T. SCHAEFERS, AND MYRA GORDON, *Botan.
Gaz.* **126**, 100–107 (1965). From *Lasallia papulosa*, but probably formed
during extraction.

RAMAUT, J. L., *Rev. Bryol. Lichénol.* **29**, 307–320 (1960). In *Parmelia
revoluta* and *P. subrudecta*, possibly an artifact of the chromatographic
method.

RAO, V. S. AND T. R. SESHADRI, *Proc. Indian Acad. Sci.* **13A**, 199–202
(1941). From *Roccella montagnei*.

SUBRAMANIAN, S. S. AND M. N. SWAMY, *J. Sci. Ind. Res.* (India) **20C**, 275–276 (1961). From *Roccella montagnei*.

PHENOLIC CARBOXYLIC ACID DERIVATIVES: DIARYL AND TRIARYL ESTERS — ORCINOL SERIES

para-DEPSIDES AND TRIDEPSIDES

anziaic acid

$C_{24}H_{30}O_7$

m.p. 124° d (ethanol-water), needles.

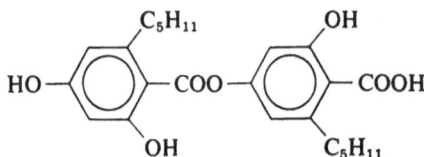

ASAHINA, Y., *J. Japan. Botany* **12**, 516–525 (1936); *ibid.* **13**, 529–536 (1937). Microcrystal tests.

ASAHINA, Y. AND M. HIRAIWA, *Chem. Ber.* **68**, 1705–1708 (1935). From *Anzia japonica* (as *A. gracilis*), first isolation.

ASAHINA, Y. AND M. HIRAIWA, *Chem. Ber.* **70**, 1826–1828 (1937). Synthesis.

CULBERSON, W. L. AND CHICITA F. CULBERSON, *Contrib. U.S. Natl. Herb.* **34**, 449–558 (1968). In the genus *Cetrelia*, microcrystal tests, thin-layer chromatography.

KUROKAWA, S. AND YOKO JINZENJI, *Bull. Natl. Sci. Mus.* (Tokyo) **8**, 369–374 (1965). Distribution in the genus *Anzia* in Japan, microcrystal tests, thin-layer chromatography.

SANTESSON, J., *Acta Chem. Scand.* **21**, 1162–1172 (1967). Thin-layer chromatography.

confluentic acid

$C_{28}H_{36}O_8$

m.p. 157° (methanol); ir (CCl_4), ν cm^{-1}: 3520, 3110, 1755, 1710, and 1685.

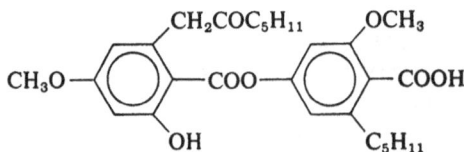

CULBERSON, CHICITA F., *Bryologist* **69**, 312–317 (1966). From *Herpothallon sanguineum*, microchemical tests.

FOLLMANN, G. AND S. HUNECK, *Willdenowia*, in press. From *Entero-grapha crassa*.

HUNECK, S., *Chem. Ber.* **95**, 328–332 (1962). From *Lecidea confluens*, structure proved.

HUNECK, S., *Naturwissenschaften* **49**, 374–375 (1962). From *Lecidea tumida*.

HUNECK, S., *Z. Naturforsch.* **20b**, 1137–1138 (1965). From *Lecidea macrocarpa*.

HUNECK, S., C. DJERASSI, D. BECHER, M. BARBER, M. VON ARDENNE, K. STEINFELDER, AND R. TÜMMLER, *Tetrahedron* **24**, 2707–2755 (1968). Mass spectrum.

SANTESSON, J., *Acta Chem. Scand.* **21**, 1162–1172 (1967). Thin-layer chromatography.

ZOPF, W., *Ann. Chem.* **306**, 282–321 (1899). From *Lecidea confluens*, first extraction, as confluentin (m.p. 147–148°).

diploschistesic acid

$C_{16}H_{14}O_8$

m.p. 174° d (acetone-water-acetic acid); NaOCl+ blue.

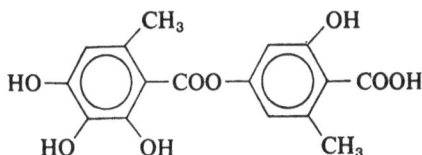

ASAHINA, Y. AND M. YASUE, *Chem. Ber.* **69**, 2327–2330 (1936). Synthesis.

KOLLER, G. AND H. HAMBURG, *Monatsh. Chem.* **65**, 367–374 (1935). From *Diploschistes scruposus*, structure studied, patellaric acid possibly identical to diploschistesic acid.

SANTESSON, J., *Acta Chem. Scand.* **21**, 1162–1172 (1967). Thin-layer chromatography.

SESHADRI, T. R. AND G. B. VENKATASUBRAMANIAN, *J. Chem. Soc.* **1959**, 1658–1659 (1959). Synthesis from lecanoric acid.

ZOPF, W., *Ann. Chem.* **346**, 82–127 (1906). From *Diploschistes scruposus*.

divaricatic acid

$C_{21}H_{24}O_7$

m.p. 137–138° (benzene), needles.

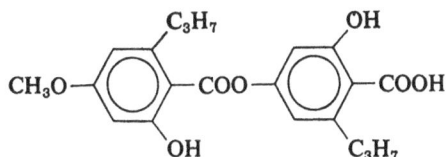

ASAHINA, Y., *J. Japan. Botany* **12**, 859–872 (1936); *ibid.* **13**, 855–861 (1937). Microchemical tests.

ASAHINA, Y. AND M. HIRAIWA, *Chem. Ber.* **68**, 1705–1708 (1935). From *Anzia japonica* (as *A. gracilis*).

ASAHINA, Y. AND M. HIRAIWA, *Chem. Ber.* **70**, 1826–1828 (1937). Synthesis.

ASAHINA, Y. AND T. HIRAKATA, *Chem. Ber.* **65**, 1665–1668 (1932). From *Evernia esorediosa*, structure proved.

CULBERSON, W. L., *Bryologist* **66**, 224–236 (1963). Distribution in the genus *Haematomma*.

FUZIKAWA, F., K. NAKAJIMA, A. TOKUOKA, Y. HITOSA, T. OMATSU, T. SUMIYAMA, AND M. KASHIWAGI, *Yakugaku Zasshi* **74**, 176–179 (1954). Antibacterial activity.

HALE, M. E., JR. AND S. KUROKAWA, *Contrib. U.S. Natl. Herb.* **36**, 121–191 (1964). Distribution in the genus *Parmelia*.

HESS, D., *Planta* **52**, 65–76 (1958). Paper chromatography.

HESSE, O., *J. Prakt. Chem.* **57**, 232–318 (1898). From *Evernia divaricata*, first isolation.

HUNECK, S., *Naturwissenschaften* **50**, 645 (1963). From *Lecidea kochiana*.

KUROKAWA, S. AND YOKO JINZENJI, *Bull. Natl. Sci. Mus.* (Tokyo) **8**, 369–374 (1965). Distribution in the genus *Anzia* in Japan, thin-layer chromatography.

MITSUNO, M., *Pharm. Bull.* (Tokyo) **1**, 170–173 (1953). Paper chromatography.

RAMAUT, J. L., *Bull. Soc. Roy. Botan. Belg.* **93**, 27–40 (1961). From *Haematomma ventosum*.

SANTESSON, J., *Acta Chem. Scand.* **21**, 1162–1172 (1967). Thin-layer chromatography.

erythrin

$C_{20}H_{22}O_{11}$

m.p. 156–157°; [α]$_D$ +8.0°; ir (KBr), ν cm^{-1}: 3448, 1639, 1625, 1504, 1462, 1399, 1250, 1245, 1075, 909, 877, and 833; uv (MeOH), λ$_{max}$ (log ε) mμ: 269 (4.0) and 303 (4.3).

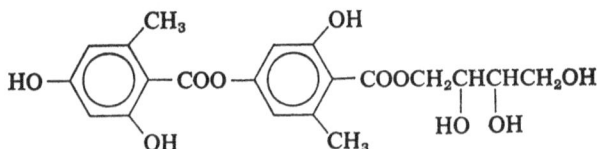

ASAHINA, Y., *J. Japan. Botany* **12**, 859–872 (1936). From *Roccella phycopsis*, microcrystal tests.

HUNECK, S. AND G. FOLLMANN, *Z. Naturforsch.* **22b**, 362–363 (1967). From *Ingaderia pulcherrima*.

HUNECK, S. AND G. FOLLMANN, *Z. Naturforsch.* **22b**, 1368–1369 (1967). From *Dirina ceratoniae* from Tunisia (1.8%).

HUNECK, S. AND G. FOLLMANN, Z. *Naturforsch.* **22b**, 1369–1370 (1967). From *Combea mollusca* (3.5%).

HUNECK, S. AND G. FOLLMANN, *Pharmazie* **23**, 156 (1968). From *Dirina approximata, D. catalinariae, D. chilena, D. limitata,* and *Dirinastrum chilenum.*

HUNECK, S. AND G. FOLLMANN, *Ber. Deut. Botan. Ges.,* in press. From numerous species of *Roccella.*

HUNECK, S. AND G. FOLLMANN, in preparation. From *Chiodecton cretaceum.*

HUNECK, S. AND G. FOLLMANN, in preparation. From *Roccella africana, R. flaccida, R. linearis* var. *primaria, R. montagnei* f. *obtusa, R. podocarpa,* and *R. tinctoria* var. *subpodicellata.*

HUNECK, S., G. FOLLMANN, AND H. ULLRICH, Z. *Naturforsch.* **23b**, 292–293 (1968). In *Roccella linearis* and from *R. teneriffensis* (2.5%) and *R. boergesenii* (2.1%).

HUNECK, S., G. FOLLMANN, W. A. WEBER, AND G. TROTET, Z. *Naturforsch.* **22b**, 671–673 (1967). From *Roccella babingtonii, R. fimbriata, R. fucoides,* and a sample of *R. portentosa* from the Galápagos Islands.

HUNECK, S., ANNICK MATHEY, AND G. TROTET, Z. *Naturforsch.* **22b**, 1367–1368 (1967). From *Roccella fuciformis* from the English Channel (1.3%) and from Morocco (2%).

HUNECK, S. AND G. TROTET, Z. *Naturforsch.* **22b**, 363 (1967). From *Dirina repanda.*

MANAKTALA, S. K., S. NEELAKANTAN, AND T. R. SESHADRI, *Tetrahedron* **22**, 2373–2376 (1966). Synthesis of (±)-erythrin.

RAO, V. S. AND T. R. SESHADRI, *Proc. Indian Acad. Sci.* **13A**, 199–202 (1941). From *Roccella montagnei.*

RAO, V. S. AND T. R. SESHADRI, *Proc. Indian Acad. Sci.* **16A**, 23–28 (1942). Structure proved.

SAKURAI, Y., *Yakugaku Zasshi* **61**, 108–115 (1941); German summary, pp. 45–46 (1941); reference from *Chem. Abstr.* **36**, 1599[5] (1942). From *Roccella montagnei,* structure studied.

SUBRAMANIAN, S. S. AND M. N. SWAMY, *J. Sci. Ind. Res.* (India), **20C**, 275–276 (1961). From *Roccella montagnei.*

ZERNER, E., *Monatsh. Chem.* **35**, 1021–1024 (1914). Structure studied.

evernic acid
$C_{17}H_{16}O_7$
m.p. 172–174° d (acetone), needles.

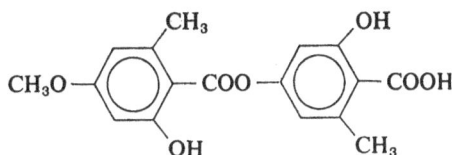

ASAHINA, Y., *J. Japan. Botany* **12**, 859–872 (1936); *ibid.* **13**, 855–861 (1937). Microchemical tests.

CULBERSON, CHICITA F., *Phytochemistry* **2**, 335–340 (1963). From *Evernia prunastri.*

CURD, F. H., A. ROBERTSON, AND R. J. STEPHENSON, *J. Chem. Soc.* **1933**, 130–133 (1933). From *Evernia prunastri.*

FISCHER, E. AND H. O. L. FISCHER, *Chem. Ber.* **47**, 505–512 (1914). Structure proved.

FUZIKAWA, F. AND K. ISHIGURO, *Yakugaku Zasshi* **56**, 837–840 (1936); German summary, pp. 149–150 (1936). Synthesis.

HALE, M. E., JR. AND S. KUROKAWA, *Contrib. U.S. Natl. Herb.* **36**, 121–191 (1964). In the genus *Parmelia.*

HESS, D., *Planta* **52**, 65–76 (1958). Paper chromatography.

HUNECK, S., C. DJERASSI, D. BECHER, M. BARBER, M. VON ARDENNE, K. STEINFELDER, AND R. TÜMMLER, *Tetrahedron* **24**, 2707–2755 (1968). Mass spectrum of the methyl ester.

HUNECK, S. AND R. TÜMMLER, *Ann. Chem.* **685**, 128–133 (1965). From *Evernia prunastri*, thin-layer chromatography.

KOLLER, G., *Monatsh. Chem.* **61**, 286–292 (1932). From *Ramalina pollinaria.*

MANAKTALA, S. K., S. NEELAKANTAN, AND T. R. SESHADRI, *Indian J. Chem.* **3**, 520–522 (1965). Synthesis of methyl ester.

MITSUNO, M., *Pharm. Bull.* (Tokyo) **1**, 170–173 (1953). Paper chromatography.

NEELAKANTAN, S., R. PADMASANI, AND T. R. SESHADRI, *J. Sci. Ind. Res.* (India) **20B**, 510–511 (1961). Synthesis of evernic acid and methyl evernate.

NEELAKANTAN, S., R. PADMASANI, AND T. R. SESHADRI, *Tetrahedron* **21**, 3531–3536 (1965). Synthesis of evernic acid and methyl evernate.

ROBERTSON, A. AND R. J. STEPHENSON, *J. Chem. Soc.* **1932**, 1388–1395 (1932). Structure studied, synthesis of methyl evernate.

SANTESSON, J., *Acta Chem. Scand.* **21**, 1162–1172 (1967). Thin-layer chromatography.

SARMA, K. G. AND S. HUNECK, *Pharmazie*, in press. From *Usnea misaminensis.*

STAHL, E. AND P. J. SCHORN, *Z. Physiol. Chem.* **325**, 263–274 (1961). Thin-layer chromatography.

glomelliferic acid
$C_{25}H_{30}O_8$
m.p. 143–144° (benzene).

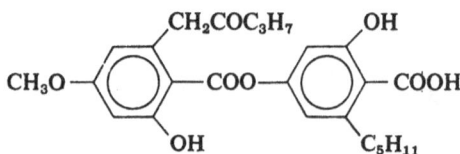

ASAHINA, Y. AND H. NOGAMI, *Chem. Ber.* **70**, 1498–1499 (1937). From *Parmelia glomellifera*, structure studied.

HUNECK, S. AND K. H. OVERTON, *Misc. Bryol. Lichenol.* (Japan), in press. From *Lecidea leucophaea*.

KROG, HILDUR, *Nytt Mag. Naturvidenskapene* **88**, 57–85 (1951). From *Parmelia isidiotyla* and in some samples of *P. pulla*, microcrystal test.

MINAMI, K., *Yakugaku Zasshi* **64**, 315–317 (1944); reference from *Chem. Abstr.* **45**, 2939F (1951). Synthesis of glomellin.

SANTESSON, J., *Acta Chem. Scand.* **21**, 1162–1172 (1967). Thin-layer chromatography.

ZOPF, W., *Ann. Chem.* **306**, 282–321 (1899). From *Parmelia glomellifera*, first isolation.

ZOPF, W., *Ann. Chem.* **321**, 37–61 (1902). From *Parmelia glomellifera*.

gyrophoric acid

$C_{24}H_{20}O_{10}$

m.p. 220–225° d (acetone); m.p. 212–214° d (acetone-methanol).

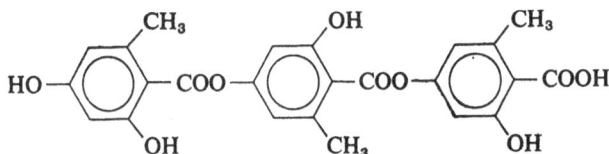

HO—(ring, CH₃, OH)—COO—(ring, OH, CH₃)—COO—(ring, CH₃, OH)—COOH

ASAHINA, Y., *J. Japan. Botany* **12**, 516–525 (1936); *ibid.* **13**, 529–536 (1937). Microchemical tests.

ASAHINA, Y., Lichens of Japan. Vol. II. Genus *Parmelia*. Tokyo. 1952. Microchemical tests.

ASAHINA, Y. AND F. FUZIKAWA, *Chem. Ber.* **65**, 983–984 (1932). Synthesis.

ASAHINA, Y. AND N. KUTANI, *Yakugaku Zasshi* **519**, 423–429 (1925). Structure studied.

ASAHINA, Y. AND M. WATANABE, *Chem. Ber.* **63**, 3044–3048 (1930). From *Umbilicaria esculenta*, structure proved.

ASAHINA, Y. AND I. YOSIOKA, *Chem. Ber.* **70**, 200–206 (1937). Synthesis.

CANTER, F. W., A. ROBERTSON, AND R. B. WATERS, *J. Chem. Soc.* **1933**, 493–495 (1933). Synthesis of methyl O-tetramethylgyrophorate.

CULBERSON, CHICITA F. AND W. L. CULBERSON, *Lloydia* **21**, 189–192 (1958). In *Lasallia papulosa* of different ages.

DHAR, M. L., S. NEELAKANTAN, S. RAMANUJAM, AND T. R. SESHADRI, *J. Sci. Ind. Res.* (India) **18B**, 111–113 (1959). From *Parmelia sanctiangelii*.

FOX, C. H. AND K. MOSBACH, *Acta Chem. Scand.* **21**, 2327–2330 (1967). Biosynthesis, incorporation of ^{14}C from $^{14}CO_2$ into gyrophoric acid in *Lasallia pustulata*.

HALE, M. E., JR., *Bull. Torrey Botan. Club* **79**, 251–259 (1952). In a strain of *Rinodina oreina*.

120   *Chemical and Botanical Guide to Lichen Products*

HALE, M. E., JR., *Contrib. U.S. Natl. Herb.* **36**, 193–358 (1965). Distribution in the genus *Parmelia*.
HALE, M. E., JR. AND S. KUROKAWA, *Contrib. U.S. Natl. Herb.* **36**, 121–191 (1964). Distribution in the genus *Parmelia*.
HESS, D., *Planta* **52**, 65–76 (1958). Paper chromatography.
HUNECK, S., *Naturwissenschaften* **49**, 396 (1962). From *Acarospora fuscata*.
HUNECK, S., *Naturwissenschaften* **16**, 477 (1965). From *Lecidea griseoatra*.
HUNECK, S., *Z. Naturforsch.* **20b**, 1137–1138 (1965). From *Lecanora fuscoatra*.
HUNECK, S.,*Z. Naturforsch.* **21b**, 80–81 (1966). From *Acarospora montana*.
HUNECK, S., C. DJERASSI, D. BECHER, M. BARBER, M. VON ARDENNE, K. STEINFELDER, AND R. TÜMMLER, *Tetrahedron* **24**, 2707–2755 (1968). Mass spectrum.
HUNECK, S. AND G. FOLLMANN, *Z. Naturforsch.* **20b**, 496 (1965). From *Dolichocarpus chilensis*.
HUNECK, S., G. FOLLMANN, AND W. A. WEBER, *Willdenowia*, in press. From *Dactylina arctica* and a sample of *D. ramulosa* from North America.
KOLLER, G., *Monatsh. Chem.* **61**, 147–161 (1932). From *Lasallia pustulata*, synthesis of methyl O-tetraacetylgyrophorate.
KOLLER, G. AND G. PFEIFFER, *Monatsh. Chem.* **62**, 241–251 (1933). From *Umbilicaria deusta*.
KOLLER, G. AND G. PFEIFFER, *Monatsh. Chem.* **62**, 359–372 (1933). From *Lasallia pustulata*.
MITSUNO, M., *Pharm. Bull.* (Tokyo) **1**, 170–173 (1953). Paper chromatography.
RUNEMARK, H., *Opera Botan.* (Lund) **2** (1), 1–152 (1956). Distribution in the genus *Rhizocarpon*, paper chromatography.
SANTESSON, J., *Acta Chem. Scand.* **21**, 1162–1172 (1967). Thin-layer chromatography.
VARTIA, K. O., *Ann. Med. Exptl. Biol. Fenniae* (Helsinki) **28** (Suppl. 7), 1–82 (1950). From *Lasallia pustulata, Lecidea granulosa*, and *Ochrolechia tartarea*.
ZELLNER, J., *Monatsh. Chem.* **66**, 81–86 (1935). From *Umbilicaria mammulata*.

hiascic acid
$C_{24}H_{20}O_{11}$
m.p. 190.5° (ethanol).

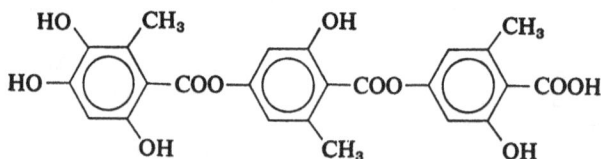

ASAHINA, Y. AND T. KUSAKA, *Bull. Chem. Soc. Japan* **17**, 152–159 (1942). From *Cetraria delisei* (as *C. hiascens*), first isolation, structure proved.
SANTESSON, J., *Acta Chem. Scand.* **21**, 1162–1172 (1967). Thin-layer chromatography.

imbricaric acid
$C_{23}H_{28}O_7$
m.p. 125.5–126.0° (benzene-petroleum ether); ir (nujol), ν cm^{-1}: 1670 (bonded C=O) and 1650 and 1615 (aryl C=C); uv (95% EtOH), λ_{max} (ε) mμ: 212 (47,100), 269.5 (19,000), and 305 (11,700); uv (95% EtOH), λ_{min} (ε) mμ: 242 (8,850) and 288.5 (9,250).

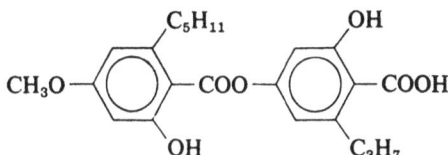

ASAHINA, Y., *J. Japan. Botany* **14**, 39–44 (1938). Microcrystal tests.
ASAHINA, Y., *J. Japan. Botany* **39**, 209–215 (1964). In *Haematomma polycarpum* and *H. puniceum* "ssp. *pacificum* Asah. *ad. interim.*"
ASAHINA, Y. AND F. FUZIKAWA, *Chem. Ber.* **68**, 634–639 (1935). From *Cetrelia cetrarioides*, structure proved.
ASAHINA, Y. AND I. YOSIOKA, *Chem. Ber.* **70**, 1823–1826 (1937). Synthesis.
CULBERSON, CHICITA F. AND W. L. CULBERSON, *Bryologist* **69**, 192–202 (1966). From *Cetrelia alaskana*, microchemical tests.
CULBERSON, W. L. AND CHICITA F. CULBERSON, *Contrib. U.S. Natl. Herb.* **34**, 449–558 (1968). Distribution in the genus *Cetrelia*.
HESS, D., *Planta* **52**, 65–76 (1958). Paper chromatography.
ZOPF, W., *Ann. Chem.* **317**, 110–145 (1901). From *Parmelia locarnensis*.
ZOPF, W., *Ann. Chem.* **321**, 37–61 (1902). From *Cetrelia cetrarioides* (as *Parmelia perlata*).

lecanoric acid
$C_{16}H_{14}O_7$
m.p. 177–187.5° d, shrinking at 183°, bubbling at 184.5° (ethanol-water); ir (nujol), ν cm^{-1}: 3600 and 3475 (aryl OH), 1660 (bonded C=O), and 1625 and 1595 (aryl C=C); uv (95% ethanol), λ_{max} (log ε) mμ: 214 (4.63), 270.5 (4.30), 303.5 (4.09); λ_{min} (log ε) mμ: 244.5 (3.89) and 290 (4.01); nmr (CDCl$_3$), δ ppm: 2.56 and 2.60 (2 aryl CH$_3$ groups) and 6.26 and 6.69 (4 aryl protons).

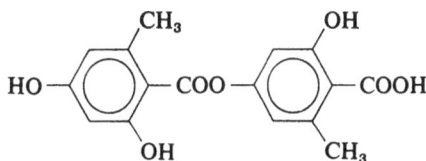

AHMANN, G. B. AND ANNICK MATHEY, *Bryologist* **70**, 93–97 (1967). From *Parmelia tinctorum* and *Pseudevernia intensa*.

ASAHINA, Y., J. *Japan. Botany* **12**, 516–525 (1936); *ibid.* **13**, 529–536 (1937). Microchemical tests.

ASAHINA, Y., Lichens of Japan. Vol. II. Genus *Parmelia*. Tokyo. 1952. Microcrystal tests.

DHAR, M. L., S. NEELAKANTAN, S. RAMANUJAM, AND T. R. SESHADRI, *J. Sci. Ind. Res.* (India) **18B**, 111–113 (1959). From *Parmelia meizospora* and *P. subargentifera*.

FISCHER, E. AND H. O. L. FISCHER, *Chem. Ber.* **46**, 1138–1148 (1913). Synthesis.

HALE, M. E., JR., *Contrib. U.S. Natl. Herb.* **36**, 193–358 (1965). Distribution in the genus *Parmelia*.

HALE, M. E., JR. AND S. KUROKAWA, *Contrib. U.S. Natl. Herb.* **36**, 121–191 (1964). Distribution in the genus *Parmelia*.

HARPER, S. H. AND R. M. LETCHER, *Proc. Trans. Rhodesian Sci. Assoc.* **51**, 156–184 (1966). Infrared spectrum, from *Dermatiscum thunbergii* (7.5%), *Parmelia dilatata* (1.25%), and *P. gossweileri* (6.6%).

HESS, D., *Planta* **52**, 65–76 (1958). Paper chromatography.

HESSE, O., J. *Prakt. Chem.* **62**, 430–480 (1900). Structure proved.

HUNECK, S., C. DJERASSI, D. BECHER, M. BARBER, M. VON ARDENNE, K. STEINFELDER, AND R. TÜMMLER, *Tetrahedron* **24**, 2707–2755 (1968). Mass spectrum.

HUNECK, S. AND G. FOLLMANN, *Z. Naturforsch.* **20b**, 1012–1013 (1965). From *Byssocaulon niveum* and *Stereocaulon corticulatum*.

HUNECK, S. AND G. FOLLMANN. *Z. Naturforsch.* **22b**, 362–363 (1967). From *Roccella arboricola*.

HUNECK, S. AND G. FOLLMANN, *Z. Naturforsch.* **22b**, 1185–1188 (1967). From *Roccellina condensata*.

HUNECK, S. AND G. FOLLMANN, *Z. Naturforsch.* **22b**, 1369–1370 (1967). From *Roccella vicentina* (20%).

HUNECK, S. AND G. FOLLMANN, *Pharmazie* **23**, 156 (1968). From *Dirina veruculosa*.

HUNECK, S. AND G. FOLLMANN, *Ber. Deut. Botan. Ges.*, in press. From *Chiodecton sphaerale* and numerous species of *Roccella*.

HUNECK, S. AND G. FOLLMANN, in preparation. From *Roccella africana*, *R. flaccida*, *R. linearis* var. *primaria*, *R. montagnei* f. *obtusa*, *R. podocarpa*, and *Dirina capensis*.

HUNECK, S., G. FOLLMANN, AND H. ULLRICH, *Z. Naturforsch.* **23b** 292–293 (1968). From *Roccella canariensis* (3.2%) and *R. immutata* (12.5%) and in *R. linearis*.

HUNECK, S., G. FOLLMANN, W. A. WEBER, AND G. TROTET, *Z. Naturforsch.* **22b**, 671–673 (1967). From *Roccella gayana* and a sample of *R. portentosa* from Chile.

KOLLER, G. AND G. PFEIFFER, *Monatsh. Chem.* **62**, 169–171 (1933). From *Parmelia glabra*, glabratic acid identical to lecanoric acid.

MITSUNO, M., *Pharm. Bull.* (Tokyo) **1**, 170–173 (1953). Paper chromatography.

MURTY, T. K., *J. Sci. Ind. Res.* (India) **19B**, 508–509 (1960). From *Parmelia tinctorum.*

NEELAKANTAN, S., R. PADMASANI, AND T. R. SESHADRI, *J. Sci. Ind. Res.* (India) **20B**, 510–511 (1961). Synthesis of methyl lecanorate.

NEELAKANTAN, S., R. PADMASANI, AND T. R. SESHADRI, *Indian J. Chem.* **2**, 478–484 (1964). Halogenation.

NEELAKANTAN, S., R. PADMASANI, AND T. R. SESHADRI, *Tetrahedron* **21**, 3531–3536 (1965). Synthesis of methyl lecanorate.

RAMAUT, J. L., *Bull. Soc. Roy. Botan. Belg.* **93**, 27–40 (1961). From *Sticta fuliginosa.*

RANGASWAMI, S. AND V. S. RAO, *Indian J. Pharm.* **17**, 49–50 (1955). From *Parmelia tinctorum.*

RAO, V. S. AND T. R. SESHADRI, *Proc. Indian Acad. Sci.* **13A**, 199–202 (1941). From *Roccella montagnei.*

SANTESSON, J., *Acta Chem. Scand.* **21**, 1162–1172 (1967). Thin-layer chromatography.

WACHTMEISTER, C. A., *Acta Chem. Scand.* **6**, 818–825 (1952). Paper chromatography.

WEBER, W. A., *Svensk Botan. Tidskr.* **59**, 59–64 (1965). In *Hubbsia lumbricoides.*

YAMAZAKI, M., M. MATSUO, AND S. SHIBATA, *Chem. Pharm. Bull.* (Tokyo) **13**, 1015–1017 (1965). Biosynthesis in *Parmelia tinctorum.*

YAMAZAKI, M. AND S. SHIBATA, *Chem. Pharm. Bull.* (Tokyo) **14**, 96–97 (1966). Biosynthesis in *Parmelia tinctorum.*

ZOPF, W., *Ann. Chem.* **313**, 317–344 (1900). From *Parmelia subrudecta* (as *P. borreri*).

methyl 3,5-dichlorolecanorate (= tumidulin)
$C_{17}H_{14}Cl_2O_7$
m.p. 177–177.5° d (benzene); m.p. 174–175° (methanol-water); ir (KBr), ν cm^{-1}: 3400, 1715, 1685, 1620, 1600, and 1575; uv (95% ethanol), λ_{max} (log ε) mμ: 259 (4.2) and 318 (4.1); λ_{min} (log ε) mμ: 244 (4.1) and 286 (3.9); nmr (CDCl$_3$-pyridine, 9:1 v/v), δ ppm: 2.53 and 2.70 (2 aryl CH$_3$) and 3.91 (1 OCH$_3$); NaOCl+ red.

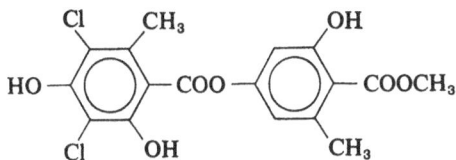

BENDZ, GERD, J. SANTESSON, AND C. A. WACHTMEISTER, *Acta Chem. Scand.* **19**, 1185–1187 (1965). From *Ramalina ceruchis.*

BENDZ, GERD, J. SANTESSON, AND C. A. WACHTMEISTER, *Acta Chem. Scand.* **19**, 1188–1190 (1965). From *Ramalina ceruchis* and *R. flaccescens*, structure proved, synthesis.

HUNECK, S., *Chem. Ber.* **99**, 1106–1110 (1966). Structure proved.

HUNECK, S., C. DJERASSI, D. BECHER, M. BARBER, M. VON ARDENNE, K. STEINFELDER, AND R. TÜMMLER, *Tetrahedron* **24**, 2707–2755 (1968). Mass spectrum.

HUNECK, S. AND G. FOLLMANN, *Bol. Univ. Chile* **7**, 56–57 (1965). From *Ramalina peruviana*.

HUNECK, S. AND G. FOLLMANN, *Z. Naturforsch.* **20b**, 611–612 (1965). From *Ramalina tumidula*, as tumidulin.

HUNECK, S. AND G. FOLLMANN, *Z. Naturforsch.* **21b**, 90–91 (1966). From *Ramalina chilensis*.

HUNECK, S. AND G. FOLLMANN, *Z. Naturforsch.* **21b**, 713–714 (1966). From *Ramalina inanis*.

HUNECK, S. AND G. FOLLMANN, *Z. Naturforsch.* **22b**, 110–111 (1967). From *Ramalina cactacearum*.

SANTESSON, J., *Acta Chem. Scand.* **21**, 1162–1172 (1967). Thin-layer chromatography.

microphyllinic acid
$C_{29}H_{36}O_9$
m.p. 116° (benzene-petroleum ether).

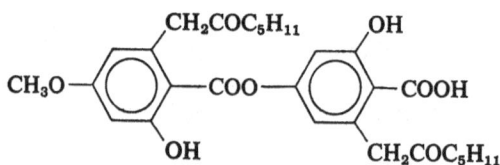

ASAHINA, Y., *J. Japan. Botany* **14**, 39–44 (1938). Microchemical tests.

ASAHINA, Y. AND F. FUZIKAWA, *Chem. Ber.* **68**, 2022–2026 (1935). From *Cetrelia japonica*, first isolation, structure proved.

CULBERSON, W. L. AND CHICITA F. CULBERSON, *Contrib. U.S. Natl. Herb.* **34**, 449–558 (1968). In *Cetrelia japonica*, microchemical tests.

SANTESSON, J. *Acta Chem. Scand.* **21**, 1162–1172 (1967). Thin-layer chromatography.

obtusatic acid
$C_{18}H_{18}O_7$
m.p. 208–209° d (acetone).

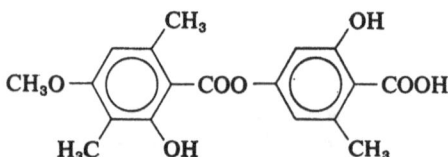

ASAHINA, Y., *J. Japan. Botany* **13**, 855–861 (1937). Microchemical tests.

ASAHINA, Y., *J. Japan. Botany* **15**, 205–223 (1939). In *Ramalina ligulata* and some other species of this genus from Japan, microchemical tests.

ASAHINA, Y. AND F. FUZIKAWA, *Chem. Ber.* **65**, 580–583 (1932). From *Ramalina commixta*, structure proved.

ASAHINA, Y. AND F. FUZIKAWA, *Yakugaku Zasshi* **53**, 846–847 (1933); reference from *Chem. Abstr.* **28**, 3397[1] (1934). Synthesis of methyl obtusatate dimethyl ether.

FUZIKAWA, F., *Yakugaku Zasshi* **56**, 237–258 (1936); German summary, pp. 25–29 (1936). Synthesis.

KOLLER, G., *Monatsh.* **61**, 286–292 (1932). From *Ramalina pollinaria*, as ramalic acid.

MITSUNO, M., *Pharm. Bull.* (Tokyo) **1**, 170–173 (1953). Paper chromatography.

SANTESSON, J., *Acta Chem. Scand.* **21**, 1162–1172 (1967). Thin-layer chromatography.

ZOPF, W., *Ann. Chem.* **352**, 1–44 (1907). From *Ramalina obtusata*.

olivetoric acid
$C_{26}H_{32}O_8$
m.p. 151° (benzene), needles.

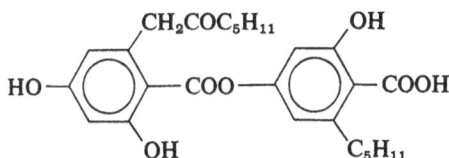

ASAHINA, Y., *J. Japan. Botany* **12**, 516–525 (1936); *ibid.* **13**, 529–536 (1937). Microcrystal tests.

ASAHINA, Y., Lichens of Japan. Vol. II. Genus *Parmelia*. Tokyo. 1952. Microcrystal tests.

ASAHINA, Y. AND J. ASANO, *Chem. Ber.* **65**, 475–482 (1932). From *Cornicularia pseudosatoana*, structure studied.

ASAHINA, Y. AND J. ASANO, *Chem. Ber.* **65**, 584–586 (1932). Structure studied.

ASAHINA, Y. AND F. FUZIKAWA, *Chem. Ber.* **67**, 163–168 (1934). Structure corrected.

ASAHINA, Y. AND F. FUZIKAWA, *Chem. Ber.* **68**, 2026–2028 (1935). From *Cetrelia olivetorum*.

CULBERSON, CHICITA F., *Science* **143**, 255–256 (1964). With physodic acid in *Cetraria ciliaris*.

CULBERSON, CHICITA F., *Bryologist* **68**, 435–439 (1965). With physodic acid in a strain of *Pseudevernia furfuracea*.

CULBERSON, W. L. AND CHICITA F. CULBERSON, *Contrib. U.S. Natl. Herb.* **34**, 449–558 (1968). Distribution in the genus *Cetrelia*, microchemical tests.

HALE, M. E., JR., *Contrib. U.S. Natl. Herb.* **36**, 193–358 (1965). Distribution in the genus *Parmelia*.

HALE, M. E., JR. AND S. KUROKAWA, *Contrib. U.S. Natl. Herb.* **36**, 121–191 (1964). Distribution in the genus *Parmelia*.

HUNECK, S., C. DJERASSI, D. BECHER, M. BARBER, M. VON ARDENNE, K. STEINFELDER, AND R. TÜMMLER, *Tetrahedron* **24**, 2707–2755 (1968). Mass spectrum.

MITSUNO, M., *Pharm. Bull.* (Tokyo) **1**, 170–173 (1953). Paper chromatography.

SANTESSON, J., *Acta Chem. Scand.* **21**, 1162–1172 (1967). Thin-layer chromatography.

ZOPF, W., *Ann. Chem.* **295**, 222–300 (1897); *ibid.* **313**, 317–344 (1900). From *Pseudevernia furfuracea*.

perlatolic acid

$C_{25}H_{32}O_7$

m.p. 107–108° (benzene-petroleum ether).

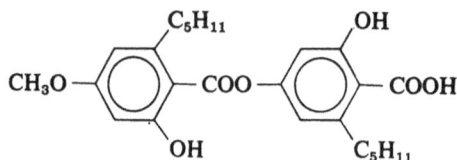

ASAHINA, Y., *J. Japan. Botany* **14**, 39–44 (1938); *ibid.* **15**, 465–472 (1939); *ibid.* **16**, 185–193 (1940). Microcrystal tests.

ASAHINA, Y., Lichens of Japan. Vol. I. Genus *Cladonia*. Tokyo. 1950. Microcrystal tests.

ASAHINA, Y., Lichens of Japan. Vol. II. Genus *Parmelia*. Tokyo. 1952. Microcrystal tests.

ASAHINA, Y. AND F. FUZIKAWA, *Chem. Ber.* **68**, 634–639 (1935). From *Cetrelia cetrarioides* (as *Parmelia perlata*).

ASAHINA, Y. AND I. YOSIOKA, *Chem. Ber.* **70**, 1823–1826 (1937). Synthesis.

BREADEN, T. W., J. KEANE, AND T. J. NOLAN, *Sci. Proc. Roy. Dublin Soc.* **23**, 6–9 (1942). From *Cladonia impexa*.

CULBERSON, CHICITA F. AND W. L. CULBERSON, *Bryologist* **69**, 192–202 (1966). Microchemical tests.

CULBERSON, W. L. AND CHICITA F. CULBERSON, *Contrib. U.S. Natl. Herb.* **34**, 449–558 (1968). Distribution in the genus *Cetrelia*, microchemical tests.

EVANS, A. W., *Rhodora* **52**, 77–123 (1950). Distribution in the genus *Cladonia* in Connecticut.

EVANS, A. W., *Rhodora* **45**, 417–438 (1943). Distribution in the genus *Cladonia*.

HALE, M. E., JR. AND S. KUROKAWA, *Contrib. U.S. Natl. Herb.* **36**, 121–191 (1964). Distribution in the genus *Parmelia*.
MITSUNO, M., *Pharm. Bull.* (Tokyo) **1**, 170–173 (1953). Paper chromatography.
SANTESSON, J., *Acta Chem. Scand.* **21**, 1162–1172 (1967). Thin-layer chromatography.

planaic acid
$C_{27}H_{36}O_7$
m.p. 110–111° (methanol); ir (CCl_4), ν cm^{-1}: 1712 (C=O) and 1760 (depside C=O).

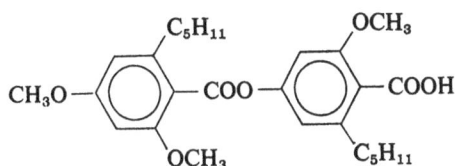

HUNECK, S., *Z. Naturforsch.* **20b**, 1119–1122 (1965). From *Lecidea plana*, structure proved.
HUNECK, S., *Z. Naturforsch.* **20b**, 1137–1138 (1965). From *Lecidea lithophila*.
HUNECK, S., C. DJERASSI, D. BECHER, M. BARBER, M. VON ARDENNE, K. STEINFELDER, AND R. TÜMMLER, *Tetrahedron* **24**, 2707–2755 (1968). Mass spectrum.

sphaerophorin
$C_{23}H_{28}O_7$
m.p. 137° (benzene).

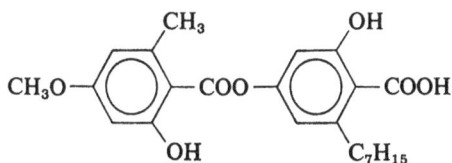

ASAHINA, Y., *J. Japan. Botany* **14**, 39–44 (1938). Microcrystal tests.
ASAHINA, Y. AND A. HASHIMOTO, *Chem. Ber.* **67**, 416–420 (1934). From *Sphaerophorus melanocarpus* and *S. meiophorus*, structure proved.
HASHIMOTO, A., *Yakugaku Zasshi* **58**, 221–223 (1938). Synthesis.
HESS, D., *Planta* **52**, 65–76 (1958). Paper chromatography.
MITSUNO, M., *J. Japan. Botany* **14**, 659–669 (1938). Distribution in the genus *Sphaerophorus*, microcrystal tests.

MITSUNO, M., *Pharm. Bull.* (Tokyo) **1**, 170–173 (1953). Paper chromatography.

RAMAUT, J. L., *Bull. Soc. Roy. Botan. Belg.* **93**, 27–40 (1961). From *Sphaerophorus fragilis*.

SANTESSON, J., *Acta Chem. Scand.* **21**, 1162–1172 (1967). Thin-layer chromatography.

ZOPF, W., *Ann. Chem.* **300**, 322–357 (1898). From *Sphaerophorus fragilis* and *S. globosus*, first extraction.

tenuiorin

$C_{26}H_{24}O_{10}$

m.p. 178–180°, solidifies and then decomposes at 238° (benzene), plates; ir (KBr), ν cm^{-1}: 1580, 1610, 1680 (C=O), and 3420 (OH).

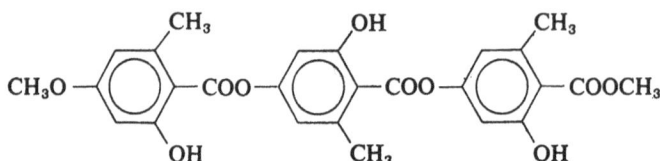

$$CH_3O-\underset{OH}{\overset{CH_3}{\bigcirc}}-COO-\underset{CH_3}{\overset{OH}{\bigcirc}}-COO-\underset{OH}{\overset{CH_3}{\bigcirc}}-COOCH_3$$

ASAHINA, Y., *J. Japan. Botany* **32**, 161–164 (1957). Microcrystal test, paper chromatography, hydrolysis.

ASAHINA, Y. AND M. YANAGITA, *Chem. Ber.* **66**, 1910–1912 (1933). From *Lobaria linita* (as *L. pulmonaria* f. *tenuior*), structure proved, synthesis.

HUNECK, S. AND R. TÜMMLER, *Ann. Chem.* **685**, 128–133 (1965). From *Peltigera polydactyla*, peltigerin identical to tenuiorin, mass spectrum, thin-layer chromatography.

KUROKAWA, S., YOKO JINZENJI, S. SHIBATA, AND HSÜCH-CHING CHIANG, *Bull. Natl. Sci. Mus.* (Tokyo) **9**, 101–114 (1966). In the genus *Peltigera* in Japan, thin-layer chromatography.

ZELLNER, J., *Monatsh. Chem.* **59**, 300–304 (1932). From *Peltigera canina*, as peltigerin.

ZOPF, W., *Ann. Chem.* **364**, 273–313 (1909). From several *Peltigera* species, first extraction, as peltigerin.

umbilicaric acid

$C_{25}H_{22}O_{10}$

m.p. 185–189° (ethanol), plates; m.p. 203° (ethanol-water).

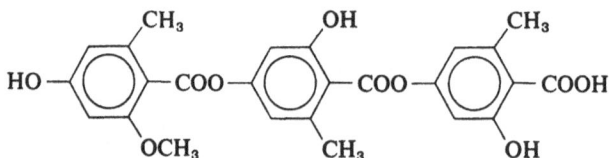

$$HO-\underset{OCH_3}{\overset{CH_3}{\bigcirc}}-COO-\underset{CH_3}{\overset{OH}{\bigcirc}}-COO-\underset{OH}{\overset{CH_3}{\bigcirc}}-COOH$$

ASAHINA, Y., *J. Japan. Botany* **14**, 39–44 (1938). Microcrystal test.
ASAHINA, Y. AND I. YOSIOKA, *Chem. Ber.* **70**, 200–206 (1937). Synthesis.
HESS, D., *Planta* **52**, 65–76 (1958). Paper chromatography.
HESSE, O., *J. Prakt. Chem.* **58**, 465–561 (1898). From *Umbilicaria polyphylla*, structure studied.
HESSE, O., *J. Prakt. Chem.* **63**, 522–553 (1901). Structure studied.
KOLLER, G. AND G. PFEIFFER, *Monatsh. Chem.* **62**, 241–251 (1933); *ibid.* **62**, 359–372 (1933). From *Umbilicaria deusta*, structure proved.
ZOPF, W. *Ann. Chem.* **300**, 322–357 (1898). From *Umbilicaria deusta*, *U. hyperborea*, and *U. polyphylla*.
ZOPF, W., *Ann. Chem.* **340**, 276–309 (1905). From *Actinogyra polyrrhiza*.

meta-DEPSIDES

boninic acid
$C_{25}H_{32}O_8$
m.p. 134.5° (benzene-petroleum ether), plates.

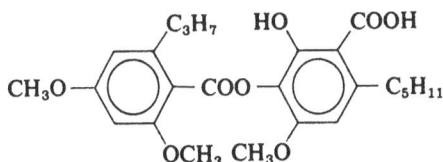

ASAHINA, Y., *J. Japan. Botany* **14**, 244–250 (1938); *ibid.* **15**, 205–223 (1939). Microcrystal test.
ASAHINA, Y. AND T. KUSAKA, *Chem. Ber.* **70**, 1815–1821 (1937). From *Ramalina boninensis*, first isolation, structure proved, synthesis.

cryptochlorophaeic acid
$C_{25}H_{32}O_8$
m.p. 182–184° (benzene); ir (nujol), ν cm^{-1}: 3400 (bonded OH), 1735 (depside C=O), 1650 (bonded COOH), and 1620 and 1580 (aryl C=C); ir (THF), ν cm^{-1}: 1650 (chelated C=O) and 1765 (depside C=O); uv (EtOH), λ_{max} mμ: 259 and 295 (shoulder).

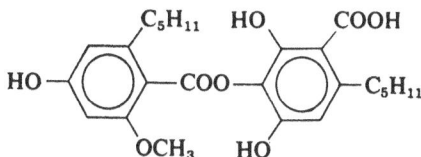

ASAHINA, Y., *J. Japan. Botany* **16**, 709–727 (1940). From *Cladonia cryptochlorophaea* (m.p. 166°), microchemical tests.

Asahina, Y., Lichens of Japan. Vol. I. Genus *Cladonia*. Tokyo. 1950. Microcrystal tests.

Culberson, Chicita F., *J. Pharm. Sci.* **54**, 1815–1816 (1965). From *Parmelia cryptochlorophaea.*

Culberson, Chicita F., *Bryologist* **70**, 397–405 (1967). From *Ramalina paludosa*, column chromatography, microchemical tests.

Evans, A. W., *Trans. Conn. Acad. Arts Sci.* **35**, 519–626 (1944). Distribution in Connecticut Cladonias.

Hale, M. E., Jr., *Contrib. U.S. Natl. Herb.* **36**, 193–358 (1965). Distribution in the genus *Parmelia.*

Huneck, S., C. Djerassi, D. Becher, M. Barber, M. von Ardenne, K. Steinfelder, and R. Tümmler, *Tetrahedron* **24**, 2707–2755 (1968). Mass spectrum.

Santesson, J., *Acta Chem. Scand.* **21**, 1162–1172 (1967). Thin-layer chromatography.

Shibata, S. and Hsüch-Ching Chiang, *Phytochemistry* **4**, 133–139 (1965). From *Cladonia cryptochlorophaea*, structure proved.

homosekikaic acid

$C_{24}H_{30}O_8$

m.p. 133–137° (benzene-petroleum ether).

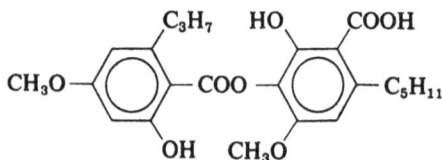

Asahina, Y., *J. Japan. Botany* **14**, 244–250 (1938). Microchemical tests.

Asahina, Y., *J. Japan. Botany* **18**, 620–625 (1942). In *Cladonia submultiformis*, microcrystal tests.

Asahina, Y., Lichens of Japan. Vol. I. Genus *Cladonia*. Tokyo. 1950. In *Cladonia dissimilis*, microchemical tests.

Asahina, Y. and T. Kusaka, *Chem. Ber.* **70**, 1815–1821 (1937). Synthesis.

Asahina, Y. and T. Kusaka, *Chem. Ber.* **70**, 1821–1823 (1937). From *Cladonia pityrea* and *C. subpityrea*, structure studied.

Dahl, E., *Medd. Grønland* **150**(2), 1–176 (1950). Microchemical tests.

Evans, A. W., *Trans. Conn. Acad. Arts Sci.* **35**, 519–626 (1944). Distribution in the genus *Cladonia.*

Santesson, J., *Acta Chem. Scand.* **21**, 1162–1172 (1967). Thin-layer chromatography.

Zopf, W., *Ber. Deut. Botan. Ges.* **36**, 51–113 (1908). From *Cladonia nemoxyna*, first extraction, as nemoxynic acid.

merochlorophaeic acid

$C_{24}H_{30}O_8$

m.p. 164–166° (benzene-hexane); ir (KBr), ν cm^{-1}: 3400 (bonded OH), 1750 (depside C=O), 1650 (bonded COOH), and 1620 and 1600 (benzenoid); uv (EtOH), λ_{max} mμ: 259 and 295 (shoulder); nmr (CDCl$_3$), τ ppm: 8.97 (terminal CH$_3$), 8.53 (intermediate CH$_2$), 7.03 (aryl substituted CH$_2$), 5.95 and 6.06 (2 OCH$_3$), and 3.35 (aryl proton).

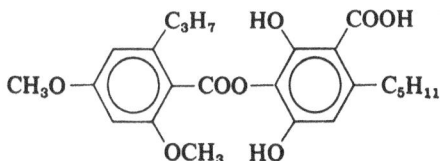

ASAHINA, Y., *J. Japan. Botany* **16**, 709–727 (1940). Microchemical tests.

ASAHINA, Y., *J. Japan. Botany* **18**, 663–683 (1942). In *Cladonia pseudorangiformis*, microchemical tests.

CULBERSON, CHICITA F., *Bryologist* **70**, 397–405 (1967). Microchemical tests.

DAHL, E., *Medd. Grønland* **150** (2), 1–176 (1950). Microchemical tests.

SHIBATA, S. AND HSÜCH-CHING CHIANG, *Phytochemistry* **4**, 133–139 (1965). From *Cladonia merochlorophaea*, structure proved, uv spectrum.

novochlorophaeic acid

Structure not known.

AHTI, T., *Ann. Botan. Fennici* **3**, 380–390 (1966). In some boreal, esorediate or sparingly sorediate, samples of *Cladonia merochloropaea*.

DAHL, E., *Medd. Grønland* **150** (2), 1–176 (1950). In *Cladonia chlorophaea sens. lat.*, first detection, microcrystal test.

paludosic acid

$C_{23}H_{28}O_8$

m.p. 158.5–159.5° (benzene), fine colorless needles; uv (95% EtOH), λ_{max} mμ: 216.5, 253, and 290; λ_{min} mμ: 239.5 and 279.5; ir (nujol), ν cm^{-1}: 3470 (OH), 1735 and 1720 (depside C=O), 1645 (bonded C=O), 1625 and 1585 (aryl C=C), and 1225 (a medium peak absent in the spectrum of cryptochlorophaeic acid).

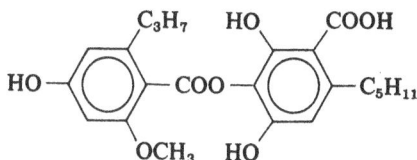

CULBERSON, CHICITA F., *Bryologist* **70**, 397–405 (1967). From *Ramalina paludosa*, first extraction, tentative structure proposed, column chromatography, microchemical tests.

ramalinolic acid

$C_{23}H_{28}O_8$

m.p. 163–164° (benzene).

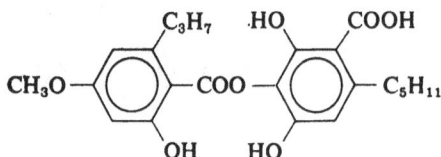

$$CH_3O-\underset{OH}{\overset{C_3H_7}{\bigcirc}}-COO-\underset{HO}{\overset{HO\quad COOH}{\bigcirc}}-C_5H_{11}$$

ASAHINA, Y., *J. Japan. Botany* **14**, 244–250 (1938). In *Ramalina calicaris, R. geniculata, R. nervulosa,* and *R. usnea,* microcrystal tests.

ASAHINA, Y. AND T. KUSAKA, *Chem. Ber.* **69**, 450–455 (1936). From *Ramalina nervulosa,* structure proved.

ASAHINA, Y. AND T. KUSAKA, *Chem. Ber.* **70**, 1821–1823 (1937). Microcrystal test.

ASAHINA, Y. and T. KUSAKA, *Chem. Ber.* **69**, 1896–1899 (1936). Synthesis.

ASAHINA, Y. and S. NONOMURA, *Chem. Ber.* **66**, 30–35 (1933). Color test evidence in *Ramalina geniculata.*

SANTESSON, J., *Acta Chem. Scand.* **21**, 1162–1172 (1967). Thin-layer chromatography.

SASAKI, I., *J. Japan. Botany* **18**, 626–632 (1942). In *Physcia aegialita.*

scrobiculin

$C_{22}H_{26}O_8$

m.p. 135.5–136° (benzene-petroleum ether); ir (nujol), ν cm^{-1}: 3420 (aryl OH), 1660 and 1650 (bonded C=O of ester groups), and 1630 and 1590 (aryl C=C); uv (95% EtOH), λ_{max} (log ε) mμ: 217.5 (4.69), 265 (4.51), and 302.5 (4.13); λ_{min} (log ε) mμ: 239 (3.95) and 286.5 (4.03); nmr (CDCl$_3$), δ ppm: 0.95 (terminal methyls), 1.65 (intermediate methylenes), 2.93 (aryl methylenes), 3.81 (aryl OCH$_3$), 3.93 (ester OCH$_3$), and 6.38–6.49 (aryl protons).

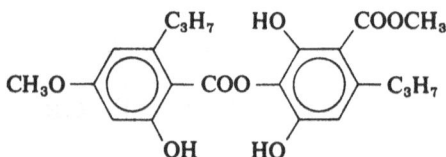

$$CH_3O-\underset{OH}{\overset{C_3H_7}{\bigcirc}}-COO-\underset{HO}{\overset{HO\quad COOCH_3}{\bigcirc}}-C_3H_7$$

ASAHINA, Y., *J. Japan. Botany* **21**, 83–86 (1947). In *Lobaria scrobiculata*.
CULBERSON, CHICITA F., *Phytochemistry* **6**, 719–725 (1967). From *Lobaria amplissima* and *L. scrobiculata*, structure proved, nmr spectrum.
CULBERSON, CHICITA F., *Bryologist* **70**, 70–75 (1967). Microchemical tests.
HALE, M. E., JR., *Bryologist* **60**, 35–39 (1957). In *Lobaria amplissima* and *L. scrobiculata*.
HALE, M. E., JR., *Lichenologist* **1**, 266–267 (1961). In *Lobaria amplissima*.

sekikaic acid
$C_{22}H_{26}O_8$
m.p. 150–151° (benzene), colorless prisms.

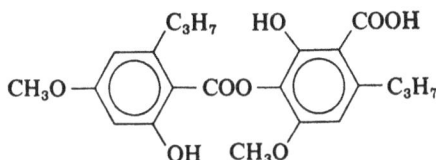

ASAHINA, Y., *J. Japan. Botany* **14**, 244–250 (1938). Microcrystal tests.
ASAHINA, Y., *J. Japan. Botany* **14**, 721–730 (1938). In the genus *Ramalina* in Japan, microcrystal tests.
ASAHINA, Y. AND T. KUSAKA, *Chem. Ber.* **69**, 450–455 (1936). From *Ramalina nervulosa*, with ramalinolic acid.
ASAHINA, Y. AND T. KUSAKA, *Chem. Ber.* **70**, 1815–1821 (1937). Microcrystal test.
ASAHINA, Y. AND S. NONOMURA, *Chem. Ber.* **66**, 30–35 (1933). From *Ramalina geniculata* and "a variety of *R. farinacea*" (probably *R. nervulosa*).
ASAHINA, Y. AND M. YASUE, *Chem. Ber.* **68**, 132–134 (1935). Synthesis of methyl dimethylsekikaiate.
ASAHINA, Y. AND M. YASUE, *Chem. Ber.* **68**, 1133–1137 (1935). Synthesis.
HUNECK, S., C. DJERASSI, D. BECHER, M. BARBER, M. VON ARDENNE, K. STEINFELDER, AND R. TÜMMLER, *Tetrahedron* **24**, 2707–2755 (1968). Mass spectrum.
HUNECK, S. AND G. FOLLMANN, *Z. Naturforsch.* **21b**, 90–91 (1966). From *Ramalina chilensis*.
HUNECK, S. AND G. TROTET, *Z. Naturforsch.* **21b**, 904 (1966). From *Ramalina boulhautiana*.
KUROKAWA, S. AND YOKO JINZENJI, *Bull. Natl. Sci. Mus.* (Tokyo) **8**, 369–374 (1965). Distribution in the genus *Anzia* in Japan, thin-layer chromatography.

134   *Chemical and Botanical Guide to Lichen Products*

MITTAL, O. P., S. NEELAKANTAN, AND T. R. SESHADRI, *J. Sci. Ind. Res.* (India) **11B**, 386–387 (1952). From *Ramalina calicaris.*

NAKAO, M., *Yakugaku Zasshi* **496**, 423–497 (1923); reference from *Chem. Abstr.* **17**, 3184[9] (1923). Extraction from a lichen mixture.

SANTESSON, J., *Acta Chem. Scand.* **21**, 1162–1172 (1967). Thin-layer chromatography.

SASAKI, I., *J. Japan. Botany* **18**, 626–632 (1942). In *Physcia aegialita.*

DEPSIDONES

alectoronic acid

$C_{28}H_{32}O_9$

m.p. 193° (benzene); hydrate (ethanol-water), m.p. 120–121°, resolidifies at 140°, melts 193°.

$$\text{HO}-\overset{\displaystyle CH_2COC_5H_{11}}{\underset{\displaystyle O}{\bigcirc}}-COO-\overset{\displaystyle OH}{\underset{\displaystyle CH_2COC_5H_{11}}{\bigcirc}}-COOH$$

ASAHINA, Y., *J. Japan. Botany* **14**, 318–323 (1938). Microcrystal tests.

ASAHINA, Y., Lichens of Japan. Vol. II. Genus *Parmelia.* Tokyo. 1952. Microcrystal tests.

ASAHINA, Y. AND F. FUZIKAWA, *Chem. Ber.* **67**, 163–168 (1934). Structure proved.

ASAHINA, Y. AND A. HASHIMOTO, *Chem. Ber.* **66**, 641–649 (1933). From *Alectoria japonica* and *A. sarmentosa.*

ASAHINA, Y., Y. KANAOKA, AND F. FUZIKAWA, *Chem. Ber.* **66**, 649–655 (1933). Structure studied.

ASANO, M. AND T. AZUMI, *Yakugaku Zasshi* **58**, 194 (1938). From *Cetraria halei* and *C. pseudocomplicata.*

CULBERSON, W. L. AND CHICITA F. CULBERSON, *Contrib. U.S. Natl. Herb.* **34**, 449–558 (1968). Distribution in the genus *Cetrelia*, microchemical tests.

HALE, M. E., JR. AND S. KUROKAWA, *Contrib. U.S. Natl. Herb.* **36**, 121–191 (1964). Distribution in the genus *Parmelia.*

HESS, D., *Planta* **52**, 65–76 (1958). Paper chromatography.

MITSUNO, M., *Pharm. Bull.* (Tokyo) **1**, 170–173 (1953). Paper chromatography.

SANTESSON, J., *Acta Chem. Scand.* **21**, 1162–1172 (1967). Thin-layer chromatography.

WACHTMEISTER, C. A., *Botan. Notiser* **109**, 313–324 (1956). Paper chromatography.

α-collatolic acid

$C_{29}H_{34}O_9$

m.p. 124–125° (benzene-petroleum ether); hydrate (ethanol-water), m.p. 90–95°.

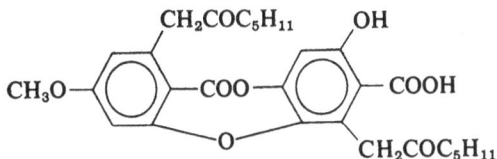

ASAHINA, Y., *J. Japan. Botany* **14**, 318–323 (1938). Microcrystal tests.

ASAHINA, Y., Y. KANAOKA, AND F. FUZIKAWA, *Chem. Ber.* **66**, 649–655 (1933). From *Cetrelia nuda*, structure proved.

ASAHINA, Y. AND F. FUZIKAWA, *Chem. Ber.* **67**, 163–168 (1934). Structure proved.

ASAHINA, Y. AND F. FUZIKAWA, *Chem. Ber.* **67**, 169–170 (1934). From *Lecanora atra*, lecanorolic acid identical to α-collatolic acid.

CULBERSON, W. L. AND CHICITA F. CULBERSON, *Brittonia* **17**, 182–190 (1965). In the genus *Asahinea*, microchemical tests.

CULBERSON, W. L. AND CHICITA F. CULBERSON, *Contrib. U.S. Natl. Herb.* **34**, 449–558 (1968). Distribution in the genus *Cetrelia*, microchemical tests.

FUZIKAWA, F., K. NAKAJIMA, A. TOKUOKA, Y. HITOSA, S. NAKAZAWA, T. OMATSU, AND T. SUMIYAMA, *Yakugaku Zasshi* **73**, 740–743 (1953). Antibacterial activity.

HESS, D., *Planta* **52**, 65–76 (1958). Paper chromatography.

MITSUNO, M., *Pharm. Bull.* (Tokyo) **1**, 170–173 (1953). Paper chromatography.

SANTESSON, J., *Acta Chem. Scand.* **21**, 1162–1172 (1967). Thin-layer chromatography.

WACHTMEISTER, C. A., *Botan. Notiser* **109**, 313–324 (1956). Paper chromatography.

ZOPF, W., *Ann. Chem.* **295**, 222–300 (1897). From *Lecanora atra*, as lecanorolic acid.

diploicin

$C_{16}H_{10}Cl_4O_5$

m.p. 232° (ethanol or benzene).

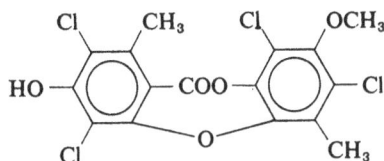

BARRY, V., *Nature* **158**, 131–132 (1946). Antitubercular activity of the monosodium salt of the hydrolysis product.

BROWN, C. J., D. E. CLARK, W. D. OLLIS, AND P. L. VEAL, *Proc. Chem. Soc.* **1960**, 393–394 (1960). Synthesis.

HUNECK, S., C. DJERASSI, D. BECHER, M. BARBER, M. VON ARDENNE, K. STEINFELDER, AND R. TÜMMLER, *Tetrahedron* **24**, 2707–2755 (1968). Mass spectrum.

NEELAKANTAN, S., R. PADMASANI, AND T. R. SESHADRI, *Current Sci.* (India) **33**, 365–366 (1964). Synthesis of the methyl ether.

NOLAN, T. J., *Chem. Ind.* (London) **12**, 512–513 (1934). From *Buellia canescens*.

NOLAN, T. J., *Sci. Proc. Roy. Dublin Soc.* **21**, 67–71 (1934). From *Buellia canescens*.

NOLAN, T. J., J. ALGAR, E. P. MCCANN, W. A. MANAHAN, AND NIALL NOLAN, *Sci. Proc. Roy. Dublin Soc.* **24**, 319–334 (1948). Structure determined.

SMITH, C. R., JR., *J. Org. Chem.* **25**, 588–591 (1960). Study toward synthesis.

SPILLANE, P. A., J. KEANE, AND T. J. NOLAN, *Sci. Proc. Roy. Dublin Soc.* **21**, 333–343 (1936). From *Buellia canescens*, structure studied.

ZOPF, W., *Ann. Chem.* **336**, 46–85 (1904). From *Buellia canescens*, first isolation.

grayanic acid

$C_{23}H_{26}O_7$

m.p. 186–189° d (50% ethanol); ir (nujol, KBr, $CHCl_3$), ν cm^{-1}: 1750 (lactone C=O), 1650 (chelated C=O), and 1570 and 1610 (aryl C=C); uv, λ_{max} (log ε) mμ: 258 (4.10) and 300–310 (3.5) shoulder; nmr ($CHCl_3$), τ ppm: 7.51 (aryl methyl), 9.10 (terminal methyl), 8.63 (intermediate methylenes), and 6.75 (aryl substituted methylene); nmr (acetone), δ ppm; 6.13, 6.66 (doublet), and 6.80 (doublet) (5', 3, and 5 protons respectively).

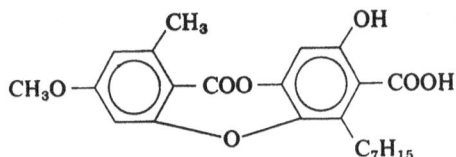

ASAHINA, Y., *J. Japan. Botany* **15**, 465–472 (1939). From *Cladonia grayi*, first isolation.

ASAHINA, Y., *J. Japan. Botany* **16**, 709–727 (1940). In *Cladonia cylindrica*, microchemical tests.

ASAHINA, Y., Lichens of Japan. Vol. I. Genus *Cladonia*. Tokyo. 1950. Microcrystal tests.

DAHL, E., *Medd. Grønland* **150** (2), 1–176 (1950). Microcrystal tests.

EVANS, A. W., *Bull. Torrey Botan. Club* **70**, 139–151 (1943). Microchemical tests.

SANTESSON, J., *Acta Chem. Scand.* **21**, 1162–1172 (1967). Thin-layer chromatography.

SHIBATA, S. AND HSÜCH-CHING CHIANG, *Chem. Pharm. Bull.* (Tokyo) **11**, 926–930 (1963). From *Cladonia grayi*, structure proved.

lividic acid
Structure not known.

CULBERSON, CHICITA F., *Phytochemistry* **5**, 815–818 (1966). From *Parmelia livida*.

CULBERSON, CHICITA F., *Bryologist* **70**, 70–75 (1967). Thin-layer chromatography.

CULBERSON, W. L., *Am. J. Botany* **48**, 168–174 (1961). In *Parmelia livida*, microcrystal test, as an unidentified substance.

HALE, M. E., JR., *Brittonia* **10**, 177–180 (1958). In *Parmelia livida*, as an unidentified substance.

HALE, M. E., JR. AND S. KUROKAWA, *Contrib. U.S. Natl. Herb.* **36**, 121–191 (1964). In *Parmelia dactylifera*, *P. immaculata*, and *P. livida*, as an unknown substance.

lobaric acid
$C_{25}H_{28}O_8$
m.p. 196–197° (acetone).

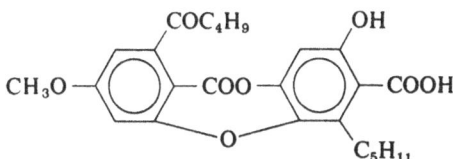

ASAHINA, Y., *J. Japan. Botany* **14**, 318–323 (1938); *ibid.* **26**, 161–165 (1951). Microcrystal tests.

ASAHINA, Y., Lichens of Japan. Vol. II. Genus *Parmelia*. Tokyo. 1952. Microcrystal test.

ASAHINA, Y. AND M. HIRAIWA, *Chem. Ber.* **68**, 1705–1708 (1935). From *Anzia hypoleucoides*.

ASAHINA, Y. AND S. NONOMURA, *Chem. Ber.* **68**, 1698–1704 (1935). From *Stereocaulon exutum*, *S. paschale*, and *S. sorediiferum*, structure studied.

ASAHINA, Y. AND M. YASUE, *Chem. Ber.* **69**, 643–649 (1936). Structure studied.

ASAHINA, Y. AND M. YASUE, *Chem. Ber.* **70**, 206–209 (1937). Structure proved.

HESS, D., *Planta* **52**, 65–76 (1958). Paper chromatography.

HUNECK, S., Z. *Naturforsch.* **21b**, 888–890 (1966). From *Lecanora badia*.

HUNECK, S., C. DJERASSI, D. BECHER, M. BARBER, M. VON ARDENNE, K. STEINFELDER, AND R. TÜMMLER, *Tetrahedron* **24**, 2707–2755 (1968). Mass spectrum.

HUNECK, S. AND G. FOLLMANN, Z. *Naturforsch.* **21b**, 714–715 (1966). From *Stereocaulon antarcticum*.

KUROKAWA, S. AND YOKO JINZENJI, *Bull. Natl. Sci. Mus.* (Tokyo) **8**, 369–374 (1965). In *Anzia hypoleucoides*, thin-layer chromatography.

MITSUNO, M., *Pharm. Bull.* (Tokyo) **1**, 170–173 (1953). Paper chromatography.

SANTESSON, J., *Acta Chem. Scand.* **21**, 1162–1172 (1967). Thin-layer chromatography.

4-O-methylphysodic acid

$C_{27}H_{32}O_8$

m.p. 151–152°; ir (nujol), ν cm^{-1}: 1740 (depsidone C=O), 1735 (ketone C=O), 1650 (bonded COOH), and 1620 and 1575 (benzenoid); uv (95% EtOH), λ_{max} (log ε) mμ: 211 (4.67) and 260 (4.15); λ_{min} (log ε) mμ: 247 (4.14); nmr (acetone-d$_6$), δ ppm: 0.90 (terminal CH$_3$), 1.50 (intermediate CH$_2$), 2.52 (—COCH$_2$CH$_2$—), 3.28 (ArCH$_2$CH$_2$—), 3.86 (OCH$_3$), 4.06 (ArCH$_2$CO—), 6.66–6.84 (aromatic protons), and 11.00 (ArOH).

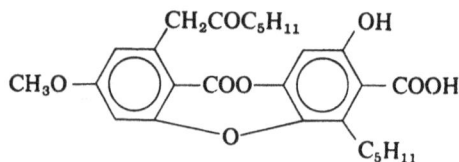

CULBERSON, CHICITA F., *Phytochemistry* **5**, 815–818 (1966). From *Parmelia livida*, structure determined.

CULBERSON, CHICITA F., *Bryologist* **70**, 70–75 (1967). Microchemical tests.

SANTESSON, J., *Acta Chem. Scand.* **21**, 1162–1172 (1967). Thin-layer chromatography.

norlobaridone

$C_{23}H_{26}O_6$

m.p. 188° (ethanol-water or benzene); m.p. 188–190° (ether-petroleum ether); ir (nujol), ν cm^{-1}: 3250 (s, broad) and 1680 (s); uv (95% EtOH), λ_{max} (log ε) mμ: 270 (3.96).

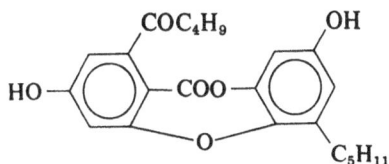

ABBAYES, H. DES, *Mém. Inst. Sci. Madagascar*, Sér. B, **10** (2), 81–121 (1961). In *Parmelia ecaperata*, paper chromatography.

ASAHINA, Y., Lichens of Japan. Vol. II. Genus *Parmelia*. Tokyo. 1952. In *Parmelia scabrosa* Tayl., as loxodic acid, microchemical tests.

BRINER, G. P., G. E. GREAM, AND N. V. RIGGS, *Australian J. Chem.* **13**, 277–284 (1960). From *Parmelia scabrosa* Tayl., as compound P.

GREAM, G. E. AND N. V. RIGGS, *Australian J. Chem.* **13**, 285–295 (1960). From *Parmelia scabrosa* Tayl., structure proved.

RAMAUT, J. L., *Rev. Bryol. Lichénol.* **33**, 587–591 (1964–1965). In *Parmelia nairobiensis*, thin-layer chromatography.

physodic acid
$C_{26}H_{30}O_8$
m.p. 205° d (acetone-carbon disulfide or methanol-water).

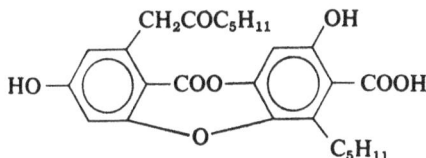

ASAHINA, Y., *J. Japan. Botany* **14**, 318–323 (1938). Microcrystal tests.

ASAHINA, Y., Lichens of Japan. Vol. II. Genus *Parmelia*. Tokyo. 1952. Microcrystal tests.

ASAHINA, Y. AND H. NOGAMI, *Chem. Ber.* **67**, 805–811 (1934); *ibid.* **68**, 77–80 (1935); *ibid.* **68**, 1500–1503 (1935). From *Hypogymnia physodes*, structure proved.

CULBERSON, CHICITA F., *Science* **143**, 255–256 (1964). With olivetoric acid in *Cetraria ciliaris*.

CULBERSON, CHICITA F., *Bryologist* **68**, 435–439 (1965). With olivetoric acid in a strain of *Pseudevernia furfuracea*.

CULBERSON, CHICITA F., *Bryologist* **70**, 70–75 (1967). From *Parmelia livida*, thin-layer chromatography.

DAHL, E., *Medd. Grønland* **150** (2), 1–176 (1950). Microcrystal test.

HALE, M. E., JR., *Science* **123**, 671 (1956). Ultraviolet spectrum.

HESS, D., *Planta* **52**, 65–76 (1958). Paper chromatography.

HESSE, O., *Chem. Ber.* **30**, 1983–1989 (1897); *J. Prakt. Chem.* **57**, 409–447 (1898). From *Hypogymnia physodes*.

HUNECK, S., C. DJERASSI, D. BECHER, M. BARBER, M. VON ARDENNE, K. STEINFELDER, AND R. TÜMMLER, *Tetrahedron* **24**, 2707–2755 (1968). Mass spectrum.

KROG, HILDUR, *Nytt Mag. Naturvidenskapene* **88**, 57–85 (1951). In species of *Cavernularia*, *Hypogymnia*, and *Pseudevernia*.

NUNO, MARIKO, *J. Japan. Botany* **39**, 97–103 (1964). Distribution in the genus *Hypogymnia*, thin-layer chromatography.

SANTESSON, J., *Acta Chem. Scand.* **21**, 1162–1172 (1967). Thin-layer chromatography.

STOLL, A., J. RENZ, AND A. BRACK, *Experientia* **3**, 111–113 (1947). From *Pseudevernia furfuracea*.

VARTIA, K. O., *Ann. Med. Exptl. Biol. Fenniae* (Helsinki) **28**, 7–19 (1950); *ibid.* **28** (Suppl. 7), 1–82 (1950). From *Hypogymnia physodes*.

WACHTMEISTER, C. A., *Botan. Notiser* **109**, 313–324 (1956). Paper chromatography.

ZELLNER, J., *Monatsh. Chem.* **64**, 6–11 (1934). From *Hypogymnia physodes*.

variolaric acid
$C_{16}H_{10}O_7$
m.p. 296° d (80% aqueous acetone).

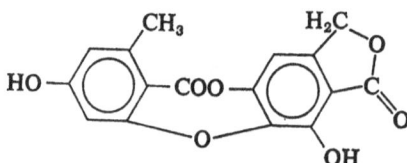

MURPHY, D., J. KEANE, AND T. J. NOLAN, *Sci. Proc. Roy. Dublin Soc.* **23**, 71–82 (1943). From *Ochrolechia parella*, structure proved.

SANTESSON, J., *Acta Chem. Scand.* **21**, 1162–1172 (1967). Thin-layer chromatography.

SCHUNCK, E., *Ann. Chem.* **54**, 257–284 (1845). From *Ochrolechia parella*, as parellic acid.

VERSEGHY, K., *Ann. Hist.-Nat. Mus. Natl. Hung.* **51**, 145–159 (1959); *Beih. Nova Hedwigia*, Heft 1, 1–146 (1962). Distribution in the genus *Ochrolechia*.

ZOPF, W., *Ann. Chem.* **321**, 37–61 (1902). From *Pertusaria lactea*.

DEPSONE

picrolichenic acid
$C_{25}H_{30}O_7$
m.p. 190° d (acetic acid–water), prisms; m.p. 184–187° d (benzene), prisms; $[\alpha]_D^{20}$ 0° (c = 5, $CHCl_3$); ir (KBr), ν cm^{-1}: 1820 (spirolactone)

and 1670 (bonded COOH and dienone); uv, λ_{max} (ε) mμ: 245 (22,000) and 270–277 (7800).

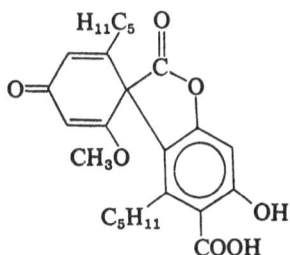

ALMS, *Ann. Chem.* **1**, 61–68 (1832). From *Pertusaria amara,* first extraction, as picrolichenin.

DAVIDSON, T. A. AND A. I. SCOTT, *Proc. Chem. Soc.* **1960**, 390–391 (1960). Total synthesis.

DAVIDSON, T. A. AND A. I. SCOTT, *J. Chem. Soc.* **1961**, 4075–4078 (1961). Total synthesis.

ERDTMAN, H. AND C. A. WACHTMEISTER, *Chem. Ind.* (London) **1957**, 1042 (1957). From *Pertusaria amara,* structure studied.

HARPER, S. H. AND R. M. LETCHER, *Proc. Trans. Rhodesian Sci. Assoc.* **51**, 156–184 (1966). Nmr spectrum, from *Pertusaria* sp.

HUNECK, S., C. DJERASSI, D. BECHER, M. BARBER, M. VON ARDENNE, K. STEINFELDER, AND R. TÜMMLER, *Tetrahedron* **24**, 2707–2755 (1968). Mass spectrum.

SANTESSON, J., *Acta Chem. Scand.* **21**, 1162–1172 (1967). Thin-layer chromatography.

WACHTMEISTER, C. A., *Acta Chem. Scand.* **12**, 147–164 (1958). From *Pertusaria amara,* structure proved, infrared and ultraviolet spectra.

WACHTMEISTER, C. A., *Svensk Kem. Tidskr.* **70**, 117–133 (1958). Review.

ZOPF, W., *Ann. Chem.* **313**, 317–344 (1900); *ibid.* **321**, 37–61 (1902). From *Pertusaria amara.*

PHENOLIC CARBOXYLIC ACID DERIVATIVES:
DIARYL ESTERS — β-ORCINOL SERIES

para-DEPSIDES

atranorin

$C_{19}H_{18}O_8$

m.p. 196° (acetone); ir (nujol), ν cm^{-1}: 1300 (not found in chloroatranorin).

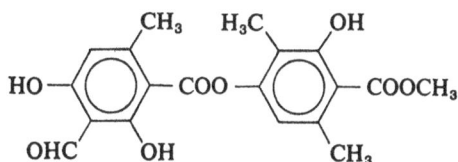

AGHORAMURTHY, K., K. G. SARMA, AND T. R. SESHADRI, *J. Sci. Ind. Res.* (India) **20B**, 166–168 (1961). From *Stereocaulon tomentosum* ssp. *myriocarpum* var. *orizabae*.

AHMANN, G. B. AND ANNICK MATHEY, *Bryologist* **70**, 93–97 (1967). From *Parmelia tinctorum* and *Pseudevernia intensa*.

ASAHINA, Y., *J. Japan. Botany* **12**, 859–872 (1936); *ibid.* **13**, 529–536 (1937); *ibid.* **16**, 185–193 (1940); *ibid.* **16**, 517–522 (1940); *ibid.* **19**, 301–311 (1943). Microcrystal tests.

ASAHINA, Y., Lichens of Japan. Vol. I. Genus *Cladonia*. Tokyo. 1950. Microcrystal test.

ASAHINA, Y., *J. Japan. Botany* **38**, 225–228 (1963). From *Parmelia galbina*.

ASAHINA, Y. AND H. AKAGI, *Chem. Ber.* **71**, 980–985 (1938). From *Parmelia leucotyliza*.

ASAHINA, Y. AND F. FUZIKAWA, *Chem. Ber.* **68**, 946–947 (1935). From *Parmelia acetabulum*.

ASAHINA, Y. AND M. HIRAIWA, *Chem. Ber.* **68**, 1705–1708 (1935). From *Anzia hypoleucoides*, *A. japonica*, and *A. opuntiella*.

ASAHINA, Y., Y. KANAOKA, AND F. FUZIKAWA, *Chem. Ber.* **66**, 649–655 (1933). From *Cetrelia nuda* (as *Cetraria collata*).

ASAHINA, Y. AND S. NONOMURA, *Chem. Ber.* **68**, 1698–1704 (1935). From *Stereocaulon exutum*, *S. paschale*, and *S. sorediiferum*.

ASAHINA, Y. AND Y. TANASE, *Chem. Ber.* **67**, 766–773 (1934). From *Parmelia zollingeri*.

ASAHINA, Y. AND Y. TANASE, *Chem. Ber.* **67**, 1434–1435 (1934). From *Parmelia saxatilis*.

ASAHINA, Y., M. YANAGITA, T. HIRAKATA, AND M. IDA, *Chem. Ber.* **66**, 1080–1086 (1933). From *Menegazzia terebrata*.

BACHMANN, O., *Österr. Botan. Z.* **110**, 103–107 (1963). Thin-layer chromatography.

CULBERSON, CHICITA F., *Phytochemistry* **4**, 951–961 (1965). From *Ramalina druidarum*.

CULBERSON, CHICITA F., *Phytochemistry* **5**, 815–818 (1966). From *Parmelia livida*.

CULBERSON, W. L., *Bryologist* **69**, 472–487 (1966). In the genus *Heterodermia* and *Anaptychia palmulata* in the Carolinas, microchemical tests.

CULBERSON, W. L., AND CHICITA F. CULBERSON, *Contrib. U.S. Natl. Herb.* **34**, 449–558 (1968). Distribution in the genera *Cetrelia* and *Platismatia*.

CURD, F. H., A. ROBERTSON, AND R. J. STEPHENSON, *J. Chem. Soc.* **1933**, 130–133 (1933). From *Evernia prunastri*.

DHAR, M. L., S. NEELAKANTAN, S. RAMANUJAM, AND T. R. SESHADRI, *J. Sci. Ind. Res.* (India) **18B**, 111–113 (1959). From *Anaptychia ciliaris*, *Parmelia sancti-angelii*, and *Physcia setosa*.

EVANS, A. W., *Rhodora* **45**, 417–438 (1943). Distribution in the genus *Cladonia*, microchemical tests.

FERNÁNDEZ, O. AND A. PIZARROSO, *Rev. Real. Acad. Cienc. Exact. Fís. Nat. Madrid* **52**, 557–563 (1958). From *Usnea canariensis*.

HALE, M. E., JR., *Contrib. U.S. Natl. Herb.* **36**, 193–358 (1965). Distribution in the genus *Parmelia*.

HARDIMAN, JOSEPHINE, J. KEANE, AND T. J. NOLAN, *Sci. Proc. Roy. Dublin Soc.* **21**, 141–145 (1935). From *Lecanora gangaleoides*.

HARPER, S. H. AND R. M. LETCHER, *Proc. Trans. Rhodesian Sci. Assoc.* **51**, 156–184 (1966). Infrared and nmr spectra, from *Parmelia dilatata* and *P. gossweileri*.

HESS, D., *Planta* **52**, 65–76 (1958). Paper chromatography.

HUNECK, S., *Naturwissenschaften* **49**, 608 (1962). From *Lepraria neglecta*.

HUNECK, S., *Z. Naturforsch.* **21b**, 199–200 (1966). From *Stereocaulon nanodes*.

HUNECK, S., C. DJERASSI, D. BECHER, M. BARBER, M. VON ARDENNE, K. STEINFELDER, AND R. TÜMMLER, *Tetrahedron* **24**, 2707–2755 (1968). Mass spectrum.

HUNECK, S. AND G. FOLLMANN, *Z. Naturforsch.* **20b**, 1012–1013 (1965). From *Stereocaulon corticulatum*.

HUNECK, S. AND G. FOLLMANN, *Z. Naturforsch.* **20b**, 1138–1139 (1965). From *Parmelia pseudoreticulata*.

HUNECK, S. AND G. FOLLMANN, *Z. Naturforsch.* **21b**, 91–92 (1966). From *Himantormia lugubris* and *Thamnolecania gerlachei*.

HUNECK, S. AND G. FOLLMANN, *Z. Naturforsch.* **21b**, 714–715 (1966). From *Stereocaulon antarcticum*.

HUNECK, S. AND G. FOLLMANN, *Z. Naturforsch.* **22b**, 461 (1967). From *Stereocaulon ramulosum* (0.4%).

HUNECK, S. AND G. FOLLMANN, *Z. Naturforsch.* **22b**, 791–792 (1967). From *Pseudocyphellaria nitida* var. *subglauca* (0.03%).

KENNEDY, G., J. BREEN, J. KEANE, AND T. J. NOLAN, *Sci. Proc. Roy. Dublin Soc.* **21**, 557–566 (1937). From *Lecanora rupicola*.

KOLLER, G. AND H. HAMBURG, *Monatsh. Chem.* **65**, 375–379 (1935). From *Pertusaria dealbata*.

KROG, HILDUR, *Nytt Mag. Naturvidenskapene* **88**, 57–85 (1951). Microcrystal test.

KUROKAWA, S., *Nova Hedwigia* **6**, 1–115 (1962). Distribution in the genus *Anaptychia*.

KUROKAWA, S. AND YOKO JINZENJI, *Bull. Natl. Sci. Mus.* (Tokyo) **8**, 369–374 (1965). Distribution in the genus *Anzia* in Japan.

MANAKTALA, S. K., S. NEELAKANTAN, AND T. R. SESHADRI, *Indian J. Chem.* **3**, 520–522 (1965). Synthesis.

MITSUNO, M., *Pharm. Bull.* (Tokyo) **1**, 170–173 (1953). Paper chromatography.

NEELAKANTAN, S., R. PADMASANI, AND T. R. SESHADRI, *Tetrahedron Letters* **1962**, 287–289 (1962). Synthesis.

NEELAKANTAN, S., R. PADMASANI, AND T. R. SESHADRI, *Indian J. Chem.* **2**, 478–484 (1964). Halogenation.

NEELAKANTAN, S., R. PADMASANI, AND T. R. SESHADRI, *Tetrahedron* **21**, 3531–3536 (1965). Synthesis.

NEELAKANTAN, S., S. RANGASWAMI, AND V. S. RAO, *Indian J. Pharm.* **16**, 173–175 (1954). From *Anaptychia speciosa, Heterodermia hypoleuca*, and *H. leucomela*.

RANGASWAMI, S. AND V. S. RAO, *Indian J. Pharm.* **17**, 50–53 (1955). From *Parmelia cirrhata*.

ST. PFAU, A., *Helv. Chim. Acta* **9**, 650–669 (1926). Structure proved.

ST. PFAU, A., *Helv. Chim. Acta* **16**, 282–286 (1933). Structure studied.

SANTESSON, J., *Acta Chem. Scand.* **19**, 2254–2256 (1965). Thin-layer chromatography.

SANTESSON, J., *Acta Chem. Scand.* **21**, 1162–1172 (1967). Thin-layer chromatography.

SARMA, K. G. AND S. HUNECK, *Pharmazie*, in press. From *Anaptychia neoleucomelaena* and *Crocynia neglecta*.

SESHADRI, T. R. AND G. B. V. SUBRAMANIAN, *J. Indian Chem. Soc.* **40**, 7–8 (1963). Synthesis of methyl homodiploschistesate, column separation of atranorin and chloroatranorin.

SHAH, L. G., *J. Indian Chem. Soc.* **31**, 253–256 (1954). From *Parmelia camtschadalis*.

SPILLANE, P. A., J. KEANE, AND T. J. NOLAN, *Sci. Proc. Roy. Dublin Soc.* **21**, 333–343 (1936). From *Buellia canescens*.

VARTIA, K. O., *Ann. Med. Exptl. Biol. Fenniae* (Helsinki) **27**, 46–54 (1949). From *Stereocaulon paschale*.

VARTIA, K. O., *Ann. Med. Exptl. Biol. Fenniae* (Helsinki) **28**, 7–19 (1950). From *Hypogymnia physodes*.

VARTIA, K. O., *Ann. Med. Exptl. Biol. Fenniae* (Helsinki) **28** (Suppl. 7), 1–82 (1950). From *Lasallia pustulata, Lecidia granulosa*, and *Ochrolechia tartarea*.

WACHTMEISTER, C. A., *Botan. Notiser* **109**, 313–324 (1956). Paper chromatography.

WETMORE, C. M., *Publ. Mus., Mich. State Univ., Biol. Ser.* **3**, 209–464 (1967). In *Lecidea insularis, L. pertingens, L. stigmatea*, and *Xylographa vitiligo*.

YAMAZAKI, M., M. MATSUO, AND S. SHIBATA, *Chem. Pharm. Bull.* (Tokyo) **13**, 1015–1017 (1965). Biosynthesis in *Parmelia tinctorum*.

YAMAZAKI, M. AND S. SHIBATA, *Chem. Pharm. Bull.* (Tokyo) **14**, 96–97 (1966). Biosynthesis in *Parmelia tinctorum*.

YOSIOKA, I., M. YAMAKI, AND I. KITAGAWA, *Chem. Pharm. Bull.* (Tokyo) **14**, 804–807 (1966). From *Parmelia entotheiochroa*.

ZOPF, W., *Ann. Chem.* **288**, 38–74 (1895). From *Haematomma coccineum* and *Lecanora melanaspis*.

Zopf, W., *Ann. Chem.* **313**, 317–344 (1900). From *Parmelia subrudecta* (as *P. borreri*).

baeomycesic acid
$C_{19}H_{18}O_8$
m.p. 222–223° d (acetone-water).

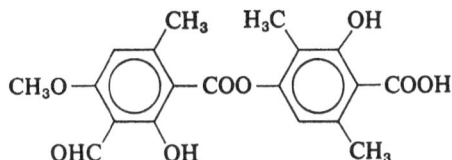

$$CH_3O-\overset{CH_3}{\underset{OHC\quad OH}{\bigcirc}}-COO-\overset{H_3C\quad OH}{\underset{CH_3}{\bigcirc}}-COOH$$

Asahina, Y., *J. Japan. Botany* **14**, 650–659 (1938); *ibid.* **18**, 620–625 (1942). Microcrystal tests.

Asahina, Y., Y. Tanase, and I. Yosioka, *Chem. Ber.* **69**, 125–127 (1936). From *Baeomyces roseus*, structure verified.

Asahina, Y. and M. Yasue, *Chem. Ber.* **70**, 1496–1497 (1937). From *Thamnolia subuliformis*.

Bendz, Gerd, J. Santesson, and C. A. Wachtmeister, *Acta Chem. Scand.* **19**, 1250–1252 (1965). In the genus *Siphula*, thin-layer chromatography.

Evans, A. W., *Trans. Conn. Acad. Arts Sci.* **35**, 519–626 (1944). In *Cladonia strepsilis*.

Koller, G. and W. Maass, *Monatsh. Chem.* **66**, 57–63 (1935). From *Baeomyces roseus*, first isolation, structure proved.

Santesson, J., *Acta Chem. Scand.* **19**, 2254–2256 (1965). Thin-layer chromatography.

Santesson, J., *Acta Chem. Scand.* **21**, 1162–1172 (1967). Thin-layer chromatography.

Satô, M., *Nova Hedwigia* **5**, 149–155 (1963). In *Thamnolia subuliformis*.

barbatic acid
$C_{19}H_{20}O_7$
m.p. 187° d (benzene).

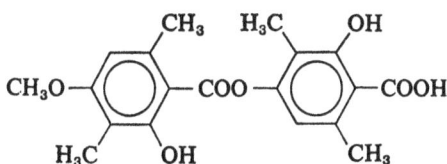

$$CH_3O-\overset{CH_3}{\underset{H_3C\quad OH}{\bigcirc}}-COO-\overset{H_3C\quad OH}{\underset{CH_3}{\bigcirc}}-COOH$$

Asahina, Y., *J. Japan. Botany* **12**, 859–872 (1936); *ibid.* **13**, 855–861 (1937). Microcrystal tests.

Asahina, Y., Lichens of Japan. Vol. I. Genus *Cladonia*. Tokyo. 1950. Microcrystal tests.

ASAHINA, Y. AND F. FUZIKAWA, *Chem. Ber.* **67**, 1793–1795 (1934). From *Cladonia amaurocraea*, coccellic acid identical to barbatic acid.

ASAHINA, Y. AND T. TUKAMOTO, *Chem. Ber.* **66**, 1255–1263 (1933). From *Usnea longissima* and *U. perplectans*.

DHAR, M. L., S. NEELAKANTAN, S. RAMANUJAM, AND T. R. SESHADRI, *J. Sci. Ind. Res.* (India) **18B**, 111–113 (1959). From *Usnea longissima* and *U. orientalis.*

FUZIKAWA, F., *Yakugaku Zasshi* **56**, 237–258 (1936); German summary, pp. 25–29 (1936). Synthesis.

HALE, M. E., JR. AND S. KUROKAWA, *Contrib. U.S. Natl. Herb.* **36**, 121–191 (1964). Distribution in the genus *Parmelia.*

HESS, D., *Planta* **52**, 65–76 (1958). Paper chromatography.

MANAKTALA, S. K., S. NEELAKANTAN, AND T. R. SESHADRI, *Indian J. Chem.* **3**, 520–522 (1965). Synthesis of the methyl ester.

MITSUNO, M., *Pharm. Bull.* (Tokyo) **1**, 170–173 (1953). Paper chromatography.

MORS, W. B., *Rev. Brasil. Biol.* **12**, 389–400 (1952). From *Usnea ludicra.*

MURTY, T. K. AND S. S. SUBRAMANIAN, *J. Sci. Ind. Res.* (India) **18B**, 394–395 (1959). From *Usnea venosa.*

ROBERTSON, A. AND R. J. STEPHENSON, *J. Chem. Soc.* **1932**, 1675–1681 (1932). From *Usnea barbata*, structure studied, synthesis of methyl barbatate.

RUNEMARK, H., *Opera Botan.* (Lund) **2**(1), 1–152 (1956). Distribution in the genus *Rhizocarpon.*

ST. PFAU, A., *Helv. Chim. Acta* **11**, 864–876 (1928). From *Usnea ceratina*, structure proved.

SANTESSON, J., *Acta Chem. Scand.* **21**, 1162–1172 (1967). Thin-layer chromatography.

SARMA, K. G. AND S. HUNECK, *Pharmazie*, in press. From *Usnea compacta* and *U. misaminensis.*

SHIBATA, S., *Acta Phytochim.* **14**, 9–38 (1944). From *Cladonia floerkeana.*

STENHOUSE, J. AND C. E. GROVES, *Ann. Chem.* **203**, 285–305 (1880). From *Usnea barbata*, first isolation.

WACHTMEISTER, C. A., *Botan. Notiser* **109**, 313–324 (1956). Paper chromatography.

chloroatranorin
$C_{19}H_{17}ClO_8$

m.p. 208–208.5° (acetone); ir (nujol), ν cm^{-1}: 1665 (bonded C=O), 1600 (aryl C=C), 1285, 1270, and no band at 1300.

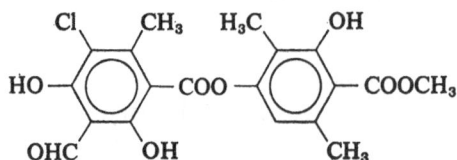

AHMANN, G. B. AND ANNICK MATHEY, *Bryologist* **70**, 93–97 (1967). From *Parmelia tinctorum* and *Pseudevernia intensa*, microchemical tests.

ASAHINA, Y. AND F. FUZIKAWA, *Chem. Ber.* **68**, 634–639 (1935). From *Cetrelia cetrarioides* (as *Parmelia perlata*).

ASAHINA, Y. AND F. FUZIKAWA, *Chem. Ber.* **68**, 2022–2026 (1935). From *Cetrelia japonica.*

ASAHINA, Y. AND H. NOGAMI, *Chem. Ber.* **68**, 1500–1503 (1935). From *Hypogymnia physodes*, structure studied.

CULBERSON, CHICITA F., *Phytochemistry* **2**, 335–340 (1963). In *Evernia prunastri.*

CULBERSON, CHICITA F., *J. Pharm. Sci.* **54**, 1815–1816 (1965). From *Parmelia cryptochlorophaea.*

CULBERSON, CHICITA F., *Phytochemistry* **4**, 951–961 (1965). From *Ramalina druidarum.*

CURD, F. H., A. ROBERTSON, AND R. J. STEPHENSON, *J. Chem. Soc.* **1933**, 130–133 (1933). From *Evernia prunastri.*

FERNÁNDEZ, O. AND A. PIZARROSO, *Rev. Real Acad. Cienc. Exact. Fís. Nat. Madrid* **52**, 557–563 (1958). From *Usnea canariensis.*

HARDIMAN, JOSEPHINE, J. KEANE, AND T. J. NOLAN, *Sci. Proc. Roy. Dublin Soc.* **21**, 141–145 (1935). From *Lecanora gangaleoides.*

HUNECK, S., C. DJERASSI, D. BECHER, M. BARBER, M. VON ARDENNE, K. STEINFELDER, AND R. TÜMMLER, *Tetrahedron* **24**, 2707–2755 (1968). Mass spectrum.

HUNECK, S. AND G. FOLLMANN, *Z. Naturforsch.* **20b**, 1138–1139 (1965). From *Parmelia pseudoreticulata.*

HUNECK, S. AND G. FOLLMANN, *Z. Naturforsch.* **21b**, 714–715 (1966). From *Anaptychia neoleucomelaena.*

KENNEDY, G., J. BREEN, J. KEANE, AND T. J. NOLAN, *Sci. Proc. Roy. Dublin Soc.* **21**, 557–566 (1937). From *Lecanora rupicola.*

KOLLER, G. AND K. PÖPL, *Monatsh. Chem.* **64**, 106–113 (1934); *ibid.* **64**, 126–130 (1934). From *Pseudevernia furfuracea*, structure studied.

NEELAKANTAN, S., R. PADMASANI, AND T. R. SESHADRI, *Indian J. Chem.* **2**, 478–484 (1964). Synthesis from atranorin.

ST. PFAU, A., *Helv. Chim. Acta* **17**, 1319–1328 (1934). From *Evernia prunastri*, structure proved.

SANTESSON, J., *Acta Chem. Scand.* **19**, 2254–2256 (1965); *ibid.* **21**, 1162–1172 (1967). Thin-layer chromatography.

SESHADRI, T. R. AND G. B. V. SUBRAMANIAN, *J. Indian Chem. Soc.* **40**, 7–8 (1963). Column separation of atranorin and chloroatranorin.

SPILLANE, P. A., J. KEANE, AND T. J. NOLAN, *Sci. Proc. Roy. Dublin Soc.* **21**, 333–343 (1936). From *Buellia canescens.*

YAMAZAKI, M., M. MATSUO, AND S. SHIBATA, *Chem. Pharm. Bull.* (Tokyo) **13**, 1015–1017 (1965). Biosynthesis in *Parmelia tinctorum.*

4-O-demethylbarbatic acid

$C_{18}H_{18}O_7$

m.p. 176–177° d (methanol-water), needles; ir (KBr), ν cm^{-1}: 1638 (acid C=O) and 1666 (depside C=O); nmr (60 MHz, DMSO-d_6): 119 (2 CH_3), 150 (2 CH_3), 384 (1 H), and 399 (1 H) Hz.

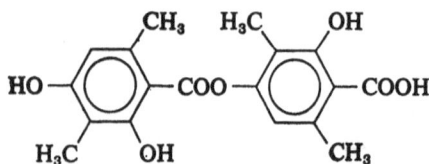

FUZIKAWA, F., *Yakugaku Zasshi* **56**, 237–258 (1936); German summary, pp. 25–29. Synthesis, as norbarbatic acid.

HUNECK, S., G. FOLLMANN, AND J. SANTESSON, *Z. Naturforsch.* **23b**, 856–860 (1968). From *Ramalina subdecipiens*, first extraction, structure proved, mass spectrum.

diffractaic acid
$C_{20}H_{22}O_7$
m.p. 189–190° (benzene).

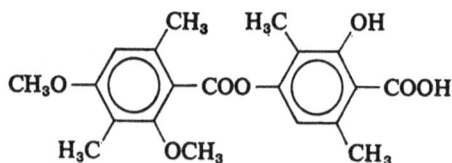

ASAHINA, Y., *J. Japan. Botany* **12**, 687–690 (1936). In *Alectoria ochroleuca*, microcrystal test.

ASAHINA, Y., *J. Japan. Botany* **12**, 859–872 (1936); *ibid.* **13**, 855–861 (1937). Microchemical tests.

ASAHINA, Y. AND F. FUZIKAWA, *Chem. Ber.* **65**, 175–178 (1932). From *Usnea diffracta*, structure proved.

ASAHINA, Y. AND F. FUZIKAWA, *Chem. Ber.* **65**, 1668 (1932). From *Usnea diffracta*, Hesse's dirhizonic acid identical to diffractaic acid.

ASAHINA, Y. AND F. FUZIKAWA, *Yakugaku Zasshi* **52**, 991–993 (1932); reference from *Chem. Abstr.* **27**, 1338[2] (1933). Synthesis.

FUZIKAWA, F., *Yakugaku Zasshi* **60**, 473–479 (1940); German summary, pp. 177–178 (1940). Antibiotic activity.

HESS, D., *Planta* **52**, 65–76 (1958). Paper chromatography.

KUROKAWA, S., *Bull. Natl. Sci. Mus.* (Tokyo) **10**, 369–376 (1967). In *Parmelia mesogenes* and *P. insueta*, microchemical tests.

MITSUNO, M., *Pharm. Bull.* (Tokyo) **1**, 170–173 (1953). Paper chromatography.

SANTESSON, J., *Acta Chem. Scand.* **21**, 1162–1172 (1967). Thin-layer chromatography.

WALL, M. E. AND K. H. DAVIS, unpublished communication. From *Usnea deminuta*.

squamatic acid
$C_{19}H_{18}O_9$
m.p. 219° d (acetic acid); m.p. 228° d (acetone).

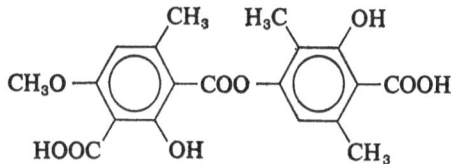

ASAHINA, Y., *J. Japan. Botany* **14**, 39–44 (1938); *ibid.* **14**, 650–659 (1938); *ibid.* **14**, 767–773 (1938). Microcrystal tests.

ASAHINA, Y., Lichens of Japan, Vol. I. Genus *Cladonia*. Tokyo. 1950. Microcrystal tests.

ASAHINA, Y. AND A. HASHIMOTO, *Chem. Ber.* **67**, 416–420 (1934). From *Sphaerophorus meiophorus*, as iso-squamatic acid.

ASAHINA, Y. AND M. HIRAIWA, *Chem. Ber.* **68**, 1708–1710 (1935). From *Thamnolia subuliformis* (as *T. vermicularis* var. *taurica*).

ASAHINA, Y. AND Y. SAKURAI, *Chem. Ber.* **70**, 64–66 (1937). Synthesis of dimethyl squamatate.

ASAHINA, Y. AND Z. SIMOSATO, *Chem. Ber.* **71**, 2561–2568 (1938). Iso-squamatic acid identical to squamatic acid.

ASAHINA, Y. AND Y. TANASE, *Chem. Ber.* **70**, 62–63 (1937). From *Cladonia bellidiflora*, structure studied.

ASAHINA, Y. AND M. YANAGITA, *Chem. Ber.* **66**, 36–39 (1933). From *Cladonia uncialis*.

ASAHINA, Y. AND M. YANAGITA, *Chem. Ber.* **66**, 393–397 (1933). From a lichen mixture, as iso-squamatic acid.

ASAHINA, Y. AND M. YASUE, *Chem. Ber.* **70**, 1496–1497 (1937). From *Cladonia squamosa* and *Thamnolia subuliformis*.

BENDZ, GERD, J. SANTESSON, AND C. A. WACHTMEISTER, *Acta Chem. Scand.* **19**, 1250–1252 (1965). Distribution in the genus *Siphula*, thin-layer chromatography.

DAHL, E., *Medd. Grønland* **150** (2), 1–176 (1950). Microcrystal tests.

EVANS, A. W., *Trans. Conn. Acad. Arts Sci.* **35**, 519–626 (1944). Distribution in Connecticut Cladonias.

HARPER, S. H. AND R. M. LETCHER, *Proc. Trans. Rhodesian Sci. Assoc.* **51**, 156–184 (1966). Infrared spectrum, from *Dermatiscum thunbergii* (1.3%).

HESS, D., *Planta* **52**, 65–76 (1958). Paper chromatography.

HESSE, O., *J. Prakt. Chem.* **62**, 430–480 (1900). From *Cladonia squamosa* f. *ventricosa*, first isolation.

HUNECK, S., C. DJERASSI, D. BECHER, M. BARBER, M. VON ARDENNE, K. STEINFELDER, AND R. TÜMMLER, *Tetrahedron* **24**, 2707–2755 (1968). Mass spectrum.

MITSUNO, M., *J. Japan. Botany* **14**, 659–669 (1938). Distribution in the genus *Sphaerophorus*, microcrystal tests.

MITSUNO, M., *Pharm. Bull.* (Tokyo) **1**, 170–173 (1953). Paper chromatography.

SANTESSON, J., *Acta Chem. Scand.* **21**, 1162–1172 (1967). Thin-layer chromatography.

SATÔ, M., *Nova Hedwigia* **5**, 149–155 (1963). In *Thamnolia subuliformis*.

SHIBATA, S., *Acta Phytochim.* **14**, 9–38 (1944). From *Cladonia pseudodidyma*.

WACHTMEISTER, C. A., *Botan. Notiser* **109**, 313–324 (1956). Paper chromatography.

ZOPF, W., *Ber. Deut. Botan. Ges.* **36**, 51–113 (1908). From *Cladonia crispata*.

meta-DEPSIDES

decarboxythamnolic acid
$C_{18}H_{16}O_9$
m.p. 215° d (acetone).
May form from thamnolic acid during extraction or chromatography.

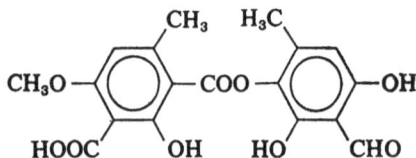

ASAHINA, Y. AND M. HIRAIWA, *Chem. Ber.* **69**, 330–333 (1936). From *Cladonia polydactyla*, possibly formed during extraction, synthesis from thamnolic acid.

ASAHINA, Y. AND MARIKO NUNO, *J. Japan. Botany* **39**, 313–317 (1964). In *Thamnolia vermicularis* and 12 other species, always with thamnolic acid, possibly of secondary origin, thin-layer chromatography.

BENDZ, GERD, J. SANTESSON, AND C. A. WACHTMEISTER, *Acta Chem. Scand.* **19**, 1250–1252 (1965). In *Siphula decumbens*, possibly an artifact.

SANTESSON, J., *Acta Chem. Scand.* **19**, 2254–2256 (1965). Thin-layer chromatography.

SOLBERG, Y. J., *Acta Chem. Scand.* **11**, 1477–1484 (1957). From *Haematomma ventosum*.

haemathamnolic acid

$C_{19}H_{16}O_{10}$

m.p. 202–204° d (acetone), pale yellow laths; uv (95% ethanol), λ_{max} (log ε) mμ: 351 (3.82), 283 (4.39), and 222 (4.50); λ_{min} (log ε) mμ: 313 (3.55) and 244 (4.18); ir (KBr), ν_{max} cm^{-1}: 3400–2400m, peaking at 3000–2800s, 1750s, and 1660–1620s.

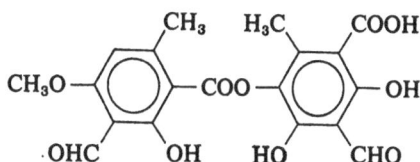

CH$_3$ H$_3$C COOH

CH$_3$O — COO — OH

·OHC OH HO CHO

HARPER, S. H. AND R. M. LETCHER, *J. Chem. Soc.* **17C**, 1603–1608 (1967). From *Pertusaria rhodesiaca*, first extraction, structure proved, nmr data for derivatives, thin-layer chromatography.

hypothamnolic acid

$C_{19}H_{18}O_{10}$

m.p. 225–227° d (80% acetone-water).

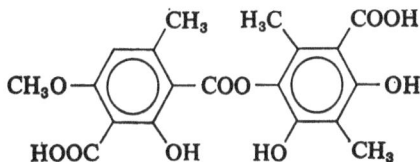

CH$_3$ H$_3$C COOH

CH$_3$O — COO — OH

HOOC OH HO CH$_3$

AGHORAMURTHY, K., T. R. SESHADRI, AND G. B. VENKATASUBRAMANIAN, *Tetrahedron* **1**, 310–316 (1957). Synthesis of dimethyl hypothamnolate.
ASAHINA, Y., *J. Japan. Botany* **18**, 489–502 (1942); *ibid.* **18**, 620–625 (1942). Microcrystal tests.
ASAHINA, Y., Lichens of Japan. Vol. I. Genus *Cladonia*. Tokyo. 1950. Microcrystal tests.
ASAHINA, Y., M. AOKI, AND F. FUZIKAWA, *Chem. Ber.* **74**, 824–831 (1941). From *Cladonia pseudostellata*, first extraction, structure proved.
BENDZ, GERD, J. SANTESSON, AND C. A. WACHTMEISTER, *Acta Chem. Scand.* **19**, 1250–1252 (1965). Distribution in the genus *Siphula*, thin-layer chromatography.
SANTESSON, J., *Acta Chem. Scand.* **21**, 1162–1172 (1967). Thin-layer chromatography.

thamnolic acid

$C_{19}H_{16}O_{11}$

m.p. 223° d (acetone); 210–212° (dioxane), pale yellow needles with solvent of crystallization, $C_{19}H_{16}O_{11} \cdot C_4H_8O_2$.

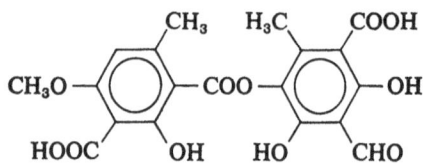

AGHORAMURTHY, K., K. G. SARMA, AND T. R. SESHADRI, *J. Sci. Ind. Res.* (India) **20B**, 166–168 (1961). From *Thamnolia vermicularis.*

AGHORAMURTHY, K., T. R. SESHADRI, AND G. B. VENKATASUBRAMANIAN, *Tetrahedron* **1**, 310–316 (1957). Synthesis of dimethyl thamnolate.

ASAHINA, Y., *J. Japan. Botany* **14**, 650–659 (1938); *ibid.* **14**, 767–773 (1938); *ibid.* **28**, 1–3 (1953). Microchemical tests.

ASAHINA, Y. AND F. FUZIKAWA, *Chem. Ber.* **65**, 58–60 (1932). Structure studied.

ASAHINA, Y. AND M. HIRAIWA, *Chem. Ber.* **69**, 330–333 (1936). From *Cladonia polydactyla*, structure proved.

ASAHINA, Y. AND M. HIRAIWA, *Chem. Ber.* **72**, 1402–1404 (1939). Hirtellic acid identical to thamnolic acid.

ASAHINA, Y. AND S. IHARA, *Chem. Ber.* **62**, 1196–1207 (1929); *ibid.* **65**, 55–57 (1932). Structure studied.

BENDZ, GERD, J. SANTESSON, AND C. A. WACHTMEISTER, *Acta Chem. Scand.* **19**, 1250–1252 (1965). In *Siphula decumbens*, thin-layer chromatography.

CULBERSON, W. L., *Madroño* **16**, 31 (1961). In some samples of *Parmeliopsis placorodia*, microcrystal tests.

EVANS, A. W., *Trans. Conn. Acad. Arts Sci.* **35**, 519–626 (1944). Distribution in Connecticut Cladonias, microcrystal tests.

HESSE, O., *J. Prakt. Chem.* **62**, 430–480 (1900). From *Thamnolia vermicularis*, structure studied.

HUNECK, S. AND G. FOLLMANN, *Z. Naturforsch.* **22b**, 689 (1967). From *Cladonia gorgonina* f. *turgidior.*

HUNECK, S. AND G. FOLLMANN, *Z. Naturforsch.* **22b**, 791–792 (1967). From *Usnea eulychniae* (1.5%).

KOLLER, G. AND H. HAMBURG, *Monatsh. Chem.* **65**, 375–379 (1935). From *Pertusaria dealbata.*

MITSUNO, M., *Pharm. Bull.* (Tokyo) **1**, 170–173 (1953). Paper chromatography.

SANTESSON, J., *Acta Chem. Scand.* **19**, 2254–2256 (1965). Thin-layer chromatography.

SANTESSON, J., *Acta Chem. Scand.* **21**, 1162–1172 (1967). Thin-layer chromatography.

SATÔ, M., *Nova Hedwigia* **5**, 149–155 (1963). In *Thamnolia subuliformis.*

SOLBERG, Y. J., *Acta Chem. Scand.* **11**, 1477–1484 (1957). From *Haematomma ventosum*, X-ray diffraction data.

WACHTMEISTER, C. A., *Acta Chem. Scand.* **9**, 1395–1396 (1955). From *Pertusaria corallina* and *Thamnolia vermicularis*, ocellatic acid identical to thamnolic acid, paper chromatography.

WACHTMEISTER, C. A., *Botan. Notiser* **109**, 313–324 (1956). Paper chromatography.

DEPSIDONES

fumarprotocetraric acid

$C_{22}H_{16}O_{12}$

m.p. 250–260° d, discolors above 230° (acetone)

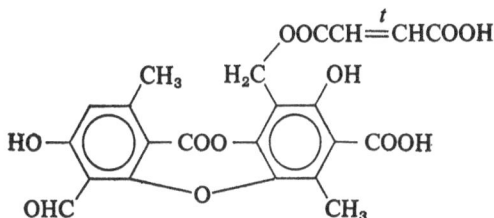

ASAHINA, Y., *J. Japan. Botany* **14**, 650–659 (1938). Microchemical tests.

ASAHINA, Y., Lichens of Japan. Vol. I. Genus *Cladonia*. Tokyo. 1950. Microcrystal tests.

ASAHINA, Y., Lichens of Japan. Vol. II. Genus *Parmelia*. Tokyo. 1952. Microcrystal tests.

ASAHINA, Y., Lichens of Japan. Vol III. Genus *Usnea*. Tokyo. 1956. Paper chromatography.

ASAHINA, Y. AND T. KUSAKA, *Chem. Ber.* **70**, 1821–1823 (1937). From *Cladonia pityrea*.

ASAHINA, Y. AND Y. TANASE, *Chem. Ber.* **66**, 700–703 (1933). From *Cetraria islandica*, structure studied.

ASAHINA, Y. AND Y. TANASE, *Chem. Ber.* **67**, 411–416 (1934); *ibid.* **67**, 766–773 (1934). Structure proved.

BACHMANN, O., *Österr. Botan. Z.* **110**, 103–107 (1963). Thin-layer chromatography.

CULBERSON, CHICITA F., *Phytochemistry* **4**, 951–961 (1965). From *Cladonia subtenuis*.

DAHL, E., *Medd. Grønland* **150** (2), 1–176 (1950). Microchemical tests.

EVANS, A. W., *Rhodora* **45**, 417–438 (1943). Distribution in the genus *Cladonia*, subgenus *Cladina*.

EVANS, A. W., *Trans. Conn. Acad. Arts Sci.* **35**, 519–626 (1944). Distribution in the *Cladoniae* of Connecticut.

GERTIG, H., *Dissertationes Pharm.* **15**, 235–240 (1963). From *Cetraria islandica*.

HESS, D., *Planta* **52**, 65–76 (1958). Paper chromatography.

HESSE, O., *J. Prakt. Chem.* **57**, 232–318 (1898). From *Cetraria islandica*, first isolation.

HUNECK, S. AND G. FOLLMANN, *Z. Naturforsch.* **22b**, 1368–1369 (1967). From *Cladonia endiviaefolia* from Spain (0.32%).

HUNECK, S. AND G. FOLLMANN, *Ber. Deut. Botan. Ges.*, in press. From *Dendrographa leucophaea* and *D. minor*.

MITSUNO, M., *Pharm. Bull.* (Tokyo) **1**, 170–173 (1953). Paper chromatography.

SANTESSON, J., *Acta Chem. Scand.* **19**, 2254–2256 (1965). Thin-layer chromatography.

SANTESSON, J., *Acta Chem. Scand.* **21**, 1162–1172 (1967). Thin-layer chromatography.

SHIBATA, S. AND HSÜCH-CHING CHIANG, *Phytochemistry* **4**, 133–139 (1965). From *Cladonia cryptochlorophaea*.

WACHTMEISTER, C. A., *Botan. Notiser* **109**, 313–324 (1956). Paper chromatography.

galbinic acid
m.p. 160–220° discolors, 260° d.
Structure unknown.

ASAHINA, Y., *J. Japan. Botany* **38**, 225–228 (1963). In *Parmelia galbina*, *P. metarevoluta*, and *P. obsessa*, thin-layer chromatography.

ASAHINA, Y., *J. Japan. Botany* **38**, 257–260 (1963). In *Usnea galbinifera*, thin-layer chromatography.

ASAHINA, Y., *J. Japan. Botany* **42**, 321–326 (1967). In *Usnea undulata*.

CULBERSON, W. L., *Am. J. Botany* **48**, 168–174 (1961). In *Parmelia galbina*, microcrystal test.

gangaleoidin
$C_{18}H_{14}Cl_2O_7$
m.p. 214–215° (ethanol-acetone).

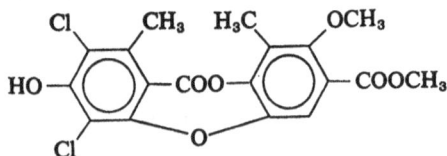

DAVIDSON, V. E., J. KEANE, AND T. J. NOLAN, *Sci. Proc. Roy. Dublin Soc.* **23**, 143–163 (1943). Structure proved.

HARDIMAN, JOSEPHINE, J. KEANE, AND T. J. NOLAN, *Sci. Proc. Roy. Dublin Soc.* **21**, 141–145 (1935). From *Lecanora gangaleoides*, first isolation.

NOLAN, T. J. AND J. KEANE, *Sci. Proc. Roy. Dublin Soc.* 22, 199–209 (1940). Structure studied.

hypoprotocetraric acid

$C_{18}H_{16}O_7$

m.p. 242–243° d, turning pink near 230° (acetone-petroleum ether); m.p. 250–251° d (methanol-water); ir (nujol), ν cm^{-1}: 3490, 1685, 1650, 1620, 1580, 1495, 1415, 1395, 1263, 1190, and 1138; for the monohydrate from ethanol-water, ir (nujol), ν cm^{-1}: 3590, 3240, 2340–2580, 1680, 1625, 1590, 1310, 1265, 1210, 1130, and 1075; nmr ((CD$_3$)$_2$SO), τ ppm: 7.90, 7.77, 7.71, and 7.49 (4 aromatic methyl groups) and 3.47 (1 aromatic proton).

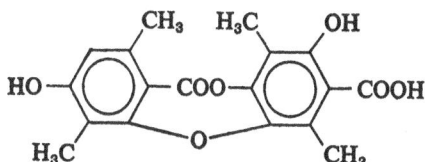

AGHORAMURTHY, K., K. G. SARMA, AND T. R. SESHADRI, *Tetrahedron* 12, 173–177 (1961). Synthesis from virensic acid.

ASAHINA, Y. AND Y. TANASE, *Chem. Ber.* 66, 700–703 (1933). Synthesis from cetraric acid.

ASAHINA, Y. AND Y. TANASE, *Chem. Ber.* 67, 411–416 (1934). Synthesis from fumarprotocetraric acid and physodalic acid.

ASAHINA, Y. AND Y. TANASE, *Chem. Ber.* 67, 766–773 (1934). Synthesis from protocetraric acid.

ASAHINA, Y. AND M. YANAGITA, *Chem. Ber.* 66, 1217–1220 (1933). Synthesis from protocetraric acid.

CULBERSON, CHICITA F., *Phytochemistry* 4, 951–961 (1965). From a strain of *Ramalina siliquosa*, first extraction, identification.

CULBERSON, CHICITA F., *Bryologist* 68, 301–304 (1965). In a strain of *Ramalina farinacea*, microchemical tests.

CULBERSON, W. L., *Rev. Bryol. Lichénol.* 34, 841–851 (1966). In *Ramalina druidarum*.

CULBERSON, W. L., *Brittonia* 19, 333–352 (1967). In *Ramalina hypoprotocetrarica*.

HUNECK, S., C. DJERASSI, D. BECHER, M. BARBER, M. VON ARDENNE, K. STEINFELDER, AND R. TÜMMLER, *Tetrahedron* 24, 2707–2755 (1968). Mass spectrum.

HUNECK, S. AND G. FOLLMANN, *Z. Naturforsch.* 20b, 611–612 (1965). From *Ramalina tumidula*, as coquimboic acid.

HUNECK, S. AND J.-M. LEHN, *Z. Naturforsch.* 21b 299 (1966). Coquimboic acid identical to hypoprotocetraric acid, infrared spectrum.

myriocarpic acid

m.p. 255–268° d; PD+, $FeCl_3+$ purple.
Structure not known.

AGHORAMURTHY, K., K. G. SARMA, AND T. R. SESHADRI, *J. Sci. Ind. Res.*
(India) **20B**, 166–168 (1961). From *Stereocaulon tomentosum* ssp.
myriocarpum var. *orizabae*.

norstictic acid

$C_{18}H_{12}O_9$
m.p. 286–287° d (acetone-water).

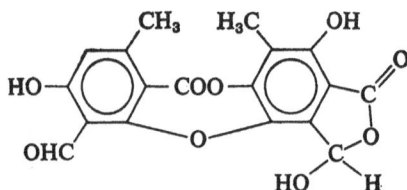

ASAHINA, Y., *J. Japan. Botany* **14**, 650–659 (1938). Microchemical tests.
ASAHINA, Y., Lichens of Japan. Vol. I. Genus *Cladonia*. Tokyo. 1950.
Microcrystal tests.
ASAHINA, Y., Lichens of Japan. Vol. II. Genus *Parmelia*. Tokyo. 1952.
Microchemical tests.
ASAHINA, Y. AND F. FUZIKAWA, *Chem. Ber.* **67**, 1789–1792 (1934).
Structure proved, uv spectrum.
ASAHINA, Y. AND F. FUZIKAWA, *Chem. Ber.* **68**, 946–947 (1935). From
Parmelia acetabulum.
ASAHINA, Y. AND M. YANAGITA, *Chem. Ber.* **67**, 799–803 (1934). From
Lobaria pulmonaria, structure studied.
ASAHINA, Y. AND M. YANAGITA, *Chem. Ber.* **67**, 1965–1969 (1934).
Synthesis from stictic acid.
ASAHINA, Y., M. YANAGITA, AND I. YOSIOKA, *Chem. Ber.* **69**, 1370–1375
(1936). From *Lobaria pulmonaria* and *L. oregana*, synthesis from stic-
tic acid.
BACHMANN, O., *Nova Hedwigia* **4**, 309–311 (1962). In *Lobaria pulmon-
aria, Stereocaulon dactylophyllum*, and *Umbilicaria torrefacta*, paper
chromatography.
BREEN, J., J. KEANE, AND T. J. NOLAN, *Sci. Proc. Roy. Dublin Soc.* **21**,
587–592 (1937). From *Pertusaria pseudocorallina*.
CULBERSON, CHICITA F., *Phytochemistry* **6**, 719–725 (1967). In *Lobaria
scrobiculata*.
CULBERSON, W. L., *Bryologist* **69**, 472–487 (1966). Distribution in the
genus *Heterodermia* in the Carolinas, microchemical tests.

EVANS, A. W., *Trans. Conn. Acad. Arts Sci.* **35**, 519–626 (1944). In *Cladonia subcariosa.*

HALE, M. E., JR., *Contrib. U.S. Natl. Herb.* **36**, 193–358 (1965). Distribution in the genus *Parmelia.*

HALE, M. E., JR. AND S. KUROKAWA, *Contrib. U.S. Natl. Herb.* **36**, 121–191 (1964). Distribution in the genus *Parmelia.*

HESS, D., *Planta* **52**, 65–76 (1958). Paper chromatography.

HUNECK, S., *Naturwissenschaften* **50**, 646 (1963). From *Lecanora radiosa.*

HUNECK, S., *Naturwissenschaften* **51**, 536 (1964). From *Lecidea pantherina.*

HUNECK, S., C. DJERASSI, D. BECHER, M. BARBER, M. VON ARDENNE, K. STEINFELDER, AND R. TÜMMLER, *Tetrahedron* **24**, 2707–2755 (1968). Mass spectrum.

HUNECK, S. AND G. FOLLMANN, *Z. Naturforsch.* **21b**, 90–91 (1966). From *Ramalina chilensis.*

HUNECK, S. AND G. FOLLMANN, *Z. Naturforsch.* **22b**, 110–111 (1967). From *Medusulina chilena.*

HUNECK, S. AND G. FOLLMANN, *Z. Naturforsch.* **22b**, 362–363 (1967). From *Roccella canariensis.*

HUNECK, S. AND G. FOLLMANN, *Z. Naturforsch.* **22b**, 1368–1369 (1967). From *Diploschistes ocellatus*, from Spain (2.7%).

HUNECK, S. AND G. FOLLMANN, in preparation. From *Roccella mossamedana.*

HUNECK, S. AND M. SIEGEL, *Naturwissenschaften* **50**, 154–155 (1963). From *Buellia sororioides.*

KUROKAWA, S., *Nova Hedwigia* **6**, 1–115 (1962). Distribution in the genus *Anaptychia.*

MITSUNO, M., *Pharm. Bull.* (Tokyo) **1**, 170–173 (1953). Paper chromatography.

MORS, W. B., *Rev. Brasil. Biol.* **12**, 389–400 (1952). From *Usnea aspera* and *U. ludicra.*

NOLAN, T. J. AND J. KEANE, *Nature* **132**, 281 (1933). From *Lobaria pulmonaria.*

RAO, P. S., K. G. SARMA, AND T. R. SESHADRI, *Current Sci.* (India) **35**, 147–148 (1966). From *Lobaria isidiosa* var. *subisidiosa.*

RUNEMARK, H., *Opera Botan.* (Lund) **2** (1), 1–152 (1956). Distribution in the genus *Rhizocarpon*, paper chromatography.

SANTESSON, J., *Acta Chem. Scand.* **19**, 2254–2256 (1965). Thin-layer chromatography.

SANTESSON, J., *Acta Chem. Scand.* **21**, 1162–1172 (1967). Thin-layer chromatography.

WACHTMEISTER, C. A., *Botan. Notiser* **109**, 313–324 (1956). Paper chromatography.

WETMORE, C. M., *Publ. Mus., Mich. State Univ., Biol. Ser.* **3**, 209–464 (1967). In *Xylographa hians* and *X. vitiligo.*

WIRTH, M. AND M. E. HALE, JR., *Contrib. U.S. Natl. Herb.* **36**, 63–119 (1963). Distribution in Mexican Graphidaceae.

pannarin
  $C_{18}H_{15}ClO_6$
  m.p. 216–217° (acetone).

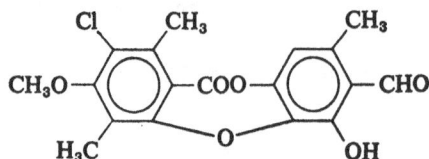

ASAHINA, Y., *J. Japan. Botany* **16**, 401–404 (1940). In *Pannaria fulvescens, P. lurida, P. pityrea*, and *P. rubiginosa*, microchemical tests.
ASAHINA, Y., *J. Japan. Botany* **20**, 129–134 (1944). In *Bombyliospora japonica*.
ASAHINA, Y. AND S. SHIBATA, Chemistry of Lichen Substances. Tokyo. 1954. Review of structural studies by Yosioka (1941).
HESS, D., *Planta* **52**, 65–76 (1958). Paper chromatography.
HUNECK, S., *Z. Naturforsch.* **21b**, 80–81 (1966). From *Lecanora hercynica*.
HUNECK, S., C. DJERASSI, D. BECHER, M. BARBER, M. VON ARDENNE, K. STEINFELDER, AND R. TÜMMLER, *Tetrahedron* **24**, 2707–2755 (1968). Mass spectrum.
SANTESSON, J., *Acta Chem. Scand.* **19**, 2254–2256 (1965). Thin-layer chromatography.
SANTESSON, J., *Acta Chem. Scand.* **21**, 1162–1172 (1967). Thin-layer chromatography.

physodalic acid
  $C_{20}H_{16}O_{10}$
  m.p. 230–260° d (acetic acid or ethyl ether).
  Confused in the literature with protocetraric acid.

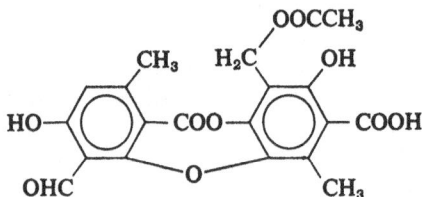

ASAHINA, Y., *J. Japan. Botany* **21**, 3–7 (1947). In *Hypogymnia physodes*, microcrystal test.

ASAHINA, Y., Lichens of Japan. Vol. II. Genus *Parmelia*. Tokyo. 1952. Microcrystal test.

ASAHINA, Y. AND M. YANAGITA, *Chem. Ber.* **66**, 1217–1220 (1933). Synthesis from protocetraric acid.

HESS, D., *Planta* **52**, 65–76 (1958). Paper chromatography.

HUNECK, S., G. FOLLMANN, AND W. A. WEBER, *Willdenowia*, in press. From *Dactylina chinensis* and *D. ramulosa*.

KOLLER, G. AND K. LOCKER, *Monatsh. Chem.* **58**, 209–212 (1931). From *Hypogymnia physodes*, capraric acid identical to physodalic acid.

KROG, HILDUR, *Nytt Mag. Naturvidenskapene* **88**, 57–85 (1951). Microcrystal tests.

KUROKAWA, S., *Bull. Natl. Sci. Mus.* (Tokyo) **10**, 369–376 (1967). In *Parmelia ferax* and *P. gerlachei*, microchemical tests.

MITSUNO, M., *Pharm. Bull.* (Tokyo) **1**, 170–173 (1953). Paper chromatography.

NUNO, MARIKO, *J. Japan. Botany* **39**, 97–103 (1964). Distribution in the genus *Hypogymnia*, thin-layer chromatography.

SANTESSON, J., *Acta Chem. Scand.* **19**, 2254–2256 (1965). Thin-layer chromatography.

SANTESSON, J., *Acta Chem. Scand.* **21**, 1162–1172 (1967). Thin-layer chromatography.

WACHTMEISTER, C. A., *Botan. Notiser* **109**, 313–324 (1956). Paper chromatography.

ZELLNER, J., *Monatsh. Chem.* **64**, 6–11 (1934). From *Hypogymnia physodes*.

ZOPF, W., *Ann. Chem.* **295**, 222–300 (1897). From *Hypogymnia physodes*, first extraction.

protocetraric acid
$C_{18}H_{14}O_9$
m.p. 245–250° d, carbonizes (acetone).

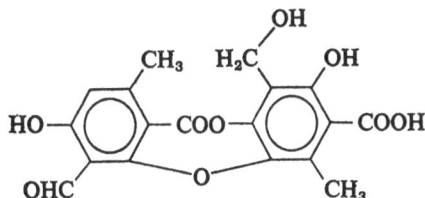

ASAHINA, Y., *J. Japan. Botany* **27**, 239–242 (1952). Microcrystal tests.

ASAHINA, Y., Lichens of Japan. Vol. II. Genus *Parmelia*. Tokyo. 1952. Microcrystal tests.

ASAHINA, Y., Lichens of Japan. Vol. III. Genus *Usnea*. Tokyo. 1956. Paper chromatography.

ASAHINA, Y. AND Y. TANASE, *Chem. Ber.* **67**, 766–773 (1934). From *Parmelia zollingeri*, structure proved.

ASAHINA, Y. AND T. TUKAMOTO, *Chem. Ber.* **66**, 1255–1263 (1933). From *Usnea perplectans*.

ASAHINA, Y. AND M. YANAGITA, *Chem. Ber.* **66**, 1217–1220 (1933). From *Parmelia caperata*.

ASANO, M. AND Z. OHTA, *Chem. Ber.* **66**, 1020–1023 (1933). From *Parmelia caperata*.

HALE, M. E., JR. AND S. KUROKAWA, *Contrib. U.S. Natl. Herb.* **36**, 121–191 (1964). Distribution in the genus *Parmelia*.

HARPER, S. H. AND R. M. LETCHER, *Proc. Trans. Rhodesian Sci. Assoc.* **51**, 156–184 (1966). From *Parmelia dilatata*, infrared and nmr spectra.

HESSE, O., *Chem. Ber.* **30**, 1983–1989 (1897). From *Parmelia caperata*, first isolation, as capraric acid.

HUNECK, S. AND G. FOLLMANN, *Z. Naturforsch.* **21b**, 715–716 (1966). From *Usnea lacerata*.

HUNECK, S. AND G. FOLLMANN, *Z. Naturforsch.* **22b**, 791–792 (1967). From *Roccellinastrum spongioideum* (1.8%).

HUNECK, S., G. FOLLMANN, W. A. WEBER, AND G. TROTET, *Z. Naturforsch.* **22b**, 671–673 (1967). From *Roccella portentosa* from the Galápagos Islands (2.6%).

KOLLER, G., E. KRAKAUER, AND K. PÖPL, *Monatsh. Chem.* **64**, 3–5 (1934). From *Ramalina farinacea*.

MORS, W. B., *Rev. Brasil. Biol.* **12**, 389–400 (1952). From *Usnea elongata*.

SHAH, L. G., *J. Indian Chem. Soc.* **31**, 253–256 (1954). From *Parmelia camtschadalis*.

psoromic acid
$C_{18}H_{14}O_8$
m.p. 265° (ethanol), needles.

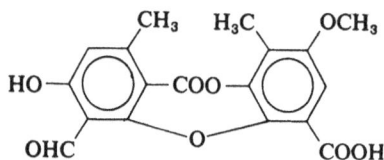

ASAHINA, Y., *J. Japan. Botany* **14**, 318–323 (1938). Microcrystal tests.

ASAHINA, Y., *J. Japan. Botany* **16**, 185–193 (1940). From *Cladonia aberrans*, microchemical tests.

ASAHINA, Y., Lichens of Japan. Vol. I. Genus *Cladonia*. Tokyo. 1950. Microcrystal tests.

ASAHINA, Y., *J. Japan. Botany* **33**, 1–5 (1958). Paper chromatography.

ASAHINA, Y. AND H. HAYASHI, *Chem. Ber.* **66**, 1023–1030 (1933). From *Alectoria sulcata*, structure studied.

ASAHINA, Y. AND H. HAYASHI, *Chem. Ber.* **70**, 810–812 (1937). Structure studied.

ASAHINA, Y. AND S. SHIBATA, *Chem. Ber.* **72**, 1399–1402 (1939). Structure proved.

BACHMANN, O., *Österr. Botan. Z.* **110**, 103–107 (1963). Thin-layer chromatography.

DAHL, E., *Medd. Grønland* **150** (2), 1–176 (1950). Microcrystal tests.

DHAR, M. L., S. NEELAKANTAN, S. RAMANUJAM, AND T. R. SESHADRI, *J. Sci. Ind. Res.* (India) **18B**, 111–113 (1959). From *Usnea aspera*.

EVANS, A. W., *Rhodora* **45**, 417–438 (1943). Distribution in the genus *Cladonia*, microchemical tests.

EVANS, A. W., *Trans. Conn. Acad. Arts Sci.* **35**, 519–626 (1944). Distribution in the genus *Cladonia*.

HAYASHI, H., *Yakugaku Zasshi* **57**, 598–602 (1937); German summary, pp. 112–114 (1937). Structure studied.

HUNECK, S., C. DJERASSI, D. BECHER, M. BARBER, M. VON ARDENNE, K. STEINFELDER, AND R. TÜMMLER, *Tetrahedron* **24**, 2707–2755 (1968). Mass spectrum.

HUNECK, S. AND G. FOLLMANN, *Z. Naturforsch.* **18b**, 991–992 (1963). From *Ingaderia pulcherrima*.

HUNECK, S. AND G. FOLLMANN, *Z. Naturforsch.* **19b**, 658–659 (1964). From *Chiodecton stalactinum*.

HUNECK, S. AND G. FOLLMANN, *Z. Naturforsch.* **21b**, 713–714 (1966). From *Ramalina tigrina*.

HUNECK, S. AND G. FOLLMANN, *Z. Naturforsch.* **22b**, 362–363 (1967). From *Darbishirella gracillima*.

HUNECK, S. AND G. FOLLMANN, *Z. Naturforsch.* **22b**, 1185–1188 (1967). From *Pentagenella fragillima* (30%).

HUNECK, S. AND G. FOLLMANN, in preparation. From *Roccellodea nigerrima* and *Roccellographa cretacea*.

RUNEMARK, H., *Opera Botan.* (Lund) **2** (1), 1–152 (1956). Distribution in the genus *Rhizocarpon*, paper chromatography.

SANTESSON, J., *Acta Chem. Scand.* **19**, 2254–2256 (1965). Thin-layer chromatography.

SANTESSON, J., *Acta Chem. Scand.* **21**, 1162–1172 (1967). Thin-layer chromatography.

SCHUMACKER, R., *Bull. Soc. Roy. Sci. Liège* **30**, 452–457 (1961). From *Everniopsis trulla*.

SPICA, G., *Gazz. Chim. Ital.* **12**, 431 (1882); reference from *Chem. Ber.* **16**, 427 (1883). From *Squamarina crassa*.

WACHTMEISTER, C. A., *Botan. Notiser* **109**, 313–324 (1956). Paper chromatography.

salazinic acid

$C_{18}H_{12}O_{10}$
m.p. 260–280° d, carbonizes (80% acetone-water), needles.

The "salazinic acid" of Zopf (*Ann. Chem.* **295**, 222–300 (1897)) is now known as norstictic acid.

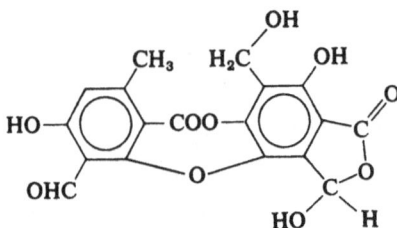

Asahina, Y., *J. Japan. Botany* **14**, 650–659 (1938). Microcrystal tests.

Asahina, Y., Lichens of Japan. Vol II. Genus *Parmelia*. Tokyo. 1952. Microcrystal test.

Asahina, Y. and J. Asano, *Chem. Ber.* **66**, 689–699 (1933). From *Parmelia cetrata* var. *sorediifera* and *P. taractica* (as *P. conspersa* var. *hypoclysta*).

Asahina, Y. and J. Asano, *Chem. Ber.* **66**, 893–897 (1933); *ibid.* **66**, 1031 (1933); *ibid.* **66**, 1215–1217 (1933). Structure proved.

Asahina, Y. and F. Fuzikawa, *Chem. Ber.* **68**, 946–947 (1935). Microcrystal test.

Asahina, Y. and Y. Tanase, *Chem. Ber.* **67**, 1434–1435 (1934). From *Parmelia saxatilis*, saxatilic acid identical to salazinic acid.

Asahina, Y. and T. Tukamoto, *Chem. Ber.* **66**, 1255–1263 (1933). From *Usnea perplectans.*

Asahina, Y. and T. Tukamoto, *Chem. Ber.* **67**, 963–971 (1934). From *Usnea japonica.*

Asahina, Y. and M. Yanagita, *Chem. Ber.* **67**, 799–803 (1934). From *Ramalina crassa* (as *R. scopulorum*).

Bachmann, O., *Österr. Botan. Z.* **110**, 103–107 (1963). Thin-layer chromatography.

Briner, G. P., G. E. Gream, and N. V. Riggs, *Australian J. Chem.* **13**, 277–284 (1960). From *Parmelia scabrosa* Tayl. (as *P. conspersa*).

Culberson, W. L., *Bryologist* **69**, 472–487 (1966). Distribution in the genus *Heterodermia* in the Carolinas.

Dhar, M. L., S. Neelakantan, S. Ramanujam, and T. R. Seshadri, *J. Sci. Ind. Res.* (India) **18B**, 111–113 (1959). From *Parmelia meizospora* and *Usnea orientalis.*

Hale, M. E., Jr., *Contrib. U.S. Natl. Herb.* **36**, 193–358 (1965). Distribution in the genus *Parmelia.*

Hale, M. E., Jr. and S. Kurokawa, *Contrib. U.S. Natl. Herb.* **36**, 121–191 (1964). Distribution in the genus *Parmelia.*

Hess, D., *Planta* **52**, 65–76 (1958). Paper chromatography.

Huneck, S. and G. Follmann, *Z. Naturforsch.* **20b**, 1138–1139 (1965). From *Parmelia pseudoreticulata.*

HUNECK, S. AND G. FOLLMANN, Z. *Naturforsch.* **21b**, 715–716 (1966). From *Usnea aureola*.

KOLLER, G. AND A. KLEIN, *Monatsh. Chem.* **64**, 80–86 (1934). Structure studied.

KOLLER, G. AND A. KLEIN, *Monatsh. Chem.* **65**, 91–92 (1934). Saxatilic acid identical to salazinic acid.

KUROKAWA, S., *Nova Hedwigia* **6**, 1–115 (1962). Distribution in the genus *Anaptychia*.

MITSUNO, M., *Pharm. Bull.* (Tokyo) **1**, 170–173 (1953). Paper chromatography.

MORS, W. B., *Rev. Brasil. Biol.* **12**, 389–400 (1952). From *Usnea ludicra*.

MURTY, T. K. AND S. S. SUBRAMANIAN, *J. Sci. Ind. Res.* (India) **18B**, 394–395 (1959). From *Usnea venosa*.

NOLAN, T. J. AND J. KEANE, *Nature* **132**, 281 (1933). From *Lobaria pulmonaria*.

RANGASWAMI, S. AND V. S. RAO, *Indian J. Pharm.* **17**, 50–53 (1955). From *Parmelia cirrhata*.

RANGASWAMI, S. AND V. S. RAO, *Indian J. Pharm.* **17**, 70 (1955). From *Usnea florida*.

SANTESSON, J., *Acta Chem. Scand.* **19**, 2254–2256 (1965). Thin-layer chromatography.

SANTESSON, J., *Acta Chem. Scand.* **21**, 1162–1172 (1967). Thin-layer chromatography.

SARMA, K. G. AND S. HUNECK, *Pharmazie*, in press. From *Usnea compacta* and *U. rubicunda* var. *ceratinella*.

SHAH, L. G., *J. Indian Chem. Soc.* **31**, 253–256 (1954). From *Parmelia camtschadalis*.

WACHTMEISTER, C. A., *Botan. Notiser* **109**, 313–324 (1956). Paper chromatography.

stictic acid
$C_{19}H_{14}O_9$
m.p. 270–272° d (acetone).

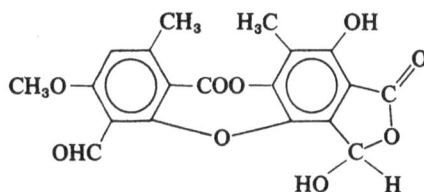

AGARWAL, S. C., K. AGHORAMURTHY, K. G. SARMA, AND T. R. SESHADRI, *J. Sci. Ind. Res.* (India) **20B**, 613–615 (1961). From *Lobaria isidiosa*.

AGHORAMURTHY, K., K. G. SARMA, AND T. R. SESHADRI, *J. Sci. Ind. Res.* (India) **20B**, 166–168 (1961). From *Stereocaulon foliolosum* and *S. tomentosum*.

ASAHINA, Y., *J. Japan. Botany* **14**, 650–659 (1938). Microcrystal tests.

ASAHINA, Y., *J. Japan. Botany* **19**, 301–311 (1943). In species of *Baeomyces*.

ASAHINA, Y., Lichens of Japan. Vol. I. Genus *Cladonia*. Tokyo. 1950. Microcrystal test.

ASAHINA, Y., Lichens of Japan. Vol. III. Genus *Usnea*. Tokyo. 1956. Microcrystal tests and paper chromatography.

ASAHINA, Y., Y. TANASE, AND I. YOSIOKA, *Chem. Ber.* **69**, 125–127 (1936). From *Baeomyces placophyllus*.

ASAHINA, Y. AND M. YANAGITA, *Chem. Ber.* **67**, 799–803 (1934). From *Lobaria pulmonaria*.

ASAHINA, Y. AND M. YANAGITA, *Chem. Ber.* **67**, 1965–1969 (1934). Conversion to norstictic acid.

ASAHINA, Y., M. YANAGITA, T. HIRAKATA, AND M. IDA, *Chem. Ber.* **66**, 1080–1086 (1933). From *Menegazzia terebrata* and *Stereocaulon nabewariense*.

ASAHINA, Y., M. YANAGITA, AND T. OMAKI, *Chem. Ber.* **66**, 943–947 (1933). From *Lobaria pulmonaria*, structure proved.

ASAHINA, Y., M. YANAGITA, AND I. YOSIOKA, *Chem. Ber.* **69**, 1370–1375 (1936). From *Lobaria oregana*.

BACHMANN, O., *Nova Hedwigia* **4**, 309–311 (1962). In *Lobaria pulmonaria* and *Umbilicaria torrefacta*, paper chromatography.

BACHMANN, O., *Österr. Botan. Z.* **110**, 103–107 (1963). Thin-layer chromatography.

CULBERSON, CHICITA F., *Phytochemistry* **6**, 719–725 (1967). From *Lobaria scrobiculata*.

CURD, F. H. AND A. ROBERTSON, *J. Chem. Soc.* **1935**, 1379–1381 (1935). From *Ramalina curnowii* (as *R. scopulorum*), stictic acid and scopuloric acid identical.

DHAR, M. L., S. NEELAKANTAN, S. RAMANUJAM, AND T. R. SESHADRI, *J. Sci. Ind. Res.* (India) **18B**, 111–113 (1959). From *Usnea orientalis*.

HALE, M. E., JR., *Contrib. U.S. Natl. Herb.* **36**, 193–358 (1965). Distribution in the genus *Parmelia*.

HALE, M. E., JR. AND S. KUROKAWA, *Contrib. U.S. Natl. Herb.* **36**, 121–191 (1964). Distribution in the genus *Parmelia*.

HESS, D., *Planta* **52**, 65–76 (1958). Paper chromatography.

HESSE, O., *J. Prakt. Chem.* **57**, 409–447 (1898). From *Lobaria pulmonaria* (as *Sticta pulmonaria*).

HUNECK, S., C. DJERASSI, D. BECHER, M. BARBER, M. VON ARDENNE, K. STEINFELDER, AND R. TÜMMLER, *Tetrahedron* **24**, 2707–2755 (1968). Mass spectrum.

HUNECK, S. AND G. FOLLMANN, *Z. Naturforsch.* **21b**, 715–716 (1966). From *Usnea rubicunda*.

HUNECK, S. AND G. FOLLMANN, *Z. Naturforsch.*, in press. From *Lecanora jussuffii*.

KUROKAWA, S., *Nova Hedwigia* **6**, 1–115 (1962). In the genus *Anaptychia*.

MOHAN, MARGARET, J. KEANE, AND T. J. NOLAN, *Sci. Proc. Roy. Dublin Soc.* **21**, 593–594 (1937). From *Parmelia conspersa* and *Lobaria pulmonaria*.

RAO, P. S., K. G. SARMA, AND T. R. SESHADRI, *Current Sci.* (India) **34**, 9–11 (1965). From *Lobaria isidiosa* and *L. retigera*.

RAO, P. S., K. G. SARMA, AND T. R. SESHADRI, *Current Sci.* (India) **35**, 147–148 (1966). From *Lobaria isidiosa* var. *subisidiosa* and *L. subretigera*.

RUNEMARK, H., *Opera Botan.* (Lund) **2** (1), 1–152 (1956). Distribution in the genus *Rhizocarpon*, paper chromatography.

SANTESSON, J., *Acta Chem. Scand.* **19**, 2254–2256 (1965). Thin-layer chromatography.

SANTESSON, J., *Acta Chem. Scand.* **21**, 1162–1172 (1967). Thin-layer chromatography.

MITSUNO, M., *Pharm. Bull.* (Tokyo) **1**, 170–173 (1953). Paper chromatography.

WACHTMEISTER, C. A., *Botan. Notiser* **109**, 313–324 (1956). Paper chromatography.

ZOPF, W., *Ann. Chem.* **352**, 1–44 (1907). From *Ramalina curnowii* (as *R. scopulorum*), as scopuloric acid.

vicanicin
$C_{18}H_{16}Cl_2O_5$
m.p. 248–250°.

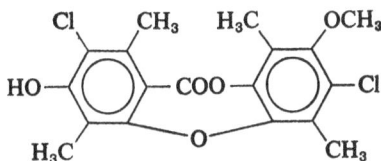

DYER, J. R., A. C. BAILLIE, V. M. BALTHIS, AND J. A. BERTRAND, Abstracts of Papers Presented at the Southeastern Regional Meeting of the American Chemical Society, Atlanta, Georgia, Nov. 1–3, 1967. Structure revised, mass spectrum, nmr, X-ray analysis of the iodoacetate.

NEELAKANTAN, S., T. R. SESHADRI, AND S. S. SUBRAMANIAN, *Tetrahedron Letters* **1959** (9), 1–4 (1959). From *Teloschistes flavicans*, structure studied.

NEELAKANTAN, S., T. R. SESHADRI, AND S. S. SUBRAMANIAN, *Tetrahedron* **18**, 597–604 (1962). Structure studied.

virensic acid
$C_{18}H_{14}O_8$
m.p. 245–247°; ir (KBr), ν cm^{-1}: 3509–3333 (w), 2985 (m), 1724 (s),

1639 (s), 1613 (s), 1543 (m), 1422 (m), 1389 (m), 1348 (m), 1305 (w), 1266 (s), 1242 (m), 1198 (m), 1149 (s), 1121 (m), 1015 (w), 881 (w), 843–833 (w), 794 (w), 781 (w), 743 (m), and 704 (w); uv (MeOH), λ_{max} (log ε) mμ: 240 (4.47) and 308 (3.68); λ_{min} (log ε) mμ: 225 (4.34) and 289 (3.48).

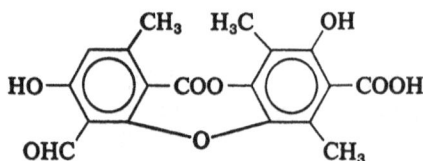

AGHORAMURTHY, K., K. G. SARMA, AND T. R. SESHADRI, *Tetrahedron* 12, 173–177 (1961). From *Alectoria tortuosa* (as *A. virens*), structure proved.

BENZYL ESTER

barbatolic acid
$C_{18}H_{14}O_{10}$
m.p. 206–207° d, softens from 190° (acetic acid or dioxane), needles.

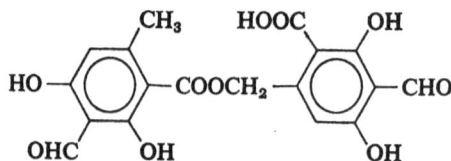

ASAHINA, Y., *J. Japan. Botany* 16, 517–522 (1940). Microcrystal test.
HESS, D., *Planta* 52, 65–76 (1958). Paper chromatography.
HUNECK, S., C. DJERASSI, D. BECHER, M. BARBER, M. VON ARDENNE, K. STEINFELDER, AND R. TÜMMLER, *Tetrahedron* 24, 2707–2755 (1968). Mass spectrum.
HUNECK, S. AND G. FOLLMANN, *Z. Naturforsch.* 21b, 91–92 (1966). From *Himantormia lugubris*.
SANTESSON, J., *Acta Chem. Scand.* 19, 2254–2256 (1965). Thin-layer chromatography.
SANTESSON, J., *Acta Chem. Scand.* 21, 1162–1172 (1967). Thin-layer chromatography.
SCHÖPF, C., K. HEUCK, AND R. DUNTZE, *Ann. Chem.* 491, 220–251 (1931). From *Usnea barbata* (probably contaminated with *Alectoria implexa*), first isolation, structure studied.
SUOMINEN, E. E., *Suomen Kemistilehti* 12B, 26–28 (1939); reference from *Chem. Zentr.* 111, 385–386 (1940). From *Alectoria implexa* f. *fuscidula*, structure proved.

PHENOLIC CARBOXYLIC ACID DERIVATIVE: DIQUINONE

pyxiferin

$C_{13}H_8O_8$

m.p. 300° (chloroform), red crystals; ir, λ μ: 2.95, 6.05, 6.20, 6.35, 6.60, 6.90, 7.40, 7.80, 8.60, 9.00, 9.35, 10.00, 10.35, 11.80, 12.40, and 12.60.

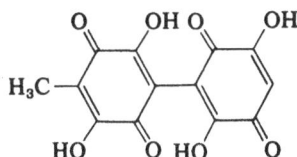

CHANDRASENAN, K., S. NEELAKANTAN, AND T. R. SESHADRI, *Bull. Natl. Inst. Sci. India* No. **28**, 92–98 (1965). From *Pyxine coccifera*, first isolation, structure proposed.

PHENOLIC CARBOXYLIC ACID DERIVATIVES: DIBENZOFURANS

dendroidin

$C_{16}H_{12}O_8$

m.p. 293°.

Structure not known.

ASAHINA, Y., *J. Japan. Botany* **36** (7), 225–232 (1961). In *Stereocaulon dendroides*, microcrystal tests.

didymic acid

$C_{22}H_{26}O_5$

m.p. 172–173° d (ligroin).

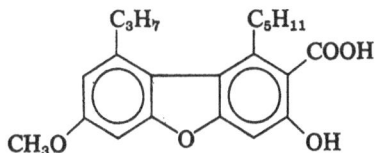

ASAHINA, Y., *J. Japan. Botany* **15**, 465–472 (1939). From *Cladonia floerkeana*, microcrystal tests.

ASAHINA, Y., *J. Japan. Botany* **19**, 301–311 (1943). Microcrystal test.

ASAHINA, Y., Lichens of Japan. Vol. I. Genus *Cladonia*. Tokyo. 1950. Microcrystal tests.

ASAHINA, Y., *J. Japan. Botany* **35**, 167–171 (1960). In *Cladonia corallifera*, microchemical tests.

EVANS, A. W., *Trans. Conn. Acad. Arts Sci.* **35**, 519–626 (1944). Distribution in Cladonias from Connecticut.

NOGAMI, H., *Yakugaku Zasshi* **64**, 47–50 (1944); reference from *Chem. Abstr.* **45**, 2929g (1951). From four species of *Cladonia*, structure studied.

SANTESSON, J., *Acta Chem. Scand.* **21**, 1162–1172 (1967). Thin-layer chromatography.

SHIBATA, S., *Acta Phytochim.* **14**, 9–38 (1944). Structure proved.

SHIBATA, S. AND Y. MIURA, *Japan. J. Med. Sci. Biol.* **2**, 22–24 (1949). Antibacterial effects.

SHIBATA, S., Y. MIURA, H. SUGIMURA, AND Y. TOYOIZUMI, *Yakugaku Zasshi* **68**, 303–305 (1948). Antibacterial activity against gram-positive bacteria.

pannaric acid

$C_{16}H_{12}O_7$

m.p. 243–245°, for anhydrous crystals (acetone-water, forms a dihydrate); ir (KBr), ν cm^{-1}: 1662 (chelated C=O) and 1690 (intermolecular hydrogen-bonded C=O); ir (THF), ν cm^{-1}: 1658 (chelated C=O and 1710 (intermolecular hydrogen-bonded C=O).

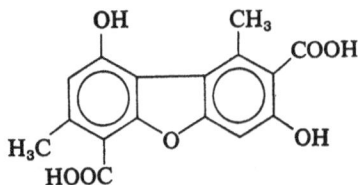

ÅKERMARK, B., *Acta Chem. Scand.* **15**, 985–990 (1961). Infrared spectra and hydrogen-bonding.

ÅKERMARK, B., H. ERDTMAN, AND C. A. WACHTMEISTER, *Acta Chem. Scand.* **13**, 1855–1862 (1959). From *Lepraria membranacea*, structure proved, paper chromatography.

SANTESSON, J., *Acta Chem. Scand.* **21**, 1162–1172 (1967). Thin-layer chromatography.

porphyrilic acid

$C_{16}H_{10}O_7$

m.p. 280–283° d, darkens after 270°.

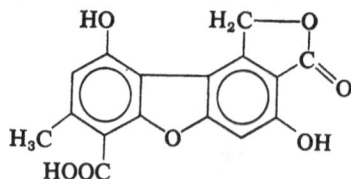

ERDTMAN, H. AND C. A. WACHTMEISTER, *Nature* **172**, 724–725 (1953). From *Haematomma coccineum*, structure studied.

ERDTMAN, H., AND C. A. WACHTMEISTER, *Chem. Ind.* (London) **1956**, 960 (1956). Structure proved.

HUNECK, S., Z. *Naturforsch.* **21b**, 80–81 (1966). From *Lecidea silacea.*

HUNECK, S., C. DJERASSI, D. BECHER, M. BARBER, M. VON ARDENNE, K. STEINFELDER, AND R. TÜMMLER, *Tetrahedron* **24**, 2707–2755 (1968). Mass spectrum of the methyl ester dimethyl ether.

SANTESSON, J., *Acta Chem. Scand.* **21**, 1162–1172 (1967). Thin-layer chromatography.

WACHTMEISTER, C. A., *Acta Chem. Scand.* **8**, 1433–1441 (1954). Structure studied.

WACHTMEISTER, C. A., *Acta Chem. Scand.* **10**, 1404–1413 (1956). Structure proved.

WACHTMEISTER, C. A., *Botan. Notiser* **109**, 313–324 (1956). Paper chromatography.

ZOPF, W., *Ann. Chem.* **346**, 82–127 (1906). From *Haematomma coccineum* and *H. porphyrium*, first extraction.

schizopeltic acid

$C_{19}H_{18}O_7$

m.p. 228–230° (acetone-water); m.p. 233–235° (methanol), colorless needles; ir (KBr), ν cm^{-1}: 3550 (CO$\underline{OH}$), 1725 (ester C=O), and 1680 (acid C=O); uv (EtOH), λ_{max} (ε) mμ: 241 (31,000), 274 (15,000), 290 (13,500), 301 (10,000), and 314 (10,500); uv (MeOH), λ_{max} (log ε) mμ: 236 (4.62), 270 (4.29), 286 (4.26), 302 (4.07), and 312 (4.12); nmr ((CD$_3$)$_2$SO), δ ppm: 2.61 and 2.70 (2 aryl CH$_3$), 3.94 (3 aryl CH$_3$), and 6.83 and 7.34 (2 aryl protons).

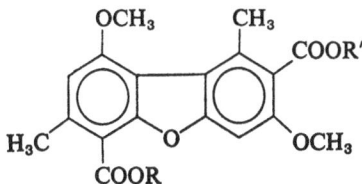

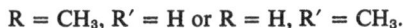

$R = CH_3$, $R' = H$ or $R = H$, $R' = CH_3$.

HUNECK, S. AND G. FOLLMANN, Z. *Naturforsch.* **22b**, 1185–1188 (1967). From *Roccellina luteola.*

SANTESSON, J., *Acta Chem. Scand.* **21**, 1111 (1967). From *Reinkella parishii* and *Schizopelte californica*, first isolation, structure studied.

SANTESSON, J., *Acta Chem. Scand.* **21**, 1162–1172 (1967). Thin-layer chromatography.

strepsilin

$C_{15}H_{10}O_5$

m.p. 324° (acetic acid).

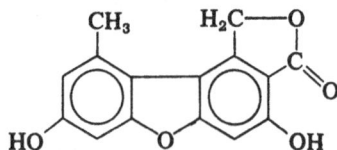

ASAHINA, Y., *J. Japan. Botany* **15**, 465–472 (1939). Microchemical tests.
ASAHINA, Y., Lichens of Japan. Vol. I. Genus *Cladonia*. Tokyo. 1950.
Microcrystal test.
EVANS, A. W., *Trans. Conn. Acad. Arts Sci.* **35**, 519–626 (1944). In
Cladonia strepsilis.
SANTESSON, J., *Acta Chem. Scand.* **21**, 1162–1172 (1967). Thin-layer
chromatography.
SHIBATA, S., *Yakugaku Zasshi* **64**(8), 20–21 (1944); see *Chem. Abstr.* **45**,
5677 (1951). Structure proved.
SHIBATA, S., *Pharm. Bull.* (Tokyo) **5**, 488–491 (1957). Structure con-
firmed, infrared spectrum.
WACHTMEISTER, C. A., *Botan. Notiser* **109**, 313–324 (1956). Paper chro-
matography.
ZOPF, W., *Ann. Chem.* **327**, 317–354 (1903). From *Cladonia strepsilis*.

USNIC ACIDS

(+)-isousnic acid

$C_{18}H_{16}O_7$

m.p. 150–152° (benzene-methanol), yellow prisms; $[\alpha]_D^{21}$ + 500° (dioxane).

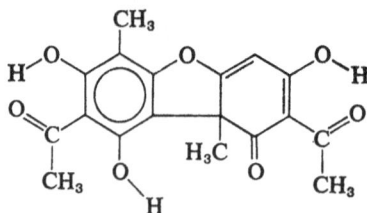

SHIBATA, S. AND H. TAGUCHI, *Tetrahedron Letters* **1967**, 4867–4871
(1967). From *Cladonia mitis*, first isolation, structure proved, nmr
spectrum, isousnic acid also found microchemically in *C. arbuscula*
and *C. submitis* along with usnic acid, not found in *Evernia*, *Parmelia*,
or *Usnea* samples tested.

(−)-isousnic acid

$C_{18}H_{16}O_7$

m.p. 150–152°; $[\alpha]_D$ −490° (dioxane).

For the structure see (+)-isousnic acid.

SHIBATA, S. AND H. TAGUCHI, *Tetrahedron Letters* **1967**, 4867–4871 (1967). With (−)-usnic acid in *Cladonia pleurota*, first report.

(+)-usnic acid

$C_{18}H_{16}O_7$

m.p. 203–204° (benzene or chloroform-ethanol), yellow prisms or needles; $[\alpha]_D^{20}$ +495° ($CHCl_3$).

For structure and references see under (−)-usnic acid.

(−)-usnic acid

$C_{18}H_{16}O_7$

m.p. 203° (benzene or chloroform-ethanol), yellow prisms or needles; $[\alpha]_D^{20}$ −495° ($CHCl_3$).

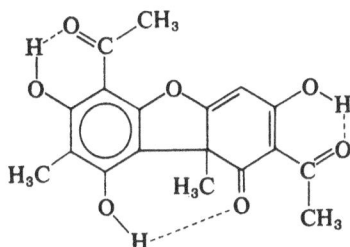

ASAHINA, Y., *J. Japan. Botany* **12**, 859–872 (1936); *ibid.* **13**, 529–536 (1937). Microchemical tests.

ASAHINA, Y., *J. Japan. Botany* **16**, 185–193 (1940). (−)-Usnic acid from *Cladonia aberrans*, microcrystal tests.

ASAHINA, Y., Lichens of Japan. Vol. I. Genus *Cladonia*. Tokyo. 1950. Microcrystal tests.

ASAHINA, Y. AND A. HASHIMOTO, *Chem. Ber.* **66**, 641–649 (1933). (−)-Usnic acid from *Alectoria japonica* and *A. sarmentosa* and (+)-usnic acid from *A. ochroleuca*.

ASAHINA, Y. AND S. SHIBATA, Chemistry of Lichen Substances. Tokyo. 1954. Review of structural studies.

ASAHINA, Y. AND M. YANAGITA, *Chem. Ber.* **69**, 1646–1649 (1936); *ibid.* **70**, 66–70 (1937); *ibid.* **70**, 1500–1505 (1937); *ibid.* **71**, 2260–2269 (1938); *ibid.* **72**, 1140–1146 (1939). Structure and reactions studied.

ASANO, M. AND T. AZUMI, *Chem. Ber.* **68**, 995–997 (1935). (−)-Usnic acid from *Cetraria stracheyi* f. *ectocarpisma*.

BARTON, D. H. R. AND T. BRUUN, *J. Chem. Soc.* **1953**, 603–609 (1953). Structure confirmed.

BARTON, D. H. R., A. M. DEFLORIN, AND O. E. EDWARDS, *Chem. Ind.* (London) **1955**, 1039–1040 (1955). Synthesis of (±)-usnic acid.

BARTON, D. H. R., A. M. DEFLORIN, AND O. E. EDWARDS, *J. Chem. Soc.* **1956**, 530–534 (1956). A two-step synthesis of (±)-usnic acid.

BARTON, D. H. R. AND G. QUINKERT, *J. Chem. Soc.* **1960**, 1–9 (1960). Racemization in dioxane solution with uv light.

BENDZ, GERD, GERD BOHMAN, AND J. SANTESSON, *Acta Chem. Scand.* **21**, 1376–1377 (1967). Thin-layer chromatographic method for separating and identifying (+)- and (−)-usnic acid depending upon preferential complex formation of (+)-usnic acid and brucine, (+)-usnic acid from *Cladonia arbuscula* (as *C. silvatica*) and (−)-usnic acid from *C. alpestris*.

BICK, I. R. C. AND D. H. S. HORN, *Australian J. Chem.* **18**, 1405–1410 (1965). Most stable tautomer deduced from nmr studies on a related natural β-triketone.

BORKOWSKI, B., WANDA WOŹNIAK, H. GERTIG, AND BOGUSLAWA WER-AKSO, *Dissertationes Pharm.* **16**, 189–194 (1964). (+)-Usnic acid from *Usnea dasypoga*.

BRINER, G. P., G. E. GREAM, AND N. V. RIGGS, *Australian J. Chem.* **13**, 277–284 (1960). (+)-Usnic acid from *Parmelia scabrosa* Tayl. (as *P. conspersa*) and *Ramalina fraxinea*.

BROCK, T. D., *J. Bacteriol.* **85**, 527–531 (1963). Inhibition of M protein synthesis.

CAPRIOTTI, A., *Giorn. Microbiol.* **7**, 187–206 (1959). Inhibitory effects of the antibiotic "USNO" on yeasts.

CAPRIOTTI, A., *Antibiot. Chemother.* **11**, 409–410 (1961). Effects of the antibiotic "USNO" on yeasts.

CHAN, W. R. AND C. H. HASSALL, *J. Chem. Soc.* **1956**, 3495–3499 (1956). Uv and ir absorption spectra.

CULBERSON, CHICITA F., *Phytochemistry* **2**, 335–340 (1963). (+)-Usnic acid from *Evernia prunastri* and (−)-usnic acid from *Cetraria nivalis*.

CULBERSON, CHICITA F., *Phytochemistry* **6**, 719–725 (1967). (+)-Usnic acid from *Lobaria scrobiculata*.

CURD, F. H. AND A. ROBERTSON, *J. Chem. Soc.* **1933**, 437–444 (1933); *ibid.* **1933**, 714–720 (1933); *ibid.* **1933**, 1173–1179 (1933); *ibid.* **1937**, 894–901 (1937). Structure proved.

DAHL, E., *Medd. Grønland* **150** (2), 1–176 (1950). Microchemical tests.

DEAN, F. M., *Sci. Progr.* (London) **40**, 635–644 (1952). Review.

DEAN, F. M., P. HALEWOOD, S. MONGKOLSUK, A. ROBERTSON, AND W. B. WHALLEY, *J. Chem. Soc.* **1953**, 1250–1261 (1953). Resolution of (±)-usnic acid.

DHAR, M. L., S. NEELAKANTAN, S. RAMANUJAM, AND T. R. SESHADRI, *J. Sci. Ind. Res.* (India) **18B**, 111–113 (1959). (+)-Usnic acid from *Parmelia taractica, Usnea aspera, U. longissima,* and *U. orientalis*.

FEDOSEEV, K. G. AND P. A. YAKIMOV, *Tr. Leningr. Khim. Farmatsevt. Inst.* **9**, 139–149 (1960). Quantitative measurements in *Cladonia* sp.

FORSÉN, S., M. NILSSON, AND C. A. WACHTMEISTER, *Acta Chem. Scand.* **16**, 583–590 (1962). Ir and nmr analysis of hydrogen-bonding.

GERTIG, H., *Acta Polon. Pharm.* **18**, 57–66 (1961). Quantitative measurements with ferric chloride or uranyl acetate.

GERTIG, H. AND Z. BANASIEWICZ, *Acta Polon. Pharm.* **18**, 67–71 (1961). Quantitative measurements in several species of *Cladonia* and *Usnea*.

HALE, M. E., JR., *Contrib. U.S. Natl. Herb.* **36**, 193–358 (1965). Distribution in the genus *Parmelia*.

HALE, M. E., JR. AND S. KUROKAWA, *Contrib. U.S. Natl. Herb.* **36**, 121–191 (1964). Distribution in the genus *Parmelia*.

HARPER, S. H. AND R. M. LETCHER, *Proc. Trans. Rhodesian Sci. Assoc.* **51**, 156–184 (1966). Infrared and nmr spectra, (+)-usnic acid from *Parmelia dilatata* and *Usnea implicita*.

HUNECK, S., *Z. Naturforsch.* **21b**, 199–200 (1966). (+)-Usnic acid from *Lecanora handelii*.

HUNECK, S., *Z. Naturforsch.* **21b**, 888–890 (1966). (+)-Usnic acid from *Lecanora badia*, *L. polytropa*, and *L. sulphurea*.

HUNECK, S., C. DJERASSI, D. BECHER, M. BARBER, M. VON ARDENNE, K. STEINFELDER, AND R. TÜMMLER, *Tetrahedron* **24**, 2707–2755 (1968). Mass spectrum.

HUNECK, S. AND G. FOLLMANN, *Naturwissenschaften* **51**, 291–292 (1964). (−)-Usnic acid from *Lecanora melanophthalma* and (+)-usnic acid from *Ramalina terebrata*.

HUNECK, S. AND G. FOLLMANN, *Z. Naturforsch.* **20b**, 611–612 (1965). (+)-Usnic acid from *Ramalina tumidula*.

HUNECK, S. AND G. FOLLMANN, *Z. Naturforsch.* **20b**, 1012–1013 (1965). (+)-Usnic acid from *Nephroma gyelnikii*.

HUNECK, S. AND G. FOLLMANN, *Z. Naturforsch.* **21b**, 90–91 (1966). (+)-Usnic acid from *Ramalina chilensis*.

HUNECK, S. AND G. FOLLMANN, *Z. Naturforsch.* **21b**, 91–92 (1966). From *Himantormia lugubris*.

HUNECK, S. AND G. FOLLMANN, *Z. Naturforsch.* **21b**, 713–714 (1966). (+)-Usnic acid from *Ramalina inanis* and *R. tigrina*.

HUNECK, S. AND G. FOLLMANN, *Z. Naturforsch.* **21b**, 715–716 (1966). (+)-Usnic acid from *Usnea aureola*, *U. lacerata*, and *U. rubicunda*.

HUNECK, S. AND G. FOLLMANN, *Z. Naturforsch.* **22b**, 461 (1967). (+)-Usnic acid from *Usnea pusilla* (2.7%).

HUNECK S. AND G. FOLLMANN, *Z. Naturforsch.* **22b**, 689 (1967). (−)-Usnic acid from *Cladonia reticulata*.

HUNECK, S. AND G. FOLLMANN, *Z. Naturforsch.* **22b**, 791–792 (1967). (+)-Usnic acid from *Usnea eulychniae* (3.4%).

HUNECK, S. AND G. FOLLMANN, *Z. Naturforsch.* **22b**, 1368–1369 (1967). (−)-Usnic acid from *Cladonia endiviaefolia*, from Spain (0.9%).

HUNECK, S., G. FOLLMANN, AND W. A. WEBER, *Willdenowia*, in press. From *Dactylina arctica, D. chinensis, D. endochrysea, D. madrepori-formis,* and *D. ramulosa.*

HUNECK, S. AND G. TROTET, *Z. Naturforsch.* **21b**, 904 (1966). (+)-Usnic acid from *Ramalina boulhautiana.*

JONES, F. T. AND K. J. PALMER, *J. Am. Chem. Soc.* **72**, 1820–1822 (1950). Optical, crystallographic, and X-ray diffraction data.

LAAKSO, P. V. AND MARGIT GUSTAFSSON, *Suomen Kemistilehti* **25B**(2), 7–10 (1952). In *Cladonia alpestris,* a colorimetric method with ferric chloride.

LAZAREV, N. V. AND V. P. SAVICZ, [The New Antibiotic Binan, or the Sodium Salt of Usnic Acid (Botanical and Medical Investigations).] Moscow-Leningrad. 1957. Review.

MACKENZIE, S., *J. Am. Chem. Soc.* **74**, 4067–4069 (1952). Uv spectrum and derivatives.

MACKENZIE, S., *J. Am. Chem. Soc.* **77**, 2214–2216 (1955). Racemization in various solvents.

MARSHAK, A., *Public Health Rept.* **62**, 3–19 (1947). From *Ramalina reticulata.*

MARSHAK, A., G. T. BARRY, AND L. C. CRAIG, *Science* **106**, 394–395 (1947). Antibiotic properties.

MITCHELL, J. C., *J. Invest. Dermat.* **47**, 167–168 (1966). Specificity for the (+)-isomer in delayed hypersensitivity.

NAITO, M., C. TAKI, A. SHIHODA, F. FUZIKAWA, K. NAKAJIMA, H. FUJII, A. TOKUOKA, AND M. OKAMOTO, *Yakugaku Zasshi* **71**, 113–115 (1951). Strong activity against the tubercule bacillus *in vitro.*

PENTTILA, A. AND H. M. FALES, *Chem. Comm.* **1966**, 656–657 (1966). Biosynthesis *in vitro* with horseradish peroxidase.

RAMAUT, J., R. SCHUMACKER, J. LAMBINON, AND C. BAUDUIN, *Bull. Jardin Botan. État Bruxelles* **36**, 399–414 (1966). Quantitative meas-urements of usnic acid in *Cladonia impexa, C. leucophaea,* and *C. tenuis.*

ROCHLEDER, F. AND W. HELDT, *Ann. Chem.* **48**, 1–18 (1843). From *Ramalina calicaris, Usnea barbata,* and other species, first isolation.

SALKOWSKI, H., *Ann. Chem.* **314**, 97–111 (1901); *ibid.* **319**, 391–399 (1901); *ibid.* **377**, 123–126 (1910). Direction of rotation of polarized light of samples from numerous species extracted by Zopf.

SANTESSON, J., *Acta Chem. Scand.* **21**, 1162–1172 (1967). Thin-layer chromatography.

SARMA, K. G. AND S. HUNECK, *Pharmazie,* in press. (+)-Usnic acid from *Ramalina linearis, R. subamplicata, Usnea compacta, U. misaminensis,* and *U. rubicunda* var. *ceratinella.*

SAVICZ, V. P., M. A. LITVINOV, AND E. N. MOISSEJEVA, *Planta Med.* **8**, 191–202 (1960). Review of Russian antibiotic studies.

SCHÖPF, C. AND K. HEUCK, *Ann. Chem.* **459**, 233–286 (1927). From *Usnea barbata,* structure studied.

SCHÖPF, C. AND F. ROSS, *Naturwissenschaften* **26**, 772–773 (1938). Structure studied.

SCHÖPF, C. AND F. ROSS, *Ann. Chem.* **546**, 1–40 (1941). Structure confirmed.

SCHUMACKER, R., *Bull. Soc. Roy. Sci. Liège* **30**, 452–457 (1961). From *Everniopsis trulla.*

SHARMA, R. K. AND P. J. JANNKE, *Indian J. Chem.* **4**, 16–18 (1966). pK values for the three hydroxyls.

SHIBATA, S., Y. MIURA, T. UKITA, AND T. TAMURA, *Yakugaku Zasshi* **68**, 298–300 (1948). Antibacterial activity due to dihydroxy α,β-unsaturated ketone structure.

SHIBATA, S. AND J. SHOJI, *Kagaku no Ryoiki* **15**, 803–810 (1961); reference from *Chem. Abstr.* **56**, 8664F (1962). Nmr data, review.

SHIBATA, S., J. SHOJI, N. TOKUTAKE, Y. KANEKO, H. SHIMIZU, AND HSÜCH-CHING CHIANG, *Chem. Pharm. Bull.* (Tokyo) **10**, 477–483 (1962). Ozonolysis of di-O-acetylusnic acid.

SHOJI, J., *Chem. Pharm. Bull.* (Tokyo) **10**, 483–491 (1962). Isodihydrousnic acid.

STARK, J. B., E. D. WALTER, AND H. S. OWENS, *J. Am. Chem. Soc.* **72**, 1819–1820 (1950). A method of isolation from *Ramalina reticulata.*

STOLL, A., A. BRACK, AND J. RENZ, *Experientia* **3**, 115–116 (1947). (+)-Usnic acid from *Cladonia mitis, Evernia divaricata, Ramalina capitata, Usnea dasypoga, U. florida,* and *U. hirta,* (−)-usnic acid from *Alectoria ochroleuca, Cetraria nivalis,* and *Cladonia deformis,* antibacterial activity.

STORK, G., *Chem. Ind.* **1955**, 915–916 (1955). Racemization.

TAGUCHI, H., U. SANKAWA, AND S. SHIBATA, *Tetrahedron Letters* **1966**, 5211–5214 (1966). Biosynthesis.

TAKAHASHI, K., A. ARAI, K. OSHIMA, Y. UEDA, AND S. MIYASHITA, *Chem. Pharm. Bull.* (Tokyo) **10**, 607–611 (1962). Methylusnic acid.

TAKAHASHI, K. AND S. MIYASHITA, *Chem. Pharm. Bull.* (Tokyo) **10**, 603–607 (1962). Methyldihydrousnic acid.

VIRTANEN, O. E. AND O. E. KILPIÖ, *Suomen Kemistilehti* **30B**, 8–9 (1957). Antifungal activity of "USNO."

WACHTMEISTER, C. A., *Acta Chem. Scand.* **8**, 1433–1441 (1954). Up to 20% of dry weight of (−)-usnic acid from *Haematomma coccineum.*

WASICKY, R., *Anal. Chem.* **34**, 1346–1347 (1962). Thin-layer chromatography on microslides.

WIDMAN, O., *Ann. Chem.* **310**, 230–301 (1900); *ibid.* **324**, 139–200 (1902). Optical isomerism observed, structure studied.

CHROMONES

lepraric acid

$C_{18}H_{18}O_8$

m.p. 155–156.5° d (chloroform-benzene), colorless crystals; m.p. 161–

162° (methanol), plates; ir, ν cm^{-1}; 2700–3300 (carboxyl), 1714 (α,β-unsaturated ester and carboxyl) 1659 and 1625 (chromone), and 1659 and 1625 (aromatic); uv, λ_{max} (log ε) mμ: 216 (4.39), 233 (4.46), 251 (4.37), and 292 (3.88); nmr (DMSO-d$_6$ and CDCl$_3$; 100 MHz), τ ppm: 3.51 and 3.89 (1 aryl proton and 1 proton of —COCH=C), 4.91 (—COOCH$_2$—), 6.93 (—COCH$_2$C=C—), 4.28 (CH$_3$C=CHCO), 7.82 (CH$_3$C=CHCO—), 6.11 (—OCH$_3$), and 7.63 (aryl—CH$_3$); mass spectrum: 317 (M − COOH).

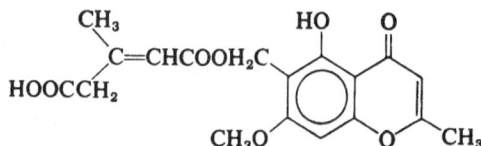

ABERHART, D. J., S. HUNECK, and K. H. OVERTON, *J. Chem. Soc.* (C), in press. Structure proved.

HUNECK, S. AND G. FOLLMANN, *Ber. Deut. Botan. Ges.*, in press. From *Sagenidium molle.*

HUNECK, S., G. FOLLMANN, AND H. ULLRICH, *Z. Naturforsch.* **23b**, 292–293 (1968). From *Roccella teneriffensis* (8.7%), as fuciformic acid.

HUNECK, S., ANNICK MATHEY, AND G. TROTET, *Z. Naturforsch.* **22b**, 1367–1368 (1967). From *Roccella fuciformis* from the English Channel (0.5%) and from Morocco (0.8%), as fuciformic acid.

SOVIAR, K., O. MOTL, Z. SAMEK, AND J. SMOLÍKOVÁ, *Tetrahedron Letters* **1967**, 2277–2279 (1967). From *Lepraria latebrarum*, structure studied.

siphulin

C$_{24}$H$_{26}$O$_7$

m.p. 180° d; ir (KBr), ν cm^{-1}: 1625; ir (THF), ν cm^{-1}: 1658, 1615, and 1595; uv (EtOH), λ_{max} (ε) mμ: 251 (29,000), 264 (21,000), and 293 (23,000); λ_{min} (ε) mμ: 248 (28,000), 259 (19,500), and 274.5 (17,500). $\lambda_{infl.}$ (ε) mμ: 242.5 (29,500).

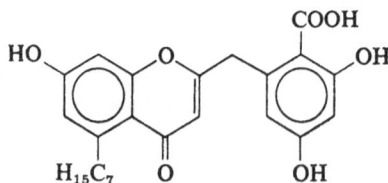

BRUUN, T., *Tetrahedron Letters* **1960**(4), 1–4 (1960). From *Siphula ceratites*, first isolation.

BRUUN, T., *Acta Chem. Scand.* **19**, 1677–1693 (1965). Structure proved, data on many derivatives.

sordidone

$C_{11}H_9ClO_4$

m.p. 260–262° cream-colored crystals; uv (EtOH), λ_{max} (ε) mμ: 263 (16,000), 296 (5,750), and 332 (3650); ir (nujol), ν cm^{-1}: 1660, 1624, and 1586; nmr (C_5D_5N), τ ppm: 7.79 and 7.58 (singlets, 2 aryl methyls), 3.87 (one aryl proton), and 0.05 and −3.60 (broad singlets, 2 OH groups).

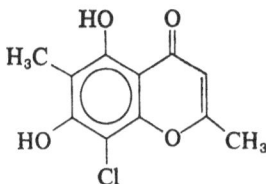

ARSHAD, M., J. P. DEVLIN, W. D. OLLIS, AND R. E. WHEELER, *Chem. Commun.* **1968**, 154–155 (1968). From *Lecanora sordida*, structure proved, synthesis from eugenitol, sordidone not identical to Hesse's "thiophanic acid" as previously suggested by Kennedy *et al.*

HUNECK, S., *Tetrahedron Letters* **1966**, 3547–3549 (1966). From *Lecanora rupicola*, as an unidentified chlorine-containing compound possibly identical to Hesse's "thiophaninic acid," m.p. 265–266°.

KENNEDY, G., J. BREEN, J. KEANE, AND T. J. NOLAN, *Sci. Proc. Roy. Dublin Soc.* **21**, 557–566 (1937). From *Lecanora rupicola*, compound ($C_{24}H_{20}Cl_2O_9$) thought to be thiophanic acid.

SANTESSON, J., *Acta Chem. Scand.* **21**, 1162–1172 (1967). Thin-layer chromatography, from *Lecanora rupicola*, as rupicolin.

XANTHONES

arthothelin

$C_{14}H_7Cl_3O_5$

m.p. 275–276° (ethyl acetate), yellow prisms; uv (MeOH), λ_{max} (log ε) mμ: 247 (4.5), 282 (3.95, shoulder), 318 (4.15), and 354 (3.98).

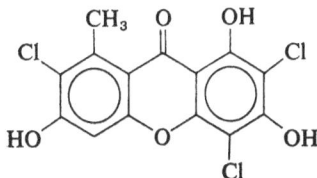

HUNECK, S. AND G. FOLLMANN, *Z. Naturforsch.* **22b**, 461 (1967). From *Arthothelium pacificum*.

SANTESSON, J., *Acta Chem. Scand.* **21**, 1162–1172 (1967). Thin-layer chromatography, as deschlorothiophanic acid, in *Lecanora straminea*.

SANTESSON, J., *Arkiv Kemi*, in press. From *Lecanora straminea*, structure proved.

SANTESSON, J., *Arkiv Kemi*, in press. From *Lecanora pinguis* and *Lecidea guernea*.

SANTESSON, J. AND G. SUNDHOLM, *Arkiv. Kemi*, in press. Synthesis.

concretin

$C_{14}H_7Cl_3O_5$

m.p. 287° d (methanol), yellow needles; no methoxyl, forms a triacetate (m.p. 220–222°) and a tri-O-methyl derivative; possibly a dechloro-thiophanic acid.

Structure not known.

BREEN, J., J. KEANE, AND T. J. NOLAN, *Sci. Proc. Roy. Dublin Soc.* **21**, 587–592 (1937). From *Pertusaria pseudocorallina* (as *P. concreta* f. *westringii*), empirical formula and derivatives studied.

2,4-dichloronorlichexanthone

$C_{14}H_8Cl_2O_5$

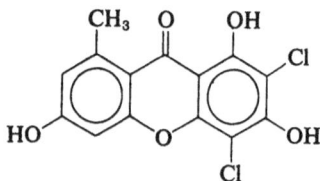

SANTESSON, J., in preparation. From *Lecanora straminea*.

SANTESSON, J. AND G. SUNDHOLM, *Arkiv Kemi*, in press. Synthesis, structure.

2,7-dichloronorlichexanthone

$C_{14}H_8Cl_2O_5$

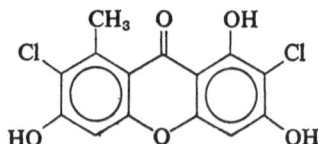

SANTESSON, J., *Arkiv Kemi*, in press. From *Lecanora straminea*, structure proved.

lichexanthone

$C_{16}H_{14}O_5$

m.p. 187–190° (acetone), yellow needles.

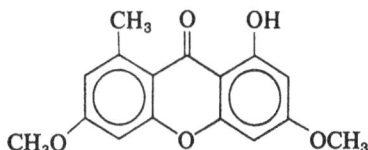

AGHORAMURTHY, K. AND T. R. SESHADRI, *J. Sci. Ind. Res.* (India) **12B**, 350–352 (1953). Synthesis.

ASAHINA, Y., Lichens of Japan. Vol II. Genus *Parmelia*. Tokyo. 1952. Microcrystal test.

ASAHINA, Y. AND H. NOGAMI, *Bull. Chem. Soc. Japan* **17**, 202–207 (1942). From *Parmelia formosana*, first isolation, structure studied, synthesis of the methyl ether.

CULBERSON, W. L. AND M. E. HALE, JR., *Bryologist* **68**, 113–116 (1965). In *Pyxine caesiopruinosa* and probably in *P. chrysanthoides*.

GROVER, P. K., G. D. SHAH, AND R. C. SHAH, *J. Sci. Ind. Res.* (India) **15B**, 629–630 (1956). A new synthesis.

HALE, M. E., JR. AND S. KUROKAWA, *Contrib. U.S. Natl. Herb.* **36**, 121–191 (1964). Distribution in the genus *Parmelia*.

HARPER, S. H. AND R. M. LETCHER, *Proc. Trans. Rhodesian Sci. Assoc.* **51**, 156–184 (1966). In an unidentified species of *Pertusaria* from Rhodesia, infrared and nmr spectra.

KRISHNAMURTI, M. AND T. R. SESHADRI, *J. Sci. Ind. Res.* (India) **14B**, 258–260 (1955). Circular-paper chromatography.

LEUCKERT, C. AND H. HERTEL, *Nova Hedwigia* **14**, 291–300 (1967). In *Lecidea stigmatea*, as unidentified substance St 1.

SANTESSON, J., *Acta Chem. Scand.* **21**, 1162–1172 (1967). Thin-layer chromatography.

WIRTH, M. AND M. E. HALE, JR., *Contrib. U.S. Natl. Herb.* **36**, 63–119 (1963). In *Graphina confluens*.

norlichexanthone

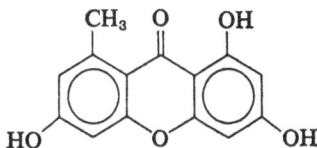

$C_{14}H_{10}O_5$

SANTESSON, J., *Acta Chem. Scand.*, in press. From *Lecanora reuteri.*
SANTESSON, J., in preparation. From *Lecanora straminea.*

thiophanic acid

$C_{14}H_6Cl_4O_5$

m.p. 243° (benzene), yellow prisms; mass spectrum, m/e: 394, 396, 398, 400, and 402; uv, λ_{max} (ε) mμ: 248 (40,000), 320 (13,600), and 360 (16,000); nmr $((CD_3)_2SO)$, τ ppm: 7.22 (aryl methyl).

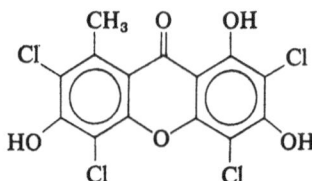

HESSE, O., *J. Prakt. Chem.* **58**, 465–561 (1898). From *Lecanora rupicola*, first isolation.

HUNECK, S., *Tetrahedron Letters* **1966**, 3547–3549 (1966). From *Lecanora rupicola*, structure proved.

KENNEDY, G., J. BREEN, J. KEANE, AND T. J. NOLAN, *Sci. Proc. Roy. Dublin Soc.* **21**, 557–566 (1937). From *Lecanora rupicola*, but see sordidone.

thiophaninic acid

$C_{15}H_{10}Cl_2O_5$

m.p. 269–271° (ethyl acetate), lemon-yellow prisms.

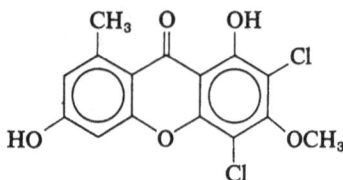

SANTESSON, J., *Acta Chem. Scand.* **21**, 1162–1172 (1967). Thin-layer chromatography, as flavicanone, from *Pertusaria flavicans.*

WACHTMEISTER, C. A. AND J. SANTESSON, unpublished. Structure proved, flavicanone identical to Zopf's thiophaninic acid.

SANTESSON, J., *Arkiv Kemi*, in press. From *Pertusaria flavicunda.*

thuringione

$C_{15}H_9Cl_3O_5$

m.p. 278–279° (ethyl acetate); uv (MeOH), λ_{max} (log ε) mμ: 246 (4.53), 314 (4.18), and 356 (4.01).

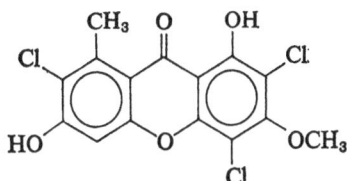

HUNECK, S. AND J. SANTESSON, unpublished. Structure proved.

SANTESSON, J., *Acta Chem. Scand.* **21**, 1162–1172 (1967). Thin-layer chromatography, as thuringion, from *Lecidea carpathica*.

SANTESSON, J., *Arkiv Kemi*, in press. From *Lecanora pinguis*.

SANTESSON, J., *Arkiv Kemi*, in press. Synthesis.

vinetorin

$C_{15}H_{11}ClO_5$

m.p. 243–245°.

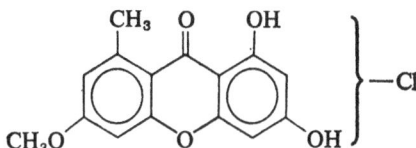

HUNECK, S. AND J. POELT, unpublished. From a new species of *Lecanora*.

ANTHRAQUINONES

chiodectonic acid

m.p. 303° (acetic acid–water), red powder.

An hydroxyanthraquinone carboxylic acid. Structure not known.

HESSE, O., *J. Prakt. Chem.* **70**, 449–502 (1904). From *Herpothallon sanguineum*, first extraction.

KOLUMBE, E., *Mikrokosmos* **21**, 53–55 (1927). From *Herpothallon sanguineum*.

RIBEIRO, O. AND W. B. MORS, *Bol. Inst. Quím. Agr.* (Rio de Janeiro) **15**, 1–14 (1949). From *Herpothallon sanguineum*.

RIBEIRO, O. AND W. B. MORS. *Anais Assoc. Quím. Brasil* **9**, 182–189 (1950). Structure studied.

chrysophanol

$C_{15}H_{10}O_4$

m.p. 193–196° (ethanol), yellow plates.

Also known in higher plants and nonlichen-forming fungi.

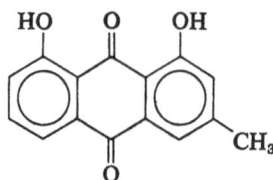

SHIBATA, S., *J. Japan. Botany*, in press. In *Acroscyphus sphaerophoroides*.

1,3-dihydroxy-8-methoxy-2-chloro-6-methylanthraquinone

$C_{16}H_{11}ClO_5$

Recrystallized from glacial acetic acid; mass spectrum, m/e: 318 and 270.

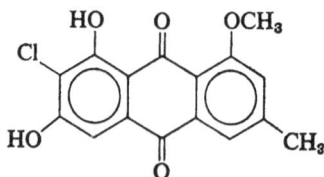

BENDZ, GERD, GERD BOHMAN, AND J. SANTESSON, *Acta Chem. Scand.* **21**, 2889–2890 (1967). From *Nephroma laevigatum*, structure proved, as 7-chloro-1-O-methylemodin.

emodin

$C_{15}H_{10}O_5$

m.p. 254–256° (ethanol), orange-red needles.

Also known in higher plants and nonlichen-forming fungi.

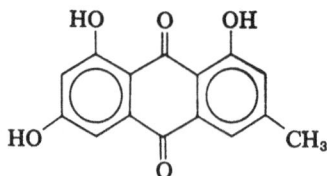

BENDZ, GERD, GERD BOHMAN, AND J. SANTESSON, *Acta Chem. Scand.* **21**, 2889–2890 (1967). From *Nephroma laevigatum*.

YOSIOKA, I., H. YAMAUCHI, K. MORIMOTO, AND I. KITAGAWA, *Tetrahedron Letters* **1968**, 1149–1152 (1968). From *Heterodermia obscurata.*

endocrocin

$C_{16}H_{10}O_7$

m.p. 318° (acetic acid or acetone-water), orange-red crystals; ir (nujol), ν cm^{-1}: 1718 (carboxylic acid C=O), 1615 (chelated anthraquinone C=O), and 1666 (unchelated anthraquinone C=O).

Also known from the fungi *Aspergillus amstelodami, Penicillium islandicum*, and *Claviceps purpurea.*

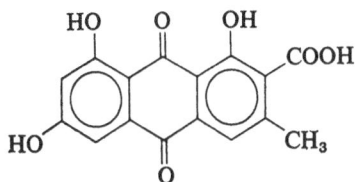

ASAHINA, Y., *J. Japan. Botany* **11**, 10–27 (1935). Description of medullary pigment in *Cetraria endocrocea.*

ASAHINA, Y. AND F. FUZIKAWA, *Chem. Ber.* **68**, 1558–1565 (1935). From *Cetraria endocrocea*, structure determined.

JOSHI, B. S., S. RAMANATHAN, AND K. VENKATARAMAN, *Tetrahedron Letters* **1962**, 951–955 (1962). Synthesis, structure confirmed.

JOSHI, B. S., S. RAMANATHAN, AND K. VENKATARAMAN, *Bull. Natl. Inst. Sci. India*, No. **28**, 122–124 (1965). Synthesis, discussion of ir and nmr spectra of the fully methylated derivative.

SANTESSON, J., *Acta Chem. Scand.* **21**, 1162–1172 (1967). Thin-layer chromatography.

TAKIDO, M., *Pharm. Bull.* (Tokyo) **4**, 45–48 (1956). Paper chromatography.

fallacinal

$C_{16}H_{10}O_6$

m.p. 250–252° (chloroform-ethanol), orange-red needles.

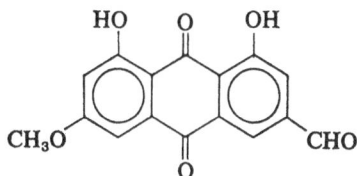

ASANO, M. AND S. FUZIWARA, *Yakugaku Zasshi* **56**, 1007–1010 (1936); German summary, p. 101 (1936). "Fallacin" from *Xanthoria fallax.*

ASANO, M. AND H. ARATA, *Yakugaku Zasshi* **60**, 521–525 (1940); German summary, pp. 206–208 (1940). "Fallacin" partially purified by removal of parietin.

MURAKAMI, T., *Pharm. Bull.* (Tokyo) **4**, 298–302 (1956). From *Xanthoria fallax*, separation from fallacinol (=teloschistin) and parietin, structure proved.

RAJAGOPALAN, T. R. AND T. R. SESHADRI, *Proc. Indian Acad. Sci.* **49A**, 1–5 (1959). From *Teloschistes flavicans*, synthesis from teloschistin.

fragilin

$C_{16}H_{11}ClO_5$

m.p. 267–268° (chloroform; vacuum sublimation), yellow; uv (CHCl$_3$), λ_{max} (ε) mμ: 271.5 (36,500), 312.5 (14,000), and 434.5 (15,000) ; λ_{min} (ε) mμ: 241.5 (12,500), 300.0 (12,500), and 335.0 (1,800); ir (KBr), ν cm^{-1}: 1680 and 1630 (C=O).

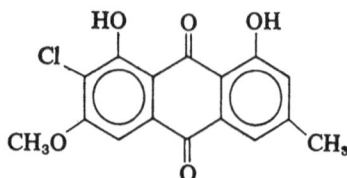

ASAHINA, Y. AND A. HASHIMOTO, *Chem. Ber.* **67**, 416–420 (1934). From *Sphaerophorus melanocarpus*.

BENDZ, GERD, GERD BOHMAN, AND J. SANTESSON, *Acta Chem. Scand.* **21**, 2889–2890 (1967). From *Nephroma laevigatum*.

BRUUN, T., D. HOLLIS, AND R. RYHAGE, *Acta Chem. Scand.* **19**, 839–844 (1965). From *Sphaerophorus fragilis* and *S. globosus*, structure proved, nmr spectrum, mass spectrum.

SANTESSON, J., *Acta Chem. Scand.* **21**, 1162–1172 (1967). Thin-layer chromatography, in *Sphaerophorus fragilis*.

ZOPF, W., *Ann. Chem.* **300**, 322–357 (1898). From *Sphaerophorus fragilis* and *S. globosus* (as *S. coralloides*).

1-hydroxy-3,8-dimethoxy-2-chloro-6-methylanthraquinone

$C_{17}H_{13}ClO_5$

Purified by sublimation and washing with petroleum ether.

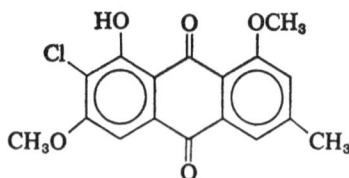

BENDZ, GERD, GERD BOHMAN, AND J. SANTESSON, *Acta Chem. Scand.* **21**, 2889–2890 (1967). From *Nephroma laevigatum*, structure proved, as 7-chloro-1,6-di-O-methylemodin.

mysaquinone

Violet-red.
Structure not known.

SANTESSON, J., *Acta Chem. Scand.* **21**, 1162–1172 (1967). Thin-layer chromatography, in *Mycoblastus sanguinarius*.

nephromin

$C_{16}H_{12}O_6$
m.p. ~196° d (acetic acid), small yellowish needles; soluble (purple-red) in alkali hydroxides.
Structure not known. But see also several pigments included in this section that have been more recently described from *Nephroma laevigatum*. These pigments are 1,3-dihydroxy-8-methoxy-2-chloro-6-methylanthraquinone, emodin, fragilin, 1-hydroxy-3,8-dimethoxy-2-chloro-6-methylanthraquinone, and 1,3,8-trihydroxy-2-chloro-6-methylanthraquinone, and nephromin could be identical to one of them.

GALUN, MARGALITH AND HANNA LAVEE, *Bryologist* **69**, 324–333 (1966). In *Nephroma laevigatum*, microcrystal test.
HESSE, O., *J. Prakt. Chem.* **57**, 409–447 (1898). From *Nephroma laevigatum*.
WETMORE, C. M., *Publ. Mus., Michigan State Univ., Biol. Ser.* **1**, 369–452 (1960). In *Nephroma laevigatum*.

norsolorinic acid

$C_{20}H_{18}O_7$
m.p. 269–270° (ethanol), red plates; ir (KBr), ν cm^{-1}: 3400, 1680, and 1625; uv (EtOH), λ_{max} (log ε) mμ: 270 (4.32), 286 (4.35), 312 (4.44), 453 (sh., 3.95), and 466 (4.00).

ANDERSON, H. A., R. H. THOMSON, AND J. W. WELLS, *J. Chem. Soc.* **1966C**, 1727–1729 (1966). From *Solorina crocea*, structure proved, synthesis from solorinic acid.

parietin (= **physcion**)

$C_{16}H_{12}O_5$

m.p. 206–207° (acetic acid), orange needles.
Also known from *Aspergillus glaucus*, *A. ruber*, *Penicillium herquei*, *Polygonum cuspidatum* (as a glucoside), *P. multiforum*, *Rheum rhaponticum*, *Rumex alpinus*, *R. crispus*, *R. ecklonianus*, *R. hymenosepalus*, *R. obtusifolius* (as a glycoside), and *Ventilago madraspatana*.

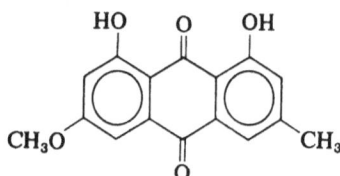

AYYANGAR, N. R., D. S. BAPAT, AND B. S. JOSHI, *J. Sci. Ind. Res.* (India) **20B**, 493–497 (1961). A new synthesis.

DHAR, M. L., S. NEELAKANTAN, S. RAMANUJAM, AND T. R. SESHADRI, *J. Sci. Ind. Res.* (India) **18B**, 111–113 (1959). From *Xanthoria parietina*.

EDER, R. AND F. HAUSER, *Helv. Chim. Acta* **8**, 126–139 (1925). Structure proved.

GALUN, MARGALITH AND HANNA LAVEE, *Bryologist* **69**, 324–333 (1966). From *Fulgensia fulgens*.

HARPER, S. H. AND R. M. LETCHER, *Proc. Trans. Rhodesian Sci. Assoc.* **51**, 156–184 (1966). Infrared spectrum, from *Caloplaca cinnabarina*.

HESS, D., *Planta* **52**, 65–76 (1958). Paper chromatography.

HESSE, O., *Ann. Chem.* **284**, 157–191 (1895). From *Xanthoria parietina*, as physcion.

HÖRHAMMER, L., H. WAGNER, AND G. BITTNER, *Pharm. Ztg.* **108**, 259–262 (1963). Thin-layer chromatography.

HUNECK, S. AND G. FOLLMANN, *Z. Naturforsch.* **20b**, 1012–1013 (1965). From *Stereocaulon corticulatum*.

HUNECK, S. AND G. FOLLMANN, *Z. Naturforsch.* **21b**, 91–92 (1966). From *Polycauliona regalis*.

MORS, W. B., *Bol. Inst. Quím. Agr.* (Rio de Janeiro) **23**, 1–16 (1951). From *Teloschistes exilis*, ultraviolet spectrum, microsublimation.

NEELAKANTAN, S. AND T. R. SESHADRI, *J. Sci. Ind. Res.* (India) **11B**, 126–127 (1952). From *Xanthoria elegans*.

RAJAGOPALAN, T. R. AND T. R. SESHADRI, *Proc. Indian Acad. Sci.* **49A**, 1–5 (1959). From *Teloschistes flavicans*, with fallacinal.

ROCHLEDER, F. AND W. HELDT, *Ann. Chem.* **48**, 1–18 (1843). From *Xanthoria parietina*, as chrysophanic acid, first extraction.

SANTESSON, J., *Acta Chem. Scand.* **21**, 1162–1172 (1967). Thin-layer chromatography.

SCHRATZ, E. AND H. J. VETHACKE, *Planta Med.* **6**, 44–69 (1958). Paper chromatography.

TAKIDO, M., *Pharm. Bull.* (Tokyo) **4**, 45–48 (1956). Paper chromatography.

THOMAS, E. A., *Ber. Schweiz. Botan. Ges.* **45**, 191–197 (1936). In cultures of the fungal component of *Caloplaca murorum* and *Xanthoria elegans*.

THOMAS, E. A., *Beitr. Kryptogamenflora Schweiz* **9**, 1–208 (1939). In the fungal cultures from several species of *Caloplaca* and *Xanthoria*.

TOMASELLI, R., *Atti Soc. Ital. Sci. Nat. Museo Civico Storia Nat.* (Milano) **97**, 357–361 (1958). Ecological factors influencing anthraquinone production by *Xanthoria parietina*.

TOMASELLI, R., *Arch. Botan. Biogeogr. Ital.* **39**, 1–20 (1963). In pure fungal cultures of *Xanthoria parietina*.

parietinic acid

$C_{16}H_{10}O_7$

m.p. ~300° with sublimation; ir (KBr), ν cm^{-1}: 2900–3000 (broad), 1700, and 1400.

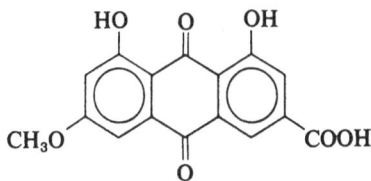

EDER, R. AND F. HAUSER, *Helv. Chim. Acta* **8**, 126–139 (1925). Synthesis.

ESCHRICH, W., *Biochem. Z.* **330**, 73–78 (1958). From *Xanthoria aureola* and in *X. contortuplicata* and *X. parietina*, first isolation from a natural source, identification, synthesis from parietin.

SANTESSON, J., *Acta Chem. Scand.* **21**, 1162–1172 (1967). Thin-layer chromatography.

rhodocladonic acid

$C_{17}H_{12}O_9$

m.p. > 360°, decomposes from 260°, brown plates (nitrobenzene), red needles (sublimation).

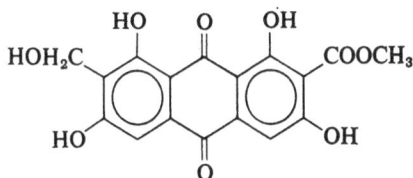

The chemical structure is not definitely proved.

HESS, D., *Planta* **52**, 65–76 (1958). Paper chromatography.

KOLLER, G. AND H. HAMBURG, *Monatsh. Chem.* **68**, 202–206 (1936). Structure studied.

MATSUURA, S. AND K. OHTA, *Yakugaku Zasshi* **82**, 963–966 (1962). Attempted synthesis, ultraviolet spectrum.

SANTESSON, J., *Acta Chem. Scand.* **21**, 1162–1172 (1967). Thin-layer chromatography.

SHIBATA, S., *Yakugaku Zasshi* **61**, 320–325 (1941); *Nippon Kagaku Soran* **15**, 1149–1150 (1941); reference from *Chem. Abstr.* **44**, 9396e (1950). Structure proposed.

SHIBATA, S., M. TAKITO, AND O. TANAKA, *J. Am. Chem. Soc.* **72**, 2789–2790 (1950). Paper chromatography.

rugulosin

$C_{30}H_{24}O_{10}$

m.p. 293° d (ethanol), yellow crystals; $[\alpha]_D^{19}$ +492° (c = 0.5, dioxane). Also known in nonlichen-forming fungi.

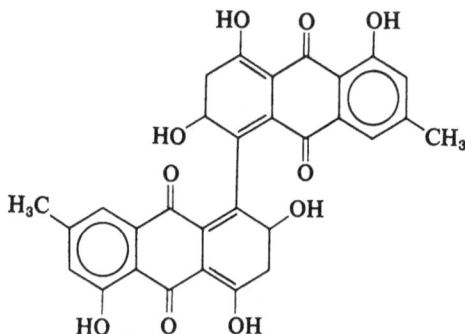

SHIBATA, S., *J. Japan. Botany*, in press. In *Acroscyphus sphaerophoroides*.

skyrin

$C_{30}H_{18}O_{10}$

m.p. 200° d (ether), yellow plates. Also known from nonlichen-forming fungi.

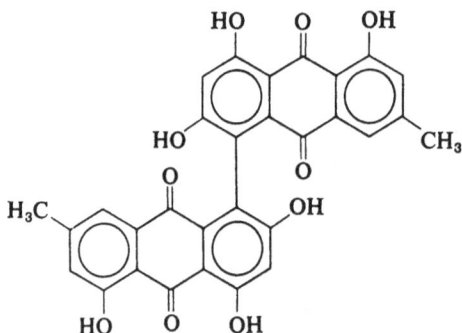

SHIBATA, S., *J. Japan. Botany*, in press. In *Acroscyphus sphaerophoroides*.

solorinic acid

$C_{21}H_{20}O_7$

m.p. 201° (acetic acid), orange-red crystals.

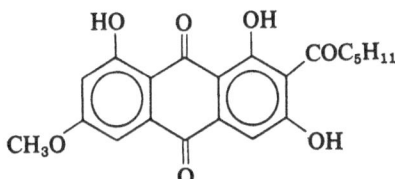

ANDERSON, H. A., J. SMITH, R. H. THOMSON, AND J. W. WELLS, *Bull. Natl. Inst. Sci. India*, No. **28**, 46–51 (1965). From *Solorina crocea*, nmr data on the tri-O-methyl derivative.

ANDERSON, H. A., R. H. THOMSON, AND J. W. WELLS, *J. Chem. Soc.* **1966C**, 1727–1729 (1966). From *Solorina crocea*, with norsolorinic acid.

HESSE, O., *J. Prakt. Chem.* **92**, 425–466 (1915). Structure studied.

KOLLER, G. AND H. RUSS, *Monatsh. Chem.* **70**, 54–72 (1937). From *Solorina crocea*, structure proved.

SANTESSON, J., *Acta Chem. Scand.* **21**, 1162–1172 (1967). Thin-layer chromatography.

ZOPF, W., *Ann. Chem.* **284**, 107–132 (1895). From *Solorina crocea*.

teloschistin

$C_{16}H_{12}O_6$

m.p. 244–246° (benzene; purified through the acetate), orange needles.

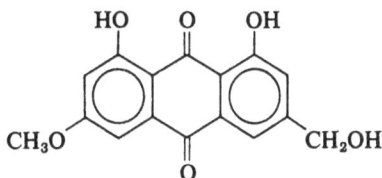

ASANO, M. AND H. ARATA, *Yakugaku Zasshi* **60**, 521–525 (1940); German summary, pp. 206–208 (1940). "Fallacin" partially purified by removal of parietin.

ASANO, M. AND S. FUZIWARA, *Yakugaku Zasshi* **56**, 1007–1010 (1936); German summary, p. 101 (1936). "Fallacin" separated from *Xanthoria fallax*.

MURAKAMI, T., *Pharm. Bull.* (Tokyo) **4**, 298–302 (1956). From *Xanthoria fallax*, separated from fallacinal and parietin, as fallacinol.

NEELAKANTAN, S., S. RANGASWAMI, T. R. SESHADRI, AND S. S. SUBRA-
MANIAN, *Proc. Indian Acad. Sci.* **33A**, 142–147 (1951). Structure proved.
NEELAKANTAN, S. AND T. R. SESHADRI, *J. Sci. Ind. Res.* (India) **13B**,
884–885 (1954). Synthesis from parietin.
NEELAKANTAN, S., T. R. SESHADRI, AND S. SUBRAMANIAN, *Proc. Indian
Acad. Sci.* **44A**, 42–45 (1956). Synthesis.
SESHADRI, T. R. AND S. S. SUBRAMANIAN, *Proc. Indian Acad. Sci.* **30A**,
67–73 (1949). From *Teloschistes flavicans*, structure proposed.
RAJAGOPALAN, T. R. AND T. R. SESHADRI, *Proc. Indian Acad. Sci.* **49A**,
1–5 (1959). From *Teloschistes flavicans*, also by partial methylation of
citreoroein.

1,3,8-trihydroxy-2-chloro-6-methylanthraquinone

$C_{15}H_9ClO_5$

m.p. 286–287° (methanol), orange needles; uv-vis (EtOH), λ_{max} (log ε)
mμ: 273 (4.35), 282 (4.35), 307 (4.18), 431 (4.05), 460 (3.96), and 521
(3.42); ir (KBr), ν cm^{-1}: 3333, 1663, and 1611; nmr (dioxane), τ ppm:
−0.76 (OH), −1.84 (OH), −2.71 (OH), 2.51 (br.s.; proton at C-5),
2.80 (s.; proton at C-4), and 2.95 (br.s.; proton at C-7).
Also known in an unidentified *Penicillium*.

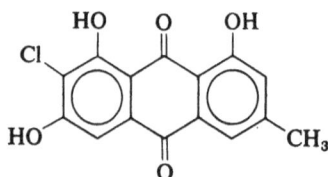

BENDZ, GERD, GERD BOHMAN, AND J. SANTESSON, *Acta Chem. Scand.* **21**,
2889–2890 (1967). From *Nephroma laevigatum*, as 7-chloroemodin.
YOSIOKA, I., H. YAMAUCHI, K. MORIMOTO, AND I. KITAGAWA, *Tetrahedron Letters* **1968**, 1149–1152 (1968). From *Heterodermia obscurata*.

1,3,8-trihydroxy-2,4-dichloro-6-methylanthraquinone

$C_{15}H_8Cl_2O_5$

m.p. 267–269° (benzene), orange needles; ir (KBr), ν cm^{-1}: 3317, 1666,
and 1623; nmr (dioxane), τ ppm: −0.74 (OH), −1.64 (OH), −3.43
(OH), 2.47 (br.s.; proton at C-5), and 2.92 (br.s.; proton at C-7).

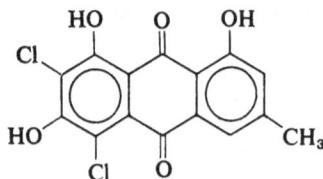

Yosioka, I., H. Yamauchi, K. Morimoto, and I. Kitagawa, *Tetrahedron Letters* **1968**, 1149–1152 (1968). From *Heterodermia obscurata*, first extraction, structure proved.

xanthorin

$C_{16}H_{12}O_6$
m.p. 253° (toluene), red.

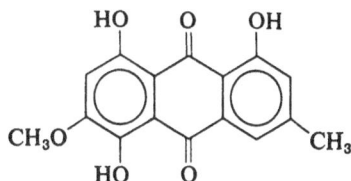

Santesson, J., *Acta Chem. Scand.* **21**, 1162–1172 (1967). Thin-layer chromatography, in *Laurera purpurina*, as lauropurpone.

Steglich, W., W. Lösel, and W. Reininger, *Tetrahedron Letters* **1967**, 4719–4721 (1967). From *Xanthoria elegans*, structure proved, synthesis.

Tanaka, O. and C. Kaneko, *Pharm. Bull.* (Tokyo) **3**, 284–286 (1955). Synthesis, as 4,5,8-trihydroxy-7-methoxy-2-methylanthraquinone.

Wachtmeister, C. A. and C. Stensiö, in preparation. From *Laurera, purpurina*, synthesis.

Secondary Products Originating by the Mevalonic Acid Pathway

CAROTENES

β-carotene

$C_{40}H_{56}$
m.p. 181–182° (benzene-methanol), violet prisms.
Widely distributed in nature.

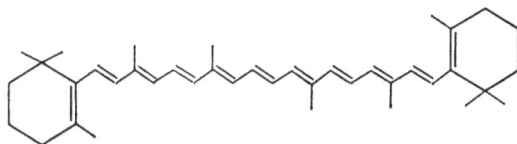

Giudici de Nicola, Marina and R. Tomaselli, *Boll. Ist. Botan. Univ. Catania*, Ser. 3, **2**, 22–28 (1961). From *Trebouxia decolorans* cultured from the lichen *Xanthoria parietina*.

GIUDICI DE NICOLA, MARINA AND G. DI BENEDETTO, *Boll. Ist. Botan. Univ. Catania*, Ser. 3, **3**, 22–33 (1962). In the algal partner of *Xanthoria parietina*.

HUNECK, S., ANNICK MATHEY, AND G. TROTET, *Z. Naturforsch.* **22b**, 1367–1368 (1967). From *Roccella fuciformis* from the English Channel (0.01%) and from Morocco (0.005%).

MURTY, T. K. AND S. S. SUBRAMANIAN, *J. Sci. Ind. Res.* (India) **17C**, 105–106 (1958); *ibid.* **18B**, 162–163 (1959). From *Roccella montagnei*.

MURTY, T. K. AND S. S. SUBRAMANIAN, *Res. Ind.* (New Delhi) **4**, 176 (1959). From *Roccella montagnei*.

γ-carotene

$C_{40}H_{56}$

m.p. 176.5–178° (benzene), violet prisms.

A rare carotene known in low concentration in plants.

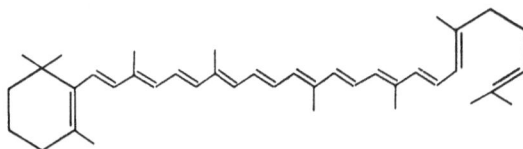

MURTY, T. K. AND S. S. SUBRAMANIAN, *J. Sci. Ind. Res.* (India) **18B**, 162–163 (1959). From *Roccella montagnei*.

MURTY, T. K. AND S. S. SUBRAMANIAN, *Res. Ind.* (New Delhi) **4**, 176 (1959). From *Roccella montagnei*.

violoxanthin

$C_{40}H_{56}O_4$

m.p. 200° (methanol), brown-yellow prisms; $[\alpha]_{Cd}^{25}$ +35° (CHCl₃).

Well known in blossoms and fruits of the higher plants.

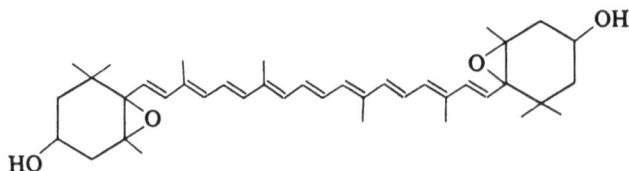

GIUDICI DE NICOLA, MARINA AND G. DI BENEDETTO, *Boll. Ist. Botan. Univ. Catania*, Ser. 3, **3**, 22–33 (1962). In the algal partner of *Xanthoria parietina*.

GIUDICI DE NICOLA, MARINA AND R. TOMASELLI, *Boll. Ist. Botan. Univ. Catania*, Ser. 3, **2**, 22–28 (1961). From *Trebouxia decolorans* cultured from the lichen *Xanthoria parietina*.

xanthophyll

$C_{40}H_{56}O_2$

m.p. 193° (methanol), violet prisms; $[\alpha]_{Cd}^{20}$ +160° ($CHCl_3$).
Common in plants.

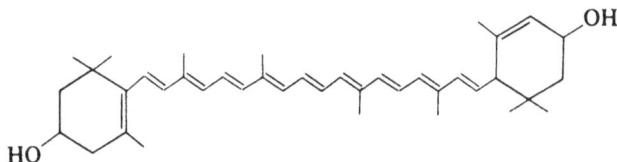

GIUDICI DE NICOLA, MARINA AND G. DI BENEDETTO, *Boll. Ist. Botan. Univ. Catania*, Ser. 3, 3, 22–33 (1962). In the algal partner of *Xanthoria parietina*.
GIUDICI DE NICOLA, MARINA AND R. TOMASELLI, *Boll. Ist. Botan. Univ. Catania*, Ser. 3, **2**, 22–28 (1961). From *Trebouxia decolorans* cultured from the lichen *Xanthoria parietina*.

unspecified carotenoid pigments

AGARWAL, S. C., K. AGHORAMURTHY, K. G. SARMA, AND T. R. SESHADRI, *J. Sci. Ind. Res.* (India) **20B**, 613–615 (1961). From *Lobaria isidiosa*.
HENRIKSSON, ELISABET, *Physiol. Plant.* **16**, 867–869 (1963). In species of the Collemataceae.
HUNECK, S. AND G. FOLLMANN, *Z. Naturforsch.* **20b**, 496 (1965). Five carotene pigments in *Dolichocarpus chilensis*, thin-layer chromatography.
MURTY, T. K. AND S. S. SUBRAMANIAN, *J. Sci. Ind. Res.* (India) **18B**, 394–395 (1959). From *Usnea venosa*.
RAO, P. S., K. G. SARMA, AND T. R. SESHADRI, *Current Sci.* (India) **34**, 9–11 (1961). From *Lobaria retigera*.
RAO, P. S., K. G. SARMA, AND T. R. SESHADRI, *Current Sci.* (India) **35**, 147–148 (1966). From *Lobaria isidiosa* var. *subisidiosa*.
SUBRAMANIAN, S. S. AND M. N. SWAMY, *J. Sci. Ind. Res.* (India) **20C**, 275–276 (1965). From *Roccella montagnei*.
ZELLNER, J., *Monatsh. Chem.* **66**, 81–86 (1935). From *Umbilicaria mammulata*.

STEROLS

ergosterol

$C_{28}H_{44}O$

m.p. 160–163° (ethanol or ether); m.p. 165°; $[\alpha]_D^{20}$ −133° ($CHCl_3$).
Widely distributed in plants.

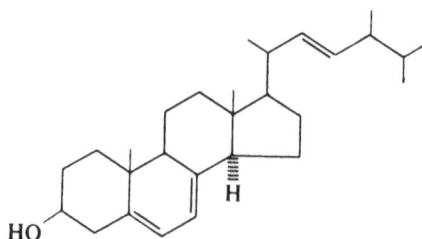

ALERTSEN, A. R., T. BRUUN, AND ELLEN HEMMER, *Acta Chem. Scand.* **16**, 541–542 (1962). From *Cornicularia muricata*.
BLIX, G. AND H. RYDIN, *Acta Soc. Med. Upsalien.* **37**, 333–340 (1932). From *Cladonia rangiferina*.
HUNECK, S. AND G. FOLLMANN, *Z. Naturforsch.* **22b**, 1182–1185 (1967). From *Pseudocyphellaria hirsuta* (0.01%).
MURTY, T. K. AND S. S. SUBRAMANIAN, *J. Sci. Ind. Res.* (India) **18B**, 91–92 (1959). From *Roccella montagnei*.
MURTY, T. K. AND S. S. SUBRAMANIAN, *J. Sci. Ind. Res.* (India) **18B**, 394–395 (1959). From *Usnea venosa*.
SARMA, K. G. AND S. HUNECK, *Pharmazie*, in press. From *Dermatocarpon vellereum*.
SNATZKE, G., *J. Chromatog.* **8**, 110–117 (1962). Paper chromatography.
SUBRAMANIAN, S. S. AND M. N. SWAMY, *J. Sci. Ind. Res.* (India) **20C**, 275–276 (1961). From *Roccella montagnei*.
ZELLNER, J., *Monatsh. Chem.* **59**, 300–304 (1932). From *Peltigera canina*.
ZELLNER, J., *Monatsh. Chem.* **66**, 81–86 (1935). From *Umbilicaria mammulata*.

fungisterol

$C_{28}H_{48}O$

m.p. 148–149° (acetone); $[\alpha]_D^{23}$ −0.2° (CHCl$_3$).
Also known from nonlichen-forming fungi.

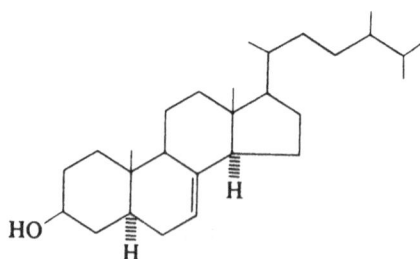

ZELLNER, J., *Monatsh. Chem.* **66**, 81–86 (1935). From *Pseudevernia furfuracea*.

β-sitosterol

$C_{29}H_{50}O$

m.p. 140–141° (methanol-methylene chloride), colorless prismatic plates; $[\alpha]_D$ −36° ($CHCl_3$).

A common plant steroid.

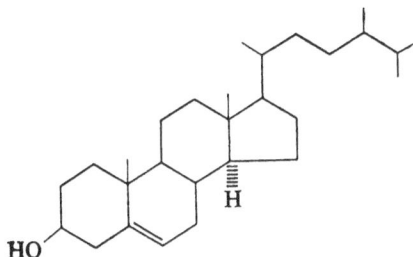

BAISTED, D. J., E. CAPSTACK, JR., AND W. R. NES, *Biochemistry* **1**, 537–541 (1962). Biosynthesis in *Pisum sativum* seeds.

HUNECK, S. AND G. FOLLMANN, *Z. Naturforsch.* **20b**, 1138–1139 (1965). From *Lecanora dispersa*.

HUNECK, S. AND G. FOLLMANN, *Z. Naturforsch.* **21b**, 714–715 (1966). From *Stereocaulon antarcticum*, thin-layer chromatography.

NICHOLAS, H. J., *J. Biol. Chem.* **237**, 1476–1480 (1962). Biosynthesis in *Salvia officinalis*.

steroid of unknown structure

m.p. 147–148°.

HUNECK, S., *Tetrahedron Letters* **1966**, 3547–3549 (1966). From *Lecanora rupicola*.

TERPENES

DITERPENES

(−)-16α-hydroxykaurane

$C_{20}H_{34}O$

m.p. 216–217° (methanol), needles; $[\alpha]_D^{20}$ −64.2° (c = 1; chloroform-ethanol, 1:1 v/v); ir (KBr), ν cm^{-1}: 3305 (broad, OH).

Also known in *Gibberella fujikuroi* and *Trachylobium verrucosum*.

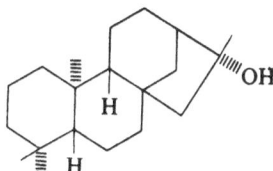

BENDZ, GERD, J. SANTESSON, AND C. A. WACHTMEISTER, *Acta Chem. Scand.* **19**, 1185–1187 (1965). From *Ramalina ceruchis, R. ceruchoides, R. combeoides,* and *R. homalea,* as ceruchdiol.

BRIGGS, L. H., B. F. CAIN, R. C. CAMBIE, B. R. DAVIS, P. S. RUTLEDGE, AND J. K. WILMSHURST, *J. Chem. Soc.* **1963**, 1345–1355 (1963). Structure studied.

HANSON, J. R., *J. Chem. Soc.* **1963**, 5061–5066 (1963). (−)-Kauranol identical to (−)-16α-hydroxykaurane.

HUGEL, GEORGETTE, LILIANE LODS, J. M. MELLOR, D. W. THEOBALD, AND G. OURISSON, *Bull. Soc. Chim. France* **1963**, 1974–1976 (1963). Structure studied.

HUNECK, S. AND G. FOLLMANN, *Z. Naturforsch.* **20b**, 611–612 (1965). From *Ramalina tumidula,* as ceruchinol.

HUNECK, S. AND G. FOLLMANN, *Z. Naturforsch.* **21b**, 713–714 (1966). From *Ramalina tigrina,* ceruchinol and ceruchdiol identical to (−)-16α-hydroxykaurane.

HUNECK, S., G. FOLLMANN, W. A. WEBER, AND G. TROTET, *Z. Naturforsch.* **22b**, 671–673 (1967). From *Roccella fimbriata.*

LEHN, J.-M. AND S. HUNECK, *Z. Naturforsch.* **20b**, 1013 (1965). Ceruchinol identified as (−)-16α-hydroxykaurane, ir spectrum.

nephrin

$C_{20}H_{32}$

Structure not known.

HESSE, O., *J. Prakt. Chem.* **58**, 465–561 (1898). From *Nephroma arcticum* and *N. laevigatum.*

WETMORE, C. M., *Publ. Mus., Michigan State Univ., Biol. Ser.* **1**, 369–452 (1960). An accessory in *Nephroma bellum, N. expallidum,* and *N. laevigatum,* microcrystal test.

ZOPF, W., *Ann. Chem.* **364**, 273–313 (1909). From *Nephroma laevigatum.*

TRITERPENES

7β-acetoxy-22-hydroxyhopane

$C_{23}H_{54}O_3$

m.p. 247–248° (benzene); $[\alpha]_D^{20}$ +26° (c = 0.71, $CHCl_3$); ir (nujol), ν cm^{-1}: 3520 (OH), 1700 (OAc), and 1270 (OAc); nmr ($CDCl_3$), δ ppm: 0.78 and 0.83 (12), 1.03 (3), 1.10 (3), 1.17 (6), 1.97 (3), and 5.10 (1 proton quartet, 5 and 11 cps splitting).
Structure II.

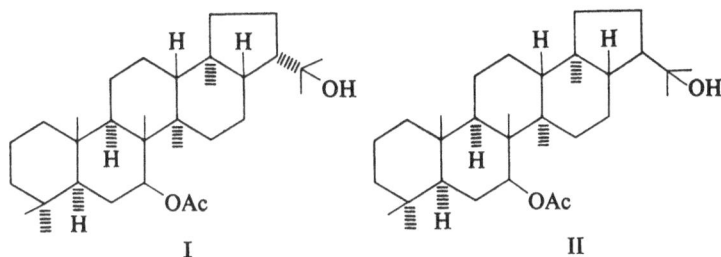

I

II

CORBETT, R. E. AND H. YOUNG, *J. Chem. Soc.* **1966C**, 1556–1563 (1966). From *Sticta billardierii*, proved to be 7β-acetoxy-22-hydroxyhopane, structure represented as I.

HUNECK, S., personal communication. 7β-Acetoxy-22-hydroxyhopane must be represented as structure II on the basis of studies by I. Yosioka, T. Nakanishi, and I. Kitagawa (*Tetrahedron Letters* **1968**, 1485–1490 (1968)) establishing a β-configuration of the side chain in hopane.

HUNECK, S. AND G. FOLLMANN, *Z. Naturforsch.* **22b**, 1182–1185 (1967). From *Pseudocyphellaria intricata* var. *thouarsii*.

16β-O-acetylleucotylic acid

$C_{32}H_{52}O_5$

Isolated as the methyl ester (m.p. 176°); $[\alpha]_D$ +95° (CHCl₃); ir (CHCl₃), ν cm⁻¹: 3530 (OH), 1740 (OAc), 1715 (COOCH₃), and 1240 ~1200; nmr (CDCl₃), τ ppm: 6.33 (COOC$\underline{H}$₃), 7.61 (OCOC$\underline{H}$₃), and 4.75 ($\underline{H}$C—OAc).

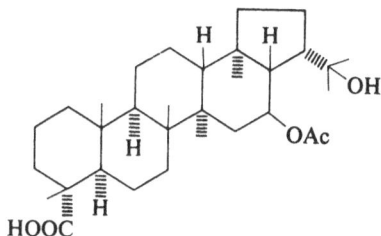

YOSIOKA, I., M. YAMAKI, AND I. KITAGAWA, *Chem. Pharm. Bull.* (Tokyo) **14**, 804–807 (1966). From *Parmelia entotheiochroa*, isolated as the methyl ester.

6α-O-acetylleucotylin

$C_{32}H_{54}O_4$

m.p. 225°; $[\alpha]_D$ +36° (CHCl$_3$); ir (CHCl$_3$), ν cm^{-1}: 3400 (OH) and
1720 and 1250 (OAc); nmr (CDCl$_3$), τ ppm: 7.95 (OCOC$\underline{H}_3$), 4.80
($\underline{H}$C—OAc), and 5.90 ($\underline{H}$C—OH).

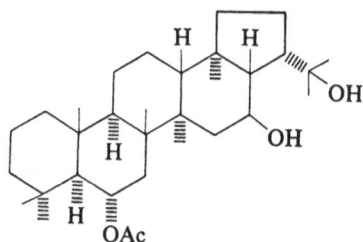

YOSIOKA, I., M. YAMAKI, AND I. KITAGAWA, *Chem. Pharm. Bull.* (Tokyo)
14, 804–807 (1966). From *Parmelia entotheiochroa*.

6-deoxy-16β-O-acetylleucotylin

$C_{32}H_{54}O_3$

m.p. 228°; $[\alpha]_D$ +52° (CHCl$_3$); ir (CHCl$_3$), ν cm^{-1}: 3550 (OH) and 1735
and 1240 ~ 1200 (OAc); nmr (CDCl$_3$), τ ppm: 7.90 (OCOC$\underline{H}_3$) and
4.75 ($\underline{H}$C—OAc).

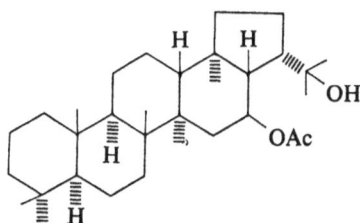

YOSIOKA, I., M. YAMAKI, AND I. KITAGAWA, *Chem. Pharm. Bull.* (Tokyo)
14, 804–807 (1966). From *Parmelia entotheiochroa*.

6-deoxyleucotylin

$C_{30}H_{52}O_2$

m.p. 268°; $[\alpha]_D$ +68° (CHCl$_3$); ir (CHCl$_3$), ν cm^{-1}: 3350 (OH).

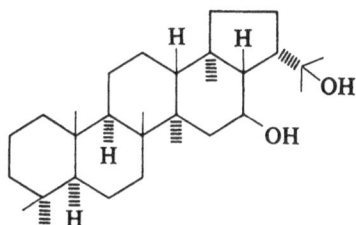

YOSIOKA, I., M. YAMAKI, AND I. KITAGAWA, *Chem. Pharm. Bull.* (Tokyo) **14**, 804–807 (1966). From *Parmelia entotheiochroa*.

6α,16β-di-O-acetylleucotylin

$C_{34}H_{56}O_5$

m.p. 232°, $[\alpha]_D$ +109°; ir (nujol), ν cm^{-1}: 3400 (OH) and 1730 and 1245 (OAc); nmr (CDCl$_3$), τ ppm: 7.90 and 7.93 (2 OCOC$\underline{H}_3$) and 4.80 (2 $\underline{H}$C—OAc).

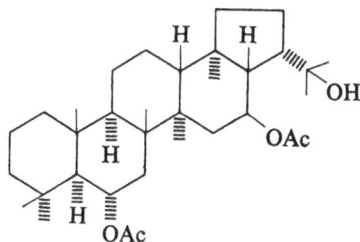

YOSIOKA, I., M. YAMAKI, AND I. KITAGAWA, *Chem. Pharm. Bull.* (Tokyo) **14**, 804–807 (1966). From *Parmelia entotheiochroa*.

15α,22-dihydroxyhopane

$C_{30}H_{52}O_2$

m.p. 249° (benzene); $[\alpha]_D^{20}$ +34° (c = 1.3, CHCl$_3$).

Structure II.

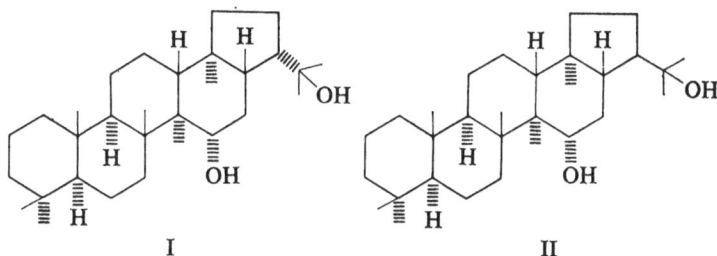

I   II

CORBETT, R. E. AND H. YOUNG, *J. Chem. Soc.* **1966C**, 1556–1563 (1967). From *Sticta billardierii*, as compound B.

CORBETT, R. E. AND H. YOUNG, *J. Chem. Soc.* **1966C**, 1564–1567 (1966). From *Sticta billardierii*, proved to be 15α,22-dihydroxyhopane, structure represented as I.

HUNECK, S., personal communication. 15α,22-Dihydroxyhopane must be represented as structure II on the basis of studies by I. Yosioka, T. Nakanishi, and I. Kitagawa (*Tetrahedron Letters* **1968**, 1485–1490 (1968)) establishing a β-configuration of the side chain in hopane.

HUNECK, S. AND G. FOLLMANN, *Z. Naturforsch.* **22b**, 1182–1185 (1967). From *Pseudocyphellaria intricata* var. *thouarsii*.

dolichorrhizin

m.p. 200°.
Structure not known.

KUROKAWA, S., YOKO JINZENJI, S. SHIBATA, AND HSÜCH-CHING CHIANG, *Bull. Natl. Sci. Mus.* (Tokyo) **9**, 101–114 (1966). Distribution in Japanese Peltigeras, thin-layer chromatography, first detection.

durvilldiol

$C_{30}H_{52}O_2$
m.p. 264–266°.
A triterpene of unknown structure.

HUNECK, S. AND G. FOLLMANN, *Z. Naturforsch.* **22b**, 1182–1185 (1967). From *Pseudocyphellaria durvillei*, oxidized by the Jones reagent to durvilldione.

durvillonol

$C_{30}H_{50}O_2$
m.p. 218–220°; ir, ν cm^{-1}: 3510 (OH) and 1700 (C=O).
A hydroxy triterpene ketone of unknown structure.

HUNECK, S. AND G. FOLLMANN, *Z. Naturforsch.* **22b**, 1182–1185 (1967). From *Pseudocyphellaria durvillei*, oxidized by the Jones reagent to durvilldione ($C_{30}H_{48}O_2$).

eulecanorol

$C_{30}H_{52}O_3$
m.p. 262–263° (ethanol-chloroform), colorless needles; acetate, m.p. 185–187°.
Structure not known.

HUNECK, S., *Z. Naturforsch.* **21b**, 888–890 (1966). From *Lecanora polytropa*, first isolation, chromatographic separation from zeorin, thin-layer chromatography.

friedelin

$C_{30}H_{50}O$

m.p. 257–265° (ethyl acetate-methylene chloride); $[\alpha]_D^{14}$ −22° to −27.8° $(CHCl_3)$.

Also known from *Balanops australiana, Ceratopetalum apetalum, Clerodendron trichotomum, Quercus suber, Rhododendron westlandii, Salix japonica, Shiia sieboldii,* and *Zelkova serrata.*

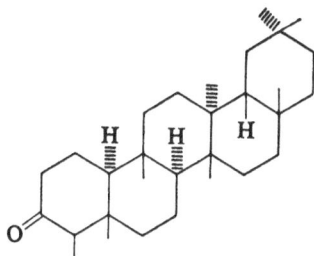

BROWNLIE, G., F. S. SPRING, R. STEVENSON, AND W. S. STRACHAN, *Chem. Ind.* (London) **1955**, 686–687 (1955); *ibid.* **1955**, 1156–1158 (1955); *J. Chem. Soc.* (London) **1956,** 2419–2427 (1956). Structure and configuration studied.

BRUUN, T., *Acta Chem. Scand.* **8**, 71–75 (1954). From *Alectoria ochroleuca, Cetraria cucullata, Stereocaulon paschale,* and several others, first extraction from lichens.

BRUUN, T. AND P. R. JEFFRIES, *Acta Chem. Scand.* **8**, 1948–1949 (1954). From *Cetraria nivalis.*

COREY, E. J., AND J. J. URSPRUNG, *J. Am. Chem. Soc.* **77**, 3667–3668 (1955); *ibid.* **77**, 3668–3669 (1955); *ibid.* **78**, 5041–5051 (1956). Structure and configuration studied.

DUTLER, H., O. JEGER, AND L. RUZICKA, *Helv. Chim. Acta* **38**, 1268–1273 (1955). Structure and configuration studied.

KLEIN, E. AND S. HUNECK, unpublished. From *Evernia prunastri.*

SHAMMA, M., R. E. GLICK, AND R. O. MUMMA, *J. Org. Chem.* **27**, 4512–4517 (1962). The nmr spectrum.

STEVENSON, R., *J. Org. Chem.* **26**, 2142–2143 (1961). A new isolation method.

epi-friedelinol

A report by Bruun (*Acta Chem. Scand.* **8**, 71–75 (1954)) of this substance in *Cetraria nivalis* was later corrected (BRUUN, T. AND P. R. JEFFRIES, *Acta Chem. Scand.* **8**, 1948–1949 (1954)) to friedelin and an unknown substance.

leucotylic acid

$C_{30}H_{50}O_4$
m.p. 260°; $[\alpha]_D$ +330° (c = 0.15, $CHCl_3$); ir (nujol), ν cm^{-1}: 3200 (OH) and 1690 (C=O).

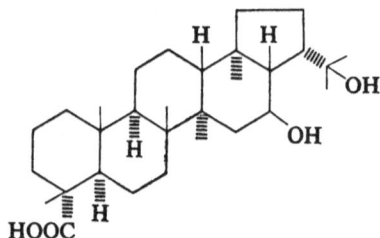

YOSIOKA, I., T. NAKANISHI, AND E. TSUDA, *Tetrahedron Letters* **1966**, 607–612 (1966). From *Parmelia leucotyliza*, structure proved.
YOSIOKA, I., M. YAMAKI, AND I. KITAGAWA, *Chem. Pharm. Bull.* (Tokyo) **14**, 804–807 (1966). From *Parmelia entotheiochroa.*

leucotylin

$C_{30}H_{52}O_3$
m.p. 335–336° (methanol); $[\alpha]_D^{20}$ +56.5° (c = 0.566, $CHCl_3$).

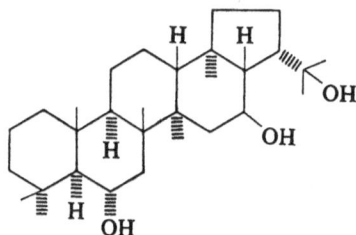

ASAHINA, Y., *J. Japan. Botany* **26**, 225–228 (1951). In *Parmelia homogenes.*
ASAHINA, Y. AND H. AKAGI, *Chem. Ber.* **71**, 980–985 (1938). From *Parmelia leucotyliza*, structure studied.
HUNECK, S., *Chem. Ber.* **94**, 614–622 (1961). From *Lecanora muralis*, structure studied.
NAKANISHI, T., T. FUJIWARA, AND K. TOMITA, *Tetrahedron Letters* **1968**, 1491–1495 (1968). Structure proved by X-ray analysis of the 16β-O-*p*-bromobenzoate of 6-ketoleucotylin.
YOSIOKA, I. AND T. NAKANISHI, *Chem. Pharm. Bull.* (Tokyo) **11**, 1468–1470 (1963). Structure studied.

YOSIOKA, I., M. YAMAKI, AND I. KITAGAWA, *Chem. Pharm. Bull.* (Tokyo) **14**, 804–807 (1966). From *Parmelia entotheiochroa*.

YOSIOKA, I., T. NAKANISHI, AND I. KITAGAWA, *Tetrahedron Letters* **1968**, 1485–1490 (1968). Structure studied.

YOSIOKA, I., M. YAMAKI, T. NAKANISHI, AND I. KITAGAWA, *Tetrahedron Letters* **1966**, 2227–2235 (1966). Structure of methyl isoleucotylate.

pyxinic acid

$C_{30}H_{50}O_4$

m.p. 254–255°; $[\alpha]_D$ +62° (c = 0.3, EtOH); ir (nujol), ν cm^{-1}: 3560, 3500, and 1705.

Structure II.

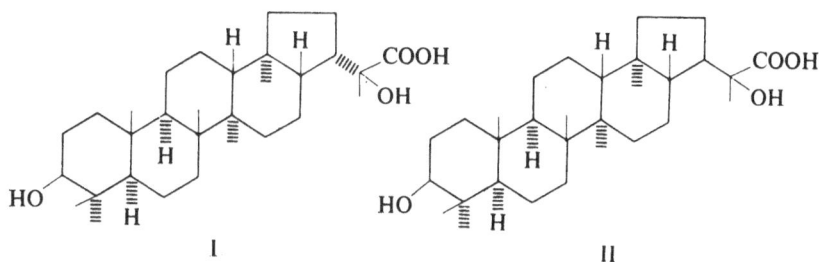

YOSIOKA, I., A. MATSUDA, AND I. KITAGAWA, *Tetrahedron Letters* **1966**, 613–616 (1966). From *Pyxine endocrysina*, structure I proposed.

HUNECK, S., personal communication. Pyxinic acid must have structure II. On the basis of studies by I. Yosioka, T. Nakanishi, and I. Kitagawa (*Tetrahedron Letters* **1968**, 1485–1490 (1968)), showing a β-configuration of the side chain in hopane, this side chain must have the β-configuration in pyxinic acid, since the structure of this compound was originally proven by conversion to the norketone of hydroxyhopane.

retigeranic acid

m.p. 218–221° (dioxane-acetone), colorless, long needles; $[\alpha]_D$ −59° (c = 1.0, CHCl$_3$); ir, ν cm^{-1}: 1667 (α,β-unsaturated COOH); uv, λ_{max} (log ε) mμ: 239 (3.92).

Structure not known.

AGARWAL, S. C., K. AGHORAMURTHY, K. G. SARMA, AND T. R. SESHADRI, *J. Sci. Ind. Res.* (India) **20B**, 613–615 (1961). From two unidentified species of *Lobaria*, as compound B (m.p. 225–227.5°), a diterpene.

RAO, P. S., K. G. SARMA, AND T. R. SESHADRI, *Current Sci.* (India) **34**, 9–11 (1965). From *Lobaria retigera*, as triterpene B.

RAO, P. S., K. G. SARMA, AND T. R. SESHADRI, *Current Sci.* (India) **35**, 147–148 (1966). From *Lobaria subretigera*.

retigerdiol

m.p. 270–273° (chloroform), colorless prisms; $[\alpha]_D$ +28° (c = 0.5, $CHCl_3$); diacetate, m.p. 276–279°; dibenzoate, m.p. 304–306°. Structure not known.

AGARWAL, S. C., K. AGHORAMURTHY, K. G. SARMA, and T. R. SESHADRI, *J. Sci. Ind. Res.* (India) **20B**, 613–615 (1961). From an unidentified species of *Lobaria*, as compound A (m.p. 283–284°).

RAO, P. S., K. G. SARMA, AND T. R. SESHADRI, *Current Sci.* (India) **34**, 9–11 (1965). From *Lobaria retigera*, as triterpene A.

RAO, P. S., K. G. SARMA, AND T. R. SESHADRI, *Current Sci.* (India) **35**, 147–148 (1966). From *Lobaria subretigera*.

taraxerene

$C_{30}H_{50}$
m.p. 238–239°; $[\alpha]_D$ +3° ($CHCl_3$).

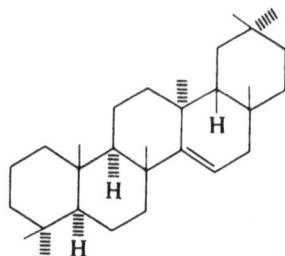

BRUUN, T., *Acta Chem. Scand.* **8**, 1291–1292 (1954). From *Cladonia deformis*, first isolation, identified with taraxerene from reduction of taraxerone.

triterpene C

m.p. 318–320°
Structure not known.

AGARWAL, S. C., K. AGHORAMURTHY, K. G. SARMA, AND T. R. SESHADRI, *J. Sci. Ind. Res.* (India) **20B**, 613–615 (1961). From *Lobaria isidiosa* and an unidentified species of *Lobaria*, a triterpene acid.

triterpene D

$C_{31}H_{50}O_5$

m.p. 289–291° (acetic acid), colorless needles; $[\alpha]_D^{25}$ +20° (c = 1.0, pyridine); acetate, m.p. 286–288°; methyl ester, 244–246°.
Structure not known.

AGARWAL, S. C., K. AGHORAMURTHY, K. G. SARMA, AND T. R. SESHADRI, *J. Sci. Ind. Res.* (India) **20B**, 613–615 (1961). From an unidentified species of *Lobaria*, m.p. 267–269°.
RAO, P. S., K. G. SARMA, AND T. R. SESHADRI, *Current Sci.* (India) **34**, 9–11 (1965). From *Lobaria retigera*, an unsaturated triterpene hydroxy acid.

triterpene N-1

$C_{30}H_{50}O$

m.p. 258°; $[\alpha]_D$ −12° (CHCl₃); ir (CHCl₃), ν cm⁻¹: 1720 (C=O).
Structure not known.

YOSIOKA, I., M. YAMAKI, AND I. KITAGAWA, *Chem. Pharm. Bull.* (Tokyo) **14**, 804–807 (1966). From *Parmelia entotheiochroa*.

ursolic acid

$C_{30}H_{48}O_3$

m.p. 285–291° (ethanol), needles; $[\alpha]_D^{20}$ +66° (pyridine).
Common in plants.

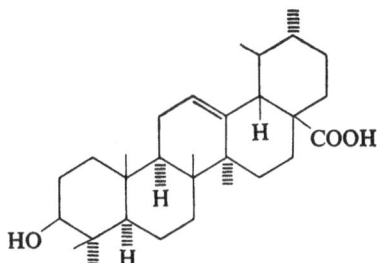

BREADEN, T. W., J. KEANE, AND T. J. NOLAN, *Sci. Proc. Roy. Dublin Soc.* **23**, 6–9 (1942); *ibid.* **23**, 197–200 (1944). From *Cladonia arbuscula* and *C. impexa*, first isolation from lichens.
EVANS, A. W., *Trans. Conn. Acad. Arts Sci.* **35**, 519–626 (1944). In *Cladonia arbuscula*, as substance E, microcrystal test.

EVANS, A. W., *Rhodora* **52**, 77–123 (1950). Microcrystal test.
GLICK, R. E., R. O. MUMMA, AND M. SHAMMA, *Chem. Ind.* (London) **1959**, 1092–1093 (1959). The nmr spectrum.
NICHOLAS, H. J., *J. Biol. Chem.* **237**, 1476–1480 (1962). Biosynthesis in *Salvia officinalis.*
PASICH, B., *Nature* **181**, 765 (1958). Chromatography on alumina-impregnated paper.
ZÜRCHER, A., O. JEGER, AND L. RUZICKA, *Helv. Chim. Acta* **37**, 2145–2152 (1954). Structure confirmation.

zeorin

$C_{30}H_{52}O_2$

m.p. (variable) ~245–253°, ~223–227°; [α] + 54° (c = 0.50, $CHCl_3$); nmr ($CDCl_3$), δ ppm: 0.87 (4α-methyl), 1.01 (4β-methyl), 1.15–1.18 (10β-methyl), 1.03 (8β-methyl), 0.97 (14α-methyl), and 0.77 (18α-methyl); nmr, τ ppm: 9.14 (4α-methyl), 9.02 (4β-methyl), 8.96 (8β-methyl), 8.83 (10β-methyl), 9.00 (14α-methyl), 9.25 (18α-methyl), and 8.83 (22α-methyl and 22β-methyl).

Structure II.

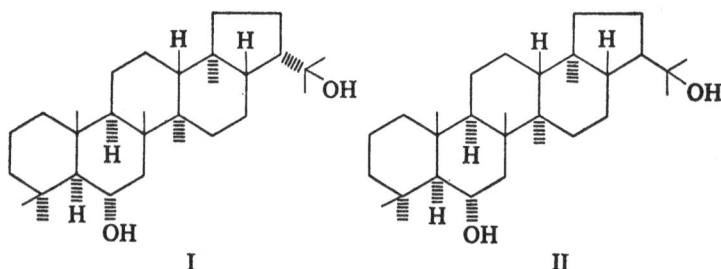

I   II

ASAHINA, Y., *J. Japan. Botany* **14**, 767–773 (1938). Microcrystal tests.
ASAHINA, Y. AND H. AKAGI, *Chem. Ber.* **71**, 980–985 (1938). From *Parmelia leucotyliza*, with leucotylin.
ASAHINA, Y. AND I. YOSIOKA, *Chem. Ber.* **73**, 742–747 (1940). From *Anaptychia speciosa, Heterodermia hypoleuca*, and *H. obscurata.*
BARTON, D. H. R. AND T. BRUUN, *J. Chem. Soc.* **1952**, 1683–1690 (1952). From *Nephroma arcticum*, structure studied.
BARTON, D. H. R., P. DE MAYO, AND J. C. ORR, *J. Chem. Soc.* **1958**, 2239–2248 (1958). Structure studied.
CULBERSON, W. L., *Am. J. Botany* **48**, 168–174 (1961). In *Parmelia galbina*, microcrystal test.

CULBERSON, W. L., *Bryologist* **69**, 472–487 (1966). Distribution in the genus *Heterodermia* in the Carolinas.

EVANS, A. W , *Trans. Conn. Acad. Arts Sci.* **35**, 519–626 (1944). Distribution in Connecticut Cladonias.

HUNECK, S., *Chem. Ber.* **94**, 614–622 (1961). From *Lecanora muralis*.

HUNECK, S., Z. *Naturforsch.* **21b**, 80–81 (1966). From *Lecanora hercynica*.

HUNECK, S., *Z. Naturforsch.* **21b**, 199–200 (1966). From *Lecanora handelii*.

HUNECK, S., *Z. Naturforsch.* **21b**, 888–890 (1966). From *Lecanora polytropa* and *L. sulphurea*.

HUNECK, S. AND G. FOLLMANN, *Z. Naturforsch.* **20b**, 1012–1013 (1965). From *Nephroma gyelnikii*.

HUNECK, S. AND G. FOLLMANN, *Z. Naturforsch.* **21b**, 714–715 (1966). From *Anaptychia neoleucomelaena*.

HUNECK, S. AND J.-M. LEHN, *Bull. Soc. Chim. France* **1963**, 1702–1706 (1963). Nmr study of configuration.

IKEKAWA, N., S. NATORI, H. AGETA, K. IWATA, AND M. MATSUI, *Chem. Pharm. Bull.* (Tokyo) **13**, 320–325 (1965). Gas chromatography of zeorin.

KUROKAWA, S., *Nova Hedwigia* **6**, 1–115 (1962). Distribution in the genus *Anaptychia*, microcrystal tests.

KUROKAWA, S., YOKO JINZENJI, S. SHIBATA, AND HSÜCH-CHING CHIANG, *Bull. Natl. Sci. Mus.* (Tokyo) **9**, 101–114 (1966). Distribution in Japanese Peltigeras, thin-layer chromatography.

NEELAKANTAN, S., S. RANGASWAMI, AND V. S. RAO, *Indian J. Pharm.* **16**, 173–175 (1954). From *Anaptychia speciosa, Heterodermia hypoleuca*, and *H. leucomela*.

RYABININ, A. A. AND L. G. MATYUKHINA, *J. Gen. Chem. USSR* (Engl. Transl.) **27**, 311–315 (1957); *ibid.* **28**, 2628–2630 (1958). From *Cladonia deformis*, structure studied.

SARMA, K. G. AND S. HUNECK, *Pharmazie*, in press. From two undetermined species of *Lepraria* and from *Anaptychia neoleucomelaena* from the Himalayas.

SHIBATA, S., T. FURUYA, AND H. IIZUKA, *Chem. Pharm. Bull.* (Tokyo) **13**, 1254–1257 (1965). In *Peltigera dolichorrhiza*, gas-liquid chromatography.

WETMORE, C. M., *Publ. Mus., Michigan State Univ., Biol. Ser.* **1**, 369–452 (1960). Distribution in North American Nephromas, microcrystal tests.

YOSIOKA, I., T. NAKANISHI, AND I. KITAGAWA, *Chem. Pharm. Bull.* (Tokyo) **15**, 353 (1967). Evidence for structure I.

YOSIOKA, I., T. NAKANISHI, AND I. KITAGAWA, *Tetrahedron Letters* **1968**, 1485–1490 (1968). Structure II proved.

YOSIOKA, I., T. NAKANISHI, AND E. TSUDA, *Tetrahedron Letters* **1966**, 607–612 (1966). From *Parmelia leucotyliza*, with leucotylic acid and leucotylin, comparison of nmr data.

YOSIOKA, I., M. YAMAKI, AND I. KITAGAWA, *Chem. Pharm. Bull.* (Tokyo) **14**, 804–807 (1966). From *Parmelia entotheiochroa*, with several related triterpenes.

ZOPF, W., *Ann. Chem.* **364**, 273–313 (1909). From *Nephroma antarcticum*, *N. arcticum*, *N. laevigatum*, and *N. parile*.

Secondary Products Originating by the Shikimic Acid Pathway

TERPHENYLQUINONES

polyporic acid

$C_{18}H_{12}O_4$

m.p. 315°, bath preheated to 310°, purple cubes (acetone), red needles (pyridine), or brown scales (xylene).

Also known from *Polyporus nidulans* and *P. rutilans*.

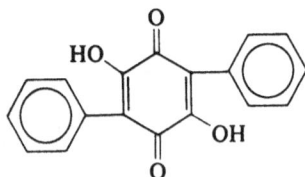

BURTON, J. F. AND B. F. CAIN, *Nature* **184**, 1326–1327 (1959). From *Sticta coronata*, antileukemic activity.

CAIN, B. F., *J. Chem. Soc.* **1961**, 936–940 (1961). From *Sticta coronata*, antitumor activity.

CAIN, B. F., *J. Chem. Soc.* **1963**, 356–359 (1963); *ibid.* **1964**, 5472–5474 (1964); *ibid.* **1966C**, 1041–1045 (1966). Antitumor activity of derivatives.

FICHTER, F., *Ann. Chem.* **361**, 363–378 (1908). Synthesis.

FRANK, R. L., G. R. CLARK, AND J. N. COKER, *J. Am. Chem. Soc.* **72**, 1824–1826 (1950). Oxidation to pulvinic dilactone.

FRANK, R. L., G. R. CLARK, AND J. N. COKER, *J. Am. Chem. Soc.* **72**, 1827–1829 (1950). Metallic complexes.

GRIPENBERG, J., *Acta Chem. Scand.* **12**, 1762–1767 (1958). Ultraviolet spectrum.

MAASS, W. S. G. AND A. C. NEISH, *Can. J. Botany* **45**, 59–72 (1967). Biosynthesis.

MOSBACH, K., *Biochem. Biophys. Res. Commun.* **17**, 363–367 (1964). Biosynthesis.

MURRAY, J., *J. Chem. Soc.* **1952**, 1345–1350 (1952). From *Sticta colensoi* and *S. coronata*.
SANTESSON, J., *Acta Chem. Scand.* **21**, 1162–1172 (1967). Thin-layer chromatography.
SHILDNECK, P. R. AND R. ADAMS, *J. Am. Chem. Soc.* **53**, 2373–2379 (1931). Synthesis.
WEISS, E., *Ann. Chem.* **361**, 378–382 (1908). Synthesis.
ZOPF, W., *Ann. Chem.* **317**, 110–145 (1901). From *Sticta origmaea*, first extraction from lichens.

thelephoric acid

$C_{20}H_{12}O_9$
m.p. >350° (pyridine), dark violet crystals.
Also known from nonlichen-forming fungi in the genera *Hydnum*, *Phlebia*, *Polyporus*, *Polyozellus*, and *Thelephora*.

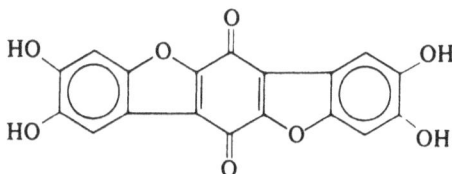

AGARWAL, S. C., K. AGHORAMURTHY, K. G. SARMA, AND T. R. SESHADRI, *J. Sci. Ind. Res.* (India) **20B**, 613–615 (1961). From *Lobaria isidiosa*.
AGHORAMURTHY, K., K. G. SARMA, AND T. R. SESHADRI, *Tetrahedron Letters* **1959**(8), 20–24 (1959); *ibid.* **1960**(16), 4–10 (1960). Structure studied.
ASAHINA, Y. AND S. SHIBATA, *Chem. Ber.* **72**, 1531–1533 (1939). From *Lobaria retigera*, first isolation from lichens, microcrystal test.
GRIPENBERG, J., *Acta Chem. Scand.* **12**, 1411–1414 (1958); *Tetrahedron* **10**, 135–143 (1960). Structure proved, synthesis.
KÖGL, F., H. ERXLEBEN, AND L. JÄNECKE, *Ann. Chem.* **482**, 105–119 (1930). Structure studied.
LOUNASMAA, M., *Acta Chem. Scand.* **19**, 540–541 (1965). Synthesis.
RAO, P. S., K. G. SARMA, AND T. R. SESHADRI, *Current Sci.* (India) **35**, 147–148 (1966). From *Lobaria isidiosa* var. *subisidiosa*.
READ, G. AND L. C. VINING, *Can. J. Chem.* **37**, 1442–1445 (1959). Structure studied.
ZOPF, W., *Botan. Ztg.* **47**, 68–82 (1889). From fungi of the genus *Thelephora*, first isolation.

PULVINIC ACID DERIVATIVES

calycin

$C_{18}H_{10}O_5$

m.p. 249–249.5° (acetic acid), red crystals; ir (KBr), ν cm^{-1}: 1715 (C=O); ir (dioxane), ν cm^{-1}: 1722; uv (cyclohexane), λ_{max} (ϵ) mμ: 208 (19,800), 237 (15,700), 253 (7,700), and 430 (24,700); uv (dioxane), λ_{max} (ϵ) mμ: 422 (17,000).

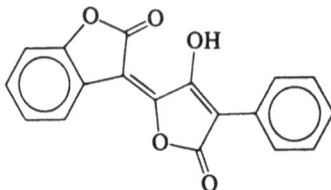

ÅKERMARK, B., *Acta Chem. Scand.* **15**, 1695–1700 (1961). Synthesis, structure revised.

ASANO, M. AND Y. KAMEDA, *Chem. Ber.* **68**, 1568–1571 (1935). From *Sticta aurata*, structure studied.

BENDZ, GERD, J. SANTESSON, AND C. A. WACHTMEISTER, *Acta Chem. Scand.* **19**, 1776–1777 (1965). Thin-layer chromatography.

GROVER, P. K. AND T. R. SESHADRI, *J. Sci. Ind. Res.* (India) **18B**, 238–240 (1959). From *Lecidea lucida*.

HARPER, S. H. AND R. M. LETCHER, *Proc. Trans. Rhodesian Sci. Assoc.* **51**, 156–184 (1966). Infrared and ultraviolet spectra, thin-layer chromatography, from *Buellia rhodesiaca*.

HESSE, O., *J. Prakt. Chem.* **58**, 465–561 (1898). Structure studied.

HUNECK, S. AND G. FOLLMANN, *Z. Naturforsch.* **22b**, 791–792 (1967). In *Pseudocyphellaria nitida* var. *subglauca*.

HUNECK, S. AND G. FOLLMANN, *Z. Naturforsch.* **22b**, 1182–1185 (1967). From *Pseudocyphellaria aurata, P. crocata, P. durvillei,* and *P. hirsuta*.

MAASS, W. S. G. AND A. C. NEISH, *Can. J. Botany* **45**, 59–72 (1967). Biosynthesis in *Pseudocyphellaria crocata*.

MITSUNO, M., *Pharm. Bull.* (Tokyo) **3**, 60–62 (1955). Paper chromatography.

MURRAY, J., *J. Chem. Soc.* **1952**, 1345–1350 (1952). From *Sticta colensoi* and *S. coronata*.

SANTESSON, J., *Acta Chem. Scand.* **21**, 1162–1172 (1967). Thin-layer chromatography.

SENFT, E., *Ber. Deut. Botan. Ges.* **34**, 592–600 (1916). From *Chrysothrix nolitangere*.

WACHTMEISTER, C. A., *Botan. Notiser* **109**, 313–324 (1956). Paper chromatography.

ZOPF, W., *Ann. Chem.* **284**, 107–132 (1895). From *Candelaria concolor, Candelariella aurella, Candelariella medians, Candelariella vitellina, Lepraria candelaris*, and *L. chlorina*.

epanorin

$C_{25}H_{25}NO_6$

m.p. 135–136° (methanol), yellow needles; $[\alpha]_D^{26}$ −1.86 ± 0.2° (c = 6.48, CHCl$_3$).

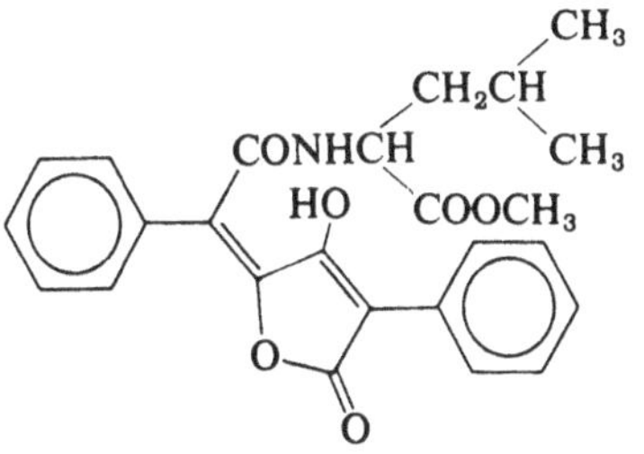

COKER, J. N., Univ. of Illinois Thesis Abstract. Urbana. 1950. Structure proved.

FRANK, R. L., S. M. COHEN, AND J. N. COKER, *J. Am. Chem. Soc.* **72**, 4454–4457 (1950). Synthesis from pulvinic anhydride and L-leucine.

JONES, M. P., J. KEANE, AND T. J. NOLAN, *Nature* **154**, 580 (1944). Structure studied.

MITSUNO, M., *Pharm. Bull.* (Tokyo) **3**, 60–62 (1955). Paper chromatography.

SANTESSON, J., *Acta Chem. Scand.* **21**, 1162–1172 (1967). Thin-layer chromatography.

ZOPF, W., *Ann. Chem.* **313**, 317–344 (1900). From *Lecanora epanora*, first extraction.

leprapinic acid

$C_{20}H_{16}O_6$

m.p. 164–165° (methanol), golden-yellow plates; ir (KBr), ν cm⁻¹: 3030 (w), 1776 (s), 1686 (s), 1621 (s), 1603 (s), 1499 (m), 1458 (m), and 1441 (s); uv (MeOH), λ_{max} (log ε) mμ: 270 (4.18) and 316 (infl., 3.92).

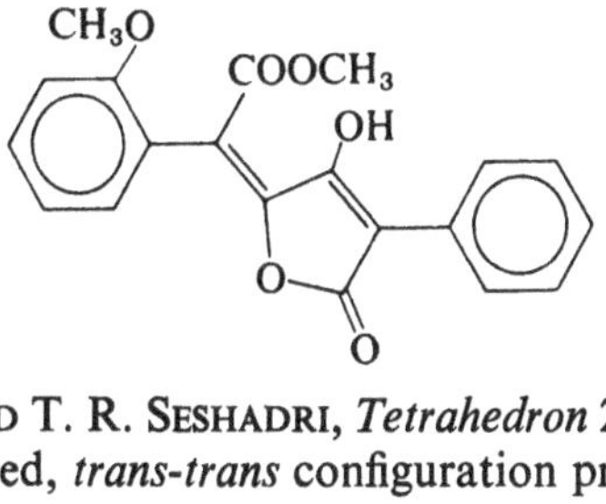

AGARWAL, S. C. AND T. R. SESHADRI, *Tetrahedron* **21**, 3205–3208 (1965). Structure confirmed, *trans-trans* configuration proved, biogenesis discussed.

GROVER, P. K. AND T. R. SESHADRI, *J. Sci. Ind. Res.* (India) **18B**, 238–240 (1959). From *Lecidea lucida*.

MITTAL, O. P. AND T. R. SESHADRI, *J. Chem. Soc.* **1955**, 3053–3055 (1955). From *Lepraria citrina*, structure studied, first isolation.

MITTAL, O. P. AND T. R. SESHADRI, *J. Chem. Soc.* **1956**, 1734–1735 (1956). Synthesis.

SARMA, K. G. AND S. HUNECK, *Pharmazie*, in press. From two unidentified Leprarias from the Himalayas.

leprapinic acid methyl ether

$C_{21}H_{18}O_6$

m.p. 150–152° (methanol), colorless needles; ir (CHCl$_3$), ν cm^{-1}: 2994 (m), 2899 (w), 1773 (s), 1634 (s), 1603 (m), 1493 (m), 1453 (m), and 1435 (m); uv (MeOH), λ_{max} (log ε) mμ: 229 (infl., 4.27), 261 (4.14), and 336 (4.44).

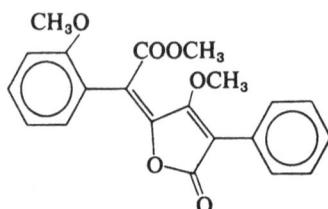

AGARWAL, S. C. AND T. R. SESHADRI, *Tetrahedron* **21**, 3205–3208 (1965). Purification, synthesis from leprapinic acid with diazomethane, *trans-trans* configuration of leprapinic acid proved.

GROVER, P. K. AND T. R. SESHADRI, *J. Sci. Ind. Res.* (India) **18B**, 238–240 (1959). From *Lepraria chlorina*, first isolation, structure proved.

pinastric acid

$C_{20}H_{16}O_6$

m.p. 202–204° (benzene), orange rectangular plates.

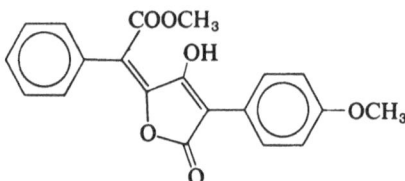

AGARWAL, S. C. AND T. R. SESHADRI, *Tetrahedron* **19**, 1965–1968 (1963). Structure studied.

AGARWAL, S. C. AND T. R. SESHADRI, *Tetrahedron* **20**, 17–24 (1964). Structure of Asano and Kameda (1935) confirmed.

AGARWAL, S. C. AND T. R. SESHADRI, *Indian J. Chem.* **2**, 17–22 (1964). Structure studied.

ASANO, M. AND Y. KAMEDA, *Chem. Ber.* **67**, 1522–1526 (1934); *ibid.* **68**, 1565–1567 (1935). From *Cetraria juniperina*, structure proved.

BENDZ, GERD, J. SANTESSON, AND C. A. WACHTMEISTER, *Acta Chem. Scand.* **19**, 1776–1777 (1965). Thin-layer chromatography.

GROVER, P. K. AND T. R. SESHADRI, *Tetrahedron* **4**, 105–110 (1958). Stereochemistry studied.

GROVER, P. K. AND T. R. SESHADRI, *Tetrahedron* **6**, 312–314 (1959). Structure studied.

HARPER, S. H. AND R. M. LETCHER, *Proc. Trans. Rhodesian Sci. Assoc.* **51**, 156–184 (1966). Infrared and ultraviolet spectra, thin-layer chromatography, from *Temnospora fulgens*.

HESS, D., *Planta* **52**, 65–76 (1958). Paper chromatography.

HUNECK, S., C. DJERASSI, D. BECHER, M. BARBER, M. VON ARDENNE, K. STEINFELDER, AND R. TÜMMLER, *Tetrahedron* **24**, 2707–2755 (1968). Mass spectrum.

KOLLER, G. AND A. KLEIN, *Monatsh. Chem.* **63**, 213–215 (1933). Synthesis.

KOLLER, G. AND G. PFEIFFER, *Monatsh. Chem.* **62**, 160–168 (1933). Structure studied.

MITTAL, O. P. AND T. R. SESHADRI, *J. Chem. Soc.* **1955**, 3053–3055 (1955); *ibid.* **1956**, 1734–1735 (1956). From *Lepraria candelaris*, structure studied.

MITSUNO, M., *Pharm. Bull.* (Tokyo) **3**, 60–62 (1955). Paper chromatography.

SANTESSON, J., *Acta Chem. Scand.* **21**, 1162–1172 (1967). Thin-layer chromatography.

VARTIA, K. O., *Ann. Med. Exptl. Biol. Fenniae* (Helsinki) **28** (Suppl. 7), 1–82 (1950). From *Cetraria pinastri*.

WACHTMEISTER, C. A., *Botan. Notiser* **109**, 313–324 (1956). Paper chromatography.

ZOPF, W., *Ann. Chem.* **284**, 107–132 (1895). From *Cetraria juniperina*, *C. pinastri*, and *Lepraria candelaris* (as *L. flava*).

pulvinic acid
$C_{18}H_{12}O_5$
m.p. 215–217° d (ethanol), yellow crystals.
Also known in the fungus *Ahlesia lichenicola*.

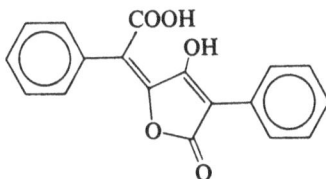

AGARWAL, S. C. AND T. R. SESHADRI, *Tetrahedron* **20**, 17–24 (1964). Reaction with *o*-phenylenediamine.

BENDZ, GERD, J. SANTESSON, AND C. A. WACHTMEISTER, *Acta Chem. Scand.* **19**, 1776–1777 (1965). Thin-layer chromatography.

BRODERSEN, R. AND A. KJAER, *Acta Pharmacol. Toxicol.* **2**, 109–220 (1946). Antibacterial activity and toxicity.

FRANK, R. L., G. R. CLARK, AND J. N. COKER, *J. Am. Chem. Soc.* **72**, 1824–1826 (1950). Synthesis from polyporic acid, ir spectrum.

HARPER, S. H. AND R. M. LETCHER, *Proc. Trans. Rhodesian Sci. Assoc.* **51**, 156–184 (1966). Infrared and ultraviolet spectra, thin-layer chromatography.

KARRER, P., K. A. GEHRCKENS, AND W. HEUSS, *Helv. Chim. Acta* **9**, 446–457 (1926). Configuration.

MAASS, W. S. G. AND A. C. NEISH, *Can. J. Botany* **45**, 59–72 (1967). Biosynthesis mentioned.

MITSUNO, M., *Pharm. Bull.* (Tokyo) **3**, 60–62 (1955). Paper chromatography.

MURRAY, J., *J. Chem. Soc.* **1952**, 1345–1350 (1952). From *Sticta coronata*.

SANTESSON, J., *Phytochemistry* **6**, 685–686 (1967). In *Thelocarpon epibolum* and *T. laureri* and in the fungus *Ahlesia lichenicola*, thin-layer chromatography.

SPIEGEL, A., *Ann. Chem.* **219**, 1–56 (1883). Synthesis from vulpinic acid.

VOLHARD, J., *Ann. Chem.* **282**, 1–21 (1894) and Schenck, R., *ibid.* **282**, 21–44 (1894). Synthesis.

pulvinic dilactone

$C_{18}H_{10}O_4$

m.p. 222–224° (glacial acetic acid), yellow needles; m.p. 227° (benzene).

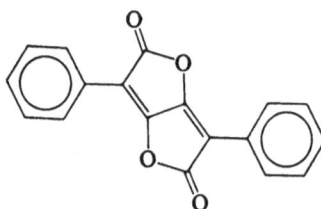

BENDZ, GERD, J. SANTESSON, AND C. A. WACHTMEISTER, *Acta Chem. Scand.* **19**, 1776–1777 (1965). Thin-layer chromatography.

CAIN, B. F., *J. Chem. Soc.* **1961**, 936–940 (1961). From *Sticta coronata*. antitumor activity.

FRANK, R. L., G. R. CLARK, AND J. N. COKER, *J. Am. Chem. Soc.* **72**, 1824–1826 (1950). Synthesis from polyporic acid.

GROVER, P. K. AND T. R. SESHADRI, *J. Sci. Ind. Res.* (India) **18B**, 238–240 (1959). From *Candelaria concolor*.

HARPER, S. H. AND R. M. LETCHER, *Proc. Trans. Rhodesian Sci. Assoc.* **51**, 156–184 (1966). Infrared and ultraviolet spectra, thin-layer chromatography.

HUNECK, S. AND G. FOLLMANN, *Z. Naturforsch.* **22b**, 791–792 (1967). In *Pseudocyphellaria nitida* var. *subglauca*.

HUNECK, S. AND G. FOLLMANN, *Z. Naturforsch.* **22b**, 1182–1185 (1967). From *Pseudocyphellaria aurata, P. crocata, P. durvillei,* and *P. hirsuta*.

MAASS, W. S. G. AND A. C. NEISH, *Can. J. Botany* **45**, 59–72 (1967). Biosynthesis in *Pseudocyphellaria crocata*, column chromatography.

MAASS, W. S. G., G. H. N. TOWERS, AND A. C. NEISH, *Ber. Deut. Botan. Ges.* **77**, 157–161 (1964). Biosynthesis in *Pseudocyphellaria crocata*.

MITSUNO, M., *Pharm. Bull.* (Tokyo) **3**, 60–62 (1955). Paper chromatography.

MURRAY, J., *J. Chem. Soc.* **1952**, 1345–1350 (1952). From *Sticta colensoi* and *S. coronata*.

RUNGE, F. AND URSULA KOCH, *Chem. Ber.* **91**, 1217–1224 (1958). Synthesis.

SANTESSON, J., *Acta Chem. Scand,* **21**, 1162–1172 (1967). Thin-layer chromatography.

SANTESSON, J., *Phytochemistry* **6**, 685–686 (1967). In *Thelocarpon epibolum* and *T. laureri*, thin-layer chromatography.

SPIEGEL, A., *Chem. Ber.* **15**, 1546–1554 (1882); *Ann. Chem.* **219**, 1–56 (1883). Synthesis from pulvinic acid.

WACHTMEISTER, C. A., *Botan. Notiser* **109**, 313–324 (1956). Paper chromatography.

WALTON, R. P., M. DE V. COTTEN, AND W. M. McCORD, *Proc. Soc. Exptl. Biol. Med.* **74**, 548–550 (1950). Action on the heart.

ZOPF, W., *Ann. Chem.* **306**, 282–321 (1899). From *Candelaria concolor, Candelariella aurella, Candelariella vitellina,* and *Pseudocyphellaria aurata*.

rhizocarpic acid

$C_{28}H_{23}NO_6$

m.p. 177–178° (ethanol), yellow needles; $[\alpha]_D^{20}$ +110.4° ± 2.1° (c = 1.22, $CHCl_3$).

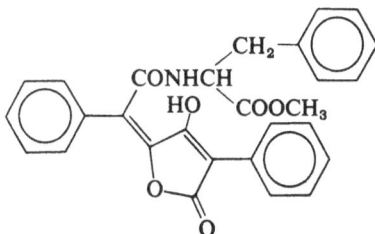

ASAHINA, Y., *J. Japan. Botany* **34**, 65–66 (1959). In *Acarospora gobiensis*, paper chromatography.

BENDZ, GERD, J. SANTESSON, AND C. A. WACHTMEISTER, *Acta Chem. Scand.* **19**, 1776–1777 (1965). From *Acarospora chlorophana* and *Coniocybe furfuracea*, thin-layer chromatography.

FRANK, R. L., S. M. COHEN, AND J. N. COKER, *J. Am. Chem. Soc.* **72**, 4454–4457 (1950). Synthesis.

HARPER, S. H. AND R. M. LETCHER, *Proc. Trans. Rhodesian Sci. Assoc.* **51**, 156–184 (1966). Infrared and ultraviolet spectra, thin-layer chromatography, from *Acarospora schleicheri* and *Dermatiscum thunbergii*.

HUNECK, S., *Z. Naturforsch.* **21b**, 80–81 (1966). From *Lecanora hercynica*.

HUNECK, S., C. DJERASSI, D. BECHER, M. BARBER, M. VON ARDENNE, K. STEINFELDER, AND R. TÜMMLER, *Tetrahedron* **24**, 2707–2755 (1968). Mass spectrum.

JONES, M. P., J. KEANE, AND T. J. NOLAN, *Nature* **154**, 580 (1944). Structure studied.

RUNEMARK, H., *Opera Botan.* (Lund) **2** (1), 1–152 (1956). Distribution in the genus *Rhizocarpon*, paper chromatography.

SANTESSON, J., *Acta Chem. Scand.* **21**, 1162–1172 (1967). Thin-layer chromatography.

SARMA, K. G. AND S. HUNECK, *Pharmazie*, in press. From *Acarospora oxytona*.

ZOPF, W., *Ann. Chem.* **284**, 107–132 (1895). From *Rhizocarpon geographicum*.

stictaurin

$C_{18}H_{10}O_5 + C_{18}H_{10}O_4$
m.p. 212°.
A 1:1 molecular complex of pulvinic dilactone and calycin.

MITSUNO, M., *Pharm. Bull.* (Tokyo) **3**, 60–62 (1955). Paper chromatography.

MURRAY, J., *J. Chem. Soc.* **1952**, 1345–1350 (1952). From *Sticta coronata*.

ZOPF, W., *Ann. Chem.* **306**, 282–321 (1899). From *Pseudocyphellaria aurata* (as *Sticta aurata*), *Candelaria concolor*, *Candelariella aurella*, and *Candelariella vitellina*, first isolation.

ZOPF, W., *Ann. Chem.* **317**, 110–145 (1901). From *Pseudocyphellaria crocata*, *Sticta flavicans*, *S. impressa*, and *S. origmaea*.

vulpinic acid

$C_{19}H_{14}O_5$
m.p. 146–148° (ethanol or ether), yellow crystals; m.p. 148–149° (benzene), yellow needles.
Also known in the fungus *Ahlesia lichenicola*.

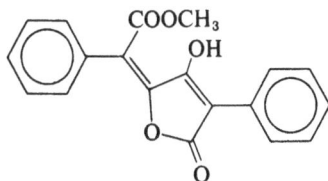

AGHORAMURTHY, K., K. G. SARMA, AND T. R. SESHADRI, *Tetrahedron* 12, 173–177 (1961). From *Alectoria tortuosa*.

ASAHINA, Y., *J. Japan. Botany* 29, 33–34 (1954). Microcrystal test.

ASANO, M. AND Y. ARATA, *Yakugaku Zasshi* 59, 679–687 (1939); German summary, pp. 286–290 (1939). Reduction with sodium amalgam, structure studied.

ASANO, M. AND Y. KAMEDA, *Chem. Ber.* 68, 1565–1567 (1935). From a mixture of *Cetraria juniperina* and *C. pinastri*.

BENDZ, GERD, J. SANTESSON, AND C. A. WACHTMEISTER, *Acta Chem. Scand.* 19, 1776–1777 (1965). Thin-layer chromatography.

BRODERSEN, R. AND A. KJAER, *Acta Pharmacol. Toxicol.* 2, 109–220 (1946). Antibacterial action.

FRANK, R. L., G. R. CLARK, AND J. N. COKER, *J. Am. Chem. Soc.* 72, 1824–1826 (1950). Synthesis from polyporic acid, ir spectrum.

FUZIKAWA, F., K. NAKAJIMA, O. WADAI, M. TORII, S. NAKAZAWA, T. OMATSU, AND T. TOYODA, *Yakugaku Zasshi* 73, 250–252 (1953). Antibacterial activity.

HALE, M. E., JR., *Contrib. U.S. Natl. Herb.* 36, 193–358 (1965). Distribution in the genus *Parmelia*.

HARPER, S. H. AND R. M. LETCHER, *Proc. Trans. Rhodesian Sci. Assoc.* 51, 156–184 (1966). Infrared and ultraviolet spectra, thin-layer chromatography.

HESS, D., *Planta* 52, 65–76 (1958). Paper chromatography.

HUNECK, S., C. DJERASSI, D. BECHER, M. BARBER, M. VON ARDENNE, K. STEINFELDER, AND R. TÜMMLER, *Tetrahedron* 24, 2707–2755 (1968). Mass spectrum.

KARRER, P., K. A. GEHRCKENS, AND W. HEUSS, *Helv. Chim. Acta* 9, 446–457 (1926). Configuration studied.

MAZZA, F. P., *Rend. Accad. Sci. Napoli* 31, 182–190 (1925); reference from *Chem. Abstr.* 21, 1110[6] (1927). Derivatives and properties.

MITSUNO, M., *Pharm. Bull.* (Tokyo) 3, 60–62 (1955). Paper chromatography.

MÖLLER, F. AND A. STRECKER, *Ann. Chem.* 113, 56–77 (1860). From *Letharia vulpina* (as *Cetraria vulpina*), structure studied.

MOSBACH, K., *Biochem. Biophys. Res. Commun.* 17, 363–367 (1964). Biosynthesis.

SANTESSON, J., *Acta Chem. Scand.* 21, 1162–1172 (1967). Thin-layer chromatography.

SANTESSON, J., *Phytochemistry* **6**, 685–686 (1967). In *Thelocarpon epibolum, T. laureri,* and the fungus *Ahlesia lichenicola,* thin-layer chromatography.

SCHÖNBERG, A. AND A. SINA, J. *Chem. Soc.* **1946**, 601–604 (1946). Thermal rearrangement.

SPIEGEL, A., *Chem. Ber.* **13**, 1629–1635 (1880); *ibid.* **13**, 2219–2221 (1880); *ibid.* **14**, 1686–1696 (1881); *ibid.* **15**, 1546–1554 (1882); *Ann. Chem.* **219**, 1–56 (1883). Structure proved.

STAHL, E. AND P. J. SCHORN, Z. *Physiol. Chem.* **325**, 263–274 (1961). Thin-layer chromatography.

VARTIA, K. O., *Ann. Med. Exptl. Biol. Fenniae* (Helsinki) **28** (Suppl. 7), 1–82 (1950). From *Cetraria pinastri.*

VOLHARD, J., *Ann. Chem.* **282**, 1–21 (1894). Synthesis.

WACHTMEISTER, C. A., *Botan. Notiser* **109**, 313–324 (1956). Paper chromatography.

ZOPF, W., *Ann. Chem.* **284**, 107–132 (1895). From *Calicium chlorinum, Chaenotheca chrysocephala,* and *Lepraria candelaris.*

Well Known or Recently Described Secondary Products of Uncertain Position

alectorialic acid

$C_{22}H_{20}O_{11}$

m.p. 175–176° d (ethanol-water), colorless needles; uv, λ_{max} mμ: 237, 260, and 335.

Structure not known.

HESS, D., *Planta* **52**, 65–76 (1953). Paper chromatography, in *Alectoria nigricans.*

LAMB, I. M., *British Antarctic Survey, Scientific Report No.* **38**, London, 1964. In *Alectoria nigricans,* microchemical tests.

SOLBERG, Y. J., Z. *Naturforsch.* **22b**, 777–783 (1967). From *Alectoria nigricans* and *Parmelia alpicola,* structure studied, uv and ir spectra.

bellidiflorin

Red to dark brown prisms (benzene); soluble in KOH (yellow solution), insoluble in $NaHCO_3$.

Structure not known.

ASAHINA, Y., J. *Japan. Botany* **14**, 767–773 (1938). In *Cladonia bellidiflora,* microchemical tests.

ASAHINA, Y., J. *Japan. Botany* **15**, 465–472 (1939). In *Cladonia corallifera.*

ASAHINA, Y., J. *Japan. Botany* **15**, 602–620, 663–671 (1939). In several species of *Cladonia.*

ASAHINA, Y., *J. Japan. Botany* **35**, 167–171 (1960). In *Cladonia coralli-fera*.
ZOPF, W., Die Flechtenstoffe. Jena. 1907 Description of properties.

constictic acid
Structure not known.

ASAHINA, Y., *J. Japan. Botany* **43**, 97–101 (1968). Accompanies stictic acid in many lichens, thin-layer chromatography and microcrystal tests.

destrictinic acid
m.p. 215° (ethyl ether), indigo-blue powder.
Structure not known.

ASAHINA, Y., *J. Japan. Botany* **18**, 489–502 (1942). In *Cladonia destricta*.
EVANS, A. W., *Trans. Conn. Acad. Arts Sci.* **35**, 519–626 (1944). In *Cladonia destricta*.
ZOPF, W., *Ann. Chem.* **327**, 317–354 (1903); *ibid.* **346**, 82–127 (1906). From *Cladonia destricta*.

echinocarpic acid
PD+ deep orange, KC−.
Structure not known.

KUROKAWA, S., *J. Japan. Botany* **40**, 264–269 (1965). In *Parmelia echino-carpa, P. planiuscula*, and *P. schizospatha*, microchemical tests.

endococcin
$C_{16}H_{12}O_5$ or $C_{15}H_{10}O_5$
Yellow needles (acetone-methanol); insoluble in Na_2CO_3, soluble in NaOH (purple-red), and soluble in H_2SO_4 (green).
Structure not known.

NOLAN, T. J. AND J. KEANE, *Sci. Proc. Roy. Dublin Soc.* **22**, 199–209 (1940). From *Lecanora gangaleoides*, converted to rhodophyscin in boiling acetic acid
ZOPF, W., *Ann. Chem.* **340**, 276–309 (1905). From *Physcia endococcinea*.

entothein
$C_{23}H_{22}O_{10}$
m.p. 240°, yellow.
Structure not known.

ABBAYES, H. DES, *Mém. Inst. Sci. Madagascar, Sér. B*, **10**(2), 81–121 (1961). In *Parmelia aurulenta*.

ASAHINA, Y., *J. Japan. Botany* **26**, 225–228 (1951). From *Parmelia entotheiochroa*, also in several other species of *Parmelia* studied, $FeCl_3+$ intense wine-red.
ASAHINA, Y., Lichens of Japan. Vol. II. Genus *Parmelia*. Tokyo. 1952. In several species of *Parmelia*.
HALE, M. E., JR., *Bryologist* **61**, 81–85 (1958). In *Parmelia aurulenta* and *P. obsessa*.
YOSIOKA, I., M. YAMAKI, AND I. KITAGAWA, *Chem. Pharm. Bull.* (Tokyo) **14**, 804–807 (1966). From *Parmelia entotheiochroa*.

eumitrin a and eumitrin b
Structures not known.

ASAHINA, Y., *J. Japan. Botany* **42**, 1–9 (1967). In *Usnea bayleyi* and several new subspecies, microchemical tests.

galapagin
$C_{21}H_{24}O_{11}$
m.p. 163–165° (methanol), very pale yellowish needles; ir (KBr), $v\ cm^{-1}$: 3340 (OH), 1738 (C=O), and 1652; uv (MeOH), λ_{max} (log ε) mμ: 242 (4.26), 248 (4.25), 256 (4.27); soluble in NaOH (yellow) and conc. H_2SO_4 (yellow); $Ca(OCl)_2-$, Gibbs reagent negative, $FeCl_3+$ ink-blue; no OCH_3.
Structure not known.

HUNECK, S., G. FOLLMANN, W. A. WEBER, AND G. TROTET, *Z. Natur-forsch.* **22b**, 671–673 (1967). From *Roccella portentosa* from the Galá-pagos Islands, first extraction.

mollin
m.p. 270–271° d (acetone), colorless needles; uv (MeOH), λ_{max} mμ: 228, 252, 284, and 318; $FeCl_3+$ violet, $NaHCO_3$ insoluble, NaOH soluble (yellow); acetyl derivative, m.p. 208–209°.
Structure not known.

HUNECK, S. AND G. FOLLMANN, *Z. Naturforsch.* **22b**, 666–670 (1967). From *Roccellaria mollis*.

pachycarpin
Insoluble in KOH; K−, C−, KC−, PD−, $FeCl_3+$ pale reddish-violet; chloroform and KOH with heat gives a violet color and a faint fluores-cence after dilution with water; no change after 12 hours in cold con-centrated H_2SO_4.
Structure not known, possibly a dibenzofuran derivative.

ASAHINA, Y., *J. Japan. Botany* **39**, 165–171 (1964). In *Haematomma pachycarpum*, microcrystal test in the GE solution, thin-layer chromatography.

phlebin A and phlebin B
Structures not known.

KUROKAWA, S., YOKO JINZENJI, S. SHIBATA, AND HSÜCH-CHING CHIANG, *Bull. Natl. Sci. Mus.* (Tokyo) **9**, 101–114 (1966). In the genus *Peltigera* in Japan, thin-layer chromatography, first detection.

rhodophyscin
$C_{15}H_{10}O_5 \cdot H_2O$ or $C_{18}H_{12}O_6 \cdot H_2O$.
Deep red crystals (acetone-methanol); soluble in Na_2CO_3 and in NaOH (purple-red); soluble in H_2SO_4 (raspberry-red becoming green); $FeCl_3 -$; no methoxyl; forms a yellow acetate.
Structure not known.

HALE, M. E., JR., *Contrib. U.S. Natl. Herb.* **36**, 193–358 (1965). Distribution in the genus *Parmelia*.
NOLAN, T. J. AND J. KEANE, *Sci. Proc. Roy. Dublin Soc.* **22**, 199–209 (1940). From *Lecanora gangaleoides*, also from endococcin in boiling acetic acid.
ZOPF, W., *Ann. Chem.* **340**, 276–309 (1905). From *Physcia endococcinea*.

roccellin
m.p. 206–207° (ethyl acetate), ir (KBr), ν cm^{-1}: 3330 (OH), 1720 (C=O), and 1652 (C=O); uv (MeOH), λ_{max} mμ: 228, 254, 286, and 316, nearly identical to mollin; $FeCl_3 +$ blue-violet, NaOH soluble (yellow); acetyl derivative, m.p. 210°.
Structure not known.

HUNECK, S. AND G. FOLLMANN, *Z. Naturforsch.* **22b**, 666–670 (1967). From *Roccellaria mollis*.

scabrosin A and scabrosin B
Structures not known.

KUROKAWA, S., YOKO JINZENJI, S. SHIBATA, AND HSÜCH-CHING CHIANG, *Bull. Natl. Sci. Mus.* (Tokyo) **9**, 101–114 (1966). In *Peltigera horizontalis* and European samples of *P. scabrosa*, thin-layer chromatography, first detection.

substances I, II, III, IV, V
Structures not known.

KUROKAWA, S., YOKO JINZENJI, S. SHIBATA, AND HSÜCH-CHING CHIANG, *Bull. Natl. Sci. Mus.* (Tokyo) **9**, 101–114 (1966). Distribution in Japanese Peltigeras, thin-layer chromatography.

unknown nitrogen-containing substance

m.p. 308–310° d (dioxane-water), colorless flakes; no coloration with $FeCl_3$, KOH, PD, or the Liebermann-Burchard reaction.

HUNECK, S. AND G. FOLLMANN, *Z. Naturforsch.* **22b**, 1369–1370 (1967). From *Roccella vicentina*.

HUNECK, S., G. FOLLMANN, AND H. ULLRICH, *Z. Naturforsch.* **23b**, 292–293 (1968). From *Roccella canariensis*.

F. A List of Species Containing Unidentified Substances

1. Colorless K+ Substances

Graphina heteroplacoides Red.; *G. interstes* Müll. Arg.; *G. obtectula* Müll. Arg.; *G. palmeri* Zahlbr.; *G. triangularis* Zahlbr.; *G. virginea* (Eschw.) Müll. Arg.
Graphis humilis Vain.
Lecidea tumida Mass.
Parmelia formosana Zahlbr.; *P. laevigata* (Sm.) Ach.; *P. regis* Lynge
Phaeographina strigops Wirth & Hale
Phaeographis exaltata (Mont. & Bosch) Müll. Arg.
Rhizocarpon cookeanum Magn.
Usnea lunaria Mot.

2. KC+ Substances

Nephroma sikkimense Asah.
Parmelia affixa Hale & Kurok.; *P. apophysata* Hale & Kurok.; *P. aptata* Kremp.; *P. confluescens* Nyl.; *P. damaziana* Zahlbr.; *P. gracilis* (Müll. Arg.) Vain.; *P. horrescens* Tayl.; *P. laevigata* (Sm.) Ach.; *P. madagascariacea* (Hue) Abb.; *P. meiosperma* (Hue) Dodge; *P. monilifera* Kurok.; *P. nodakensis* Asah.; *P. palmarum* Lynge; *P. paulensis* Zahlbr.; *P. regis* Lynge; *P. subfatiscens* Kurok.; *P. texana* Tuck.; *P. virginica* Hale; *P. xanthina* (Müll. Arg.) Vain.

3. C+ Substances

Lecidea tumida Mass.
Parmelia bostrychodes Zahlbr.; *P. echinocarpa* Kurok.; *P. explanata* Hale; *P. exsecta* Tayl.; *P. imbricatula* Zahlbr.; *P. planiuscula* Kurok.; *P. prolongata* Kurok.; *P. schizospatha* Kurok.
Peltigera aphthosa (L.) Willd.; *P. malacea* (Ach.) Funck; *P. venosa* (L.) Baumg.
Solorina saccata (L.) Ach.

4. PD+ Substances

Alectoria bicolor (Ehrh.) Nyl.; *A. chalybeiformis* (L.) S. Gray; *A. lanea* (Ehrh.) Vain.
Anaptychia dissecta Kurok.; *A. pandurata* Kurok.; *A. spinulosa* Kurok.
Cladonia acuminata var. *norrlinii* (Vain.) Magn.
Hypogymnia alpicola (Th. Fr.) Hav.; *H. intestiniformis* (Vill.) Räs.
Lecanora albula (Nyl.) Hue
Parmelia chapadensis Lynge; *P. cryptoxantha* Abb.; *P. dentella* Hale & Kurok.; *P. direagens* Hale; *P. echinocarpa* Kurok.; *P. gracilis* (Müll. Arg.) Vain.; *P. imperfecta* Kurok.; *P. olivacea* (L.) Ach.; *P. planiuscula* Kurok.; *P. ramosissima* Kurok.; *P. schizospatha* Kurok.; *P. scytodes* Kurok.; *P. sublimbata* Nyl.

5. FeCl₃+ Substances

Cetraria rhytidocarpa f. *nipponensis* Asah.; *C. rugosa* (Asah.) Satô
Hypogymnia enteromorpha (Ach.) Nyl.; *H. fragillima* (Hillm.) Rass.; "*H. lugubris*"; *H. mundata* f. *sorediosa* (Bitt.) Rass.; *H. vittata* (Ach.) Gas.
Pseudevernia furfuracea (L.) Zopf

6. Pigments

Acarospora fuscata (Schrad.) Arn. (brownish-purple)
Anaptychia albicans Kurok. (yellow, K+ yellow); *A. comosa* (Eschw.) Mass.; *A. corallophora* (Tayl.) Lynge (pale yellow, K+ yellow); *A. fauriei* Kurok. (two pigments: K+ violet; yellow, K+ yellow); *A. firmula* (Nyl.) Dodge & Awas. (yellow, K+ purple); *A. flabellata* (Fée) Mass. (yellow, K+ purple); *A. fragilissima* Kurok. (brownish); *A. hypocaesia* Yas. (yellow, K+ purple); *A. hypochraea* Vain. (yellow, K+ purple); *A. lamelligera* (Tayl.) Kurok. (yellow, K+ yellow); *A. loriformis* Kurok. (yellow, K+ purple); *A. lutescens* Kurok. (yellow, K+ yellow); *A. obesa* (Pers.) Zahlbr. (orange, K+ violet); *A. rugulosa* Kurok. (yellow, K+ purple); *A. subascendens* Asah. (yellow, K+ purple); *A. usambarensis* Kurok. (brown, K−); *A. vulgaris* (Vain.) Kurok. (K+ purple).
Asahinea chrysantha (Tuck.) W. Culb. & C. Culb. (purple or lavender); *A. kurodakensis* (Asah.) W. Culb. & C. Culb. (purple or lavender); *A. scholanderi* (Llano) W. Culb. & C. Culb. (purple or lavender)
Baeomyces sanguineus Asah. (red)
Cetraria ornata Müll. Arg. (yellow)
Cladonia bacillaris f. *reagens* Evans (anthraquinone derivative); *C. bacillaris* f. *tingens* Asah. (anthraquinone derivative); *C. floerkeana* f. *tingens* Asah. (K+ purple); *C. metacorallifera* f. *tingens* Asah. (K+ violet)
Haematomma coccineum (Dicks.) Körb. (red); *H. lapponicum* Räs. (red, acetone soluble); *H. porphyrium* (Pers.) Zopf (red, acetone insoluble); *H. puniceum* (Sm. *ex* Ach.) Mass. (red, acetone insoluble); *H. subpuniceum* (Müll. Arg.) B. de Lesd. (red, acetone insoluble); *H. ventosum* (L.) Mass.

Heterodermia appalachensis (Kurok.) W. Culb. (yellow, K+ yellow); *H. casarettiana* (Mass.) Trev. (yellow, not K+ purple); *H. dendritica* (Pers.) Poelt (yellow, K+ purple); *H. obscurata* (Nyl.) Trev. (yellow-orange, K+ purple); *H. squamulosa* (Degel.) W. Culb. (whitish-brown, K−)

Lasallia asiae-orientalis Asah. (anthraquinone derivative); *L. mayebarae* (Satô) Asah. (anthraquinone derivative); *L. papulosa* (Ach.) Llano (red, anthraquinone derivative)

Lecidea handelii Zahlbr. (anthraquinone derivative)

Maronella laricina M. Stein. (K+)

Oropogon asiaticus Asah. (anthraquinone derivative); *O. loxensis* (Fée) Th. Fr.

Parmelia amagiensis Asah. (yellow, FeCl$_3$−, K+ violet); *P. appendiculata* Fée (yellow, K−); *P. araucariarum* Zahlbr. (yellow, K−); *P. arcana* Kurok. (yellow); *P. bahiana* Nyl. (K+ purple); *P. chapadensis* Lynge (pale yellow); *P. congruens* Ach. (yellow); *P. consimilis* Vain. (yellow); *P. crocoides* Hale (orange-red, K−); *P. crustacea* Lynge (K+ purple); *P. cryptoxantha* Abb. (probably entothein); *P. cyphellata* (Lynge) Sant. (yellow); *P. denegans* Nyl. (K+ purple); *P. ebulliens* Hale (yellow); *P. endomiltoides* Nyl. (two pigments, one K+); *P. endosulphurea* (Hillm.) Hale (K−); *P. erasmia* Hale (orange-red, K+); *P. galbina* Ach. (yellow); *P. gracilescens* Vain. (yellow, possibly entothein); *P. heterochroa* Hale & Kurok. (anthraquinone derivative); *P. huei* Asah. (yellow); *P. hypomilta* Fée (K+ purple-black); *P. hypomiltoides* Vain. (orange-red, K+); *P. lecanoracea* Müll. Arg. (K+ purple-black); *P. lythogoeana* Dodge (K+ purple); *P. malesiana* Hale (K+); *P. malmei* Lynge (K+ purple); *P. mesogenes* Nyl. (orange-red, K+); *P. michoacanensis* B. de Lesd. (yellow); *P. minima* Lynge (anthraquinone); *P. myelochroa* Hale (yellow-orange); *P. obessa* Ach. (yellow, possibly entothein); *P. osteoleuca* Nyl. (K+ purple); *P. permutata* Stirt. (yellow); *P. peruviana* Nyl. (yellow, K−); *P. prolongata* Kurok. (pale yellow); *P. regnellii* Lynge (anthraquinone derivative); *P. rigidula* Kurok. (pale yellow); *P. rutidota* Hook. f. & Tayl. (K+ purple-black); *P. sancti-angelii* Lynge (yellow, acetone soluble); *P. silvatica* Lynge (K+ purple); *P. subcolorata* Hale (K−); *P. violacea* Kurok. (K+); *P. xantholepis* Mont. & Bosch

Platismatia erosa W. Culb. & C. Culb. (yellow)

Pseudevernia olivetorina Zopf (yellow)

Pyxine caesiopruinosa (Nyl.) Imsh. (yellow)

Ramalina boulhautiana Mah. & Gillet (red)

Sticta coronata Müll. Arg. (two anthraquinone derivatives); *S. gilva* (Ach.) Ach.

Trypetheliopsis boninensis Asah. (yellow, K+ purple)

Umbilicaria mammulata (Ach.) Tuck.

Usnea bakongoensis Duvign.; *U. canariensis* (Ach.) Du Rietz; *U. contorta* Jatta (K+ purple); *U. creberrima* Vain. (red); *U. croceorubescens* Vain. (red); *U. dorogawensis* Asah. (red); *U. eizanensis* Asah. (red); *U. hossei* f. *subtrichodea* Asah. (yellow, K+purple); *U. imlpicita* (Stirt.) Zahlbr. (red); *U. indigena* Mot. (acetone soluble, K+

purple); *U. misaminensis* (Vain.) Mot. (K+ purple-black); *U. mutabilis* Stirt. (red); *U. pectinata* Tayl. (yellow, K+ purple); *U. roseola* Vain. (red); *U. shikokiana* Asah. (yellow, acetone soluble, K+ purple); *U. strigosa* (Ach.) A. Eat. (red)
Xanthoria parietina (L.) Th. Fr. (anthraquinone derivative)

7. Terpenes

Lobaria isidiosa (Müll. Arg.) Vain.; *L. retigera* (Bory) Trev.; *L. subretigera* Inum.
Parmelia entotheiochroa Hue
Sticta coronata Müll. Arg.

8. Saccharides

Hypogymnia physodes (L.) Nyl.
Parmelia centrifuga (L.) Ach.
Pseudevernia furfuracea (L.) Zopf
Roccella fuciformis (L.) DC.; *R. hypomecha* (Ach.) Bory; *R. linearis* (Ach.) Vain.
Umbilicaria mammulata (Ach.) Tuck.

9. Neutral Substances

Cetraria endocrocea (Asah.) Satô
Sticta colensoi Bab.; *S. gilva* (Ach.) Ach.; *S. glaucolurida* Nyl.
Teloschistes flavicans (Sw.) Norm.
Usnea venosa Mot.

10. Miscellaneous Aliphatic Substances

Cetraria ambigua Bab.; *C. cucullata* (Bell.) Ach. (proto-α-lichesterinic acid); *C. islandica* (L.) Ach. (paralichesterinic acid and unsaturated fatty acids); *C. rhytidocarpa* Mont. & Bosch; *C. rhytidocarpa* f. *nipponensis* Asah.; *C. togashii* Asah.; *C. ulophylloides* Asah.
Cladonia alpicola (Flot.) Vain.
Cornicularia aculeata (Schreb.) Ach. (proto-α-lichesterinic acid); *C. muricata* Ach. (proto-α-lichesterinic acid and dilichesterinic acid)
Hypogymnia physodes (L.) Nyl. (unsaturated fatty acids)
Lecidea handelii Zahlbr.
Lepraria farinosa (Hoffm.) Ach. (hydroxyroccellic acid); *L. latebrarum* Ach. (hydroxyroccellic acid)
Parmelia arcana Kurok.; *P. centrifuga* (L.) Ach.; *P. hypoleucites* Nyl.; *P. keitauensis* Asah.; *P. lophogena* Abb.; *P. lythgoeana* Dodge; *P. macrocarpoides* f. *subcomparata* Vain.; *P. mesotropa* Müll. Arg.; *P. panniformis* (Nyl.) Vain.; *P. reddenda* Stirt.; *P. saxatilis* (L.) Ach.; *P. sorocheila* Vain.; *P. subarnoldii* Abb.; *P. subrugata* Kremp.; *P. tropica* Vain.
Peltigera canina (L.) Willd.
Pseudevernia furfuracea (L.) Zopf
Ramalina curnowii Cromb. *ex* Nyl.
Stereocaulon pseudoarbuscula Asah.
Umbilicaria mammulata (Ach.) Tuck.
Usnea barbata (L.) Wigg. (unsaturated fatty acids)

11. Miscellaneous Substances

Anaptychia diademata (Tayl.) Kurok.; *A. dissecta* var. *koyana* Kurok.; *A. isidiophora* (Nyl.) Vain.; *A. speciosa* (Wulf.) Mass.

Anzia hypoleucoides Müll. Arg.

Baeomyces aggregatus Asah.

Calicium japonicum Asah.

Cetraria islandica (L.) Ach.; *C. ulophylloides* Asah.

Cladonia atlantica Evans; *C. botryocarpa* Merr.; *C. cariosa* (Ach.) Spreng.; *C. conista* (Ach.) Robb.; *C. cristatella* Tuck.; *C. cubana* (Vain.) Evans; *C. degenerans* var. *haplothea* Ach.; *C. delessertii* (Nyl.) Vain.; *C. floerkeana* var. *alpina* Asah.; *C. gymnopoda* Vain.; *C. linearis* Evans

Enterographa crassa (DC.) Fée

Graphina albostriata (Vain.) Zahlbr.; *G. collosporella* (Vain.) Zahlbr.; *G. confluens* (Fée) Müll. Arg. (a chemical strain); *G. pseudosophistica* (Vain.) Zahlbr.

Graphis abaphoides Nyl.

Haematomma lapponicum Räs.; *H. puniceum* (Sm. *ex* Ach.) Mass.; *H. ventosum* (L.) Mass.

Hendrickxia alcicornis Duvign.

Heterodermia hypoleuca (Ach.) Trev.; *H. pseudospeciosa* (Kurok.) W. Culb.; *H. tremulans* (Müll. Arg.) W. Culb.

Hypogymnia alpicola (Th. Fr.) Hav.; *H. tubulosa* (Schaer.) Hav.

Lecanora gangaleoides Nyl.; *L. pruinosa* Chaub.; *L. radiosa* (Hoffm.) Schaer.; *L. reuteri* Schaer.; *L. straminea* (Wahlenb.) Ach.

Lepraria candelaris (L.) Fr.

Nephroma bellum (Spreng.) Tuck.; *N. expallidum* (Nyl.) Nyl.; *N. gyelnikii* (Räs.) Lamb; *N. helveticum* Ach.; *N. isidiosum* (Nyl.) Gyeln.; *N. laevigatum* Ach.; *N. sinense* Zahlbr.

Ochrolechia spp.

Oropogon loxensis (Fée) Th. Fr.; *O. tanakae* Asah.

Parmelia appendiculata Fée; *P. araucariarum* Zahlbr.; *P. cirrhata* Fr.; *P. ebulliens* Hale; *P. endosulphurea* (Hillm.) Hale; *P. formosana* Zahlbr.; *P. ikomae* Asah.; *P. keitauensis* Asah.; *P. laevigata* ssp. *extremi-orientalis* Asah.; *P. livida* Tayl.; *P. mesogenes* Nyl.; *P. michoacanensis* B. de Lesd.; *P. novella* Vain.; *P. perlata* (Huds.) Ach.; *P. regnellii* Lynge; *P. sorediosa* Almb.; *P. subaurulenta* Nyl.; *P. subcrinita* Nyl.; *P. substygia* Räs.; *P. tiliacea* (Hoffm.) Ach.; *P. tortula* Kurok.

Peltigera canina (L.) Willd., *P. dolichorrhiza* (Nyl.) Nyl.; *P. horizontalis* (Huds.) Baumg.; *P. microphylla* (And.) Gyeln.; *P. nigripunctata* Bitt.; *P. polydactyla* (Neck.) Hoffm.; *P. pruinosa* (Gyeln.) Inum.; *P. scabrosa* Th. Fr.; *P. scutata* (Dicks.) Duby; *P. subscutata* Gyeln.; *P. venosa* (L.) Baumg.

Pertusaria pertusa (L.) Tuck.

Phaeographina elliptica Wirth & Hale

Phaeographis inustoides (Fink) Red.

Pseudevernia olivetorina (Zopf) Zopf

Pyxine caesiopruinosa (Nyl.) Imsh.

Ramalina druidarum W. Culb.

Rhizocarpon saanaënse Räs.

Squamarina crassa var. *platyloba* (Matt.) Poelt; *S. crassa* f. *pseudocrassa* (Matt.) Poelt; *S. nivalis* Frey & Poelt; *S. periculosa* (Duf.) Poelt

Stereocaulon albicans Th. Fr. (Chem. strain 1); *S. arbuscula* var. *aberrans* Asah.; *S. botryosum* f. *congestum* (Magn.) Frey; *S. coralligerum* Meyer; *S. curtatum* Nyl.; *S. foliolosum* var. *strictum* (Bab.) Lamb; *S. japonicum* var. *commixtum* Asah.; *S. japonicum* ssp. *etigoense* Asah.; *S. mixtum* Nyl.; *S. nesaeum* Nyl.; *S. proximum* Nyl.; *S. ramulosum* (Sw.) Räusch.; *S. tennesseense* Magn.

Sticta coronata Müll. Arg.; *S. flavicans* Hook. f. & Tayl.

Usnea arguta Mot.; *U. asahinai* Mot.; *U. croceorubescens* Vain.; *U. diffracta* ssp. *subdiffracta* Asah.; *U. eizanensis* Asah.; *U. elongata* Mot.; *U. mutabilis* Stirt.; *U. nipparensis* Asah.; *U. pseudintumescens* Asah.

Chapter IV

Chemical Summary of the Genera

Class Ascomycetes

SUBCLASS ASCOMYCETIDAE

ORDER LECANORALES

Lichinaceae

Lichina (1 species)

arabinose, fructose, galactose, glucosan, glucose, glycerol, glycogen, mannosido-mannitol, mannitol, sucrose, xylose, complex polysaccharides

Collemataceae

Collema (13 taxa)

carotenes

Leptogium (3 species)

carotenes

Pannariaceae

Pannaria (5 species)

atranorin, pannarin

Peltigeraceae

Nephroma (14 taxa)

(+)-usnic acid

1,3,-dihydroxy-8-methoxy-2-chloro-6-methylanthraquinone, emodin, fragilin, 1-hydroxy-3,8-dimethoxy-2-chloro-6-methylanthraquinone, nephromin, 1,3,8-trihydroxy-2-chloro-6-methylanthraquinone
nephrin, zeorin
arabitol, mannitol
ascorbic acid

Peltigera (23 taxa)

tenuiorin

unidentified fatty acids
ergosterol, zeorin
3-O-β-D-galactofuranosyl-D-mannitol, 3-O-β-D-glucopyrano-syl-D-mannitol, D-mannitol, sucrose
ascorbic acid, chlorophyll *a*, dolichorrhizin, *myo*-inositol, phlebin A, phlebin B, scabrosin A, scabrosin B

Solorina (2 species)

norsolorinic acid, solorinic acid
D-mannitol

Stictaceae

Lobaria (14 taxa)

gyrophoric acid, scrobiculin, tenuiorin
norstictic acid, stictic acid
(+)-usnic acid, (−)-usnic acid

carotenes, ergosterol, retigeradiol, retigeranic acid
thelephoric acid
D-arabitol, D-mannitol

Pseudocyphellaria (2 species)

atranorin

7β-acetoxy-22-hydroxyhopane, 15α,22-dihydroxyhopane, durvilldiol, durvillonol, ergosterol
calycin, pulvinic acid, pulvinic dilactone

Sticta (12 species)

lecanoric acid
norstictic acid, stictic acid

unidentified anthraquinone pigments
7β-acetoxy-22-hydroxyhopane, 15α,22-dihydroxyhopane, unidentified terpenes
calycin, pulvinic dilactone, polyporic acid
ammonia, methylamine, trimethylamine

Graphidaceae

Graphina (29 taxa)

atranorin, protocetraric acid, norstictic acid, salazinic acid,
stictic acid

lichexanthone
zeorin

Graphis (17 species)

lecanoric acid
atranorin, protocetraric acid, norstictic acid, salazinic acid,
stictic acid

Medusulina (2 species)

norstictic acid, stictic acid

Phaeographina (6 species)

norstictic acid, stictic acid

Phaeographis (5 taxa)

norstictic acid

Xylographa

atranorin, norstictic acid

Cladoniaceae

Baeomyces (9 species)

atranorin, baeomycesic acid, norstictic acid, stictic acid
didymic acid

Cladonia (227 taxa)

cryptochlorophaeic acid, divaricatic acid, grayanic acid,
homosekikaic acid, merochlorophaeic acid, perlatolic acid

atranorin, baeomycesic acid, barbatic acid, decarboxy-thamnolic acid, fumarprotocetraric acid, hypothamnolic acid, norstictic acid, psoromic acid, squamatic acid, stictic acid, thamnolic acid
didymic acid, strepsilin
(+)-isousnic acid, (−)-isousnic acid, (+)-usnic acid, (−)-usnic acid

norrangiformic acid, protolichesterinic acid, pseudonor-rangiformic acid, rangiformic acid, tetrahydroxy fatty acids
rhodocladonic acid
ergosterol, friedelin, taraxerene, ursolic acid, zeorin
D-arabitol, lichenin, mannitol, sucrose, trehalose, unidentified polysaccharides
ascorbic acid, bellidiflorin, cervicornic acid, destrictinic acid, dimethyl sulfone, ethanolamine, succinic acid

Glossodium (1 species)

thamnolic acid

Pycnothelia (1 species)

atranorin

(+)-protolichesterinic acid

Thysanothecium (2 taxa)

divaricatic acid

Lecideaceae

Bombyliospora (1 species)

pannarin

zeorin

Catillaria (2 species)

(−)-usnic acid

Lecidea (51 taxa)

confluentic acid, divaricatic acid, glomelliferic acid, gyrophoric acid, lecanoric acid, planaic acid
atranorin, fumarprotocetraric acid, norstictic acid
porphyrilic acid

roccellic acid
arthothelin, lichexanthone, thuringione
parietin, unidentified hydroxyanthraquinone
calycin, leprapinic acid, rhizocarpic acid

Megalospora (4 species)

usnic acid

zeorin

Mycoblastus (1 species)

atranorin

caperatic acid

Rhizocarpon (40 taxa)

gyrophoric acid
barbatic acid, norstictic acid, psoromic acid, stictic acid

rhizocarpic acid

Temnospora

pinastric acid, vulpinic acid

Stereocaulaceae

Stereocaulon (76 taxa)

divaricatic acid, lecanoric acid, lobaric acid
atranorin, fumarprotocetraric acid, (α-methylethersalazinic
acid), norstictic acid, salazinic acid, squamatic acid, stictic
acid, thamnolic acid
usnic acid

ventosic acid
parietin
ergosterol, friedelin, β-sitosterol, zeorin
D-arabitol
ascorbic acid, dendroidin

Umbilicariaceae

Actinogyra (2 species)

gyrophoric acid, lecanoric acid, umbilicaric acid

arabitol, mannitol

Agyrophora (3 species)

 gyrophoric acid

 arabitol, mannitol, sucrose, trehalose, umbilicin

Dermatiscum (1 species)

 lecanoric acid
 squamatic acid

 rhizocarpic acid
 arabitol, D-mannitol
 choline sulfate ester

Lasallia (5 species)

 gyrophoric acid

 unidentified anthraquinone
 arabitol, D-mannitol, pustulan, sucrose, trehalose, umbilicin

Omphalodiscus (4 species)

 gyrophoric acid

 arabitol, mannitol

Umbilicaria (16 species)

 gyrophoric acid, umbilicaric acid
 norstictic acid, stictic acid

 unidentified fatty acid
 ergosterol
 arabitol, galactose, glucose, mannitol, pustulan, sucrose,
  volemitol, polysaccharide like lichenin

Diploschistaceae

Diploschistes (5 taxa)

 diploschistesic acid, lecanoric acid
 norstictic acid

Pertusariaceae

Perforaria (1 species)
 norstictic acid, stictic acid

Pertusaria (21 taxa)

> gyrophoric acid, lecanoric acid
> atranorin, haemathamnolic acid, norstictic acid, salazinic acid, thamnolic acid
> picrolichenic acid
>
> concretin, lichexanthone, thiophaninic acid
> arabitol, mannitol

Acarosporaceae

Acarospora (9 taxa)

> gyrophoric acid
> usnic acid
>
> acaranoic acid, acarenoic acid
> rhizocarpic acid

Thelocarpon (2 species)

> pulvinic acid, pulvinic dilactone, vulpinic acid

Lecanoraceae

Candelariella (4 species)

> calycin, pulvinic dilactone
> D-mannitol

Haematomma (16 taxa)

> divaricatic acid, imbricaric acid
> atranorin, decarboxythamnolic acid, psoromic acid, thamnolic acid
> porphyrilic acid
> (+)-usnic acid, (−)-usnic acid
>
> tetrahydroxy fatty acids, ventosic acid
> zeorin
> arabitol, mannitol, sucrose, trehalose, umbilicin, pachycarpin

Icmadophila (1 species)

> thamnolic acid

Lecanora (80 taxa)

α-collatolic acid, erythrin, gyrophoric acid, lobaric acid
atranorin, chloroatranorin, fumarprotocetraric acid, gangaleoidin, (α-methylethersalazinic acid), norstictic acid, pannarin, physodalic acid, protocetraric acid, psoromic acid,
stictic acid
(+)-usnic acid, (−)-usnic acid

aspicilin, rangiformic acid, roccellic acid
sordidone
arthothelin, 2,4-dichloronorlichexanthone, 2,7-dichloronor-
lichexanthone, norlichexanthone, thiophanic acid, thuringione, vinetorin
eulecanorol, leucotylin, β-sitosterol, zeorin
epanorin, rhizocarpic acid
erythritol, lichenin, mannitol, sucrose, trehalose
endococcin, placodin, rhodophyscin

Ochrolechia (54 taxa)

erythrin, gyrophoric acid, lecanoric acid, variolaric acid
atranorin

arabitol, mannitol

Phlyctis (2 species)

norstictic acid

Placopsis (1 species)

gyrophoric acid

Squamarina (10 taxa)

psoromic acid
(−)-usnic acid

Thamnolecania (1 species)

atranorin

Parmeliaceae

Anzia (8 species)

anziaic acid, divaricatic acid, lobaric acid, sekikaic acid
atranorin, chloroatranorin

Asahinea (3 species)

alectoronic acid, α-collatolic acid
atranorin
usnic acid

Candelaria (2 species)

calycin, pulvinic dilactone

Cavernularia (2 species)

physodic acid
atranorin

Cetraria (35 taxa)

alectoronic acid, α-collatolic acid, gyrophoric acid, hiascic
    acid, microphyllinic acid, olivetoric acid, physodic acid
atranorin, fumarprotocetraric acid, stictic acid
(+)-usnic acid, (−)-usnic acid

caperatic acid, (−)-lichesterinic acid, (−)-*allo*-protoliches-
    terinic acid, (+)-protolichesterinic acid, (−)-protoliches-
    terinic acid, nephromopsinic acid, nephrosteranic acid,
    nephrosterinic acid, tetrahydroxy fatty acids
endocrocin
ergosterol, friedelin
pinastric acid, vulpinic acid
arabitol, cellulose, hemicellulose, isolichenin, lichenin, man-
    nitol, sucrose, trehalose, umbilicin
ascorbic acid

Cetrelia (14 species)

alectoronic acid, anziaic acid, α-collatolic acid, glomelliferic
    acid, imbricaric acid, microphyllinic acid, olivetoric acid,
    perlatolic acid
atranorin, chloroatranorin

Hendrickxia (1 species)

atranorin
usnic acid

Hypogymnia (30 taxa)

olivetoric acid, physodic acid
atranorin, chloroatranorin, physodalic acid, protocetraric
    acid
usnic acid

tetrahydroxy fatty acids, unsaturated fatty acids
β-carotene, ergosterol, xanthophyll
arabitol, lichenin, mannitol
ascorbic acid, chlorophyll *a*, chlorophyll *b*

Menegazzia (2 species)

atranorin, stictic acid

Parmelia (476 taxa)

alectoronic acid, α-collatolic acid, cryptochlorophaeic acid, divaricatic acid, evernic acid, glomelliferic acid, gyrophoric acid, imbricaric acid, lecanoric acid, lobaric acid, 4-O-methylphysodic acid, norlobaridone, olivetoric acid, perlatolic acid, physodic acid
alectorialic acid, atranorin, barbatic acid, chloroatranorin, diffractaic acid, fumarprotocetraric acid, norbarbatic acid, norstictic acid, physodalic acid, protocetraric acid, salazinic acid, stictic acid
(+)-usnic acid

caperatic acid, (+)-protolichesterinic acid
methyl β-orcinolcarboxylate
lichexanthone
16-β-O-acetylleucotylic acid, 6α-O-acetylleucotylin, 6-deoxy-16β-O-acetylleucotylin, 6-deoxyleucotylin, 6α,16β-di-O-acetylleucotylin, leucotylic acid, leucotylin, zeorin
vulpinic acid
arabitol, fructose, galactose, glucose, isolichenin, lichenin, D-mannitol, sucrose, volemitol
ascorbic acid, echinocarpic acid, entothein, galbinic acid, lividic acid, rhodophyscin, subauriferin

Parmeliopsis (4 species)

divaricatic acid
atranorin, thamnolic acid
usnic acid

Platismatia (10 species)

atranorin, fumarprotocetraric acid

caperatic acid
arabitol, mannitol

Pseudevernia (5 species)

lecanoric acid, olivetoric acid, physodic acid
atranorin, chloroatranorin

unidentified fatty acids
ergosterol, fungisterol
arabitol, erythritol, lichenin, mannitol

Usneaceae

Alectoria (18 species)

alectoronic acid, olivetoric acid
alectorialic acid, atranorin, diffractaic acid, fumarproto-
    cetraric acid, psoromic acid, virensic acid
barbatolic acid
(−)-usnic acid

friedelin
vulpinic acid
arabitol, bryopogin, hemicellulose, lichenin, mannitol
ascorbic acid

Cornicularia (7 species)

olivetoric acid, physodic acid

protolichesterinic acid
ergosterol

Dactylina (1 species)

gyrophoric acid
physodalic acid
usnic acid

dufourein, endochrysin

Endocena (1 species)

thamnolic acid

Evernia (7 species)

divaricatic acid, evernic acid
atranorin, chloroatranorin
(+)-usnic acid
friedelin
arabitol, mannitol

Everniopsis (1 species)

atranorin, psoromic acid
usnic acid

Himantormia (1 species)

    atranorin

    barbatolic acid
    usnic acid

Letharia (2 species)

    atranorin

    vulpinic acid
    arabitol, lichenin, mannitol

Neuropogon (2 species)

    usnic acid

Oropogon (4 species)

    protocetraric acid, psoromic acid

Ramalina (63 taxa)

    boninic acid, cryptochlorophaeic acid, divaricatic acid, evernic acid, methyl 3,5-dichlorolecanorate, paludosic acid, ramalinolic acid, sekikaic acid
    atranorin, chloroatranorin, hypoprotocetraric acid, norbarbatic acid, norstictic acid, obtusatic acid, protocetraric acid, psoromic acid, salazinic acid, stictic acid
    (+)-usnic acid

    aspicilin, tetrahydroxy fatty acids
    (−)-16α-hydroxykaurane
    D-arabitol, isolichenin, lichenin, mannitol
    ascorbic acid, riboflavin

Siphula (8 species)

    atranorin, baeomycesic acid, barbatic acid, chloroatranorin, decarboxythamnolic acid, hypothamnolic acid, norstictic acid, squamatic acid, thamnolic acid
    porphyrilic acid, strepsilin

    siphulin
    siphulitol

Thamnolia (2 species)

    baeomycesic acid, decarboxythamnolic acid, squamatic acid, thamnolic acid

    arabitol, mannitol

Usnea (117 taxa)

evernic acid
atranorin, barbatic acid, chloroatranorin, decarboxythamnolic acid, diffractaic acid, fumarprotocetraric acid,
norstictic acid, obtusatic acid, protocetraric acid, psoromic
acid, salazinic acid, squamatic acid, stictic acid, thamnolic
acid
barbatolic acid
(+)-usnic acid

caperatic acid, tetrahydroxy fatty acids, unsaturated fatty
acids
unidentified anthraquinones
carotene, ergosterol
D-arabitol, isolichenin, lichenin, mannitol
ascorbic acid, galbinic acid, riboflavin

Physciaceae

Anaptychia (9 species)

atranorin

arabitol, mannitol

Buellia (8 species)

diploicin
atranorin, chloroatranorin, norstictic acid
usnic acid

calycin, rhizocarpic acid
mannitol

Heterodermia (84 taxa)

atranorin, chloroatranorin, norstictic acid, salazinic acid

emodin, 1,3,8-trihydroxy-2-chloro-6-methylanthraquinone,
1,3,8-trihydroxy-2,4-dichloro-6-methylanthraquinone
zeorin
D-arabitol

Physcia (24 taxa)

divaricatic acid, ramalinolic acid, sekikaic acid
atranorin

zeorin
volemitol
endococcin, rhodophyscin

Pyxine (4 species)

lichexanthone
pyxinic acid
pyxiferin

Rinodina (3 species)

gyrophoric acid
atranorin, fumarprotocetraric acid, norstictic acid
(+)-usnic acid

zeorin

Tornabenia (3 species)

zeorin

Teloschistaceae

Caloplaca (13 species)

parietin
calycin, pulvinic dilactone

Fulgensia (1 species)

parietin

Polycauliona (1 species)

parietin

Teloschistes (5 taxa)

vicanicin

fallacinal, parietin, teloschistin

Xanthoria (7 taxa)

atranorin

fallacinal, parietin, parietinic acid, teloschistin, xanthorin
β-carotene, violoxanthin, xanthophyll
isolichenin, lichenin, D-mannitol
chlorophyll *a*, chlorophyll *b*, glyceric acid

ORDER SPHAERIALES

Verrucariaceae

Dermatocarpon (5 species)

ergosterol
D-mannitol, sucrose, trehalose, volemitol

Endocarpon (1 species)

mannitol, volemitol

ORDER CALICIALES

Caliciaceae

Calicium (3 species)

rhizocarpic acid, vulpinic acid

Chaenotheca (2 species)

calycin, vulpinic acid

Coniocybe (1 species)

rhizocarpic acid, vulpinic acid

Cypheliaceae

Cyphelium (2 species)

rhizocarpic acid

Sphaerophoraceae

Acrocyphus

gyrophoric acid, lecanoric acid
atranorin
usnic acid

chrysophanol, rugulosin, skyrin
calycin

Sphaerophorus (6 species)

sphaerophorin
squamatic acid

fragilin
arabitol, mannitol
ascorbic acid

SUBCLASS LOCULOASCOMYCETIDAE

ORDER MYRANGIALES

Arthoniaceae

Arthonia (6 species)

usnic acid

Arthothelium

arthothelin

ORDER PLEOSPORALES

Arthopyreniaceae

Laurera

xanthorin

ORDER HYSTERIALES

Opegraphaceae

Chiodecton (2 species)

erythrin, gyrophoric acid, lecanoric acid
psoromic acid

Enterographa

confluentic acid

Roccellaceae

Combea

erythrin

Darbishirella (1 species)

psoromic acid

Dendrographa (1 species)

fumarprotocetraric acid

Dirina (2 species)

> erythrin, lecanoric acid
>
> roccellic acid
> erythritol
> portentol

Dirinastrum

> erythrin

Dolichocarpus (1 species)

> gyrophoric acid
>
> carotenes

Hubbsia (1 species)

> lecanoric acid

Ingaderia (1 species)

> erythrin, lecanoric acid
>
> erythritol

Pentagenella

> psoromic acid

Reinkella

> schizopeltic acid

Roccella (14 species)

> erythrin, lecanoric acid
> norstictic acid, protocetraric acid, psoromic acid
>
> (+)-montagnetol, (±)-montagnetol, orcinol
> aspicilin, roccellic acid, tetrahydroxy fatty acids
> β-carotene, γ-carotene, ergosterol
> arabitol, erythritol, galactose, isolichenin, lichenin, mannitol,
>   tagatose, unidentified di- and trisaccharides
> acetylportentol, ascorbic acid, choline sulfate ester, lepraric
>   acid, picroroccellin, portentol, riboflavin, unknown
>   compound (m.p. 129–130°), unknown nitrogen-containing
>   compound (m.p. 308–310°)

Roccellaria (1 species)

roccellaric acid
zeorin˙
erythritol
roccellin, mollin

Roccellina

lecanoric acid
schizopeltic acid

(+)-roccellic acid
portentol, unknown compounds (m.p. 177–178°; m.p. 295°)

Roccellinastrum

protocetraric acid

Roccellodea

psoromic acid

Roccellographa

psoromic acid

Sagenidium

lepraric acid

Schizopelte

schizopeltic acid

Simonyella

simonyellin

Lepraria (7 species)

atranorin
pannaric acid

roccellic acid
zeorin
calycin, leprapinic acid, leprapinic acid methyl ether, pinastric acid, vulpinic acid
arabitol, mannitol
unknown substances (m.p. 176–177° and m.p. 239–240°),
unknown substances (m.p. 276–277° and m.p. 286–288°)

Genera of Uncertain Position

Byssocaulon (1 species)

> lecanoric acid

Chrysothrix (1 species)

> calycin

Herpothallon (1 species)

> confluentic acid

> chiodectonic acid

Chapter V

Botanical Guide to Lichen Products

For an explanation of the organization of material in this chapter, see pages 8–12. For the abbreviations of the names of the chemical tests, see page 10.

Acarospora chlorophana (Wahlenb. *ex* Ach.) Mass. Acaranoic acid, acarenoic acid; rhizocarpic acid.

Acaranoic acid, acarenoic acid. Bendz, Santesson, and Tibell (1966), by preparative-layer chromatography, mixed acids identical to pleopsidic acid; Salkowski (1910), as pleopsidic acid.

Rhizocarpic acid. Asahina (1959*a*); Bendz, Santesson, and Wachtmeister (1965*d*), used as an authentic source of this substance.

REVIEWS. Acaranoic acid and acarenoic acid (as pleopsidic acid), rhizocarpic acid: W. Brieger (1923), and as *Pleopsidium chlorophanum*; Hesse (1912), and as *P. chlorophanum*; Thies (1932), as *P. chlorophanum*; Zopf (1907), and as *P. chlorophanum*.

Acarospora chrysops (Tuck.) Magn. Rhizocarpic acid. Hale (1957*b*), mention only.

Acarospora fuscata (Schrad.) Arn. Gyrophoric acid, unknown brownish-purple pigment.

Gyrophoric acid. Huneck (1962*b*), from Germany, by extraction.

Unknown brownish-purple pigment. Hale (1958*e*), mention only.

OTHER REPORT. Water-soluble pigments produced by fungus in pure culture: Diner, Ahmadjian, and Rosenkrantz (1964).

Acarospora gobiensis Magn. Rhizocarpic acid. Asahina (1959*a*), from the Gobi Desert, China.

Acarospora montana Magn. Gyrophoric acid; copper, iron.

Gyrophoric acid. Huneck (1966*a*), from Germany, 0.4% by extraction of 15.0 g. of lichen.

Copper, iron. Lange and Ziegler (1963), a quantitative study.

Acarospora oxytona (Ach.) Mass. Rhizocarpic acid. Hale (1957*b*), mention only.

Acarospora sinopica (Wahlenb. *ex* Ach.) Körb. Copper, iron. Lange and Ziegler (1963), a quantitative study.

Acarospora smaragdula (Wahlenb. *ex* *A*ch.) Mass. Usnic acid. J. C. Mitchell (1965), mention only.

Acarospora smaragdula var. *lesdainii* f. *subochracea* Magn. Copper, iron. Lange and Ziegler (1963), a quantitative study.

Acolium tigillare (Ach.) S. Gray, see *Cyphelium tigillare* (Ach.) Ach.

Actinogyra muehlenbergii (Ach.) Schol. Gyrophoric acid. Hale (1956*e*), from North America, as *Umbilicaria muehlenbergii*; Hale (1957*c*), from West Virginia, an exsiccati specimen.
    REVIEW. Parmelia-brown: Thies (1932), as *Gyrophora muehlenbergii*.

Actinogyra polyrrhiza (L.) Schol. Gyrophoric acid, lecanoric acid, umbilicaric acid; arabitol, mannitol.
    Gyrophoric acid. Hale (1956*e*), from North America, as *Umbilicaria polyrrhiza*.
    Arabitol, mannitol. Lindberg, Misiorny, and Wachtmeister (1953), by paper chromatography, no volemitol, as *Umbilicaria polyrrhiza*.
    REVIEWS. Gyrophoric acid, lecanoric acid, umbilicaric acid: Thies (1932); Zopf (1907); both reports as *Gyrophora polyrrhiza*. — Gyrophoric acid, umbilicaric acid: W. Brieger (1923); Hesse (1912); both reports as *G. polyrrhiza*.

Agyrophora lyngei (Schol.) Llano Gyrophoric acid. Hale (1956*e*), from North America, as *Umbilicaria lyngei*.

Agyrophora microphylla (Laur.) Llano
    REVIEW. Parmelia-brown: Thies (1932), as *Gyrophora microphylla*.

Agyrophora rigida (Du Rietz) Llano Gyrophoric acid; arabitol, mannitol, sucrose, trehalose, umbilicin.
    Gyrophoric acid. Hale (1956*e*), from North America, as *Umbilicaria rigida*.
    Arabitol, mannitol, sucrose, trehalose, umbilicin. Lindberg, Misiorny, and Wachtmeister (1953), by paper chromatography and column chromatography, no volemitol, as *Umbilicaria rigida*.
    REVIEWS. Parmelia-brown: Thies (1932), as *Gyrophora anthracina*. — Sucrose, trehalose, umbilicin: Shibata (1958), as *Umbilicaria rigida*. — Umbilicin: Shibata (1963), as *U. rigida*.

Alectoria articulata (L.) Link, see *Usnea articulata* (L.) Hoffm.

Alectoria bicolor (Ehrh.) Nyl.   Unidentified substance (PD+ red). Dahl (1950), from Greenland, PD+ red substance also found in *Alectoria lanea*.

Alectoria cana (Ach.) Leight.   Barbatolic acid (sometimes as alectoric acid). Asahina (1940*d*), as *Alectoria implexa* var. *cana*; Kurokawa (1959*c*), from Japan, as *A. implexa* var. *cana*, by microcrystal tests.
   OTHER REPORT. Alectoronic acid, atranorin, psoromic acid, salazinic acid: Solberg (1956), mention only.
   REVIEWS. Barbatolic acid, bryopogonic acid: W. Brieger (1923), and as *Alectoria jubata* var. *cana*; Hesse (1912), and as *A. jubata* var. *cana*; Thies (1932); Zopf (1907), and as *Bryopogon canus*.

Alectoria canariensis Ach., see *Usnea canariensis* (Ach.) Du Rietz

Alectoria chalybeiformis (L.) S. Gray   Unidentified substance (PD+ red); D-arabitol; ascorbic acid; enzymes.
   Unidentified substance (PD+ red). Lamb (1964), from Antarctica, soredia PD+.
   D-Arabitol. Solberg (1955), from Norway, as *Alectoria jubata* var. *chalybeiformis*.
   Ascorbic acid. Karev and Kochevykh (1962).
   Enzymes. Moissejeva (1961), enzymes tabulated, as *Bryopogon chalybeiformis*.

Alectoria crinalis Ach., see *Ramalina crinalis* (Ach.) Gyeln.

Alectoria divergens (Ach.) Nyl., see *Cornicularia divergens* Ach.

Alectoria divergens var. *satoana* (Gyeln.) Asah., see *Cornicularia pseudo-satoana* Asah.

Alectoria fremontii Tuck.   Vulpinic acid. Hale (1957*b*), mention only.

Alectoria implexa (Hoffm.) Nyl.   Barbatolic acid; arabitol, mannitol; enzymes.
   Barbatolic acid. Asahina (1940*d*), as *Alectoria implexa* f. *fuscidula*; J. Santesson (1965); Suominen (1939), from Karelo-Finnish S.S.R., 3% by extraction, as *A. implexa* f. *fuscidula*.
   Arabitol, mannitol. Lindberg, Misiorny, and Wachtmeister (1953), by paper chromatography, no volemitol.
   Enzymes. Moissejeva (1961), enzymes tabulated.

OTHER REPORTS, Atranorin: Vartia (1949), from Finland. — Barbatolic acid, bryopogonic acid: D. Hess (1958). — Psoromic acid: Asahina (1936*b*) (see *Alectoria zopfii*).
REVIEWS. Atranorin, psoromic acid (as parellic acid): W. Brieger (1923); Hesse (1912); Thies (1932); Zopf (1907), and as *Bryopogon implexus*. — Barbatolic acid: Asahina (1951*i*), and also *Alectoria implexa* f. *fuscidula*; Asahina and Shibata (1954). — Bryopogonic acid: Zopf (1907), as *A. jubata* var. *implexa*. — Psoromic acid: Asahina and Shibata (1954) (see *Alectoria zopfii*).

Alectoria implexa var. *cana* (Ach.) Flag., see *Alectoria cana* (Ach.) Leight.

Alectoria japonica Tuck., see *Alectoria laeta* (Tayl.) Linds.

Alectoria jubata (L.) Ach.   Bryopogin, lichenin, hemicellulose, unidentified reducing sugar; protein; folic acid-, folinic acid-, and vitamin B_{12}-group factors; calcium, phosphorus.
Bryopogin. Votoček and Burda (1926), a polysaccharide of D-glucose, galactose, and mannose, other polysaccharide fractions discussed, as *Bryopogon jubatus*.
Lichenin, hemicellulose, unidentified reducing sugar. Yanovsky and Kingsbury (1938).
Protein, fat, calcium, phosphorus. Scotter (1965), from northern Canada.
Folic acid-, folinic acid-, and vitamin B_{12}-group factors. Sjöström and Ericson (1953).

Alectoria jubata var. *cana* (Ach.) Flag., see *Alectoria cana* (Ach.) Leight.

Alectoria jubata var. *chalybeiformis* (L.) G. Web., see *Alectoria chalybeiformis* (L.) S. Gray.

Alectoria jubata var. *implexa* (Hoffm.) Ach., see *Alectoria implexa* (Hoffm.) Nyl.

Alectoria jubata var. *subcana* Nyl., see *Alectoria subcana* (Nyl.) Gyeln.

Alectoria laeta (Tayl.) Linds.   Alectoronic acid, (−)-usnic acid. Asahina (1936*b*); Asahina and Hashimoto (1933), from Japan, by extraction, as *Alectoria japonica*.
Alectoronic acid, usnic acid. Asahina (1938*d*), mention in description of microcrystal tests; Satô (1934), from Sakhalin, as *Alectoria laeta* f. *subfibrillosa*.
Usnic acid. Asahina (1935*c*), as *Alectoria japonica*.

REVIEWS. Alectoronic acid, usnic acid: Asahina and Shibata (1954), from Japan. — Alectoronic acid: Shibata (1958, 1963). All review reports are as *Alectoria japonica*.

Alectoria lanea (Ehrh.) Vain. Unidentified substance (PD+ red). Dahl (1950), from Greenland and Iceland, PD+ red substance also found in *Alectoria bicolor*.

Alectoria minuscula Nyl.
Probably no extracellular lichen substances. Lamb (1964).

Alectoria nidulifera Norrl. Fumarprotocetraric acid. Kurokawa (1959*c*), from Japan.

Alectoria nigricans (Ach.) Nyl. Unidentified substance (=alectorialic acid?). D. Hess (1958), from Germany; Lamb (1964), by microchemical tests.
REVIEWS. Alectorialic acid: W. Brieger (1923); Hesse (1912); Thies (1932); Zopf (1907).

Alectoria ochroleuca (Hoffm.) Mass. Diffractaic acid, (−)-usnic acid; tetrahydroxy fatty acids; friedelin; D-arabitol; enzymes.
Diffractaic acid, (−)-usnic acid. Stoll, Brack, and Renz (1950).
Diffractaic acid, usnic acid. D. Hess (1958), from Germany.
(−)-Usnic acid. Moissejeva (1957), 4.0%; Moissejeva (1961), 4.0%, enzymes tabulated; Savicz, Kuprevicz, Litvinov, Moissejeva, and Rassadina (1956), 4% by extraction; Stoll, Brack, and Renz (1947), 8% by extraction; Stoll, Renz, and Brack (1947), from Switzerland.
Usnic acid. Asahina (1935*c*); Steiner (1955); Tomaselli (1957), from Italy.
Tetrahydroxy fatty acids. Solberg (1960).
Friedelin. Bruun (1954*a*), from Norway, by extraction.
OTHER REPORTS. Barbatic acid, usnic acid, linoleic acid, oleic acid, erythritol (probably really D-arabitol according to Asahina [1939*b*] and Pueyo [1960, 1965]), D-glucose, choline, an aliphatic carboxylic acid: Klima (1933), by extraction (but see also Solberg [1957, 1960] and Steiner [1957]). — Barbatic acid: Schulte (1905), by a microcrystal test on sections of strands. — Diffractaic acid, (+)-usnic acid: Asahina (1936*b*); Asahina (1936*e*); Asahina and Hashimoto (1933), from Hokkaido, by extraction; Satô (1934).
REVIEWS. Diffractaic acid, D-arabitol: Asahina and Shibata (1954). — Diffractaic acid, friedelin: Shibata (1958, 1963). — Barbatic acid, (−)-usnic acid: W. Brieger (1923), as *Alectoria rigida*; Hesse (1912), as *A. rigida*; Savicz, Litvinov, and Moissejeva (1960); Thies (1932), as *A. rigida*; Zopf (1907), and as *A. rigida*.

Alectoria pubescens (L.) R. H. Howe
No substances known. Lamb (1964).

Alectoria rigida (Fr.) Dalla Torre & Sarnth., see *Alectoria ochroleuca* (Hoffm.) Mass.

Alectoria sarmentosa (Ach.) Ach.   Alectoronic acid, (−)-usnic acid; arabitol, mannitol.
Alectoronic acid, (−)-usnic acid. Asahina and Hashimoto (1933), from Europe, by extraction.
Alectoronic acid, usnic acid. Asahina (1938*d*), mention in description of microcrystal tests.
Alectoronic acid. Satô (1934), from Japan.
Usnic acid. Hale (1957*b*); Tomaselli (1957), from Italy; Zopf (1900), 2.9% by extraction.
Arabitol, mannitol. Lindberg, Misiorny, and Wachtmeister (1953), by paper chromatography, no volemitol.
REVIEWS. Alectoronic acid: Asahina and Shibata (1954), from Europe; Shibata (1958, 1963). — (−)-Usnic acid: W. Brieger (1923); Hesse (1912); Thies (1932); Zopf (1907).

Alectoria satoana Gyeln., see *Cornicularia divergens* Ach., *Cornicularia pseudosatoana* Asah., and *Cornicularia satoana* (Gyeln.) Asah.

Alectoria subcana (Nyl.) Gyeln.   Fumarprotocetraric acid.
Asahina (1940*d*), as *Alectoria jubata* var. *subcana*.

Alectoria sulcata (Lév.) Nyl.   Atranorin, psoromic acid (sometimes as sulcatic acid). Asahina (1938*e*), by microcrystal tests; Asahina and Hayashi (1928); Asahina and Hayashi (1933), mention only.
Psoromic acid. Kurokawa (1952); Mitsuno (1953), by paper chromatography; Kurokawa (1964), mention only.
REVIEWS. Atranorin, psoromic acid: Asahina and Shibata (1954); Shibata (1958, 1963); Thies (1932).

Alectoria tortuosa Merr.   Virensic acid; vulpinic acid; D-arabitol. Aghoramurthy, Sarma, and Seshadri (1961*a*), from India, structural analysis of the new depsidone named virensic acid.
Vulpinic acid. Asahina (1955*a*), by microchemical tests.
OTHER REPORT. Atranorin, psoromic acid: Dhar, Neelakantan, Ramanujam, and Seshadri (1959), from India, by extraction of 20 g. of lichen.
REVIEW. Virensic acid: Neelakantan (1965).
All reports are as *Alectoria virens*.

Alectoria virens Tayl., see *Alectoria tortuosa* Merr.

Alectoria zopfi Asah.   Psoromic acid. Asahina (1940*d*).
REVIEWS. Psoromic acid: Asahina and Shibata (1954); Shibata (1958, 1963).

Allarthonia caesia Flot., see *Arthonia caesia* (Flot.) Körb.

Anaptychia albicans* Kurok.   Atranorin; zeorin; unidentified yellow pigment (K−). Kurokawa (1962*a*), from South America, type description, yellow pigment on undersurface also found in *Anaptychia lutescens.*

Anaptychia albidiflava* Kurok.   Atranorin, unidentified substance (PD + yellow); unidentified anthraquinone; zeorin. Kurokawa (1962*a*), from India, type description, unidentified PD + substance also found in *Anaptychia dissecta.*

Anaptychia albopruinosa* Kurok.   Atranorin; zeorin. Kurokawa (1962*a*), from Africa, type description.

Anaptychia allardii* Kurok.   Atranorin, norstictic acid, salazinic acid; zeorin. Kurokawa (1962*a*), from Central and South America and the West Indies.

Anaptychia angustiloba* (Müll. Arg.) Kurok.   Atranorin, norstictic acid, salazinic acid; zeorin. Kurokawa (1962*a*), from Asia and Australia.

Anaptychia appalachensis Kurok., see *Heterodermia appalachensis* (Kurok.) W. Culb.

Anaptychia appendiculata* Kurok.   Atranorin; zeorin. Kurokawa (1962*a*), from Africa, type description.

Anaptychia arsenei* Kurok.   Atranorin, salazinic acid; zeorin. Kurokawa (1962*a*), from Mexico, type description.

Anaptychia awasthii* Kurok.   Atranorin, norstictic acid, salazinic acid; zeorin. Kurokawa (1962*a*), from Asia, type description.

Anaptychia barbifera* (Nyl.) Trev.   Atranorin, norstictic acid, salazinic acid; zeorin. Kurokawa (1962*a*), from Central and South America.

Anaptychia casarettiana Mass., see *Heterodermia casarettiana* (Mass.) Trev.

* Species of the genus *Heterodermia* which have not yet been formally transferred from *Anaptychia* are listed here under *Anaptychia* and marked with an asterisk. Those species for which the combinations in *Heterodermia* have been published are cross-referenced to that genus.

Anaptychia chilensis* Kurok. Atranorin; zeorin. Kurokawa (1962*a*), from Chile, type description.

Anaptychia ciliaris (L.) Körb.   Atranorin ($\pm$); arabitol, mannitol.
    Atranorin. Dhar, Neelakantan, Ramanujam, and Seshadri (1959), from Kashmir, by extraction; Vartia (1949), from Finland.
    Arabitol, mannitol. Lindberg, Misiorny, and Wachtmeister (1953), by paper chromatography, no volemitol.
    No lichen substances found. Galun and Lavee (1966), from Israel; Kurokawa (1962*a*), from Europe, as *Anaptychia ciliaris* f. *ciliaris, A. c. f. agriopa, A. c. f. melanosticta, A. c. f. nigrescens,* and *A. c. f. verrucosa.*
    REVIEWS. Atranorin: W. Brieger (1923); Hesse (1912); both reports as *Physcia ciliaris*; Zopf (1907). — Atranorin, parmelia-brown: Thies (1932), as *P. ciliaris.*

Anaptychia comosa* (Eschw.) Mass. Atranorin; zeorin; unidentified pigment ($\pm$). Kurokawa (1962*a*), from Asia and Central and South America.
    See also *Heterodermia galactophylla.*

Anaptychia congoensis* Kurok. Atranorin; zeorin. Kurokawa (1962*a*), from Africa, type description.

Anaptychia corallophora* (Tayl.) Lynge   Atranorin; zeorin; unidentified pale yellow pigment (K + yellow). Kurokawa (1962*a*), from Central and South America and the West Indies, pale yellow pigment on under-surface.
    Atranorin, zeorin. Hale (1957*b*), mention only.

Anaptychia coronata* Kurok. Atranorin, salazinic acid; zeorin. Kurokawa (1962*a*), from India, type description.

Anaptychia cubensis* Kurok. Atranorin, norstictic acid, salazinic acid; zeorin. Kurokawa (1962*a*), from Central and South America and Cuba, type description.

Anaptychia dactyliza* (Nyl.) Zahlbr.   Atranorin; zeorin. Kurokawa (1962*a*), from Africa and South America, also *Anaptychia dactyliza* f. *serpens* with an "almost entirely blackened thallus."

Anaptychia dendritica (Pers.) Vain., see *Heterodermia dendritica* (Pers.) Poelt

Anaptychia dendritica var. *propagulifera* Vain., see *Heterodermia dendritica* (Vain.) Poelt

Anaptychia diademata* (Tayl.) Kurok. Atranorin; zeorin. Kurokawa (1959*b*), from Africa, Asia, and Central and South America, as *Anaptychia esorediata*, and an unidentified substance also found in *Heterodermia hypoleuca;* Kurokawa (1962*a*), from Africa, Asia, Mexico, and South America, and also *A. diademata* f. *angustata* from Japan, *A. d.* f. *brachyloba* from Asia, and *A. d.* f. *condensata* from Formosa, Japan, and the Philippines.

Atranorin. Asahina (1939*d*), as *Anaptychia speciosa* f. *compactior* from Japan and as *Chaudhuria indica*, atranorin apparently in cortex and medulla.

Anaptychia dissecta* Kurok. Atranorin, norstictic acid, salazinic acid, unidentified substance (PD + deep yellow); zeorin. Kurokawa (1962*a*), from India and Japan, unidentified substance determined by microcrystal tests.

OTHER REPORT. Atranorin, norstictic acid, salazinic acid, unidentified substance A, unidentified substance B (PD + deep yellow): Kurokawa (1959*b*), the substance A also found in *Heterodermia hypoleuca.*

Anaptychia dissecta var. *koyana** Kurok. Atranorin; zeorin; unidentified substance. Kurokawa (1959*b*), from Formosa and Japan; Kurokawa (1962*a*), from Central America, Formosa, India, and Japan.

Anaptychia domingensis (Ach.) Mass., see *Heterodermia domingensis* (Ach.) Trev.

Anaptychia echinata (Tayl.) Kurok., see *Heterodermia echinata* (Tayl.) W. Culb.

Anaptychia echinata var. *pterocarpa* Kurok., see *Heterodermia echinata* (Tayl.) W. Culb.

Anaptychia erinacea* (Ach.) Trev. Atranorin; zeorin. Kurokawa (1962*a*), from Baja California and coastal California, and also *Anaptychia erinacea* f. *cilatomarginata.*

Anaptychia esorediata (Vain.) Du Rietz & Lynge, see *Anaptychia diademata* (Tayl.) Kurok.

Anaptychia fauriei* Kurok. Atranorin; zeorin; two unidentified pigments. Kurokawa (1962*a*), from Hawaii, type description, yellow pigment on undersurface K + yellow, a second pigment K + violet.

Anaptychia firmula* (Nyl.) Dodge & Awas. Atranorin; unidentified yellow pigment (K + purple); zeorin. Kurokawa (1962*a*), from India and southwestern China, unidentified yellow pigment probably an anthraquinone.

Anaptychia flabellata* (Fée) Mass.   Atranorin; unidentified yellow pigment (K + purple); zeroin.   Kurokawa (1960*a*), from Central and South America, Ceylon, Fiji, and the Philippines, yellow pigment on undersurface also found in *Heterodermia obscurata*, as *Anaptychia fulvescens* and as *A. fulvescens* var. *rottbollii* from Formosa, the Philippines, South America, and Yunnan; Kurokawa (1962*a*), from Africa, Asia, and Central and South America, and also *A. flabellata* var. *rottbollii* from Asia and South America, differing by the distribution of pigment.

Anaptychia flabellata var. *corcovadoensis** Kurok.   Atranorin, norstictic acid, salazinic; zeorin; unidentified yellow pigment. Kurokawa (1962*a*), from Brazil, type description.

Anaptychia fragilissima* Kurok.   Atranorin; zeorin; unidentified brownish pigment (±). Kurokawa (1962*a*), from eastern Asia, type description.

Anaptychia fulvescens (Vain.) Kurok., see *Anaptychia flabellata* (Fée) Mass.

Anaptychia fulvescens var. *rottbollii* (Vain.) Kurok., see *Anaptychia flabellata* (Fée) Mass.

Anaptychia fusca (Huds.) Vain.   Atranorin; arabitol, mannitol.
    Atranorin. Kurokawa (1962*a*), from Europe, a small amount found microchemically.
    Arabitol, mannitol. Lindberg, Misiorny, and Wachtmeister (1953), by paper chromatography, no volemitol.

Anaptychia galactophylla (Tuck.) Trev., see *Heterodermia galactophylla* (Tuck.) W. Culb.

Anaptychia granulifera (Ach.) Mass., see *Heterodermia granulifera* (Ach.) W. Culb.

Anaptychia heterochroa Vain., see *Heterodermia obscurata* (Nyl.) Trev.

Anaptychia himalayensis* Awas.   Atranorin, norstictic acid, salazinic acid; zeorin. Kurokawa (1962*a*), from Asia.

Anaptychia hypocaesia* Yas.   Atranorin, salazinic acid; zeorin; unidentified yellow substance. Kurokawa (1960*a*), from India, Japan, and the Philippines, yellow pigment on undersurface also found in *Heterodermia obscurata*.
    Atranorin, salazinic acid, zeorin. Kurokawa (1962*a*), from India, Japan, and Hawaii.

Anaptychia hypochraea* Vain. Atranorin; unidentified yellow pigment (K + purple); zeorin. Kurokawa (1962*a*), from Asia and South America, unidentified pigment probably also found in *Heterodermia obscurata* and *Anaptychia flabellata*.

Anaptychia hypoleuca (Ach.) Mass., see *Heterodermia hypoleuca* (Ach.) Trev.

Anaptychia hypoleuca var. *colorata* Zahlbr., see *Heterodermia obscurata* (Nyl.) Trev.

Anaptychia hypoleuca var. *sorediifera* (Müll. Arg.) Vain., see *Heterodermia obscurata* (Nyl.) Trev.

Anaptychia incana* (Stirt.) Zahlbr. Atranorin; zeorin. Kurokawa (1962*a*), from Asia.

Anaptychia indica* Magn. Atranorin; zeorin. Kurokawa (1962*a*), from India.

Anaptychia intricata (Desf.) Mass., see *Tornabenia intricata* (Desf.) Kurok.

Anaptychia isidiophora* (Nyl.) Vain. Atranorin; zeorin; unidentified substance. Kurokawa (1959*b*), from Africa, Japan, and Manchuria, unidentified substance also found in *Heterodermia hypoleuca* by micro-crystal tests.
Atranorin, zeorin. Kurokawa (1962*a*), from Africa, Asia, and Central and South America.

Anaptychia isidiza Kurok. No lichen substances found. Kurokawa (1962*a*), from Asia.

Anaptychia japonica* (Satô) Kurok. Atranorin; zeorin. Kurokawa (1960*c*), from Formosa and Japan; Kurokawa (1962*a*), from Africa and eastern Asia.

Anaptychia japonica var. *reagens** Kurok. Atranorin, norstictic acid, salazinic acid; zeorin. Kurokawa (1960*c*), from Japan, salazinic acid as an accessory substance; Kurokawa (1962*a*), from Africa and Japan.

Anaptychia kaspica Gyeln. Calcium oxalate. Kurokawa (1962*a*), from Asia, Europe, and North America, no other substances found.

Anaptychia lamelligera* (Tayl.) Kurok. Atranorin; zeorin; unidentified yellow pigment (K + yellow). Kurokawa (1962*a*), from Central America and the West Indies, yellow pigment on undersurface.

Anaptychia "leucomelaena" auct., see *Heterodermia leucomela* (L.) Poelt

Anaptychia leucomelaena f. *albociliata* (Nyl.) Hue, see *Heterodermia leucomela* (L.) Poelt

Anaptychia leucomelaena var. *angustifolia* Meyen & Flot., see *Heterodermia leucomela* (L.) Poelt

Anaptychia leucomelaena f. *palmiformis* Kurok., see *Heterodermia leucomela* (L.) Poelt

Anaptychia leucomelaena f. *verrucifera* Kurok., see *Heterodermia leucomela* (L.) Poelt

Anaptychia loriformis* Kurok.   Atranorin; unidentified yellow pigment (K + purple); zeorin. Kurokawa (1962a), from East Africa, type description, yellow pigment found on undersurface.

Anaptychia lutescens* Kurok.   Atranorin; zeorin; unidentified yellow pigment (K + yellow). Kurokawa (1961), from Africa, Central and South America, China, and Formosa, yellow pigment on undersurface, type description. Kurokawa (1962a), from Africa, Central and South America, China, Formosa, Jamaica, and Hawaii, unidentified yellow pigment on undersurface probably a pulvinic acid derivative.

Anaptychia magellanica* Zahlbr.   Atranorin; zeorin. Kurokawa (1962a), and also *Anaptychia magellanica* var. *pectinata* from Mexico and *A. magellanica* var. *pectinata* f. *subisidiosa* from Mexico.

Anaptychia microphylla* (Kurok.) Kurok.   Atranorin, norstictic acid (±), salazinic acid; zeorin. Kurokawa (1962a), from Japan, and also *Anaptychia microphylla* f. *granulosa*, from Japan and Korea.

Anaptychia multiciliata* Kurok.   Atranorin; zeorin. Kurokawa (1962a), from South America, type description.

Anaptychia neoleucomelaena* Kurok.   Chloroatranorin; zeorin. Huneck and Follmann (1966d), from Chile, 1.6% chloroatranorin, 0.18% zeorin, by extraction of 27.0 g. of lichen.
OTHER REPORTS. Atranorin, zeorin: Kurokawa (1961), from Africa Asia, Fiji, Mexico, and South America, type description; Kurokawa (1962a), from Africa, Burma, Fiji, Formosa, Japan, Korea, Mexico, the Philippines, and South America, and also *Anaptychia neoleucomelaena* f. *circinalis* from Ecuador and Jamaica, *A. n.* f. *sorediosa* from the Congo, India, and Japan, and *A. n.* f. *squarrosa* from Africa, Asia, Costa Rica, Mexico, and Hawaii.

Anaptychia obesa* (Pers.) Zahlbr. Atranorin; unidentified orange pigment (K + violet); zeorin. Kurokawa (1962*a*), from Hawaii, orange pigment on undersurface.

Anaptychia obscurata (Nyl.) Vain., see *Heterodermia obscurata* (Nyl.) Trev.

Anaptychia ophioglossa (Tayl.) Kurok., see *Heterodermia leucomela* (L.) Poelt

Anaptychia ophioglossa f. *albociliata* (Nyl.) Kurok., see *Heterodermia leucomela* (L.) Poelt

Anaptychia palmulata (Michx.) Vain. Atranorin ($\pm$). W. L. Culberson (1966*a*), from North Carolina and South Carolina, by microchemical tests on 36 samples.
No lichen substances found. Hale (1956*h*), from the eastern U.S.A.; Hale (1957*a*), from Arkansas and Oklahoma; Kurokawa (1962*a*), from Asia and North America.

Anaptychia palpebrata* (Tayl.) Vain. Atranorin; zeorin. Kurokawa (1962*a*), from South America.

Anaptychia pandurata* Kurok. Atranorin, unidentified substance (PD + deep yellow), unidentified yellow pigment (K + violet); zeorin. Kurokawa (1962*a*), from Asia, type description, PD + unknown also in *Anaptychia dissecta*, yellow pigment on undersurface.

Anaptychia pellucida* Awas. Atranorin; zeorin. Kurokawa (1962*a*), from Asia.

Anaptychia podocarpa* (Bél.) Mass. Atranorin, norstictic acid, salazinic acid; zeorin. Kurokawa (1962*a*), from Africa, Asia, the Americas, the West Indies, and Hawaii, reddish color of lichen possibly caused by decomposition of salazinic acid.

Anaptychia polyrhiza* Kurok. Atranorin; zeorin. Kurokawa (1962*a*), from Central America, type description.

Anaptychia pseudospeciosa Kurok., see *Heterodermia pseudospeciosa* (Kurok.) W. Culb.

Anaptychia pseudospeciosa var. *inactiva* Kurok., see *Heterodermia tremulans* (Müll. Arg.) W. Culb.

Anaptychia pseudospeciosa f. *tagawae* Kurok., see *Heterodermia pseudospeciosa* (Kurok.) W. Culb.

Anaptychia pseudospeciosa var. *tremulans* (Müll. Arg.) Kurok., see *Heterodermia tremulans* (Müll. Arg.) W. Culb.

*Anaptychia punctifera** Kurok.   Atranorin, norstictic acid, salazinic acid; zeorin. Kurokawa (1962a), from India, type description.

Anaptychia ravenelii (Tuck.) Zahlbr., see *Heterodermia domingensis* (Ach.) Trev.

*Anaptychia rubescens** (Räs.) Kurok.   Atranorin, norstictic acid, salazinic acid; zeorin. Kurokawa (1962a), from India.

*Anaptychia rugulosa** Kurok.   Atranorin; unidentified yellow pigment (K + purple); zeorin. Kurokawa (1962a), from Mexico, yellow pigment in the medulla, and also *Anaptychia rugulosa* var. *isidiosa* from Mexico, type descriptions.

Anaptychia sorediifera (Müll. Arg.) Du Rietz & Lynge, see *Heterodermia obscurata* (Nyl.) Trev.

*Anaptychia speciosa** (Wulf.) Mass.   Atranorin; zeorin. Asahina (1938f), mention in description of microcrystal tests; Asahina and Yosioka (1940); Kurokawa (1959b), from Europe, and an unidentified substance also in *Heterodermia hypoleuca*; Kurokawa (1962a), from Europe, specimens outside Europe referred to *H. pseudospeciosa*; Neelakantan, Rangaswami, and Rao (1954), from Sikkim, 3.3% atranorin, 0.75% zeorin, by extraction.

OTHER REPORT. Atranorin, stictic acid, zeorin: Hale (1957a), from Arkansas, Missouri, and Oklahoma, by microcrystal tests.

REVIEWS. Atranorin, zeorin: Asahina and Shibata (1954); W. Brieger (1923), and as *Physcia speciosa*; Hesse (1912), and as *P. speciosa;* Shibata (1958, 1963); Thies (1932), as *P. speciosa;* Zopf (1907).

Anaptychia speciosa f. *compactior* Zahlbr., see *Anaptychia diademata* (Tayl.) Kurok.

Anaptychia speciosa var. *microspora* Kurok., see *Heterodermia tremulans* (Müll. Arg.) W. Culb.

*Anaptychia spinigera** Kurok.   Atranorin; zeorin. Kurokawa (1962a), from Peru, type description.

*Anaptychia spinulosa** Kurok.   Atranorin, unidentified substance (PD + deep yellow); zeorin. Kurokawa (1962a), from Asia, type description, unidentified PD + substance also found in *Anaptychia dissecta*.

Anaptychia squamulosa Degel., see *Heterodermia squamulosa* (Degel.) W. Culb.

Anaptychia stellata* (Vain.) Kurok. Atranorin; zeorin. Kurokawa (1962*a*), from South America and the West Indies, and also *Anaptychia stellata* var. *squarrolosa* from Brazil.

Anaptychia stippaea (Ach.) Nádv. No lichen substances found. Kurokawa (1962*a*), from Europe, microchemical tests negative.

Anaptychia subaquila (Nyl.) Kurok. No lichen substances found. Kurokawa (1962*a*), from Europe, microchemical tests negative.

Anaptychia subascendens* Asah. Atranorin; unidentified yellow pigment (K + purple); zeorin. Asahina (1958*f*), from Japan, type description; Kurokawa (1962*a*), from Formosa and Japan, yellow pigment on undersurface.

Anaptychia subcomosa* (Nyl.) Trev. Atranorin; zeorin. Kurokawa (1962*a*), from Central and South America.

Anaptychia subheterochroa Kurok., see *Heterodermia dendritica* (Pers.) Poelt

Anaptychia subheterochroa var. *propagulifera* (Vain.) Kurok., see *Heterodermia dendritica* (Pers.) Poelt

Anaptychia tentaculata (Zahlbr.) Kurok. No lichen substances found. Kurokawa (1962*a*), from Formosa, microchemical tests negative.

Anaptychia togashii* Kurok. Atranorin; zeorin. Kurokawa (1962*a*), from Asia, type description.

Anaptychia trichophora* Kurok. Atranorin; zeorin. Kurokawa (1962*a*), from Bolivia, type description.

Anaptychia trichophoroides* Kurok. Atranorin, norstictic acid, salazinic acid; zeorin. Kurokawa (1962*a*), from Mexico, type description.

Anaptychia tropica* Kurok. Atranorin, salazinic acid; zeorin. Kurokawa (1962*a*), from Costa Rica and Mexico, type description, and also *Anaptychia tropica* var. *antillarum* from Mexico and the West Indies.

Anaptychia ulothricoides (Vain.) Vain. Calcium oxalate. Kurokawa (1962*a*), from Asia, no other substances found, and also *Anaptychia ulothricoides* f. *tenuior*, microchemical tests negative, calcium oxalate crystals observed, especially in *A. ulothricoides* f. *tenuior*.

Anaptychia usambarensis* Kurok.   Atranorin; zeorin; unidentified brown pigment (K −). Kurokawa (1962a), type description.

Anaptychia vulgaris* (Vain.) Kurok.   Unidentified pigment (K + purple). Kurokawa (1962a), pigment on undersurface probably an anthraquinone.

Anzia colpodes (Ach.) Stizenb.   Atranorin (±), divaricatic acid. W. L. Culberson (1961a), from North America.
   Divaricatic acid. Hale (1954a), from Arkansas, an exsiccati specimen; Hale (1955a), from North America, only divaricatic acid in all 105 samples tested; Hale (1956a); Hale (1956f); Hale (1957a).

Anzia colpota Vain.   Atranorin (±), divaricatic acid, sekikaic acid. Kurokawa and Jinzenji (1965), from Japan, by thin-layer chromatography and microcrystal tests.

Anzia gracilis Asah., see *Anzia japonica* (Tuck.) Müll. Arg.

Anzia hypoleucoides Müll. Arg.   Atranorin, lobaric acid; two unidentified substances. Kurokawa and Jinzenji (1965), from Japan, by thin-layer chromatography.
   Atranorin, lobaric acid. Asahina and Hiraiwa (1935a), from Japan, by extraction.
   Lobaric acid. Asahina (1938d); Satô (1954), from Japan.
   REVIEWS. Lobaric acid: Asahina and Shibata (1954); Shibata (1958, 1963).

Anzia japonica (Tuck.) Müll. Arg.   Anziaic acid, atranorin (±). Kurokawa and Jinzenji (1965), from Japan, by thin-layer chromatography.
   Anziaic acid, atranorin. Kurokawa (1956), from Japan, as *Anzia opuntiella* var. *ryogamiensis*.
   Anziaic acid. Asahina (1936d), mention in description of microcrystal tests, as *Anzia gracilis*; Asahina (1937c), by microcrystal tests, as *A. opuntiella*; Asahina and Hiraiwa (1935a), from Japan, by extraction, as *A. gracilis*; Satô (1954), from Japan, as *A. opuntiella*.
   REVIEWS. Anziaic acid: Asahina and Shibata (1954); Shibata (1958, 1963); all review reports as *Anzia opuntiella*.
   See also *Anzia opuntiella*.

Anzia japonica f. *robusta* Asah., see *Anzia opuntiella* Müll. Arg.

Anzia japonica f. *sublinearis* Asah., see *Anzia opuntiella* Müll. Arg.

Anzia japonica f. *typica* Asah., see *Anzia opuntiella* Müll. Arg.

Anzia leucobatoides (Nyl.) Zahlbr. Anziaic acid. Asahina (1936*d*), mention in description of microcrystal tests, as *Anzia leucobatoides* f. *hypomelaena*; Satô (1954), from China, mention only.

REVIEWS. Anziaic acid: Asahina and Shibata (1954); Shibata (1958, 1963); all review reports as *Anzia leucobatoides* f. *hypomelaena*.

Anzia opuntiella Müll. Arg. Atranorin (±), chloroatranorin (±), divaricatic acid and/or sekikaic acid.

Atranorin (±), divaricatic acid and/or sekikaic acid. Kurokawa and Jinzenji (1965), from Japan, by thin-layer chromatography, a taxonomic revision.

Atranorin, chloroatranorin, divaricatic acid. Asahina and Hiraiwa (1935*a*), from Japan, by extraction, as *Anzia japonica*; Seshadri (1944), mention only, as *A. japonica*.

Divaricatic acid. Asahina (1937*a*), as *Anzia japonica* f. *typica*, *A. j.* f. *robusta*, and *A. j.* f. *sublinearis*; W. L. Culberson (1961*a*), mention only, as *A. japonica*.

OTHER REPORTS. For reports of anziaic acid in this species see *Anzia japonica*.

REVIEW. Chloroatranorin: Asahina and Shibata (1954), as *Anzia japonica*.

Anzia opuntiella var. *ryogamiensis* Kurok., see *Anzia japonica* (Tuck.) Müll. Arg.

Anzia ornata (Zahlbr.) Asah. Atranorin, divaricatic acid, sekikaic acid. Kurokawa and Jinzenji (1965), from Japan, by thin-layer chromatography.

Atranorin, divaricatic acid. W. L. Culberson (1961*a*), from North Carolina.

Sekikaic acid. Asahina (1937*a*), *Anzia japonica* var. *ornata* raised to the rank of species.

Anzia stenophylla Asah. Atranorin, divaricatic acid; calcium oxalate.

Atranorin, divaricatic acid. Kurokawa and Jinzenji (1965), from Japan, by thin-layer chromatography.

Calcium oxalate. Satô (1954), crystals in medulla.

Arthonia caesia (Flot.) Körb. Usnic acid. Hale (1957*b*), mention only, as *Allarthonia caesia*.

Arthonia cinnabarina (DC.) Wallr.

REVIEW. Arthonia-violet: Thies (1932), as *Arthonia gregaria*.

Arthonia gregaria (Weig.) Körb., see *Arthonia cinnabarina* (DC.) Wallr.

Arthonia impolita (Ehrh. *ex* Hoffm.) Borr.
 Lepranthin. Salkowski (1910).
 REVIEWS. Lecanoric acid, lepranthaic acid (sometimes as lepranthic acid), lepranthin: W. Brieger (1923); Hesse (1912); Thies (1932); Zopf (1907); all review reports as *Leprantha impolita*.

Arthonia lobata f. *decussata* (Flot.) Szat.
 REVIEWS. Atranorin, lecanoric acid: W. Brieger (1923); Hesse (1912); Thies (1932); Zopf (1907); all reports as *Pachnolepia decussata*.

Arthonia obscura Ach., see *Arthonia reniformis* (Pers.) Ach.

Arthonia radiata (Pers.) Ach.
 REVIEW. Lecanora-red: Thies (1932), as *Arthonia vulgaris*.

Arthonia reniformis (Pers.) Ach.
 REVIEW. Lecanora-red: Thies (1932), as *Arthonia obscura*.

Arthonia vulgaris Schaer., see *Arthonia radiata* (Pers.) Ach.

Asahinea chrysantha (Tuck.) W. Culb. & C. Culb. Alectoronic acid, atranorin, α-collatolic acid (±), usnic acid; unidentified purple or lavender pigments (±). W. L. Culberson and Culberson (1965), from Japan, the Northwest Territories, the U.S.S.R., and Alaska, by microchemical tests.
 Alectoronic acid, atranorin, usnic acid. Mitsuno (1953), no α-collatolic acid by paper chromatography, as *Cetraria chrysantha*.
 Alectoronic acid, atranorin. Dahl (1952), as *Platysma chrysanthum*; Tavares (1954), mention only, as *P. chrysanthum*.
 Alectoronic acid, usnic acid. Hale (1957b), as *Cetraria chrysantha*.
 Alectoronic acid. Asahina (1938d), mention in description of microchemical tests, as *Cetraria chrysantha*.
 Usnic acid. Asahina (1953b), mention only, as *Cetraria chrysantha*.

Asahinea kurodakensis (Asah.) W. Culb. & C. Culb. Alectoronic acid, atranorin, α-collatolic acid; unidentified purple or lavender pigments. W. L. Culberson and Culberson (1965), from Japan, by microchemical tests.
 Alectoronic acid, atranorin, α-collatolic acid. Asahina (1953b), from Hokkaido, as *Cetraria kurodakensis*, type description; Mitsuno (1953), by paper chromatography, trace alectoronic acid, no usnic acid, as *C. kurodakensis*.

Asahinea scholanderi (Llano) W. Culb & C. Culb. Alectoronic acid, atranorin, α-collatolic acid; unidentified purple or lavender pigments. W. L. Culberson and Culberson (1965), from the region from Siberia to the Bering Strait and from northern and central Alaska, by micro-chemical tests.

Atranorin, alectoronic acid. Krog (1962), from Alaska, as *Cetraria scholanderi.*

Aspicilia adunans (Nyl.) Arn., see *Lecanora adunans* Nyl.

Aspicilia adunans f. *glacialis* Arn., see *Lecanora adunans* Nyl.

Aspicilia affinis (Eversm.) Meresch., see *Lecanora affinis* Eversm.

Aspicilia caesiocinerea (Nyl.) Arn., see *Lecanora caesiocinerea* Nyl.

Aspicilia calcarea (L.) Mudd, see *Lecanora calcarea* (L.) Somm.

Aspicilia calcarea var. *farinosa* (Flörke) Hazsl., see *Lecanora farinosa* (Flörke) Nyl.

Aspicilia candida (Anzi) Hue, see *Lecanora candida* (Anzi) Nyl.

Aspicilia cinerea (L.) Körb., see *Lecanora cinerea* (L.) Somm.

Aspicilia esculenta (Pall.) Flag., see *Lecanora esculenta* (Pall.) Eversm.

Aspicilia fruticulosa (Eversm.) Flag., see *Lecanora fruticulosa* Eversm.

Aspicilia gibbosa (Ach.) Körb., see *Lecanora gibbosula* Magn.

Aspicilia laevata (Ach.) Arn., see *Lecanora laevata* (Ach.) Nyl.

Aspicilia laevata f. *albicans* Arn., see *Lecanora laevata* f. *albicans* (Arn.) Zsch.

Aspicilia sylvatica Arn., see *Lecanora sylvatica* (Arn.) Sandst.

Aspicilia verrucosa (Ach.) Körb., see *Lecanora verrucosa* Ach.

Bacidia acclinis (Mass.) Zahlbr.
REVIEW. Bacidia-green: Thies (1932).

Bacidia citrinella (Ach.) Branth & Rostr.
REVIEWS. Rhizocarpic acid: W. Brieger (1923); Hesse (1912); Thies (1932); Zopf. (1907); all reports as *Raphiospora flavovirescens.*

Bacidia fuscorubella (Hoffm.) Bausch
REVIEW. Lecanora-red: Thies (1932).

Bacidia muscorum (Sw.) Mudd
REVIEW. Bacidia-green: Thies (1932).

Baeomyces absolutus Tuck. Baeomycesic acid. Asahina (1943*f*), from Japan, and also *Baeomyces absolutus* var. *stipitatus.*

Baeomyces aggregatus Asah. Atranorin, didymic acid; unidentified substance. Asahina (1943*f*), unidentified substance by microcrystals in the GE solution, type description; Asahina (1957*f*), from Japan.

Baeomyces botryophorus Zahlbr. Norstictic acid. Asahina (1943*f*), from Japan.

Baeomyces brevis Zahlbr. Stictic acid. Asahina (1943*f*), from Japan.

Baeomyces fungoides (Sw.) Ach. Baeomycesic acid. Asahina (1942*d*), description of microchemical test; Asahina (1943*f*), from Japan.
REVIEWS. Baeomycesic acid: Asahina and Shibata (1954); Shibata (1958, 1963).

Baeomyces placophyllus Ach. Stictic acid. Asahina (1943*f*), from Japan, by microchemical tests; Asahina, Tanase, and Yosioka (1936); Asahina, Yanagita, and Yosioka (1936), from Formosa, mention only.
OTHER REPORT. Norstictic acid: Hale (1957*b*), mention only.
REVIEWS. Stictic acid: Asahina and Shibata (1954); Shibata (1958, 1963). — Atranorin: W. Brieger (1923); Hesse (1912); Zopf (1907); all three reports as *Sphyridium placophyllum.*

Baeomyces roseus Pers. Atranorin, baeomycesic acid. Hale (1957*b*), mention only.
Baeomycesic acid. Asahina (1938*e*), by microcrystal tests; Asahina (1943*f*), from Japan; Asahina, Tanase, and Yosioka (1936), from Japan, by extraction; Koller and Maass (1935), from Germany, by extraction; Schatz (1963), from Massachusetts.
REVIEWS. Baeomycesic acid: Asahina and Shibata (1954); Shibata (1958, 1963).

Baeomyces rufus (Huds.) Rebent. Stictic acid. Asahina (1943*f*), from Japan.
OTHER REPORT. Atranorin, norstictic acid: D. Hess (1958), from Germany.

Baeomyces sanguineus Asah. Norstictic acid; unidentified red pigment. Asahina (1943*f*), from Formosa, and also *Baeomyces sangineus* var. *ablutum*, type descriptions.

Biatora atrofusca Flot., see *Lecidea templetonii* Tayl.

Biatora granulosa (Ehrh.) Flot., see *Lecidea granulosa* (Ehrh.) Ach.

Biatora infidula (Nyl.) Dalla Torre & Sarnth., see *Lecidea sylvicola* Flot.

Biatora lightfootii "f. *commutata*," see *Catillaria lightfootii* (Sm.) Oliv.

Biatora lucida (Ach.) Fr., see *Lecidea lucida* (Ach.) Ach.

Biatora mollis (Wahlenb.) Arn., see *Lecidea mollis* (Wahlenb.) Nyl.

Biatora ochracea Hepp, see *Protoblastenia ochracea* (Hepp) Zahlbr.

Biatora turgidula (Fr.) Nyl., see *Lecidea turgidula* Fr.

Biatora viridescens (Schrad.) Mann, see *Lecidea viridescens* (Schrad.) Ach.

Bilimbia melaena (Nyl.) Arn., see *Micarea melaena* (Nyl.) Hedl.

Blastenia arenaria (Pers.) Mass., see *Caloplaca arenaria* (Pers.) Müll. Arg.

Blastenia arenaria var. *teicholytum* Ach., see *Caloplaca teicholyta* (Ach.) J. Stein.

Blastenia percrocata Arn., see *Caloplaca percrocata* (Arn.) J. Stein.

Bombyliospora japonica Zahlbr.   Pannarin; zeorin. Asahina (1944), from Japan, and also *Bombyliospora japonica* f. *asahinae* and *B. japonica* f. *purpurascens* from Japan.

Bryopogon canus (Ach.) Choisy, see *Alectoria cana* (Ach.) Leight.

Bryopogon chalybeiformis (L.) Link, see *Alectoria chalybeiformis* (L.) S. Gray.

Bryopogon implexus (Hoffm.) Elenk., see *Alectoria implexa* (Hoffm.) Nyl.

Bryopogon jubatus (L.) Link, see *Alectoria jubata* (L.) Ach.

Buellia atrata (Sm.) Anzi
 REVIEW. Lecidea-green: Thies (1932).

Buellia canescens (Dicks.) De Not.   Atranorin, chloroatranorin, diploicin; mannitol.

Atranorin, chloroatranorin, diploicin. Spillane, Keane, and Nolan (1936), from Ireland, and a high-melting unidentified substance, by extraction.

Diploicin, mannitol. Asahina (1951*i*); Nolan (1934*b*), from Ireland, by extraction, also a substance similar to atranorin but with no methoxyl and an unidentified water-soluble substance, Zopf's "catolechin" probably the same as diploicin.

Diploicin. Barry (1946); Brown, Clark, Ollis, and Veal (1960), mention only, synthesis of diploicin; Neelakantan and Seshadri (1960*b*), mention only; Nolan (1934*a*); Nolan, Algar, McCann, Manahan, and Nolan (1948), structure study; Ollis (1955–1957), mention only; Yoshida (1951), mention only.

REVIEWS. Atranorin, catolechin, diploicin: W. Brieger (1923); Hesse (1912); Thies (1932); Zopf (1907); all four reports as *Diploicia canescens*. — Atranorin, chloroatranorin, diploicin: Asahina and Shibata (1954). — Diploicin: Neelakantan (1965); Shibata (1958, 1963).

Buellia parasema De Not.
REVIEW. Lecanora-red: Thies (1932).

Buellia pulchella (Schrad.) Tuck., see *Buellia wahlenbergii* (Ach.) Sheard

Buellia punctata (Hoffm.) Mass.
REVIEW. Lecanora-red: Thies (1932).

Buellia sororia Th. Fr.   Norstictic acid. Schindler (1936).

Buellia sororioides Erichs.   Norstictic acid. Huneck and Siegel (1963), from Germany, 6% norstictic acid by extraction of 0.9 g. of lichen.

Buellia stillingiana J. Stein.   Atranorin, norstictic acid. Hale (1958*e*).

Buellia wahlenbergii (Ach.) Sheard   Usnic acid; rhizocarpic acid. Hale (1957*b*), as *Buellia pulchella*, mention only.

Byssocaulon niveum Mont.   Lecanoric acid. Huneck and Follman (1965*e*), from Chile, 3.1% by extraction of 7.0 g. of lichen.

Calicium chlorinum (Ach.) Schaer.   Vulpinic acid. Mittal and Seshadri (1957), mention only.
REVIEWS. Vulpinic acid: Asahina and Shibata (1954); W. Brieger (1923); Hesse (1912); Shibata (1958, 1963). — Calycin, lepraric acid,

leprarin, lobaric acid (as stereocaulic acid), vulpinic acid: Thies (1932). — Calycin, lobaric acid (as stereocaulic acid), vulpinic acid: Zopf (1907).

Calicium chrysocephalum (Turn.) Ach., see *Chaenotheca chrysocephala* (Turn.) Th. Fr.

Calicium hyperellum (Ach.) Ach., see *Calicium viride* Pers.

Calicium japonicum Asah.   Unidentified substance. Asahina (1958c), from Japan, by microchemical tests, unidentified substance shows a permanent yellow color with alcoholic Ca(OCl)$_2$, type description.

Calicium trichiale Ach., see *Chaenotheca trichialis* (Ach.) Hellb.

Calicium viride Pers.   Rhizocarpic acid.
    REVIEWS. Rhizocarpic acid: Asahina and Shibata (1954); W. Brieger (1923); Hesse (1912); Shibata (1958, 1963); Thies (1932); Zopf (1907); all reports as *Calicium hyperellum*.

Callopisma arenarium (Pers.) Trev., see *Caloplaca arenaria* (Pers.) Müll. Arg.

Callopisma aurantiacum (Lightf.) Mass., see *Caloplaca aurantiaca* (Lightf.) Th. Fr.

Callopisma erythrocarpum (Pers.) J. Stein., see *Caloplaca teicholyta* (Ach.) J. Stein.

Callopisma exsecutum (Nyl.) Arn., see *Caloplaca exsecuta* (Nyl.) Dalla Torre & Sarnth.

Callopisma flavovirescens (Wulf.) Mass., see *Caloplaca flavovirescens* (Wulf.) Dalla Torre & Sarnth.

Callopisma jungermanniae (Vahl) Räs., see *Caloplaca jungermanniae* (Vahl) Th. Fr.

Callopisma percrocatum (Arn.) Jatta, see *Caloplaca percrocata* (Arn.) J. Stein.

Callopisma teicholytum (Ach.) Müll. Arg., see *Caloplaca teicholyta* (Ach.) J. Stein.

Callopisma vitellinum (Ehrh.) Mudd, see *Candelariella vitellina* (Ehrh.) Müll. Arg.

Caloplaca arenaria (Pers.) Müll. Arg.
    REVIEWS. Blastenin: W. Brieger (1923), as *Blastenia arenaria;* Hesse
(1912), as *B. arenaria;* Thies (1932), as *B. arenaria;* Zopf (1907), as *B.
arenaria* and *Callopisma arenarium.*

Caloplaca aurantia (Pers.) Hellb.   Parietin (sometimes as physcion).
    REVIEWS. Parietin: W. Brieger (1923), as *Gasparrinia sympageum;*
Hesse (1912), as *G. sympageum;* Thies (1932), as *Placodium sympageum;*
Zopf (1907), as *P. sympageum.*

Caloplaca aurantiaca (Lightf.) Th. Fr.   Parietin.
    REVIEWS. Parietin: W. Brieger (1923); Hesse (1912); Thies (1932), as
Callopisma aurantiacum.

Caloplaca cirrochroa (Ach.) Th. Fr.   Parietin (sometimes as physcion).
    REVIEWS. Parietin: W. Brieger (1923), as *Gasparrinia cirrochroa;*
Hesse (1912), as *G. cirrochroa;* Thies (1932), as *Placodium cirrochroum;*
Zopf (1907), as *P. cirrochroum.*

Caloplaca decipiens (Arn.) J. Stein.   Parietin (as physcion).
    REVIEWS. Parietin: W. Brieger (1923); Hesse (1912); both reports as
Gasparrinia decipiens.

Caloplaca elegans (Link) Th. Fr., see *Xanthoria elegans* (Link) Th. Fr.

Caloplaca exsecuta (Nyl.) Dalla Torre & Sarnth.
    REVIEW. Lecidia-green: Thies (1932), as *Callopisma exsecutum.*

Caloplaca flavovirescens (Wulf.) Dalla Torre & Sarnth.   Parietin (sometimes as physcion).
    REVIEWS. Parietin: W. Brieger (1923), as *Callopisma aurantiacum*
β *flavovirescens;* Hesse (1912), as *Callopisma aurantiacum* β *flavovirescens;* Thies (1932), as *Callopisma aurantiacum* β *flavovirescens;* Zopf
(1907), as *Callopisma flavovirescens.*

Caloplaca jungermanniae (Vahl) Th. Fr.   Parietin.
    REVIEWS. Parietin: W. Brieger (1923); Hesse (1912); Thies (1932);
Zopf (1907); all reports as *Callopisma jungermanniae.*

Caloplaca lallaveri Flag.   Stictaurin (=calycin and pulvinic dilactone).
Galun and Lavee (1966), from Israel, by a microcrystal test.

Caloplaca murorum (Hoffm.) Th. Fr.   Parietin (sometimes as physcion).
Thomas (1936), parietin also produced by the lichen fungus in culture.
    REVIEWS. Parietin: W. Brieger (1923), as *Gasparrinia murorum;* Hesse
(1912), as *G. murorum;* Thies (1932), as *Placodium murorum;* Zopf
(1907), as *P. murorum.*

Caloplaca percrocata (Arn.) J. Stein.
REVIEWS. Blastenin: W. Brieger (1923); Hesse (1912); Thies (1932); all three reports as *Blastenia percrocata;* Zopf (1907), as *Callopisma percrocatum.*

Caloplaca teicholyta (Ach.) J. Stein.
REVIEWS. Atranorin, fumarprotocetraric acid, gyrophoric acid, unidentified anthraquinone pigment: Zopf (1907), as *Blastenia arenaria* var. *teicholytum* and as *Callopisma teicholytum*, unidentified pigment found in the apothecia. — Atranorin, gyrophoric acid: W. Brieger (1923); Hesse (1912); as *B. arenaria* var. *teicholytum* and as *Callopisma teicholytum.* — Fumarprotocetraric acid: Thies (1932), as *Callopisma teicholytum.*

Candelaria concolor (Dicks.) B. Stein   Stictaurin (=calycin and pulvinic dilactone). Hale (1957a), from Arkansas, Missouri, and Oklahoma.
Pulvinic dilactone (as pulvinic lactone). Grover and Seshadri (1959), from India, by extraction.
REVIEWS. Stictaurin: W. Brieger (1923); Hesse (1912); Thies (1932), and as *Candelaria laciniosa;* Zopf (1907), and as *C. laciniosa.*

Candelaria fibrosa (Fr.) Müll. Arg.   Calycin, pulvinic dilactone (as pulvic anhydride). Hale (1957a), from Arkansas, Missouri, and Oklahoma.

Candelaria laciniosa (Duf.) Kieff., see *Candelaria concolor* (Dicks.) B. Stein

Candelaria medians (Nyl.) Flag., see *Candelariella medians* (Nyl.) A. L. Sm.

Candelaria vitellina (Ehrh.) Mass., see *Candelariella vitellina* (Ehrh.) Müll Arg.

Candelariella aurella (Hoffm.) Zahlbr.   Stictaurin.
REVIEWS. Stictaurin (=calycin and pulvinic dilactone): W. Brieger (1923); Hesse (1912); Thies (1932); Zopf (1907); all reports as *Gyalolechia aurella.*

Candelariella medians (Nyl.) A. L. Sm.   Calycin, pulvinic dilactone (as pulvinic acid anhydride). Eigler and Poelt (1965), from Sweden, by paper chromatography.
REVIEWS. Rhizocarpic acid, stictaurin (=calycin and pulvinic dilactone): W. Brieger (1923); Hesse (1912); Thies (1932); all three reports as *Candelaria medians* and *Gasparrinia medians;* Zopf (1907), as *Candelaria medians.*

Candelariella reflexa (Nyl.) Lett.   Stictaurin.
REVIEWS. Stictaurin (=calycin and pulvinic dilactone): Thies (1932); Zopf (1907); both reports as *Gyalolechia reflexa.*

Candelariella vitellina (Ehrh.) Müll. Arg.   Stictaurin; D-mannitol.
Mannitol. Asahina (1934*b*), mention only, as *Callopisma vitellinum;*
Pueyo (1960), mention only, as *Callopisma vitellinum;* Pueyo (1965),
mention only.
REVIEWS. D-Mannitol, stictaurin: Thies (1932); Zopf (1907); both
reports as *Candelaria vitellina*. — D-Mannitol: Asahina and Shibata
(1954), as *Callopisma vitellinum*. — Stictaurin: W. Brieger (1923); Hesse
(1912); both reports as *Candelaria vitellina*.

Catillaria athallina (Hepp) Hellb.
REVIEW. Thalloidima-green: Thies (1932).

Catillaria lightfootii (Sm.) Oliv.   (−)-Usnic acid. Salkowski (1910).
REVIEWS. (−)-Usnic acid: W. Brieger (1923); Hesse (1912); Thies
(1932); Zopf (1907); all review reports as *Biatora lightfootii* f. *commutata*.

Catocarpus oreites (Vain.) Eitn., see *Rhizocarpon oreites* (Vain.) Zahlbr.

Cavernularia hultenii Degel.   Atranorin, physodic acid. Krog (1951*b*),
from Norway.

Cavernularia lophyrea (Ach.) Degel.   Atranorin, physodic acid. Dahl
(1952), from Norway, as *Hypogymnia lophyrea*; Krog (1951*b*), from
Washington.

Cenomyce "*destricta*," see *Cladonia destricta* (Nyl.) Nyl.

Cetraria aculeata (Schreb.) Fr., see *Cornicularia aculeata* (Schreb.) Ach.

Cetraria alaskana C. Culb. & W. Culb., see *Cetrelia alaskana* (C. Culb. &
W. Culb.) W. Culb. & C. Culb.

Cetraria alvarensis (Wahlenb.) Vain.
Pinastric acid, vulpinic acid. Mittal and Seshadri (1957), mention
only, as *Cetraria tubulosa*.
(−)-Usnic acid. Salkowski (1910), as *Cetraria tubulosa*.
REVIEWS. Pinastric acid (as chrysocetraric acid), (−)-usnic acid, vulpinic acid: W. Brieger (1923); Hesse (1912); both reports as *Cetraria
juniperina* var. *tubulosa*; Thies (1932), and cetrarialic acid, as *C. juniperina* var. *tubulosa* and as *C. tubulosa*; Zopf (1907), and cetrarialic acid,
as *Platysma tubulosa*. — Pinastric acid, vulpinic acid: Asahina and

Shibata (1954), as *C. juniperina* var. *tubulosa* and as *C. tubulosa*; Shibata (1958, 1963), as *C. juniperina* var. *tubulosa* and *C. tubulosa*; Zopf (1907), as *C. tubulosa*. — Pinastric acid, (−)-usnic acid: Thies (1932), as *C. juniperina* var. *alvarensis*.

Cetraria ambigua Bab.   Usnic acid; unidentified fatty acid. Asahina (1955*a*), from India.

Cetraria aurescens Tuck.   Usnic acid. Hale (1957*b*), mention only.

Cetraria caperata Vain., see *Cetraria pinastri* (Scop.) S. Gray

Cetraria chicitae W. Culb., see *Cetrelia chicitae* (W. Culb.) W. Culb. & C. Culb.

Cetraria chlorophylla (Willd.) Vain., see *Cetraria scutata* (Wulf.) Poetsch

Cetraria chrysantha Tuck., see *Asahinea chrysantha* (Tuck.) W. Culb. & C. Culb.

Cetraria ciliaris Ach.   Atranorin, olivetoric acid, physodic acid. C. F. Culberson (1964), from North Carolina, separation by column chromatography, mention of a strain with alectoronic acid and atranorin (see *Cetraria halei*); C. F. Culberson (1965*a*), mention only; W. L. Culberson and Culberson (1967*a*), from southern Finland and North America, by microchemical tests on 21 samples; Mosbach (1964*c*), mention in discussion of biosynthesis; Shibata (1965), mention in discussion of biosynthesis; Vězda (1965), from North Carolina, an exsiccati specimen.
   Atranorin, olivetoric acid. Hale (1963*a*), other strains with alectoronic acid and atranorin (see *Cetraria halei*) and with atranorin and protolichesterinic acid (see *Cetraria orbata*).
   Olivetoric acid. Hale (1957*c*), from Virginia, an exsiccati specimen. See also *Cetraria halei*.

Cetraria collata (Nyl.) Müll. Arg., see *Cetrelia collata* (Nyl.) W. Culb. & C. Culb. and *Cetrelia nuda* (Hue) W. Culb. & C. Culb.

Cetraria collata f. *isidiata* Asah., see *Cetrelia braunsiana* (Müll. Arg.) W. Culb. & C. Culb.

Cetraria collata f. *microphyllina* Zahlbr., see *Cetrelia japonica* (Zahlbr.) W. Culb. & C. Culb.

Cetraria collata f. *nuda* (Hue) Zahlbr., see *Cetrelia nuda* (Hue) W. Culb. & C. Culb.

Cetraria commixta (Nyl.) Th. Fr.
    Stictic acid. Dahl (1952), mention only; Tavares (1954), mention only.
    Enzymes. Moissejeva (1961), as *Cetraria fahlunensis*, enzymes tabulated.
    REVIEWS. Atranorin, fumarprotocetraric acid: W. Brieger (1923), as *Cetraria fahlunensis*; Hesse (1912), as *C. fahlunensis*; Thies (1932), as *Parmelia fahlunensis* and *Platysma fahlunensis*; Zopf (1907), as *C. fahlunensis* and *Platysma fahlunensis*.

Cetraria crispa (Ach.) Nyl., see *Cetraria ericetorum* Opiz

Cetraria crispa f. *subtubulosa* (Fr.) Hue, see *Cetraria ericetorum* Opiz

Cetraria crispa var. *tenuifolia* (Retz.) Degel., see *Cetraria ericetorum* Opiz

Cetraria cucullata (Bell.) Ach.   (−)-Usnic acid; protolichesterinic acid; friedelin; enzymes; ascorbic acid.
    Usnic acid, protolichesterinic acid. Dahl (1952), mention only.
    (−)-Usnic acid. Moissejeva (1957), 0.6%; Moissejeva (1961), enzymes tabulated; Savicz, Kuprevicz, Litvinov, Moissejeva, and Rassadina (1956), 0.6%; Stoll, Brack, and Renz (1947), 0.70% by extraction; Stoll, Renz, and Brack (1947), from Switzerland.
    Usnic acid. D. Hess (1958); Tomaselli (1957), from Italy, mention only.
    Friedelin. Bruun (1954a), from Norway, 0.002% by extraction, a detailed chemical study.
    Ascorbic acid. Karev and Kochevykh (1962).
    OTHER REPORT. No usnic acid: Gertig and Banasiewicz (1961), from Poland.
    REVIEWS. Protolichesterinic acid, proto-α-lichesterinic acid, (−)-usnic acid: Thies (1932), no fumarprotocetraric acid. — Protolichesterinic acid, (−)-usnic acid: W. Brieger (1923); Hesse (1912); Zopf (1907), and as *Platysma cucullatum*. — (−)-Usnic acid: Savicz, Litvinov, and Moissejeva (1960). — Friedelin: Shibata (1958, 1963).

Cetraria delisei (Bory) Th. Fr.   Gyrophoric acid, hiascic acid; friedelin; ascorbic acid, nicotinic acid, riboflavin, thiamine.
    Gyrophoric acid, hiascic acid. Asahina (1951i), as *Cetraria hiascens*; Dahl (1952), mention only; Erdtman and Wachtmeister (1957a), as *C. hiascens*, mention in discussion of biosynthesis; Tavares (1954), mention only.
    Gyrophoric acid. Hale (1956e).
    Hiascic acid. Asahina and Kusaka (1942), a detailed chemical study, as *Cetraria hiascens*.

Friedelin. Bruun (1954*a*), from Norway.

Ascorbic acid, nicotinic acid, riboflavin, thiamine. Gustafson (1954), a quantitative study.

Ascorbic acid. Karev and Kochevykh (1962), as *Cetraria hiascens*.

REVIEWS. Gyrophoric acid, hiascic acid: Asahina and Shibata (1954), as *Cetraria hiascens*. — Hiascic acid, friedelin: Shibata (1958, 1963), and as *C. hiascens*.

Cetraria diffusa (G. Web.) Lynge, see *Parmeliopsis hyperopta* (Ach.) Vain.

Cetraria endocrocea (Asah.) Satô   Endocrocin; nephrosteranic acid, nephrosterinic acid; unidentified neutral substance. Asahina and Yanagita (1937), unidentified neutral substance possibly identical to caperin, as *Nephromopsis endocrocea*.

Endocrocin, unidentified neutral substance. Asahina and Fuzikawa (1935*d*), from Japan, and an unidentified fatty acid, as *Nephromopsis endocrocea*.

Endocrocin. Franck and Reschke (1960), sample of endocrocin used to prove identity to pigment from ergot, infrared and ultraviolet spectra, analyses and paper chromatographic comparisons, as *Nephromopsis endocrocea*; Shibata (1965), mentioned in discussion of biosynthesis.

REVIEWS. Endocrocin, nephrosteranic acid, nephrosterinic acid, caperin: Asahina and Shibata (1954), and as *Nephromopsis endocrocea*. — Endocrocin, nephrosteranic acid (as nephrosteraic acid), nephrosterinic acid (as nephrosteric acid): Shibata (1958, 1963); and as *N. endocrocea*. — Endocrocin: Neelakantan (1965), as *N. endocrocea*.

Cetraria endocrocea f. *clarkii* (Asah.) Satô   No endocrocin. Asahina (1935*b*), as *Nephromopsis endocrocea* f. *clarkii*.

Cetraria ericetorum Opiz   Fumarprotocetraric acid (±); (−)-lichesterinic acid (±), (−)-*allo*-protolichesterinic acid (±), (+)-protolichesterinic acid (±), (−)-protolichesterinic acid; friedelin; iron, manganese, zinc.

Fumarprotocetraric acid, (−)-protolichesterinic acid (from Japan). (−)-*allo*-Protolichesterinic acid (±), (−)-lichesterinic acid (±), (+)-protolichesterinic acid (±) (from Europe). Asahina (1937*e*).

Fumarprotocetraric acid, (−)-protolichesterinic acid (±) (from Japan). (−)-*allo*-Protolichesterinic acid (from Norway) with (−)-lichesterinic acid (from northern Germany) or (+)-protolichesterinic acid (from Baden). Asahina and Yasue (1937*a*), as *Cetraria islandica* var. *tenuifolia*.

Protolichesterinic acid, lichesterinic acid. Dahl (1952).

(−)-Protolichesterinic acid. Asano (1927), from Japan, 1.3% by extraction, as *Cetraria islandica* f. *angustifolia*; Asahina and Yanagita (1936), from Japan, as *C. tenuifolia*; Asano (1927), misdetermined as *C.*

islandica f. *angustifolia* according to Asahina and Yanagita (1936) and Asahina (1939*b*).

(+)-Protolichesterinic acid. Stoll, Renz, and Brack (1947), from Switzerland, and (±)-usnic acid (apparently an error), as *Cetraria islandica* var. *crispa*.

Friedelin. Bruun (1954*a*), from Norway, 0.002% by extraction, a detailed chemical study.

Iron, manganese, zinc. Lounamaa (1965), a quantitative study.

OTHER REPORT. Atranorin: Vartia (1949), from Finland, as *Cetraria tenuifolia*.

REVIEWS. Fumarprotocetraric acid, proto-α-lichesterinic acid, protolichesterinic acid: Thies (1932), as *Cetraria subtubulosa*. — Fumarprotocetraric acid, protolichesterinic acid: Zopf (1907), as *C. subtubulosa*. — (−)-Lichesterinic acid, (−)-protolichesterinic acid: Asahina and Shibata (1954), and as *C. islandica* f. *tenuifolia*. — (−)-Lichesterinic acid, friedelin: Shibata (1958, 1963).

Except where indicated otherwise, all reports are as *Cetraria crispa*.

Cetraria fahlunensis (L.) Schreb., see *Cetraria commixta* (Nyl.) Th. Fr.

Cetraria fallax (G. Web.) And., see *Platismatia glauca* (L.) W. Culb. & C. Culb.

Cetraria gilva Asah.   Alectoronic acid. Asahina (1953*b*), type description.

Cetraria glauca (L.) Ach., see *Platismatia glauca* (L.) W. Culb. & C. Culb.

Cetraria halei W. Culb. & C. Culb.   Alectoronic acid, atranorin (±), α-collatolic acid (±). W. L. Culberson and Culberson (1967*a*), from North America and southern Finland east through central Asia to Kamchatka and Japan, by microcrystal tests and paper chromatography on 65 samples.

Alectoronic acid, atranorin. Hale (1963*a*), as a chemical strain of *Cetraria ciliaris*.

Alectoronic acid, α-collatolic acid. Mitsuno (1953), no atranorin, no usnic acid, by paper chromatography, as *Cetraria ciliaris*.

Alectoronic acid. Asano and Azumi (1938), by extraction, as *Nephromopsis ciliaris*.

α-Collatolic acid. Asahina (1938*d*), as *Nephromopsis ciliaris*.

REVIEWS. Alectoronic acid: Asahina and Shibata (1954); Shibata (1958, 1963); all review reports as *Nephromopsis ciliaris*.

Cetraria hiascens (Fr.) Th. Fr., see *Cetraria delisei* (*Bory*) Th. Fr.

Cetraria islandica (L.) Ach.   (−)-Lichesterinic acid (±), (−)-*allo*-protolichesterinic acid (±), (+)-protolichesterinic acid (±), (from Japan). Fumarprotocetraric acid; (+)-protolichesterinic acid (from Europe).

Unidentified unsaturated fatty acids; friedelin; ergosterol; arabitol, cellulose, hemicellulose, isolichenin, lichenin, mannitol, sucrose, trehalose, umbilicin; ribonuclease, other enzymes; ascorbic acid, folic acid-, folinic acid-, and vitamin B_{12}-group factors, vitamin B_1; unidentified substance; iron, manganese, zinc.

Lichesterinic acid, (−)-*allo*-protolichesterinic acid or (+)-protolichesterinic acid. Asahina (1937e), from Japan.

(−)-*allo*-Protolichesterinic acid, (+)-protolichesterinic acid. Asahina and Yanagita (1936), from Japan.

(−)-*allo*-Protolichesterinic acid and/or (+)-protolichesterinic acid (Japan). Fumarprotocetraric acid, (+)-protolichesterinic acid (Europe). Asahina and Yasue (1937a).

Protolichesterinic acid. Bendz, Santesson, and Tibell (1966), source of an authentic sample of this substance; Hale (1957c), from West Virginia, an exsiccati specimen.

Fumarprotocetraric acid, (+)-protolichesterinic acid, unidentified substance (m.p. 159–162°). Gertig (1963), from Poland, by extraction, no usnic acid.

Fumarprotocetraric acid, (+)-protolichesterinic acid. Vartia (1950a); Vartia (1950b), from Finland, another sample with only protolichesterinic acid.

Fumarprotocetraric acid, lichesterinic acid, protolichesterinic acid. Dahl (1952); Solberg (1956), mention only.

Fumarprotocetraric acid, protolichesterinic acid. Borkowski, Woźniak, Gertig, and Werakso (1964), from Poland; Sticher (1965a), no usnic acid, cetraric acid and protocetraric acid from alcohol and water extracts respectively; Sticher (1965b), no usnic acid.

Fumarprotocetraric acid. Asahina and Tanase (1933); Asahina and Tanase (1934a), from Sakhalin; Mitsuno (1953), by paper chromatography; J. Santesson (1965), by thin-layer chromatography.

Unidentified fatty acids. Wagner and Friedrich (1965), two C_{16} acids with one double bond and one with three double bonds, three C_{18} acids with one, two, and four double bonds, by thin-layer chromatography.

Friedelin. Bruun (1954a), from Poland, 30 mg. extracted from 10 kg. of lichen.

Ergosterol. Blix and Rydin (1932).

Arabitol, mannitol, sucrose, trehalose, umbilicin. Lindberg, Misiorny, and Wachtmeister (1953), no volemitol.

Cellulose, hemicellulose, isolichenin, lichenin. Buston and Chambers (1933), no pectin, pentosans or lignin.

Hemicellulose, isolichenin, lichenin. Chanda, Hirst, and Manners (1957).

Hemicellulose. Granichstädten and Percival (1943), a chemical study.

Isolichenin, lichenin. Fleming and Manners (1966b), mention only, structural study on isolichenin, as Iceland moss; K. Hess and Lauridsen (1940); Meyer and Gürtler (1947b), by extraction; Peat, Whelan, and

Roberts (1957), structural study on lichenin; Peat, Whelan, Turvey, and Morgan (1961), structural study on isolichenin.

Lichenin. Acker, Diemair, and Samhammer (1955*a*), as Icelandic moss; Acker, Diemair, and Samhammer (1955*b*), as Icelandic moss; Drake (1943); Fleming and Manners (1966*a*), mention only, as Icelandic moss, enzymatic hydrolysis study, lichenin not identical to barley glucan; K. Hess and Fries (1927); Jewell and Lewis (1918); Karrer, Staub, and Staub (1924); Meyer and Gürtler (1947*a*), by extraction; Perlin and Suzuki (1962), structural study, comparison with a similar product from cereal grains.

Ribonuclease. Rennert and Gubanski (1960).

Enzymes. Moissejeva (1961), enzymes tabulated.

Ascorbic acid. Karev and Kochevykh (1962).

Folic acid-, folinic acid-, and vitamin B_{12}-group factors. Sjöström and Ericson (1953).

Iron, manganese, zinc. Lounamaa (1965), a quantitative study.

No usnic acid. Gertig and Banasiewicz (1961), from Poland.

OTHER REPORTS. (+)-Protolichesterinic acid, (±)-usnic acid: Stoll, Brack, and Renz (1950); Stoll, Renz, and Brack (1947), from Switzerland. — (−)-Protolichesterinic acid: Asano and Kanematsu (1932), from Japan, but see *Cetraria ericetorum*; Cavaltito, Fruehauf, and Bailey (1948). — (±)-Usnic acid: Stoll, Brack, and Renz (1947), 0.04% by extraction. — Usnic acid: Klosa (1951*a*); Tomaselli (1957), from Italy, listed as an Italian lichen containing usnic acid. — (Chloro)atranorin, fumarprotocetraric acid: D. Hess (1958), from Germany, by paper chromatography. — α-Ethyletherprotocetraric acid (as cetraric acid): Koller and Krakauer (1929), as *Cladonia islandica*, (compound apparently formed by decomposition of fumarprotocetraric acid during alcohol extraction).

REVIEWS. (−)-Lichesterinic acid (Japan), (−)-*allo*-protolichesterinic acid (Japan), (+)-protolichesterinic acid (Europe), fumarprotocetraric acid, umbilicin, friedelin: Shibata (1958, 1963), and as *Cetraria islandica* var. *orientalis*. — (−)-*allo*-Protolichesterinic acid, (+)-protolichesterinic acid, fumarprotocetraric acid, isolichenin, lichenin, vitamin B_1: Schindler (1956/1957). — (−)-*allo*-Protolichesterinic acid, (−)-lichesterinic acid: Asahina and Shibata (1954), from Japan, reported for *C. islandica* var. *orientalis*. — (−)-Lichesterinic acid, (+)-protolichesterinic acid, fumarprotocetraric acid: Asahina and Shibata (1954). — Paralichesterinic acid, protolichesterinic acid, proto-α-lichesterinic acid, α-ethyletherprotocetraric acid (as cetrarin, probably formed during extraction), fumarprotocetraric acid: W. Brieger (1923); Hesse (1912); Thies (1932), and as *C. platyna*. — Paralichesterinic acid, protolichesterinic acid, fumarprotocetraric acid: Zopf (1907), and as *C. islandica* var. *vulgaris* and *C. platyna*. — Isolichenin, lichenin, sucrose, trehalose: Shibata (1958). — Lichenin: Neelakantan (1965). — Crude fat: Steiner (1957), 0.40% according to one study and 4.30% according to another study.

Cetraria islandica f. *angustifolia* Mats. & Miyoshi, see *Cetraria ericetorum* Opiz

Cetraria islandica var. *crispa* Ach., see *Cetraria ericetorum* Opiz

Cetraria islandica var. *tenuifolia* (Retz.) Vain., see *Cetraria ericetorum* Opiz

Cetraria islandica var. *vulgaris* Schaer., see *Cetraria islandica* (L.) Ach.

Cetraria japonica Zahlbr., see *Cetrelia japonica* (Zahlbr.) W. Culb. & C. Culb.

Cetraria juniperina (L.) Ach. (+)-Usnic acid; pinastric acid, vulpinic acid. Asano and Kameda (1935*b*), from southern Sakhalin, by extraction, lichen mixed with *Cetraria pinastri*.
   Usnic acid, pinastric acid, vulpinic acid. Koller and Pfeiffer (1933*b*).
   Usnic acid, pinastric acid. Dahl (1952).
   Pinastric acid, vulpinic acid. Mittal and Seshadri (1957), mention only.
   Pinastric acid. Bendz, Santesson, and Wachtmeister (1965*d*), used as an authentic source of this substance; Koller and Klein (1934*c*), mention only; Mitsuno (1955).
   Vulpinic acid. Hale (1957*a*), from Arkansas and Missouri, reported for *Cetraria juniperina* var. *virescens*.
   REVIEWS. (−)-Usnic acid, pinastric acid (as chrysocetraric acid): W. Brieger (1923); Hesse (1912); Thies (1932), pinastric acid questionable. — (−)-Usnic acid, pinastric acid: Thies (1932), as *Cetraria juniperina* var. *genuina*. — Usnic acid: Zopf (1907), as *Platysma juniperinum*. — Pinastric acid, vulpinic acid: Shibata (1958, 1963). — Pinastric acid: Asahina and Shibata (1954).

Cetraria juniperina var. *alvarensis* (Wahlenb.) Torss., see *Cetraria alvarensis* (Wahlenb.) Vain.

Cetraria juniperina var. *genuina* Körb., see *Cetraria juniperina* (L.) Ach.

Cetraria juniperina var. *terrestris* Schaer., see *Cetraria tilesii* Ach.

Cetraria juniperina var. *tubulosa* Schaer., see *Cetraria alvarensis* (Wahlenb.) Vain.

Cetraria kurodakensis Asah., see *Asahinea kurodakensis* (Asah.) W. Culb. & C. Culb.

Cetraria lacunosa Ach., see *Platismatia lacunosa* (Ach.) W. Culb. & C. Culb.

Cetraria lacunosa var. *acharii* Du Rietz, see *Platismatia lacunosa* (Ach.) W. Culb. & C. Culb.

Cetraria lacunosa var. *macounii* Du Rietz, see *Platismatia lacunosa* (Ach.) W. Culb. & C. Culb.

Cetraria laureri Kremp. (−)-Usnic acid; protolichesterinic acid.
    Usnic acid, protolichesterinic acid. Asahina (1959c), from Japan, by microchemical tests; Dahl (1952), mention only, and an unidentified substance (±); Tavares (1954), mention only, and unidentified substances (±).
    (−)-Usnic acid. Salkowski (1910), as *Cetraria complicata.*
    OTHER REPORTS. Atranorin, fumarprotocetraric acid, (−)-usnic acid, protolichesterinic acid: Asahina (1936c), from Europe, review of Zopf's work, also listed as *Platysma complicatum* but without atranorin; Klosa (1951b), mention, as *Cetraria complitata* [sic].
    REVIEWS. Atranorin, fumarprotocetraric acid, (−)-usnic acid, proto-α-lichesterinic acid, protolichesterinic acid: Thies (1932), as *Cetraria complicata* and *Parmelia complicata.* — Atranorin, fumarprotocetraric acid, (−)-usnic acid, protolichesterinic acid: Zopf (1907), as *C. complicata* and *Platysma complicatum.* — Atranorin, (−)-usnic acid, protolichesterinic acid: W. Brieger (1923); Hesse (1912); both reports as *C. complicata.*

Cetraria microphyllica W. Culb. & C. Culb. Atranorin, microphyllinic acid. W. L. Culberson and Culberson (1967a), from Japan, by microcrystal tests and paper chromatography, type description.

Cetraria nigricans (Retz.) Nyl. Protolichesterinic acid. Dahl (1952), mention only; Tavares (1954), mention only.

Cetraria nivalis (L.) Ach. (−)-Usnic acid; tetrahydroxy fatty acids; friedelin; ergosterol; enzymes, protein; ascorbic acid; unidentified substance; calcium, phosphorus.
    (−)-Usnic acid. C. F. Culberson (1963), by extraction; Moissejeva (1957), 1.1%; Moissejeva (1961), 1.1%, enzymes tabulated; Savicz, Kuprevicz, Litvinov, Moissejeva, and Rassadina (1956), 1.1%; Stoll, Brack, and Renz (1947), 2.75% by extraction; Stoll, Renz, and Brack (1947), from Switzerland.
    Usnic acid. Dahl (1952); Gertig and Banasiewicz (1961), from Poland, 2.48%; Tomaselli (1957), from Italy, mention only.
    Tetrahydroxy fatty acids. Solberg (1957); Solberg (1960).

Friedelin. Bruun (1954*a*), from Norway, *epi*-friedelinol also reported but later found to be a mixture of friedelin and an unidentified substance, see Bruun and Jefferies (1954).

Ergosterol. Blix and Rydin (1932).

Ascorbic acid. Karev and Kochevykh (1962).

Protein, fat, calcium, phosphorus. Scotter (1965), from northern Canada.

OTHER REPORT. Pinastric acid, usnic acid, vulpinic acid: D. Hess (1958), from Germany.

REVIEWS. Crude fat: Steiner (1957), 0.99%. — Nivalic acid, (−)-usnic acid: W. Brieger (1923); Thies (1932). — (−)-Usnic acid: Hesse (1912); Savicz, Litvinov, and Moissejeva (1960); Zopf (1907), and as *Platysma nivale*. — Friedelin, *epi*-friedelinol: Shibata (1958, 1963).

Cetraria oakesiana Tuck.   Usnic acid; caperatic acid. Hale (1956*h*), from the Appalachians, tentative identification of caperatic acid; Hale (1957*c*), from West Virginia, an exsiccati specimen.

REVIEWS. (+)-Usnic acid: W. Brieger (1923); Hesse (1912); Thies (1932); Zopf (1907); all review reports as *Platysma oakesianum*.

Cetraria orbata (Nyl.) Fink   Protolichesterinic acid. W. L. Culberson and Culberson (1967*a*), from North America, no atranorin, by microcrystal tests.

OTHER REPORT. Atranorin, protolichesterinic acid: Hale (1963*a*), as a chemical strain of *Cetraria ciliaris*.

Cetraria ornata Müll. Arg.   Fumarprotocetraric acid; unidentified yellow pigment.

Fumarprotocetraric acid. Asahina (1937*e*).

Unidentified yellow pigment. Asahina (1935*b*), yellow pigment in the medulla, as *Nephromopsis endoxantha*.

Cetraria pinastri (Scop.) S. Gray   (−)-Usnic acid; pinastric acid, vulpinic acid. Klosa (1952*c*), mention only.

Usnic acid, pinastric acid, vulpinic acid. Dahl (1952); Koller and Pfeiffer (1933*b*); Vartia (1950*b*), from Finland, by extraction.

(−)-Usnic acid, vulpinic acid. Stoll, Renz, and Brack (1947), from Switzerland.

(−)-Usnic acid. Stoll, Brack, and Renz (1947), 0.55% by extraction.

Pinastric acid, vulpinic acid. D. Hess (1958), from Germany; Mitsuno (1955); Mittal and Seshadri (1957), mention only.

Pinastric acid. Koller and Klein (1934*c*), mention only; Wachtmeister (1956*a*), used as an authentic source of this substance.

Vulpinic acid. Stoll, Brack, and Renz (1950); Vartia (1949), from Finland.

OTHER REPORTS. (+)-Usnic acid, pinastric acid, vulpinic acid: Asano and Kameda (1935*b*), from southern Sakhalin, sample mixed with *Cetraria juniperina*. — Usnic acid, parietin, vulpinic acid: Krog (1951*a*), as *C. caperata*, mention only. — Atranorin, vulpinic acid: Steiner (1955). REVIEWS. Vulpinic acid, pinastric acid (as chrysocetraric acid), (−)-usnic acid: W. Brieger (1923); Hesse (1912); Thies (1932). — Vulpinic acid, pinastric acid, usnic acid: Zopf (1907), and as *Platysma pinastri*. — Vulpinic acid, pinastric acid: Asahina and Shibata (1954); Shibata (1958, 1963).

Cetraria platyna Ach., see *Cetraria islandica* (L.) Ach.

Cetraria pseudocomplicata Asah.   Alectoronic acid, usnic acid (a trace). Asahina (1936*c*), from Japan, type description, no protolichesterinic acid.
   Alectoronic acid. Asahina (1938*d*), description of microcrystal tests; Asahina (1954*b*), from Japan; Asano and Azumi (1938), by extraction; Mitsuno (1953), by paper chromatography.
   REVIEWS. Alectoronic acid: Asahina and Shibata (1954); Shibata (1958, 1963).

Cetraria rhytidocarpa Mont. & Bosch   Unidentified protolichesterinic acid-type fatty acid. Asahina (1954*b*); Asahina (1955*a*).

Cetraria rhytidocarpa f. *nipponensis* Asah.   Physodic acid, unidentified substance (FeCl$_3$+ bluish). Nuno (1964), from Japan, by thin-layer chromatography.
   OTHER REPORT. Unidentified phenolic substance (±), a fatty acid: Asahina (1954*b*), type description.

Cetraria richardsonii Hook.   Alectoronic acid. Hale (1956*f*).

Cetraria rugosa (Asah.) Satô   Physodic acid, usnic acid, unidentified substance (FeCl$_3$+ bluish). Nuno (1964), from Japan, by thin-layer chromatography.
   Usnic acid. Mitsuno (1953), no alectoronic acid, atranorin, or α-collatolic acid, by paper chromatography.

Cetraria sanguinea Schaer., see *Cetrelia sanguinea* (Schaer.) W. Culb. & C. Culb.

Cetraria scholanderi Llano, see *Asahinea scholanderi* (Llano) W. Culb. & C. Culb.

Cetraria scutata (Wulf.) Poetsch   Atranorin; protolichesterinic acid. Vartia (1950*b*), from Finland.

Atranorin. Vartia (1949), from Finland.

OTHER REPORT. Atranorin, physodic acid, protolichesterinic acid: Klosa (1951*b*).

REVIEWS. Atranorin, protolichesterinic acid: W. Brieger (1923); Hesse (1912); Zopf (1907), and as *Platysma chlorophyllum.* — Atranorin, proto-α-lichesterinic acid, protolichesterinic acid: Thies (1932).

Except where indicated otherwise, all reports are as *Cetraria chlorophylla.*

Cetraria sepincola (Ehrh.) Ach. Protolichesterinic acid. Dahl (1952), mention only; Tavares (1954), mention only.

Cetraria stracheyi Bab. Caperatic acid. Kurokawa (1964), mention only.

Cetraria stracheyi f. *ectocarpisma* (Hue) Satô (−)-Usnic acid; caperatic acid, (−)-lichesterinic acid, nephromopsinic acid, (−)-protolichesterinic acid. Asano and Azumi (1935), unknown substance A identical to nephromopsinic acid and unknown substance B identical to (−)-protolichesterinic acid according to Asano and Taniguti (1939), as *Nephromopsis stracheyi* f. *ectocarpisma.*

Nephromopsinic acid. Bendz, Santesson, and Tibell (1966), used as an authentic source of this substance, isolation by preparative-layer chromatography, as *Nephromopsis stracheyi* f. *ectocarpisma.*

REVIEWS. (−)-Usnic acid, caperatic acid, (−)-lichesterinic acid, nephromopsinic acid, (−)-protolichesterinic acid: Asahina and Shibata (1954), as *Nephromopsis stracheyi* f. *ectocarpisma.* — Caperatic acid, (−)-lichesterinic acid, nephromopsinic acid: Shibata (1958, 1963), as *N. stracheyi* f. *ectocarpisma.*

Cetraria stuppea (Flot.) Sandst., see *Cornicularia muricata* Ach.

Cetraria "subtubulosa," see *Cetraria ericetorum* Opiz

Cetraria tenuifolia (Retz.) R. H. Howe, see *Cetraria ericetorum* Opiz

Cetraria terrestris (Schaer.) Fink, see *Cetraria tilesii* Ach.

Cetraria thomsonii Stirt., see *Parmelia thomsonii* (Stirt.) W. Culb.

Cetraria tilesii Ach.
Usnic acid, vulpinic acid. Hale (1957*b*), mention only.

REVIEWS. Pinastric acid (as chrysocetraric acid): W. Brieger (1923); Hesse (1912); as *Cetraria juniperina* var. *terrestris*, probably misidentified. — Terrestrin, vulpinic acid: W. Brieger (1923); Thies (1932); both reports as *C. terrestris.*

Cetraria togashii Asah.   Usnic acid; protolichesterinic acid-type fatty acid. Asahina (1953*b*), from Japan, type description.

Cetraria tubulosa (Schaer.) B. de Lesd., see *Cetraria alvarensis* (Wahlenb.) Vain.

Cetraria tuckermanii Oakes, see *Platismatia tuckermanii* (Oakes) W. Culb. & C. Culb.

Cetraria ulophylloides Asah.   Unidentified substance; protolichesterinic acid-type fatty acid. Asahina (1953*b*), from Japan, by microcrystal tests, type description.

Cetraria wallichiana (Tayl.) Müll. Arg.   Alectoronic acid. Mitsuno (1953), by paper chromatography, no atranorin, α-collatolic acid, or usnic acid.

Cetrelia alaskana (C. Culb. & W. Culb.) W. Culb. & C. Culb.   Atranorin, imbricaric acid, unidentified substance (KC + red). C. F. Culberson and Culberson (1966), from Alaska, 6.3% imbricaric acid, 0.37% atranorin, by extraction of 2.0 g. of lichen, type description of *Cetraria alaskana*; W. L. Culberson and Culberson (1968), from Alaska, by microchemical tests on 14 samples.

Cetrelia braunsiana (Müll. Arg.) W. Culb. & C. Culb.   Alectoronic acid, atranorin, α-collatolic acid. W. L. Culberson and Culberson (1968), from Formosa, Japan, the Himalayas, Luzon, and Yunnan, by microchemical tests on 32 samples; Mitsuno (1953), by paper chromatography, as *Cetraria collata* f. *isidiata*.

α-Collatolic acid. Asahina (1938*b*), description of microcrystal tests, as *Cetraria collata* f. *isidiata*.

Cetrelia cetrarioides (Del. *ex* Duby) W. Culb. & C. Culb.   Atranorin, chloroatranorin ($\pm$), perlatolic acid ($\pm$), imbricaric acid ($\pm$), unidentified substance (KC + red) ($\pm$).

Atranorin, chloroatranorin ($\pm$), perlatolic acid and/or imbricaric acid. Asahina (1938*b*), description of microcrystal tests; Asahina and Fuzikawa (1935*b*), from Austria and Japan, as *Parmelia perlata*, taxonomic problems and work of Zopf and Hesse discussed, see also comment on this work in Asahina and Yosioka (1937).

Atranorin, imbricaric acid or perlatolic acid, unidentified substance (KC + red) ($\pm$). W. L. Culberson and Culberson (1968), from Austria, France, Germany, Poland, Sikkim, Switzerland, Wales, Yunnan, West Virginia, and Washington, by microchemical tests on 69 samples of which 34 contained imbricaric acid and 35 contained perlatolic acid, unidentified KC + substance possibly also C +.

Atranorin, perlatolic acid and/or imbricaric acid. Asahina (1952*a*), from Japan and Sakhalin; Asahina (1952*e*), from Japan; W. L. Culberson (1962*a*), from eastern Asia, Europe, and North America, more common in Europe than in North America, previously referred to as the "perlatolic acid strain"; Dahl (1952), from Norway, mention only; Krog (1951*b*), from Norway, the lichen found to be KC+ red due to an unidentified substance; Shibata (1965), mention in description of biosynthesis, as Chem. strain 1, see also *Cetrelia olivetorum* (Chem. strain 2, with atranorin and olivetoric acid) and *C. chicitae* (Chem. strain 3, with alectoronic acid and α-collatolic acid).

Atranorin, perlatolic acid. W. L. Culberson (1958), from Asia, Europe, the Appalachians, and Washington, as the "perlatolic acid strain," some samples with atranorin and olivetoric acid from Asia, Europe, and eastern North America (see *Cetrelia olivetorum*) and some samples with alectoronic acid, atranorin, and α-collatolic acid from North America (see *Cetrelia chicitae*); Hale (1960*b*), from the Alps, by microchemical tests on the lectotype specimen; M. E. Mitchell (1960), from Ireland, sample tested by Asahina who found also a KC+ substance not demonstrable by microchemical tests; Ramaut (1961*a*), from Yugoslavia; Ramaut (1961*b*), from Yugoslavia.

Perlatolic acid or imbricaric acid. Asahina (1937*e*).

Perlatolic acid and imbricaric acid. Asahina (1939*b*), mention only; Hale (1957*b*), mention only.

Perlatolic acid. Asahina (1940*f*), description of microcrystal tests; W. L. Culberson (1965), individuals with this acid previously called a chemical strain; Krog (1962), from Alaska.

OTHER REPORTS. Olivetoric acid: Hale (1956*e*), from North America, as a chemical strain (see *Cetrelia olivetorum*); Hale (1959*b*), mentions confusion with *Parmelia austrosinensis*. — Atranorin, lecanoric acid: Rangaswami and Rao (1955*b*), from India, as *P. perlata*. — Alectoronic acid: Hale (1957*b*), mention only, as *P. perlata* (see *Cetrelia chicitae*).

REVIEWS. Atranorin, chloroatranorin (±), perlatolic acid and/or imbricaric acid: Asahina and Shibata (1954), and as *Parmelia perlata*. — Atranorin, imbricaric acid, perlatolic acid: W. Brieger (1923); Hesse (1912); Thies (1932), and parmelia-brown; Zopf (1907); all four reports as *P. perlata*. — Atranorin, chloroatranorin, perlatolic acid: Shibata (1958, 1963).

Unless indicated otherwise, all reports are as *Parmelia cetrarioides*.

Cetrelia chicitae (W. Culb.) W. Culb. & C. Culb. Alectoronic acid, atranorin, α-collatolic acid. W. L. Culberson (1958), from North America, as one of the three chemical strains of *Parmelia cetrarioides* (= *Cetrelia cetrarioides*); W. L. Culberson (1965), from Canada, France, Japan, Korea, the U.S.S.R., Connecticut, Massachusetts, Maryland, Michigan, Minnesota, North Carolina, Pennsylvania, Tennessee,

Virginia, and West Virginia, as *Cetraria chicitae*; W. L. Culberson and Culberson (1968), from Canada, France, Formosa, Japan, Java, Korea, the Philippines, Sabah, the U.S.A., and the U.S.S.R., by microchemical tests on 24 samples.

Cetrelia collata (Nyl.) W. Culb. & C. Culb.   Atranorin, imbricaric acid, unidentified substance (KC + red) (±). W. L. Culberson and Culberson (1968), from Nepal, Szechwan, and Yunnan, by microchemical tests on 17 samples.
See also *Cetrelia nuda*.

Cetrelia davidiana W. Culb. & C. Culb.   Atranorin, olivetoric acid. W. L. Culberson and Culberson (1968), from Szechwan and Yunnan, by microchemical tests on seven samples, type description.

Cetrelia delavayana W. Culb. & C. Culb.   Atranorin, perlatolic acid, unidentified substance (KC+) (±). W. L. Culberson and Culberson (1968), from Szechwan and Yunnan, by microchemical tests on eight samples, type description.

Cetrelia isidiata (Asah.) W. Culb. & C. Culb.   Anziaic acid, atranorin. W. L. Culberson and Culberson (1968), from Formosa and Japan, by microcrystal tests and paper chromatography on four samples.

Cetrelia japonica (Zahlbr.) W. Culb. & C. Culb.   Atranorin, chloroatranorin, microphyllinic acid. Asahina and Fuzikawa (1935e), by extraction, as *Cetraria collata* f. *microphyllina*.
   Atranorin, microphyllinic acid. Asahina (1938b), description of microcrystal tests, as *Cetraria japonica;* W. L. Culberson and Culberson (1968), from Formosa, Japan, Java, northern Sabah, and South Korea, by microcrystal tests and paper chromatography on 22 samples; Mitsuno (1953), by paper chromatography, as *Cetraria japonica*.
   Microphyllinic acid. Asahina and Fuzikawa (1935a), and an unidentified neutral substance (probably atranorin), as *Cetraria collata* f. *microphyllina;* Fuzikawa (1939), as *Cetraria collata* f. *microphyllina;* Satô (1935), as *Cetraria japonica*.
   OTHER REPORT. Olivetoric acid: Asahina (1939b), as *Cetraria japonica*.
   REVIEWS. Microphyllinic acid: Asahina and Shibata (1954), and as *Cetraria collata* f. *microphyllina*; Shibata (1958, 1963), as *Cetraria japonica*.

Cetrelia nuda (Hue) W. Culb. & C. Culb.   Alectoronic acid, atranorin, α-collatolic acid. W. L. Culberson and Culberson (1968), from Formosa, Japan, and Yunnan, by microchemical tests on 22 samples; Mitsuno (1953), by paper chromatography, no usnic acid, as *Cetraria collata* and *Cetraria collata* f. *nuda*.

Atranorin, α-collatolic acid. Asahina, Kanaoka, and Fuzikawa (1933), from Japan, by extraction, as *Cetraria collata* but later identified as *Cetraria collata* f. *nuda* by Asahina and Fuzikawa (1935*a*); Dahl (1952), mention only, as *Platysma collatum*.

α-Collatolic acid. Asahina (1937*e*), as *Cetraria collata*; Asahina (1938*d*), discussion of microcrystal tests, as *Cetraria collata*; Kurokawa (1964), mention only, as *Cetraria collata*.

REVIEWS. α-Collatolic acid: Asahina and Shibata (1954); Shibata (1958, 1963); all review reports as *Cetraria collata*.

Cetrelia olivetorum (Nyl.) W. Culb. & C. Culb.   Atranorin, chloro-atranorin, olivetoric acid.

Atranorin, olivetoric acid. Asahina (1952*a*), from Japan and Korea; Asahina (1952*e*), from Japan and Korea; W. L. Culberson (1962*a*), from eastern Asia, Europe, eastern North America, and Hawaii; W. L. Culberson and Culberson (1968), from Asia, Europe, and the U.S.A., by microchemical tests on 19 samples, as *Cetrelia olivetorum*; Krog (1951*b*), from Norway, as *Parmelia cetrarioides* var. *rubescens*; Shibata (1965), mention in discussion of biosynthesis.

Chloroatranorin, olivetoric acid. Asahina and Fuzikawa (1935*f*), from Japan, by extraction.

Olivetoric acid. Asahina (1936*d*), description of microcrystal tests; Asahina and Asano (1932*a*), mention only; Asahina and Fuzikawa (1935*b*), mention only.

OTHER REPORTS. Atranorin, glomelliferic acid: Dahl (1952), from Norway, mention only, as *Parmelia cetrarioides* var. *rubescens*. — Erythrin: Manaktala, Neelakantan, and Seshadri (1966), mention only; V. S. Rao and Seshadri (1942*b*), mention only.

REVIEWS. Olivetoric acid: Asahina and Shibata (1954); Shibata (1958).

Unless indicated otherwise, all reports are as *Parmelia olivetorum*.

Cetrelia pseudolivetorum (Asah.) W. Culb. & C. Culb.   Atranorin, olivetoric acid. Asahina (1952*a*), from Japan, type description of *Parmelia pseudolivetorum*; Asahina (1952*e*), from Japan, as *P. pseudolivetorum*; W. L. Culberson and Culberson (1968), from Formosa, the Himalayas, Japan, and Yunnan, by microchemical tests on 31 samples.

Cetrelia sanguinea (Schaer.) W. Culb. & C. Culb.   Anziaic acid, atranorin. W. L. Culberson (1966*c*), from the Himalayas, by microchemical tests on the holotype specimen of *Aspidelia wattii*, as *Cetraria sanguinea*; W. L. Culberson and Culberson (1968), from southwestern China, Japan, Java, and Sumatra, by microcrystal tests and paper chromatography on 67 samples.

Anziaic acid. Asahina (1936*d*), from Java and Sumatra, description of microcrystal tests, as *Cetraria sanguinea*; Asahina (1939*b*), mention only, as *Cetraria sanguinea*.

REVIEWS. Anziaic acid: Asahina and Shibata (1954); Shibata (1958, 1963); all review reports as *Cetraria sanguinea*.

Cetrelia sinensis W. Culb. & C. Culb. Atranorin, imbricaric acid. W. L. Culberson and Culberson (1968), from Formosa and Yunnan, by microcrystal tests and paper chromatography on six samples, type description.

Chaenotheca chrysocephala (Turn.) Th. Fr. Vulpinic acid. Hale (1957*b*), mention only; Mittal and Seshadri (1957), as *Cyphelium chrysocephalum*.

REVIEWS. Calycin, vulpinic acid: W. Brieger (1923); Hesse (1912); Zopf (1907). — Vulpinic acid: Asahina and Shibata (1954); Shibata (1958, 1963); Thies (1932). All review reports are as *Cyphelium chrysocephalum* and as *Calicium chrysocephalum*.

Chaenotheca trichialis (Ach.) Hellb. Calycin.

REVIEWS. Calycin: W. Brieger (1923); Hesse (1912); Thies (1932); Zopf (1907); all reports as *Calicium trichiale*.

Chaudhuria indica Zahlbr., see *Anaptychia diademata* (Tayl.) Kurok.

Chiodecton japonicum Zahlbr. Gyrophoric acid. W. L. Culberson (1963*a*), from Japan.

Chiodecton sanguineum (Sw.) Vain., see *Herpothallon sanguineum* (Sw.) Tobl.

Chiodecton stalactinum Nyl. Psoromic acid. Huneck and Follmann (1964*b*), from Chile, by extraction of 30 g. of lichen.

Chlorea vulpina (L.) Nyl., see *Letharia vulpina* (L.) Hue

Chrysothrix nolitangere (Mont.) Mont. Calycin. Bendz, Santesson, and Wachtmeister (1965*d*), used as an authentic source of this substance; Senft (1916).

REVIEW. Calycin: Thies (1932).

Cladia retipora (La Bill.) Nyl., see *Cladonia retipora* (La Bill.) Fr.

Cladia sullivanii (Müll. Arg.) Martin, see *Cladonia sullivanii* Müll. Arg.

Cladina alpestris (L.) Harm., see *Cladonia alpestris* (L.) Rabenh.

"*Cladina alpestris* var. *laxiuscula*," see *Cladonia impexa* Harm.

Cladina alpestris f. *spumosa* (Flörke) Norrl., see *Cladonia spumosa* (Flörke) Sandst.

Cladina amaurocraea (Flörke) Oliv., see *Cladonia amaurocraea* (Flörke) Schaer.

"*Cladina condensata*," see *Cladonia impexa* Harm.

"*Cladina destricta*," see *Cladonia destricta* (Nyl.) Nyl.

"*Cladina laxiuscula*," see *Cladonia impexa* Harm.

Cladina rangiferina (L.) Nyl., see *Cladonia rangiferina* (L.) G. Web. *ex* Wigg.

"*Cladina spumosa*," see *Cladonia spumosa* (Flörke) Sandst.

Cladina sylvatica (L.) Leight., see *Cladonia arbuscula* (Wallr.) Rabenh.

"*Cladina sylvatica* var. *silvestris*," see *Cladonia impexa* Harm.

Cladina uncialis (L.) Nyl., see *Cladonia uncialis* (L.) Wigg.

Cladonia abbreviatula Merr.   Didymic acid, thamnolic acid. Evans (1952*b*), from Florida, no usnic acid.

Cladonia aberrans (Abb.) Stuck.   Perlatolic acid, psoromic acid, (−)-usnic acid; rangiformic acid. Asahina (1950*c*), from Japan, as *Cladonia alpestris* f. *aberrans*, rangiformic acid determination not certain.
   Perlatolic acid, psoromic acid. Evans (1943*b*), from Japan and North America, by microchemical tests, and unidentified substances A (±) and B (±) (later found to be crystal forms of usnic acid in the GE solution), as *Cladonia alpestris* f. *aberrans*.
   Psoromic acid, (−)-usnic acid. Asahina (1940*f*), from Japan, as *Cladonia alpestris* f. *aberrans*, by extraction.
   Psoromic acid. Evans (1943*a*), mention in discussion of microcrystal tests, as *Cladonia alpestris* f. *aberrans*.
   No fumarprotocetraric acid. Evans (1944*b*).
   OTHER REPORT. Fumarprotocetraric acid, usnic acid, rhodocladonic acid: Dahl (1952), mention only.

Cladonia acuminata (Ach.) Norrl.   Atranorin, norstictic acid. Ahti (1964);
Asahina (1943*a*), from Vermont; Dahl (1952), mention only; Evans
(1947*b*), from Vermont; Kurokawa (1959*c*), from Japan.
     OTHER REPORT. Norstictic acid or α-methylethersalazinic acid: Asa-
hina (1936*c*).

Cladonia acuminata var. *norrlinii* (Vain.) Magn.   Atranorin, unidentified
substance (PD +). Ahti (1964), unidentified substance may be psoromic
acid; Dahl (1950), PD + reaction often faint.

Cladonia aggregata (Sw.) Ach.   Barbatic acid only (from Australia,
Bolivia, India, Japan, Mexico, Puerto Rico, and New Zealand). Fumarprotocetraric acid only (from Chile and New Zealand). Barbatic acid,
fumarprotocetraric acid (from Colombia, New Zealand, and Peru).
Barbatic acid ($\pm$), norstictic acid ($\pm$), stictic acid (from Mexico and
New Zealand). Nuno (1962), by microchemical tests on many samples.
     Barbatic acid. Asahina (1942*a*), from Bolivia, India, and Japan, a
sample from Chile found to be PD + ; Asahina (1950*c*), from Formosa
and Japan; Asahina (1955*a*), from India; Dahl (1952), mention only;
Fuzikawa and Ishiguro (1936); Fuzikawa, Shinamura, and Tarui (1941).
     Fumarprotocetraric acid. Follmann (1965).

Cladonia alaskana Evans
     Fumarprotocetraric acid, usnic acid. Evans (1947*a*), from Alaska,
fumarprotocetraric acid determination by color test only, no atranorin,
barbatic acid, or squamatic acid; Dahl (1952), mention only.

Cladonia alcicornis (Lightf.) Flörke   ($-$)-Usnic acid. Salkowski (1910).
     REVIEWS. (+)-Usnic acid: W. Brieger (1923); Hesse (1912); Thies
(1932); Zopf (1907).

Cladonia alpestris (L.) Rabenh.   Perlatolic acid ($\pm$), ($-$)-usnic acid;
pseudonorrangiformic acid (sometimes as Evans' substance C), rangiformic acid, tetrahydroxy fatty acid like ventosic acid; friedelin; ergosterol; arabitol, mannitol, unidentified polysaccharides; protein;
ascorbic acid; calcium, iron, manganese, phosphorus, zinc, ^{137}Cs, ^{40}K,
^{54}Mn, ^{210}Pb, ^{210}Po, ^{226}Ra, ^{106}Ru, ^{125}Sb.
     Perlatolic acid, ($-$)-usnic acid, rangiformic acid. Asahina (1950*c*),
from Japan and Korea, tentative identification of rangiformic acid.
     Perlatolic acid ($\pm$), usnic acid, unidentified substances ($\pm$). Dahl
(1952), mention only, and psoromic acid ($\pm$) but see *Cladonia aberrans*;
Evans (1943*b*), from Europe and North America, A, B, and C as accessories (substances A and B later found to be crystal forms of usnic
acid in the GE solution), specimens PD + yellow (psoromic acid) are
C. alpestris f. *aberrans* (see *C. aberrans*); Evans (1944*b*); Evans (1950),
from Connecticut, unidentified substances A, B, and C as accessories, all
samples PD −, no *C. alpestris* f. *aberrans*.

(−)-Usnic acid. Moissejeva (1957), 0.6%; Moissejeva (1961), 0.6%, enzymes tabulated; Pätiälä, Pätiälä, Siintola, and Heilala (1948); Salkowski (1910), and *Cladonia alpestris* var. *laxiuscula*; Savicz, Kuprevicz, Litvinov, Moissejeva, and Rassadina (1956), 0.6%; Siintola, Heilala, Pätiälä, and Pätiälä (1948); Vartia (1949), from Finland, by extraction; Virtanen, Viitanen, and Kortekangas (1954), from Finland, by extraction.

Usnic acid. Ark, Bottini, and Thompson (1960), source of usnic acid for preparing sodium usnate; Curd and Robertson (1933), from Norway, a detailed chemical study; Dahl (1950), from Greenland, and several other lichen acids; Gertig and Banasiewicz (1961), from Poland, 1.08%; Hale (1954b), from Baffin Island; Laakso and Gustafsson (1952), from Finland, a quantitative study on the distribution of this pigment in the lichen; Taguchi, Sankawa, and Shibata (1966), ^{14}C study on the biosynthesis of usnic acid; Vartia (1950b), from Finland; Virtanen and Kärki (1956), mention only.

Pseudonorrangiformic acid. Evans (1955a).

Tetrahydroxy fatty acid like ventosic acid. Solberg (1957); Solberg (1960), mention only.

Friedelin. Bruun (1954a), from Norway.

Ergosterol. Blix and Rydin (1932).

Arabitol, mannitol. Lindberg, Misiorny, and Wachtmeister (1953), by paper chromatography, no volemitol.

A branched polysaccharide of D-galactose, D-glucose, and D-mannose. Aspinall, Hirst, and Warburton (1955); Chanda, Hirst, and Manners (1957), mention only.

Lignin-like substance, crude protein. Salo (1957).

Protein, fiber, ash, calcium, phosphorus, crude fat analyses. Scotter (1965), from northern Canada.

Ascorbic acid. Karev and Kochevykh (1962).

Iron, manganese, zinc. Lounamaa (1965), quantitative study.

^{137}Cs, ^{40}K, ^{54}Mn, ^{106}Ru, ^{125}Sb. Häsänen and Miettinen (1966), from Finnish Lapland.

^{210}Pb, ^{210}Po, ^{226}Ra. Holtzman (1966), from Finland.

Unidentified substances. Asahina (1940f), a general discussion and microchemical tests, see also *Cladonia aberrans*.

REVIEWS. Friedelin, lichenin: Shibata (1958). — (−)-Usnic acid: W. Brieger (1923), as *Cladina alpestris*; Hesse (1912), as *Cladina alpestris*; Savicz, Litvinov, and Moissejeva (1960); Thies (1932); Zopf (1907), as *Cladina alpestris*.

Cladonia alpestris f. *aberrans* Abb., see *Cladonia aberrans* (Abb.) Stuck.

Cladonia alpicola (Flot.) Vain. Psoromic acid; aliphatic acid similar to rangiformic acid (±). Asahina (1943a), no atranorin, and also *Cetraria alpicola* var. *karelica*; Asahina (1950c).

Psoromic acid. Asahina (1936*c*); Asahina (1938*e*), description of microcrystal tests; Dahl (1950), from Greenland; Dahl (1952), mention only; Evans (1944*b*); Evans (1950), mention only, no atranorin; Hale (1954*b*), from Baffin Island; Kurokawa (1964), mention only; Zopf (1908), from Tirol, as *Cladonia alpicola* var. *foliosa* f. *macrophylla*.

REVIEWS. Psoromic acid: Asahina and Shibata (1954); Shibata (1958, 1963).

Cladonia alpicola var. *macrophylla* (Schaer.) Oliv., see *Cladonia alpicola* (Flot.) Vain.

Cladonia amaurocraea (Flörke) Schaer.   Barbatic acid (also as coccellic acid and cenomycin), (−)-usnic acid. Asahina and Fuzikawa (1934*b*), from North Korea, by extraction; Asahina (1950*c*); Zopf (1908), from Tirol.

Barbatic acid, usnic acid. Asahina (1942*a*), by microchemical tests of specimens from Sandstede's exsiccati; Dahl (1950), from Greenland, microchemical tests discussed; Dahl (1952), mention only; Evans (1943*a*), mention in discussion of microcrystal tests; Evans (1944*b*), from Connecticut; Evans (1947*b*), from North America; Hale (1954*b*), from Baffin Island; Pišút (1961).

Barbatic acid. Fuzikawa and Ishiguro (1936); Kurokawa (1964), mention only.

(−)-Usnic acid. Stoll, Brack, and Renz (1947), 0.22% by extraction; Stoll, Renz, and Brack (1947), from Switzerland.

Usnic acid. Vartia (1950*b*), from Finland.

Ascorbic acid. Karev and Kochevykh (1962).

REVIEWS. Barbatic acid, (−)-usnic acid: W. Brieger (1923); Hesse (1912); Thies (1932); Zopf (1907); all four reports as *Cladina amaurocraea*. — Barbatic acid: Asahina and Shibata (1954); Shibata (1958, 1963).

Cladonia angustata Nyl.   Barbatic acid, didymic acid, usnic acid. Asahina (1950*c*), from Japan.

Cladonia apodocarpa Robb.   Atranorin, fumarprotocetraric acid. Asahina (1942*b*), from Connecticut, Kentucky, Massachusetts, New Jersey, North Carolina, and Vermont; Dahl (1952), mention only; Evans (1944*b*), from Connecticut, atranorin by microcrystal test and fumarprotocetraric acid by PD+ reaction; Evans (1947*a*), from North America, probably fumarprotocetraric acid; Evans (1950), from Connecticut.

OTHER REPORT. No atranorin: Mozingo (1961), from Tennessee.

Cladonia arbuscula (Wallr.) Rabenh.   Fumarprotocetraric acid, (+)-usnic acid; ursolic acid (±) (sometimes as Evans' substance E); alanine, α-aminobutyric acid, γ-aminobutyric acid, arginine (?), asparagine, aspartic acid, ethanolamine, glutamic acid, glutamine, glycine, isoleucine, lysine, phenylalanine, proline, sarcosine, serine, threonine, tyrosine, valine, enzymes; ascorbic acid, folic acid-, folinic acid-, and vitamin B_{12}-group factors; copper, iron, manganese, zinc, ^{210}Pb, ^{210}Po, ^{226}Ra.

Fumarprotocetraric acid, usnic acid, ursolic acid ($\pm$). Evans (1944*b*), from Connecticut, ursolic acid reported as substance E, also substance A ($\pm$) (later found to be a crystal form of usnic acid in the GE solution); Evans (1950), from Connecticut, microchemical determination of ursolic acid (as substance E) described.

Fumarprotocetraric acid, usnic acid, ursolic acid. Hale (1956*b*), from Connecticut; Hale (1956*g*), from Connecticut, an exsiccati specimen; Schatz (1963), from New Jersey, mention only.

Fumarprotocetraric acid, (+)-usnic acid. Asahina (1950*c*), from Japan.

Fumarprotocetraric acid, usnic acid. Asahina (1941*b*); Bustinza (1951); Duvigneaud (1939); Evans (1943*a*), from North America, mention in discussion of microcrystal tests; Evans (1943*b*), and substance A ($\pm$) (later found to be a crystal form of usnic acid in the GE solution); Hale (1954*b*), from Baffin Island; D. Hess (1958), from Germany, by paper chromatography.

(+)-Usnic acid. Moissejeva (1957), 0.2%; Moissejeva (1961), enzymes tabulated; Salkowski (1910); Savicz, Kuprevicz, Litvinov, Moissejeva, and Rassadina (1956), 0.2%; Vartia (1950*a*), mention only.

Usnic acid. Dahl (1950); Duvigneaud (1940); Gertig (1961); Gertig and Banasiewicz (1961), from Poland, 0.85%; Marshak, Barry, and Craig (1947), used as a source of an authentic sample of this substance; Mozingo (1961), from Tennessee, mentioned in key; Tomaselli (1957), from Italy; Vartia (1950*b*), from Finland.

Ursolic acid. Asahina (1951*i*), mention only.

Alanine, α-aminobutyric acid, γ-aminobutyric acid, arginine (?), asparagine, aspartic acid, ethanolamine, glutamic acid, glutamine, glycine, isoleucine, lysine, phenylalanine, proline, sarcosine, serine, threonine, tyrosine, valine. Linko, Alfthan, Miettinen, and Virtanen (1953).

Ascorbic acid. Karev and Kochevykh (1962).

Folic acid-, folinic acid-, and vitamin B_{12}-group factors. Sjöström and Ericson (1953).

Copper, iron. Lange and Ziegler (1963), a quantitative study, as *Cladonia arbuscula.*

Iron, manganese, zinc. Lounamaa (1965), a quantitative study, as *Cladonia arbuscula.*

Zinc. Lambinon, Maquinay, and Ramaut (1964), from Belgium, a quantitative study, as *Cladonia arbuscula.*

^{210}Pb, ^{210}Po, ^{226}Ra. Holtzman (1966), from Alaska.

OTHER REPORTS. Fumarprotocetraric acid, usnic acid, unidentified substance (±): Dahl (1952), mention only. — Fumarprotocetraric acid, perlatolic acid, usnic acid (weakly levorotatory), ursolic acid: Breaden, Keane, and Nolan (1944), from Ireland, by extraction (apparently a mixture of species).

REVIEWS. Fumarprotocetraric acid, perlatolic acid, ursolic acid, usnic acid: Asahina and Shibata (1954). — Fumarprotocetraric acid, ursolic acid: Shibata (1958, 1963). — Fumarprotocetraric acid, silvatic acid, (+)-usnic acid: W. Brieger (1923); Hesse (1912); Thies (1932). — Fumarprotocetraric acid, (+)-usnic acid: Zopf (1907). — (+)-Usnic acid: Savicz, Litvinov, and Moissejeva (1960).

Except where indicated, all reports are as *Cladonia sylvatica* or *Cladina sylvatica*.

Cladonia atlantica Evans   Baeomycesic acid, squamatic acid. Dahl (1952), mention only; Evans (1944*b*), from Connecticut, and unidentified substance F (±) (which may be barbatic acid), type description; Evans (1950), from Connecticut; Evans (1952*b*), from Massachusetts south to North Carolina.

Cladonia aueri Räs.   Thamnolic acid. Dahl (1952), possibly also barbatic acid, mention only.

Cladonia bacillaris (Ach.) Nyl.   Barbatic acid (sometimes as coccellic acid or cenomycin), (−)-usnic acid (±); rhodocladonic acid. Dahl (1952), mention only, and didymic acid (±) (but see *Cladonia bacillaris* var. *pacifica*).

Barbatic acid, (−)-usnic acid, rhodocladonic acid. Zopf (1908), from Germany, no thamnolic acid.

Barbatic acid, usnic acid (±). Asahina (1939*g*); Evans (1943*a*), mention in discussions of microchemical tests; Evans (1944*b*), from Connecticut, and substance F (which may be barbatic acid), by microchemical tests on 269 samples, usnic acid relatively rare, some samples with didymic acid (but see *Cladonia bacillaris* var. *pacifica*).

Barbatic acid. Asahina (1936*e*), mention in description of microcrystal tests; Asahina (1937*d*), and as *Cladonia bacillaris* var. *clavata*; Kurokawa (1952), as *C. bacillaris* var. *clavata*; Mitsuno (1953).

(−)-Usnic acid. Salkowski (1910), as *Cladonia bacillaris* var. *clavata*.

OTHER REPORTS. Barbatic acid, didymic acid, usnic acid: Asahina (1950*c*) (but see *Cladonia bacillaris* var. *pacifica*). — Barbatic acid, didymic acid: Asahina (1939*e*), mention in description of microcrystal tests (but see *C. bacillaris* var. *pacifica*). — Didymic acid: Nogami (1944), mention in study on chemical structure of this substance; Asahina (1952*b*), from Japan, detection with $Ca(OCl)_2$ (but see *C. bacillaris* var. *pacifica*). — Cladestin, squamatic acid, usnic acid, zeorin: Klosa (1951*b*). — No usnic acid: Mozingo (1961), from Tennessee.

REVIEWS. Barbatic acid (as coccellic acid and cenomycin), (−)-usnic acid, rhodocladonic acid: Thies (1932), and as *Cladonia bacillaris* var. *clavata*; Zopf (1907), and as *C. bacillaris* var. *clavata*. — Bellidiflorin, barbatic acid: Shibata (1958). — Barbatic acid: Shibata (1963). — Rhodocladonic acid: W. Brieger (1923); Hesse (1912); both reports as *C. bacillaris* var. *clavata*. — Barbatic acid, didymic acid, rhodocladonic acid: Asahina and Shibata (1954), and as *C. bacillaris* var. *clavata*.

Cladonia bacillaris var. *pacifica* Asah. Barbatic acid, didymic acid, usnic acid (±). Asahina (1939c), from Japan, Formosa, and Sakhalin, type description; Evans (1944b), from Japan and Connecticut, not in Europe, suggests dropping this variety and considering didymic acid an accessory in *Cladonia bacillaris*.
Barbatic acid. Shibata (1934), by extraction.
REVIEWS. Didymic acid: Shibata (1958, 1963).

Cladonia bacillaris f. *reagens* Evans Barbatic acid, didymic acid (±); unidentified anthraquinone pigment. Evans (1944b), from Europe and Connecticut, suggests dropping *Cladonia bacillaris* f. *tingens* previously separated by Asahina for didymic acid-containing plants.

Cladonia bacillaris f. *tingens* Asah. Barbatic acid, didymic acid; unidentified anthraquinone pigment. Asahina (1939c), type description.

Cladonia bacilliformis (Nyl.) Vain. Barbatic acid, usnic acid. Asahina (1943b); Dahl (1952), mention only.
Usnic acid. Dahl (1950).

Cladonia balfourii Cromb. Fumarprotocetraric acid. Asahina (1940a), from North Carolina; Asahina (1943b); Dahl (1952), mention only; Evans (1952b), from Florida.

Cladonia beaumontii (Tuck.) Vain. Baeomycesic acid, squamatic acid. Dahl (1952), mention only; Evans (1950), from Connecticut, Massachusetts, New Jersey, New York, and North Carolina; Evans (1952b), from Massachusetts south to Florida.
Baeomycesic acid. Abbayes (1963), mention only.

Cladonia bellidiflora (Ach.) Schaer. Bellidiflorin, squamatic acid, (−)-usnic acid; rhodocladonic acid.
Bellidiflorin, squamatic acid, usnic acid, rhodocladonic acid. Dahl (1952), mention only.
Bellidiflorin, squamatic acid, usnic acid. Asahina (1938f), mention in description of microcrystal tests; Asahina (1939c), from Europe and Japan, and also *Cladonia bellidiflora* f. *tubaeformis* from Norway, no

zeorin; Asahina (1939g); Asahina (1950c), from Japan; Evans (1943a), from North America, mention in discussion of microchemical tests.

Bellidiflorin, squamatic acid. Hale (1954b).

Bellidiflorin, usnic acid. Dahl (1950).

Squamatic acid, (−)-usnic acid. Asahina and Tanase (1937), from Japan, by extraction.

Squamatic acid, usnic acid. Asahina (1938b), mention in description of microcrystal tests; Asahina (1938e), mention in description of microcrystal tests; D. Hess (1958), from Germany.

Squamatic acid. Evans (1951).

OTHER REPORT. No usnic acid: Gertig and Banasiewicz (1961), from Poland.

REVIEWS. Bellidiflorin, rhodocladonic acid, squamatic acid, (−)-usnic acid, zeorin: Thies (1932). — Rhodocladonic acid, (−)-usnic acid: W. Brieger (1923); Hesse (1912); Zopf (1907). — Rhodocladonic acid: Asahina and Shibata (1954).

Cladonia bellidiflora var. *coccocephala* (Ach.) Vain.   Bellidiflorin, squamatic acid, (−)-usnic acid; rhodocladonic acid; zeorin.

Bellidiflorin, squamatic acid, usnic acid, rhodocladonic acid, zeorin. Zopf (1908).

Bellidiflorin, squamatic acid, usnic acid. Asahina (1938f), mention in description of microcrystal tests.

REVIEWS. Bellidiflorin, rhodocladonic acid, squamatic acid, (−)-usnic acid, zeorin: Thies (1932). — Bellidiflorin, rhodocladonic acid, squamatic acid, usnic acid, zeorin: W. Brieger (1923); Zopf (1907). — Bellidiflorin, squamatic acid, usnic acid, zeorin: Hesse (1912). — Squamatic acid, zeorin: Asahina and Shibata (1954); Shibata (1958, 1963).

Cladonia borbonica (Del.) Nyl.   Fumarprotocetraric acid. Evans (1950), from Réunion, no grayanic acid as in *Cladonia cylindrica*.

Cladonia borbonica (Del.) Nyl. from North America, see *Cladonia cylindrica* (Evans) Evans

Cladonia borbonica f. *cylindrica* Evans, see *Cladonia cylindrica* (Evans) Evans

Cladonia boryi Tuck.   Usnic acid only. Asahina and Simosato (1938), lichen from south Sakhalin extracted by Asahina and Yanagita (1933b) probably a mixture, no depsides found by microchemical tests; Asahina (1942a), from Japan; Dahl (1952), mention only; Evans (1944b), from Connecticut; Evans (1950), mention only; Evans (1955a).

OTHER REPORT. Squamatic acid (as iso-squamatic acid), (−)-usnic acid, zeorin: Asahina and Yanagita (1933b).

Cladonia botryocarpa Merr.   Barbatic acid. Asahina (1942*b*), one sample with baeomycesic acid and squamatic acid possibly a different species; Dahl (1952), and an unidentified substance (±), mention only; Evans (1947*a*), from Florida, and unidentified substance F (which may be barbatic acid), by microcrystal tests; Evans (1952*b*), from Florida; Evans (1955*b*), from Florida.

Cladonia botrytes (Hag.) Willd.   Barbatic acid, usnic acid. Asahina (1943*b*); Asahina (1950*c*), from Japan; Dahl (1952), mention only; Mozingo (1961), mentioned in key.

Cladonia bouillennei Duvign.   Thamnolic acid, usnic acid. Duvigneaud (1939), type description; Evans (1951), from Belgium.

Cladonia brevis Sandst.   Psoromic acid. Dahl (1952), mention only; Evans (1944*b*), from Connecticut; Evans (1950), from Connecticut.
No atranorin or norstictic acid. Mozingo (1961).

Cladonia caespiticia (Pers.) Flörke   Fumarprotocetraric acid. Asahina (1942*b*), tested samples from Sandstede's exsiccati, previous report by Zopf of squamatic acid and atranorin probably based on contaminated material; Asahina (1950*c*), from Japan; Dahl (1952), mention only; Evans (1944*b*); Evans (1950), from Connecticut.
OTHER REPORTS. Atranorin, squamatic acid: Zopf (1908). — Atranorin, fumarprotocetraric acid, squamatic acid: Evans (1943*a*), mention in discussion of microchemical tests.
REVIEWS. Atranorin: W. Brieger (1923); Hesse (1912); Thies (1932).

Cladonia calycantha (Del.) Nyl.   Fumarprotocetraric acid only. Asahina (1940*b*), from Japan, no atranorin; Asahina (1943*b*); Asahina (1950*c*), from Formosa and Japan; Asahina (1956*a*), from Japan, as *Cladonia calycantha* f. *recurvans*; Asahina (1956*b*), from Formosa and Japan, and also *C. calycantha* var. *gracilior* and *C. calycantha* f. *recurvans* from Japan; Dahl (1952), mention only; Evans (1943*a*), mention in discussion of microcrystal tests.

Cladonia capitata (Michx.) Spreng.   Fumarprotocetraric acid. Asahina (1943*a*), from Alabama, New Jersey, North Carolina, and South Carolina, as *Cladonia mitrula*, also contains calcium oxalate, samples tested from Sandstede's exisccati including *C. mitrula* f. *abbreviata*, *C. mitrula* f. *imbricatula*, and *C. mitrula* f. *pallida*; Asahina (1950*c*), from Japan; Evans (1944*b*), from Connecticut; Evans (1950), from Connecticut; Evans (1952*b*), from Costa Rica, the U.S.A., and the West Indies; Pišút (1961).
OTHER REPORT. Atranorin (±), fumarprotocetraric acid: Dahl (1952), mention only.

Cladonia carassensis Vain.   Decarboxythamnolic acid, thamnolic acid. Asahina and Nuno (1964), by thin-layer chromatography.

Thamnolic acid, unidentified substances ($\pm$). Evans (1944*b*), from Connecticut, unknown substances F and G as accessories, by microcrystal tests; Evans (1950), from Connecticut and Oregon.

Thamnolic acid. Asahina (1942*b*), from Massachusetts and Oregon; Asahina (1959*d*), from Japan, as *Cladonia carassensis* ssp. *japonica*; Dahl (1952), mention only; Pišút (1961).

Cladonia cariosa (Ach.) Spreng.   Atranorin; unidentified substance. Asahina (1940*d*), unidentified substance could be rangiformic acid, by microcrystal tests; Asahina (1943*a*), possibly rangiformic acid, tested samples from Sandstede's exisccati; Asahina (1950*c*); Dahl (1952), mention only; Evans (1944*b*), from Connecticut; Evans (1950), from Connecticut, as *Cladonia cariosa* f. *cribosa*.

Atranorin. Hale (1954*b*), from Baffin Island.

OTHER REPORTS. Atranorin, methylethersalazinic acid ($\pm$), psoromic acid ($\pm$), unidentified substance ($\pm$): Dahl (1950). — Atranorin, bryopogonic acid: Zopf (1908), as *Cladonia cariosa* f. *squamulosa*.

REVIEWS. Atranorin: W. Brieger (1923); Hesse (1912).

Cladonia carneola (Fr.) Fr.   Usnic acid; zeorin. Asahina (1943*b*), microchemical tests on samples in Sandstede's exsiccati; Dahl (1952), mention only.

Usnic acid. Dahl (1950); Vartia (1949), from Finland, mention only.

Cladonia caroliniana (Schwein.) Tuck.   Usnic acid only. Asahina (1942*a*), from North America; Evans (1943*a*), from North America, mention in discussion of microchemical tests; Evans (1952*b*), from Florida, mention only; Evans (1955*a*), from North America; Hale (1957*c*), from West Virginia, as *Cladonia caroliniana* f. *dilatata*, an exsiccati specimen.

Usnic acid. Evans (1944*b*), from Connecticut, as *Cladonia caroliniana* f. *dilatata* with unidentified substances A ($\pm$) and B or only A ($\pm$), *C. caroliniana* f. *fibrillosa*, *C. caroliniana* f. *prolifera* with unidentified substances A ($\pm$) and B, and *C. caroliniana* f. *tenuiramea* with unidentified substance A ($\pm$); Evans (1950), from Connecticut, and unidentified substance A, (substances A and B later found to be crystal forms of usnic acid in the GE solution); Mozingo (1961).

Cladonia cenotea (Ach.) Schaer.   Squamatic acid (sometimes as uncinatic acid). Asahina (1942*b*), by microcrystal tests on samples from Sandstede's exsiccati; Asahina (1950*c*), from Japan and Korea; Dahl (1950), from Greenland; Dahl (1952), mention only; Evans (1943*a*), mention in discussion of microchemical tests; Evans (1944*b*), from Connecticut; Evans (1950), from Connecticut; Zopf (1908), reported as uncinatic acid.

OTHER REPORT. Squamatic acid, thamnolic acid: D. Hess (1958), from Germany.

REVIEWS. Squamatic acid: W. Brieger (1923), as *Cladonia uncinata*; Hesse (1912), as *C. uncinata*; Thies (1932), as *C. uncinata*; Zopf (1907), and as *C. uncinata*.

Cladonia ceratophylla (Sw.) Spreng.
Fumarprotocetraric acid. Asahina (1942*b*), tentative identification, could not repeat KC+ reaction reported by Sandstede; Dahl (1952), mention only.

Cladonia ceratophyllina Vain. Atranorin (±), fumarprotocetraric acid.
Atranorin (±). Asahina (1955*a*), from India and Japan.
Fumarprotocetraric acid. Asahina (1950*c*), from Japan.

Cladonia cervicornis Ach., see *Cladonia verticillata* var. *cervicornis* (Ach.) Flörke

Cladonia chlorophaea (Flörke) Spreng. Fumarprotocetraric acid. Asahina (1940*a*), from Japan and North America; Asahina (1943*b*); Asahina (1950*c*), from Japan; Evans (1943*a*), in discussion of microcrystal tests; Evans (1952*b*), from Florida, as *Cladonia chlorophaea* f. *simplex*, mention only; Hale (1954*b*), from Baffin Island; D. Hess (1958), from Germany; Mozingo (1961), as the fumarprotocetraric acid strain; Shibata (1965), mention in review on biosynthesis; Shibata and Chiang (1965), mention, chemical study on cryptochlorophaeic acid and merochlorophaeic acid.
OTHER REPORTS. Fumarprotocetraric acid, chlorophaeic acid: Zopf (1908), from Germany, by extraction, no atranorin. — Fumarproto-cetraric acid, chlorophaeic acid (±): Asahina (1939*e*), description of microcrystal tests and discussion. — Fumarprotocetraric acid, unidenti-fied substance (±): Dahl (1952), mention only. — Atranorin, fumar-protocetraric acid: Asahina (1941*a*), as *Cladonia chlorophaea* f. *conistea*. — Atranorin (±), fumarprotocetraric acid: Evans (1944*b*), discusses taxonomic history of *C. grayi*, *C. cryptochorophaea*, and *C. chlorophaea* (pp. 593–596). — Many chemical races containing chlorophaeic acid, fumarprotocetraric acid, grayanic acid, merochlorophaeic acid, novo-chlorophaeic acid, usnic acid, and an unidentified substance in various combinations: Dahl (1950), as *C. pyxidata* ssp. *chlorophaea*. — Fumar-protocetraric acid (Chem. strain 1) or grayanic acid (Chem. strain 2) or cryptochlorophaeic acid (Chem. strain 3) or merochlorophaeic acid (Chem. strain 4) or novochlorophaeic acid (Chem. strain 5) or usnic acid (Chem. strain 6) or none of these substances ("acid free phase"): Lamb (1951*a*), suggested taxonomic treatment; Lamb (1951*b*). — Fumarpro-tocetraric acid (±) and cryptochlorophaeic acid (cryptochlorophaeic acid strain) or fumarprotocetraric acid (±) and grayanic acid (grayanic acid strain): Mozingo (1961). — Usnic acid: Tomaselli (1957), from Italy, mention only.

REVIEWS. Fumarprotocetraric acid: Asahina and Shibata (1954); Neelakantan (1965); Shibata (1958, 1963).— Chlorophaeic acid, fumarprotocetraric acid: W. Brieger (1923); Hesse (1912); Thies (1932); Zopf (1907).

Cladonia chondrotypa Vain. Decarboxythamnolic acid, thamnolic acid. Asahina and Nuno (1964), by thin-layer chromatography. Thamnolic acid. Asahina (1941*d*), from Formosa; Asahina (1942*b*).

Cladonia clavulifera Vain. Fumarprotocetraric acid. Asahina (1943*a*), from Japan, Connecticut, North Carolina, and Vermont, and also *Cladonia clavulifera* f. *subfastigiata* and *C. clavulifera* f. *subvestita*, microchemical tests on specimens from Sandstede's exsiccati; Asahina (1950*c*), from Japan; Dahl (1952), mention only; Evans (1944*a*); Evans (1944*b*), from Connecticut and Maine; Evans (1950), from Connecticut; Evans (1952*b*).
OTHER REPORT. No atranorin or norstictic acid: Mozingo (1961), mention only.

Cladonia coccifera (L.) Willd. Barbatic acid (sometimes as cenomycin or coccellic acid), (−)-usnic acid; rhodocladonic acid. Zopf (1908), from Tirol, no zeorin, as *Cladonia coccifera* var. *stemmatina*.
Barbatic acid, usnic acid, rhodocladonic acid. Klosa (1951*b*), mention only.
Barbatic acid, usnic acid. Asahina (1939*g*); Evans (1943*a*), mention only, in discussion of microchemical tests; Evans (1944*b*), from Labrador and Quebec; Evans (1955*a*), mention only; Hale (1954*b*), from Baffin Island.
Barbatic acid. Asahina (1937*d*), as *Cladonia coccifera* var. *stemmatina*, description of microcrystal tests; Evans (1955*b*).
(−)-Usnic acid. Salkowski (1910), as *Cladonia coccifera* var. *stemmatina*.
Usnic acid. Gertig and Banasiewicz (1961), from Poland, 0.78%; D. Hess (1958), from Germany; D. Hess (1960), mention only; Vartia (1949), from Finland, mention only; Vartia (1950*b*), from Finland.
OTHER REPORTS. Barbatic acid (in *Cladonia coccifera* var. *stemmatina*), squamatic acid (±), unidentified substances (±): Dahl (1950), many samples tested microchemically, zeorin-containing samples referred to *C. pleurota*. — Barbatic acid (±), rhodocladonic acid, usnic acid, zeorin (±): Dahl (1952), see also *C. pleurota*.
REVIEWS. Barbatic acid, rhodocladonic acid, (−)-usnic acid: Thies (1932), as *Cladonia coccifera* f. *extensa*, *C. coccifera* var. *minuta*, and *C. coccifera* var. *stemmatina*. — Barbatic acid, rhodocladonic acid, (−)-usnic acid, zeorin: W. Brieger (1923); Hesse (1912); Zopf (1907), and as *C. coccifera* var. *stemmatina*. — Barbatic acid: Asahina and Shibata (1954),

and as *C. coccifera* var. *stemmatina*; Shibata (1958, 1963), as *C. coccifera* var. *stemmatina*. — Lecanora-red: Thies (1932), in the apothecia. — Rhodocladonic acid: W. Brieger (1923); Hesse (1912); both reports as *C. coccifera* var. *extensa* and as *C. coccifera* f. *minuta*.

Cladonia coccifera var. *pleurota* (Flörke) Schaer., see *Cladonia pleurota* (Flörke) Schaer.

Cladonia confusa Sant. Perlatolic acid (±), usnic acid. Evans (1955*a*), from Jamaica and Mexico.

Cladonia coniocraea (Flörke) Spreng. Fumarprotocetraric acid only. Asahina (1943*b*), from Jamaica, Connecticut, New Jersey, Pennsylvania, and Vermont, microchemical tests on samples from Sandstede's exsiccati, previous report (by Zopf) of atranorin not confirmed microchemically; Asahina (1950*c*); Dahl (1950), from Greenland, discussion; Dahl (1952), mention only; Evans (1944*b*), from Connecticut, discussion, samples with atranorin referred to *Cladonia ochrochlora*.

OTHER REPORT. Atranorin, fumarprotocetraric acid: Zopf (1908), from Germany, as *Cladonia fimbriata* var. *apolepta* f. *coniocraea*.

REVIEWS. Atranorin, fumarprotocetraric acid: W. Brieger (1923); Hesse (1912); both reports as *Cladonia fimbriata* var. *apolepta* f. *coniocraea*. — Atranorin: Thies (1932), as *C. fimbriata* var. *coniocraea* and *C. fimbriata* var. *apolepta* f. *coniocraea*.

Cladonia conista (Ach.) Robb. Fumarprotocetraric acid; unidentified substance. Asahina (1943*b*); Dahl (1952), as *Cladonia fimbriata* var. *ambigua*; Evans (1944*b*), from Connecticut, unidentified substance H, and *C. conista* f. *simplex*; Evans (1950), from Connecticut, as *C. conista* f. *simplex*, unidentified substance H.

Fumarprotocetraric acid. Dahl (1952), mention only.

OTHER REPORTS. Atranorin (±), fumarprotocetraric acid, unidentified substance: Asahina (1941*a*), from Asia, Europe, and North America, and *Cladonia fimbriata* var. *ambigua* and *C. conista* f. *simplex*, and as *C. fimbriata* f. *conista*, unidentified substance found by microcrystal tests (GE and GAW solutions). — Atranorin: Asahina (1940*a*).

Cladonia conistea (Del.) Asah. Atranorin, fumarprotocetraric acid. Asahina (1943*b*); Asahina (1950*c*), from Japan; Dahl (1952), mention only.

Cladonia convoluta (Lam.) Cout., see *Cladonia endiviaefolia* Fr.

Cladonia corallifera (Kunze) Nyl. Bellidiflorin, didymic acid, usnic acid; rhodocladonic acid. Dahl (1952), mention only.

Bellidiflorin, didymic acid, usnic acid. Asahina (1939c), from South America; Asahina (1960b), discussion, plants with no didymic acid not considered as *Cladonia corallifera*, by microcrystal tests.

Bellidiflorin, usnic acid. Asahina (1939e), from North America, but see also Asahina (1960b); Asahina (1939g).

OTHER REPORT. Squamatic acid, usnic acid, bellidiflorin: Evans (1944b), a misquote of Asahina.

Cladonia corallifera* var. *gracilescens (Nyl.) Vain.   Squamatic acid, usnic acid. Asahina (1960b), no didymic acid.

Cladonia corallifera* var. *kunzeana Vain.   Thamnolic acid ($\pm$), usnic acid. Asahina (1960b), no bellidiflorin, didymic acid, or squamatic acid.

OTHER REPORTS. Bellidiflorin, squamatic acid, usnic acid, zeorin: Asahina (1939g). — Bellidiflorin, squamatic acid, usnic acid: Evans (1951), mention only.

Cladonia cornuta (L.) Hoffm.   Fumarprotocetraric acid; ascorbic acid.

Fumarprotocetraric acid. Asahina (1943b), from Japan, and *Cladonia cornuta* f. *cylindrica*, *C. cornuta* f. *scyphosa*, and *C. cornuta* f. *subdilata*, also some samples from Sandstede's exsiccati; Asahina (1950c), from Japan; Dahl (1950); Hale (1954b), from Baffin Island; Zopf (1908), from Germany, no atranorin.

Ascorbic acid. Karev and Kochevykh (1962).

Cladonia cornutoradiata (Coem.) Vain., see *Cladonia subulata* (L.) Wigg.

Cladonia corymbosula Nyl.   Barbatic acid. Asahina (1943a), from Cuba and Honduras.

Cladonia crenulata (Ach.) Flörke, see *Cladonia deformis* (L.) Hoffm.

Cladonia crispata (Ach.) Flot.   Squamatic acid; ascorbic acid.

Squamatic acid only. Asahina (1942b), from Japan, and *Cladonia crispata* var. *cetrariaeformis*, *C. c.* var. *dilacerata*, *C. c.* f. *divulsa*, *C. c.* var. *elegans*, *C. c.* var. *gracilescens*, *C. c.* var. *infundibulifera*, *C. c.* f. *jesoensis*, *C. c.* f. *rigidula*, *C. c.* var. *virgata*, also tested microchemically all varieties and forms in Sandstede's exsiccati; Asahina (1950c), from Japan; Dahl (1950), from Greenland; Dahl (1952), mention only; Evans (1943a), from North America, mention in discussion of microchemical tests; Evans (1944b); Evans (1955b); Zopf (1908), by extraction of two samples, as *C. crispata* var. *gracilescens* from Germany, and as *C. crispata* var. *virgata* from Austria.

Ascorbic acid. Karev and Kochevykh (1962).

OTHER REPORT. Fumarprotocetraric acid, squamatic acid: D. Hess (1958), from Germany.

REVIEWS. Cladonin, squamatic acid: W. Brieger (1923); Hesse (1912); Thies (1932); as *Cladonia crispata* var. *gracilescens*. — Squamatic acid: Zopf (1907), as *C. crispata* var. *gracilescens*.

Cladonia cristatella Tuck. Barbatic acid, didymic acid, usnic acid; rhodocladonic acid. Dahl (1952), mention only.

Barbatic acid, didymic acid, usnic acid. Asahina (1939*g*); Evans (1943*a*), mention in discussion of microcrystal tests; Evans (1944*b*), from Connecticut, and substance F (which may be barbatic acid).

Barbatic acid, didymic acid, rhodocladonic acid. Hale (1956*b*), as *Cladonia cristatella* f. *beauvoisii*; Hale (1956*g*), from West Virginia, an exsiccati specimen.

Barbatic acid, didymic acid, usnic acid ($\pm$). Evans (1952*b*), usnic acid present in samples from Massachusetts, Minnesota, Ohio, Vermont, and West Virginia and absent in samples from Alabama, Arkansas, Florida, and Georgia.

Barbatic acid. Robinson (1959), from North Carolina.

Didymic acid, usnic acid, rhodocladonic acid. Ahmadjian (1964), mention only, red and yellow pigments in cultures of fungal component not rhodocladonic and usnic acids; Castle and Kubsch (1949), also produced by the lichen fungus in culture.

Didymic acid. Asahina (1939*e*), description of microcrystal tests.

Cladonia cryptochlorophaea Asah. Cryptochlorophaeic acid, fumarprotocetraric acid. Asahina (1940*a*), from Japan and North America, type description; Asahina (1943*b*); Asahina (1950*c*), from Japan and Korea; Dahl (1952), mention only; Evans (1943*a*), discussion of microchemical tests; Evans (1950), from Connecticut.

Crytochlorophaeic acid, fumarprotocetraric acid ($\pm$) (see also *Cladonia cryptochlorophaea* f. *inactiva*). Evans (1944*b*), from Connecticut; Shibata and Chiang (1965), from Japan, 0.16% cryptochlorophaeic acid by extraction, a detailed chemical study on cryptochlorophaeic acid.

Cryptochlorophaeic acid. Shibata (1965), mentioned in review on biosynthesis.

REVIEW. Cryptochlorophaeic acid, fumarprotocetraric acid: Neelakantan (1965).

Cladonia cryptochlorophaea f. *inactiva* Asah. Cryptochlorophaeic acid. Asahina (1940*a*), from North America, not in Japan, no fumarprotocetraric acid; Asahina (1950*c*), from North America.

Cladonia cubana (Vain.) Evans

No constant lichen substance: baeomycesic acid or barbatic acid and substance F (which may be barbatic acid), possibly a trace of fumarprotocetraric acid in some samples (by color test), no atranorin, no squamatic acid. Evans (1947*a*), from British Guiana, Cuba, Jamaica, and Puerto Rico.

Cladonia cyanipes (Somm.) Nyl.   Barbatic acid, (−)-usnic acid. Asahina 1950*c*).
   Barbatic acid, usnic acid. Asahina (1943*b*), microchemical tests on specimens from Sandstede's exsiccati, also *Cladonia cyanipes* f. *connectans*; Dahl (1952), mention only; Evans (1947*b*), from Vermont; Hale (1954*b*), from Baffin Island.
   (−)-Usnic acid. Zopf (1908), mention only.
   Usnic acid. Dahl (1950); Evans (1943*a*), mention in discussion of microchemical tests; Zopf (1900), from Tirol, by extraction.
   REVIEWS. (−)-Usnic acid: W. Brieger (1923); Hesse (1912); Thies (1932); Zopf (1907).

Cladonia cylindrica (Evans) Evans   Fumarprotocetraric acid, grayanic acid. Asahina (1940*a*), from North America, as *Cladonia borbonica* f. *cylindrica* ; Asahina (1943*b*), as *C. borbonica*; Dahl (1952), mention only, as *C. borbonica*; Evans (1943*a*), as *C. borbonica*, mention in discussion of microchemical tests; Evans (1944*b*), from Connecticut, as *C. borbonica*; Evans (1950), from Connecticut, *C. borbonica* f. *cylindrica* containing grayanic acid raised to the rank of species, see *C. borbonica*.
   Grayanic acid. Evans (1952*b*); Mozingo (1961), from Tennessee, mentioned in key.

Cladonia decorticata (Flörke) Spreng.   Perlatolic acid. Asahina (1943*a*); Dahl (1952), mention only; Evans (1950), from Connecticut; Hale (1956*f*).
   OTHER REPORTS. Unidentified substance: Dahl (1950); Evans (1944*b*).
— (−)-Usnic acid: Stoll, Renz, and Brack (1947), from Switzerland.

Cladonia deformis (L.) Hoffm.   Bellidiflorin, (−)-usnic acid; rhodocladonic acid; taraxerene, zeorin; enzymes; dimethyl sulfone.
   Bellidiflorin (±), usnic acid, rhodocladonic acid, zeorin. Dahl (1952), mention only as *Cladonia deformis* f. *crenulata*.
   Bellidiflorin, usnic acid, rhodocladonic acid. Pišút (1961), mention only.
   Bellidiflorin, usnic acid, zeorin. Asahina (1950*c*), from Japan.
   Bellidiflorin (±), usnic acid, zeorin. Asahina (1939*c*), from Japan and Sakhalin, as *Cladonia crenulata*; Evans (1943*a*), mention in description of microcrystal tests; Evans (1944*b*), from North America, discussion of taxonomy and chemistry of this and related species; Thomson (1963), mention only.
   (−)-Usnic acid, rhodocladonic acid, zeorin. Zopf (1908), and two colorless acids, mention only.

Usnic acid, zeorin. Asahina (1939*g*), as *Cladonia deformis* f. *crenulata*; Ryabinin and Matyukhina (1957), 0.25% zeorin by extraction; Zopf (1900), from western Austria, and two colorless substances, 3.3% usnic acid by extraction.

(−)-Usnic acid. Moissejeva (1957), 3.0%; Moissejeva (1961), 3.0%, enzymes tabulated; Savicz, Kuprevicz, Litvinov, Moissejeva, and Rassadina (1956), 3.0% by extraction; Stoll, Brack, and Renz (1947), 1.0% by extraction; Stoll, Renz, and Brack (1947), from Switzerland.

Usnic acid. D. Hess (1958), from Germany, by paper chromatography; Vartia (1950*b*), from Finland.

Taraxerene. Bruun (1954*b*), from Norway, a detailed chemical study; Shibata (1965), mentioned in discussion of biosynthesis.

Dimethyl sulfone. Bruun and Sörensen (1954), from Norway, a detailed chemical study.

OTHER REPORTS. Usnic acid, zeorin, cladonic acid: Klosa (1951*b*). For discussions of the relationship of this species to *Cladonia gonecha* see Asahina (1939*c*), Dahl (1950), Evans (1944*b*), and Pišút (1961).

REVIEWS. Rhodocladonic acid, (−)-usnic acid, zeorin: W. Brieger (1923); Hesse (1912); Thies (1932); Zopf (1907), and two colorless acidic substances. — Rhodocladonic acid: Asahina and Shibata (1954); W. Brieger (1923), as *Cladonia deformis* f. *alpestris*, *C. d.* f. *crenulata*, and *C. d.* f. *phyllocephala*; Thies (1932), as *C. deformis* f. *alpestris*, *C. d.* f. *crenulata*, and *C. d.* f. *phyllocephala*. — (−)-Usnic acid: Savicz, Litvinov, and Moissejeva (1960). — Taraxerene, dimethyl sulfone: Shibata (1963). — Dimethyl sulfone: Neelakantan (1965); Shibata (1958).

See also *Cladonia gonecha*.

Cladonia deformis f. *gonecha* Ach., see *Cladonia gonecha* (Ach.) Asah.

Cladonia degenerans (Flörke) Spreng. Fumarprotocetraric acid. Asahina (1943*b*); Asahina (1950*c*), from Japan; Dahl (1950), no atranorin; Dahl (1952), mention only; Hale (1954*b*), from Baffin Island.

Cladonia degenerans var. *haplothea* Ach. Unidentified colorless acid. Zopf (1908), from western Austria, no fumarprotocetraric acid.

Cladonia delessertii (Nyl.) Vain. Squamatic acid. Asahina (1942*b*), and a substance which hinders the microcrystal test in the GAAn solution; Asahina (1950*c*), from Japan; Dahl (1950); Dahl (1952), mention only.

Cladonia delicata (Ehrh.) Flörke, see *Cladonia parasitica* (Hoffm.) Hoffm.

Cladonia denticollis Hoffm., see *Cladonia squamulosa* (Scop.) Hoffm.

Cladonia destricta (Nyl.) Nyl. Usnic acid; destrictinic acid. Asahina (1942*a*), tested samples from Sandstede's exsiccati, no squamatic acid;

Dahl (1952), mention only; Evans (1944*b*), and substances A (±) and B (±) (later found to be crystal forms of usnic acid in the GE solution), no squamatic acid.

OTHER REPORTS. Squamatic acid, usnic acid, cladestin, destrictinic acid: Klosa (1951*b*). — Squamatic acid, (−)-usnic acid, cladestin, destrictinic acid: Zopf (1908), from Germany.

REVIEWS. Squamatic acid, (−)-usnic acid, cladestin, cladestinic acid, destrictinic acid: W. Brieger (1923), and as *Cladina destricta*; Hesse (1912), and as *Cladina destricta*; Thies (1932), and also destrictaic acid, and as *Cenomyce destricta* and *Cladina destricta*. — Squamatic acid, (−)-usnic acid, cladestin, destrictinic acid: Zopf (1907), and as *Cladina destricta*.

Cladonia didyma (Fée) Vain.   Barbatic acid, didymic acid; rhodocladonic acid. Dahl (1952), mention only.

Barbatic acid, didymic acid. Asahina (1939*c*), from South America; Asahina (1939*g*); Evans (1943*a*), mentioned in discussion of micro-chemical tests; Mozingo (1961), from Tennessee, mentioned in key.

Barbatic acid (±), didymic acid (±). Evans (1944*b*), and substance F (which may be barbatic acid); Evans (1947*a*); Evans (1952*b*), from Florida.

Didymic acid. Asahina (1939*e*), from Japan and North America, description of microcrystal tests; Asahina (1951*i*); Nogami (1944), mention only.

OTHER REPORT. Barbatic acid, didymic acid, thamnolic acid: Mozingo (1961), from Tennessee, as the "thamnolic acid strain (*Cladonia vulcanica*)."

REVIEWS. Didymic acid: Asahina and Shibata (1954).—Rhodocla-donic acid: W. Brieger (1923); Thies (1932); both reports as *Cladonia didyma* var. *muscigena*.

Cladonia didyma var. *vulcanica* (Zoll.) Vain., see *Cladonia vulcanica* Zoll.

Cladonia digitata (L.) Hoffm.   Bellidiflorin, decarboxythamnolic acid, thamnolic acid, rhodocladonic acid.

Bellidiflorin, thamnolic acid, rhodocladonic acid. Dahl (1952), mention only.

Bellidiflorin, thamnolic acid. Asahina (1939*c*), from Europe, Japan, and Sakhalin, as *Cladonia digitata* f. *brachytes*, *C. d.* var. *glabrata*, and *C. d.* var. *monstrosa*; Asahina (1939*g*); Asahina (1950*c*), from Japan, Korea, and the Kurile Islands.

Bellidiflorin (±), thamnolic acid. Evans (1943*a*), mention only; Evans (1944*b*), from Connecticut, and unidentified substance G.

Decarboxythamnolic acid, thamnolic acid. Asahina and Nuno (1964), by thin-layer chromatography.

Thamnolic acid, rhodocladonic acid. Zopf (1908), as *Cladonia digitata* f. *brachytes* and *C. digitata* var. *monstrosa*, no usnic acid, no zeorin, no barbatic acid (as coccellic acid and cenomycin).

Thamnolic acid. Asahina (1938*e*), description of microcrystal tests; Dahl (1950); Duvigneaud (1939), no usnic acid; D. Hess (1958), from Germany; Kurokawa (1964), mention only; Vartia (1950*b*), from Finland.

No usnic acid. Mozingo (1961), mentioned in key.

OTHER REPORTS. Thamnolic acid, usnic acid, zeorin: Klosa (1951*b*). — (−)-Usnic acid: Stoll, Renz, and Brack (1947), from Switzerland. — Usnic acid: Tomaselli (1957), from Italy, mention only.

REVIEWS. Rhodocladonic acid, thamnolic acid, usnic acid: Zopf (1907). — Rhodocladonic acid, thamnolic acid: Asahina and Shibata (1954); Thies (1932). — Rhodocladonic acid: W. Brieger (1923), as *Cladonia digitata* f. *brachytes*, *C. d.* var. *ceruchoides*, *C. d.* var. *glabrata*, and *C. d.* var. *monstrosa*; Hesse (1912), as *C. digitata* var. *monstrosa*; Thies (1932), as *C. digitata* f. *brachytes*, *C. d.* var. *ceruchoides*, *C. d.* var. *glabrata*, and *C. d.* var. *monstrosa*. — Thamnolic acid: W. Brieger (1923); Hesse (1912); Shibata (1963).

Cladonia diplotypa Nyl. Thamnolic acid. Abbayes (1962), from Africa.

Cladonia dissimilis (Asah.) Asah. Atranorin, fumarprotocetraric acid (±), homosekikaic acid. Asahina (1950*c*), from Japan, fumarprotocetraric acid by color reaction; Asahina (1956*b*), from Japan, fumarprotocetraric acid by color reaction.

Atranorin, homosekikaic acid. Dahl (1952), mention only; Kurokawa (1957), from Japan.

Cladonia divaricata Nyl. Fumarprotocetraric acid, usnic acid. Evans (1952*b*), from Brazil and Cuba.

Cladonia ecmocyna (Ach.) Nyl. Atranorin, fumarprotocetraric acid. Asahina (1943*b*), tested samples from Sandstede's exsiccati; Asahina (1950*c*), from Japan; Dahl (1950), from Greenland; Dahl (1952), mention only; Evans (1952*a*), from northern North America; Pišút (1961), mention only.

Cladonia ecmocyna f. *foveata* Dahl
No atranorin (?). Dahl (1950), from Greenland, discussion difficult to interpret.

Cladonia elongata (Jacq.) Hoffm. Atranorin (±), fumarprotocetraric acid; ascorbic acid.
Atranorin (±), fumarprotocetraric acid. Asahina (1943*b*), as *Cladonia gracilis* var. *elongata*.

Atranorin, fumarprotocetraric acid. Zopf (1908), from Tirol, as *Cladonia gracilis* var. *elongata*.

Atranorin. Evans (1943*a*), mention in discussion of microcrystal tests. No atranorin. Černohorský (1959), as *Cladonia gracilis* var. *elongata*. Fumarprotocetraric acid. Hale (1954*b*), from Baffin Island. Ascorbic acid. Karev and Kochevykh (1962).

REVIEWS. Atranorin, fumarprotocetraric acid: W. Brieger (1923); Hesse (1912); Thies (1932); Zopf (1907). All review reports are as *Cladonia gracilis* var. *elongata*.

Cladonia endiviaefolia Fr.   Fumarprotocetraric acid, (−)-usnic acid; mannitol.

Fumarprotocetraric acid, (−)-usnic acid. Zopf (1908), from Capri, as *Cladonia foliacea* var. *convoluta*.

Fumarprotocetraric acid, usnic acid. Evans (1952*b*), as *Cladonia convoluta*.

(−)-Usnic acid. Salkowski (1910), as *Cladonia foliacea* var. *convoluta*. Mannitol. Pueyo (1964–1965*a*), from France, 2.3% isolated.

REVIEWS. Fumarprotocetraric acid: W. Brieger (1923); Hesse (1912); Thies (1932); all review reports as *Cladonia foliacea* var. *convoluta*.

Cladonia endoxantha Vain., see *Cladonia hypoxantha* Tuck.

Cladonia erythrosperma Vain.   Thamnolic acid. Abbayes (1962), mention only.

Cladonia evansii Abb.   Atranorin, perlatolic acid. Asahina (1940*f*), description of microcrystal tests; Dahl (1952), mention only; Evans (1943*b*); Evans (1952*b*), from Cuba and southeastern U.S.A., usnic acid possibly an accessory substance according to a private communication from Asahina; Evans (1955*a*); Hale (1956*b*); Hale (1956*g*), from Alabama, an exsiccati specimen.

Atranorin. C. F. Culberson (1964), used as an authentic source of this substance.

OTHER REPORT. Atranorin, unidentified substance: Duvigneaud and Bleret (1940), by microchemical tests, (description of the unknown suggests atranorin).

REVIEWS. Perlatolic acid: Asahina and Shibata (1954). — Imbricaric acid: Shibata (1958, 1963).

Cladonia fallax Abb., see *Cladonia pycnoclada* (Gaud.) Nyl.

Cladonia farinacea (Vain.) Evans   Fumarprotocetraric acid; ursolic acid (±). Evans (1950), from eastern U.S.A.

Fumarprotocetraric acid. Dahl (1952), mention only; Hale (1954*b*), from Baffin Island.

Cladonia fauriei Abb. Thamnolic acid, usnic acid. Evans (1951), from Hawaii.

Cladonia fimbriata (L.) Fr. Fumarprotocetraric acid; zinc.
Fumarprotocetraric acid. Asahina (1943*b*); Dahl (1950), from Greenland, no atranorin, discussion of early work by Zopf; Dahl (1952), mention only; Evans (1944*b*), from Connecticut; Hale (1954*b*), from Baffin Island.
Zinc. Lambinon, Maquinay, and Ramaut (1964), from Belgium, a quantitative study.
OTHER REPORT. Atranorin, fumarprotocetraric acid (and as fimbriatic acid), usnic acid: Klosa (1951*b*).
REVIEWS. Fumarprotocetraric acid: Asahina and Shibata (1954); Shibata (1958, 1963).

Cladonia fimbriata var. *ambigua* Asah., see *Cladonia conista* (Ach.) Robb.

Cladonia fimbriata var. *apolepta* f. *coniocraea* (Flörke) Sandst., see *Cladonia coniocraea* (Flörke) Spreng.

Cladonia fimbriata var. *coniocraea* (Flörke) Vain., see *Cladonia coniocraea* (Flörke) Spreng.

Cladonia fimbriata f. *conista* (Ach.) Nyl., see *Cladonia conista* (Ach.) Robb.

Cladonia fimbriata var. *cornutoradiata* (Coem.) Coem., see *Cladonia subulata* (L.) Wigg.

Cladonia fimbriata var. *cornutoradiata* f. *nemoxyna* (Ach.) Parrique, see *Cladonia nemoxyna* (Ach.) Nyl.

Cladonia fimbriata f. *fibula* (Ach.) Vain.
REVIEWS. Rhodocladonic acid, thamnolic acid, (−)-usnic acid: W. Brieger (1923); Hesse (1912). — Rhodocladonic acid, thamnolic acid. Thies (1932).

Cladonia fimbriata var. *simplex* (Weiss) Flot.
REVIEWS. Atranorin, thamnolic acid: Thies (1932), as *Cladonia fimbriata* var. *tubaeformis*. — Thamnolic acid, (−)-usnic acid: W. Brieger (1923); Hesse (1912); both reports as *C. fimbriata* var. *tubaeformis*. — Fumarprotocetraric acid (as fimbriatic acid): W. Brieger (1923); Hesse (1912); Zopf (1907).

Cladonia fimbriata var. *simplex* f. *major* (Hag.) Vain., see *Cladonia major* (Hag.) Sandst.

Cladonia fimbriata var. *simplex* f. *minor* (Hag.) Magn. Fumarprotocetraric acid (sometimes as fimbriatic acid). Asahina (1943*b*); Zopf (1908), from Germany, no atranorin.
REVIEWS. Fumarprotocetraric acid: W. Brieger (1923); Hesse (1912); Thies (1932); Zopf (1907), no atranorin.

Cladonia fimbriata var. *tubaeformis* Ach., see *Cladonia fimbriata* var. *simplex* (Weiss) Flot.

Cladonia flabelliformis (Flörke) Vain. Thamnolic acid. Duvigneaud (1939), no usnic acid; Koller and Hamburg (1935*b*), mention only.
OTHER REPORT. Thamnolic acid, usnic acid: Klosa (1951*b*), mention only.

Cladonia flabelliformis var. *polydactyla* (Flörke) Vain. Thamnolic acid, (−)-usnic acid. Asahina and Ihara (1929).
REVIEWS. Rhodocladonic acid, thamnolic acid, (−)-usnic acid: Thies (1932). — Rhodocladonic acid: W. Brieger (1923).

Cladonia floerkeana (Fr.) Somm. Barbatic acid (sometimes as coccellic acid), didymic acid (±), usnic acid; rhodocladonic acid. Dahl (1952), mention only.
Barbatic acid, didymic acid, usnic acid (±). Asahina (1950*c*), usnic acid rare.
Barbatic acid, didymic acid (±), usnic acid (±). Evans (1944*b*), from Connecticut, and also unidentified substance F (possibly identical to barbatic acid), by microcrystal tests, discussion of the occurrence of didymic acid; Evans (1952*b*), from Florida.
Barbatic acid, didymic acid. Asahina (1939*e*), description of microcrystal tests.
Barbatic acid. Asahina (1936*e*), description of microcrystal tests; Asahina (1937*d*), description of microcrystal tests; Asahina (1939*g*); Asahina (1942*b*), as *Cladonia floerkeana* var. *trachypoda*; Dahl (1950), from Greenland; Evans (1955*a*); D. Hess (1958), from Germany; Kurokawa (1964), mention only.
Didymic acid. Nogami (1944), mention, study on chemical structure.
REVIEWS. Barbatic acid, didymic acid, rhodocladonic acid: Asahina and Shibata (1954). — Barbatic acid, didymic acid: Shibata (1958, 1963). — Barbatic acid, rhodocladonic acid: Zopf (1907). — Barbatic acid, thamnolic acid, rhodocladonic acid: W. Brieger (1923); Hesse (1912); Thies (1932), and *Cladonia floerkeana* f. *carcata*.

Cladonia floerkeana var. *alpina* Asah. Barbatic acid; unidentified substance. Asahina (1939*c*), from Japan, by microcrystal tests, type description.

Cladonia floerkeana "f. *carcata*," see *Cladonia floerkeana* (Fr.) Somm.

Cladonia floerkeana var. *intermedia* Hepp   Barbatic acid (sometimes as coccellic acid or coenomycin); rhodocladonic acid. Zopf (1908), from Germany, no thamnolic acid or usnic acid.

Barbatic acid. Asahina (1937*d*), description of microcrystal tests; Asahina (1939*c*), from Europe and southern Sakhalin.

OTHER REPORT. Didymic acid: Evans (1944*b*), from Connecticut.

REVIEWS. Barbatic acid, rhodocladonic acid, thamnolic acid: W. Brieger (1923); Hesse (1912); Thies (1932). — Barbatic acid, rhodocladonic acid: Zopf (1907).

Cladonia floerkeana var. *suboceanica* Asah.   Barbatic acid, didymic acid, usnic acid ($\pm$). Asahina (1939*c*), from Formosa and Japan, usnic acid rare.

Barbatic acid. Shibata (1934), by extraction.

REVIEWS. Didymic acid: Asahina and Shibata (1954). — Barbatic acid, didymic acid: Shibata (1958, 1963).

Cladonia floerkeana f. *tingens* Asah.   Unidentified pigment (K + purple). Asahina (1939*c*).

Cladonia foliacea (Huds.) Schaer.   Fumarprotocetraric acid, ($-$)-usnic acid. Asahina (1943*b*); Zopf (1908), from Langeoog, as *Cladonia foliacea* var. *alcicornis*.

Fumarprotocetraric acid, usnic acid. Evans (1944*b*); Evans (1950); Evans (1952*b*); Pišút (1961), mention only.

OTHER REPORTS. Atranorin, fumarprotocetraric acid, usnic acid: Dahl (1952), mention only. — No usnic acid: Gertig and Banasiewicz (1961), from Poland, as *Cladonia foliacea* var. *alcicornis*.

REVIEWS. Fumarprotocetraric acid: W. Brieger (1923); Hesse (1912); Thies (1932); all review reports as *Cladonia foliacea* var. *alcicornis*.

Cladonia foliacea var. *convoluta* (Lam.) Vain., see *Cladonia endiviaefolia* Fr.

Cladonia formosana Asah.   Psoromic acid; protolichesterinic acid. Abbayes (1962), as *Cladonia formosana* var. *decaryana*; Asahina (1941*e*), from Formosa, type description; Asahina (1943*b*); Asahina (1950*c*), from Formosa and Japan, rarely with usnic acid (see *C. formosana* f. *aberrans*); Dahl (1952), mention only; Evans (1955*a*), from the Philippines.

Cladonia formosana f. *aberrans* Asah.   Psoromic acid, usnic acid; protolichesterinic acid. Asahina (1941*e*); Evans (1955*a*).

Usnic acid. Evans (1955*b*), from eastern Asia; Abbayes (1962), mention only.

Cladonia furcata (Huds.) Schrad. Atranorin ($\pm$), fumarprotocetraric acid; ascorbic acid; zinc.

Atranorin ($\pm$), fumarprotocetraric acid. Asahina (1942*b*), microchemical tests on 65 samples from Sandstede's exsiccati, including *Cladonia furcata* f. *fissa*, *C. f.* var. *foliosa*, *C. f.* f. *paradoxa*, *C. f.* var. *pinnata*, *C. f.* var. *racemosa*, *C. f.* var. *racemosa* f. *corymbosa* (= *C. f.* var. *corymbosa*) from Connecticut and Pennsylvania, *C. f.* var. *racemosa* f. *flaccida*, *C. f.* var. *racemosa* f. *regalis*, and *C. f.* var. *subulata*, atranorin present in two samples of *C. furcata*, two samples of *C. f.* var. *pinnata* f. *turgida* and one sample of *C. f.* var. *racemosa* f. *furcatosubulata*; Dahl (1952), mention only; Evans (1944*b*), from Connecticut, fumarprotocetraric acid by the PD reaction, atranorin by microcrystal tests, and also *C. furcata* var. *racemosa* f. *furcatosubulata*; Evans (1950), from Connecticut and Massachusetts, microchemical tests on many samples; Evans (1952*b*), from Canada and the U.S.A.

Atranorin, fumarprotocetraric acid. D. Hess (1958), from Germany; Klosa (1951*b*); Pišút (1961), and an unknown substance ($\pm$), mention only; Zopf (1908), from Bavaria and Tirol, as *Cladonia furcata* var. *racemosa* and *C. furcata* var. *pinnata*.

Fumarprotocetraric acid. Asahina (1938*e*), description of microcrystal tests; Asahina (1950*c*); Hale (1954*a*), from Arkansas, an exsiccati specimen, as *Cladonia furcata* var. *racemosa*; Hale (1956*a*), as *C. furcata* var. *racemosa*; Schatz (1963), mention only.

Ascorbic acid. Karev and Kochevykh (1962), as *Cladonia furcata* var. *pinnata*.

Zinc. Lambinon, Maquinay and Ramaut (1964), from Belgium, a quantitative study.

No usnic acid. Duvigneaud (1940); Gertig and Banasiewicz (1961), from Poland, as *Cladonia furcata* var. *racemosa*.

OTHER REPORT. Usnic acid: Tomaselli (1957), from Italy, mention only.

REVIEWS. Atranorin, fumarprotocetraric acid: W. Brieger (1923); Hesse (1912); Thies (1932); all three reports as *Cladonia furcata* var. *pinnata* and as *C. furcata* var. *racemosa*. — Fumarprotocetraric acid: Shibata (1958, 1963).

Cladonia furcata var. *palamaea* (Ach.) Nyl., see *Cladonia multiformis* Merr.

Cladonia furcata var. *racemosa* f. *corymbosa* (Ach.) Oliv., see *Cladonia furcata* (Huds.) Schrad.

Cladonia furcata var. **racemosa** f. **recurva** Jatta
REVIEW. One colorless acid. Zopf (1907).

Cladonia furfuracea Vain. Fumarprotocetraric acid only. Asahina (1943*b*); Dahl (1952), mention only.

Cladonia floridana Vain. Barbatic acid (±); thamnolic acid. Asahina (1942*b*), from North Carolina, New Jersey, New York, and South Carolina, by microchemical tests on specimens from Sandstede's exsiccati; Dahl (1952), mention only; Evans (1952*b*), from Florida.

Cladonia glauca Flörke Squamatic acid. Asahina (1942*b*), and also *Cladonia glauca* f. *capreolata, C. g.* f. *fastigiata, C. g.* f. *fruticulosa, C. g.* f. *muricelloides,* and *C. g.* f. *rigida,* by microchemical tests on specimens from Sandstede's exsiccati; Asahina (1950*c*), from Japan, and also *C. glauca* f. *subacuta;* Dahl (1952), mention only; Evans (1943*a*), from North America, mention in discussion of microchemical tests; Evans (1944*b*), from Connecticut and Massachusetts; Zopf (1908), from Germany.

No usnic acid. Gertig and Banasiewicz (1961), from Poland.

OTHER REPORTS. Fumarprotocetraric acid, squamatic acid: D. Hess (1958), from Germany. — Squamatic acid, usnic acid: Klosa (1951*b*).

REVIEWS. Squamatic acid: W. Brieger (1923); Hesse (1912); Thies (1932); Zopf (1907).

Cladonia gonecha (Ach.) Asah. Bellidiflorin (±), squamatic acid, usnic acid; rhodocladonic acid. Dahl (1952), mention only, as *Cladonia deformis* f. *gonecha.*

Squamatic acid, usnic acid, bellidiflorin. Asahina (1939*c*), from Japan and Sakhalin; Asahina (1939*g*), as *Cladonia deformis* f. *gonecha;* Asahina (1950*c*), from Japan; Evans (1943*a*), from North America, mention in discussion of microchemical tests; Evans (1944*b*), from Greenland, Labrador, and northeastern U.S.A.; Thomson (1963), mention only, bellidiflorin (±).

Squamatic acid, usnic acid. Evans (1955*a*); Hale (1954*b*), from Baffin Island.

Cladonia gorgonina (Bory) Vain. Thamnolic acid. Abbayes (1962), mention only; Asahina (1942*b*), from Brazil; Dahl (1952), mention only.

Cladonia gracilescens (Flörke) Vain., see *Cladonia lepidota* Nyl.

Cladonia graciliformis Zahlbr. Bellidiflorin, squamatic acid, usnic acid; rhodocladonic acid. Dahl (1952), mention only.
Bellidiflorin, squamatic acid, usnic acid. Asahina (1939*c*), from Japan; Asahina (1950*c*), from Japan.
Squamatic acid, usnic acid. Evans (1955*a*).

Cladonia gracilis (L.) Willd. Atranorin (±), fumarprotocetraric acid. Atranorin, fumarprotocetraric acid. D. Hess (1958), from Germany. Atranorin. Vartia (1949), from Finland.

Fumarprotocetraric acid. Dahl (1952), mention only; Klosa (1951b), mention only.

Fumarprotocetraric acid only. Asahina (1943b); Evans (1944b), from Connecticut.

OTHER REPORT. Usnic acid: Tomaselli (1957), from Italy.

Cladonia gracilis* var. *chordalis (Flörke) Schaer.　Fumarprotocetraric acid; amino acids, protein.

Fumarprotocetraric acid. Asahina (1943b); Hale (1954b), from Baffin Island; Zopf (1908), from Germany, no atranorin.

Amino acids, protein. Ramakrishnan and Subramanian (1965), from the Himalayas, alanine, arginine, glutamic acid, glycine, leucine, serine, threonine, tryptophan (trace), tyrosine, and valine as free amino acids, these and aspartic acid, isoleucine, methionine, phenylalanine, and an unidentified substance as combined amino acids, by paper chromatography, as *Cladonia gracilis* var. *chordalis* f. *foliosa*.

No usnic acid. Gertig and Banasiewicz (1961), from Poland.

REVIEWS. Fumarprotocetraric acid: W. Brieger (1923); Hesse (1912); Thies (1932); Zopf (1907).

Cladonia gracilis* var. *dilatata Flörke　Fumarprotocetraric acid only. Asahina (1943b).

Cladonia gracilis var. *elongata* (Jacq.) Flörke, see *Cladonia elongata* (Jacq.) Hoffm

Cladonia granulans Vain.　Bellidiflorin (±), squamatic acid, usnic acid; rhodocladonic acid. Dahl (1952), mention only.

Bellidiflorin (±), squamatic acid, usnic acid. Asahina (1939c), from Japan, and also *Cladonia granulans* f. *leucocarpa*; Asahina (1939g); Asahina (1950c), from Japan.

Bellidiflorin, squamatic acid, usnic acid. Asahina (1954c), type description of *Cladonia granulans* f. *sorediascens*.

Squamatic acid, usnic acid. Evans (1955a).

Cladonia grayi Merr. *ex* Sandst.　Fumarprotocetraric acid (±), grayanic acid (sometimes as chlorophaeic acid). Dahl (1952), mention only; Evans (1943a), mention in discussion of microcrystal tests; Evans (1944b), from Connecticut, PD reaction given for samples of many forms of the species; Evans (1950), from Connecticut, as *Cladonia grayi* f. *carpophora*, *C. g.* f. *cyathiformis*, *C. g.* f. *lacerata*, *C. g.* f. *prolifera*, *C. g.* f. *squamulosa*; Evans (1952b); Shibata and Chiang (1965), mention only.

Grayanic acid. Asahina (1939e), by extraction, description of microcrystal tests; Asahina (1940a), as *Cladonia grayi* f. *squamulosa*, from

Japan, reserves *C. grayi* f. *aberrans* for plants also containing fumar-protocetraric acid; Asahina (1943*b*); Asahina (1950*c*); Robinson (1959), from North Carolina; Shibata (1965), mentioned in discussions of bio-synthesis; Shibata and Chiang (1963), from Japan, a detailed chemical study.

REVIEW. Fumarprotocetraric acid, grayanic acid: Neelakantan (1965).

Cladonia grayi f. *aberrans* Asah. Fumarprotocetraric acid, grayanic acid. Asahina (1940*a*), from North America, not in Japan, type description; Asahina (1943*b*); Asahina (1950*c*), from North America.

Cladonia gymnopoda Vain. Fumarprotocetraric acid; unidentified substance ($\pm$). Asahina (1943*b*), tested samples from Sandstede's exsiccati; Dahl (1952), mention only.

Cladonia heteroclada Asah. Squamatic acid, usnic acid. Asahina (1959*d*), from Japan, type description.

Cladonia hitatiensis Asah. Decarboxythamnolic acid, thamnolic acid, usnic acid. Asahina and Nuno (1964), by thin-layer chromatography.

Cladonia hondoensis Asah. Barbatic acid; succinic acid. Asahina (1942*b*); Asahina (1942*c*), type description; Asahina (1950*c*); Dahl, (1952), mention only.

Barbatic acid. Evans (1955*b*).

Cladonia hypoxantha Tuck. Bellidiflorin, thamnolic acid. Asahina (1939*g*), as *Cladonia endoxantha*; Evans (1943*a*), from North America, mention in discussion of microcrystal tests, as *C. endoxantha*; Evans (1952*b*), from Florida.

Cladonia impexa Harm. Perlatolic acid, ($-$)-usnic acid ($\pm$); ursolic acid; D-arabitol, mannitol; zinc.

Perlatolic acid, ($-$)-usnic acid, ursolic acid, D-arabitol. Breaden, Keane, and Nolan (1942), from Ireland, a detailed chemical study; Breaden, Keane, and Nolan (1944), identification of ursolic acid. For comments on the work of Breaden, Keane, and Nolan see Asahina (1951*i*).

Perlatolic acid, ($-$)-usnic acid. Asahina (1940*f*), description of micro-crystal tests; Asahina (1950*c*), from Kamchatka, as *Cladonia impexa* f. *subpelleucida*.

Perlatolic acid, usnic acid. Dahl (1952), and an unknown substance ($\pm$), mention only; Duvigneaud (1939), samples without usnic acid as *Cladonia subimpexa*, perlatolic acid as "érinaceine"; Duvigneaud and Bleret (1940), from Brazil and Croatia, perlatolic acid as "érinaceine," and also *C. impexa* f. *exalbescens* from Costa Rica, samples without usnic acid as *C. subimpexa* from Belgium; Evans (1943*a*), mention in discussion of microchemical tests; Evans (1943*b*), discussion, *C. subimpexa* with no usnic acid possibly a shade form.

(–)-Usnic acid. Salkowski (1910), as *Cladonia sylvatica* var. *erinacea*. Usnic acid. Bustinza (1951), no fumarprotocetraric acid; Duvigneaud (1940); Gertig and Banasiewicz (1961), from Poland, 1.6% and 1.5% from *Cladonia impexa* f. *laxiuscula*.

Arabitol, mannitol. Pueyo (1960).

Zinc. Lambinon, Maquinay, and Ramaut (1964), from Belgium, a quantitative study.

OTHER REPORTS. Cornicularin, (–)-usnic acid: Klosa (1951*b*). — Imbricaric acid, usnic acid: D. Hess (1958), from Germany. — (+)-Usnic acid: Salkowski (1910), as *Cladina sylvatica* var. *silvestris*.

REVIEWS. Cornicularin, (–)-usnic acid: W. Brieger (1923), as *Cladina condensata*; Hesse (1912), as *Cladina condensata*; Thies (1932). — Perlatolic acid, (–)-usnic acid, ursolic acid, D-arabitol: Asahina and Shibata (1954). — Perlatolic acid, ursolic acid, D-arabitol: Shibata (1958, 1963). — (–)-Usnic acid: W. Brieger (1923), as *Cladina alpestris* var. *laxiuscula*; Hesse (1912), as *Cladina alpestris* var. *laxiuscula*; Zopf (1907), and a colorless acid, as *Cladina alpestris* var. *laxiuscula*, *Cladina condensata*, and *Cladina laxiuscula*.

For reports on North American plants see *Cladonia pacifica* and *Cladonia terrae-novae*.

Cladonia impexa "f. *condensata*," see *Cladonia pacifica* Ahti and *Cladonia terrae-novae* Ahti

Cladonia impexa f. *laxiuscula* (Del.) Mig., see *Cladonia impexa* Harm., *Cladonia pacifica* Ahti, and *Cladonia terrae-novae* Ahti

Cladonia impexa f. *spumosa* (Flörke) Mig., see *Cladonia terrae-novae* Ahti

Cladonia incrassata Flörke   Bellidiflorin (±), didymic acid (±) (sometimes as incrassatic acid), squamatic acid, (–)-usnic acid; rhodocladonic acid.

Bellidiflorin (±), didymic acid (±), squamatic acid, usnic acid, rhodocladonic acid. Dahl (1952), mention only.

Bellidiflorin (±), didymic acid (±), usnic acid. Evans (1944*b*), from Connecticut.

Bellidiflorin, usnic acid, rhodocladonic acid. Pišút (1961), from Czechoslovakia, mention only.

Bellidiflorin (±), squamatic acid, usnic acid. Asahina (1939*c*), from Japan; Asahina (1950*c*), from Japan.

Didymic acid, squamatic acid, (–)-usnic acid. Asahina (1939*e*), description of microchemical tests.

Didymic acid (±), squamatic acid, usnic acid. Asahina (1939*c*), from

Europe; Evans (1943*a*), mention in discussion of microchemical tests. Squamatic acid, usnic acid. Asahina (1939*g*); Evans (1955*a*).

REVIEWS. Didymic acid, rhodocladonic acid: Asahina and Shibata (1954). — Didymic acid: Shibata (1958, 1963). — Rhodocladonic acid, (−)-usnic acid: W. Brieger (1923); Hesse (1912); Thies (1932); Zopf (1907).

Cladonia incrassata f. *epiphylla* (Fr.) Vain. Usnic acid only. Asahina (1939*e*), description of microchemical tests.

Cladonia inobeana Asah. Thamnolic acid. Asahina (1963*a*), from Japan, no other lichen substances found by microchemical tests, type description.

Cladonia japonica Vain. Thamnolic acid. Asahina (1938*f*), description of microchemical tests, and as *Cladonia subsquamosa* f. *subulata*; Asahina (1942*b*), from Japan; Asahina (1950*c*), from Japan.

REVIEWS. Thamnolic acid: Asahina and Shibata (1954); Shibata (1963).

Cladonia koyaensis Asah. Barbatic acid, usnic acid. Asahina (1953*e*), from Japan, type description.

Cladonia krempelhuberi (Vain.) Vain. Atranorin, fumarprotocetraric acid. Asahina (1956*b*), from Japan; Dahl (1952), mention only.

Cladonia laxiuscula Del. (−)-Usnic acid.

REVIEW. (−)-Usnic acid: Thies (1932).

Cladonia lepidota Nyl. Atranorin, fumarprotocetraric acid. Asahina (1943*b*), tested a sample from Sandstede's exsiccati; Asahina (1950*c*), from Japan; Dahl (1950), from Greenland; Dahl (1952), mention only; Hale (1954*b*), from Baffin Island.

Atranorin. Asahina (1935*c*), no usnic acid, mention only; Asahina (1955*a*), mention, as *Cladonia lepidota* var. *gracilescens*; Dahl (1953).

Cladonia leporina Fr. Baeomycesic acid, bellidiflorin (±), didymic acid (±), squamatic acid, usnic acid; rhodocladonic acid.

Baeomycesic acid, bellidiflorin (±), didymic acid (±), squamatic acid. Evans (1947*a*), from Cuba and eastern and south-central U.S.A., by microchemical tests.

Baeomycesic acid, squamatic acid, usnic acid, rhodocladonic acid. Dahl (1952), mention only; Hale (1956*b*); Hale (1956*g*), from Alabama, an exsiccati specimen.

Didymic acid, usnic acid. Asahina (1939*g*).

Didymic acid. Asahina (1939*e*), description of microcrystal tests; Asahina (1952*b*), mention only.
Squamatic acid. Evans (1952*b*), from Florida.
Usnic acid. Mozingo (1961), from Tennessee, mentioned in key.

Cladonia leptoclada Abb.   Perlatolic acid (±), usnic acid. Evans (1955*a*), from Africa, New Caledonia, and New Zealand.

Cladonia leptophylla (Ach.) Flörke   Atranorin (±), fumarprotocetraric acid. Asahina (1943*a*), tested samples from Sandstede's exsiccati, atranorin rare.

Cladonia leucophaea Abb.   Fumarprotocetraric acid. Asahina (1941*b*), not in Japan, 0.3%, no usnic acid; Duvigneaud (1940); Evans (1943*b*), from Europe.
OTHER REPORTS. Fumarprotocetraric acid, unidentified substance (±): Dahl (1952), mention only. — Atranorin, fumarprotocetraric acid: Duvigneaud (1939).

Cladonia linearis Evans   Atranorin; unidentified substance (±). Evans (1947*a*), from North Carolina and Tennessee, type description; Dahl (1952), mention only.

Cladonia macilenta Hoffm.
Barbatic acid, bellidiflorin (±), didymic acid, thamnolic acid, rhodocladonic acid. Dahl (1952), mention only.
Barbatic acid, bellidiflorin, thamnolic acid. Asahina (1939*g*), as *Cladonia macilenta* var. *squamigera;* Evans (1943*a*), mention in discussion of microchemical tests.
Barbatic acid (±), bellidiflorin (±), thamnolic acid (±). Evans (1944*b*), from Connecticut, and substance F (±) (possibly identical to barbatic acid) and substance G (±), discussion of chemistry and taxonomy.
Barbatic acid, didymic acid, thamnolic acid. Asahina (1939*c*), from Japan, as *Cladonia macilenta* var. *ostreata*; Asahina (1950*c*), from Japan, as *C. macilenta* var. *ostreata* and *C. macilenta* var. *squamigera.*
Barbatic acid (as coccellic acid and cenomycin), thamnolic acid, rhodocladonic acid. Zopf (1908), from Germany, as *Cladonia macilenta* var. *styracella.*
Barbaric acid, thamnolic acid. Asahina (1939*g*), as *Cladonia macilenta* var. *styracella.*
Barbatic acid. Asahina (1936*e*), description of microcrystal tests; Asahina (1937*d*), as *Cladonia macilenta* var. *styracella*; D. Hess (1958), from Germany.
Thamnolic acid. Asahina (1938*e*), description of microcrystal tests; Duvigneaud (1939), no usnic acid; Evans (1955*a*).

Zinc. Lambinon, Maquinay, and Ramaut (1964), from Belgium, a quantitative study.
No usnic acid. Mozingo (1961), from Tennessee, mentioned in key.
REVIEWS. Barbatic acid (as coccellic acid), rhodocladonic acid, thamnolic acid: Asahina and Shibata (1954); W. Brieger (1923), and *Cladonia macilenta* var. *styracella*; Hesse (1912), and *C. macilenta* var. *styracella*; Thies (1932), and *C. macilenta* var. *styracella*; Zopf (1907), and *C. macilenta* var. *styracella*. — Barbatic acid, thamnolic acid: Shibata (1963). — Barbatic acid: Shibata (1958).

Cladonia macrophyllodes Nyl. Atranorin fumarprotocetaric acid. Asahina (1943*b*), tested samples from Sandstede's exsiccati; Dahl (1950); Dahl (1952), mention only; Hale (1954*b*).

Cladonia macroptera Räs., see *Cladonia scabriuscula* (Del. *ex* Duby) Leight.

Cladonia magyarica Vain. Atranorin, fumarprotocetraric acid. Pišút (1961).

Cladonia major (Hag.) Sandst. Fumarprotocetraric acid only (sometimes also as fimbriatic acid). Asahina (1943*b*), as *Cladonia fimbriata* var. *simplex* f. *major*; Evans (1944*b*), from Connecticut.
OTHER REPORTS. Atranorin, fumarprotocetraric acid: Evans (1943*b*); Zopf (1908), from Germany, as *Cladonia fimbriata* var. *simplex* f. *major*. — Fumarprotocetraric acid, usnic acid: D. Hess (1968), from Germany.
REVIEWS. Atranorin, fumarprotocetraric acid: Thies (1932); Zopf (1907); both reports as *Cladonia fimbriata* var. *simplex* f. *major*. — Fumarprotocetraric acid: W. Brieger (1923); Hesse (1912); both reports as *C. fimbriata* var. *simplex* f. *major*.

Cladonia mateocyatha Robb. Fumarprotocetraric acid. Asahina (1943*b*), from Connecticut and North Carolina; Dahl (1952), mention only; Evans (1943*a*), mention in discussion of microcrystal tests; Evans (1944*b*), from Connecticut; Hale (1957*c*), from West Virginia, an exsiccati specimen.

Cladonia medusina (Bory) Nyl. Thamnolic acid, usnic acid. Asahina (1942*a*), from Madagascar; Dahl (1952), mention only.

Cladonia merochlorophaea Asah. Fumarprotocetraric acid, merochlorophaeic acid. Asahina (1940*a*), from Japan and North America, type description; Asahina (1943*b*); Dahl (1952), mention only; Evans (1944*b*), from northeastern U.S.A., fumarprotocetraric acid not always present (see *Cladonia merochlorophaea* f. *inactiva*); Shibata and Chiang (1965), from Japan, 0.28% merochlorophaeic acid by extraction, detailed chemical study, fumarprotocetraric acid not always present (see *C. merochlorophaea* f. *inactiva*).

Merochlorophaeic acid. Shibata (1965), mentioned in discussion of biosynthesis.

REVIEW. Fumarprotocetraric acid, merochlorophaeic acid: Neelakantan (1965).

Cladonia merochlorophaea f. *inactiva* Asah.   Merochlorophaeic acid.
Asahina (1940*a*), from Japan; Asahina (1943*b*); Evans (1944*b*), from northeastern U.S.A., as *Cladonia merochlorophaea*.

Cladonia metacorallifera Asah.   Bellidiflorin (±), didymic acid, squamatic acid, usnic acid; rhodocladonic acid.
Bellidiflorin (±), didymic acid, squamatic acid, usnic acid. Asahina (1939*c*), from Japan, and also *Cladonia metacorallifera* f. *squamosa* from Japan.
Bellidiflorin, didymic acid, squamatic acid, usnic acid. Asahina (1950*c*), from Japan.
Didymic acid, squamatic acid, usnic acid, rhodocladonic acid. Dahl (1952), mention only.
Squamatic acid, usnic acid. Evans (1955*a*), from Japan and Alaska.
Squamatic acid. Evans (1955*b*), from Chile.

Cladonia metacorallifera var. *reagens* Asah.   Bellidiflorin (±), decarboxythamnolic acid, didymic acid, thamnolic acid, usnic acid; rhodocladonic acid.
Bellidiflorin (±), didymic acid, thamnolic acid, usnic acid, rhodocladonic acid. Dahl (1952), mention only.
Bellidiflorin (±), didymic acid, thamnolic acid, usnic acid. Asahina (1939*c*), from Japan, type description.
Bellidiflorin, didymic acid, thamnolic acid, usnic acid. Asahina (1950*c*), from Japan.
Decarboxythamnolic acid, didymic acid, thamnolic acid, usnic acid. Asahina and Nuno (1964), by thin-layer chromatography.
Thamnolic acid, usnic acid. Evans (1951), from Japan.

Cladonia metacorallifera f. *tingens* Asah.   Bellidiflorin, didymic acid, squamatic acid, usnic acid; unidentified yellow pigment (K+ violet). Asahina (1939*c*), type description.

Cladonia miniata Meyer   Rhodocladonic acid.
REVIEWS. Rhodocladonic acid: W. Brieger (1923); Thies (1932).

Cladonia mitis Sandst.   Fumarprotocetraric acid (±), psoromic acid (±), (+)-usnic acid; norrangiformic acid (±), rangiformic acid (±) (sometimes as Evans' substance D); enzymes.

Fumarprotocetraric acid ($\pm$), (+)-usnic acid, norrangiformic acid ($\pm$), rangiformic acid ($\pm$). Asahina (1941*b*), and mention of an unknown neutral substance (cladonin).

Fumarprotocetraric acid ($\pm$), usnic acid, rangiformic acid ($\pm$). Asahina (1950*c*), from Japan and Korea, and *Cladonia mitis* f. *divaricata* (with and without rangiformic acid), *C. mitis* f. *prolifera* (with rangiformic acid), *C. mitis* f. *tenuis* (without rangiformic acid); Evans (1950), from Europe and North America, discussion of *C. arbuscula*, *C. mitis*, and *C. submitis*, microchemical tests on many samples from Sandstede's exsiccati, substances A and B probably just crystal forms of usnic acid in the GE solution, substance D identical to rangiformic acid.

Psoromic acid ($\pm$), usnic acid, rangiformic acid ($\pm$). Dahl (1952), mention only; Evans (1943*b*), from arctic America, Europe, Alaska, Oregon, Virginia, Wisconsin, and Wyoming, psoromic acid in some European samples, also substances A ($\pm$) and B ($\pm$) (probably crystal forms of usnic acid in the GE solution), see also *Cladonia submitis*, specimens with perlatolic acid referred to *C. impexa*; Evans (1944*b*), from Massachusetts and New Jersey, psoromic acid rare, also substances A and B (probably crystal forms of usnic acid in the GE solution); Thomson (1963), mention only.

(+)-Usnic acid, rangiformic acid. Asahina and Sasaki (1942), from Japan, by extraction.

(+)-Usnic acid, rangiformic acid ($\pm$). Asahina (1958*a*), from Japan, Evans' substance D found to be rangiformic acid, some Japanese samples lacking rangiformic acid and pseudonorrangiformic acid (see *Cladonia submitis*).

Usnic acid, rangiformic acid. Asahina (1941*b*), as *Cladonia sylvatica* f. *inactiva*.

(+)-Usnic acid. Stoll, Brack, and Renz (1947), 0.60% by extraction; Stoll, Renz, and Brack (1947), from Switzerland.

Usnic acid. Burkholder and Evans (1945); Dahl (1950); Duvigneaud (1939), no perlatolic acid (as "érinaceine"); Gertig and Banasiewicz (1961), from Poland, 0.85%; Hale (1954*b*), from Baffin Island; Kurokawa (1952); Taguchi, Sankawa, and Shibata (1966), ^{14}C and ^{3}H study on the biosynthesis of this substance.

Rangiformic acid ($\pm$). Evans (1955*a*), identical to substance D.

Enzymes. Moissejeva (1961), enzymes tabulated.

REVIEWS. Rangiformic acid, (+)-usnic acid, an unidentified neutral substance: Asahina and Shibata (1954). — Rangiformic acid: Shibata (1958, 1963).

Cladonia mitrula Tuck., see *Cladonia capitata* (Michx.) Spreng.

Cladonia multiformis Merr. Fumarprotocetraric acid; ursolic acid ($\pm$) (sometimes as Evans' substance E). Evans (1944*b*), from Connecticut.

Fumarprotocetraric acid. Asahina (1942*b*), from Nova Scotia, Connecticut, Pennsylvania, and Vermont, no atranorin, also as *Cladonia furcata* var. *palamacea;* Dahl (1952), mention only; Evans (1943*a*), mention in discussion of microchemical tests; Evans (1944*b*), from Connecticut, as *C. furcata* var. *palamacea*; Evans (1950), no atranorin.

Cladonia nemoxyna (Ach.) Nyl.   Fumarprotocetraric acid (±), homosekikaic acid (sometimes as nemoxynic acid). Asahina (1943*b*), from Connecticut, New Jersey, Oregon, Pennsylvania, and Vermont; Asahina (1950*c*), from Formosa, Japan, and Korea; Evans (1944*b*), from Connecticut; Pišút (1961).

Homosekikaic acid. Asahina (1938*a*); Evans (1950), from Connecticut; Mozingo (1961), from Tennessee, mentioned in key; Zopf (1908), from Germany, by extraction, no fumarprotocetraric acid or atranorin, as *Cladonia fimbriata* var. *cornutoradiata* f. *nemoxyna*.

OTHER REPORTS. Homosekikaic acid, unknown substance: Dahl (1950), microcrystal tests described; Dahl (1952), mention only.

REVIEWS. Homosekikaic acid: Asahina and Shibata (1954); W. Brieger (1923), as *Cladonia fimbriata* var. *cornutoradiata* f. *nemoxyna*; Hesse (1912), as *C. fimbriata* var. *cornutoradiata* f. *nemoxyna*; Shibata (1958, 1963); Thies (1932), as *C. fimbriata* var. *cornutoradiata* f. *nemoxyna*.

Cladonia nipponica Asah.   Usnic acid. Asahina (1950*c*), from Japan; Asahina (1957*e*); Evans (1955*a*), from Alaska.

Cladonia norrlinii Vain.   Atranorin, psoromic acid (±). Asahina (1943*a*), one sample from Vermont lacking psoromic acid; Evans (1950), from Connecticut, New Hampshire, and Vermont.

OTHER REPORT. Atranorin, unidentified PD + substance: Evans (1944*b*).

Cladonia nylanderi Cout.   Atranorin, fumarprotocetraric acid. Asahina (1943*b*); Dahl (1952), mention only; Evans (1944*b*), no usnic acid; Evans (1952*b*).

Cladonia oceanica Vain.   Barbatic acid, squamatic acid, usnic acid. Asahina (1953*d*), from Hawaii.

Cladonia ochracea Scriba   Fumarprotocetraric acid. Asahina (1943*b*); Dahl (1952), mention only.

Cladonia ochrochlora Flörke   Fumarprotocetraric acid. Asahina (1943*b*); Asahina (1950*c*), from Japan; Evans (1944*b*).

Cladonia pachycladodes Vain.   Usnic acid only. Asahina (1942*a*), from North Carolina, also one sample from Sandstede's exsiccati; Dahl

(1952), mention only; Evans (1947*a*), from Florida, North Carolina, and South Carolina; Evans (1952*b*), from Florida.

Cladonia pacifica Ahti   Perlatolic acid, usnic acid. Evans (1955*a*), from Alaska and the Aleutian Islands, as *Cladonia impexa* f. *condensata* and *Cladonia impexa* f. *laxiuscula*.

Cladonia papillaria (Ehrh.) Hoffm., see *Pycnothelia papillaria* (Ehrh.) Duf.

Cladonia parasitica (Hoffm.) Hoffm.   Decarboxythamnolic acid, thamnolic acid. Asahina and Nuno (1964), by thin-layer chromatography.
   Thamnolic acid. Asahina (1942*b*), tested samples from Sandstede's exsiccati; Asahina (1950*c*); Dahl (1952), mention only; Evans (1944*b*), from Connecticut, also with substances F (±) and G; Evans (1947*a*); Evans (1950), from Connecticut; Evans (1952*b*), from Florida; Zopf (1908), from Germany.
   OTHER REPORT. Squamatic acid: Evans (1943*a*), mention in discussion of microcrystal tests, (apparently an error).
   All reports are as *Cladonia delicata*.

Cladonia patagonica Evans   Barbatic acid. Evans (1955*b*), from Patagonia, type description.

Cladonia peltasta (Ach.) Spreng.   Barbatic acid, usnic acid. Asahina (1942*a*), from Réunion; Dahl (1952), mention only.

Cladonia perforata Evans   Squamatic acid, usnic acid. Evans (1952*b*), from Florida, type description.

Cladonia persquamulosa Merr., see *Cladonia santensis* Tuck.

Cladonia piedmontensis Merr.   Usnic acid only. Asahina (1943*b*), from Connecticut, New Jersey, and North Carolina, microchemical tests on samples from Sandstede's exsiccati; Dahl (1952), mention only; Evans (1944*b*), from Connecticut; Evans (1950), from Connecticut, and also *Cladonia piedmontensis* f. *epiphylla, C. p.* f. *lepidifera, C. p.* f. *obconica, C. p.* f. *squamosissima,* and *C. p.* f. *squamulosa*; Mozingo (1961), from Tennessee, mentioned in key; Robinson (1959), from North Carolina.

Cladonia pityrea (Flörke) Fr.   Fumarprotocetraric acid, homosekikaic acid (±). Asahina and Kusaka (1937*b*), from Japan, 1.2% fumarprotocetraric acid, 0.13% homosekikaic acid, by extraction, as *Cladonia pityrea* var. *zwackhii* f. *phyllophora*; Asahina (1943*b*), samples with fumarprotocetraric acid only cited from Costa Rica, Connecticut, New Jersey, North Carolina, and Pennsylvania; Asahina (1950*c*), and very rarely atranorin [?], fumarprotocetraric acid only in samples from

Europe and North America, usually also homosekikaic acid in samples from Japan and tropical Asia; Dahl (1952), mention only; Evans (1943*a*), mention in discussion of microchemical tests; Evans (1944*b*), from Connecticut, no homosekikaic acid in forms of *C. pityrea* var. *zwackhii*.

Fumarprotocetraric acid, homosekikaic acid. Asahina (1937*e*), from Japan, as *Cladonia pityrea* f. *phyllophora*; Asahina (1955*b*), from Aogashima, as *C. pityrea* var. *zwackhii* f. *phyllophora*.

Fumarprotocetraric acid. Evans (1952*b*), from Europe and North America; Evans (1955*a*); Zopf (1908), from Germany, as *Cladonia pityrea* var. *zwackhii* and as *C. pityrea* var. *cladomorpha* (= *C. pityrea* var. *zwackhii* f. *cladomorpha*).

Homosekikaic acid. Asahina (1937*e*), as *Cladonia pityrea* (var. *zwackhii*) f. *crassiuscula*; Asahina (1938*c*), from Japan, description of microcrystal tests; Asahina (1939*b*), from Japan, mention only.

No grayanic acid. Mozingo (1961), from Tennessee, mentioned in key.

REVIEWS. Fumarprotocetraric acid: W. Brieger (1923); Hesse (1912); Thies (1932); Zopf (1907); all four reports as *Cladonia pityrea* var. *cladomorpha* and *C. pityrea* var. *zwackhii*. — Homosekikaic acid: Asahina and Shibata (1954), and as *C. pityrea* f. *phyllophora*; Shibata (1958, 1963), as *C. pityrea* f. *phyllophora*.

Cladonia pleurota (Flörke) Schaer.   Bellidiflorin (±), (−)-usnic acid; rhodocladonic acid; zeorin.

Bellidiflorin (±), usnic acid, zeorin. Asahina (1939*c*), from Japan, Norway, and Sahkalin, bellidiflorin very rare; Asahina (1950*c*), from Japan and Korea, and as *Cladonia pleurota* (var. *hygrophila*) f. *denticulata*; Evans (1943*a*), mention in discussion of microchemical tests.

(−)-Usnic acid, rhodocladonic acid, zeorin. Zopf (1908), from Germany, no barbatic acid (as coccellic acid and cenomycin).

Usnic acid, zeorin. Asahina (1938*f*), description of microcrystal tests; Asahina (1939*c*), as *Cladonia pleurota* var. *hygrophila*, from Japan and Sakhalin, and as *C. pleurota* f. *frondescens* from Japan; Asahina (1939*g*); Asahina (1950*c*), from Japan, as *C. pleurota* var. *esorediata*, no barbatic acid; Evans (1944*b*), from Connecticut; Evans (1955*a*); Hale (1954*b*).

(−)-Usnic acid. Salkowski (1910).

Usnic acid. Tomaselli (1957), from Italy, mention only; Vartia (1950*b*).

OTHER REPORT. Bellidiflorin (±), rhodocladonic acid, squamatic acid, usnic acid, zeorin, unidentified substance: Dahl (1952), mention only.

REVIEWS. Rhodocladonic acid, (−)-usnic acid, zeorin: W. Brieger (1923); Hesse (1912); Thies (1932); Zopf (1907); all four reports also as *Cladonia coccifera* var. *pleurota*. — Rhodocladonic acid: Asahina and Shibata (1954).

Cladonia polycarpia Merr.   Atranorin, stictic acid (±), norstictic acid (±).
Evans (1944*a*), from Alabama, Florida, Mississippi, North Carolina,

South Carolina, and Virginia, numerous samples tested microchemically; Evans (1952*b*), from Florida, discussion of chemical variation, plants with atranorin, stictic acid, and norstictic acid referred to *Cladonia sub-clavulifera*; Robinson (1959), from North Carolina.
Atranorin, norstictic acid (±). Mozingo (1961), mention only.

Cladonia polydactyla (Flörke) Spreng. Bellidiflorin (±), thamnolic acid; rhodocladonic acid.
Bellidiflorin, thamnolic acid, rhodocladonic acid. Dahl (1952), mention only.
Bellidiflorin (±), thamnolic acid. Asahina (1950*c*).
Bellidiflorin, thamnolic acid. Asahina (1939*c*), from Japan; Asahina (1939*g*); Evans (1943*a*), mention in discussion of microchemical tests.
Thamnolic acid. Asahina (1938*e*), description of microcrystal tests; Evans (1955*a*), no usnic acid.
OTHER REPORT. Thamnolic acid, usnic acid: Asahina and Hiraiwa (1936), from Japan, but see *Cladonia polydactyla* f. *perplexans* and *C. polydactyla* var. *theiophila*.
REVIEWS. Rhodocladonic acid, thamnolic acid: Asahina and Shibata (1954). — Thamnolic acid: Shibata (1958, 1963).

Cladonia polydactyla var. *perplexans* Asah., see *Cladonia polydactyla* f. *perplexans* (Asah.) Asah.

Cladonia polydactyla f. *perplexans* (Asah.) Asah. Bellidiflorin (±), thamnolic acid, usnic acid. Asahina (1950*c*), from Japan.
Thamnolic acid, usnic acid. Asahina (1939*c*), from Japan, type description for *Cladonia polydactyla* var. *perplexans*, distinguished from *C. incrassata* by containing thamnolic acid; Evans (1951), from Japan.

Cladonia polydactyla var. *theiophila* (Asah.) Asah. Decarboxythamnolic acid, thamnolic acid, usnic acid.
Decarboxythamnolic acid, thamnolic acid. Asahina and Nuno (1964), by thin-layer chromatography.
Thamnolic acid, usnic acid. Asahina (1939*c*), from Japan, type description of *Cladonia theiophila*; Evans (1951), from Japan, as *C. theiophila*.
Thamnolic acid. Kurokawa (1964), as *Cladonia macilenta* ssp. *theiophylla*.
Usnic acid. Asahina (1950*c*), from Japan, other substances of *Cladonia polydactyla* implied.

Cladonia prostrata Evans Atranorin, fumarprotocetraric acid. Evans (1952*b*), from Alabama, Florida, and Georgia.
OTHER REPORT. Squamatic acid. Hale (1957*b*).

Cladonia pseudalcicornis Asah., see *Cladonia verticillata* (Hoffm.) Schaer.

Cladonia pseudodidyma Asah.   Bellidiflorin (±), didymic acid, squamatic acid; rhodocladonic acid.
Bellidiflorin (±), didymic acid, squamatic acid. Asahina (1939*c*), from Formosa and Japan, type description; Asahina (1950*c*), from Formosa and Japan, rarely bellidiflorin.
Didymic acid, squamatic acid, rhodocladonic acid. Dahl (1952), mention only.
Didymic acid. Nogami (1944), mention in study of chemical structure.
Squamatic acid. Shibata (1934), from Japan, by extraction.
REVIEWS. Didymic acid, rhodocladonic acid: Asahina and Shibata (1954). — Didymic acid: Shibata (1958, 1963).

Cladonia pseudodidyma var. **subpygmaea** Asah.   Didymic acid, squamatic acid. Asahina (1939*c*), from Japan, type description; Asahina (1950*c*), from Japan.

Cladonia pseudoevansii Asah.   Perlatolic acid, usnic acid. Asahina (1940*f*), description of microchemical tests; Asahina (1950*c*), from Japan and Korea; Dahl (1952), mention only; Evans (1943*b*), from Japan.
Perlatolic acid. Mitsuno (1953), by paper chromatography.
REVIEWS. Imbricaric acid, perlatolic acid: Shibata (1958, 1963). — Perlatolic acid: Asahina and Shibata (1954).

Cladonia pseudohondoensis Asah.   Decarboxythamnolic acid, thamnolic acid. Asahina and Nuno (1964), by thin-layer chromatography.
Thamnolic acid. Asahina (1959*d*), from Japan, type description.

Cladonia pseudomacilenta Asah.   Squamatic acid, usnic acid; rhodocladonic acid.
Squamatic acid, rhodocladonic acid. Dahl (1952), mention only.
Squamatic acid, usnic acid. Asahina (1943*d*), from Japan, and *Cladonia pseudomacilenta* var. *subsquamigera*, type descriptions; Asahina (1950*c*), from Japan; Asahina (1953*d*), and *C. pseudomacilenta* var. *subsquamigera*; Evans (1955*a*).

Cladonia pseudorangiformis Asah.   Atranorin, merochlorophaeic acid, psoromic acid. Asahina (1942*b*); Asahina (1942*c*), type description; Asahina (1950*c*), from Japan and Korea; Dahl (1952), mention only.
OTHER REPORT. Atranorin, merochlorophaeic acid, perhaps psoromic acid: Evans (1955*a*), from North America, microcrystal test for psoromic acid failed.

Cladonia pseudostellata Asah. Hypothamnolic acid, (−)-usnic acid. Asahina (1950*c*), from Japan; Asahina, Aoki, and Fuzikawa (1941), from Japan, by extraction, as *Cladonia uncialis* f. *obtusata.*

Hypothamnolic acid, usnic acid. Asahina (1942*d*), type description, differs chemically from *Cladonia uncialis*; Asahina (1952*b*); Dahl (1952), mention only.

Hypothamnolic acid. Asahina (1951*i*); Bendz, Santesson, and Wachtmeister (1965*c*), mention only.

REVIEWS. Hypothamnolic acid, usnic acid: Asahina and Shibata (1954), from Japan. — Hypothamnolic acid: Shibata (1958, 1963).

Cladonia pungens (Ach.) Flörke, see *Cladonia rangiformis* Hoffm.

Cladonia pycnoclada (Gaud.) Nyl. Fumarprotocetraric acid, usnic acid. Asahina (1941*b*), from Formosa; Asahina (1950*c*), from Formosa and perhaps Japan; Dahl (1952), mention only; Evans (1943*b*). All reports are as *Cladonia fallax.*

Cladonia pyxidata (L.) Hoffm. Atranorin (±), fumarprotocetraric acid; zinc.

Atranorin (±), fumarprotocetraric acid. Dahl (1950), as *Cladonia pyxidata* ssp. *neglecta* and *C. pyxidata* ssp. *pocillum*; Dahl (1952), mention only; Pišút (1961).

Atranorin, fumarprotocetraric acid. Hale (1954*b*), from Baffin Island, as *Cladonia pyxidata* var. *neglecta* and *C. pyxidata* var. *pocillum.*

Fumarprotocetraric acid. Asahina (1943*b*), and as *Cladonia pyxidata* var. *neglecta*; Asahina (1950*c*), from Japan, the Kurile Islands, and Manchuria; Evans (1943*a*), mention in discussion of microchemical tests; Evans (1944*b*); Zopf (1908), as *C. pyxidata* var. *neglecta* from western Austria and *C. pyxidata* f. *cerina* from Tirol.

Zinc. Lambinon, Maquinay, and Ramaut (1964), from Belgium, a quantitative study, as *Cladonia pyxidata* var. *neglecta.*

OTHER REPORTS. Psoromic acid (as parellic acid and stereocaulic acid): Klosa (1951*b*). — Usnic acid: Tomaselli (1957), from Italy, mention only.

REVIEWS. Fumarprotocetraric acid: W. Brieger (1923), as *Cladonia pyxidata* var. *cerina* and *C. pyxidata* var. *neglecta*; Hesse (1912), as *C. pyxidata* var. *cerina* and *C. pyxidata* var. *neglecta*; Thies (1932), as *C. pyxidata* var. *cerina* and *C. pyxidata* var. *neglecta*; Zopf (1907), no psoromic acid. — Psoromic acid (as parellic acid): W. Brieger (1923); Hesse (1912); Thies (1932).

Cladonia pyxidata var. *chlorophaea* Flörke, see *Cladonia chlorophaea* (Flörke) Spreng.

Cladonia rangiferina (L.) G. Web. *ex* Wigg. Atranorin, fumarprotocetraric acid; fat; ergosterol; arabitol, lichenin, mannitol, sucrose, trehalose; amino acids; enzymes, protein; ascorbic acid; calcium, iron, manganese, phosphorus, zinc.

Atranorin, fumarprotocetraric acid. Asahina (1941*b*), from Japan and Korea; Asahina (1950*c*), from Japan and Korea; Bustinza (1951); Dahl (1952), mention only; Duvigneaud (1939); Evans (1943*a*), mention in discussion of microchemical tests; Evans (1943*b*); Evans (1944*b*), from Connecticut, no usnic acid; Evans (1950), from Connecticut; Evans (1952*b*), from Florida; Hale (1954*b*), from Baffin Island; Hale (1956*b*); Hale (1956*g*), from Connecticut, an exsiccati specimen; D. Hess (1958), from Germany; Kurokawa (1952).

Atranorin. Asahina (1935*c*), no usnic acid; Asahina (1936*e*), description of microcrystal tests; Dahl (1950); Mozingo (1961), from Tennessee, mentioned in key; Vartia (1949), mention only.

Ergosterol. Aghoramurthy and Seshadri (1954), mention only.

Arabitol, mannitol, sucrose, trehalose. Lindberg, Misiorny, and Wachtmeister (1953), by paper chromatography and charcoal column chromatography.

Arabitol, mannitol. Pueyo (1965), from France, by column chromatography.

Fat, protein, calcium, phosphorus. Scotter (1965), from northern Canada.

Lichenin, amino acids, protein. Ramakrishnan and Subramanian (1965), from the Himalayas, alanine, arginine, leucine, serine, tyrosine, and valine as free amino acids, these and aspartic acid, glutamic acid, glycine, isoleucine, lysine, methionine, phenylalanine, threonine, tryptophan (trace), and an unidentified substance as combined amino acids, by paper chromatography.

Enzymes. Moissejeva (1961), enzymes tabulated.

Ascorbic acid. Karev and Kochevykh (1962).

Iron, manganese, zinc. Lounamaa (1965), a quantitative study.

OTHER REPORTS. Ergosterol, vitamin D: Blix and Rydin (1932), from Sweden. — (+)-Usnic acid: Borkowski, Gertig, and Jeljaszewicz (1958), from Poland, 0.76%; Stoll, Brack, and Renz (1947), 0.55% by extraction; Stoll, Renz, and Brack (1947), from Switzerland. — Usnic acid: Gertig and Banasiewicz (1961), from Poland; Tomaselli (1957), from Italy, mention only.

REVIEWS. Atranorin, fumarprotocetraric acid, sucrose, trehalose: Shibata (1958). — Atranorin, fumarprotocetraric acid: Asahina and Shibata (1954); W. Brieger (1923), and as *Cladina rangiferina*; Hesse (1912), and as *Cladina rangiferina*; Shibata (1963); Thies (1932); Zopf (1907), and as *Cladina rangiferina*. — Fat: Steiner (1957). — No usnic acid: Savicz, Kuprevicz, Litvinov, Moissejeva, and Rassadina (1956).

Cladonia rangiferina "var. *spumosa*," see *Cladonia spumosa* (Flörke) Sandst.

Cladonia rangiformis Hoffm. Atranorin, fumarprotocetraric acid ($\pm$); norrangiformic acid, rangiformic acid ($\pm$); zinc.

Atranorin, fumarprotocetraric acid ($\pm$), rangiformic acid. Asahina (1942*b*), samples of *Cladonia rangiformis* var. *foliosa, C. r.* var. *muricata,* and *C. r.* var. *sorediophora* with or without fumarprotocetraric acid and samples of *C. r.* var. *pungens* without fumarprotocetraric acid.

Atranorin, rangiformic acid. Evans (1943*a*), mention in discussion of microchemical tests.

Atranorin, rangiformic acid ($\pm$). Evans (1955*a*); Zopf (1908), from Germany, a sample of *Cladonia rangiformis* var. *pungens* with no rangiformic acid.

Rangiformic acid. Asahina (1951*i*), mention only; Bendz, Santesson, and Tibell (1966), used as an authentic source of this substance; Dahl (1952), mention only; Paternò (1882), by extraction.

Norrangiformic acid. Asahina and Sakurai (1951), mention only.

Zinc. Lambinon, Maquinay, and Ramaut (1964), from Belgium, a quantitative study.

OTHER REPORTS. Atranorin, atranorinic acid (=atrinic acid), rangiformic acid: Klosa (1951*b*), mention. — No atranorin, no fumarprotocetraric acid: Černohorský (1959). — Usnic acid: Tomaselli (1957), from Italy, as *Cladonia pungens.*

REVIEWS. Atranorin, atranorinic acid, rangiformic acid: W. Brieger (1923); Hesse (1912); Thies (1932). — Atranorin, rangiformic acid: Zopf (1907). — Rangiformic acid: Asahina and Shibata (1954); Shibata (1958, 1963).

Cladonia rappii Evans Psoromic acid. Asahina (1956*b*), from Europe and North America; Evans (1952*b*), from Denmark, Germany, Scotland, and Florida, type description.

Cladonia ravenelii Tuck. Didymic acid ($\pm$), thamnolic acid; rhodocladonic acid.

Didymic acid ($\pm$), thamnolic acid. Evans (1952*b*), one sample lacking didymic acid.

Thamnolic acid, rhodocladonic acid. Dahl (1952), mention only.

Thamnolic acid. Asahina (1939*g*); Evans (1943*a*), mention in discussion of microchemical tests.

Cladonia retipora (La Bill.) Fr. Atranorin, usnic acid ($\pm$); rangiformic acid. Nuno (1962), from Australia and New Zealand.

Atranorin, usnic acid, rangiformic acid. Martin (1965), as *Cladia retipora.*

Atranorin. Dahl (1952), mention only.

Cladonia rhodoleuca Vain. Thamnolic acid. Abbayes (1963).

Cladonia robbinsii Evans   Barbatic acid, usnic acid. Dahl (1952), mention only; Evans (1944*b*), from Connecticut, and substance F (which may be barbatic acid); Evans (1950), from Connecticut, and substance F; Hale (1956*g*), from Kansas, an exsiccati specimen.
    Barbatic acid. Hale (1956*f*).
    OTHER REPORT. Atranorin, barbatic acid: Hale (1956*b*).

Cladonia sandstedei Abb.   Atranorin, fumarprotocetraric acid. Dahl (1952), mention only; Evans (1943*b*); Evans (1955*a*), from Colombia, Jamaica, Mexico, Peru, Puerto Rico, and Venezuela.

Cladonia santensis Tuck.   Thamnolic acid. Asahina (1938*f*), as *Cladonia persquamulosa*, description of microcrystal tests; Asahina (1942*b*), tested samples from Sandstede's exsiccati; Dahl (1952), mention only; Evans (1943*a*), from North America, mention in discussion of micro-chemical tests; Evans (1952*b*), from New Jersey to Florida.
    REVIEWS. Thamnolic acid: Asahina and Shibata (1954); Shibata (1963); both reports as *Cladonia persquamulosa*.

Cladonia scabriuscula (Del. *ex* Duby) Leight.   Atranorin (±), fumarpro-tocetraric acid. Asahina (1942*b*), from Japan and North America, tested samples from Sandstede's exsiccati, four of 20 samples with atranorin, and as *Cladonia scabriuscula* f. *adspera* from Japan, *C. s.* f. *cancellata* from Japan, *C. s.* f. *elegans* from Canada, Japan, and Vermont, *C. s.* f. *farinacea* from Japan, Connecticut, and Pennsylvania, and *C. s.* f. *sublaevis* from Japan; Dahl (1952), mention only.
    Fumarprotocetraric acid. Asahina (1940*c*), from Japan, as *Cladonia macroptera* (= *C. scabriuscula* f. *elegans*); Asahina (1941*d*), mention only; Asahina (1950*c*); Dahl (1950); Dahl (1952), as *C. macroptera*, mention only; Evans (1943*a*), mention in discussion of microchemical tests.
    OTHER REPORTS. Atranorin, squamatic acid: D. Hess (1958), from Germany. — Fumarprotocetraric acid, ursolic acid (±) (as unidentified substance E): Evans (1944*b*).

Cladonia scabriuscula f. *surrecta* Sandst.   Fumarprotocetraric acid; ursolic acid (±) (sometimes as Evans' substance E). Evans (1944*b*), from Connecticut; Evans (1950), from Connecticut.
    Fumarprotocetraric acid. Asahina (1942*b*), from Japan.

Cladonia scholanderi Abb.   Thamnolic acid, usnic acid. Dahl (1952), mention only.

Cladonia shikokiana Asah.   Atranorin, fumarprotocetraric acid. Asahina (1957*f*), from Japan, type description.

Cladonia "silvatica," see *Cladonia arbuscula* (Wallr.) Rabenh.

Cladonia simulata Robb. Fumarprotocetraric acid. Evans (1952*b*), from Florida, Georgia, Massachusetts, and North Carolina, no usnic acid.

Cladonia solida Vain. Fumarprotocetraric acid. Asahina (1943*a*), no atranorin, tested one sample from Sandstede's exsiccati; Dahl (1952), mention only.

Cladonia spumosa (Flörke) Sandst. (−)-Usnic acid. Salkowski (1910), as *Cladina alpestris* var. *spumosa*.
REVIEWS. (−)-Usnic acid: W. Brieger (1923), as *Cladina alpestris* var. *spumosa*; Hesse (1912), as *Cladina alpestris* var. *spumosa*; Thies (1932), as *Cladonia rangiferina* var. *spumosa*. — (−)-Usnic acid, unidentified substance: Zopf (1907), as *Cladina alpestris* var. *spumosa* and as *Cladina spumosa*.

Cladonia squamosa (Scop.) Hoffm. Squamatic acid. Asahina (1942*b*), from Japan, and as *Cladonia squamosa* var. *denticollis, C. s.* f. *levicorticata, C. s.* var. *muricella, C. s.* var. *muricella* f. *plumosa, C. s.* var. *phyllocoma, C. s.* var. *phyllocoma* f. *myosurioides, C. s.* f. *pityrea, C. s.* var. *polychonia, C. s.* f. *pseudocrispata, C. s.* f. *squamosissima,* and *C. s.* f. *subulata;* Asahina (1950*c*); Asahina and Yasue (1937*b*), from Europe, as *C. squamosa* var. *denticollis;* Dahl (1950), mention only; Dahl (1952), mention only; Evans (1943*a*), mention in discussion of microchemical tests; Evans (1944*b*), from Connecticut, samples PD+ yellow with baeomycesic acid (see *C. atlantica*), also substances F (±) and G (±); Evans (1947*a*), mention only; Evans (1950), mention only; Hale (1954*b*), from Baffin Island; Hale (1956*f*); D. Hess (1958), from Germany; Zopf (1908), as *C. squamosa* var. *denticollis, C. s.* f. *frondosa, C. s.* var. *multibrachiata, C. s.* var. *turfacea,* and *C. s.* f. *ventricosa.*
No usnic acid. Gertig and Banasiewicz (1961), from Poland.
REVIEWS. Squamatic acid: Asahina and Shibata (1954); W. Brieger (1923), as *Cladonia squamosa* var. *denticollis, C. s.* f. *frondosa,* and *C. s.* f. *ventricosa;* Hesse (1912), as *C. squamosa* var. *denticollis, C. s.* f. *frondosa, C. s.* f. *ventricosa;* Shibata (1958); Thies (1932), and as *C. squamosa* var. *denticollis, C. s.* f. *frondosa, C. s.* var. *multibrachiata, C. s.* f. *pseudocrispata, C. s.* var. *turfacea,* and *C. s.* f. *ventricosa;* Zopf (1907), as *C. squamosa* var. *denticollis, C. s.* f. *frondosa, C. s.* var. *multibrachiata* f. *pseudocrispata, C. s.* var. *multibrachiata* f. *turfacea,* and *C. s.* f. *ventricosa.*

Cladonia squamulosa (Scop.) Hoffm. Squamatic acid.
REVIEW. Squamatic acid: Zopf (1907), as *Cladonia denticollis.*

Cladonia strepsilis (Ach.) Vain. Baeomycesic acid, squamatic acid, strepsilin. Asahina (1939*e*), description of microcrystal tests; Asahina (1943*b*);

Asahina (1950*c*); Dahl (1952), mention only; Evans (1943*a*), mention in discussion of microchemical tests; Evans (1944*b*), from Connecticut, and unidentified substance F ($\pm$).

Baeomycesic acid, strepsilin. Hale (1956*b*), as *Cladonia strepsilis* f. *subsessilis*; Hale (1956*g*), from Connecticut, an exsiccati specimen.

Strepsilin. Asahina (1951*i*); Wachtmeister (1956*a*), used as an authentic source of this substance.

OTHER REPORTS. Strepsilin, thamnolic acid: Zopf (1908), from Germany. — Thamnolic acid: Asahina and Ihara (1929), mention only.

REVIEWS. Baeomycesic acid, squamatic acid, strepsilin: Asahina and Shibata (1954). — Strepsilin, thamnolic acid: W. Brieger (1923); Hesse (1912); Thies (1932); Zopf (1907). — Strepsilin: Neelakantan (1965); Shibata (1958, 1963).

Cladonia subcariosa Nyl.   Norstictic acid. Asahina (1938*e*), description of microcrystal tests; Asahina (1943*a*), one sample from Sandstede's exsiccati also containing atranorin thought to be *Cladonia symphicarpia*; Asahina (1950*c*), from the Bonin Islands and Japan; Asahina (1963*d*), by thin-layer chromatography, used as an authentic source of this substance; Dahl (1950), mention only; Evans (1943*a*), mention in discussion of microchemical tests; Evans (1944*a*); Evans (1944*b*), from Connecticut; Evans (1950), from Connecticut; Evans (1952*b*), from Alabama and Georgia; Kurokawa (1964), mention only; Robinson (1959), from North Carolina, atranorin and stictic acid absent.

OTHER REPORTS. Atranorin, norstictic acid: Dahl (1952), mention only; Mozingo (1961), from Tennessee, mention only.

REVIEWS. Norstictic acid: Asahina and Shibata (1954); Shibata (1958, 1963).

Cladonia subcervicornis (Vain.) Kernst.   Atranorin, fumarprotocetraric acid. Asahina (1943*b*); Dahl (1952), mention only; Hale (1954*b*), from Baffin Island.

Atranorin. Dahl (1950); Dahl (1953).

OTHER REPORT. Atranorin, fumarprotocetraric acid, cervicornin: Zopf (1908), as *Cladonia verticillata* var. *subcervicornis*, cervicornin in the apothecia.

REVIEWS. Atranorin, cervicornin, fumarprotocetraric acid: Thies (1932); Zopf (1907). — Fumarprotocetraric acid: W. Brieger (1923); Hesse (1912).

Cladonia subchordalis Evans   Usnic acid. Evans (1955*b*), from Patagonia, type description.

Cladonia subclavulifera Asah.   Atranorin, norstictic acid, stictic acid. Asahina (1943*a*), type description; Dahl (1952), mention only.

Cladonia subconistea Asah. Atranorin, psoromic acid. Asahina (1941*a*); Asahina (1943*b*); Asahina (1950*c*), from Formosa and Japan; Dahl (1952), mention only.

Cladonia subdelicatula Vain. *ex* Sandst. Thamnolic acid. Asahina (1963*a*), from Brazil, by a microchemical test on the lectotype specimen.

Cladonia subimpexa Duvign., see *Cladonia impexa* Harm.

Cladonia submedusina Müll. Arg. Squamatic acid, usnic acid. Asahina (1942*a*), from Madagascar and Réunion; Dahl (1952), mention only.

Cladonia submitis Evans Usnic acid; pseudonorrangiformic acid (sometimes as Evans' unidentified substance C). Asahina and Sakurai (1951).
Usnic acid, pseudonorrangiformic acid. Evans (1943*b*), from Connecticut, Delaware, Massachusetts, New York, Rhode Island, Vermont, Virginia, and West Virginia, and unknown substance A ($\pm$) (later found to be a crystal form of usnic acid in the GE solution), also as *Cladonia submitis* f. *prolifera* from Connecticut; Evans (1944*b*), from Connecticut, and unknown substance A ($\pm$) (=usnic acid), and as *C. submitis* f. *prolifera* from Connecticut; Evans (1950), from Connecticut; Evans (1955*a*); Thomson (1963), mention only.
Pseudonorrangiformic acid. Asahina (1958*a*), from Japan, the Kurile Islands, and Sakhalin.
OTHER REPORTS. Usnic acid, unknown substance: Dahl (1952), mention only. — Usnic acid, rangiformic acid: Hale (1957*b*), mention only.

Cladonia submultiformis Asah. Atranorin, fumarprotocetraric acid, homosekikaic acid. Asahina (1942*b*); Asahina (1942*d*), type description, species differs chemically from *Cladonia multiformis*.
Fumarprotocetraric acid, homosekikaic acid. Asahina (1950*c*), from Formosa; Dahl (1952), mention only.

Cladonia subpityrea Sandst.
Fumarprotocetraric acid. Asahina (1937*e*).
Homosekikaic acid. Asahina and Kusaka (1937*b*), from Japan; Asahina (1938*c*), description of microcrystal tests.

Cladonia subrangiformis Sandst. Atranorin ($\pm$), fumarprotocetraric acid. Asahina (1942*b*), checked samples in Sandstede's exsiccati, one sample with no atranorin.
Atranorin, fumarprotocetraric acid. Dahl (1952), mention only.

Cladonia subsetacea Robb. *ex* Evans Baeomycesic acid ($\pm$), squamatic acid, usnic acid. Evans (1952*b*), from Florida, North Carolina, and South Carolina, no baeomycesic acid in the sample from Florida.

Baeomycesic acid, squamatic acid, usnic acid. Evans (1947*a*); Dahl (1952), mention only.

Cladonia subsquamosa (Nyl.) Vain.   Decarboxythamnolic acid, thamnolic acid. Asahina and Nuno (1964), by thin-layer chromatography.

Thamnolic acid. Asahina (1938*f*), description of microchemical tests, and *Cladonia subsquamosa* f. *nuda*; Asahina (1942*b*); Asahina (1950*c*); Dahl (1952), mention only; Evans (1943*a*), mention in discussion of microchemical tests.

REVIEWS. Thamnolic acid: Asahina and Shibata (1954); Shibata (1963).

Cladonia subsquamosa f. *subulata* Sandst., see *Cladonia japonica* Vain.

Cladonia substrepsilis Sandst.   Fumarprotocetraric acid. Asahina (1943*b*); Asahina (1950*c*), from Japan; Dahl (1952), mention only.

Cladonia subtenuis (Abb.) Evans   Fumarprotocetraric acid, usnic acid; ursolic acid ($\pm$) (sometimes as Evans' substance E). Dahl (1952), mention only; Evans (1944*b*), from eastern and south-central U.S.A., also substance A ($\pm$) (later found to be a crystal form of usnic acid in the GE solution); Evans (1950), from Connecticut; Evans (1952*b*), from Florida.

Fumarprotocetraric acid, ursolic acid. Evans (1955*a*).

Fumarprotocetraric acid. C. F. Culberson (1965*c*), by extraction.

Cladonia subtenuis f. *cinerascens* (Abb.) Evans

Usnic acid frequently absent. Evans (1944*b*), from Connecticut, no usnic acid in samples tested.

Cladonia subulata (L.) Wigg.   Fumarprotocetraric acid; zinc.

Fumarprotocetraric acid. Asahina (1938*a*); Asahina (1943*b*); Dahl (1952), mention only; Evans (1944*b*), from Connecticut; D. Hess (1958), from Germany; Zopf (1908), from Germany, no atranorin, no fimbriatic acid (?), as *Cladonia fimbriata* var. *cornutoradiata*.

Zinc. Lambinon, Maquinay, and Ramaut (1964), from Belgium, a quantitative study.

REVIEWS. Fumarprotocetraric acid: W. Brieger (1923), as *Cladonia fimbriata* var. *cornutoradiata*; Hesse (1912), as *C. fimbriata* var. *cornutoradiata*; Zopf (1907), no atranorin, no fimbriatic acid (?), and as *C. fimbriata* var. *cornutoradiata*.

Except where indicated otherwise, all reports are as *Cladonia cornutoradiata*.

Cladonia sullivanii Müll. Arg. Didymic acid (±), divaricatic acid. Martin (1965), from Australia, New Zealand, and Tasmania, as *Cladia sullivanii*.
Divaricatic acid. Nuno (1962), from New Zealand.

Cladonia sylvatica (L.) Hoffm., see *Cladonia arbuscula* (Wallr.) Rabenh.

Cladonia sylvatica var. *erinacea* (Desm.) Kickx, see *Cladonia impexa* Harm.

Cladonia sylvatica f. *inactiva* Asah., see *Cladonia mitis* Sandst.

Cladonia symphicarpia Sandst. Atranorin, norstictic acid; calcium oxalate. Asahina (1943*a*), tested samples in Sandstede's exsiccati.
Atranorin, norstictic acid. Asahina (1942*b*), from North Carolina; Dahl (1952), mention only.
Atranorin. Evans (1943*a*), mention in discussion of microcrystal tests. Norstictic acid. Schindler (1936).
OTHER REPORT. Atranorin, methylethersalazinic acid, norstictic acid: Dahl (1950).

Cladonia tenuis (Flörke) Harm. Fumarprotocetraric acid, (+)-usnic acid; zinc.
Fumarprotocetraric acid, (+)-usnic acid. Asahina (1941*b*), no rangiformic acid; Asahina (1950*c*), from Japan.
Fumarprotocetraric acid, usnic acid. Dahl (1952), mention only; Duvigneaud (1939); Duvigneaud (1940); Evans (1943*a*), from North America, mention in discussion of microchemical tests; Evans (1943*b*), and unknown A (±) (later found to be a crystal form of usnic acid in the GE solution); Klosa (1951*b*).
Zinc. Lambinon, Maquinay, and Ramaut (1964), from Belgium, a quantitative study.
OTHER REPORT. Fumarprotocetraric acid, usnic acid (and as unknown A (±)), ursolic acid (and as unknown E (±)): Evans (1944*b*), not in the U.S.A. where *Cladonia subtenuis* occurs instead.
REVIEWS. Fumarprotocetraric acid, (+)-usnic acid: W. Brieger (1923); Thies (1932).

Cladonia terrae-novae Ahti Perlatolic acid, usnic acid. Evans (1944*b*), from Massachusetts and New Jersey, and substance B (±) (later found to be a crystal form of usnic acid in the GE solution), as *Cladonia impexa*; Evans (1955*a*), from Nova Scotia, as *C. impexa* f. *condensata*, *C. impexa* f. *laxiuscula*, and *C. impexa* f. *spumosa* from Massachusetts.

Cladonia theiophila Asah., see *Cladonia polydactyla* var. *theiophila* (Asah.) Asah.

Cladonia tixieri Abb.   Fumarprotocetraric acid. Abbayes (1963), from Vietnam, type description.

Cladonia transcendens (Vain.) Vain.   Thamnolic acid, usnic acid ($\pm$). Evans (1951), from British Columbia, California, Oregon, and Washington.
Thamnolic acid. Evans (1951), from British Columbia, Alaska, and Oregon, as *Cladonia transcendens* f. *squamulosa.*
OTHER REPORT. Squamatic acid, usnic acid, rhodocladonic acid, bellidiflorin: Dahl (1952), mention only, but see *Cladonia yunnana.*

Cladonia transcendens var. *yunnana* Vain., see *Cladonia yunnana* (Vain.) Abb.

Cladonia turgida (Ehrh.) Hoffm.   Atranorin, fumarprotocetraric acid. Asahina (1942*b*), from Michigan and Vermont, also tested samples from Sandstede's exsiccati; Asahina (1950*c*), from Japan; Dahl (1950); Dahl (1952), mention only; Evans (1950), from Connecticut.
OTHER REPORTS. Atranorin, fumarprotocetraric acid ($\pm$): Evans (1944*b*), from Connecticut; Evans (1947*a*), mention only.

Cladonia uncialis (L.) Wigg.   Squamatic acid ($\pm$), ($-$)-usnic acid; arabitol, mannitol; ascorbic acid; iron, manganese, zinc.
Squamatic acid ($\pm$), usnic acid. Evans (1944*b*), from Canada, Europe, Russia, Scandinavia, and northeastern U.S.A., and unknown substance A ($\pm$) (later found to be a crystal form of usnic acid in the GE solution); Evans (1950), from North America, and unknown substances A ($\pm$) and B ($\pm$) (later found to be crystal forms of usnic acid in the GE solution); Evans (1952*b*), from Florida.
Squamatic acid, ($-$)-usnic acid. Asahina (1950*c*), from Japan and Korea; Asahina and Yanagita (1933*a*), from southern Sakhalin.
Squamatic acid, usnic acid. Asahina (1942*a*), microchemical tests on samples from Sandstede's exsiccati, in Japan some samples with thamnolic acid or hypothamnolic acid or usnic acid only (see under OTHER REPORTS); Asahina (1952*b*), as *Cladonia uncialis* f. *obtusata*; Asahina, Aoki, and Fuzikawa (1941), from southern Sakhalin, mention only; Dahl (1950); Hale (1954*b*), from Baffin Island; Hale (1956*b*); Hale (1956*g*), from Connecticut, an exsiccati specimen; Klosa (1951*b*).
Squamatic acid. Asahina and Hiraiwa (1935*b*), mention only; Asahina and Simosato (1938), mention only; Mitsuno (1953), by paper chromatography.
Usnic acid. Asahina (1935*c*), suggested by KC+ (yellow) reaction; Evans (1943*a*), from North America, mention in discussion of microchemical tests; Gertig and Banasiewicz (1961), from Poland, 2.38%; Mozingo (1961), from Tennessee, mentioned in key; Vartia (1949), from Finland, mention only.

Arabitol, mannitol. Lindberg, Misiorny, and Wachtmeister (1953), by paper chromatography, no volemitol.

Ascorbic acid. Karev and Kochevykh (1962).

Iron, manganese, zinc. Lounamaa (1965), a quantitative study.

OTHER REPORTS. Barbatic acid, usnic acid: Hale (1957*b*), mention only. — Thamnolic acid, (–)-usnic acid: Vartia (1950*a*); Vartia (1950*b*), from Finland, by extraction; Zopf (1908), mention only. — Hypothamnolic acid, usnic acid: Asahina, Aoki, and Fuzikawa (1941), from Japan, as *Cladonia uncialis* f. *obtusata*, sample considered a physiological strain (see *C. pseudostellata*).

REVIEWS. Squamatic acid, (–)-usnic acid: Asahina and Shibata (1954). — Squamatic acid: Shibata (1958, 1963). — Thamnolic acid, (–)-usnic acid: W. Brieger (1923), as *Cladina uncialis*; Hesse (1912), as *Cladina uncialis*; Thies (1932); Zopf (1907), and as *Cladina uncialis*.

Cladonia uncinata Hoffm., see *Cladonia cenotea* (Ach.) Schaer.

Cladonia verticillaris (Raddi) Fr. Fumarprotocetraric acid. Asahina (1943*b*), from Brazil, as *Cladonia verticillaris* f. *penicilata*; Dahl (1952), mention only.

Cladonia verticillata (Hoffm.) Schaer. Fumarprotocetraric acid. Asahina (1943*b*), and as *Cladonia pseudalcicornis* (= *C. verticillata* f. *sobolifera*); Asahina (1943*d*), type description of *C. pseudalcicornis*; Asahina (1950*c*), from Japan, as *C. pseudalcicornis*; Asahina (1956*b*), from Japan; Dahl (1952), mention only; Evans (1943*a*), mention in discussion of microchemical tests; Evans (1944*b*), from Connecticut (see also OTHER REPORTS); Hale (1957*b*), mention only; D. Hess (1958), from Germany; Zopf (1908), as *C. verticillata* var. *evoluta*, mention only, no atranorin.

No usnic acid. Gertig and Banasiewicz (1961), from Poland.

OTHER REPORTS. Atranorin, fumarprotocetraric acid: Dahl (1952), mention only, as *Cladonia pseudalcicornis*. — Atranorin (±), fumarprotocetraric acid: Asahina (1950*c*). In Evans (1944*b*), the literature reports mentioned in the text are tabulated as atranorin (±), fumarprotocetraric acid (±), homosekikaic acid (±), and psoromic acid (±).

REVIEWS. Fumarprotocetraric acid: W. Brieger (1923); Hesse (1912); Thies (1932); all review reports as *Cladonia verticillata* var. *evoluta*.

Cladonia verticillata var. *cervicornis* (Ach.) Flörke Fumarprotocetraric acid; cervicornic acid; zinc.

Fumarprotocetraric acid, cervicornic acid (in the apothecia). Zopf (1908), from Germany, no atranorin.

No atranorin. Dahl (1950), as *Cladonia cervicornis*; Dahl (1953), as *C. cervicornis*.

Zinc. Lambinon, Maquinay, and Ramaut (1964), from Belgium, a quantitative study.

REVIEWS. Cervicornic acid, fumarprotocetraric acid: W. Brieger (1923); Hesse (1912); Thies (1932), and as *Cladonia verticillata* var. *cervicornis* f. *phyllophora*.

Cladonia verticillata ssp. *dissimilis* Asah. Atranorin, homosekikaic acid. Asahina (1940*b*), from Japan, type description; Asahina (1943*b*).

Cladonia verticillata var. *subcervicornis* Vain., see *Cladonia subcervicornis* (Vain.) Kernst.

Cladonia verticillata var. *subevoluta* Asah. Atranorin, fumarprotocetraric acid. Asahina (1940*b*), from Japan, type description.

Cladonia verticillata var. *subsobolifera* Asah. Atranorin, fumarproto-cetraric acid. Asahina (1940*b*), from Japan, type description.

Cladonia vulcanica Zoll. Barbatic acid, didymic acid, thamnolic acid; rhodocladonic acid.
  Barbatic acid, didymic acid, thamnolic acid. Asahina (1950*c*), from Formosa; Evans (1952*b*), from Florida.
  Didymic acid. Evans (1943*a*), mention in discussion of microcrystal tests.
  REVIEWS. Rhodocladonic acid: W. Brieger (1923); Hesse (1912); Thies (1932); all review reports as *Cladonia didyma* var. *vulcanica*.

Cladonia wrightii Evans Thamnolic acid. Evans (1955*a*), from Cuba, Jamaica, Puerto Rico, and the West Indies.

Cladonia yunnana (Vain.) Abb. Bellidiflorin ($\pm$), squamatic acid, usnic acid. Asahina (1939*c*), one sample from Japan with bellidiflorin; Asahina (1950*c*), from Japan; Evans (1951), from China and Japan.
  All reports are as *Cladonia transcendens* var. *yunnana*.

Coccocarpia cronia (Tuck.) Vain. No lichen substances found. Hale (1957*a*), from Arkansas and Oklahoma.

Coccocarpia parmelioides (Hook.) Tuck. *ex* Curt. No lichen substances found. Hale (1957*a*), from Arkansas and Oklahoma.

Collema auriculatum Hoffm. Carotene pigments. Henriksson (1963).

Collema bachmanianum (Fink) Degel., see *Collema tenax* var. *bachmanianum* (Fink) Degel.

Collema conglomeratum var. *crassiusculum* (Malme) Degel. No lichen substances found. Hale (1957*a*), from Arkansas, Missouri, and Oklahoma.

Collema cristatum (L.) Wigg. Carotene pigments. Henriksson (1963).

Collema fasciculare (L.) Wigg. Carotene pigments. Henriksson (1963).

Collema flaccidum (Ach.) Ach. Carotene pigments. Henriksson (1963).

Collema furfuraceum (Arn.) Du Rietz Carotene pigments. Henriksson (1963).

Collema multipartitum Sm. Carotene pigments. Henriksson (1963).

Collema polycarpon Hoffm. No lichen substances; carotene pigments. Carotene pigments. Henriksson (1963).
REVIEW. No lichen substances: Zopf (1907).

Collema subfurvum (Müll. Arg.) Degel. No lichen substances found. Hale (1957*a*), from Arkansas, Missouri, and Oklahoma.

Collema subnigrescens Degel. Carotene pigments. Henriksson (1963).

Collema tenax var. *bachmanianum* (Fink) Degel. Carotene pigments. Henriksson (1963), as *Collema bachmanianum*.

Collema tenax var. *vulgare* (Schaer.) Degel. Carotene pigments. Henriksson (1963).

Collema tunaeforme (Ach.) Ach. Carotene pigments. Henriksson (1963).

Coniocybe furfuracea (L.) Ach. Rhizocarpic acid, vulpinic acid. Bendz, Santesson, and Wachtmeister (1965*d*), separated and purified by thin-layer chromatography.
Vulpinic acid. Hale (1957*b*), mention only.
REVIEWS. Coniocybic acid: W. Brieger (1923); Hesse (1912); Thies (1932); Zopf (1907).

Cornicularia aculeata (Schreb.) Ach. Protolichesterinic acid; copper, iron.
Protolichesterinic acid. Dahl (1952), mention only; Vartia (1950*b*), from Finland; Walker (1965).
No usnic acid. Gertig and Banasiewicz (1961), from Poland, as *Cetraria aculeata*.
An aliphatic acid. Lamb (1964), mention only.

Copper, iron. Lange and Ziegler (1963), a quantitative study.
OTHER REPORT. Protolichesterinic acid, cornicularin: Klosa (1951*b*).
REVIEWS. Acanthellin, proto-α-lichesterinic acid, protolichesterinic acid, rangiformic acid: Thies (1932), as *Cetraria aculeata*, no fumarprotocetraric acid. — Acanthellin, protolichesterinic acid, rangiformic acid: W. Brieger (1923), and as *Cetraria aculeata*. — Protolichesterinic acid, rangiformic acid: Hesse (1912); Zopf (1907), no fumarprotocetraric acid; both reports as *Cetraria aculeata*. — Rangiformic acid: Asahina and Shibata (1954), as *Cetraria aculeata*.

Cornicularia divergens Ach.   Olivetoric acid. Asahina (1939*a*), from Japan; Dahl (1952), mention only; Hale (1956*e*), from North America; Mitsuno (1953), by paper chromatography, as *Alectoria divergens*; Walker (1965).
REVIEWS. Olivetoric acid: Asahina and Shibata (1954), and as *Alectoria divergens*; Shibata (1958, 1963), and as *A. divergens*; Thies (1932), as *A. divergens*.

Cornicularia epiphorella (Nyl.) Du Rietz
Chemical constituents unknown. Lamb (1964), mention only.

Cornicularia muricata Ach.   Protolichesterinic acid. Klosa (1951*b*), mention only, as *Cornicularia alpina*.
REVIEWS. Cornicularin, dilichesterinic acid, proto-α-lichesterinic acid, protolichesterinic acid, stuppeaic acid: W. Brieger (1923); Thies (1932). — Cornicularin, proto-α-lichesterinic acid, protolichesterinic acid: Hesse (1912). — Protolichesterinic acid, unidentified substance (insoluble in KOH): Zopf (1907). All review reports are as *Cetraria stuppea*.

Cornicularia normoerica (Gunn.) Du Rietz   Ergosterol. Alertsen, Bruun, and Hemmer (1962), from Norway, by extraction.
OTHER REPORT. Unidentified substance: Dahl (1952), mention only.
REVIEW. Parmelia-brown: Thies (1932), as *Cornicularia tristis*.

Cornicularia odontella (Ach.) Röhl.   Protolichesterinic acid. Dahl (1952), mention only.

Cornicularia pseudosatoana Asah.   Olivetoric acid. Asahina (1936*d*), by microcrystal tests, as *Alectoria satoana*; Asahina (1937*c*), as *A. divergens* var. *satoana*; Asahina (1939*a*), from Japan, type description; Asahina and Asano (1932*a*), from Japan, by extraction, as *Cornicularia divergens* (=misidentified *A. satoana* according to Asahina and Fuzikawa (1935*f*)); Asahina and Fuzikawa (1935*f*), from Japan, as *A. satoana*.
REVIEWS. Olivetoric acid: Asahina and Shibata (1954); Shibata (1958, 1963).
See also *Cornicularia divergens* and *C. satoana*.

Cornicularia satoana (Gyeln.) Asah. Physodic acid. Asahina (1939*a*).

Cornicularia tristis (Web.) Hoffm., see *Cornicularia normoerica* (Gunn.) Du Rietz

Crocynia membranacea (Dicks.) Zahlbr., see *Lepraria membranacea* (Dicks.) Vain.

Crocynia neglecta (Nyl.) Hue, see *Lepraria neglecta* (Nyl.) Lett.

Cyphelium chrysocephalum (Turn.) Chev., see *Chaenotheca chrysocephala* (Turn.) Th. Fr.

Cyphelium notarisii (Tul.) Blombg. & Forss. Rhizocarpic acid. Hale (1957*b*), mention only.

Cyphelium tigillare (Ach.) Ach. Rhizocarpic acid. Hale (1957*b*), mention only.
REVIEWS. Acolic acid, rhizocarpic acid: W. Brieger (1923); Hesse (1912); Thies (1932); Zopf (1907); all review reports as *Acolium tigillare*.

Dactylina arctica (Hook.) Nyl. Gyrophoric acid. Hale (1956*e*), from North America; Hale (1957*b*), mention only.

Darbishirella gracillima (Kremp.) Zahlbr. Psoromic acid (sometimes as parellic acid). Follmann and Villagrán (1964), mention only; Huneck and Follmann (1963), mention only; Huneck and Follmann (1967*b*), from Chile, 14.2% by extraction of 28.0 g. of lichen.
OTHER REPORT. Psoromic acid, roccellic acid: Follmann (1964), mention only.
REVIEWS. Psoromic acid: W. Brieger (1923); Hesse (1912); Thies (1932); Zopf (1907).
See also *Ingaderia pulcherrima*.

Dendrographa leucophaea (Tuck.) Darb.
Protocetraric acid or fumarprotocetraric acid. Asahina (1936*e*).
Erythrin. Manaktala, Neelakantan, and Seshadri (1966), mention only; V. S. Rao and Seshadri (1942*b*), mention only.
REVIEW. Fumarprotocetraric acid: W. Brieger (1923); Hesse (1912); Thies (1932).

Dermatiscum thunbergii (Ach.) Nyl. Lecanoric acid, squamatic acid; rhizocarpic acid; arabitol, D-mannitol; choline sulfate ester. Harper and Letcher (1966).

Dermatocarpon fluviatile (G. Web.) Th. Fr.   Mannitol, volemitol. Lindberg, Misiorny, and Wachtmeister (1953), volemitol isolated, mannitol found by paper chromatography, no arabitol.
REVIEWS. D-Volemitol: Shibata (1958).

Dermatocarpon lachneum (Ach.) A. L. Sm.   Mannitol, volemitol. Lindberg, Misiorny, and Wachtmeister (1953), volemitol isolated, mannitol found by paper chromatography, no arabitol.
REVIEWS. D-Volemitol: Shibata (1958, 1963).

Dermatocarpon miniatum (L.) Mann   D-Mannitol, sucrose, trehalose, D-volemitol. Lindberg (1955).
    Mannitol, volemitol. Lindberg, Misiorny, and Wachtmeister (1953), volemitol isolated, mannitol found by paper chromatography, no arabitol and also *Dermatocarpon miniatum* var. *complicatum*.
    D-Volemitol. Asahina and Kagitani (1934).
    Volemitol. Asahina (1934*b*); Asahina (1939*b*), mention only; Wachtmeister (1958*a*), mention only.
    REVIEWS. D-Volemitol: Asahina and Shibata (1954); Shibata (1958), and also *Dermatocarpon miniatum* var. *complicatum*.

Dermatocarpon moulinsii (Mont.) Zahlbr.   Carbohydrate; amino acids, protein; calcium, iron, phosphorus. Ramakrishnan and Subramanian (1966*a*), from India, alanine, arginine, aspartic acid, glutamic acid, glycine, isoleucine, leucine, lysine, methionine, phenylalanine, proline, serine, threonine, tyrosine, and valine as free amino acids, and these with tryptophan and an unidentified amino acid as combined amino acids.

Dermatocarpon tuckermanii (Rav.) Zahlbr.   No lichen substances found. Hale (1956*h*), from eastern U.S.A.; Hale (1957*a*), from Arkansas, Missouri, and Oklahoma.

Dimelaena oreina (Ach.) Beltr., see *Rinodina oreina* (Ach.) Mass.

Diploicia canescens (Dicks.) Mass., see *Buellia canescens* (Dicks.) De Not.

Diploschistes albissimus (Ach.) Dalla Torre & Sarnth., see *Diploschistes gypsaceus* (Ach.) Nyl.

" *Diploschistes arenarius* Schaer.," see *Diploschistes scruposus* var. *arenarius* (Schaer.) Müll. Arg.

Diploschistes bryophilus (Ehrh.) Zahlbr.   Diploschistesic acid; zinc.
    Diploschistesic acid. Koller and Hamburg (1935*a*), mention early work of Zopf and Hesse.
    Zinc. Lambinon, Maquinay, and Ramaut (1964), from Belgium, a quantitative study, as *Diploschistes scruposus* var. *bryophilus*.

REVIEWS. Diploschistesic acid, lecanoric acid: Asahina and Shibata (1954); Neelakantan (1965). — Diploschistesic acid: Shibata (1958, 1963); Zopf (1907), questions occurrence of lecanoric acid. — Atranorin, lecanoric acid, patellaric acid: W. Brieger (1923); Hesse (1912); as *Urceolaria scruposa* var. *bryophila*. — Lecanoric acid, patellaric acid: Thies (1932), as *U. scruposa* var. *bryophila*.

Diploschistes cretaceus (Ach.) Lett.
REVIEWS. Atranorin, lecanoric acid, zeorin: Zopf (1907). — Atranorin, zeorin: Thies (1932).

Diploschistes gypsaceus (Ach.) Nyl.
REVIEWS. Atranorin, lecanoric acid, zeorin: W. Brieger (1923), as *Urceolaria scruposa* var. *cretacea*; Hesse (1912), as *U. scruposa* var. *cretacea*; Thies (1932), as *Diploschistes albissimus*; Zopf (1907), as *D. albissimus*. — Patellaric acid: W. Brieger (1923); Thies (1932); both reports as *U. albissima*.

Diploschistes scruposus (Schreb.) Norm. Diploschistesic acid, lecanoric acid. Asahina (1939*b*), mention only; Asahina and Yasue (1936), mention only; Erdtman and Wachtmeister (1957*a*), mention in discussion of biosynthesis; Koller and Hamburg (1935*a*), from Germany, by extraction.
Diploschistesic acid. Seshadri (1944), mention only; Vartia (1950*b*), from Finland.
Lecanoric acid. Hale (1956*e*), from North America.
REVIEWS. Atranorin, diploschistesic acid, lecanoric acid, patellaric acid: Zopf (1907), and as *Urceolaria scruposa*. — Atranorin, diploschistesic acid, patellaric acid: Thies (1932), and as *U. scruposa*. — Atranorin, lecanoric acid, patellaric acid: W. Brieger (1923); Hesse (1912); both reports as *U. scruposa* var. *vulgaris*. — Diploschistesic acid, lecanoric acid: Neelakantan (1965). — Diploschistesic acid: Asahina and Shibata (1954); Shibata (1958, 1963).

Diploschistes scruposus var. *arenarius* (Schaer.) Müll. Arg. Lecanoric acid.
REVIEWS. Lecanoric acid: W. Brieger (1923), as *Urceolaria scruposa* var. *arenaria*; Hesse (1912), as *U. scruposa* var. *arenaria*; Zopf (1907), as *Diploschistes arenarius*.

Diploschistes scruposus var. *bryophilus* (Ehrh.) Müll. Arg., see *Diploschistes bryophilus* (Ehrh.) Zahlbr.

Dirina lutosa Zahlbr. Roccellic acid. Huneck and Follmann (1964*b*), from Chile, by extraction of 40 g. of lichen.

Dirina repanda (Nyl.) Fr.   Erythrin, lecanoric acid; portentol; *meso*-erythritol. Huneck and Trotet (1967), 2.5% erythrin, 1.2% portentol, and 2.8% *meso*-erythritol, by extraction of 36.0 g. of lichen, lecanoric acid by thin-layer chromatography.

Dolichocarpus chilensis Sant.   Gyrophoric acid; five carotene pigments; Na$^+$, K$^+$, Mg^{++}, Cl$^-$, SO$_4$$^=$. Huneck and Follmann (1965a), from Chile, 10% crude gyrophoric acid by extraction, carotene pigments by thin-layer chromatography.

Endocarpon adscendens (Anzi) Müll. Arg.   Mannitol, volemitol. Lindberg, Misiorny, and Wachtmeister (1953), by paper chromatography, no arabitol.
REVIEWS. D-Volemitol. Shibata (1958, 1963).

Endocena informis (L.) Zahlbr.   Thamnolic acid. Bendz, Santesson, and Wachtmeister (1965c).

Enterographa crassa (DC.) Fée
REVIEWS. Stigmatidin: W. Brieger (1923); Hesse (1912); Thies (1932); Zopf (1907), and an unidentified colorless lichen acid; all reports as *Stigmatidium venosum*.

Evernia divaricata (L.) Ach.   Divaricatic acid, (+)-usnic acid.
Divaricatic acid, usnic acid. C. F. Culberson (1963), by microchemical tests; Zahlbruckner (1904), no atranorin.
Divaricatic acid. Asahina (1936e), mention in discussion of microchemical tests; Asahina and Hirakata (1932), mentioning early work of Zopf and Hesse.
(+)-Usnic acid. Moissejeva (1957), 0.4%; Moissejeva (1961), 0.4%; Stoll, Brack, and Renz (1947), 0.50% by extraction; Stoll, Renz, and Brack (1947), from Switzerland.
OTHER REPORT. Atranorin, divaricatic acid: Hale (1957b), mention only.
REVIEWS. Divaricatic acid, (+)-usnic acid: W. Brieger (1923); Hesse (1912); Thies (1932), and as *Letharia divaricata*; Zopf (1907). — Divaricatic acid: Asahina and Shibata (1954); Shibata (1958, 1963). — (+)-Usnic acid: Savicz, Litvinov, and Moissejeva (1960), 0.4%.

Evernia esorediosa (Müll. Arg.) Du Rietz   Divaricatic acid, (+)-usnic acid. Asahina and Hirakata (1932), from Japan, by extraction, as *Evernia mesomorpha* f. *esorediosa*.
Divaricatic acid, usnic acid. Asahina (1952d), trace of usnic acid, as *Evernia mesomorpha* f. *esorediosa*.
(+)-Usnic acid. Moissejeva (1957), 0.5%; Moissejeva (1961), 0.5%.

REVIEWS. Divaricatic acid: Asahina and Shibata (1954); Shibata (1958, 1963); all three reports as *Evernia mesomorpha* f. *esorediosa.* — (+)-Usnic acid: Savicz, Litvinov, and Moissejeva (1960), 0.5%.

Evernia furfuracea (L.) Mann, see *Pseudevernia furfuracea* (L.) Zopf

Evernia furfuracea var. *olivetorina* (Zopf) Räs., see *Pseudevernia olivetorina* (Zopf) Zopf

Evernia herinii Duvign., see *Evernia prunastri* (L.) Ach.

Evernia illyrica (Zahlbr.) Zahlbr. Atranorin, divaricatic acid. C. F. Culberson (1963), from the Mediterranean region, by microchemical tests; Zahlbruckner (1904), from Yugoslavia and Central Europe, no trace of usnic acid.

Divaricatic acid. Asahina and Hirakata (1932), mention only.

REVIEWS. Atranorin, divaricatic acid: W. Brieger (1923); Hesse (1912); Thies (1932); Zopf (1907).

Evernia isidiophora Zopf, see *Pseudevernia furfuracea* (L.) Zopf

Evernia mesomorpha Nyl. Divaricatic acid, (+)-usnic acid; fat; enzymes, protein; calcium, phosphorus.

Divaricatic acid, usnic acid. Asahina (1952*d*), from Japan, only a trace of usnic acid, by microchemical tests; C. F. Culberson (1963), by microchemical tests; Hale (1957*c*), from Michigan, an exsiccati specimen.

Divaricatic acid. Asahina (1936*e*), description of microcrystal tests; Asahina and Hirakata (1932), mention only, as *Evernia thamnodes*; Kurokawa (1964), mention only; Mitsuno (1953), paper by chromatography.

(+)-Usnic acid. Moissejeva (1957), 0.8%, as *Evernia thamnodes*; Moissejeva (1961), 0.8%, enzymes tabulated, as *E. thamnodes*; Savicz, Kuprevicz, Litvinov, Moissejeva, and Rassadina (1956), 0.8%, as *E. thamnodes*.

Usnic acid. Taguchi, Sankawa, and Shibata (1966), ^{14}C study on biosynthesis.

Fat, protein, calcium, phosphorus. Scotter (1965), from northern Canada.

OTHER REPORT. Evernic acid: Hale (1956*f*).

REVIEWS. Divaricatic acid, (+)-usnic acid: W. Brieger (1923); Hesse (1912); Savicz, Litvinov, and Moissejeva (1960); Thies (1932); Zopf (1907); all five reports as *Evernia thamnodes*. — Divaricatic acid, usnic acid: Asahina and Shibata (1954). — Divaricatic acid: Shibata (1958, 1963).

Evernia mesomorpha f. *esorediosa* (Müll. Arg.) Zahlbr., see *Evernia esorediosa* (Müll. Arg.) Du Rietz

Evernia olivetorina Zopf, see *Pseudevernia olivetorina* (Zopf) Zopf

Evernia perfragilis Llano   Divaricatic acid, usnic acid. C. F. Culberson (1963), by microchemical tests.
Divaricatic acid. Hale (1957b), mention only.

Evernia prunastri (L.) Ach.   Atranorin (sometimes as atranoric acid), chloroatranorin, evernic acid, (+)-usnic acid (±); arabitol, mannitol; enzymes; folic acid-, folinic acid-, and vitamin B_{12}-group factors; iron, manganese, zinc.
Atranorin, chloroatranorin, evernic acid. St. Pfau (1934), in a chemical study of chloroatranorin, by extraction of a sample contaminated with *Pseudevernia furfuracea*.
Atranorin, evernic acid, (+)-usnic acid. Asahina (1952d), from Japan; Moissejeva (1961), 0.5% (+)-usnic acid, enzymes tabulated.
Atranorin, chloroatranorin. Seshadri and Subramanian (1963), mention with a method for chromatographic separation; Spillane, Keane, and Nolan (1936), mention only.
Atranorin, evernic acid, usnic acid. Curd, Robertson, and Stephenson (1933), by extraction, authors question natural occurrence of chloroatranorin in this species.
Atranorin, evernic acid, usnic acid (±). Duvigneaud (1940), atranorin misprinted as thamnolic acid in this reference, usnic acid lacking in samples referred to as *Evernia herinii*; St. Pfau (1926).
Atranorin, evernic acid. Asahina (1936e), description of microcrystal tests; D. Hess (1958), from Germany; Ramaut, Lambinon, and Targé (1962), from Belgium, France, and the Netherlands, no usnic acid in any samples examined by paper chromatography or by extraction; Ramaut (1963b), by thin-layer chromatography; Seshadri (1944), mention in discussion of biogenesis; Vartia (1950a), by extraction; Vartia (1950b), by extraction.
Atranorin, usnic acid. Vartia (1949), from Finland.
Chloroatranorin, evernic acid, (+)-usnic acid (±). C. F. Culberson (1963), from France, by extraction and by microchemical tests, atranorin or chloroatranorin indicated by microchemical test, chloroatranorin possibly containing a trace of atranorin by extraction.
Chloroatranorin. Neelakantan and Seshadri (1960b), mention only; Yoshida (1951), mention only.
Evernic acid. E. Fischer and Fischer (1914), source of evernic acid for comparison with synthetic material; Huneck and Tümmler (1965), by extraction; Koller (1932b); Koller and Pfeiffer (1933d); Ramaut (1961c); Wachtmeister (1952), source of evernic acid for studies of chromatographic methods.

(+)-Usnic acid. Moissejeva (1957), 0.5%.
Usnic acid. Asahina (1935*c*), suggested by KC+ (yellow) reaction; Tomaselli (1957), from Italy, mention only.
Usnic acid ($\pm$). Ramaut (1965).
Arabitol, mannitol. Pueyo (1964–1965*b*), from France.
Folic acid-, folinic acid-, and vitamin B_{12}-group factors. Sjöström and Ericson (1953).
Iron, manganese, zinc. Lounamaa (1965), a quantitative study.
No usnic acid. Gertig and Banasiewicz (1961), from Poland.
OTHER REPORTS. Atranorin, barbatic acid, evernic acid, usnic acid: Solberg (1956), mention only. — Everniin: Drake (1943). — Atranorin, evernic acid, everninic acid: Klosa (1951*b*).
REVIEWS. Atranorin, chloroatranorin, evernic acid, (+)-usnic acid: Asahina and Shibata (1954); Schindler (1956/1957). — Atranorin, evernic acid, everninic acid, (+)-usnic acid: Thies (1932), and as *Evernia prunastri* f. *gracilis* and *E. prunastri* var. *vulgaris.* — Atranorin, evernic acid, (+)-usnic acid: W. Brieger (1923); Hesse (1912); Zopf (1907). — Atranorin, evernic acid: W. Brieger (1923); Hesse (1912); both reports as *E. prunastri* f. *gracilis* and *E. prunastri* var. *vulgaris.* — Atranorin: Zopf (1907), 5.3%, as *E. prunastri* var. *sorediifera.* — Chloroatranorin, evernic acid: Shibata (1958, 1963). — (+)-Usnic acid: Savicz, Litvinov, and Moissejeva (1960), 0.5%.

Evernia thamnodes (Flot.) Arn., see *Evernia mesomorpha* Nyl.

Evernia vulpina (L.) Ach., see *Letharia vulpina* (L.) Hue

Everniopsis pseudoreticulata (Duvign.) Dodge, see *Everniopsis trulla* (Ach.) Nyl.

Everniopsis trulla (Ach.) Nyl.   Atranorin, psoromic acid, usnic acid. Schumacker (1961), from Africa, by extraction and paper chromatography, as *Everniopsis pseudoreticulata.*
   Atranorin, usnic acid, unidentified substance. Duvigneaud (1942*b*), from the Republic of Congo, unidentified substance may be psoromic acid, type description of *Hendrickxia pseudoreticulata.*

Fulgensia fulgens (Sw.) Elenk.   Parietin (sometimes as physcion). Galun and Lavee (1966), from Israel, by microcrystal tests.
   REVIEWS. Parietin: W. Brieger (1923); Hesse (1912); Thies (1932); Zopf (1907).

Gasparrinia cirrochroa (Ach.) B. Stein, see *Caloplaca cirrochroa* (Ach.) Th. Fr.

Gasparrinia decipiens (Arn.) Syd., see *Caloplaca decipiens* (Arn.) J. Stein.

Gasparrinia elegans (Link) B. Stein, see *Xanthoria elegans* (Link) Th. Fr.

Gasparrinia medians (Nyl.) Syd., see *Candelariella medians* (Nyl.) A. L. Sm.

Gasparrinia muorum (Hoffm.) Torn., see *Caloplaca murorum* (Hoffm.) Th. Fr.

"*Gasparrinia sympageum,*" see *Caloplaca aurantia* (Pers.) Hellb.

Glossodium japonicum Zahlbr. Thamnolic acid. Asahina (1956a), by microchemical tests, no fumarprotocetraric acid.

Graphina aibonitensis Fink   Salazinic acid. Wirth and Hale (1963), by microchemical tests on the holotype specimen.

Graphina albostriata (Vain.) Zahlbr.   Unidentified acid. Wirth and Hale (1963), by microchemical tests on the holotype specimen.

Graphina bipartita Müll. Arg.   Norstictic acid. Wirth and Hale (1963), from Mexico, by microchemical tests on the holotype specimen.

Graphina cladophora (Vain.) Zahlbr.   Norstictic acid. Wirth and Hale (1963), by microchemical tests on an isotype of *Graphis cladophora*.

Graphina collosporella (Vain.) Zahlbr.   Unidentified acid. Wirth and Hale (1963), by microchemical tests on syntypes of *Graphis collosporella*.

Graphina confluens (Fée) Müll. Arg.   Stictic acid; lichexanthone; unidentified fatty substance (from Brazil, Costa Rica, Cuba, Jamaica, Mexico, and Puerto Rico). Stictic acid; unidentified substance (from Puerto Rico). Stictic acid only (from the Philippines). Atranorin, lichexanthone; zeorin (from Brazil). Lichexanthone only (from Brazil and Cuba). Norstictic acid (from Brazil and Cuba). Wirth and Hale (1963), by microchemical tests.

Graphina corcovadensis Red.   Stictic acid. Wirth and Hale (1963), by microchemical tests on the holotype specimen.

Graphina diorygmatoides (Vain.) Zahlbr.   Norstictic acid. Wirth and Hale (1963), by microchemical tests on an isotype of *Graphis diorygmatoides*.

Graphina elongata (Vain.) Zahlbr.   Norstictic acid. Wirth and Hale (1963), from Mexico, by microchemical tests on the holotype of *Graphis elongata*, no lichen substances found by microcrystal test in the GAo-T solution on the holotype of *Graphina elongatoradians* previously thought to be synonymous.

Graphina elongatoradians Fink   No substances found. Wirth and Hale (1963), from Puerto Rico, by a microcrystal test in the GA*o*-T solution on the holotype specimen.

Graphina heteroplacoides Red.   Unidentified acid (K + red). Wirth and Hale (1963), by a microcrystal test in the GA*o*-T solution, holotype specimen tested.

Graphina hiascens (Fée) Müll. Arg.   Norstictic acid. Wirth and Hale (1963), from Mexico, by microchemical tests.

Graphina hiascens var. *clausior* (Vain.) Zahlbr.   Stictic acid. Wirth and Hale (1963), by microchemical tests on an isotype of *Graphis hiascens* var. *clausior*.

Graphina incerta Red.   Protocetraric acid. Wirth and Hale (1963), microchemical tests on a syntype specimen.

Graphina indita var. *pularensis* (Vain.) Zahlbr.   Stictic acid. Wirth and Hale (1963), by microchemical tests on an isotype of *Graphis indita* var. *pularensis*.

Graphina insignis (Vain.) Zahlbr.   No substances found. Wirth and Hale (1963), from Mexico, lichen thallus K + red.

Graphina insignis var. *imperfecta* Red.   Atranorin. Wirth and Hale (1963), from Brazil, by microchemical tests on the holotype specimen.

Graphina interstes Müll. Arg.   Unidentified acid (K + red). Wirth and Hale (1963), by microchemical tests on the holotype specimen.

Graphina obtectula Müll. Arg.   Unidentified acid (K + yellow). Wirth and Hale (1963), by microchemical tests on the holotype specimen.

Graphina palmeri Zahlbr.   Unidentified substance (K + red). Wirth and Hale (1963), from Mexico, microcrystals the same as for *Graphina triangularis*, *G. virginea*, and *Phaeographis exaltata*, by microcrystal tests on an isotype specimen.

Graphina parilis (Kremp.) Müll. Arg.   Atranorin (±), stictic acid. Wirth and Hale (1963), from Mexico.

Graphina platicarpina Zahlbr.   Protocetraric acid. Wirth and Hale (1963), by microchemical tests on the holotype specimen.

Graphina plittii Zahlbr.   Salazinic acid. Wirth and Hale (1963), by microchemical tests on an isotype specimen.

Graphina pseudosophistica (Vain.) Zahlbr.   Unidentified acid (±). Wirth and Hale (1963), by microchemical tests on a syntype specimen.

Graphina reniformis var. *subastroidea* Red.   Protocetraric acid; lichexanthone. Wirth and Hale (1963), by microchemical tests on the holotype specimen.

Graphina riopiedrensis Fink   Stictic acid. Wirth and Hale (1963), by microchemical tests on the holotype specimen.

Graphina sulcata Fink   Norstictic acid. Wirth and Hale (1963), from Mexico, by microchemical tests.

Graphina triangularis Zahlbr.   Unidentified substance (K + red). Wirth and Hale (1963), by microchemical tests on an isotype specimen, unidentified substance also found in several other species listed under *Graphina palmeri.*

Graphina virginea (Eschw.) Müll. Arg.   Unidentified substance (K + red). Wirth and Hale (1963), from Mexico, by microchemical tests, unidentified substance also found in several other species listed under *Graphina palmeri.*

Graphis abaphoides Nyl.   Protocetraric acid; unidentified substance. Wirth and Hale (1963), by microchemical tests on an isosyntype specimen.

Graphis afzelii Ach.   Lecanoric acid. Wirth and Hale (1963), from Mexico, by microchemical tests on an isotype specimen.

Graphis albescens Vain.   Stictic acid. Wirth and Hale (1963), by microchemical tests on the holotype specimen.

Graphis albida Fink   Protocetraric acid. Wirth and Hale (1963), by microchemical tests on the holotype specimen.

Graphis arecae Vain.   Stictic acid. Wirth and Hale (1963), by microchemical tests on an isotype specimen.

Graphis caesiella Vain.   Norstictic acid. Wirth and Hale (1963), from Mexico, by microchemical tests on the holotype specimen.

Graphis desquamescens Fée   Norstictic acid. Wirth and Hale (1963), from Mexico, by microchemical tests on the holotype of *Graphis compulsa.*

Graphis diaphoroides Müll. Arg. Norstictic acid. Wirth and Hale (1963), by microchemical tests on an isosyntype specimen.

Graphis eugeniae Vain. Salazinic acid. Wirth and Hale (1963), by microchemical tests on an isosyntype specimen.

Graphis floridana Tuck. Norstictic acid. Wirth and Hale (1963), by microchemical tests on an isotype specimen.

Graphis grammatica Nyl. Stictic acid ($\pm$). Wirth and Hale (1963), from Cuba and Mexico, by microchemical tests.

Graphis humilis Vain. Unidentified acid (K + red). Wirth and Hale (1963), by microchemical tests on an isotype specimen.

Graphis implicata Fée Norstictic acid. Wirth and Hale (1963), from Mexico, by microchemical tests.

Graphis scripta (L.) Ach. Norstictic acid. Hale (1957*b*), mention only.
REVIEWS. Norstictic acid: Asahina and Shibata (1954); Shibata (1958, 1963). — Salazinic acid: W. Brieger (1923); Hesse (1912); Thies (1932); Zopf (1907).

Graphis subamylacea Zahlbr. Stictic acid. Wirth and Hale (1963), from Mexico, by microchemical tests on an isotype specimen.

Graphis tenella Ach. Atranorin, norstictic acid. Ahmadjian (1964), cultured fungal component found to produce unidentified pigments.

Graphis turbulenta Nyl. Norstictic acid. Wirth and Hale (1963), by microchemical tests on an isotype specimen.

Gyalecta ulmi (Sw.) Zahlbr.
REVIEWS. Rhizoiden-green: Thies (1932), as *Phialopsis rubra*.

Gyalolechia aurella (Hoffm.) Körb., see *Candelariella aurella* (Hoffm.) Zahlbr.

Gyalolechia reflexa (Nyl.) Dalla Torre & Sarnth., see *Candelariella reflexa* (Nyl.) Lett.

Gyrophora anthracina (Wulf.) Körb., see *Agyrophora rigida* (Du Rietz) Llano

Gyrophora crustulosa Ach., see *Omphalodiscus crustulosus* (Ach.) Schol.

Gyrophora cylindrica (L.) Ach., see *Umbilicaria cylindrica* (L.) Del.

Gyrophora deusta (L.) Ach., see *Umbilicaria deusta* (L.) Baumg.

Gyrophora dillenii (Tuck.) Müll. Arg., see *Umbilicaria mammulata* (Ach.) Tuck.

Gyrophora esculenta Miyoshi, see *Umbilicaria esculenta* (Miyoshi) Minks

Gyrophora hirsuta (Sw.) Ach., see *Umbilicaria hirsuta* (Sw. *ex* Westr.) Ach.

Gyrophora hyperborea Ach., see *Umbilicaria hyperborea* (Ach.) Ach.

Gyrophora microphylla (Laur.) Arn., see *Agyrophora microphylla* (Laur.) Llano

Gyrophora muehlenbergii Ach., see *Actinogyra muehlenbergii* (Ach.) Schol.

Gyrophora murina (Ach.) Ach., see *Umbilicaria murina* (Ach.) DC.

Gyrophora polyphylla (L.) Fink, see *Umbilicaria polyphylla* (L.) Baumg.

Gyrophora polyrrhiza (L.) Körb., see *Actinogyra polyrrhiza* (L.) Schol.

Gyrophora proboscidea (L.) Ach., see *Umbilicaria proboscidea* (L.) Schrad.

Gyrophora pustulata (L.) Ach., see *Lasallia pustulata* (L.) Mér.

Gyrophora spodochroa (Hoffm.) Ach., see *Omphalodiscus spodochrous* (Hoffm.) Schol.

Gyrophora spodochroa "var. *depressa* Ach." (= *Gyrophora depressa* (Ach.) Röhl.), see *Omphalodiscus crustulosus* (Ach.) Schol.

Gyrophora vellea (L.) Ach., see *Umbilicaria vellea* (L.) Ach.

"*Haematomma abortivum* Hepp," see *Haematomma coccineum* (Dicks.) Körb.

Haematomma cismonicum Beltr.   Thamnolic acid. W. L. Culberson (1963*a*), from Austria, Canada, France, Germany, Hungary, Switzerland, Maine, Massachusetts, New Hampshire, and North Carolina, by microchemical tests.

Haematomma coccineum (Dicks.) Körb.   Atranorin, porphyrilic acid, (−)-usnic acid; zeorin; unidentified red pigment.

Atranorin, porphyrilic acid, (−)-usnic acid. Wachtmeister (1954), a detailed chemical study of porphyrilic acid, mentions early work of Hesse and of Zopf, extraction of lichen gives very high yield of (−)-usnic acid (up to 20%).

Atranorin, usnic acid, zeorin, unidentified red pigment. W. L. Culberson (1963a), from England, Germany, and Sweden, not in North America, like *Haematomma porphyrium* but containing usnic acid.

Atranorin, usnic acid. Klosa (1951b).

Porphyrilic acid, (−)-usnic acid. Erdtman and Wachtmeister (1953); Wachtmeister (1956a), mention in a chemical study of porphyrilic acid.

Porphyrilic acid. Åkermark, Erdtman, and Wachtmeister (1959), mention only; Erdtman and Wachtmeister (1956), mention only; Wachtmeister (1956b); Wachtmeister (1958a), mention only.

(−)-Usnic acid. Salkowski (1910).

Zeorin. Hale (1957b), mention only.

REVIEWS. Atranorin, coccinic acid, haematommidin: W. Brieger (1923); Hesse (1912); Thies (1932), haematommidin possibly the same as leiphaemin; Zopf (1907); all four reports as *Haematomma coccineum* f. *abortivum* and *H. abortivum*. — Atranorin, haematommin, hydrohaematommin, lecanoric acid, leiphaemin, (−)-usnic acid, zeorin: W. Brieger (1923); Hesse (1912). — Atranorin, haematommin, hydrohaematommin, hymenorhodin, leiphaemin, porphyrilic acid, (−)-usnic acid, zeorin: Thies (1932), one variety with lecanoric acid; Zopf (1907). — Atranorin, zeorin: Asahina and Shibata (1954). — Atranorin, porphyrilic acid, zeorin: Shibata (1958, 1963). — Porphyrilic acid: Neelakantan (1965).

Haematomma elatinum (Ach.) Mass. Thamnolic acid. W. L. Culberson (1963a), from Austria, Hungary, Norway, Poland, Sweden, Maine, Massachusetts, New York, North Carolina, and Virginia.

Haematomma fauriei Zahlbr. Atranorin, psoromic acid. Asahina (1964a), from Japan.

Haematomma fuliginosum Asah. Thamnolic acid; calcium oxalate. Asahina (1964a), type description.

Haematomma lapponicum Räs. Divaricatic acid, usnic acid; unidentified substance; unidentified red pigment (acetone soluble).

Divaricatic acid, usnic acid, unidentified substance. Asahina (1964a), from Japan, as *Haematomma ventosum* var. *lapponicum*, unidentified substance present in minor quantity.

Divaricatic acid, usnic acid, unidentified red pigment (acetone soluble). W. L. Culberson (1963a), from the U.S.S.R., New Hampshire, Virginia, and Washington.

OTHER REPORTS. Atranorin, divaricatic acid: Satô (1940), from Japan, as *Haematomma ventosum* var. *lapponicum*. — Divaricatic acid, thamnolic acid, (+)-usnic acid, ventosic acid: Solberg (1957), as *H. ventosum* var. *lapponicum.* (See *H. ventosum.*)

REVIEW. Ventosic acid: Neelakantan (1965), as *Haematomma ventosum* var. *lapponicum.*

Haematomma leiphaemum (Ach.) Zopf   Atranorin; zeorin. W. L. Culberson (1963*a*), from Denmark, Germany, and Sweden, PD + thallus reaction not understood.

OTHER REPORTS. Atranorin, a fatty acid: Solberg (1956), mention only. — Atranorin, leiphaemic acid, leiphaemin, zeorin: Klosa (1951*b*), mention only.

REVIEWS. Atranorin, leiphaemic acid, leiphaemin, zeorin: W. Brieger (1923); Hesse (1912); Thies (1932); Zopf (1907).

Haematomma ochrophaeum (Tuck.) Mass.   Thamnolic acid; zeorin. W. L. Culberson (1963*a*), from Canada, Japan, Maine, Massachusetts, Michigan, New Hampshire, New York, North Carolina, and Vermont. Thamnolic acid. Asahina (1964*a*), from Japan.

Haematomma pachycarpum (Müll. Arg.) Zahlbr.   Pachycarpin. Asahina (1964*a*), from Japan, the new substance pachycarpin detected microchemically.

Haematomma polycarpum Zahlbr.   Imbricaric acid. Asahina (1964*b*), from North Carolina, species identification uncertain.

Haematomma porphyrium (Pers.) Zopf   Atranorin, porphyrilic acid; zeorin; unidentified red pigment (acetone insoluble).

Atranorin, zeorin, unidentified red pigment (acetone insoluble). W. L. Culberson (1963*a*), from Austria, France, North America, and Norway.

Atranorin, porphyrilic acid. Wachtmeister (1954), mentions Zopf's study of this species from southern Sweden.

Porphyrilic acid. Erdtman and Wachtmeister (1953); Wachtmeister (1958*a*).

OTHER REPORTS. Atranorin, porphyrilic acid, hymenorhodin, leiphaemin, zeorin: Thies (1932); Zopf (1907). — Hymenorhodin, leiphaemin, porphyrilic acid, zeorin: W. Brieger (1923); Hesse (1912). — Porphyrilic acid: Neelakantan (1965); Shibata (1958, 1963).

Haematomma puniceum (Sm. *ex* Ach.) Mass.   Atranorin; unidentified substance; unidentified red pigment. Asahina (1964*b*), from Africa and North and South America, by thin-layer chromatography and by microcrystal tests, red pigment in apothecia.

Atranorin, unidentified red pigment (acetone insoluble). W. L. Culberson (1963*a*), from North Carolina, South Carolina, Texas, and

Virginia, PD+ yellow thallus reaction not understood, red pigment in hymenium.

OTHER REPORT. The PD– thallus reaction distinguishes this species from *Haematomma fauriei*: Asahina (1964*a*), from Formosa and Japan.

Haematomma puniceum "ssp. *magellanicum* Asah. *ad interim*" Atranorin; unidentified red pigment. Asahina (1964*b*), from South America, red pigment in apothecia.

Haematomma puniceum "ssp. *pacificum* Asah. *ad interim*" Atranorin, imbricaric acid; unidentified red pigment. Asahina (1964*b*), from Ceylon, Colombia, Cuba, Formosa, India, Japan, and the Philippines, red pigment in apothecia.

Haematomma subpuniceum (Müll. Arg.) B. de Lesd. Atranorin; unidentified red pigment (acetone insoluble). W. L. Culberson (1963*a*), from Mexico and Texas, PD+ reaction of thallus not understood, red pigment in hymenium.

Haematomma ventosum (L.) Mass. Decarboxythamnolic acid ($\pm$), divaricatic acid, thamnolic acid ($\pm$) (sometimes as ventosaric acid), (+)-usnic acid; unidentified red pigment (acetone soluble); 9,10,12,13-tetrahydroxyheneicosanoic acid, ventosic acid; arabitol, mannitol, sucrose, trehalose, umbilicin.

Decarboxythamnolic acid, divaricatic acid, thamnolic acid, (+)-usnic acid, ventosic acid. Solberg (1957), from Norway, as *Haematomma ventosum* var. *lapponicum*.

Decarboxythamnolic acid, divaricatic acid, thamnolic acid, usnic acid. Asahina and Nuno (1964), from Europe, by thin-layer chromatography.

Divaricatic acid, thamnolic acid, usnic acid, unidentified substance. Asahina (1964*a*), from Europe, unidentified substance present in minor quantity.

Divaricatic acid, usnic acid, unidentified red pigment (acetone soluble). W. L. Culberson (1963*a*), from Czechoslovakia, England, France, Portugal, Sweden, Switzerland, and the U.S.S.R., probably not in North America, and also fumarprotocetraric acid (apparently a misidentification of thamnolic acid by chromatography, see Asahina, 1964*a*).

Divaricatic acid, usnic acid. Hale (1957*b*), mention only; Klosa (1951*b*), mention only; Solberg (1956), mention only; Vartia (1949), from Finland, by extraction; Vartia (1950*b*), from Finland, by extraction.

Divaricatic acid. Asahina (1936*e*), description of microcrystal tests; Kurokawa (1964), mention only; Ramaut (1961*c*), by extraction; Vartia (1950*a*), by extraction; Wachtmeister (1952).

Unidentified red pigment. Asahina (1954c), suggests this pigment is an alkali salt of an anthraquinone which is converted to the free anthraquinone in the GE solution.

9,10,12,13-Tetrahydroxyheneicosanoic acid, ventosic acid. Solberg (1960).

Arabitol, mannitol, sucrose, trehalose, umbilicin. Lindberg, Misiorny, and Wachtmeister (1953), by paper chromatography and column chromatography, trace of mannitol, no volemitol.

Umbilicin. Pueyo (1960), mention only.

OTHER REPORTS. Atranorin, divaricatic acid, usnic acid: D. Hess (1958), from Germany. — Atranorin, divaricatic acid, fumarprotocetraric acid: Satô (1940) (but see Asahina, 1964a). — Divaricatic acid, fumarprotocetraric acid, usnic acid, unidentified red pigment (acetone soluble): W. L. Culberson (1963a) (but see Asahina, 1964a).

REVIEWS. Divaricatic acid, thamnolic acid, (+)-usnic acid: Thies (1932); Zopf (1907). — Divaricatic acid, (+)-usnic acid: W. Brieger (1923); Hesse (1912). — Sucrose, trehalose, umbilicin: Shibata (1958). — Umbilicin: Shibata (1963).

Haematomma ventosum var. *lapponicum* (Räs.) Lynge, see *Haematomma lapponicum* Räs.

Hendrickxia alcicornis Duvign. Atranorin, usnic acid; unidentified substance. Duvigneaud (1942b), from Peru, type description, by microchemical tests.

Hendrickxia pseudoreticulata Duvign., see *Everniopsis trulla* (Ach.) Nyl.

Herpothallon sanguineum (Sw.) Tobl. Confluentic acid; chiodectonic acid. C. F. Culberson (1966a), from Florida, 0.65% confluentic acid by extraction, chiodectin not found, description of microchemical tests.

Chiodectonic acid. Kolumbe (1927), from Florida, as an unidentified red pigment; Ribeiro and Mors (1949), 0.9% by extraction, found to be a hydroxyanthraquinone carboxylic acid (m.p. 303°); Ribeiro and Mors (1950).

REVIEWS. Chiodectin, chiodectonic acid: Thies (1932); Zopf (1907). — Chiodectonic acid: W. Brieger (1923); Hesse (1912). All review reports are as *Chiodecton sanguineum*.

*Heterodermia appalachensis** (Kurok.) W. Culb. Atranorin; zeorin (±); unidentified yellow pigment (K + yellow). W. L. Culberson (1966a), from North Carolina, by microchemical tests on five samples; Kurokawa (1962a), from North America, type description, yellow pigment on undersurface, as *Anaptychia appalachensis*.

* Most species of *Heterodermia* have not yet been formally transferred from *Anaptychia* and they will be found under that genus.

Heterodermia casarettiana (Mass.) Trev. Atranorin, norstictic acid; zeorin ($\pm$); unidentified yellow pigment (not K+ purple). W. L. Culberson (1966*a*), from North Carolina and South Carolina, no salazinic acid by microcrystal tests and thin-layer chromatography, yellow pigment on undersurface.

OTHER REPORT. Atranorin, norstictic acid, salazinic acid, zeorin, unidentified yellow pigment (K+ yellow or brownish yellow): Kurokawa (1962*a*), from Central and South America, the West Indies, and southeastern U.S.A., yellow pigment on undersurface may be a pulvinic acid derivative, as *Anaptychia casarettiana*.

Heterodermia dendritica (Pers.) Poelt Atranorin, norstictic acid, salazinic acid; unidentified yellow pigment (K+ purple); zeorin ($\pm$). W. L. Culberson (1966*a*), from North Carolina, by microchemical tests on three samples, yellow pigment on undersurface.

Atranorin, norstictic acid, salazinic acid, unidentified yellow pigment, zeorin. Kurokawa (1960*b*), as *Anaptychia subheterochroa* from Formosa, Japan, and the Philippines and *A. subheterochroa* var. *propagulifera* from Japan, Java, and the Philippines, unidentified yellow pigment also found in *A. fulvescens* and *A. heterochroa* (= *Heterodermia obscurata*); Kurokawa (1962*a*), from Asia, as *A. dendritica*, *A. dendritica* var. *dissecta* from Japan, and *A. dendritica* var. *propagulifera* from Africa, Asia, Hawaii, and North and South America.

Norstictic acid. Kurokawa (1964), mention only, as *Anaptychia dendritica*.

Heterodermia domingensis (Ach.) Trev. Atranorin, salazinic acid; zeorin ($\pm$). W. L. Culberson (1966*a*), from North Carolina and South Carolina, by microchemical tests on 31 samples.

Atranorin, salazinic acid, zeorin. Hale (1957*a*), as *Anaptychia ravenellii*; Kurokawa (1962*a*), from Mexico, North and South America, and the West Indies, as *A. domingensis*.

Heterodermia echinata (Tayl.) W. Culb. Atranorin; zeorin. W. L. Culberson (1966*a*), from North Carolina, by microchemical tests on three samples; Kurokawa (1962*a*), from Mexico and the U.S.A. from Massachussetts to Florida and Texas, as *Anaptychia echinata* and *A. echinata* var. *pterocarpa*.

Heterodermia galactophylla (Tuck.) W. Culb. Atranorin; zeorin. W. L. Culberson (1966*a*), from North Carolina, by microchemical tests on two specimens; Kurokawa (1962*a*), from North, Central, and South America and the West Indies, as *Anaptychia galactophylla*.

Atranorin. Hale (1954*a*), from Arkansas, an exsiccati specimen, as *Anaptychia comosa*; Hale (1957*a*), from Arkansas, Missouri, and Oklahoma, as *A. galactophylla*.

Heterodermia granulifera (Ach.) W. Culb.   Atranorin, salazinic acid; zeorin ($\pm$). W. L. Culberson (1966*a*), from North Carolina and South Carolina, by microchemical tests on eight samples.

Atranorin, salazinic acid, zeorin. Hale (1957*a*), from Arkansas, Missouri, and Oklahoma, as *Anaptychia granulifera*; Hale (1957*c*) from West Virginia, an exsiccati specimen, as *A. granulifera*; Kurokawa (1962*a*), from east-central North America and from Mexico, as *A. granulifera*.

Heterodermia hypoleuca (Ach.) Trev.   Atranorin, chloroatranorin, norstictic acid ($\pm$), salazinic acid ($\pm$), unidentified substance; zeorin.

Atranorin, norstictic acid ($\pm$), salazinic acid ($\pm$), unidentified substance, zeorin. Kurokawa (1959*a*), from Japan and North America, unidentified substance by microcrystals in the GAAn solution, a report of stictic acid by Hale (1956*h*) probably based on abnormal crystal forms of zeorin in the GAo-T solution, as *Anaptychia hypoleuca*, *A. hypoleuca* var. *microphylla*, and *A. hypoleuca* var. *microphylla* f. *granulosa*.

Atranorin, norstictic acid ($\pm$), salazinic acid ($\pm$), zeorin. Kurokawa (1962*a*), from Asia and North America, as *Anaptychia hypoleuca*.

Atranorin, chloroatranorin, zeorin. Asahina and Yosioka (1940), as *Anaptychia hypoleuca*.

Atranorin, zeorin. Neelakantan, Rangaswami, and Rao (1954), from Kashmir, 0.2% atranorin and 0.04% zeorin, by extraction, as *Anaptychia hypoleuca*.

Atranorin, zeorin ($\pm$). W. L. Culberson (1966*a*), from North Carolina and South Carolina, by microchemical tests on nine samples.

Atranorin. Asahina (1937*c*), by microcrystal tests, as *Anaptychia hypoleuca*.

OTHER REPORTS. Atranorin, stictic acid, zeorin: Hale (1954*a*), from Arkansas, an exsiccati specimen, as *Anaptychia hypoleuca*; Hale (1956*a*), as *A. hypoleuca*; Hale (1957*a*), from Arkansas, Missouri, and Oklahoma, as *A. hypoleuca*; Hale (1957*b*), as *A. hypoleuca*. — Atranorin, stictic acid ($\pm$), zeorin ($\pm$): Hale (1956*h*), from eastern U.S.A., distribution mapped, as *A. hypoleuca*.

REVIEWS. Atranorin, chloroatranorin, zeorin: Asahina and Shibata (1954); Shibata (1958, 1963); all review reports as *Anaptychia hypoleuca*.

Heterodermia leucomela (L.) Poelt   Atranorin, salazinic acid; zeorin ($\pm$). W. L. Culberson (1966*a*), from North Carolina, by microchemical tests on eight samples.

Atranorin, salazinic acid, zeorin. Kurokawa (1960*c*), from Australia, Colombia, Mexico, and North America, as *Anaptychia ophioglossa* and as *A. ophioglossa* f. *albociliata* which differs by the color of the cilia, red coloration of laciniae possibly caused by decomposition of salazinic acid, see Asahina (1953c) for other examples of such coloration; Kuro-kawa (1962a), from Africa, Asia, Central and South America, and Europe, as *A. leucomelaena* and also *A, leucomelaena* f. *albociliata* from Asia, Central and South America, and Florida, *A. leucomelaena* f. *palmiformis* from Mexico with an unidentified violet pigment in the rhizines, and *A. leucomela* f. *verrucifera* from Colombia, Jamaica, and Mexico.

Atranorin, zeorin. Neelakantan, Rangaswami, and Rao (1954), from India, as *Anaptychia leucomelaena* var. *angustifolia*, 4.0% atranorin and 1.3% zeorin by extraction of a sample from Sikkim, 2.4% atranorin and 1.0% zeorin by extraction of a sample from Kodaikanal.

REVIEWS. Atranorin, zeorin: Shibata (1958, 1963), as *Anaptychia leucomelaena* var. *angustifolia*.

Heterodermia obscurata (Nyl.) Trev. Atranorin; unidentified yellow-orange pigment (K+ purple); zeorin (±). W. L. Culberson (1966a), from North Carolina and South Carolina, by microchemical tests on 52 samples, yellow-orange pigment on undersurface.

Atranorin, unidentified yellow pigment (K+ purple), zeorin. Kuro-kawa (1960a), from Africa, Formosa, Japan, Korea, North and South America, and New Zealand, as *Anaptychia heterochroa*; Kurokawa (1962a), from Africa, Asia, Europe, North and South America, and Hawaii, yellow pigment on undersurface, as *A. obscurata*.

Atranorin, zeorin. Neelakantan, Rangasawami, and Rao (1954), mention only, as *Anaptychia heterochroa*.

Unidentified yellow pigment. Hale (1956h), undersurface with an unknown anthraquinone related to parietin, as *Anaptychia hypoleuca* var. *colorata*; Kurokawa (1960c), mention only, as *A. heterochroa*.

OTHER REPORTS. Norstictic acid (±), salazinic acid (±), unidentified yellow pigment (undersurface K+ yellow): Kurokawa (1960a), as *Anaptychia sorediifera*. — Atranorin, stictic acid, unidentified anthra-quinone derivative in lower medulla: Hale (1957a), from Arkansas, Missouri, and Oklahoma, as *A. heterochroa*. — Orange-yellow hydroxy-anthraquinone in medulla: Asahina and Yosioka (1940), pigment (m.p. 278°) may be the same as "blastenin," as *A. heterochroa*. — Atran-orin, norstictic acid (±), stictic acid, zeorin, unidentified red pigment: Hale (1957b), as *A. heterochroa*. — Purple-black undersurface: Hale (1956h), as *A. hypoleuca* var. *sorediifera*.

REVIEWS. Zeorin: Asahina and Shibata (1954); Shibata (1958, 1963); all review reports as *Anaptychia heterochroa*.

Heterodermia pseudospeciosa (Kurok.) W. Culb. Atranorin, norstictic acid; zeorin (±). W. L. Culberson (1966a), from North Carolina and South Carolina, no salazinic acid by microcrystal tests and thin-layer chromatography.

OTHER REPORTS. Atranorin, norstictic acid, salazinic acid, unidentified substance (PD + yellow?), zeorin: Kurokawa (1959*b*), from Formosa, India, and Japan, type description, unidentified PD + substance also found in *Anaptychia hypoleuca* (= *Heterodermia hypoleuca*) by microcrystals in the GAAn solution, as *A. pseudospeciosa*. — Atranorin, norstictic acid, salazinic acid, zeorin: Kurokawa (1962*a*), from Africa, Asia, Hawaii, Mexico, and South America, as *A. pseudospeciosa*, and also *A. pseudospeciosa* f. *tagawae*.

Heterodermia squamulosa (Degel.) W. Culb.   Atranorin; zeorin. W. L. Culberson (1966*a*), from North Carolina and South Carolina; Kurokawa (1960*b*), from Japan and North America, as *Anaptychia squamulosa*, often also a whitish-brown compound (K −); Kurokawa (1962*a*), from Central and North America, as *A. squamulosa*.

Heterodermia tremulans (Müll. Arg.) W. Culb.   Atranorin, unidentified substance (PD + yellow?); zeorin. Kurokawa (1959*b*), from Central and North America, Japan, Java, and Hawaii, as *Anaptychia pseudospeciosa* var. *inactiva*, the unidentified PD + substance also found in *A. hypoleuca* (= *Heterodermia hypoleuca*) by microcrystals in the GAAn solution, and as *A. speciosa* var. *microspora* (with no mention of the unidentified substance) from Japan, Korea, Manchuria, and Sakhalin.

Atranorin, zeorin. Kurokawa (1962*a*), from Central and South America, China, Costa Rica, India, Japan, Manchuria, Sakhalin, and Hawaii, as *Anaptychia pseudospeciosa* var. *tremulans*.

Atranorin, zeorin (±). W. L. Culberson (1966*a*), from North Carolina and South Carolina, by microchemical tests on 32 samples.

Himantormia lugubris (Hue) Lamb   Atranorin, barbatolic acid, usnic acid. Huneck and Follmann (1966*b*), from Chile, 0.005% atranorin, 0.45% barbatolic acid, 0.0025% usnic acid, by extraction.

Atranorin, barbatolic acid. Lamb (1964), from Graham Land, no usnic acid, barbatolic acid as an unidentified (PD + yellow) β-orcinol derivative, microcrystal test described.

Hubbsia lumbricoides W. Web.   Lecanoric acid. Weber (1965), from Baja California and Mexico, by microcrystal test.

Hypogymnia almquistii (Vain.) Rass.   Olivetoric acid. Krog (1963), from Alaska, Canada, and Siberia, by microchemical tests, as *Parmelia almquistii*.

Hypogymnia alpicola (Th. Fr.) Hav.   Two unidentified substances. Krog (1951*b*), by microchemical tests.

Unidentified acid. Dahl (1952), from Norway, mention only; Krog (1963), unidentified acid PD +.

All reports are as *Parmelia alpicola*.

Hypogymnia austerodes (Nyl.) Räs. Atranorin, physodic acid. Dahl (1950), by microchemical tests, as *Parmelia austerodes*; Dahl (1952), from Norway, mention only; Krog (1951*b*), from Norway, as *P. austerodes*.

Atranorin. Hale (1954*b*), from Baffin Island, no usnic acid, as *Parmelia austerodes*.

Hypogymnia bitteriana (Zahlbr.) Räs. Atranorin (sometimes as atranoric acid), physodic acid (sometimes as farinacinic acid). Dahl (1950), as *Parmelia bitteriana*; Dahl (1952), from Norway, mention only; Krog (1951*b*), from Norway.

Physodic acid. Asahina (1947*a*), Zopf's farinacinic acid found to be identical to physodic acid, as *Parmelia farinacea* Bitt.

REVIEWS. Atranorin, physodic acid: W. Brieger (1923); Hesse (1912); Thies (1932); all three reports as *Parmelia farinacea*; Zopf (1907), as *Hypogymnia farinacea*.

Hypogymnia encausta (Sm.) W. Wats. Atranorin, protocetraric acid; tetrahydroxy fatty acids.

Atranorin, protocetraric acid. Dahl (1952), from Norway, mention only; Krog (1951*b*), from Norway, protocetraric acid and/or fumarprotocetraric acid, and unidentified acids, as *Parmelia encausta*.

Tetrahydroxy fatty acids. Solberg (1960), as *Parmelia encausta*.

No usnic acid. Gertig and Banasiewicz (1961), from Poland, as *Parmelia encausta*.

REVIEWS. Atranorin: W. Brieger (1923); Hesse (1912); both reports as *Parmelia encausta*; Thies (1932); Zopf (1907).

Hypogymnia enteromorpha (Ach.) Nyl. Atranorin, physodalic acid, physodic acid, unidentified substance ($FeCl_3$ + bluish) ($\pm$). Nuno (1964), from Formosa, Japan, and the U.S.A., unidentified accessory absent in the U.S.A., by thin-layer chromatography.

Atranorin, physodalic acid, physodic acid. Asahina (1951*a*), from Formosa; Asahina (1952*e*), from Formosa; Hale (1957*c*), from West Virginia, an exsiccati specimen.

All reports are as *Parmelia enteromorpha*.

"*Hypogymnia enteromorpha* f. *inactiva*" Atranorin ($\pm$?), physodic acid, unidentified substance ($FeCl_3$ + bluish) ($\pm$), unidentified substance ($FeCl_3$ + violet) ($\pm$). Nuno (1964), from Formosa, Japan, the Kurile Islands, Sakhalin, and the U.S.A., by thin-layer chromatography.

Atranorin, physodic acid. Asahina (1952*e*), from Formosa and Japan.

All reports are as *Parmelia enteromorpha* f. *inactiva* Asah. The combination of the name of the forma *inactiva* and the generic name *Hypogymnia* has not been made.

"*Hypogymnia farinacea* Bitt.," see *Hypogymnia bitteriana* (Zahlbr.) Räs.

Hypogymnia fragillima (Hillm.) Rass.   Atranorin ($\pm$?), physodic acid, unidentified substance ($FeCl_3$+ bluish). Nuno (1964), from Japan, Korea, and Rebun, by thin-layer chromatography.
   Atranorin, physodic acid. Asahina (1951*a*), from Japan, Korea, and Sakhalin; Asahina (1952*e*), from Japan and Korea.
   All reports are as *Parmelia fragillima.*

"**Hypogymnia fujisanensis**"   Atranorin, physodalic acid, physodic acid, unidentified substance ($FeCl_3$+ bluish). Nuno (1964), from Japan, by thin-layer chromatography.
   Atranorin, physodalic acid, physodic acid. Asahina (1952*e*), from Japan.
   All reports are as *Parmelia fujisanensis* Asah. The combination of the specific name *fujisanensis* and the generic name *Hypogymnia* has not been made.

Hypogymnia furfuracea var. *olivetorina* (Zopf) Krog, see *Pseudevernia olivetorina* (Zopf) Zopf

Hypogymnia hultenii Degel.   Atranorin, physodic acid. Dahl (1952), from Norway, mention only.

Hypogymnia hypotrypa (Nyl.) Rass.   Physodalic acid, usnic acid. Nuno (1964), from China and Formosa, by thin-layer chromatography, as *Parmelia hypotrypa*, and also *P. hypotrypa* f. *balteata* from China.

Hypogymnia hypotrypella (Asah.) Rass.   Physodalic acid, usnic acid. Asahina (1950*a*), from Japan, Korea, and Sakhalin, type description of *Parmelia hypotrypella*; Nuno (1964), from Formosa, Japan, Korea, and Sikkim, by thin-layer chromatography.
   OTHER REPORT. Physodalic acid, physodic acid, usnic acid: Asahina (1952*e*), from Formosa, Japan, and Korea.
   REVIEWS. Physodalic acid: Asahina and Shibata (1954); Shibata (1958, 1963).
   All reports are as *Parmelia hypotrypella.*

Hypogymnia intestiniformis (Vill.) Räs.   Atranorin, physodic acid, unidentified substances (PD+); iron, manganese, zinc.
   Atranorin, physodic acid, unidentified substance (PD+). Krog (1962), from Alaska, unidentified substance not fumarprotocetraric acid, as *Parmelia intestiniformis.*
   Atranorin, physodic acid. Dahl (1952), from Norway, mention only; Krog (1951*b*), from Norway, as *Parmelia intestiniformis.*

Atranorin. Hale (1957*b*), mention only, as *Parmelia intestiniformis*.
Iron, manganese, zinc. Lounamaa (1965), a quantitative study.

"*Hypogymnia lugubris*" Atranorin, physodic acid, unidentified sub-stance (FeCl₃ + bluish). Nuno (1964), from the Aleutian Islands, by thin-layer chromatography, as *Parmelia lugubris* Pers. The combination of the specific name *lugubris* with the generic name *Hypogymnia* has not been made.

"*Hypogymnia lugubris* var. *sikkimensis*" Atranorin, physodalic acid, physodic acid. Nuno (1964), from India, by thin-layer chromatography, as *Parmelia lugubris* var. *s[i]kkimensis*.

"*Hypogymnia lugubris* f. *tenuis*" Atranorin, physodalic acid, physodic acid. Nuno (1964), from New Zealand, possibly also an unidentified substance, by thin-layer chromatography, as *Parmelia lugubris* f. *tenuis* Bitt. The combination of the name of the forma *tenuis* and the generic name *Hypogymnia* has not been made.

Hypogymnia metaphysodes (Asah.) Rass. Atranorin, physodic acid, unidentified substance (FeCl₃ + bluish). Nuno (1964), from Japan, by thin-layer chromatography.
　　Atranorin, physodic acid. Asahina (1950*a*), from Japan and Sakhalin, type description of *Parmelia metaphysodes*; Asahina (1952*e*), from Japan.
　　All reports are as *Parmelia metaphysodes*.

Hypogymnia mundata f. *sorediosa* (Bitt.) Rass. Atranorin, physodalic acid, physodic acid, unidentified substance (FeCl₃ + bluish). Nuno (1964), from Japan, by thin-layer chromatography.
　　Atranorin, physodalic acid, physodic acid. Asahina (1951*a*), from Japan and Sakhalin; Asahina (1952*e*), from Japan.
　　All reports are as *Parmelia mundata* f. *sorediosa*.

Hypogymnia nikkoensis (Zahlbr.) Rass. Atranorin, physodic acid, un-identified substance (FeCl₃ + bluish). Nuno (1964), from Japan, by thin-layer chromatography.
　　Atranorin, physodic acid. Asahina (1952*e*), from Japan.
　　All reports are as *Parmelia nikkoensis*.

Hypogymnia obscurata (Bitt.) Räs. Atranorin, physodic acid. Dahl (1950), as *Parmelia obscurata*; Dahl (1952), from Norway, mention only; Krog (1951*b*), from Norway, as *P. obscurata*.
　　OTHER REPORT. Thamnolic acid: Asahina (1938*f*), description of microcrystal tests.

REVIEWS. Thamnolic acid (as hirtellic acid): W. Brieger (1923); Hesse (1912); Zopf (1907).

Hypogymnia pertusa (Schrank) Nyl., see *Menegazzia terebrata* (Hoffm.) Mass.

Hypogymnia physodes (L.) Nyl.   Atranorin, chloroatranorin, physodalic acid (sometimes as monoacetylprotocetraric acid), physodic acid, unidentified substance (FeCl$_3$+ bluish) ($\pm$); unsaturated fatty acids; ergosterol; β-carotene, xanthophyll; arabitol, lichenin, mannitol; chlorophyll *a*, chlorophyll *b*, enzymes; iron, manganese, zinc.

Atranorin, chloroatranorin, physodalic acid, physodic acid. Klosa (1953*b*), from Europe, variation in quantity of physodalic acid in autumn and spring reported, as *Parmelia physodes*.

Atranorin, chloroatranorin, physodalic acid. J. Santesson (1965), as *Parmelia physodes*.

Atranorin, physodalic acid, physodic acid, unidentified substance (FeCl$_3$+ bluish) ($\pm$). Nuno (1964), from India (unidentified accessory substance absent), Japan, Switzerland, and the U.S.A., by thin-layer chromatography, as *Parmelia physodes*, and also *P. physodes* f. *labrosa* from Japan and which lacks the unidentified accessory substance, *P. physodes* f. *subcrustacea* from Japan, and *P. physodes* f. *vittatoides* from New Zealand.

Atranorin, physodalic acid, physodic acid. Asahina (1952*e*), from Japan, as *Parmelia physodes*; Dahl (1952), from Norway, mention only; Hale (1956*b*), as *P. physodes*; Hale (1956*g*), from Connecticut, an exsiccati specimen, as *P. physodes*; D. Hess (1958), from Germany, by paper chromatography, as *P. physodes*; Krog (1951*b*), from Norway, as *P. physodes*; Tichý and Rypáček (1952), and "amorphous lichen acids," as *P. physodes*; Vartia (1949), from Finland, as *P. physodes*; Zellner (1934), from Austria, as *P. physodes*, by extraction of 3.5 kg. yielding also an unidentified carboxylic acid, hypogymnol I, hypogymnol II (hypogymnol possibly identical to ventosic acid according to Solberg (1957)), ergosterol, an unidentified polysaccharide, lichenin, erythritol (probably misidentified D-arabitol according to Asahina (1934*b*, 1939*b*)).

Atranorin, physodalic acid. Koller and Locker (1931), as *Parmelia physodes*.

Atranorin, physodic acid. Asahina and Nogami (1934), from Japan and Sakhalin, as *Parmelia physodes*; Dahl (1950), from Greenland, as *P. physodes*, and a PD+ substance (probably physodalic acid according to Dahl (1953)); Vartia (1950*a*), from Finland, by extraction, as *P. physodes*; Vartia (1950*b*), from Finland.

Atranorin. Hutchinson and Reynolds (1960), as *Parmelia physodes*.

Chloroatranorin. Asahina and Nogami (1935), from Japan, by extraction, as *Parmelia physodes*.

Physodalic acid, physodic acid. Klosa (1950), as *Parmelia physodes*; Wachtmeister (1956a), used as an authentic source of these compounds, as *P. physodes*.

Physodalic acid. Asahina (1947a), as *Parmelia physodes*; C. F. Culberson (1965c), used as an authentic source of this substance; Klosa (1952a); Mitsuno (1953), by paper chromatography, as *P. physodes*; Steiner (1955), as *P. physodes*.

Physodic acid. Asahina (1938d), description of microcrystal tests, as *P. physodes*; C. F. Culberson (1964), used as an authentic source of this substance, as *P. physodes*; Fuzikawa and Ishiguro (1936), as *P. physodes*; Moissejeva (1961), enzymes tabulated, as *P. physodes*; Stoll, Renz, and Brack (1947), from Switzerland, as *P. physodes*.

Unsaturated fatty acids. Wagner and Friedrich (1965), by thin-layer chromatography, a C_{16} acid with one double bond, a C_{16} acid with three double bonds, a C_{18} acid with two double bonds, traces of unsaturated C_{20} and C_{22} acids, as *Parmelia physodes*.

β-Carotene, xanthophyll, chlorophyll *a*, chlorophyll *b*. Godnev, Khodasevich, and Arnautova (1966), from Belorussian S.S.R., ^{14}C incorporation at low temperatures, as *Parmelia physodes*.

Arabitol, mannitol. Lindberg, Misiorny, and Wachtmeister (1953), by paper chromatography, trace mannitol, no volemitol, as *Parmelia physodes*.

Chlorophyll *a*, chlorophyll *b*. Wilhelmsen (1959), as *Parmelia physodes*.

Iron, manganese, zinc. Lounamaa (1965), a quantitative study.

No usnic acid. Gertig and Banasiewicz (1961), from Poland, as *Parmelia physodes*.

OTHER REPORT. Polyose of glucose, mannose, and galactose, hemicelluloses, true cellulose: Votoček and Burda (1926), as *Parmelia physodes*.

REVIEWS. Atranorin, capraric acid, evernuric acid, physodic acid, physodin, physol: W. Brieger (1923); Hesse (1912); Thies (1932), also physodalic acid and physodylic acid; all three reports as *Parmelia physodes*. — Atranorin, chloroatranorin, physodalic acid, physodic acid, D-arabitol: Asahina and Shibata (1954), as *P. physodes*. — Atranorin, physodalic acid, physodic acid: Zopf (1907), and as *P. physodes* and *P. physodes* f. *labrosa*. — Chloroatranorin, physodalic acid, physodic acid: Shibata (1958, 1963). — Fat, fatty acids, ergosterol: Steiner (1957), as *P. physodes*.

"*Hypogymnia pseudohypotrypa*" Atranorin, physodic acid, unidentified substance ($FeCl_3$ + bluish). Nuno (1964), from Sikkim, by thin-layer chromatography, type description of *Parmelia pseudohypotrypa* Asah. The combination of the specific name *pseudohypotrypa* and the generic name *Hypogymnia* has not been made.

Hypogymnia pseudophysodes (Asah.) Rass. Atranorin, physodic acid; unidentified substance (K+ brownish-red). Asahina (1951a), from

Formosa, Japan, and Sakhalin, type description of *Parmelia pseudophysodes*; Asahina (1952*e*), from Formosa and Japan; Nuno (1964), from Japan, unidentified substance (FeCl₃+ bluish) present but K reaction not given, by thin-layer chromatography.

All reports are as *Parmelia pseudophysodes*.

"Hypogymnia pseudophysodes f. *reagens"* Atranorin, physodalic acid, physodic acid, unidentified substance (FeCl₃+ bluish). Nuno (1964), from Japan, by thin-layer chromatography, as *Parmelia pseudophysodes* f. *reagens*.

Differs from the species by a PD + reaction of the medulla. Asahina (1951*a*), from Japan, type description of *Parmelia pseudophysodes* f. *reagens* Asah. The combination of the name of the forma *reagens* and the generic name *Hypogymnia* has not been made.

Hypogymnia submundata (Oksn.) Rass. Atranorin, physodic acid, unidentified substance. Asahina (1951*a*), from Japan and Sakhalin, unidentified substance K + red; Asahina (1952*e*), from Japan; Nuno (1964), from Japan, Sakhalin, and Sikkim, unidentified substance FeCl₃+ bluish, by thin-layer chromatography; all reports as *Parmelia submundata*.

Hypogymnia submundata f. *colorans* (Asah.) Rass. Atranorin, physodalic acid, physodic acid, unidentified substance. Nuno (1964), from Japan, by thin-layer chromatography, unidentified substance FeCl₃+ bluish, as *Parmelia submundata* f. *colorans*.

Atranorin, physodic acid, two unidentified substances. Asahina (1951*a*), type description, one unidentified substance K + red, one unidentified substance PD + yellow (probably identical to physodalic acid).

Hypogymnia subobscura (Vain.) Poelt Physodic acid. Dahl (1950), no atranorin, as *Parmelia subobscura*.

Hypogymnia tubulosa (Schaer.) Hav. Atranorin, physodic acid, unidentified substance. Nuno (1964), from Norway and Switzerland, by thin-layer chromatography, as *Parmelia tubulosa*.

Atranorin, physodic acid. Dahl (1950), as *Parmelia tubulosa*; Dahl (1952), from Norway, mention only; D. Hess (1958), from Germany, by paper chromatography, as *P. tubulosa*; Krog (1951*b*), from Norway, as *P. tubulosa*; Vartia (1950*b*), from Finland, by extraction, as *P. tubulosa*.

OTHER REPORTS. Atranorin, physodic acid or physodalic acid: Vartia (1949), from Finland, as *Parmelia tubulosa*. — Lobaric acid: Hale (1957*b*), mention only, as *P. tubulosa*.

Hypogymnia vittata (Ach.) Gas. Atranorin, physodic acid, unidentified substance ($FeCl_3 +$ bluish) ($\pm$), two unidentified substances ($FeCl_3 +$ violet) ($\pm$); ascorbic acid.

Atranorin, physodic acid, unidentified substance ($FeCl_3 +$ bluish) ($\pm$), two unidentified substances ($FeCl_3 +$ violet) ($\pm$). Nuno (1964), from Formosa, India, Japan, and the U.S.A., as *Parmelia vittata* and also *P. vittata* var. *hypotrypanea*, from China and India and *P. vittata* f. *stricta* from Formosa, India, and Japan, by thin-layer chromatography.

Atranorin, physodic acid. Asahina (1951*a*), from Japan, as *Parmelia vittata* f. *stricta*; Asahina (1952*e*), from Formosa and Japan, as *P. vittata*; Dahl (1950), as *P. vittata*; Dahl (1952), from Norway, mention only; Krog (1951*b*), from Norway, as *P. vittata*.

Physodic acid. Hale (1957*b*), mention as *Parmelia vittata*; Fuzikawa and Ishiguro (1936), as *P. vittata*.

Ascorbic acid. Karev and Kochevykh (1962), as *Parmelia vittata*.

"*Hypogymnia vittata* f. *reagens*" Atranorin, physodalic acid. Nuno (1964), from India, by thin-layer chromatography, as *Parmelia vittata* f. *reagens*.

Icmadophila ericetorum (L.) Zahlbr. Thamnolic acid. Bendz, Santesson, and Wachtmeister (1965*c*); D. Hess (1958), from Germany, by paper chromatography.

REVIEW. Icmadophilaic acid: Thies (1932), red pigment in apothecia.

Ingaderia pulcherrima Darb. Erythrin, lecanoric acid; *meso*-erythritol. Huneck and Follmann (1967*b*), from Chile, 6.3% erythrin and 1% *meso*-erythritol by extraction of 22.0 g. of lichen, lecanoric acid by thin-layer chromatography, discussion of relationship to *Darbishirella gracillima*.

OTHER REPORTS. Psoromic acid: Huneck and Follmann (1963), from Chile, by extraction; Huneck and Follmann (1964*a*), sample used for comparison with psoromic acid from *Chiodecton stalactinum*.

See also *Darbishirella gracillima*.

Lasallia asiae-orientalis Asah. Gyrophoric acid; unidentified anthraquinone derivative. Asahina (1960*a*), from Formosa and Japan, type description.

Lasallia mayebarae (Satô) Asah. Gyrophoric acid; unidentified anthraquinone derivative. Asahina (1960*a*), from Formosa.

Lasallia papulosa (Ach.) Llano Gyrophoric acid; unidentified red pigment; arabitol, mannitol.

Gyrophoric acid, unidentified red pigment. C. F. Culberson and Culberson (1958), from North Carolina, by microchemical tests.

Gyrophoric acid. Hale (1956e), from North America, as *Umbilicaria papulosa*.

Unidentified pigment. Asahina (1960a), the pigment probably a hydroxyanthraquinone derivative; Hale (1957b), mention only, as *Umbilicaria papulosa*.

Arabitol, mannitol. Lindberg, Misiorny, and Wachtmeister (1953), by paper chromatography, no volemitol, as *Umbilicaria papulosa*.

OTHER REPORT. Ethyl orsellinate, orcinol: Miller, Griffin, Schaefers, and Gordon (1965), probably formed during extraction by decomposition of gyrophoric acid, as *Umbilicaria papulosa*.

Lasallia pensylvanica (Hoffm.) Llano   Gyrophoric acid; arabitol, mannitol.

Gyrophoric acid. Hale (1956e), from North America, as *Umbilicaria pensylvanica*.

Arabitol, mannitol. Lindberg, Misiorny, and Wachtmeister (1953), by paper chromatography, as *Umbilicaria pensylvanica*, no volemitol.

Lasallia pustulata (L.) Mér.   Gyrophoric acid; arabitol, D-mannitol, pustulan (sometimes as pustulin), sucrose, trehalose, umbilicin; amino acids; enzymes; folic acid-, folinic acid-, and vitamin B_{12}-group factors; iron, manganese, zinc.

Gyrophoric acid, mannitol, umbilicin. Wachtmeister (1958a), mention only, as *Umbilicaria pustulata*.

Gyrophoric acid. Koller (1932b), from Germany, by extraction, as *Umbilicaria pustulata*; Koller and Pfeiffer (1933d), by extraction, as *U. pustulata*; Měrka (1951), as *U. pustulata*; Moissejeva (1961), mention only, enzymes tabulated, as *U. pustulata*; Mosbach (1964a), biosynthesis study, as *U. pustulata*; Mosbach (1964c), mention only, as *U. pustulata*; Schindler (1956/1957), as *U. pustulata*; Vartia (1950b), from Finland, by extraction, as *U. pustulata*.

Arabitol, mannitol, trehalose, sucrose, umbilicin. Lindberg, Misiorny, and Wachtmeister (1953), no volemitol, as *Umbilicaria pustulata*; Lindberg, Wachtmeister, and Wickberg (1953), separation by column chromatography, as *U. pustulata*.

Arabitol, mannitol, umbilicin. Pueyo (1960), mention only, as *Umbilicaria pustulata*.

D-Mannitol, pustulan, umbilicin. Lindberg, Wachtmeister, and Wickberg (1952), first isolation of umbilicin, as *Umbilicaria pustulata*.

Pustulan, umbilicin. Lindberg and McPherson (1954), pustulan extracted, as *Umbilicaria pustulata*.

Pustulan. Chanda, Hirst, and Manners (1957), mention only, as *Umbilicaria pustulata*; Drake (1943), first isolation, as *U. pustulata*.

Umbilicin. Lindberg and Wickberg (1954), a detailed chemical study, as *Umbilicaria pustulata*; Lindberg and Wickberg (1962), a chemical study, mention only, as *U. pustulata*; Pueyo (1959), mention only, as *U. pustulata*.

Amino acids. Ramakrishnan and Subramanian (1966*b*), from India, alanine, glutamic acid, glycine, isoleucine, leucine, lysine, methionine, phenylalanine, and threonine as free amino acids, these and arginine, aspartic acid, proline, serine, tryptophan, tyrosine, valine, and an unidentified substance as combined amino acids, as *Umbilicaria pustulata*.

Folic acid-, folinic acid-, and vitamin B_{12}-group factors. Sjöström and Ericson (1953), as *Umbilicaria pustulata*.

Iron, manganese, zinc. Lounamaa (1965), a quantitative study.

OTHER REPORTS. Gyrophoric acid, umbilicaric acid: D. Hess (1958), from Germany, by paper chromatography, as *Umbilicaria pustulata*. — No anthraquinone pigment as in *Lasallia papulosa*: Asahina (1960*a*). — Polyoses of galactose, glucose, and mannose, and a cellulose of glucose: Votoček and Burda (1926), as *U. pustulata*.

REVIEWS. Gyrophoric acid, parmelia-brown: Thies (1932), as *Umbilicaria pustulata*. — Gyrophoric acid, umbilicaric acid: Zopf (1907), as *U. pustulata*. — Gyrophoric acid, D-mannitol, umbilicin: Asahina and Shibata (1954), as *U. pustulata* and *Gyrophora pustulata*. — Gyrophoric acid, pustulan, umbilicin: Neelakantan (1965), as *U. pustulata*. — Gyrophoric acid, sucrose, trehalose, umbilicin: Shibata (1958), as *U. pustulata*. — Gyrophoric acid, umbilicin: Shibata (1963), as *U. pustulata*. — Gyrophoric acid: W. Brieger (1923); Hesse (1912); as *U. pustulata*.

Lecanora achariana A. L. Sm.   Usnic acid. Eigler and Poelt (1965), from Södermanland, by paper chromatography.

Lecanora admontensis Zahlbr.   Psoromic acid, usnic acid. Eigler and Poelt (1965), from Spain, by paper chromatography.

Lecanora adunans Nyl.
REVIEW. Aspicilia-green: Thies (1932), as *Aspicilia adunans* and *A. adunans* f. *glacialis*.

Lecanora affinis Eversm.   Enzymes. Moissejeva (1961), enzymes tabulated, as *Aspicilia affinis*.

Lecanora agardhiana Ach.
REVIEW. Lecidea-green: Thies (1932).

Lecanora albula (Nyl.) Hue   Unidentified substance (PD + yellow), usnic acid. Eigler and Poelt (1965), from the Alps, by paper chromatography, unidentified substance B.

Lecanora allophana (Ach.) Röhl.   Atranorin.
REVIEWS. Atranorin: W. Brieger (1923), as *Lecanora subfusca* var. *allophana*; Hesse (1912), as *L. subfusca* var. *allophana*; Thies (1932); Zopf (1907).

Lecanora alphoplaca (Wahlenb.) Ach.   Norstictic acid. Asahina (1958*b*), from Japan; Hale (1957*b*), mention only.

OTHER REPORT. α-Methylethersalazinic acid: Eigler and Poelt (1965), from Tirol, by paper chromatography.

REVIEWS. Salazinic acid, two unidentified colorless substances: Zopf (1907), as *Squamaria alphoplaca*. — Salazinic acid: W. Brieger (1923); Hesse (1912); Thies (1932); Zopf (1907); all four reports as *Placodium alphoplacum*.

Lecanora atra (Huds.) Ach.   Atranorin, α-collatolic acid, (sometimes as lecanorolic acid); sucrose, trehalose.

Atranorin, α-collatolic acid. Asahina and Fuzikawa (1934*a*), from Japan, lecanorolic acid of Zopf found to be identical to α-collatolic acid; Seshadri (1944), mention only.

α-Collatolic acid. Asahina (1938*d*), description of microcrystal tests; D. Hess (1958), from Germany, by paper chromatography; Salkowski (1910); Wachtmeister (1956*a*), source of an authentic sample of this substance.

Sucrose, trehalose. Lindberg, Misiorny, and Wachtmeister (1953), by separation on a charcoal column, no umbilicin.

REVIEWS. Atranorin, α-collatolic acid, sucrose, trehalose: Shibata (1958), and as *Lecanora grumosa* and *L. atra* var. *grumosa*. — Atranorin, α-collatolic acid: Asahina and Shibata (1954); W. Brieger (1923), and as *L. grumosa*; Hesse (1912), and as *L. grumosa*; Shibata (1963); Thies (1932), and also (+)-usnic acid and lecanora-red (in the hymenium), and as *L. grumosa*; Zopf (1907), and as *L. grumosa*.

Lecanora atra "var. *panormitata*"

REVIEWS. Atraic acid: W. Brieger (1923); Hesse (1912); Thies (1932), atraic acid identical to atralinic acid ($C_{16}H_{18}O_5$).

Lecanora badia (Hoffm.) Ach.   Lobaric acid (sometimes as stereocaulic acid or usnetinic acid), (+)-usnic acid. Huneck (1966*e*), 2.7% lobaric acid, 0.03% (+)-usnic acid, by extraction of 168 g. of lichen, as *Lecanora badia* var. *milvina*.

REVIEWS. Lobaric acid: W. Brieger (1923); Hesse (1912); Thies (1932); Zopf (1907).

Lecanora bolcana (Pollini) Poelt   Usnic acid. Eigler and Poelt (1965), from France, by paper chromatography.

Lecanora bruneri Imsh. & Brodo   Atranorin, psoromic acid. Imshaug and Brodo (1966), from Mexico, type description.

Lecanora caesiocinerea Nyl.

REVIEW. Aspicilia-green: Thies (1932), as *Aspicilia caesiocinerea*.

Lecanora caesiorubella Ach. ssp. *caesiorubella*   Atranorin, physodalic acid (as monoacetylprotocetraric acid). Imshaug and Brodo (1966), from Canada, Connecticut, Illinois, Iowa, Maine, Massachusetts, Michigan, Minnesota, New Jersey, New York, North Carolina, Ohio, Oregon, Pennsylvania, Vermont, Virginia, West Virginia, and Wisconsin.

Lecanora caesiorubella ssp. *glaucomodes* (Nyl.) Imsh. & Brodo   Atranorin, protocetraric acid. Imshaug and Brodo (1966), from Bermuda, Brazil, Colombia, Cuba, Mexico, Nicaragua, Uruguay, Alabama, Arizona, Florida, Louisiana, New Jersey, North Carolina, South Carolina, and Texas.

Lecanora caesiorubella ssp. *lathamii* Imsh. & Brodo   Atranorin, norstictic acid, protocetraric acid, unidentified substance (C+ yellow). Imshaug and Brodo (1966), from Canada and the eastern U.S.A., type description.

Lecanora caesiorubella ssp. *merrillii* Imsh. & Brodo   Atranorin, norstictic acid, protocetraric acid. Imshaug and Brodo (1966), from coastal southern California and Baja California, type description.

Lecanora caesiorubella ssp. *saximontana* Imsh. & Brodo   Atranorin, norstictic acid. Imshaug and Brodo (1966), from the southern Rocky Mountains and Minnesota, type description.

Lecanora calcarea (L.) Somm.   Erythrin (sometimes as erythrinic acid); erythritol.
    Erythrin. Manaktala, Neelakantan, and Seshadri (1966), mention only, (±)-erythrin and (±)-montagnetol synthesized; Wehmer, Thies, and Hadders (1932).
    REVIEWS. Aspicilic acid, aspicilin, aspicilia-green, erythrin: Thies (1932). — Aspicilin, erythrin: Zopf (1907). — Aspicilin: W. Brieger (1923); Hesse (1912).
    All reports are as *Aspicilia calcarea*.

Lecanora campestris (Schaer.) Hue   Atranorin.
    REVIEWS. Atranorin: W. Brieger (1923), as *Lecanora subfusca* var. *campestris*; Hesse (1912), as *L. subfusca* var. *campestris*; Thies (1932); Zopf (1907).

Lecanora candida (Anzi) Nyl.
    REVIEW. Aspicilia-green: Thies (1932), as *Aspicilia candida*.

Lecanora cateilea (Ach.) Mass.   Atranorin, psoromic acid. Imshaug and Brodo (1966), from Canada, Michigan, and Minnesota.

Lecanora cenisea Ach.   Atranorin; roccellic acid. Vartia (1950*b*), from Finland, by extraction.
Atranorin. Vartia (1949), from Finland.
REVIEWS. Atranorin, roccellic acid: W. Brieger (1923); Hesse (1912); Thies (1932); Zopf (1907). — Roccellic acid: Shibata (1958, 1963).

Lecanora chondroderma Zahlbr.   Usnic acid. Eigler and Poelt (1965), from Sikkim, by paper chromatography.

Lecanora cinerea (L.) Somm.   Norstictic acid. Schindler (1936), as *Aspicilia cinerea*.
REVIEW. Aspicilia-green: Thies (1932), as *Aspicilia cinerea*.

Lecanora circinata (Pers.) Ach., see *Lecanora radiosa* (Hoffm.) Schaer.

Lecanora concolor Ram.   Psoromic acid, usnic acid. Eigler and Poelt (1965), from Spain, by paper chromatography.

Lecanora conizaeoides Nyl. *ex* Cromb.   Fumarprotocetraric acid. Brightman (1964), from Ireland, may have been deduced from a PD + reaction.

Lecanora crassa (Huds.) Ach., see *Squamarina crassa* (Huds.) Poelt

Lecanora dispersa (Pers.) Somm.   β-Sitosterol. Huneck and Follmann (1965*d*), from Chile, 0.004% by extraction.

Lecanora dispersoareolata (Schaer.) Lamy   Usnic acid. Eigler and Poelt (1965), from the Alps, by paper chromatography.

Lecanora effusa (Pers.) Ach.   Atranorin.
REVIEW. Atranorin: Zopf (1907).

Lecanora epanora (Ach.) Ach.   Zeorin; epanorin, rhizocarpic acid; copper, iron.
Zeorin, epanorin. Zopf (1900), from Tirol.
Epanorin, rhizocarpic acid. Coker (1950); Jones, Keane, and Nolan (1944).
Epanorin. Asahina (1951*i*), mention only; Frank, Cohen, and Coker (1950), a detailed study of the chemical structure, mention early work of Zopf, as "*Epanora lecanora*"; Mittal and Seshadri (1957), mention only.
REVIEWS. Epanorin, zeorin: Asahina and Shibata (1954); W. Brieger (1923); Hesse (1912); Thies (1932); Zopf (1907). — Epanorin: Shibata (1958, 1963).

Lecanora esculenta (Pall.) Eversm. Enzymes. Moissejeva (1961), enzymes tabulated, as *Aspicilia esculenta*.

Lecanora excludens Malme Atranorin. Imshaug and Brodo (1966), from Sweden, by paper chromatography of an isotype specimen.

Lecanora farinacea Fée Atranorin, norstictic acid, protocetraric acid. Imshaug and Brodo (1966), from Brazil, Colombia, Cuba, the Dominican Republic, and Jamaica.

Lecanora farinosa (Flörke) Nyl. Erythrin.
REVIEWS. Erythrin (sometimes as erythrinic acid): W. Brieger (1923); Hesse (1912); Zopf (1907); all reports as *Aspicilia calcarea* var. *farinosa*.

Lecanora frustulosa var. *thiodes* (Spreng.) Link Atranorin; zeorin.
Zeorin. Neelakantan, Rangaswami, and Rao (1954), mention only.
REVIEWS. Atranorin, zeorin: W. Brieger (1923); Hesse (1912); Thies (1932); Zopf (1907).
All reports are as *Lecanora thiodes*.

Lecanora fruticulosa Eversm. Enzymes. Moissejeva (1961), enzymes tabulated.

Lecanora gangaleoides Nyl. Atranorin, chloroatranorin, gangaleoidin. Hardiman, Keane, and Nolan (1935), from Ireland, by extraction, also a small amount of a high-melting substance not identified; Spillane, Keane, and Nolan (1936), from Ireland.
Chloroatranorin, gangaleoidin. Neelakantan and Seshadri (1960b), mention only.
Gangaleoidin. Asahina (1951i), mention only; V. E. Davidson, Keane, and Nolan (1943), detailed study of chemical structure; Yoshida (1951), mention only.
OTHER REPORT. Atranorin, chloroatranorin, gangaleoidin, endococcin, rhodophyscin, arabitol, high-melting unknown substance similar to gangaleoidin: Nolan and Keane (1940), from Ireland, by extraction.
REVIEWS. Gangaleoidin: Asahina and Shibata (1954); Shibata (1958, 1963).

Lecanora garovaglii (Körb.) Zahlbr. Usnic acid. Eigler and Poelt (1965), from Tirol, by paper chromatography.

Lecanora gelida (L.) Ach., see *Placopsis gelida* (L.) Linds.

Lecanora gibbosula Magn.
REVIEWS. Aspicilic acid, aspicilin: W. Brieger (1923); Hesse (1912); Thies (1932), and aspicilia-green; Zopf (1907); all reports as *Aspicilia gibbosa*.

Lecanora glaucoma (Ach.) Ach., see *Lecanora rupicola* (L.) Zahlbr.

Lecanora graeca J. Stein. Psoromic acid, usnic acid. Eigler and Poelt (1965), from Greece, by paper chromatography.

Lecanora grumosa (Pers.) Röhl., see *Lecanora atra* (Huds.) Ach.

Lecanora gypsacea (Sm.) Müll. Arg., see *Squamarina gypsacea* (Sm.) Poelt

Lecanora handelii J. Stein. (+)-Usnic acid; zeorin. Huneck (1966c), from Germany, 0.15% and 0.26% zeorin, 3.8% and 2.2% (+)-usnic acid, by extraction of two samples of lichen.

Lecanora hercynica Poelt & Ullrich   Pannarin; rhizocarpic acid; zeorin. Huneck (1966a), from Germany, 0.5% pannarin, 1% rhizocarpic acid, 0.21% zeorin, by extraction of 60 g. of lichen.

Lecanora intumescens (Rebent.) Rabenh.   Atranorin, psoromic acid. Imshaug and Brodo (1966), by paper chromatography.

Lecanora laevata (Ach.) Nyl.
    REVIEW. Aspicilia-green: Thies (1932), as *Aspicilia laevata*.

Lecanora laevata f. *albicans* (Arn.) Zsch.
    REVIEW. Aspicilia-green: Thies (1932), as *Aspicilia laevata* f. *albicans*.

Lecanora leptyrodes auct.   Atranorin only. Imshaug and Brodo (1966), from Sweden, by paper chromatography.

Lecanora macrocyclos (Magn.) Degel.   Usnic acid. Eigler and Poelt (1965), from Öland, by paper chromatography.

Lecanora melanaspis (Ach.) Ach.
    Atranorin, placodin. Eigler and Poelt (1965), mention only.
    REVIEWS. Atranorin, placodin: W. Brieger (1923); Hesse (1912); Thies (1932); all three reports as *Placodium melanaspis*; Zopf (1907), as *P. melanaspis* and *Squamaria melanaspis*.

Lecanora melanophthalma (Ram.) Ram.   Psoromic acid, (−)-usnic acid.
    Psoromic acid, usnic acid. Eigler and Poelt (1965), from the Alps, and *Lecanora melanophthalma* var. *obscura*, from Spain, by paper chromatography.
    (−)-Usnic acid. Huneck and Follmann (1964a), from Chile, approximately 1% by extraction.
    REVIEWS. Placodiolic acid, rhizoplacic acid, (−)-usnic acid: W. Brieger (1923), as *Placodium opacum*; Hesse (1912), as *P. opacum*; Thies

(1932), as *P. opacum*; Zopf (1907), as *P. opacum, Rhizoplaca opaca*, and *Squamaria opaca*.

Lecanora microbola Lamb   Gyrophoric acid. Eigler and Poelt (1965), from Canada, by paper chromatography; Hale (1956*e*), from North America.

Lecanora muralis (Schreb.) Rabenh.   Usnic acid; leucotylin, zeorin; lichenin. Huneck (1961), from Germany, by extraction and column chromatography.

Leucotylin. Yosioka and Nakanishi (1963), mention only.

Usnic acid. D. Hess (1958), from Germany, by paper chromatography, as *Placodium saxicolum*; Vartia (1950*b*), from Finland, as *Parmularia muralis*.

Zeorin. Kennedy, Breen, Keane, and Nolan (1937), mention only.

OTHER REPORT. Psoromic acid, usnic acid: Eigler and Poelt (1965), from Germany, by paper chromatography.

REVIEWS. Atranorin, usnic acid, zeorin: W. Brieger (1923), as *Placodium saxicolum*, and *Squamaria saxicola*; Hesse (1912), as *P. saxicolum* and *S. saxicola*; Thies (1932), as *Lecanora saxicola, P. saxicolum*, and *S. saxicola*; Zopf (1907), as *P. saxicolum* and *S. saxicola*. — Atranorin, an acid soluble in KHCO$_3$ solution: Zopf (1907), as *S. saxicola* var. *compacta*.

Lecanora muralis var. **dubyi** (Müll. Arg.) Poelt   Usnic acid. Eigler and Poelt (1965), from the Alps, by paper chromatography.

Lecanora nemoralis Mak.   Atranorin, psoromic acid. Imshaug and Brodo (1966), from France, Germany, and Poland.

Lecanora novomexicana Magn.   Psoromic acid, usnic acid. Eigler and Poelt (1965), from New Mexico, by paper chromatography.

Lecanora orbicularis (Schaer.) Vain.   Psoromic acid, usnic acid. Eigler and Poelt (1965), from the Alps, by paper chromatography, trace of psoromic acid.

Lecanora pallida (Schreb.) Rabenh.   Atranorin, norstictic acid, protocetraric acid. Imshaug and Brodo (1966), from Canada, Mexico, Michigan, New York, North Carolina, and Washington, and also *Lecanora pallida* var. *rubescens* from Canada, Michigan, New York, Vermont, and Wisconsin.

Lecanora parella (L.) Ach., see *Ochrolechia parella* (L.) Mass.

Lecanora peltata (Ram.) Steud.   Psoromic acid, usnic acid. Eigler and Poelt (1965), from Uzbekistan, by paper chromatography, trace of usnic acid.

Lecanora pinguis Tuck.   Unidentified substances, usnic acid. Eigler and Poelt (1965), from California, by paper chromatography, unidentified substances A_1, A_2, A_8, and D.

Lecanora polytropa (Ehrh.) Rabenh.   (+)-Usnic acid; rangiformic acid; eulecanorol, zeorin. Huneck (1966e), from Germany, 0.67% (+)-usnic acid, 0.36% rangiformic acid, 0.045% zeorin, by extraction of 218 g. of lichen, first report of the new triterpene eulecanorol, mass spectrographic study on rangiformic acid.

Lecanora praeradiosa Nyl.   α-Methylethersalazinic acid. Eigler and Poelt (1965), from Upper Adige, by paper chromatography.

Lecanora pruinosa Chaub.   Unidentified substances. Eigler and Poelt (1965), from France and Sicily, by paper chromatography, unidentified substances A_1, A_2, A_3, A_4, and A_7.

Lecanora pseudopallida Gyeln.   Atranorin. Imshaug and Brodo (1966), possibly also psoromic acid.

Lecanora radiosa (Hoffm.) Schaer.   Unidentified substance. Eigler and Poelt (1965), by paper chromatography, unidentified substance C.
  REVIEWS. Psoromic acid (sometimes as parellic acid): W. Brieger (1923), as *Placodium circinatum*; Hesse (1912), as *P. circinatum*; Thies (1932), as *Lecanora circinata*; Zopf (1907), as *Squamaria circinata*. — Salazinic acid: W. Brieger (1923); Hesse (1912); Thies (1932); all three reports as *P. circinatum* var. *radiosum*; Zopf (1907), as *S. radiosa* and as *P. radiosum*.

Lecanora radiosa var. **subcircinata** (Nyl.) Zahlbr.   Norstictic acid. Huneck (1963b), from Germany, 0.25% by extraction; Schindler (1936), as *Placodium circinatum* var. *subcircinatum*.
  OTHER REPORT. Salazinic acid: Galun and Lavee (1966), from Israel, by a microcrystal test.

Lecanora reuteri Schaer.   Unidentified substances. Eigler and Poelt (1965), from Wawóz Soboczansky, by paper chromatography, unidentified substances A_1, A_2, A_3, A_4, A_5, and A_6.

Lecanora rubina (Vill.) Ach.   Usnic acid. D. Hess (1958), from Germany, by paper chromatography, as *Placodium rubinum*; D. Hess (1960), as *P. rubinum*.

REVIEWS. Placodiolic acid, (−)-usnic acid: W. Brieger (1923); Hesse (1912); Thies (1932); all three reports as *Placodium chrysoleucum*; Zopf (1907), as *P. chrysoleucum, Rhizoplaca chrysoleuca*, and *Squamaria chrysoleuca*.

Lecanora rupicola (L.) Zahlbr. Atranorin, chloroatranorin; sordidone; thiophanic acid; roccellic acid; unidentified neutral substance; mannitol.

Atranorin, chloroatranorin, sordidone, thiophanic acid, roccellic acid, unidentified neutral substance (m.p. 147–148°). Huneck (1966*d*), from Germany, 0.46% thiophanic acid, chemical structure of thiophanic acid determined, sordidone as an unidentified chlorine-containing substance (m.p. 265–266°). (See also the Chemical Guide, Chapter III, for more information on sordidone and the unidentified neutral substance.)

Atranorin, chloroatranorin, thiophanic acid, roccellic acid, mannitol. Kennedy, Breen, Keane, and Nolan (1937), from Ireland, by extraction, as *Lecanora sordida*, good review of early literature included.

Atranorin, chloroatranorin, roccellic acid. Spillane, Keane, and Nolan (1936), as *Lecanora sordida*.

Roccellic acid. Barry and McNally (1945).

Mannitol. Asahina (1951*i*), mention, as *Lecanora sordida*.

OTHER REPORTS. Atranorin, lecasterid, lecasterinic acid, roccellic acid, thiophanic acid: Klosa (1951*b*), as *Lecanora sordida.* — Atranorin, zeorin: Hale (1957*b*), mention only, as *L. sordida.* — Zeorin: Neelakantan, Rangaswami, and Rao (1954), mention only, as *L. sordida*.

REVIEWS. Atranorin, chloroatranorin, mannitol, roccellic acid, thiophanic acid: Asahina and Shibata (1954). — Atranorin (as atranoric acid), lecasterid, lecasterinic acid, roccellic acid, thiophanic acid: Thies (1932), as *Lecanora swartzii* and *L. sordida* var. *swartzii*; Zopf (1907), as *L. swartzii*, and also zeorin (as *L. sordida* var. *swartzii*). — Atranorin, lecasterid, roccellic acid, thiophanic acid: W. Brieger (1923); Hesse (1912); both reports as *L. swartzii* and *L. sordida* var. *swartzii*. — Atranorin, roccellic acid, thiophanic acid: W. Brieger (1923), as *L. glaucoma* and *L. sordida* var. *glaucoma*; Hesse (1912), as *L. glaucoma* and *L. sordida* var. *glaucoma;* Thies (1932), as *L. glaucoma*; Zopf (1907), as *L. glaucoma*, another sample with different results (see below). — Atranorin, sordidin, zeoric acid, zeorin: W. Brieger (1923); Hesse (1912); Thies (1932), and rangiformic acid; all three reports as *L. sordida.* — Atranorin, psoromic acid (sometimes as parellic acid): Thies (1932), and thiophanic acid, as *L. sordida* var. *glaucoma*; Zopf (1907), as *L. glaucoma.* — Atranorin, zeoric acid, an unidentified colorless substance: Zopf (1907), as *L. sordida.* — Roccellic acid, zeorin: Shibata (1958, 1963), as *L. glaucoma, L. sordida*, and *L. sordida* var. *swartzii.* — Roccellic acid: Neelakantan (1965), as *L. sordida*.

Lecanora saxicola (Poll.) Ach., see *Lecanora muralis* (Schreb.) Rabenh.

Lecanora sordida (Pers.) Th. Fr., see *Lecanora rupicola* (L.) Zahlbr.

Lecanora sordida var. *glaucoma* (Hoffm.) Th. Fr., see *Lecanora rupicola* (L.) Zahlbr.

Lecanora sordida var. *swartzii* (Ach.) Rabenh., see *Lecanora rupicola* (L.) Zahlbr.

Lecanora straminea (Wahlenb.) Ach.   Unidentified substances. Eigler and Poelt (1965), from Finland, by paper chromatography, unidentified substances A_1, A_2, A_3, and A_4.

Lecanora subcircinata Nyl.   α-Methylethersalazinic acid. Eigler and Poelt (1965), from Germany, by paper chromatography.

Lecanora subfusca (L.) Ach.   Atranorin (sometimes as atranoric acid). Vartia (1949), from Finland.
    REVIEWS. Atranorin: W. Brieger (1923); Hesse (1912); Thies (1932); Zopf (1907).

Lecanora subfusca var. *allophana* Ach., see *Lecanora allophana* (Ach.) Röhl.

Lecanora subfusca var. *campestris* (Schaer.) Rabenh., see *Lecanora campestris* (Schaer.) Hue

Lecanora sulphurea (Hoffm.) Ach.   (+)-Usnic acid; zeorin. Huneck (1966e), from Germany, 5.6% (+)-usnic acid, 0.0015% zeorin, by extraction of 69.0 g. of lichen.
    (+)-Usnic acid. Salkowski (1910).
    OTHER REPORTS. α-Collatolic acid (or as lecanorolic acid): Asahina and Fuzikawa (1934a), mention of early work of Paternò and Crosa (1894); Rangaswami and Rao (1955b), mention only. — (+)-Usnic acid, zeorin, sordidin: Kennedy, Breen, Keane, and Nolan (1937), mention early work of Zopf; Klosa (1951b), mention only.
    REVIEWS. α-Collatolic acid (sometimes as lecanorolic acid), (+)-usnic acid, zeorin, sordidin: W. Brieger (1923); Hesse (1912); Thies (1932); Zopf (1907).

Lecanora swartzii (Ach.) Ach., see *Lecanora rupicola* (L.) Zahlbr.

Lecanora sylvatica (Arn.) Sandst.
    REVIEW. Aspicilia-green: Thies (1932), as *Aspicilia sylvatica*.

Lecanora tartarea (L.) Ach., see *Ochrolechia tartarea* (L.) Mass.

Lecanora thiodes Spreng., see *Lecanora frustulosa* var. *thiodes* (Spreng.) Link

Lecanora valesiaca (Müll. Arg.) Stitzenb. Usnic acid. Eigler and Poelt (1965), from Upper Adige, by paper chromatography.

Lecanora varia (Ehrh.) Ach. Psoromic acid (sometimes as parellic acid), (−)-usnic acid.
(−)-Usnic acid. Salkowski (1910).
Usnic acid. Klosa (1951*b*); Vartia (1949), from Finland, mention only; Vartia (1950*b*), from Finland.
REVIEWS. Psoromic acid, (−)-usnic acid: W. Brieger (1923); Hesse (1912); Thies (1932); Zopf (1907).

Lecanora verrucosa Ach.
REVIEW. Aspicilia-green: Thies (1932), as *Aspicilia verrucosa*.

Lecanora viridula (Flörke) Hillm. Aspicilin. Huneck (1966*e*), from Germany, by extraction of 94 g. of lichen and chromatographic separation, aspicilin found to be an aliphatic lactone of unknown structure related to the fatty acids.

Lecanora wisconsinensis Magn. Atranorin. Hale (1957*b*), mention only.

Lecidea aglaea Somm.
REVIEW. Lecidea-green: Thies (1932).

Lecidea aglaeotera Nyl., see *Lecidea armeniaca* f. *algaeotera* (Nyl.) Müll. Arg.

Lecidea armeniaca f. ***algaeotera*** (Nyl.) Müll. Arg. Roccellic acid.
REVIEWS. Roccellic acid: W. Brieger (1923); Hesse (1912); Thies (1932); Zopf (1907), and α-ethyletherprotocetraric acid (as cetraric acid); all reports as *Lecidea aglaeotera*.

Lecidea athroocarpa (Ach.) Ach.
REVIEW. Lecidea-green: Thies (1932).

Lecidea cinereoatra Ach.
REVIEWS. Lecidic acid, lecidol: W. Brieger (1923); Hesse (1912); Thies (1932); Zopf (1907).

Lecidea confluens (G. Web.) Ach. Confluentic acid. Huneck (1962*c*), mention only; Huneck (1962*d*), from Germany, a detailed chemical study; Huneck (1965*a*), mention only.

REVIEWS. Confluentic acid (as confluentin): W. Brieger (1923); Hesse (1912); Thies (1932); Zopf (1907).

Lecidea crustulata (Ach.) Spreng.
REVIEW. Lecidea-green, lecanora-red: Thies (1932).

Lecidea cuprea Somm.   Fumarprotocetraric acid. Hale (1957*b*), mention only.

Lecidea decipiens (Hedw.) Ach.   Parietin.
REVIEWS. Parietin: Thies (1932); Zopf (1907); both reports as *Placodium decipiens.*

Lecidea enteroleuca Ach.
REVIEW. Rhizoiden-green: Thies (1932).

Lecidea friesii Ach.   Lecanoric acid. Hale (1956*e*), from North America; Hale (1957*b*), mention only.

Lecidea fuscoatra (L.) Ach.   Gyrophoric acid. Huneck (1965*a*), from Germany, 2.4% by extraction and 1.6% by extraction of *Lecidea fuscoatra* var. *grisella.*
REVIEW. Lecidea-green: Thies (1932), and as *Lecidea fuscoatra* var. *subcontigua* (=*L. fuscoatra* f. *mosigii*).

Lecidea granulata Magn.
REVIEW. Lecanora-red: Thies (1932).

Lecidea granulosa (Ehrh.) Ach.   Gyrophoric acid. Huneck (1965*b*), mention only; Vartia (1950*b*), from Finland, by extraction, as *Biatora granulosa.*
REVIEWS. Gyrophoric acid: W. Brieger (1923); Hesse (1912); Zopf (1907); all review reports as *Biatora granulosa.*

Lecidea grisella Flörke   Gyrophoric acid. Huneck (1965*b*), mention only.
REVIEWS. Gyrophoric acid: W. Brieger (1923); Hesse (1912); Zopf (1907).

Lecidea griseoatra (Hoffm.) Flot.   Gyrophoric acid. Hale (1956*e*), from North America; Huneck (1965*b*), from Germany, by extraction of 4 g. of lichen, as *Lecidea tenebrosa.*
REVIEW. Lecidea-green: Thies (1932).

Lecidea handelii Zahlbr.   Unidentified hydroxyanthraquinone pigment; two unidentified fatty acids. Asahina (1954*c*), by microchemical tests.
    Two unidentified fatty acids. Asahina (1954*b*), from Japan, type description of *Lecidea pseudohaematomma.*

Lecidea icterica (Mont.) Tayl.   Rhizocarpic acid. Hale (1957*b*), mention only.

Lecidea incongrua Nyl.
REVIEW. Parmelia-brown: Thies (1932).

Lecidea kochiana Hepp   Divaricatic acid. Huneck (1963*a*), from Germany, 2.5% by extraction of 16 g. of lichen.

Lecidea lactea Flörke, see *Lecidea pantherina* (Hoffm.) Th. Fr.

Lecidea latypea Ach.
REVIEW. Lecidea-green: Thies (1932).

Lecidea limprichtii (B. Stein) Zahlbr.
REVIEW. No lichen substances: Zopf (1907), as *Psora limprichtii*.

Lecidea lithophila (Ach.) Ach.   Planaic acid. Huneck (1965*a*), from Germany, 1.5% by extraction of 196 g. of lichen.

Lecidea lithyrga Ach.
REVIEW. Lecanora-red: Thies (1932).

Lecidea lucida (Ach.) Ach.   Calycin, leprapinic acid. Agarwal and Seshadri (1965), mention only; Grover and Seshadri (1959), from India, by extraction.
Calycin. Wachtmeister (1956*a*).
REVIEWS. Calycin: Shibata (1963). — Rhizocarpic acid: W. Brieger (1923); Hesse (1912); Thies (1932); Zopf (1907).
All reports are as *Biatora lucida*.

Lecidea lurida (Dill.) Ach.
REVIEW. A greenish amorphous substance: Zopf (1907), as *Psora lurida*.

Lecidea macrocarpa (DC.) Steud.   Confluentic acid; copper, iron.
Confluentic acid. Huneck (1965*a*), from Germany, 2.5% by extraction.
Copper, iron. Lange and Ziegler (1963), a quantitative study.
REVIEW. Lecidea-green, Parmelia-brown, rhizoiden-green: Thies (1932), and as *Lecidea platycarpa* and *L. platycarpa* f. *flavicunda*.

Lecidea marginata Schaer.
REVIEW. Lecidea-green: Thies (1932).

Lecidea mollis (Wahlenb.) Nyl.

REVIEWS. Diffusinic acid: W. Brieger (1923); Hesse (1912); Thies (1932); Zopf (1907); all reports as *Biatora mollis*.

Lecidea novomexicana (B. de Lesd.) W. Web.   Gyrophoric acid. Anderson (1962), from Arizona, California, Colorado, New Mexico, Oregon, and Washington, some apparently C— samples also with gyrophoric acid.

Lecidea pantherina (Hoffm.) Th. Fr.   Norstictic acid. Huneck (1964), from Austria, by extraction of 19 g. of lichen; Schindler (1936).
REVIEW. Lecidea-green: Thies (1932), as *Lecidea lactea*.

Lecidea pantherina var. *achariana* f. *sudetica* (Körb.) Magn.
REVIEWS. Salazinic acid: W. Brieger (1923); Hesse (1912); Thies (1932); Zopf (1907); all reports as *Lecidea sudetica*.

Lecidea parasema Ach.
REVIEW. Lecidea-green: Thies (1932).

Lecidea pelobotrion (Wahlenb. *ex* Ach.) Leight.   Gyrophoric acid. Hale (1956e), from North America; Hale (1957b), mention only.

Lecidea pilatii (Hepp) Körb.
REVIEW. Lecidea-green: Thies (1932).

Lecidea plana Lahm *ex* Körb.   Planaic acid. Huneck (1965a), mention only; Huneck (1965c), 0.74% by extraction, chemical structure determined, thin-layer chromatography described.

Lecidea platycarpa Ach., see *Lecidea macrocarpa* (DC.) Steud.

Lecidea platycarpa var. *flavicunda* Arn., see *Lecidea macrocarpa* (DC.) Steud.

Lecidea promiscens Nyl.
REVIEW. Lecidea-green: Thies (1932).

Lecidea pseudohandelii Asah.   Lecanoric acid. Asahina (1957f), from Japan, type description.

Lecidea pseudohaematomma Asah., see *Lecidea handelii* Zahlbr.

Lecidea scalaris (Ach.) Ach.   Lecanoric acid. Asahina (1936d); Asahina (1953e), from Europe and Japan.
REVIEWS. Lecanoric acid: W. Brieger (1923); Hesse (1912); Thies (1932); Zopf (1907).
All reports are as *Psora ostreata*.

Lecidea silacea (Ach.) Ach. Porphyrilic acid. Huneck (1966*a*), from Germany, 0.12% by extraction of 2.5 g. of lichen.

Lecidea sublutescens Nyl.
REVIEW. Lecidea-green: Thies (1932).

Lecidea sudetica Körb., see *Lecidea pantherina* var. *achariana* f. *sudetica* (Körb.) Magn.

Lecidea sylvicola Flot.
REVIEW. Lecidea-green: Thies (1932), as *Biatora infidula* (= *Lecidea sylvicola* var. *aphana*).

Lecidea templetonii Tayl.
REVIEW. Parmelia-brown, rhizoiden-green: Thies (1932), as *Biatora atrofusca*.

Lecidea tenebrosa Flot., see *Lecidea griseoatra* (Hoffm.) Flot.

Lecidea tumida Mass. Confluentic acid; unknown substance (m.p. 201–203°, K+ yellow, C+ red). Huneck (1962*c*), from Germany, by extraction.
Confluentic acid. Huneck (1965*a*), mention only.

Lecidea turgidula Fr.
REVIEW. Rhizoiden-green: Thies (1932), as *Biatora turgidula*.

Lecidea viridescens (Schrad.) Ach.
REVIEW. Aspicilia-green: Thies (1932), as *Biatora viridescens*.

Lecidea vorticosa (Flörke) Körb.
REVIEW. Lecidea-green, parmelia-brown: Thies (1932), and as *Lecidea* "*verticosa.*"

Leprantha impolita (Ehrh.) Körb., see *Arthonia impolita* (Ehrh. *ex* Hoffm.) Borr.

Lepraria candelaris (L.) Fr. Calycin, pinastric acid (sometimes as chrysocetraric acid), vulpinic acid. Klosa (1952*c*), from Germany, as *Lepraria flava*, and an unidentified substance.
Calycin. Åkermark (1961), chemical structure of calycin studied and revised, synthetic calycin compared to a sample from *Lepraria candelaris*.
Pinastric acid. Grover and Seshadri (1959), from India; Mittal and Seshadri (1955), from India, as *Lepraria flava*; Mittal and Seshadri (1957), mention only, as *L. flava*; Vartia (1950*b*), from Finland, as *L. flava*.

OTHER REPORT. Calycin, vulpinic acid, lobaric acid (as stereocaulic acid): Klosa (1951*b*), mention only, as "*Letharia* [sic] *flava*."

REVIEWS. Calyciarin, calycin, pinastric acid: Thies (1932); Zopf (1907); both reports as *Lepraria flava* "f. *quercina*." — Calyciarin (±), calycin: W. Brieger (1923); Hesse (1912); both reports as *L. flava* "f. *quercina*." — Calycin, pinastric acid: W. Brieger (1923); Hesse (1912); Thies (1932), as *L. flava*, another collection with no calycin; Zopf (1907), as *L. flava*, another collection with no calycin. — Calycin, pinastric acid: Asahina and Shibata (1954); Shibata (1958, 1963); all three reports as *L. flava* "f. *quercina*." — Calycin: Neelakantan (1965).

Lepraria chlorina (Ach.) Ach. *ex* Sm.   Calycin, leprapinic acid, leprapinic acid methyl ether; arabitol, mannitol.

Calycin, leprapinic acid, leprapinic acid methyl ether. Agarwal and Seshadri (1965), mention only.

Calycin, leprapinic acid methyl ether. Grover and Seshadri (1959), from India, a detailed chemical study.

Arabitol, mannitol. Lindberg, Misiorny, and Wachtmeister (1953), by paper chromatography, no volemitol.

OTHER REPORTS. Calycin, vulpinic acid: Klosa (1952*c*). — Vulpinic acid: Klosa (1951*b*), mention only; Mittal and Seshadri (1957), mention only.

REVIEWS. Calycin, lepraric acid, lobaric acid (as usnetic acid): W. Brieger (1923); Hesse (1912); Zopf (1907), and vulpinic acid (10.5%). — Calycin, lobaric acid (as usnetic acid), vulpinic acid: Thies (1932). — Calycin, leprapinic acid, leprapinic acid methyl ether: Shibata (1963). — Leprapinic acid methyl ether: Neelakantan (1965).

Lepraria citrina (Schaer.) Schaer.   Leprapinic acid. Agarwal and Seshadri (1965), 5% by extraction, a chemical study on this pigment; Grover and Seshadri (1959), from India; Mittal and Seshadri (1955), from India, first report of the pigment leprapinic acid; Mittal and Seshadri (1956), leprapinic acid synthesized and compared to a sample from this lichen; Mittal and Seshadri (1957), mention only.

REVIEWS. Leprapinic acid: Neelakantan (1965); Shibata (1963).

Lepraria farinosa (Hoffm.) Ach.

REVIEWS. Atranorin, (+)-usnic acid, hydroxyroccellic acid, roccellic acid, pulveraric acid: Thies (1932). — Hydroxyroccellic acid, pulveraric acid: W. Brieger (1923); Hesse (1912); Zopf (1907).

Lepraria flava (Willd.) Ach., see *Lepraria candelaris* (L.) Fr.

Lepraria flava "f. *quercina*," see *Lepraria candelaris* (L.) Fr.

Lepraria latebrarum Ach.
Atranorin, roccellic acid, leprarin. Zopf (1900).
REVIEWS. Atranorin, hydroxyroccellic acid, latebrid, leprariaic acid, leprarin, psoromic acid (sometimes as parellic acid), pulverin, roccellic acid, talebraric acid, (+)-usnic acid: W. Brieger (1923); Hesse (1912); Thies (1932), substances listed in various combinations based on numerous early reports; Zopf (1907).

Lepraria membranacea (Dicks.) Vain.   Pannaric acid; roccellic acid.
Åkermark, Erdtman, and Wachtmeister (1959), by extraction, about 1.5% pannaric acid; Huneck (1962*a*), mention only.
Roccellic acid. Åkermark (1962), a detailed study of this substance.
REVIEWS. Pannaric acid: Neelakantan (1965); Shibata (1963).
All reports are as *Crocynia membranacea*.

Lepraria neglecta (Nyl.) Lett.   Atranorin. Huneck (1962*a*), from Germany, 0.5% by extraction, no other crystalline product found, as *Crocynia neglecta*.

Leptogium corticola (Tayl.) Tuck.   No substances found. Hale (1957*a*), from Arkansas and Oklahoma, by microchemical tests.

Leptogium hildenbrandii (Garov.) Nyl.
REVIEW. No substances found: Zopf (1907), as *Mallotium hildenbrandii*.

Leptogium saturninum (Dicks.) Nyl.   Carotene pigments. Henriksson (1963).

Letharia divaricata (L.) Hue, see *Evernia divaricata* (L.) Ach.

Letharia [sic] *flava* (= *Lepraria flava* (Willd.) Ach.), see *Lepraria candelaris* (L.) Fr.

Letharia togashii Asah.   Atranorin; vulpinic acid. Asahina (1954*a*), by microchemical tests which are described.
Atranorin, unidentified yellow substance. Asahina (1952*d*), from Japan, type description, yellow substance found to be benzene-soluble but noncrystalline.

Letharia vulpina (L.) Hue   Atranorin, vulpinic acid; arabitol, lichenin, mannitol.
Atranorin, vulpinic acid. D. Hess (1958), from Germany, by paper chromatography; Klosa (1952*c*), mention only, as *Evernia vulpina*; Solberg (1956), mention only, as *E. vulpina*.

Vulpinic acid, lichenin. Karrer, Staub, and Staub (1924), as *Evernia vulpina*.

Vulpinic acid. Hale (1957*b*), mention only; Klosa (1953*a*), from Tirol, mention only, as *Chlorea vulpina*; Mitsuno (1955), by paper chromatography, as *Evernia vulpina*; Mittal and Seshadri (1957), mention only; Mosbach (1964*b*), 4.8% isolated in study of biosynthesis, as *E. vulpina*; Mosbach (1964*c*), mention only, as *E. vulpina*; Schatz (1963), from California; Schindler (1956/1957), mention only; Stoll, Brack, and Renz (1950); Stoll, Renz, and Brack (1947), from Switzerland; Wachtmeister (1956*a*), used as an authentic source of this substance, as *E. vulpina*.

Arabitol, mannitol. Lindberg, Misiorny, and Wachtmeister (1953), by paper chromatography, no volemitol.

Lichenin. Mittal and Seshadri (1954*a*), mention only, as *Evernia vulpina*; Mittal and Seshadri (1954*b*), mention only, as *E. vulpina*.

REVIEWS. Atranorin, vulpinic acid: W. Brieger (1923); Hesse (1912); Thies (1932); Zopf (1907), and as *Chlorea vulpina*. — Vulpinic acid: Asahina and Shibata (1954); Shibata (1958, 1963). All review reports are as *Evernia vulpina*.

Lichina pygmaea (Lightf.) Ag.   Fructose, glucose, mannitol, sucrose, volemitol (possibly).

Fructose, glucose, sucrose. Pueyo (1963*a*), from France.

Mannitol, volemitol. Pueyo (1963*b*), from France.

Lithographa cyclocarpa Anzi, see *Sarcogyne cyclocarpa* (Anzi) J. Stein.

Lobaria adusta Gärnt., see *Parmelia omphalodes* (L.) Ach.

Lobaria amplissima (Scop.) Forss.   Scrobiculin. C. F. Culberson (1967*a*), from Austria, Czechoslovakia, France, Hungary, Norway, Russia, Scotland, and Sweden, by thin-layer chromatography on thirteen samples, microchemical tests described; C. F. Culberson (1967*b*), from France, 0.76% by extraction, a chemical study on this substance; Hale (1957*e*), from Europe, by microchemical tests, scrobiculin reported as an unidentified KC + red substance, also found in *Lobaria scrobiculata*; Hale (1961), from Great Britain, Haiti, Italy, Mexico, Spain, and western Europe.

Lobaria erosa (Eschw.) Nyl.   Gyrophoric acid. Hale (1956*e*), from North America; Hale (1957*a*), from Arkansas.

Lobaria isidiosa (Müll. Arg.) Vain.   Stictic acid; unidentified triterpenes; carotenes; D-arabitol ($\pm$), D-mannitol ($\pm$); thelephoric acid; amino acids, protein.

Stictic acid, unidentified triterpene, carotenes, D-arabitol (±), D-mannitol (±), thelephoric acid. P. S. Rao, Sarma, and Seshadri (1965), from the Himalayas, unidentified triterpene D (m.p. 289–291°) identical to compound D of Agarwal, Aghoramurthy, Sarma, and Seshadri (1961), by extraction.

Unidentified triterpene, thelephoric acid. Agarwal, Aghoramurthy, Sarma, and Seshadri (1961), from India, by extraction, unidentified triterpene C (m.p. 318–320°).

Thelephoric acid. Neelakantan and Seshadri (1959a), mention in discussion of biosynthesis.

Amino acids, protein, Ramakrishnan and Subramanian (1965), alanine, glutamic acid, leucine, serine, threonine, tryptophan, tyrosine, and valine, as free amino acids, the same in addition to arginine, aspartic acid, glycine, histidine, isoleucine, lysine, methionine, phenylalanine, proline, and an unidentified substance, as combined amino acids, by paper chromatography, 22% crude protein.

REVIEWS. Thelephoric acid: Neelakantan (1965); Shibata (1958, 1963).

Lobaria isidiosa var. *subisidiosa* Asah.   Norstictic acid, stictic acid; carotenes; D-arabitol; thelephoric acid; amino acids.

Norstictic acid, stictic acid, carotenes, D-arabitol, thelephoric acid. P. S. Rao, Sarma, and Seshadri (1966), from the western Himalayas, traces of norstictic acid and thelephoric acid, 0.70% stictic acid, by extraction, as *Lobaria subisidiosa*.

Norstictic acid. Asahina (1947b), type description.

Amino acids. Ramakrishnan and Subramanian (1966b), from India, alanine, glycine, leucine, methionine, and threonine, as free amino acids, the same in addition to arginine, aspartic acid, glutamic acid, histidine, isoleucine, lysine, phenylalanine, proline, serine, tryptophan, tyrosine, valine, and an unidentified amino acid, as combined amino acids.

Lobaria linita (Ach.) Rabenh.   Tenuiorin; D-mannitol. Asahina and Yanagita (1933d), a detailed chemical study, as *Lobaria pulmonaria* f. *tenuior*.

Tenuiorin. Asahina (1935a), mention only, as *Lobaria pulmonaria* f. *tenuior*; Asahina (1939b), mention only, as *L. pulmonaria* f. *tenuior*; Asahina (1949b), from Europe and Japan, as *L. linita* var. *tenuior*; Asahina (1957d), from Europe, by paper chromatography; Hale (1958b), from North America, by microchemical tests; Huneck and Tümmler (1965), as *L. pulmonaria*, "peltigerin" identified as tenuiorin.

D-Mannitol. Asahina (1934b), mention only as *Lobaria pulmonaria* f. *tenuior*; Pueyo (1960), mention only, as *L. pulmonaria* f. *tenuior*.

REVIEWS. Tenuiorin, D-mannitol: Asahina and Shibata (1954). — Tenuiorin: Shibata (1958, 1963). All review reports are as *Lobaria pulmonaria* f. *tenuior*.

Lobaria oregana (Tuck.) Müll. Arg. Norstictic acid, stictic acid, (−)-usnic acid. Asahina, Yanagita, and Yosioka (1936), from Oregon, trace norstictic.

Stictic acid, usnic acid. Hale (1958*b*), from northwestern North America.

REVIEWS. Stictic acid: Shibata (1958, 1963).

Lobaria pulmonaria (L.) Hoffm. Gyrophoric acid and norstictic acid (in plants from Sakhalin). Gyrophoric acid and stictic acid (in plants from Japan). Stictic acid and/or norstictic acid (in plants from Europe). D-Arabitol, D-mannitol; ergosterol.

Gyrophoric acid (±), norstictic acid, stictic acid, D-arabitol, D-mannitol. Schindler (1956/1957), mention only, gyrophoric acid in Japanese plants only.

Gyrophoric acid, norstictic acid, stictic acid. Shibata (1965), mention in discussion of biosynthesis.

Gyrophoric acid, norstictic acid (no stictic acid), D-arabitol. Asahina and Yanagita (1934), from Sakhalin, by extraction.

Gyrophoric acid, stictic acid, D-arabitol. Asahina and Yanagita (1934), from central Japan, by extraction.

Gyrophoric acid, stictic acid. Asahina (1937*c*), description of microcrystal tests; Asahina, Yanagita, and Omaki (1933), from Japan, one sample with an unidentified substance and no stictic acid; Seshadri (1944), mention only.

Gyrophoric acid, D-arabitol. Asahina (1939*b*), mention only.

Gyrophoric acid. Asahina (1933); Asahina (1935*a*), from Japan, mention only; Asahina (1936*d*), description of microcrystal tests, as *Lobaria pulmonaria* var. *meridionalis*; Asahina (1937*e*), from Japan, mention only, as "lung lichen"; Asahina and Kutani (1925), by extraction; Asahina and Watanabe (1930), mention only.

Norstictic acid, stictic acid. Asahina (1937*a*), as *Lobaria pulmonaria* var. *meridionalis*; Asahina (1937*e*), from Europe, mention only, as "lung lichen"; Asahina, Yanagita, and Yosioka (1936), from Oregon, by extraction, a trace of norstictic acid.

Norstictic acid (±), stictic acid (±). Schindler (1936), from Africa, Asia, Europe, and North America, finds variable amounts, by microcrystal tests, 40% of the samples with norstictic acid.

Norstictic acid, arabitol. Asahina and Yanagita (1934), suggest that the unknown substance isolated from Irish lichen by Nolan and Keane (1933) may have been norstictic acid; Nolan and Keane (1933), from Ireland, by extraction, as *Sticta pulmonaria*.

Norstictic acid. Asahina and Fuzikawa (1935*c*); Seshadri and Subramanian (1949*c*), mention only.

Stictic acid. Asahina (1938*e*), description of microcrystal tests; Curd and Robertson (1935); Hale (1957*c*), from West Virginia, an exsiccati specimen; Mohan, Keane, and Nolan (1937), from Ireland, by extraction, as *Sticta pulmonaria*; Solberg (1956), mention only; Vartia (1950*b*), mention only.

Arabitol, mannitol. Lindberg, Misiorny, and Wachtmeister (1953), by paper chromatography, no volemitol.

D-Arabitol. Asahina (1934*b*), mention only; Mittal, Neelakantan, and Seshadri (1952), mention only.

Arabitol. Pueyo (1960), mention only; Pueyo (1965), mention only.

Enzymes. Moissejeva (1961), enzymes tabulated.

OTHER REPORTS. α-Methylethersalazinic acid, norstictic acid, stictic acid: Bachmann (1962), by paper chromatography; D. Hess (1958), from Germany, by paper chromatography. — Thelephoric acid: Asahina (1951*i*), but see also *Lobaria pulmonaria* f. *hypomela*.

REVIEWS. Gyrophoric acid, stictic acid, D-arabitol: Asahina and Shibata (1954), as *Lobaria pulmonaria* var. *meridionalis*. — Gyrophoric acid: Shibata (1958, 1963), as *L. pulmonaria* var. *meridionalis*. — Norstictic acid, stictic acid, D-arabitol: Shibata (1958, 1963). — Stictic acid (sometimes as stictaic acid): W. Brieger (1923); Hesse (1912); Thies (1932); Zopf (1907); all four reports as *Sticta pulmonaria*. — Ergosterol: Steiner (1957).

Lobaria pulmonaria f. *hypomela* (Del.) Cromb.   Thelephoric acid. Asahina and Shibata (1939), determined microchemically.

REVIEWS. Thelephoric acid: Asahina and Shibata (1954); Shibata (1958, 1963).

Lobaria pulmonaria f. *tenuior* Hue, see *Lobaria linita* (Ach.) Rabenh.

Lobaria quercizans Michx.   Gyrophoric acid. Hale (1956*e*), from North America; Hale (1957*c*), an exsiccati specimen; Hale (1957*e*), from North America, by microcrystal tests and by paper chromatography.

Lobaria retigera (Bory) Trev.   Stictic acid; unidentified triterpene, two other unidentified triterpenes ($\pm$); carotenes ($\pm$); D-arabitol, D-mannitol; thelephoric acid.

Stictic acid, unidentified triterpene, two other unidentified triterpenes ($\pm$), carotenes, D-arabitol, D-mannitol. P. S. Rao, Sarma, and Seshadri (1965), from the Himalayas, 0.043–1.53% stictic acid, 0.18%–0.35% thelephoric acid, 0.011–0.29% triterpene A (m.p. 270–273°), 0–2.75% triterpene B (m.p. 218–221°), 0–0.46% triterpene D (m.p. 289–291°), and waxes, by extractions of four collections.

Stictic acid, unidentified triterpene, carotenes, D-arabitol, D-mannitol, thelephoric acid. P. S. Rao, Sarma, and Seshadri (1966), from the western Himalayas, 0.106% stictic acid, 1.36% triterpene D (m.p. 289–291°), 0.018% D-arabitol, waxes, and traces of D-mannitol.

Unidentified triterpene, thelephoric acid. Asahina and Shibata (1939), unidentified triterpene gives positive Liebermann test.

Unidentified triterpene. Asahina (1955*a*), gives positive Liebermann test and microcrystals in the GE and GAAn solutions.

Thelephoric acid. Aghoramurthy, Sarma, and Seshadri (1959), a detailed chemical study; Asahina (1951*i*), mention only; Shibata (1965), mention in discussion of biosynthesis.

REVIEWS. Thelephoric acid: Asahina and Shibata (1954); Shibata (1958, 1963).

Lobaria retigera var. *sanguinolenta* Asah.   Norstictic acid. Asahina (1947*b*), from Japan, type description.

Lobaria sachalinensis Asah.   A colorless substance (Liebermann test +). Asahina (1957*d*), and *Lobaria sachalinensis* var. *kazawaensis*, a correction of the original description which indicated tenuiorin, microchemical tests discussed.

OTHER REPORT. Tenuiorin: Asahina (1949*b*), from Sakhalin, and *Lobaria sachalinensis* var. *kazawaensis*, type descriptions.

Lobaria scrobiculata (Scop.) DC.   Norstictic acid, scrobiculin, stictic acid, (+)-usnic acid, two unidentified substances.

Norstictic acid, scrobiculin, stictic acid, (+)-usnic acid. C. F. Culberson (1967*b*), from Ireland and Scotland, 0.30–0.78% of mixed norstictic acid and stictic acid, 1.11–1.63% scrobiculin, 0.33–0.74% (+)-usnic acid, by four extractions, a structural study on the new *meta*-depside scrobiculin.

Norstictic acid, scrobiculin, stictic acid, usnic acid, two unidentified substances. C. F. Culberson (1967*a*), from Australia, Canada, Canary Islands, Czechoslovakia, Finland, France, Ireland, Norway, Russia, Scotland, Spain, Switzerland, Wales, and the U.S.A., by microchemical tests on 53 samples, usnic acid possibly accessory, description of microchemical tests.

Stictic acid, scrobiculin (as an unidentified KC+ red substance). Asahina (1947*b*), as *Lobaria verrucosa*; Hale (1957*b*), as *L. verrucosa*, mention only.

"*Lobaria subisidiosa*," see *Lobaria isidiosa* var. *subisidiosa* Asah.

Lobaria subretigera Inum.   Norstictic acid, stictic acid; retigeradiol, retigeranic acid, unidentified triterpene; D-arabitol (±), D-mannitol (±). P. S. Rao, Sarma, and Seshadri (1966), from the western Himalayas, norstictic acid by thin-layer chromatography only, 0.16–0.33% stictic acid, 0.01–0.196% retigerdiol (previously reported in other species as triterpene A), 0.03–2.84% retigeranic acid (previously reported in other species as triterpene B), 0.11–0.26% triterpene D, 0–0.056% D-arabitol, 0–0.014% D-mannitol, by extraction of three collections.

Norstictic acid, stictic acid. Shibata (1965), mention in discussion of biosynthesis.
Stictic acid. Asahina (1947*b*).

Lobaria verrucosa (Huds.) Hoffm., see *Lobaria scrobiculata* (Scop.) DC.

Mallotium hildenbrandii (Garov.) Körb., see *Leptogium hildenbrandii* (Garov.) Nyl.

Maronea constans (Nyl.) Hepp
REVIEW. Parmelia-brown: Thies (1932), as *Maronea kemmleri*.

Maronea kemmleri Körb., see *Maronea constans* (Nyl.) Hepp

Maronella laricina M. Stein. Unidentified pigment (K +). Steiner (1959), from Tirol.

Medusulina chilena Dodge Norstictic acid. Huneck and Follmann (1967*a*), from Chile, 4.8% by extraction of 25.0 g. of lichen.

Medusulina texana Müll. Arg. Stictic acid. Wirth and Hale (1963).

Megalospora marginiflexa (Hook. f. & Tayl.) Zahlbr. Usnic acid; zeorin. Asahina (1944), from Japan.

Megalospora nipponensis Asah. Usnic acid; zeorin. Asahina (1957*b*), from Japan, type description.

Megalospora submarginiflexa (Vain.) Zahlbr. Usnic acid; zeorin. Asahina (1944), from Japan.

Megalospora sulphurata Meyen & Flot. Usnic acid; zeorin. Asahina (1944), from Formosa and Japan.

Menegazzia asahinae (Yas. *ex* Zahlbr.) Sant. Atranorin, stictic acid. Asahina (1950*a*), from Formosa and Japan; Asahina (1952*e*), from Formosa and Japan; both reports as *Parmelia asahinae*.

Menegazzia pertusa (Schrank) B. Stein, see *Menegazzia terebrata* (Hoffm.) Mass.

Menegazzia terebrata (Hoffm.) Mass. Atranorin, stictic acid. Abbayes (1961), from Madagascar, as *Parmelia pertusa*; Asahina (1938*e*), description of microcrystal tests, as *P. pertusa*; Asahina (1952*e*), from Formosa and Japan, as *P. pertusa*; Asahina, Yanagita, Hirakata, and Ida (1933), from Japan, as *P. pertusa*, by extraction, early literature reviewed; Dahl (1952), from Norway, mention only, as *Menegazzia pertusa*; Hale

(1955*a*), from North America, a microchemical survey of 33 samples, as *P. pertusa*; Krog (1951*b*), from Ireland, as *P. pertusa*; Solberg (1956), mention only, as *P. pertusa*.

Stictic acid. Asahina, Yanagita, and Yosioka (1936), from Japan, mention only, as *Parmelia pertusa*.

REVIEWS. Atranorin, capraric acid (from the decomposition of physodalic acid): W. Brieger (1923); Hesse (1912); both reports as *Parmelia pertusa*. — Atranorin, evernuric acid, menegazziaic acid: Thies (1932). — Atranorin, stictic acid: Asahina and Shibata (1954), as *P. pertusa*. — Evernuric acid, menegazziaic acid: W. Brieger (1923); Hesse (1912). — Atranorin, menegazziaic acid: Zopf (1907), and as *Hypogymnia pertusa*. — Physodalic acid: Thies (1932); Zopf (1907); both reports as *P. pertusa*. — Stictic acid: Shibata (1958, 1963), as *P. pertusa*.

Micarea melaena (Nyl.) Hedl.
    REVIEW. Rhizoiden-green: Thies (1932), as *Bilimbia melaena*.

Mycoblastus sanguinarius (L.) Norm.   Atranorin; caperatic acid.
    REVIEWS. Atranorin, caperatic acid: W. Brieger (1923); Hesse (1912); Thies (1932); Zopf (1907).

Nephroma antarcticum (Wulf.) Nyl.   (+)-Usnic acid; zeorin.
    Usnic acid, zeorin. Huneck and Follmann (1965*e*), mention only; Zopf (1909), no nephrin.
    (+)-Usnic acid. Salkowski (1910).
    Zeorin. Wetmore (1960), mention only.
    REVIEWS. (+)-Usnic acid, zeorin: W. Brieger (1923); Hesse (1912); Thies (1932).

Nephroma arcticum (L.) Torss.   (+)-Usnic acid; nephrin ($\pm$), zeorin ($\pm$); unidentified pigment; arabitol, mannitol; protein; ascorbic acid; calcium, phosphorus.
    (+)-Usnic acid, nephrin, zeorin. Zopf (1909), from Sweden.
    Usnic acid, nephrin ($\pm$), zeorin ($\pm$); unidentified pigment. Wetmore (1960), includes survey of early literature, unidentified pigment probably a carotene, microchemical tests described.
    Usnic acid, zeorin. Barton and Bruun (1952), from Norway, by extraction; Huneck and Follmann (1965*e*), mention only.
    Fat, protein, fiber, ash, calcium, phosphorus. Scotter (1965), from northern Canada.
    Arabitol, mannitol. Lindberg, Misiorny, and Wachtmeister (1953), arabitol by paper chromatography, mannitol isolated, no volemitol.
    Ascorbic acid. Karev and Kochevykh (1962).
    REVIEWS. Nephrin, (+)-usnic acid, zeorin: W. Brieger (1923); Hesse (1912); Thies (1932). — Nephrin, usnic acid. Zopf (1907), as *Nephromium arcticum*. — Zeorin: Asahina and Shibata (1954); Shibata (1958, 1963).

Nephroma bellum (Spreng.) Tuck. Nephrin ($\pm$), zeorin ($\pm$), two unidentified substances ($\pm$). Wetmore (1960), from Canada, Greenland, Alaska, Colorado, Connecticut, Maine, Massachusetts, Michigan, Minnesota, Montana, New Hampshire, New Mexico, New York, North Carolina, Oregon, Pennsylvania, Vermont, Virginia, Washington, West Virginia, and Wisconsin, 1% of samples with no substances found, unknown substances A (in 36% of the samples) and D (in 31% of the samples).

Nephroma expallidum (Nyl.) Nyl. Nephrin ($\pm$), zeorin ($\pm$), unidentified substance. Wetmore (1960), from Canada, Greenland, and Alaska, unknown substance B in all samples tested.

Nephroma gyelnikii (Räs.) Lamb Unidentified substance (m.p. 190°), ($+$)-usnic acid; zeorin. Huneck and Follmann (1965e), from Chile, 1.2% ($+$)-usnic acid, 0.1% zeorin, 0.01% unidentified substance, by extraction of 16 g. of lichen.

Nephroma helveticum Ach. Nephrin ($\pm$), three unidentified substances ($\pm$). Wetmore (1960), from Canada, Costa Rica, Dominican Republic, Haiti, Mexico, Panama, Arkansas, Colorado, Connecticut, Georgia, Idaho, Illinois, Indiana, Kentucky, Louisiana, Maine, Maryland, Massachusetts, Michigan, Minnesota, Montana, New Hampshire, New Jersey, New Mexico, New York, North Carolina, Ohio, Oklahoma, Pennsylvania, South Carolina, South Dakota, Tennessee, Texas, Vermont, Virginia, West Virginia, and Wisconsin, unknown substances C, F, and G, no substances found in 1% of the samples tested.
    Unidentified substance. Asahina (1962), from Japan, description of microchemical tests, refers to unknown described by Wetmore (1960); Nuno (1963), from Europe, India, Japan, Malaya, and North America, by thin-layer chromatography, some samples with other indistinct spots, one accessory with a KC+ reaction, and also *Nephroma helveticum* f. *caespitosum* from China and Japan.

Nephroma helveticum var. *sipeanum* (Gyeln.) Wetm. Unidentified substance. Nuno (1963), from North America, by thin-layer chromatography, and also two indistinct spots.
    OTHER REPORTS. Nephrin ($\pm$), three unidentified substances ($\pm$): Wetmore (1960), from Canada, Mexico, Alaska, California, Idaho, Montana, Oregon, South Dakota, and Washington, unknown substances C ($\pm$), F ($\pm$), and G, no substances found in 6% of the samples tested.

Nephroma isidiosum (Nyl.) Gyeln. Three unidentified substances. Wetmore (1960), from Alaska, unknown substances C, F, and G.

Nephroma laevigatum Ach.　Nephromin; nephrin ($\pm$), zeorin ($\pm$); unidentified substance ($\pm$); mannitol.

Nephromin, nephrin ($\pm$), zeorin ($\pm$), unidentified substance. Wetmore (1960), from Canada and the U.S.A., older literature reviewed, determinations by microchemical tests, unknown substance D.

Nephrin, zeorin, mannitol. Zopf (1909), from southern Tirol, by extraction.

Nephromin. Galun and Lavee (1966), from Israel, by a microcrystal test.

REVIEWS. Nephrin, nephromin, zeorin: W. Brieger (1923); Hessè (1912); Thies (1932), and parmelia-brown; all three reports also as *Nephroma lusitanicum.* — Nephrin, nephromin: Zopf (1907), as *Nephromium lusitanicum.* — No lichen acids found: Zopf (1907), as *Nephromium laevigatum.*

Nephroma lusitanicum Schaer., see *Nephroma laevigatum* Ach.

Nephroma parile (Ach.) Ach.　Nephrin ($\pm$), zeorin ($\pm$); five unidentified substances ($\pm$); mannitol.

Nephrin ($\pm$), zeorin ($\pm$); five unidentified substances ($\pm$). Wetmore (1960), from Canada, Greenland, and the U.S.A., unknown substances A, C, E, F, and G, 0.7% of the samples showed none of these compounds.

Zeorin, mannitol. Zopf (1909), from the Alps, by extraction.

REVIEWS. Zeorin: W. Brieger (1923); Hesse (1912); Thies (1932).

Nephroma resupinatum (L.) Ach.　Mannitol. Zopf (1909), from southern Tirol, no zeorin, no nephrin, by extraction.

No substances found. Wetmore (1960), Canada, Alaska, California, Maine, Massachusetts, Michigan, Minnesota, Montana, New Hampshire, North Carolina, Oregon, Vermont, and Washington.

REVIEW. No lichen substances: Zopf (1907), as *Nephromium tomentosum.*

Nephroma sikkimense Asah.　Two unidentified substances. Asahina (1963*b*), type description; Nuno (1963), from Sikkim, by thin-layer chromatography, also a third indistinct spot, one unidentified substance found to be KC+ red, a substance present in several other *Nephroma* spp. tested not found here.

Nephroma sinense Zahlbr.　Unidentified substance. Nuno (1963), from Yunnan, by thin-layer chromatography, substance found in several other *Nephroma* spp. tested.

Nephroma tropicum (Müll. Arg.) Zahlbr.　Unidentified substance. Asahina (1962), from Japan, the same compound found in *Nephroma helveticum*; Nuno (1963), from Japan, North America, Sikkim, and Yunnan, by thin-layer chromatography, also found in several other *Nephroma* spp. tested, also sometimes other indistinct spots, samples from Sikkim and Yunnan as *N. cf. tropicum* with a second distinct spot (KC + red), a sample from Nepal as *N. cf. tropicum* with two different distinct spots.

Nephromium arcticum (L.) Hav., see *Nephroma arcticum* (L.) Torss.

Nephromium laevigatum (Ach.) Nyl., see *Nephroma laevigatum* Ach.

Nephromium lusitanicum (Ach.) Nyl., see *Nephroma laevigatum* Ach.

Nephromium tomentosum (Hoffm.) Nyl., see *Nephroma resupinatum* (L.) Ach.

Nephromopsis ciliaris (Ach.) Hue, see *Cetraria halei* W. Culb. & C. Culb.

Nephromopsis endocrocea Asah., see *Cetraria endocrocea* (Asah.) Satô

Nephromopsis endocrocea f. *clarkii* Asah., see *Cetraria endocrocea* f. *clarkii* (Asah.) Satô

Nephromopsis endoxantha Hue, see *Cetraria ornata* Müll. Arg.

Nephromopsis stracheyi f. *ectocarpisma* Hue, see *Cetraria stracheyi* f. *ectocarpisma* (Hue) Satô

Neuropogon antarcticus (Du Rietz) Lamb   Enzymes. Moissejeva (1961), enzymes tabulated.

Neuropogon sulphureus (Kön.) Hellb.   Usnic acid; enzymes.
Usnic acid. Hale (1957*b*), mention only.
Enzymes. Moissejeva (1961), enzymes tabulated.

Ochrolechia africana Zahlbr.   Lecanoric acid, unidentified acid. Verseghy (1962), unidentified acid also in *Ochrolechia harmandii* f. *granulosa*.

Ochrolechia alboflavescens (Wulf.) Zahlbr.   Variolaric acid, unidentified acid. Verseghy (1959), the unidentified substance also found in *Ochrolechia microstictoides* and other species listed under *O. microstictoides*, possibly also an unknown substance also found in *O. antarctica* and other species listed under *O. antarctica*; Verseghy (1962), variolaric acid questioned.

Ochrolechia androgyna (Hoffm.) Arn.   Gyrophoric acid. Hale (1956*e*), from North America; Vartia (1950*b*), from Finland.

OTHER REPORTS. Lecanoric acid: Verseghy (1959), no gyrophoric acid or calyciarin; Verseghy (1962), mention only.

REVIEWS. Calyciarin (= pertusarin?), gyrophoric acid: W. Brieger (1923); Hesse (1912); Zopf (1907). — Calyciarin: Thies (1932).

Ochrolechia androgyna f. *granulosa* (Räs.) Vers. Lecanoric acid, unidentified acid. Verseghy (1959), from Finland; Verseghy (1962).

Ochrolechia androgyna f. *leprosa* (Nyl.) Vers. Lecanoric acid, two unidentified acids. Verseghy (1959); Verseghy (1962).

Ochrolechia androgyna var. *saxorum* (Oeder) Vers. Lecanoric acid, two unidentified substances. Verseghy (1959), one unidentified substance also in *Ochrolechia frigida* var. *alaskana*; Verseghy (1962).

Ochrolechia androgyna f. *tatrica* (Gyeln.) Vers. Lecanoric acid. Verseghy (1959); Verseghy (1962).

Ochrolechia antarctica (Müll. Arg.) Darb. Lecanoric acid, unidentified substance. Verseghy (1959), unidentified substance also found in *Ochrolechia microstictoides* and other species listed under *O. microstictoides*.

Ochrolechia apiculata Vers. Lecanoric acid, two unidentified substances. Verseghy (1962), type description, one unidentified substance also in *Ochrolechia rosella*, *O. balcanica*, *O. harmandii* f. *plicata*, *O. pergranulosa*, and *O. peruensis*.

Ochrolechia austroamericana (Malme) Vers. Unidentified substance. Verseghy (1962), unidentified substance also in *Ochrolechia pulvinata*.

Ochrolechia balcanica Vers. Erythrin (as erythrinic acid), gyrophoric acid, lecanoric acid, three unidentified acids. Verseghy (1962), one unidentified acid also found in *Ochrolechia apiculata*, *O. harmandii* f. *plicata*, *O. pergranulosa*, *O. peruensis*, and *O. rosella*, one unidentified substance also in *O. gallica* and *O. pulvinata*, type description.

Ochrolechia californica Vers. Erythrin, lecanoric acid, unidentified substance. Verseghy (1962), the unidentified substance also found in *Ochrolechia californica* f. *crenata*, *O. szatalaënsis*, *O. upsaliensis*, and possibly *O. parellula*.

Ochrolechia californica f. *crenata* Vers. Erythrin (as erythrinic acid), unidentified acid. Verseghy (1962), unidentified acid also found in *Ochrolechia californica* and other species listed under *O. californica*.

Ochrolechia elisabethae-kolae Vers. Lecanoric acid, possibly also an unidentified substance. Verseghy (1959), from Alaska; Verseghy (1962).

Ochrolechia frigida (Sw.) Lynge Gyrophoric acid. Hale (1954*b*), from Baffin Island; Hale (1956*e*), from North America.
OTHER REPORTS. Lecanoric acid, unidentified substance: Verseghy (1959), from Norway; Verseghy (1962).

Ochrolechia frigida var. *alaskana* Vers. Lecanoric acid, unidentified substance. Verseghy (1959), from Alaska, unidentified substance also found in *Ochrolechia androgyna* var. *saxorum*; Verseghy (1962).

Ochrolechia gallica Vers. Lecanoric acid, unidentified acid. Verseghy (1959), unidentified substance also found in *Ochrolechia pulvinata*; Verseghy (1962), unidentified substance also found in *O. balanica* and *O. pulvinata*.

Ochrolechia gonatodes (Ach.) Räs. Gyrophoric acid, four unidentified substances. Veresghy (1959), from Switzerland, possibly also lecanoric acid and a fifth unidentified substance also found in several other species listed under *Ochrolechia antarctica*; Verseghy (1962), possibly also lecanoric acid and the unidentified substance mentioned above.

Ochrolechia harmandii f. *albidocinerea* Vers. Lecanoric acid. Verseghy (1962).

Ochrolechia harmandii f. *granulosa* Vers. Lecanoric acid, unidentified acid. Verseghy (1962), the unidentified acid also found in *Ochrolechia africana*.

Ochrolechia harmandii var. *oceanica* (Räs.) Vers. Lecanoric acid. Verseghy (1962).

Ochrolechia harmandii f. *plicata* Vers. Lecanoric acid, unidentified acid. Verseghy (1962), the unidentified acid also found in *Ochrolechia apiculata* and other species listed under *O. apiculata*.

Ochrolechia havaiensis (Räs.) Vers. Lecanoric acid (?). Verseghy (1962).

Ochrolechia madeirensis Vers. Variolaric acid (?). Verseghy (1959), from Madeira; Verseghy (1962).

Ochrolechia mahluensis Räs. Erythrin (as erythrinic acid), gyrophoric acid(?), lecanoric acid. Verseghy (1962).

Ochrolechia microstictoides Räs.   Two unidentified acids. Verseghy (1959), from Finland, possibly a third unidentified substance also found in *Ochrolechia alboflavescens, O. antarctica, O. gonatodes* (?), *O. pallescens* f. *crenularia, O. pallescens* f. *nivea*, and *O. upsaliensis*; Verseghy (1962), one unidentified acid also in *O. alboflavescens*, possibly a third unidentified substance also in *O. alboflavescens* (?), *O. antarctica, O. gonatodes* (?), *O. pallescens* f. *crenularia, O. pallescens* f. *nivea*, and *O. upsaliensis* (?).

Ochrolechia pallescens (L.) Mass.   Gyrophoric acid. Asahina (1936*d*), description of microcrystal tests.

OTHER REPORTS. Lecanoric acid: Verseghy (1959), possibly also gyrophoric acid, no variolaric acid; Verseghy (1962).

REVIEWS. Gyrophoric acid: Asahina and Shibata (1954); Shibata (1958, 1963). — Variolaric acid: W. Brieger (1923); Hesse (1912); Thies (1932); Zopf (1907); all four reports as *Ochrolechia parella* var. *pallescens.*

Ochrolechia pallescens f. *crenularia* (Cromb.) Vers.   Gyrophoric acid (?), lecanoric acid, unidentified substance. Verseghy (1959), from England, the unidentified substance also in *Ochrolechia microstictoides* and several other species listed under *O. microstictoides*; Verseghy (1962).

Ochrolechia pallescens f. *nivea* (Cromb.) Vers.   Lecanoric acid, unidentified acid. Verseghy (1959), from Finland, the unidentified acid also found in *Ochrolechia microstictoides* and several other species listed under *O. microstictoides*; Verseghy (1962).

Ochrolechia parella (L.) Mass.   Variolaric acid (sometimes as ochrolechic acid or parellic acid); mannitol. Murphy, Keane, and Nolan (1943), from Ireland, by extraction, review of early literature included, as *Lecanora parella.*

Variolaric acid. Allison and Newbold (1959), mention only, as *Lecanora parella*; Asahina and Hayashi (1933), as *L. parella*; Erdtman and Wachtmeister (1956), mention only; Seshadri (1944), mention only, as *L. parella*; Wachtmeister (1956*b*).

Mannitol. Asahina (1951*i*), as *Lecanora parella.*

OTHER REPORTS. Lecanoric acid, variolaric acid: Verseghy (1959), from Germany; Verseghy (1962).

REVIEWS. Lecanoric acid, variolaric acid (sometimes as parellic acid): Thies (1932); Zopf (1907). — No lecanoric acid: W. Brieger (1923); Hesse (1912); see also Murphy, Keane, and Nolan (1943). — Variolaric acid, mannitol: Asahina and Shibata (1954), as *Lecanora parella.* — Variolaric acid: Shibata (1958, 1963), as *L. parella.*

Ochrolechia parella var. *pallescens* (L.) Rabenh., see *Ochrolechia pallescens* (L.) Mass.

Ochrolechia parellula (Müll. Arg.) Zahlbr. Lecanoric acid, unidentified substance (?). Verseghy (1962), the unidentified substance also in *Ochrolechia californica* and several other species listed under *O. californica.*

Ochrolechia pergranulosa (Räs.) Vers. Erythrin (as erythrinic acid), lecanoric acid, unidentified acid. Verseghy (1962), the unidentified acid also in *Ochrolechia apiculata* and several other species listed under *O. apiculata.*

Ochrolechia peruensis Vers. Erythrin (?) (as erythrinic acid), lecanoric acid, unidentified acid. Verseghy (1962), the unidentified acid also in *Ochrolechia apiculata* and several other species listed under *O. apiculata,* type description.

Ochrolechia pulvinata Vers. Two unidentified acids. Verseghy (1959), from Sweden, one of the unidentified substances also in *Ochrolechia gallica,* the other substance also in *O. austroamericana;* Verseghy (1962), one of the unidentified acids also in *O. gallica,* the other one also in *O. austroamericana* and *O. balancia.*

Ochrolechia rosella (Müll. Arg.) Vers. Lecanoric acid, unidentified substance. Verseghy (1962), the unidentified substance also in several other species listed under *Ochrolechia apiculata.*

Ochrolechia subpallescens Vers. Lecanoric acid. Verseghy (1962), possibly also erythrin.

Ochrolechia subpallescens f. *uruguayense* Vers. Lecanoric acid. Verseghy (1962).

Ochrolechia szatalaënsis Vers. Variolaric acid (?), two unidentified acids. Verseghy (1959), from Bulgaria, one of the unidentified acids also in *Ochrolechia upsaliensis;* Verseghy (1962), one of the unidentified acids also in *O. californica* and several other species listed under *O. californica.*

Ochrolechia tartarea (L.) Mass. Atranorin, gyrophoric acid, lecanoric acid; arabitol, mannitol.
  Atranorin, gyrophoric acid. Klosa (1951*b*).
  Gyrophoric acid, lecanoric acid. Verseghy (1959); Verseghy (1962).
  Gyrophoric acid. Fuzikawa and Ishiguro (1936), from Japan; Vartia (1950*b*), from Finland, no roccellic acid, by extraction.
  Arabitol, mannitol. Lindberg, Misiorny, and Wachtmeister (1953), no volemitol.
  OTHER REPORTS. Erythrin: Manaktala, Neelakantan, and Seshadri (1966), mention only, as *Lecanora tartarea;* V. S. Rao and Seshadri (1942*b*), mention only, as *L. tartarea.*

REVIEWS. Gyrophoric acid, roccellic acid, old report of lecanoric acid not confirmed: W. Brieger (1923); Hesse (1912); Thies (1932); Zopf (1907); all review reports as *Lecanora tartarea*.

Ochrolechia tartarea f. *effigurata* Vers.   Lecanoric acid. Verseghy (1959), from Finland; Verseghy (1962).

Ochrolechia tartarea var. *pycnidiifera* Vers.   Lecanoric acid. Verseghy (1959), from England; Verseghy (1962).

Ochrolechia upsaliensis (L.) Mass.   Three unidentified substances. Verseghy (1959), from Norway, Germany, and Sweden, one unidentified substance also in *Ochrolechia szatalaensis*, possibly also a fourth substance in other species listed under *O. antarctica*; Verseghy (1962).

Ochrolechia yasudae Vain.   Gyrophoric acid. Hale (1956e), from North America.

Omphalodiscus crustulosus (Ach.) Schol.   Gyrophoric acid; arabitol, mannitol.
    Gyrophoric acid. Zopf (1900), from Tirol, and an unidentified substance, as *Gyrophora spodochroa* var. *depressa*.
    Arabitol, mannitol. Lindberg, Misiorny, and Wachtmeister (1953), by paper chromatography, no volemitol, as *Umbilicaria crustulosa*.
    REVIEWS. Gyrophoric acid: W. Brieger (1923); Hesse (1912); Thies (1932); Zopf (1907); all four reports as *Gyrophora spodochroa* var. *depressa*. — Parmelia-brown: Thies (1932), as *G. crustulosa*.

Omphalodiscus decussatus (Vill.) Schol.   Gyrophoric acid; arabitol, mannitol.
    Gyrophoric acid. Hale (1956e), from North America.
    Arabitol, mannitol. Lindberg, Misiorny, and Wachtmeister (1953), by paper chromatography, no volemitol.
    All reports are as *Umbilicaria decussatus*.

Omphalodiscus spodochrous (Hoffm.) Schol.   Arabitol, mannitol. Lindberg, Misiorny, and Wachtmeister (1953), by paper chromatography, no volemitol, as *Umbilicaria spodochroa*.
    Mannitol. Pueyo (1960), mention only, as *Umbilicaria spodochroa*.
    REVIEW. Parmelia-brown: Thies (1932), as *Gyrophora spodochroa*.

Omphalodiscus virginis (Schaer.) Schol.   Gyrophoric acid; arabitol, mannitol.
    Gyrophoric acid. Hale (1956e), from North America.
    Arabitol, mannitol. Lindberg, Misiorny, and Wachtmeister (1953), by paper chromatography, no volemitol.
    All reports are as *Umbilicaria virginis*.

Opegrapha atra Pers.
REVIEW. Lecanora-red: Thies (1932).

Opegrapha saxicola Ach.
REVIEW. Lecanora-red: Thies (1932).

Oropogon asiaticus Asah. Psoromic acid, unidentified hydroxyanthraquinone pigment. Asahina (1952c), from Japan.
Psoromic acid. Satô (1937), from Formosa, type description; Satô (1957), mention only.

Oropogon formosana Asah. Protocetraric acid. Asahina (1952c), from Formosa, type description.
OTHER REPORT. Fumarprotocetraric acid: Satô (1957), from Formosa.

Oropogon loxensis (Fée) Th. Fr. Unidentified substance; unidentified hydroxyanthraquinone pigment.
Unidentified substance. Asahina (1952c), from Peru, micro-crystal tests described, PD + orange-red.
Unidentified hydroxyanthraquinone pigment. Asahina (1936a).
OTHER REPORT. Protocetraric acid: Satô (1937), from Formosa.

Oropogon tanakae Asah. Unidentified substance. Asahina (1952c), microcrystal test in the GE solution described; Satô (1957), unidentified substance PD −.

Pachnolepia decussata (Flot.) Körb., see *Arthonia lobata* f. *decussata* (Flot.) Szat.

Pannaria coeruleobadia (Schleich.) Mass., see *Pannaria pityrea* (DC.) Degel.

Pannaria fulvescens (Mont.) Nyl. Pannarin. Asahina (1940c), PD + orange-red; Asahina (1951i); J. Santesson (1965), by thin-layer chromatography; Yoshida (1951), mention only.
REVIEW. Pannarin: Asahina and Shibata (1954).

Pannaria lanuginosa (Hoffm.) Szat., see *Pannaria pityrea* (DC.) Degel.

Pannaria lurida (Mont.) Nyl. Pannarin. Asahina (1940c), PD + orange-red, and as *Pannaria sublurida*; Asahina (1951i); Hale (1957a), from Arkansas, Missouri, and Oklahoma; J. Santesson (1965), by thin-layer chromatography; Yosioka (1951), mention only.
REVIEW. Pannarin: Asahina and Shibata (1954).

Pannaria microphylla (Sw.) Mass.
REVIEW. Lecidea-green: Thies (1932).

Pannaria pityrea (DC.) Degel.   Atranorin, pannarin.   D. Hess (1958), from Germany, by paper chromatography, atranorin reported as chloro-atranorin.

Pannarin. Asahina (1940*c*), as *Pannaria coeruleobadia*; Asahina (1951*i*); Yoshida (1951), mention only.

REVIEWS. Pannaric acid, hydroxyroccellic acid: W. Brieger (1923); Hesse (1912); Thies (1932); Zopf (1907). — Pannarin: Asahina and Shibata (1954); Shibata (1958, 1963).

Except where otherwise indicated, all reports are as *Pannaria lanuginosa*.

Pannaria rubiginosa (Thunb. *ex* Ach.) Del.   Pannarin. Asahina (1940*c*), PD + orange-red.

Pannaria sublurida Nyl., see *Pannaria lurida* (Mont.) Nyl.

Parmelia aberrans (Vain.) Abb.   Atranorin, gyrophoric acid, usnic acid. Hale (1965*a*), from Mexico and South America, C— populations referred to *Parmelia xanthina*.

OTHER REPORT. An aliphatic acid, usnic acid: Abbayes (1961), from Madagascar.

Parmelia abessinica Kremp.   Atranorin, cryptochlorophaeic acid (±); protolichesterinic acid. Hale (1965*a*), with cryptochlorophaeic acid from Mexico and Africa, without cryptochlorophaeic acid from Africa.

OTHER REPORTS. Atranorin, lecanoric acid, salazinic acid, a reducing sugar, lichenin, isolichenin: Sastry and Seshadri (1942), from India by extraction. — Atranorin, lecanoric acid, salazinic acid: Neelakantan and Seshadri (1952*b*), mention only; Seshadri (1944), mention only; Seshadri and Subramanian (1949), from India, mention only. — Atranorin, lecanoric acid: Sastry and Seshadri (1940), mention only; Sastry and Rao (1941), from India, 3.3% lecanoric acid and 1.1% atranorin by extraction. — Lecanoric acid: Venkateswarlu (1947), species used as a source of this substance in quantity. — Salazinic acid: Neelakantan, Seshadri, and Subramanian (1951), used as an authentic source of this substance.

REVIEWS. Atranorin, salazinic acid: Asahina and Shibata (1954); Shibata (1958, 1963).

Parmelia abnuens Nyl.   Atranorin, olivetoric acid. Hale (1965*a*), from Brazil, by chromatography.

OTHER REPORTS. Atranorin, lecanoric acid: Abbayes (1961), from Brazil, type specimen tested (?); Targé (1964-1965), mention only.

Parmelia abstrusa Vain.   Norstictic acid, usnic acid. Hale (1960*a*), from Brazil, Colombia, Japan, and the West Indies; Kurokawa (1965), from Japan and Taiwan.

Norstictic acid. Hale and Kurokawa (1964), mentioned in key.

OTHER REPORTS. Norstictic acid ($\pm$), stictic acid ($\pm$), usnic acid: Asahina (1951*b*); Asahina (1952*e*), from Formosa and Japan. — Atranorin, salazinic acid, usnic acid: Targé (1964–1965), from Africa, by paper chromatography.

Parmelia abstrusa f. *laevigata* Lynge, see *Parmelia subabstrusa* Gyeln.

Parmelia abstrusoides Abb.   Usnic acid. Hale and Kurokawa (1964), mentioned in key.

Parmelia acariospora Zahlbr.   Atranorin, gyrophoric acid. Hale (1960*a*), from Brazil.

Parmelia acetabulum (Neck.) Duby   Atranorin, norstictic acid. Asahina and Fuzikawa (1935*c*), from Germany, by extraction.

Norstictic acid. Asahina (1936*d*, 1938*e*), description of microcrystal tests; Evans (1943*a*), mention only; Krog (1951*b*), from Norway; J. Santesson (1965), by thin-layer chromatography; Seshadri and Subramanian (1949*c*), also mention Hesse found atranorin and salazinic acid; Schindler (1936); Solberg (1956), mention only; Wachtmeister (1956*a*), used as an authentic source of this substance.

OTHER REPORTS. (Chloro)atranorin, α-methylethersalazinic acid: D. Hess (1958), from Germany, by paper chromatography. — Atranorin, chloroatranorin, α-methylethersalazinic acid, salazinic acid: Clauzade and Poelt (1961), mention only. — α-Methylethersalazinic acid: Ramaut (1961*b*), mentions work of D. Hess; Ramaut (1962*b*), mentions work of D. Hess; see also Ramaut (1960), for a paper chromatogram diagram. — Stictic acid: Dahl (1952), from Norway, mention only.

REVIEWS. Atranorin, salazinic acid: W. Brieger (1923); Hesse (1912); Thies (1932); Zopf (1907). — Norstictic acid: Asahina and Shibata (1954); Shibata (1958, 1963).

Parmelia acrobotrys Kurok.   Atranorin, salazinic acid, usnic acid. Hale and Kurokawa (1964), from Java, type description.

Parmelia adaugescens Nyl.   Atranorin, salazinic acid. Asahina (1951*g*), from Japan; Asahina (1952*e*), from Formosa and Japan.

Parmelia affinis Vain., see *Parmelia subaffinis* Zahlbr.

Parmelia affixa Hale & Kurok.   Atranorin, unidentified substance (KC+). Hale and Kurokawa (1964), from southwestern Africa.

Parmelia africana Müll. Arg., see *Parmelia andina* Müll. Arg.

Parmelia aldabrensis Dodge   Atranorin, norstictic acid, stictic acid ($\pm$). Hale (1965*a*), from Aldabra, Madagascar, and Tanganyika.

Parmelia aleurites Ach., see *Parmeliopsis aleurites* (Ach.) Nyl.

Parmelia allardii Hale, see *Parmelia mellissii* Dodge

Parmelia almquistii Vain., see *Hypogymnia almquistii* (Vain.) Rass.

Parmelia alpicola Th. Fr., see *Hypogymnia alpicola* (Th. Fr.) Hav.

Parmelia amagiensis Asah.   Atranorin; zeorin; a yellow substance. Asahina (1951*c*), yellow pigment (FeCl$_3$−) not entothein, type description; Asahina (1952*e*), from Japan, yellow pigment K + violet.

Parmelia amaniensis J. Stein. & Zahlbr.   Atranorin, norstictic acid, protocetraric acid. Targé (1964–1965).

Parmelia amazonica Nyl.   Atranorin, protocetraric acid. Hale (1959*b*), from Central and South America, the West Indies, and Florida, color tests on the holotype specimen; Hale (1960*a*), from Brazil, Central America, the West Indies, and Florida.
    Protocetraric acid. Hale and Kurokawa (1964), no usnic acid, mentioned in key.

Parmelia amboimensis Dodge   Atranorin, olivetoric acid. Hale (1965*a*), from Angola, by chromatography.

Parmelia ambositrana Abb.   Fumarprotocetraric acid, protocetraric acid, usnic acid. Abbayes (1961), from Madagascar, type description.

Parmelia americana (Meyen & Flot.) Mont., see *Parmelia cirrhata* Fr.

Parmelia anaptychioides Kurok.   Atranorin, protocetraric acid. Hale and Kurokawa (1964), from the Dominican Republic, Haiti, and Jamaica, type description.

Parmelia andina Müll. Arg.   Atranorin, lecanoric acid. Aghoramurthy, Neelakantan, and Seshadri (1954), from India, as *Parmelia hyporysalea*; Asahina (1952*e*), from Formosa and Japan, as *P. hyporysalea*; Hale (1958*a*), from Paraguay, as *P. paraguariensis*; Hale (1960*a*), from Paraguay, as *P. africana*; Hale (1965*a*), from Asia, central and south-central Africa, and tropical South America; Ramaut (1961*a*), from the Belgian Congo, as *P. africana*; Ramaut (1961*b*), as *P. africana*; Targé (1964–1965), from Africa, by chromatography, as *P. africana*.
    Lecanoric acid. Asahina (1926), as *Parmelia hyporysalea* var. *cinerascens*.

Parmelia andreana Müll. Arg., see *Parmelia flaventior* Stirt.

Parmelia angolensis (Vain.) Vain. *ex* Van der Byl   Atranorin, norstictic acid, perlatolic acid (traces). Targé (1964–1965), from Katanga, by chromatography.

Parmelia annexa Kurok.   Atranorin, lecanoric acid. Hale and Kurokawa (1964), from South Africa, type description.

Parmelia antillensis Nyl.   Norstictic acid. Hale and Kurokawa (1964), mentioned in key.

Parmelia apophysata Hale & Kurok.   Atranorin, unidentified substance (KC+, UV+). Hale and Kurokawa (1964), from the Dominican Republic, type description.

Parmelia appalachensis W. Culb.   Atranorin, protolichesterinic acid. W. L. Culberson (1962*a*), from Nova Scotia and the Appalachian Mountains from New Hampshire south to North Carolina, type description.
Protolichesterinic acid. Hale (1965*c*), from the Appalachian Mountains. Fatty acid. Hale and Kurokawa (1964), mention only.

Parmelia appendiculata Fée   Atranorin, unidentified colorless substances; unidentified yellow pigment (K−). Hale (1965*a*), from Réunion, unidentified pigment also in *Parmelia araucariarum* and other species.
Atranorin, unidentified orange-yellow medullary pigment. Hale (1960*a*), from Brazil, pigment also in *Parmelia endosulphurea*, as *P. merrillii* and as *P. cornuta* var. *crocea*.

Parmelia aptata Kremp.   Unidentified substance (KC+). Hale and Kurokawa (1964), no usnic acid, mentioned in key.

Parmelia araucariarum Zahlbr.   Atranorin, unidentified substances, unidentified orange-yellow medullary pigment. Hale (1965*a*), from South America.

Parmelia arcana Kurok.   Atranorin; unidentified fatty acid; unidentified pale yellow pigment. Hale and Kurokawa (1964), from South Africa, type description.

Parmelia argentina Kremp.   Alectoronic acid, atranorin. Hale (1958*a*), from Brazil, and as *Parmelia lângii* and *P. subproboscidea*; Hale (1960*a*), from Brazil and Paraguay, as *P. lângii*, *P. melanothrix* f. *microspora*, and *P. subproboscidea*; Hale (1965*a*), from Central and South America and Mexico, cilia of apothecia may react K + violet.
Alectoronic acid. Hale (1960*a*), mention only.

Parmelia arnoldii Du Rietz   Alectoronic acid (±), atranorin, α-collatolic acid; rhodophyscin (±).

Alectoronic acid, atranorin, α-collatolic acid. Mitsuno (1953), by paper chromatography.

Alectoronic acid (±), atranorin, α-collatolic acid. Abbayes (1961), from Madagascar; Abbayes (1962), from Africa; Abbayes (1963), from Vietnam.

Alectoronic acid, atranorin, rhodophyscin. Hale (1965*a*), from Canada, Central and South America, Europe, Mexico, the West Indies, California, Kentucky, Maine, Michigan, Minnesota, Oregon, Virginia, Washington, and West Virginia.

Alectoronic acid, atranorin. Dahl (1952), from Norway, mention only; Krog (1951*b*), from Norway.

Alectoronic acid. Dahl (1953).

Atranorin, α-collatolic acid. Asahina (1940*e*), from Formosa and Japan, cilia color blue-violet with alkali; Asahina (1952*e*), from Formosa and Japan.

OTHER REPORT. Atranorin, lecanoric acid, salazinic acid: Shah (1954), from India, by extraction.

Parmelia arnoldii f. *pallescens* Asah., see *Parmelia melissii* Dodge

Parmelia arseneana Gyeln.   Atranorin, norstictic acid (±), stictic acid, usnic acid. Hale (1955*b*), from North America, as *Parmelia conspersa* Chem. strain 1, plants with a blackened undersurface referred to *P. hyposila.*

Parmelia asahinae Yas. *ex* Zahlbr., see *Menegazzia asahinae* (Yas. *ex* Zahlbr.) Sant.

Parmelia aspera Mass.
No substances found. Dahl (1952), mention only; Krog (1951*b*).
REVIEW. Parmelia-brown: Thies (1932).
All reports are as *Parmelia aspidota.*

Parmelia aspidota (Ach.) Poetsch, see *Parmelia aspera* Mass.

Parmelia atrichella Nyl.   Gyrophoric acid. Hale and Kurokawa (1964), mentioned in key.

Parmelia aurulenta Tuck.   Atranorin; zeorin (±); entothein. Asahina (1951*c*), from China and Japan; Asahina (1952*e*), from Japan; Hale (1957*c*), from West Virginia, an exsiccati specimen; Hale (1958*b*), from eastern U.S.A., zeorin accessory.

Atranorin, entothein. Abbayes (1961), from Madagascar, no zeorin.

Atranorin, zeorin. Hale (1957*a*), from Arkansas, Missouri, and Oklahoma.

Parmelia austerodes Nyl., see *Hypogymnia austerodes* (Nyl.) Räs.

Parmelia austrosinensis Zahlbr. Atranorin, lecanoric acid. Abbayes (1961), from Madagascar; Abbayes (1962), from Africa; Hale (1965*a*), from Africa, Asia, Australia, Central and South America, the West Indies, Louisiana, Missouri, and Texas; Ramaut (1964–1965), by thin-layer chromatography.

Lecanoric acid. Hale (1959*b*), from Africa, China, Japan, South America, Mississippi, and Texas, a sample designated lectotype; Hale 1965*d*), from Japan.

Parmelia azulensis B. de Lesd., see *Parmelia hypoleucites* Nyl.

Parmelia bahiana Nyl. Protocetraric acid; unidentified medullary pigment (K + purple). Hale and Kurokawa (1964), no usnic acid, no lichexanthone, mentioned in key.

OTHER REPORT. Alectoronic acid and atranorin (in a sample determined by Lynge), or atranorin and an unidentified substance (in a syntype from Brazil): Hale (1960*a*).

Parmelia balansae Müll. Arg. Atranorin only. Hale (1960*a*).

Parmelia bangii Vain. Atranorin, stictic acid. Hale (1965*a*), from Bolivia and Colombia.

Parmelia benyovszkyana Gyeln. Norstictic acid (trace), stictic acid, usnic acid. Abbayes (1961), from Madagascar.

Parmelia bicornuta Müll. Arg. Lecanoric acid. Hale (1960*a*), from Brazil, mention only; Hale and Kurokawa (1964), mentioned in key.

Parmelia bitteriana Zahlbr., see *Hypogymnia bitteriana* (Zahlbr.) Räs.

Parmelia blanchetii Hue Atranorin, protocetraric acid. Hale (1965*a*), from Brazil and Colombia.

Parmelia blastica Vain. Norstictic acid. Hale (1958*a*), from the West Indies, type specimen tested microchemically; Hale (1959*a*), mention only.

Parmelia bogotensis Vain. Evernic acid, lecanoric acid. Hale and Kurokawa (1964), KC + reaction observed on thallus.

Parmelia boliviana Nyl. Barbatic acid. Hale and Kurokawa (1964), no usnic acid, mentioned in key.

Parmelia bolliana Müll. Arg.   Atranorin; protolichesterinic acid. W. L.
Culberson (1962*a*), from east-central U.S.A. south to south Texas and
Mexico, as *Parmelia frondifera*; W. L. Culberson and Culberson (1956),
from east-central U.S.A., as Chem. strain 4 ; Hale (1957*a*), as Chem.
strain 4; Hale (1957*c*), from West Virginia, an exsiccati specimen, as
Chem. strain 4.
   Protolichesterinic acid. W. L. Culberson (1960*a*), mention only, as
Parmelia frondifera; Hale (1957*b*), mention only, as a chemical strain;
Hale (1965*c*), from North America, distribution described.
   Fatty acids. Hale and Kurokawa (1964), mention only.

Parmelia bolliana Müll. Arg. *sensu* W. Culb. & C. Culb., see *Parmelia
hypoleucites* Nyl.

Parmelia bolliana Müll. Arg. Chem. strain 2, see *Parmelia praesignis* Nyl.

Parmelia bolliana Müll. Arg. Chem. strain 3, see *Parmelia subpraesignis*
Nyl.

Parmelia borreri (Sm.) Turn.   Atranorin, gyrophoric acid. Asahina (1951*d*),
from Japan, Korea, and Manchuria, type description of *Parmelia
pseudoborreri*; Asahina (1952*e*), from China, Japan, and Korea, as *P.
pseudoborreri*; W. L. Culberson (1960*a*), from Belgium and France, as
P. pseudoborreri; W. L. Culberson (1962*a*), as *P. pseudoborreri*; W. L.
Culberson and Culberson (1956), from the U.S.A., as *P. dubia* Chem.
strain 3; Hale (1965*c*), European distribution mapped, exsiccati samples
listed; Targé and Lambinon (1965), from Europe, by thin-layer chromatography, as *P. borreri* "chemovar." *pseudoborreri*; Thomson (1963),
from Asia, Europe, North America, mention in chemotaxonomic discussion, as *P. pseudoborreri*.
   Gyrophoric acid. Anderson (1962), from Colorado; Hale and Kurokawa (1964), mention only.

Parmelia borrerina Nyl.   Gyrophoric acid. Hale and Kurokawa (1964),
mention only.

Parmelia bostrychodes Zahlbr.   Barbatic acid, unidentified substance
(C+). Hale (1965*b*), the unidentified substance (also in *Parmelia
explanata*) found by chromatography.
   Barbatic acid. Hale and Kurokawa (1964), from Java, no usnic acid,
mentioned in key.

Parmelia brasiliana Nyl.   Lichexanthone, protocetraric acid. Hale and
Kurokawa (1964), no usnic acid, no rhodophyscin.
   Lichexanthone. Hale (1958*c*), mention only; Hale (1960*a*), from South
America.

Parmelia breviciliata Hale   Atranorin, alectoronic acid. Hale (1960*a*), from Brazil, as *Parmelia latissima* f. *microspora*; Hale (1965*a*), from Africa, Brazil, Madagascar, and Thailand, type description.

Parmelia brevirhiza Kurok.   Atranorin, salazinic acid. Hale and Kuro-kawa (1964), from Argentina, Chile, and Java, type description.

Parmelia bulbochaeta Hale   Atranorin only. Hale and Kurokawa (1964), from India, type description.

Parmelia camtschadalis (Ach.) Eschw.
    Atranorin, protocetraric acid, salazinic acid. Shah (1954), from India, by extraction, as *Parmelia kamtschadalis*.
    OTHER REPORT. Kamtschadalic acid: Salkowski (1910).
    REVIEWS. Atranorin, salazinic acid: Thies (1932); Zopf (1907); both reports as *Parmelia kamtschadalis*. — Atranorin: W. Brieger (1923); Hesse (1912); both reports as *P. kamtschadalis*. — Chloroatranorin: Asahina and Shibata (1954); Shibata (1958, 1963).

Parmelia canaliculata Lynge   Atranorin; protolichesterinic acid. Hale (1960*a*), from Argentina, Brazil, and Uruguay.
    Fatty acid. Hale and Kurokawa (1964), mention only.

Parmelia canescens Kurok.   Atranorin, norstictic acid. Hale and Kuro-kawa (1964), from Chile, type description.

Parmelia caperata (L.) Ach.   Atranorin, protocetraric acid (sometimes as caprararic acid or protocapraric acid), (+)-usnic acid; caperatic acid; arabitol, mannitol (trace).
    Atranorin, protocetraric acid, usnic acid, caperatic acid. Dahl (1952), from Norway, mention only.
    Atranorin, protocetraric acid, usnic acid. D. Hess (1958), from Germany, by paper chromatography.
    Atranorin, usnic acid. D. N. Rao and LeBlanc (1965).
    Protocetraric acid, (+)-usnic acid, caperatic acid. Asano and Ohta (1933), from Japan, by extraction, protocetraric acid reported as capraric acid (see Asahina and Tanase, 1934*b*).
    Protocetraric acid, usnic acid, caperatic acid. Asahina (1937*e*), from Japan, no caperin or caperidin, mention only; Asahina (1951*h*); Asa-hina (1952*e*), from Japan; Hale (1954*a*), from Arkansas, an exsiccati specimen; Hale (1956*a*); Hale (1957*a*), from Arkansas, Missouri, and Oklahoma.
    Protocetraric acid, (+)-usnic acid. Ramaut and Schumacker (1961), from Belgium, by paper chromatography, usnic acid extracted from the chromatogram found to be dextrorotary.

Protocetraric acid, usnic acid. W. L. Culberson (1955), from Wisconsin; Hale (1960*a*), from South America; Hale and Kurokawa (1964), mention in key; Hutchinson and Reynolds (1960); Ramaut (1961*a*), mention only; Ramaut (1961*b*), from Belgium; Targé (1964–1965), from Africa, by chromatography.

Protocetraric acid, caperatic acid. Kurokawa (1964), mention only.

Protocetraric acid. Asahina and Yanagita (1933*c*), from Japan, by extraction; C. F. Culberson (1965*c*), by extraction; Klosa (1952*a*), mention only; Koller, Klein, and Pöpl (1933), as protocapraric acid.

Usnic acid, caperatic acid. Klosa (1952*b*), mention only; Taguchi, Sankawa, and Shibata (1966), ^{14}C and ^{3}H labeling study on the biosynthesis of usnic acid.

Caperatic acid. Bendz, Santesson, and Tibell (1966), used as an authentic source of this substance.

Arabitol, mannitol (trace). Lindberg, Misiorny, and Wachtmeister (1953), by paper chromatography, no volemitol.

Arabitol. Pueyo (1960), by extraction.

OTHER REPORT. Protocetraric acid and/or fumarprotocetraric acid, usnic acid, caperatic acid: Krog (1951*b*), from Portugal.

REVIEWS. Caperatic acid, caperidin, caperin, (+)-usnic acid: W. Brieger (1923); Hesse (1912); Thies (1932); Zopf (1907). — Caperatic acid: Asahina and Shibata (1954); Shibata (1958, 1963).

Parmelia capitata Lynge, see *Parmelia praesorediosa* Nyl.

Parmelia caraccensis Tayl. Norstictic acid, usnic acid. Hale and Kurokawa (1964), mention only.

Parmelia caribaea Hale Atranorin, protocetraric acid. Hale and Kurokawa (1964), from the West Indies, type description.

Parmelia carneopruinata Zahlbr. Stictic acid. Hale and Kurokawa (1964), no usnic acid, mention in key.

Parmelia caroliniana Nyl. Atranorin, perlatolic acid. W. L. Culberson (1957), from southeastern U.S.A., by microchemical tests of specimens throughout the North American range; Hale (1957*a*), from Arkansas; Hale (1959*b*), from Brazil, Central America, the West Indies, Kentucky, and West Virginia; Hale (1960*a*), from southeastern U.S.A. south to Brazil.

Perlatolic acid. Hale and Kurokawa (1964), mentioned in key, no usnic acid.

Parmelia carporrhizans Tayl. Lecanoric acid. Hale and Kurokawa (1964), mentioned in key.

Parmelia centrifuga (L.) Ach. Alectoronic acid, atranorin, usnic acid; unidentified aliphatic acid; unidentified sugar alcohol; enzymes; iron, manganese, zinc.

Alectoronic acid, atranorin, usnic acid. Dahl (1952), from Norway, mention only; Krog (1951*b*), from Norway; Mitsuno (1953), by paper chromatography, no α-collatolic acid; Solberg (1956), mention only.

Alectoronic acid, usnic acid. Asahina (1951*b*), from Japan; Asahina (1952*e*), from Japan.

Alectoronic acid. Hale (1956*f*), mention only; Wachtmeister (1956*a*), used as an authentic source of this substance.

Usnic acid. Hale (1954*b*), from Baffin Island.

Unidentified aliphatic acid. Solberg (1957), acid similar to ventosic acid.

Ventosic acid. Aghoramurthy, Sarma, and Seshadri (1961*b*).

Sugar alcohol. Solberg (1960), from Norway.

Enzymes. Moissejeva (1961), enzymes tabulated.

Iron, manganese, zinc. Lounamaa (1965), a quantitative study.

Parmelia cervicornis Tuck. Alectoronic acid, atranorin. Hale and Kurokawa (1964), from Hawaii.

Parmelia cetrarioides (Del. *ex* Duby) Nyl., see *Cetrelia cetrarioides* (Del. *ex* Duby) W. Culb. & C. Culb.

Parmelia cetrarioides var. *rubescens* (Th. Fr.) Du Rietz, see *Cetrelia olivetorum* (Nyl.) W. Culb. & C. Culb.

Parmelia cetrata Ach. Atranorin, salazinic acid. Abbayes (1961), from Madagascar; Asahina (1938*e*), description of microcrystal tests; Asahina (1940*e*), from Formosa and Japan; Asahina (1951*g*), from Formosa and Japan; Asahina (1952*e*), from Formosa and Japan; Hale (1957*a*), from Arkansas; Hale (1960*a*), from temperate and tropical areas.

Salazinic acid. Fuzikawa and Ishiguro (1936); Hale (1959*a*).

REVIEWS. Atranorin, salazinic acid: Asahina and Shibata (1954). — Cetrataic acid: W. Brieger (1923); Hesse (1912); Thies (1932), from Java; Zopf (1907). — Salazinic acid: Shibata (1958, 1963).

Parmelia cetrata f. *granularis* Asah., see *Parmelia subisidiosa* (Müll. Arg.) Dodge

Parmelia cetrata var. *sorediifera* Vain., see *Parmelia reticulata* Tayl.

Parmelia chapadensis Lynge Usnic acid, unidentified substance (PD+); unidentified pale yellow medullary pigment. Hale (1960*a*), from Brazil, PD+ substance similar to protocetraric acid.

Parmelia chiapensis Hale   Alectoronic acid, atranorin; rhodophyscin. Hale (1965*a*), from Guatemala and Mexico, type description.

Parmelia chiricahuensis R. Anderson & W. Web.   Norstictic acid, stictic acid. Anderson and Weber (1962), from Arizona, by microcrystal tests, type description.

Parmelia cinerascens Lynge   Atranorin, salazinic acid. Hale (1960*a*), from Paraguay.
   Salazinic acid. Hale and Kurokawa (1964), no usnic acid, mentioned in key.

Parmelia circumnodata Nyl.   Protocetraric acid. Hale and Kurokawa (1964), mentioned in key.

Parmelia cirrhata Fr.   Atranorin, salazinic acid, unidentified substance; (+)-protolichesterinic acid; isolichenin, lichenin; amino acids; ascorbic acid, riboflavin; calcium, iron, nitrogen, phosphorus.
   Atranorin, salazinic acid, fatty acid. Abbayes (1961), from Madagascar, as *Parmelia americana*.
   Atranorin, salazinic acid, unidentified substance. Asahina (1951*a*), Asahina (1951*b*), from Formosa, as *Parmelia vexans* (=*P. cirrhata* f. *americana*); Asahina (1952*e*), from Formosa.
   Atranorin, salazinic acid, (+)-protolichesterinic acid. Aghoramurthy, Neelakantan, and Seshadri (1954), from India; Rangaswami and Rao (1955*b*), from India, by extraction.
   Atranorin, salazinic acid. Hale (1960*a*), from Brazil, as *Parmelia americana*.
   Isolichenin, lichenin, ascorbic acid, riboflavin, calcium, iron, nitrogen, phosphorus. Lal and Rao (1959), from India.
   Amino acids. Ramakrishnan and Subramanian (1966*b*), from India, as *Parmelia nepalensis*, alanine, aspartic acid, glutamic acid, glycine, isoleucine, leucine, methionine, and serine as free amino acids, these as well as arginine, lysine, phenylalanine, threonine, tryptophan, tyrosine, valine and an unidentified substance as combined amino acids, by paper chromatography.
   REVIEWS. Salazinic acid, (+)-protolichesterinic acid: Shibata (1958, 1963).

Parmelia citrella Kurok.   Salazinic acid, usnic acid. Hale and Kurokawa (1964), from Colombia, Java, and Panama, type description.

Parmelia cladonia (Tuck.) Du Rietz, see *Pseudevernia cladonia* (Tuck.) Hale & W. Culb.

Parmelia claudelii (Harm.) Vain., see *Parmelia stuppea* Tayl.

Parmelia clavulifera Räs.   Salazinic acid. Kurokawa (1964), mention in description of microcrystal tests.

Parmelia cochleata Zahlbr.   Atranorin, salazinic acid. Asahina (1951*g*), from Japan; Asahina (1952*e*), from Japan.

Parmelia concrescens Vain.   Divaricatic acid. Hale and Kurokawa (1964), no usnic acid, mentioned in key; Ramaut (1963*b*), by thin-layer chromatography.

Parmelia confluescens Nyl.   Atranorin, divaricatic acid, unidentified substance (KC+) (±). Abbayes (1961), from Madagascar, mention only.

Parmelia confoederata W. Culb.   Atranorin, lecanoric acid. W. L. Culberson (1961*c*), from southeastern U.S.A., type description.
   Lecanoric acid. Hale and Kurokawa (1964), mentioned in key.

Parmelia conformata Vain.   Atranorin, fumarprotocetraric acid, protocetraric acid, usnic acid. Hale (1965*a*), from Angola, Brazil, Mexico, Panama, and the West Indies.
   Atranorin protocetraric acid, usnic acid. Abbayes (1961), mention only.

Parmelia congensis B. Stein   Stictic acid, usnic acid. Hale (1960*a*), from Brazil, specimen determined by Lynge as *Parmelia protoalegrensis.*

Parmelia congruens Ach.   Barbatic acid or stictic acid, usnic acid; unidentified yellowish pigment. Hale (1959*a*), from Brazil, Cuba, Honduras, Madagascar, Alabama, Florida, Georgia, Louisiana, Mississippi, and South Carolina, yellowish pigment in the medulla. Hale (1960*a*), from Africa, Central and South America, southern U.S.A., and the West Indies, yellowish pigment in the medulla.
   OTHER REPORTS. Usnic acid, unidentified substance: Abbayes (1961), from Madagascar and Réunion. — Usnic acid: Hale (1957*b*), mention only, as *Parmelia leucochlora.*
   Except where indicated otherwise, all reports are as *Parmelia sphaerospora.*

Parmelia connivens Kurok.   Atranorin, usnic acid; protolichesterinic acid. Hale and Kurokawa (1964), from Guam, the Moluccas, New Guinea, and the Tanimbar Islands, type description.

Parmelia consimilis Vain.   Atranorin, protocetraric acid; unidentified yellow pigment in lower medulla.

Atranorin, protocetraric acid. Hale (1960*a*), from Brazil and the West Indies.

Protocetraric acid, unidentified pigment. Hale and Kurokawa (1964), mentioned in key.

Parmelia consors Nyl.   Atranorin. Hale (1960*a*), from Brazil.

Parmelia conspersa (Ach.) Ach.   Atranorin ($\pm$), norstictic acid ($\pm$), stictic acid, usnic acid; arabitol, mannitol.

Atranorin ($\pm$), norstictic acid ($\pm$), stictic acid, usnic acid. Hale (1955*b*), from North America, as *Parmelia isidiata* Chem. strain 1 and as *P. isidiosa*, a detailed microchemical study; Hale (1964), from Europe and North America, taxonomy clarified; see also Hale (1956*d*), a microchemical study.

Atranorin, stictic acid, usnic acid. Solberg (1956), mention only.

Norstictic acid, stictic acid, usnic acid. Asahina (1959*c*), from Japan, as *Parmelia lusitana*.

Stictic acid, usnic acid. Asahina (1949*a*), from Japan, as *Parmelia lusitana*; Hale (1957*b*), as *P. isidiosa*; Hale (1957*c*), from Virginia, as *P. isidiata* Chem. strain 1, an exsiccati specimen; Hale (1960*a*), from South America, as *P. isidiata*; Mohan, Keane, and Nolan (1937), from Ireland, by extraction, review of older chemical literature included.

Norstictic acid. Asahina (1936*d*).

Stictic acid. Asahina (1937*e*), from Ireland; Hale (1957*b*), and as *Parmelia isidiata* Chem. strain 1; J. Santesson (1965); Wachtmeister (1956*a*), used as an authentic source of this substance.

Usnic acid. D. Hess (1960), mention only; Vartia (1949), from Finland; Vartia (1950*b*), from Finland.

Arabitol, mannitol. Lindberg, Misiorny, and Wachtmeister (1953), by paper chromatography, no volemitol.

Arabitol. Pueyo (1960), mention only.

OTHER REPORTS. Atranorin, methylethersalazinic acid, stictic acid, usnic acid: Dahl (1952), from Norway, mention only; D. Hess (1958), from Germany, by paper chromatography; Krog (1951*b*), from Norway; Ramaut (1961*b*), mention only; Ramaut (1962*b*), mention only. — Salazinic acid, usnic acid: Asahina (1949*a*); Asahina (1952*e*); Asahina (1959*c*); Klosa (1952*b*); Schatz (1963), from Arkansas. — Salazinic acid: Asahina and Fuzikawa (1935*c*), mention only; Asahina (1937*e*); Krog (1962), from Alaska; Mitsuno (1953), by paper chromatography; Schatz (1963), from Pennsylvania; Schindler (1936). — Lobaric acid, salazinic acid, (+)-usnic acid: Asahina and Asano (1933), mention work of Zopf. — Norlobaridone, salazinic acid, (+)-usnic acid, D-mannitol: Briner, Gream, and Riggs (1960), from Australia, by extraction (but see *P. scabrosa* Tayl., norlobaridone identical to loxodic acid); Gream and Riggs (1960), from Australia.

Nonisidiate plants listed as *Parmelia conspersa* Chem. strain 1 (containing atranorin, norstictic acid (±), stictic acid, and usnic acid), *P. conspersa* Chem. strain 2 (containing atranorin (±), salazinic acid, and usnic acid), *P. conspersa* Chem. strain 3 (containing fumarproto-cetraric acid and usnic acid), *P. conspersa* Chem. strain 4 (containing norstictic acid and usnic acid), *P. conspersa* Chem. strain 5 (containing usnic acid only), and *P. conspersa* Chem. strain 6 (containing usnic acid and an unidentified substance). Hale (1955*b*), from North America.

REVIEWS. Conspersaic acid, salazinic acid, (+)-usnic acid: W. Brieger (1923); Hesse (1912); Thies (1932), and parmelia-brown. — Norlo-baridone, salazinic acid: Shibata (1963), see *Parmelia scabrosa* Tayl. — Norlobaridone: Neelakantan (1965), see *P. scabrosa* Tayl. — Salazinic acid, (+)-usnic acid: Zopf (1907). — Salazinic acid: Asahina and Shibata (1954); Shibata (1958).

See also *Parmelia arseneana, P. dierythra, P. hypomelaena, P. hyposila, P. lineola, P. mexicana, P. novomexicana, P. piedmontensis, P. plittii, P. scabrosa* Tayl., *P. subramigera,* and *P. tinctina.*

Parmelia conspersa (Ach.) Ach. Chem. strain 1, see *Parmelia hypopsila* Müll. Arg. (black below) and *Parmelia arseneana* Gyeln. (pale below)

Parmelia conspersa (Ach.) Ach. Chem. strain 2, see *Parmelia lineola* Berry

Parmelia conspersa (Ach.) Ach. Chem. strain 3, see *Parmelia novomexicana* Gyeln. (pale below) and *Parmelia hypomelaena* Hale (black below)

Parmelia conspersa f. *corticola* Räs. (*nom. nud.*) Usnic acid. Vartia (1950*b*), from Finland.

Parmelia conspersa var. *hypoclysta* Nyl., see *Parmelia taractica* Kremp.

Parmelia conspersa var. *hypoclysta* f. *isidiosa* Müll. Arg., see *Parmelia subramigera* Gyeln. See also *Parmelia mexicana* Gyeln.

Parmelia constrictans Nyl. Salazinic acid, usnic acid. Abbayes (1961), from Madagascar.

Parmelia continentalis Lynge Atranorin. Hale (1960*a*), from Brazil and Paraguay.

Parmelia continua Lynge Atranorin, salazinic acid. Hale (1960*a*), from Brazil.
Salazinic acid. Hale and Kurokawa (1964), from South America, mentioned in key.

Parmelia cooperi J. Stein. & Zahlbr.   Atranorin, lecanoric acid. Hale (1965*a*), from eastern Africa.

Parmelia coralliformis Hale   Atranorin, salazinic acid. Hale (1965*a*), from Mexico, type description.

Parmelia coralloidea (Meyen & Flot.) Vain., see *Parmelia tinctorum* Nyl.

Parmelia corniculans Nyl.   Alectoronic acid, atranorin, α-collatolic acid; rhodophyscin (±). Hale (1965*a*), from Java, Laos, and the Philippines.
    OTHER REPORT. Atranorin, lecanoric acid, salazinic acid: Dhar, Neelakantan, Ramanujam, and Seshadri (1959), from India, sample apparently misdetermined.

Parmelia cornuta Lynge   Atranorin; vulpinic acid. Hale (1960*a*), from Brazil; Hale (1965*a*), from Brazil.

Parmelia cornuta var. *crocea* Lynge, see *Parmelia appendiculata* Fée

Parmelia coronata Fée   Atranorin, gyrophoric acid. W. L. Culberson (1961*c*), from Mexico; Hale (1960*a*), from tropical America.
    Gyrophoric acid. Hale and Kurokawa (1964), mentioned in key.

Parmelia costaricensis Nyl.   Caperatic acid. Hale and Kurokawa (1964), from America, mention only.

Parmelia crassescens Stirt.   Atranorin, norstictic acid. Hale (1965*a*), from Brazil and British Guiana.

Parmelia crenata Kurok.   Atranorin, stictic acid. Hale and Kurokawa (1964), from Japan, no usnic acid, type description.

Parmelia crinita Ach.   Atranorin, stictic acid. Abbayes (1961), from Madagascar and Réunion; Asahina (1940*e*), from Formosa and Japan; Asahina (1952*e*), from Formosa and Japan; Hale (1954*a*), from Arkansas, an exsiccati specimen; Hale (1956*a*); Hale (1957*a*), from Arkansas, Missouri, and Oklahoma; Hale (1958*a*), from Ireland, by micro-chemical tests on the type specimen of *Parmelia proboscidea*; Hale (1960*a*), from Ireland; Hale (1965*a*), from Africa, Asia, Canada, Central and South America, Europe, the U.S.A., and the West Indies; Krog (1951*b*), from Ireland; Ramaut (1961*a*), from France, as *P. pilosella*; Ramaut (1961*b*), from France, as *P. pilosella*.
    Atranorin. Dahl (1952), from Norway, mention only.
    Stictic acid. Asahina (1952*b*); Hale (1960*b*); Hale (1965*d*), from Japan.

REVIEWS. Atranorin: Zopf (1907), as *Parmelia excrescens.* — Atranorin, pilosellic acid: Thies (1932); Zopf (1907), and one other colorless substance; both reports as *P. pilosella.* — Atranorin, salazinic acid: W. Brieger (1923); Hesse (1912); Thies (1932); all three reports as *P. excrescens.* — Pilosellic acid: W. Brieger (1923); Hesse (1912); both reports as *P. pilosella.*

Parmelia cristifera Tayl.   Atranorin, salazinic acid. Abbayes (1961), from Madagascar and Réunion; Hale (1960*a*), from the tropics; Hale (1965*a*), from the tropics.
Salazinic acid. Hale (1959*b*), mention only.

Parmelia croceopustulata Kurok.   Atranorin, protocetraric acid; rhodophyscin. Hale and Kurokawa (1964), from the Dominican Republic, Haiti, Jamaica, North Carolina, and Virginia, type description.

Parmelia crocoides Hale   Atranorin; protolichesterinic acid (in low concentration); unidentified medullary orange-red pigment (K −). Hale (1965*a*), from Costa Rica, type description.

Parmelia crozalsiana B. de Lesd.   Stictic acid. Hale (1959*b*), from Italy, Georgia, Illinois, Indiana, Kentucky, North Carolina, Ohio, and Virginia; Hale and Kurokawa (1964), mentioned in key.

Parmelia crustacea Lynge   Lichexanthone, protocetraric acid; unidentified anthraquinone pigment (K + purple). Hale (1960*a*), from Brazil.

Parmelia cryptochlorophaea Hale   Chloroatranorin, cryptochlorophaeic acid; caperatic acid. C. F. Culberson (1965*b*), from the Dominican Republic, 0.16% chloroatranorin, 2.5% cryptochlorophaeic acid and 1.7% caperatic acid, by extraction of 9.3 g. of lichen.
Atranorin, cryptochlorophaeic acid, unidentified fatty acid. Hale (1959*b*), from Brazil, Dominican Republic, Honduras, the West Indies, Georgia, and Florida, type description.
No usnic acid. Hale and Kurokawa (1964), mentioned in key.

Parmelia cryptoxantha Abb.   Atranorin, unidentified acids (PD + pale orange-yellow); protolichesterinic acid; unidentified pale yellow pigment. Hale (1965*a*), from Natal.
Atranorin, unidentified fatty acid, unidentified pigment. Abbayes (1961), from Madagascar, type description, fatty acid near protolichesterinic acid, medullary pigment probably entothein.

Parmelia crystallorum Lynge   Atranorin. Hale (1960*a*), from Brazil, notes large colorless crystals in the medulla when treated with aqueous KOH.

Parmelia culmigena Zahlbr. Evernic, lecanoric acid. Hale and Kurokawa (1964), thallus KC+ rose.

Parmelia cyphellata (Lynge) Sant. Unidentified yellow pigment. Hale and Kurokawa (1964), mentioned in key, unidentified yellow pigment in the medulla.

Parmelia dactylifera Vain. Unidentified substance (KC+). Hale and Kurokawa (1964), also found in *Parmelia livida* and *P. immaculata*.

Parmelia damaziana Zahlbr. Unidentified substance (KC+). Hale and Kurokawa (1964), no usnic acid, mentioned in key.

Parmelia darrowii Thoms. Usnic acid. Hale (1957b), mention only.

Parmelia decurtata Kurok. Atranorin, salazinic acid. Hale and Kurokawa (1964), from South Africa, type description.

Parmelia defecta Hale Atranorin, lecanoric acid. Hale (1965a), from Africa, type description.

Parmelia degelii Hale Alectoronic acid, atranorin, α-collatolic acid. Hale and Kurokawa (1964), from Angola, type description.

Parmelia delicatula Vain. Atranorin, salazinic acid, usnic acid. Hale (1958a), from Brazil, microchemical test on type specimen; Hale (1960a), from Brazil, and as *Parmelia magna* and *P. microdactyla* (atranorin not detected); Hale (1965a), from Brazil and Uruguay.

Parmelia denegans Nyl. Unidentified pigment (K+ purple). Hale and Kurokawa (1964), mentioned in key.

Parmelia densirhizinata Kurok. Alectoronic acid, atranorin. Hale and Kurokawa (1964), from Bolivia, Chile, Colombia, Dominican Republic, Equador, Guatemala, Haiti, Panama, and Peru, type description.

Parmelia dentella Hale & Kurok. Atranorin, unidentified substance (PD+ pale orange-red). Hale and Kurokawa (1964), from Alabama, type description.

Parmelia diacidula Hale Atranorin, gyrophoric acid; protolichesterinic acid (and caperatic acid?). Hale (1965a), from Natal, type description.

Parmelia dierythra Hale Norstictic acid, usnic acid. Hale (1964), from Minnesota, South Dakota, and Wisconsin, no stictic acid, type description.
See also *Parmelia conspersa*.

Parmelia diffugiens Zahlbr. Alectoronic acid, atranorin, usnic acid. Asahina (1951*b*); Asahina (1952*e*), from Japan; Mitsuno (1953), no α-collatolic acid, by paper chromatography.

REVIEWS. Alectoronic acid: Asahina and Shibata (1954); Shibata (1958, 1963).

Parmelia digitata Lynge Atranorin, barbatic acid. Hale (1960*a*), from Brazil to the West Indies.

Barbatic acid. Hale and Kurokawa (1964), mentioned in key.

Parmelia dilatata Vain. Atranorin, protocetraric acid, usnic acid (±). Hale (1965*a*), from the tropics, many localities listed.

Atranorin, protocetraric acid. Abbayes (1961), from Madagascar and Réunion.

Protocetraric acid. Hale (1959*b*), from Florida, Georgia, and South Carolina, as *Parmelia robusta*.

OTHER REPORTS. Atranorin, protocetraric acid, norstictic acid: Ramaut (1961*a*), from France; Ramaut (1961*b*), from France; Targé (1964–1965), from Africa, by paper chromatography. — Atranorin, norstictic acid, salazinic acid, three unidentified substances: Bachmann (1963), by thin-layer chromatography, as *Parmelia robusta*. — Protocetraric acid, norstictic acid: Ramaut (1962*b*), from France and Katanga; Ramaut (1963*a*), by thin-layer chromatography, as *P. robusta*.

Parmelia direagens Hale Alectoronic acid, atranorin, unidentified acid (PD +). Hale (1965*a*), from the Union of South Africa, type description.

Parmelia discordans Nyl. Atranorin, lobaric acid, protocetraric acid. Asahina (1951*h*), from Denmark, Ellesmere Island, France, Greenland, Iceland, Norway, and Spain.

Parmelia disjuncta Erichs., see *Parmelia substygia* Räs.

Parmelia disparilis Nyl. Atranorin, protocetraric acid. Hale (1965*a*), from Congo, Madagascar, and Mexico.

Parmelia dissecta Nyl. Atranorin, gyrophoric acid. Hale (1957*b*), mention only, as a chemical strain; Hale and Kurokawa (1964), mention only.

Gyrophoric acid. Hale (1956*e*), from North America; Hale and Kurokawa (1962), mention only.

OTHER REPORTS. Atranorin, physodic acid: Asahina (1951*d*), from Japan; Asahina (1952*e*), from Japan. — Atranorin, unidentified substance: Hale (1957*a*), from Arkansas, as *Parmelia hubrichtii*.

Parmelia dolosa Abb.   Atranorin, gyrophoric acid; unidentified fatty acid. Abbayes (1961), from Madagascar, unidentified fatty acid near caperatic acid, type description.

Parmelia dominicana Vain.   Atranorin, protocetraric acid, usnic acid. Hale (1958*a*), from the West Indies, microchemical tests on type specimen; Hale (1965*a*), from Africa, tropical America, and Hawaii.
OTHER REPORT. Atranorin, protocetraric acid, norstictic acid (trace): Ramaut (1964–1965), from Africa, by thin-layer chromatography.

Parmelia dubia (Wulf. *in* Jacq.) Schaer. *sens. str.*, see *Parmelia subrudecta* Nyl.

Parmelia eborina Hale   Atranorin, protocetraric acid. Hale (1965*a*), from Colombia, Dominican Republic, and Honduras, type description.

Parmelia ebulliens Hale   Atranorin; unidentified colorless acid; unidentified yellow pigment. Hale (1965*a*), from Jamaica and Mexico, by microcrystal tests, type description.

Parmelia ecaperata Müll. Arg.   Divaricatic acid, norlobaridone (sometimes as loxodic acid), usnic acid. Abbayes (1961), from Madagascar, by paper chromatography.
Divaricatic acid, norlobaridone. Targé (1964–1965), by paper chromatography.
Usnic acid. Hale and Kurokawa (1964), mentioned in key.

Parmelia echinocarpa Kurok.   Usnic acid, unidentified substances. Kurokawa (1965), from Japan, unidentified C+ substance (needles in the GE solution) also in *Parmelia planiuscula* and *P. schizospatha*, unidentified PD+ substance named "echinocarpic acid," type description.

Parmelia eciliata (Nyl.) Nyl.   Atranorin, stictic acid. Hale (1960*a*), from Argentina, Brazil, Mexico, and the West Indies; Hale (1965*a*), from Africa, Central and South America, Mexico, and the West Indies.
Stictic acid. Hale (1959*a*), from tropical America, mention only.

Parmelia elegantula (Zahlbr.) Szat.   No substances found. Dahl (1952), from Norway, mention only; Krog (1951*b*), from Norway.

Parmelia encausta (Sm.) Ach., see *Hypogymnia encausta* (Sm.) W. Wats.

Parmelia enderythraea Zahlbr.   Norstictic acid, usnic acid. Hale and Kurokawa (1964), mentioned in key.

Parmelia endochlora Leight.   Entothein. Asahina (1951*c*), sulfur-yellow medulla FeCl$_3$ + wine-red.
No usnic acid. Hale and Kurokawa (1964), mentioned in key.

Parmelia endomiltoides Nyl.   Usnic acid; two unidentified pigments. Hale and Kurokawa (1964), from South Africa, one pigment also found in *Parmelia violacea.*

Parmelia endosulphurea (Hillm.) Hale   Atranorin; unidentified pigment; unidentified substances. Hale (1965*a*), from Central and South America, Fiji, Ivory Coast, Java, Mexico, the West Indies, Alabama, Florida, and Texas, pigment also in *Parmelia appendiculata* and *P. araucariarum.*
   Atranorin, unidentified pale orange-yellow pigment. Hale (1960*a*), from Africa, Central and South America, southern U.S.A., and the West Indies, as *Parmelia lindmanii.*
   Unidentified yellow-orange pigment. Hale (1959*b*), from Central and South America, Florida, and Louisiana, as *Parmelia lindmanii,* yellow-orange pigment in the medulla.

Parmelia ensifolia Kurok.   Alectoronic acid, atranorin. Hale and Kurokawa (1964), from Dominican Republic, Haiti, Jamaica, Mexico, Panama, and Venezuela, type description.

Parmelia enteromorpha Ach., see *Hypogymnia enteromorpha* (Ach.) Nyl.

Parmelia enteromorpha f. *inactiva* Asah., see " *Hypogymnia enteromorpha* f. *inactiva.*"

Parmelia entotheiochroa Hue   Atranorin; 16β-O-acetylleucotylic acid, 6α-O-acetylleucotylin, 6-deoxy-16β-O-acetylleucotylin, 6-deoxyleucotylin, 6α,16β-di-O-acetylleucotylin, leucotylic acid, leucotylin, zeorin, unidentified triterpene; entothein. Yosioka, Yamaki, and Kitagawa (1966), from Japan, by extraction, a detailed chemical study.
   Atranorin, leucotylin, zeorin, unidentified colorless substance (C$_{41}$H$_{66}$O$_{10}$, m.p. 293°), entothein (C$_{23}$H$_{22}$O$_{11}$, m.p. 240°). Asahina (1952*e*), from Japan and Korea, only atranorin and zeorin detected microchemically.
   Atranorin, zeorin. Kurokawa (1959*a*), mention only.
   Entothein. Asahina (1951*c*), from Japan and Korea, FeCl$_3$ + wine-red reaction due to entothein.

Parmelia epiclada Hale, see *Parmelia michauxiana* Zahlbr.

Parmelia eradicata (Nyl.) Gyeln.   Salazinic acid. Abbayes (1961).

Parmelia erasmia Hale   Alectoronic acid, atranorin; unidentified orange-red medullary pigment. Hale (1965*a*), from Honduras and Mexico, unidentified pigment also found in *Parmelia hypomiltoides* and *P. mesogenes*, type description.

Parmelia erecta Berry, see *Parmelia perforata* (Jacq.) Ach.

Parmelia eruptens Kurok.   Atranorin, divaricatic acid. Hale and Kurokawa (1964), from South Africa, type description.

Parmelia erythrodes (Zahlbr.) Hale & Kurok.   Atranorin, lichexanthone, unidentified substance (KC+); rhodophyscin. Hale and Kurokawa (1964), from Brazil and Peru, unidentified substance probably identical to one in *Parmelia livida*, species resembles *P. brasiliana* except for chemistry.

Parmelia euneta Stirt.   Atranorin, gyrophoric acid. Hale (1965*a*), from Africa, Ceylon, and Haiti.

Parmelia eurysaca Hue   Atranorin, salazinic acid. Hale (1965*a*), from Mexico, Arizona, and Texas.

Parmelia exasperatula Nyl.   Atranorin ($\pm$). Krog (1951*b*), from Norway.

Parmelia "*excrescens*," see *Parmelia crinita* Ach.

Parmelia eximbricata (Gyeln.) Hale & Kurok.   Usnic acid, fumarproto-cetraric acid. Hale and Kurokawa (1964), from Cuba, Dominican Republic, Jamaica, and Florida.
     Fumarprotocetraric acid. Hale (1965*b*), mention only.

Parmelia explanata Hale   Atranorin, barbatic acid, unidentified substance (C+). Hale (1965*b*), from Thailand, unidentified C+ substance also found by chromatography in *Parmelia bostrychodes*, *P. exsecta*, and *P. imbricatula*.

Parmelia exporrecta Kurok.   Atranorin, gyrophoric acid. Hale and Kurokawa (1964), from Mexico, type description.

Parmelia exsecta Tayl.   Barbatic acid, unidentified substance (C+). Hale (1965*b*), unidentified C+ substance also found by chromatography in *Parmelia bostrychodes*, *P. explanata*, and *P. imbriculata*.
     Barbatic acid. Hale and Kurokawa (1964), from Asia, mentioned in key; Kurokawa (1964), mention only.

Parmelia exsplendens Hale   Alectoronic acid, atranorin. Hale and Kuro-kawa (1964), from Guatemala, Jamaica, Mexico, and South Africa, type description.

Parmelia fahlunensis (L.) Ach., see *Cetraria commixta* (Nyl.) Th. Fr.

Parmelia farinacea Bitt. see *Hypogymnia bitteriana* (Zahlbr.) Krog

Parmelia fasciculata Vain.   Atranorin, protocetraric acid, usnic acid (trace).   Hale (1960a), from Brazil and Colombia, and as *Parmelia fatiscens*; Hale (1965a), from Colombia, Jamaica, and Liberia.

Parmelia fatiscens Lynge, see *Parmelia fasciculata* Vain.

Parmelia finkii Zahlbr., see *Parmelia obsessa* Ach.

Parmelia fissicarpa Kurok.   Atranorin, protocetraric acid. Hale and Kuro-kawa (1964), from South Africa, type description.

Parmelia fistulata Tayl.   Atranorin, protocetraric acid. Hale (1960a), from Argentina and Uruguay.

Parmelia flava Kremp., see *Parmelia flavida* Zahlbr.

Parmelia flava var. *stellata* Lynge, see *Parmelia flavida* Zahlbr.

Parmelia flava var. *subdichotoma* Lynge, see *Parmelia flavida* Zahlbr.

Parmelia flaventior Stirt.   Atranorin ($\pm$), lecanoric acid, usnic acid. W. L. Culberson (1962a); W. L. Culberson and Culberson (1956), as *Parmelia dubia* Chem. strain 2.
    Atranorin, lecanoric acid, usnic acid. Hale (1959a), from Europe and North and South America.
    Atranorin, lecanoric acid. Aghoramurthy, Neelakantan, and Seshadri (1954), from India, as *Parmelia soredica*.
    Lecanoric acid, usnic acid. Asahina (1941c), mention only, as *Parmelia kernstockii*; Asahina (1955a), as *P. himalayensis*; W. L. Culberson (1955), from Wisconsin, no protocetraric acid, as *P. andreana*; W. L. Culberson (1960b), mention only, as *P. andreana*; Hale (1954a), from Colorado, an exsiccati specimen, as *P. andreana*; Hale (1956a), from the Rocky Mountains, Great Lakes region, and the Appalachian Mountains south to Alabama, as *P. andreana*; Ramaut (1961a), mention only, as *P. andreana*; Ramaut (1961b), from Belgium, mention only, as *P. andreana*.
    Lecanoric acid. Hale (1956e), from North America, as *Parmelia andreana*; Ramaut (1962b), by chromatography, as *P. andreana*.

OTHER REPORT. Lecanoric acid, usnic acid, *o*-orsellinic acid: Ramaut and Schumacker (1961), from Belgium, by paper chromatography, as *Parmelia andreana.*

Parmelia flavescens (Kremp.) Nyl. Atranorin, salazinic acid, usnic acid. Hale (1965*a*), from Central and South America and Mexico.
Salazinic acid, usnic acid. Hale (1960*a*), mentioned in key.

Parmelia flavida (Kremp.) Zahlbr. Protocetraric acid, usnic acid. Hale (1960*a*), from Brazil, as *Parmelia flava, P. flava* var. *stellata,* and *P. flava* var. *subdichotoma.*
Usnic acid. Hale and Kurokawa (1964), mentioned in key.
OTHER REPORT. Fumarprotocetraric acid, usnic acid: Abbayes (1961), from Madagascar.

Parmelia flavotincta Hale Atranorin. Hale (1965*a*), from Colombia, atranorin in cortex and medulla, type description.

Parmelia flavovirens Kurok. Usnic acid, protocetraric acid. Hale and Kurokawa (1964), from Chile, type description.

Parmelia fluorescens Hale Alectoronic acid, usnic acid. Hale (1965*b*), from Sabah, type description.

Parmelia formosana Zahlbr. Lichexanthone, unidentified acid ($C_{25}H_{32}O_9$), unidentified phenol (K + wine-red); rhodophyscin.
Lichexanthone, unidentified acid ($C_{25}H_{32}O_9$), unidentified phenol (K + wine-red). Asahina (1951*e*), from Formosa and Japan; Asahina and Nogami (1942), from Formosa and Japan, first isolation of lichexanthone, unknown phenolic acid (mp. 200°) found to be C — and $FeCl_3$ + violet-red, phenolic substance never crystallized; Hale (1958*c*), according to Asahina.
Lichexanthone, unknown phenolic compound (K + violet). Abbayes (1961), from Madagascar and Réunion, unidentified yellow pigment in acetone extract; Asahina (1952*e*), from Formosa and Japan, microchemistry discussed, some extracts yellow possibly due to an accessory pigment.
Lichexanthone, unidentified substance. Aghoramurthy and Seshadri (1953*b*), mention only; Asahina and Nuno (1964), unknown substance colors blue with the Barton-Evans-Gardner reagent.
Lichexanthone, rhodophyscin. Hale and Kurokawa (1964).
Lichexanthone. Asahina (1951*i*), mention only; W. L. Culberson and Hale (1965); Grover, Shah, and Shah (1956), mention in study on the synthesis of lichexanthone; Hale (1960*a*), mention only; Neelakantan and Seshadri (1961), mention only.

REVIEWS. Lichexanthone: Asahina and Shibata (1954); Neelakantan (1965); Shibata (1958, 1963).

Parmelia fracta Hale   Atranorin, protocetraric acid, usnic acid. Hale (1965*a*), from Colombia, type description.

Parmelia fragillima Hillm., see *Hypogymnia fragillima* (Hillm.) Rass.

Parmelia fraudans Nyl.   Atranorin, salazinic acid, usnic acid; proto-lichesterinic acid. Dahl (1952), from Norway, mention only; Krog (1951*b*), from Norway.
   Protolichesterinic acid. Krog (1951*a*), with photograph of crystals in the GE solution.

Parmelia frondifera Merr., see *Parmelia bolliana* Müll. Arg.

Parmelia fujisanensis Asah., see "*Hypogymnia fujisanensis.*"

Parmelia fuliginosa (Dicks.) Schaer., see *Sticta fuliginosa* (Dicks.) Ach.

Parmelia fuliginosa f. *ferruginascens* Zopf, see *Sticta fuliginosa* (Dicks.) Ach.

Parmelia fungicola Lynge   Atranorin, gyrophoric acid. Hale (1960*a*), from Brazil, gyrophoric acid not proved.
   Gyrophoric acid. Hale and Kurokawa (1964), mentioned in key.

Parmelia furfuracea (L.) Ach., see *Pseudevernia furfuracea* (L.) Zopf and *Pseudevernia olivetorina* (Zopf) Zopf

Parmelia furfuracea var. *olivetorina* (Zopf) Zahlbr., see *Pseudevernia olivetorina* (Zopf) Zopf

Parmelia galbina Ach.   Atranorin, galbinic acid; zeorin; unidentified yellow pigment. W. L. Culberson (1961*c*), from eastern North America, pigment in upper medulla.
   Atranorin, unidentified substance (K+, PD+), zeorin. Asahina (1951*e*), from Japan, microchemical tests described for the unidentified substance (= galbinic acid), lichen apparently misdetermined as *Parmelia sublaevigata*, and also *P. sublaevigata* f. *rugosa*, and *P. sublaevigata* f. *subradiata*; Asahina (1952*e*), from Japan, apparently misdetermined as *P. sublaevigata*.
   Atranorin, galbinic acid. Asahina (1963*c*), by microchemical tests, by thin-layer chromatography and by extraction of 20 g. of lichen.
   Atranorin, unidentified substances. Hale (1957*a*), from Arkansas, Missouri, and Oklahoma, as *Parmelia subquercifolia*.

Unidentified substance (PD+), unidentified pigment. Hale and Kurokawa (1964), mentioned in key.

OTHER REPORT. Atranorin, barbatic acid, zeorin: Hale (1957*b*), mention only.

Parmelia gigas Kurok. Alectoronic acid, atranorin. Hale and Kurokawa (1964), from Colombia, Equador, Mexico, Panama, and Venezuela, type description.

Parmelia glabra (Schaer.) Nyl. Lecanoric acid (sometimes as glabratic acid). Ahti (1966); Anderson (1962), from California, Colorado, New Mexico, and Nevada; Asahina (1936*d*); Hale (1956*e*), from North America, as *Parmelia olivacea* var. *glabra*; Koller and Pfeiffer (1933*a*), by extraction of 300 g., glabratic acid found to be identical to lecanoric acid.

REVIEWS. Lecanoric acid: W. Brieger (1923); Hesse (1912); Thies (1932); Zopf (1907).

Parmelia glomellifera (Nyl.) Nyl. Glomelliferic acid (sometimes as glomelliferin). Asahina (1939*b*), mention only; Asahina and Nogami (1937), from Europe, by extraction, earlier work of Zopf described; Krog (1951*b*), mention only; Minami (1944), mention only, synthesis of glomellin.

REVIEWS. Glomellic acid, glomelliferic acid, sphaerophorin: Thies (1932), and lecanora-red; Zopf (1907). — Glomellic acid, glomelliferic acid: W. Brieger (1923); Hesse (1912). — Glomelliferic acid: Asahina and Shibata (1954); Shibata (1958, 1963).

Parmelia gracilescens Vain. Alectoronic acid, atranorin, α-collatolic acid; unidentified yellow pigment. Abbayes (1961), from Madagascar, pigment may be entothein.

OTHER REPORT. Usnic acid, unidentified substance (KC+): Hale and Kurokawa (1964), mentioned in key.

Parmelia gracilescens var. *angolensis* Vain. Atranorin, salazinic acid. Abbayes (1961), from Madagascar.

Parmelia gracilis (Müll. Arg.) Vain. Atranorin, unidentified substances. Hale (1960*a*), Brazil, medulla K+ reddish, C−, KC+ red, PD+ pale orange.

Parmelia granatensis Nyl. Atranorin, gyrophoric acid. Hale (1958*a*), by microchemical tests on the type specimen.

Parmelia grayana Hue Atranorin; (+)-protolichesterinic acid. Asahina (1941*c*), from Japan, type description of *Parmelia simodensis*; Asahina

(1952*e*), from Japan, as *P. simodensis*; Hale (1965*a*), from Japan and the Union of South Africa.

REVIEWS. (+)-Protolichesterinic acid: Asahina and Shibata (1954); Shibata (1958, 1963); all review reports as *Parmelia simodensis*.

Parmelia hababiana Gyeln. Atranorin, cryptochlorophaeic acid; protolichesterinic acid (Chem. strain 1). Atranorin, protolichesterinic acid (Chem. strain 2). Hale (1965*a*), Chem. strain 1 from Africa, Argentina, Central America, China, India, Mexico, and Arizona and Chem. strain 2 from Africa and Mexico.

Parmelia haitiensis Hale Atranorin, cryptochlorophaeic acid; protolichesterinic acid. Hale (1965*a*), from Haiti, Jamaica, Mexico, Arkansas, Mississippi, North Carolina, and Virginia, as *Parmelia subtinctoria* Chem. strain 1.
Atranorin, unidentified substance (= cryptochlorophaeic acid), unidentified fatty acid. Hale (1959*b*), from Jamaica, the West Indies, Florida, North Carolina, and South Carolina.

Parmelia halei Ahti Atranorin, fumarprotocetraric acid. Ahti (1966), type description.

Parmelia hanningtoniana Müll. Arg. Atranorin, gyrophoric acid. Hale (1965*a*), from Angola and the Congo.
OTHER REPORT. Atranorin, lecanoric acid, norstictic acid, perlatolic acid: Targé (1964–1965), by paper chromatography.

Parmelia helenae B. de Lesd. Lecanoric acid. Hale (1965*c*), from Italy.

Parmelia heterochroa Hale & Kurok. Atranorin, protocetraric acid; unidentified anthraquinone pigment. Hale and Kurokawa (1964), from Australia.

Parmelia himalayensis Nyl., see *Parmelia flaventior* Stirt.

Parmelia hololoba Hale Atranorin, lecanoric acid. Hale (1965*a*), from Uganda, type description.

Parmelia homogenes Nyl. Atranorin; leucotylin, zeorin; entothein. Asahina (1951*c*), from India and Japan.
Atranorin, zeorin, entothein. Asahina (1952*e*), from Japan.
Atranorin, zeorin. Kurokawa (1959*a*), mention only.

Parmelia hookeri Tayl. Atranorin, lecanoric acid. Hale (1958*a*), by microchemical tests on the type specimen.
Lecanoric acid. Hale (1960*a*), mention only.

Parmelia horrescens Tayl.   An unidentified substance (KC+). Hale and Kurokawa (1962), from Ireland, crystals in the GE solution; Hale and Kurokawa (1964), mentioned in key.

Parmelia hubrichtii Berry, see *Parmelia dissecta* Nyl.

Parmelia huei Asah.   Lecanoric acid; unidentified yellow pigment ($\pm$); parmelia-brown. Asahina (1951*b*), from Japan, type description; Asahina (1952*e*), from Japan.
Lecanoric acid, yellow pigment ($\pm$). Athi (1966).

Parmelia hyperopta Ach., see *Parmeliopsis hyperopta* (Ach.) Vain.

Parmelia hypochraea Vain.   Salazinic acid. Hale and Kurokawa (1964), mentioned in key.

Parmelia hypoleucites Nyl.   Atranorin, lecanoric acid. W. L. Culberson (1962*a*), from North America, distribution described, as *Parmelia bolliana* Chem. strain 1; W. L. Culberson and Culberson (1956), from the U.S.A., as *P. bolliana* Chem. strain 1; Hale (1956*b*), as *P. bolliana*; Hale (1956*g*), from Oklahoma, an exsiccati specimen, as *P. bolliana*; Hale (1957*a*), as *P. bolliana* Chem. strain 1; Hale (1958*a*), from Mexico, and two unidentified fatty acids, by microchemical tests on the type specimen of *P. azulensis*.
Lecanoric acid. W. L. Culberson (1960*a*), mention only, as *Parmelia bolliana* Chem. strain 1; Hale (1956*e*), from North America, as a chemical strain of *P. bolliana*; Hale (1965*c*), North American distribution described.

Parmelia hypomelaena Hale   Fumarprotocetraric acid, usnic acid. Hale (1955*b*), from North America, as *Parmelia conspersa* Chem. strain 3, for plants pale below see *P. novomexicana*.

Parmelia hypomilta Fée   An unidentified pigment (K+ purple-black). Hale and Kurokawa (1964), pigment in lower half of medulla, mentioned in key.

Parmelia hypomiltoides Vain.   Alectoronic acid, atranorin; unidentified anthraquinone pigment (K+ orange-red). Hale (1965*a*), from southern Brazil, K+ pigment in the medulla also found in *Parmelia erasmia* and *P. mesogenes*.

Parmelia hyporysalea (Vain.) Vain., see *Parmelia andina* Müll. Arg.

Parmelia hyporysalea var. *cinerascens*, see *Parmelia andina* Müll. Arg.

Parmelia hyposila Müll. Arg.   Atranorin, norstictic acid ($\pm$), stictic acid, usnic acid. Hale (1955*b*), from North America, as *Parmelia conspersa* Chem. strain 1, for plants pale below see *P. arseneana*.

Parmelia hypotropa Nyl.   Atranorin, norstictic acid, stictic acid ($\pm$). Hale (1965*a*), from eastern North America and Mexico with norstictic acid and no stictic acid, from Africa, Europe, and California with stictic acid and little or possibly no norstictic acid.

Parmelia hypotropoides Nyl. *ex* Will., see *Parmelia perforata* (Jacq.) Ach.

Parmelia hypotrypa Nyl., see *Hypogymnia hypotrypa* (Nyl.) Rass.

Parmelia hypotrypa f. *balteata* Nyl., see *Hypogymnia hypotrypa* (Nyl.) Rass.

Parmelia hypotrypella Asah., see *Hypogymnia hypotrypella* (Asah.) Rass.

Parmelia ikomae Asah.   Atranorin, unidentified substance. Asahina (1953*b*), from Japan, crystals in the GE solution (photograph) described as similar to rangiformic acid, type description.

Parmelia imbricatula Zahlbr.   Barbatic acid, unidentified substance (C+). Hale (1965*b*), unidentified substance also found chromatographically in *Parmelia explanata* and other species listed under *P. explanata*.
    Barbatic acid. Hale and Kurokawa (1964), mentioned in key.

Parmelia immaculata Kurok.   Atranorin, unidentified substance (KC+); rhodophyscin. Hale and Kurokawa (1964), from Africa, type description, unidentified KC+ substance also found in *Parmelia dactylifera* and *P. livida*.

Parmelia imperfecta Kurok.   Atranorin, unidentified substance (PD+ orange). Hale and Kurokawa (1964), from South Africa, unidentified substance by crystals in the GA*o*-T solution, type description.

Parmelia incurva (Pers.) Fr.   Alectoronic acid, atranorin ($\pm$), usnic acid.
    Alectoronic acid, atranorin, usnic acid. Dahl (1952), from Norway, mention only; Krog (1951*b*), from Norway.
    Alectoronic acid, usnic acid. Asahina (1951*b*), from Japan; Asahina (1952*e*), from Japan; Mitsuno (1953), no atranorin, no α-collatolic acid, by paper chromatography.
    Usnic acid. Hale (1954*b*), from Baffin Island; Vartia (1949), from Finland; Vartia (1950*b*), from Finland; Zopf (1900).
    REVIEWS. (+)-Usnic acid: W. Brieger (1923); Hesse (1912); Thies (1932); Zopf (1907).

Parmelia inexspectata Abb. Atranorin, unidentified substance (K+ orange-red). Hale (1965*a*), from the Ivory Coast, unidentified K+ substance in the medulla.

Parmelia infirma Kurok. Atranorin; caperatic acid, protolichesterinic acid. Hale and Kurokawa (1964), from India, type description.

Parmelia infumata Nyl. Atranorin (±). Dahl (1952), mention only; Krog (1951*b*), from Norway.

Parmelia insinuans Nyl. Protocetraric acid; rhodophyscin (±). Hale and Kurokawa (1964), no usnic acid.

Parmelia intercalanda Vain. Atranorin, olivetoric acid. Hale (1960*a*), from Brazil.
  Olivetoric acid. Hale and Kurokawa (1964), no usnic acid, mentioned in key.

Parmelia internexa Nyl. Stictic acid. Hale and Kurokawa (1964), mentioned in key.

Parmelia intertexta Mont. & Bosch Protocetraric acid, usnic acid. Hale and Kurokawa (1964), mentioned in key.

Parmelia intestiniformis (Vill.) Ach., see *Hypogymnia intestiniformis* (Vill.) Räs.

Parmelia ischnoides Kurok. Atranorin, stictic acid. Hale and Kurokawa (1964), from South Africa, type description.

Parmelia isidiata (Anzi) Gyeln., see *Parmelia conspersa* (Ach.) Ach.

Parmelia isidiata (Anzi) Gyeln. Chem. strain 1, see *Parmelia conspersa* (Ach.) Ach. and *Parmelia plittii* Gyeln.

Parmelia isidiata (Anzi) Gyeln. Chem. strain 2, see *Parmelia mexicana* Gyeln.

Parmelia isidiata (Anzi) Gyeln. Chem. strain 3, see *Parmelia piedmontensis* Hale and *Parmelia subramigera* Gyeln.

Parmelia isidiosa (Müll. Arg.) Hale, see *Parmelia conspersa* (Ach.) Ach.

Parmelia isidiotyla Nyl. Glomelliferic acid. Dahl (1952), from Norway, mention only; Krog (1951*b*), from Norway.

OTHER REPORT. Most specimens KC +, a few KC − : Dahl (1950), from Greenland.

Parmelia isidiza Nyl. Atranorin, salazinic acid. Abbayes (1961), from Madagascar; Targé (1964–1965), from Africa, by chromatography. Salazinic acid. Hale and Kurokawa (1964), mentioned in key.

Parmelia jamaicensis Vain. *non* (Ach.) Spreng., see *Parmelia microblasta* Vain.

Parmelia "kamtschadalis," see *Parmelia camtschadalis* (Ach.) Eschw.

Parmelia keitauensis Asah. Atranorin, protocetraric acid; unidentified substance. Asahina (1951e), from Formosa, type description.
Protocetraric acid, unidentified fatty acid. Hale and Kurokawa (1964), from Asia, no usnic acid.

Parmelia kernstockii (Lynge) Zahlbr., see *Parmelia flaventior* Stirt.

Parmelia kinabalensis Hale Atranorin, norstictic acid. Hale (1965b), from Sabah, type description.

Parmelia koflerae Poelt Atranorin or chloroatranorin, salazinic acid. Clauzade and Poelt (1961), by microchemical tests, type description.

Parmelia koyaensis Asah. Atranorin, protocetraric acid; caperatic acid. Asahina (1959b), from Formosa and Japan, by microchemical tests, caperatic acid confirmed by chromatography.
Atranorin, protocetraric acid, unidentified acid. Asahina (1953a), from Japan, type description.
Protocetraric acid, unidentified fatty acids. Hale and Kurokawa (1964), mention only.

Parmelia koyaensis f. *inactiva* Asah., see *Parmelia nodakensis* Asah.

Parmelia laciniatula (Flag.) Zahlbr. No substances found. Dahl (1952), mention only; Krog (1951b), from Germany.

Parmelia laevigata (Sm.) Ach. Atranorin, barbatic acid, unidentified substance. Dahl (1952), from Norway, mention only; Hale and Kurokawa (1962), from Europe, South America, and Wales, unidentified medullary substance K + yellow and KC + deep orange; Krog (1951b), from Ireland, barbatic acid not demonstrated, photograph of crystals in the GAQ solution.
Barbatic acid. Hale and Kurokawa (1964), mentioned in key; Krog (1951a).

OTHER REPORTS. Atranorin, lecanoric acid: Hale (1957*b*), mention only, as a chemical strain. — Lecanoric acid: Hale (1956*e*), from North America, reported as a chemical strain (but see *Parmelia rockii*).

Parmelia laevigata ssp. ***extremi-orientalis*** Asah.   Atranorin, barbatic acid, physodic acid; unidentified substance. Asahina (1951*e*), from Formosa and Japan, type description; Asahina (1952*e*), from Formosa and Japan.

Parmelia laevigatula Nyl.   Lecanoric acid. Hale and Kurokawa (1964), mentioned in key.

Parmelia laevior Nyl.   Atranorin, salazinic acid. Asahina (1951*f*), from Japan; Asahina (1952*e*), from Japan; Asahina (1954*d*).

Parmelia laevior f. ***microphyllina*** Hue   Atranorin, chloroatranorin, salazinic acid. Fuzikawa and Ishiguro (1936).

Parmelia långii Lynge, see *Parmelia argentina* Kremp.

Parmelia latissima Fée   Atranorin, salazinic acid. Hale (1960*a*), from the tropics, by microchemical tests; Hale (1965*a*), from Central and South America, India, Mexico, and the West Indies.
  OTHER REPORTS. Atranorin, chloroatranorin, lecanoric acid, salazinic acid, D-arabitol: Nolan, Keane, and Davidson (1940), from Kenya, by extraction. — D-Arabitol: Asahina (1951*i*), mention only; Mittal, Neelakantan, and Seshadri (1952), mention only.
  REVIEWS. Chloroatranorin, lecanoric acid: Asahina and Shibata (1954); Shibata (1958, 1963).

Parmelia latissima var. *corniculata* Kremp., see *Parmelia zollingeri* Hepp

Parmelia latissima f. *microspora* Lynge, see *Parmelia breviciliata* Hale

Parmelia latissima var. *minima* Lynge, see *Parmelia zollingeri* Hepp

Parmelia lecanoracea Müll. Arg.   Unidentified pigment (K + purple-black). Hale and Kurokawa (1964), no usnic acid, unidentified K + pigment in medulla, mentioned in key.

Parmelia leiophylla Kurok.   Atranorin, olivetoric acid; rhodophyscin. Hale and Kurokawa (1964), from South Africa, type description.

Parmelia leucochlora Tuck., see *Parmelia congruens* Ach.

Parmelia leucopis Kremp.   Protocetraric acid. Hale and Kurokawa (1964), from South America, no usnic acid, mentioned in key.

Parmelia leucosemotheta Hue   Atranorin, salazinic acid. Hale (1960*a*), from Brazil; Hale (1965*a*), from Africa, Central and South America, and the West Indies.

Parmelia leucotyliza Nyl.   Atranorin; leucotylic acid, leucotylin, zeorin. Atranorin, leucotylin, zeorin. Asahina and Akagi (1938), from Japan, by extraction and chromatographic separation of the triterpenes; Asahina (1951*d*), from Japan; Asahina (1952*e*), from Japan.
   Atranorin, zeorin. Kurokawa (1959*a*), mention only.
   Leucotylic acid, leucotylin, zeorin. Yosioka, Nakanishi, and Tsuda (1966), from Japan, a detailed chemical study on leucotylic acid.
   Leucotylin, zeorin. Asahina (1939*b*), from Japan, mention only; Schaffner, Caglioti, Arigoni, and Jeger (1958), mention only; Yosioka and Nakanishi (1963), a chemical study on leucotylin.
   Zeorin. Neelakantan, Rangaswami, and Rao (1954), mention only.
   Leucotylin. Asahina (1951*i*), mention only.
   REVIEWS. Leucotylin, zeorin: Asahina and Shibata (1954); Shibata (1958, 1963). — Leucotylin: Neelakantan (1965).

Parmelia leucoxantha Müll. Arg.   Atranorin, protocetraric acid, usnic acid. Hale (1960*a*), from Brazil.

Parmelia limbata Laur.   Stictic acid. Hale and Kurokawa (1964), mentioned in key.
   OTHER REPORTS. Norstictic acid, usnic acid, probably also fumarprotocetraric acid: Asahina (1951*b*), from Formosa. — Fumarprotocetraric acid, calcium oxalate: Asahina (1952*e*), from Formosa.

Parmelia lindmanii Lynge (incorrect determination), see *Parmelia endosulphurea* (Hillm.) Hale

Parmelia lineariloba Kurok.   Alectoronic acid, atranorin. Hale and Kurokawa (1964), from the Dominican Republic, type description.

Parmelia lineola Berry   Atranorin ($\pm$), salazinic acid, usnic acid. Hale (1955*b*), from North America, as *Parmelia conspersa* Chem. strain 2.

Parmelia livida Tayl.   Atranorin, lividic acid, 4-O-methylphysodic acid, two unidentified substances ($\pm$), unidentified substance. C. F. Culberson (1967*a*), from Arkansas, Georgia, New Jersey, New York, North Carolina, South Carolina, Tennessee, and Virginia, by thin-layer chromatography of 39 specimens.
   Atranorin, lividic acid, 4-O-methylphysodic acid. C. F. Culberson (1966*b*), from North Carolina, 0.35% atranorin, 1.3% lividic acid, 0.35% 4-O-methylphysodic acid, by extraction and column chromatography.

Atranorin, lividic acid (reported as an unidentified substance). W. L. Culberson (1961c), from eastern and south-central U.S.A.; Hale (1958a), from Louisiana, by microchemical tests on the type specimen.

Lividic acid (reported as an unidentified substance). Hale and Kurokawa (1964), no usnic acid, mentioned in key.

Parmelia lividescens Kurok.   Atranorin, olivetoric acid. Hale and Kurokawa (1964), from South Africa, type description.

Parmelia lobulascens J. Stein.   Atranorin, gyrophoric acid. Hale (1965a), from the Cameroons, Ethiopia, and Kenya.

Parmelia lobulifera Degel.   Olivetoric acid. Hale (1956e), from North America.

Parmelia locarnensis Zopf   Gyrophoric acid, imbricaric acid.
REVIEWS. Gyrophoric acid, imbricaric acid: W. Brieger (1923); Hesse (1912); Thies (1932); Zopf (1907).

Parmelia lophogena Abb.   Atranorin, gyrophoric acid; unidentified fatty acid. Abbayes (1961), from Africa.

Atranorin, gyrophoric acid. Hale (1965a), from Africa.

Parmelia loxodes Nyl., see *Parmelia scabrosa* Tayl.

Parmelia lugubris Pers., see "*Hypogymnia lugubris.*"

"*Parmelia lugubris* var. *sikkimensis,*" see "*Hypogymnia lugubris* var. *sikkimensis.*"

Parmelia lugubris f. *tenuis* Bitt., see "*Hypogymnia lugubris* f. *tenuis.*"

Parmelia lusitana Nyl., see *Parmelia conspersa* (Ach.) Ach.

Parmelia luteoviridis Kurok.   Atranorin, gyrophoric acid, usnic acid. Hale and Kurokawa (1964), from Borneo, type description.

Parmelia lythgoeana Dodge   Unidentified pigment (K+ purple); unidentified fatty acid. Hale and Kurokawa (1964), no usnic acid or lichexanthone, unidentified K+ pigment in the medulla, mentioned in key.

Parmelia maclayana Müll. Arg.   Alectoronic acid, atranorin. Hale (1965a), from Africa; Ramaut (1964–1965), by thin-layer chromatography.

Parmelia macrocarpoides f. *subcomparata* Vain. Atranorin; unidentified fatty substance. Hale (1959*a*), from Brazil, the Dominican Republic, Guatemala, and Tennessee.

Parmelia madagascariacea (Hue) Abb. Unidentified substance (KC+), usnic acid. Abbayes (1961), from Madagascar; Hale (1959*a*), from Central America, Madagascar, and southern U.S.A.
   Usnic acid. Ramaut (1964–1965), from Tennessee, by thin-layer chromatography.

Parmelia magna Lynge, see *Parmelia delicatula* Vain.

Parmelia majoris Vain. Salazinic acid; zeorin. Hale and Kurokawa (1964), from Asia and Madagascar, no usnic acid.

Parmelia malaccensis Nyl. Protocetraric acid, usnic acid. Hale and Kurokawa (1964), mentioned in key.

Parmelia malesiana Hale   Fumarprotocetraric acid, usnic acid; unidentified pigment (±) (K+). Hale (1965*b*), from the Philippines, Taiwan, and Sabah; Kurokawa (1965), from Taiwan.

Parmelia malmei Lynge   Unidentified pigment (K+ purple). Hale and Kurokawa (1964), mentioned in key.
   See also *Parmelia minima*.

Parmelia manilensis Vain.   Protocetraric acid. Hale and Kurokawa (1964), from Asia, no fatty acids or usnic acid.

Parmelia manshurica Asah., see *Parmelia ulophyllodes* (Vain.) Sav.

Parmelia margaritata Hue   Atranorin, salazinic acid. Hale (1965*a*), from Arkansas, Iowa, Kentucky, Maryland, Massachusetts, North Carolina, Ohio, Pennsylvania, Tennessee, Virginia, West Virginia, and Wisconsin.

Parmelia marginalis Lynge   Atranorin, gyrophoric acid. Hale (1960*a*), from Brazil.

Parmelia marmariza Nyl.   Atranorin, chloroatranorin, salazinic acid.
   Atranorin, salazinic acid. Asahina (1951*g*), from Japan; Asahina (1952*e*).
   REVIEWS. Chloroatranorin, salazinic acid: Asahina and Shibata (1954); Shibata (1958, 1963).

Parmelia martinicana Nyl.   Protocetraric acid. Hale and Kurokawa (1964), no usnic acid, mentioned in key.

Parmelia maxima Hue, see *Parmelia stuppea* Tayl.

Parmelia meiosperma (Hue) Dodge   Unidentified substance (KC+). Hale and Kurokawa (1964), no usnic acid, mentioned in key.

Parmelia meizospora (Nyl.) Nyl.   Lecanoric acid, salazinic acid. Dhar, Neelakantan, Ramanujam, and Seshadri (1959), from India, by extraction of 100 g. of lichen.
  Salazinic acid. Hale and Kurokawa (1964), from eastern Asia, mentioned in key.

Parmelia melanochaeta Kurok.   Atranorin, gyrophoric acid. Hale and Kurokawa (1964), from Brazil and Paraguay, type description.

Parmelia melanothrix (Mont.) Vain.   Atranorin; protolichesterinic acid. Hale (1965a), from Brazil and Réunion.
  Atranorin, unidentified fatty substance. Abbayes (1961), from Madagascar; Hale (1960a), from Brazil and other tropical regions.

Parmelia melanothrix f. *microspora* Lynge, see *Parmelia argentina* Kremp.

Parmelia mellissii Dodge   Alectoronic acid, atranorin, α-collatolic acid; rhodophyscin (±).
  Alectoronic acid, atranorin, α-collatolic acid. Ramaut (1964–1965), from Georgia, by thin-layer chromatography.
  Alectoronic acid, atranorin, rhodophyscin (±). Hale (1965a), pantropical distribution northward into temperate U.S.A. and Japan described.
  Alectoronic acid, atranorin. Hale (1959a), from Cuba, Formosa, Jamaica, Venezuela, Alabama, Florida, Georgia, Tennessee, and Virginia, type description of *Parmelia allardii*.
  Alectoronic acid. Hale (1965d), from Japan.
  Atranorin, α-collatolic acid. Asahina (1940e), type description of *Parmelia arnoldii* f. *pallescens*.

Parmelia merrillii Lynge *non* Vain., see *Parmelia appendiculata* Fée

Parmelia merrillii Vain.   Atranorin, protocetraric acid. Abbayes (1963), from Vietnam; Hale (1965a), from Bolivia, Celebes, Formosa, and Sumatra.

Parmelia mesogenes Nyl.   Atranorin, unidentified substance; unidentified pigment (K+). Hale (1965a), from Mexico, unidentified substance by large colorless needle clusters observed in the GE solution, K+ pigment in medulla also found in *Parmelia erasmia* and *P. hypomiltoides*.

Parmelia mesotropa Müll. Arg. Atranorin; caperatic acid. Hale (1965*a*), from Guatemala, Mexico, and South America, caperatic acid possibly mixed with protolichesterinic acid.

Atranorin, unidentified fatty substance similar to caperatic acid. Hale (1960*a*), from South America.

Parmelia metaphysodes Asah., see *Hypogymnia metaphysodes* (Asah.) Rass.

Parmelia metarevoluta Asah. Atranorin, galbinic acid. Asahina (1960*a*), from Japan, microchemical tests described, galbinic acid as an unidentified, PD + substance, type description.

Galbinic acid. Asahina (1963*c*).

Parmelia mexicana Gyeln. Atranorin (±), salazinic acid, usnic acid. Hale (1955*b*), North American distribution described, as *Parmelia isidiata* Chem. strain 2.

Salazinic acid, usnic acid. Hale (1954*a*), from Arkansas, an exsiccati specimen, as *Parmelia isidiata*; Hale (1956*a*), as *P. isidiata* Chem. strain 2; Hale (1964), from the U.S.A.

Salazinic acid. Asahina (1951*b*), as *Parmelia conspersa* var. *hypoclysta* f. *isidiosa* (probably *P. mexicana* misidentified according to a private communication from M. E. Hale, Jr.).

Other reports probably based on samples of this species determined as *Parmelia conspersa* are listed under *P. conspersa*.

Parmelia meyeri Zahlbr. Usnic acid. Hale and Kurokawa (1964), mentioned in key.

Parmelia michauxiana Zahlbr. Atranorin, protocetraric acid. Hale (1959*a*), from southeastern U.S.A., as *Parmelia epiclada* and *P. epiclada* var. *lacinata*, type descriptions; Ramaut (1964–1965), by thin-layer chromatography.

Protocetraric acid. Hale and Kurokawa (1964), mentioned in key.

Parmelia michoacanensis B. de Lesd. Unidentified substance; unidentified yellow pigment. Hale (1958*a*), from Mexico, microchemical tests on type specimen, unidentified yellow pigment in medulla.

Parmelia microblasta Vain. Norstictic acid, usnic acid. Hale (1959*a*), from Cuba, Jamaica, Panama, and the West Indies, as *Parmelia norsticta*; Hale (1960*a*), as *P. jamaicensis*; Hale and Kurokawa (1964), mentioned in key.

Norstictic acid. Hale (1958*a*), from Brazil, by microchemical tests on type specimen (which also contains a trace of usnic acid according to a private communication from M. E. Hale, Jr.).

OTHER REPORTS. Norstictic acid only: Hale (1959*a*), no usnic acid; Hale (1960*a*), no usnic acid.

Parmelia microdactyla Hale, see *Parmelia delicatula* Vain.

Parmelia microsticta Müll. Arg.   Atranorin; protolichesterinic acid. Hale (1960*a*), from Uruguay.
  Fatty acid. Hale and Kurokawa (1964), mention only.

Parmelia minarum Vain.   Atranorin, gyrophoric acid. Hale (1958*a*), from Brazil, by microchemical tests on type specimen; Hale (1960*a*), from Africa and Central and South America.

Parmelia minima Lynge   Lichexanthone, protocetraric acid; unidentified anthraquinone. Hale (1960*a*), from Brazil, probably identical to *Parmelia malmei.*

Parmelia minuscula (Nyl.) Nyl.   No substances found. Krog (1951*b*), from Norway.

Parmelia miranda Hale   Salazinic acid, usnic acid. Hale (1965*a*), from Mexico, type description.

Parmelia molliuscula Ach.   Norstictic acid, stictic acid, usnic acid. Hale (private communication), by microchemical tests on the type specimen.
  Norstictic acid, usnic acid. Hale (1957*b*), mention only.
  Usnic acid. F. T. Jones and Palmer (1950).
  REVIEW. Usnic acid: Asahina and Shibata (1954).

Parmelia molybdiza Nyl.   Lecanoric acid. Hale and Kurokawa (1964), mentioned in key.

Parmelia monilifera Kurok.   Atranorin, barbatic acid, unidentified substance (KC+). Hale and Kurokawa (1964), from Venezuela, type description, lichen thallus also C+ pale orange.

Parmelia mougeotii Schaer.   Norstictic acid, stictic acid, usnic acid. Asahina (1961*c*), from Germany.
  (+)-Usnic acid. Salkowski (1910).
  OTHER REPORTS. Atranorin, α-methylethersalazinic acid, stictic acid, usnic acid: Dahl (1952), from Norway; Krog (1951*b*), from Norway.
  REVIEWS. (+)-Usnic acid: W. Brieger (1923); Hesse (1912); Thies (1932); Zopf (1907).

Parmelia muelleri Vain.
No usnic acid. Hale and Kurokawa (1964), mentioned in key.

Parmelia multispora Schneid.   Unidentified substances. Ahti (1966).

Parmelia mundata f. *sorediosa* Bitt., see *Hypogymnia mundata* f. *sorediosa* (Bitt.) Rass.

Parmelia mutata Vain.   Salazinic acid. Hale and Kurokawa (1964), mentioned in key.

Parmelia myelochroa Hale   Atranorin, barbatic acid (?); unidentified yellow-orange pigment. Hale (1965*a*), from Brazil, Honduras, Mexico, and Peru, yellow-orange pigment in medulla, type description.

Parmelia nairobiensis J. Stein. & Zahlbr.   Atranorin, divaricatic acid, nor-lobaridone (or as loxodic acid). Ramaut (1964–1965), from Africa, by thin-layer chromatography.
Divaricatic acid. Hale and Kurokawa (1964), mentioned in key.

Parmelia natalensis J. Stein. & Zahlbr.   Alectoronic acid, atranorin. Hale (1965*a*), from Africa.

Parmelia neglecta Asah., see *Parmelia praesorediosa* Nyl.

Parmelia nepalensis Tayl. *ex* Hook. f., see *Parmelia cirrhata* Fr.

Parmelia niitakana Asah.   Atranorin, salazinic acid. Asahina (1951*f*), from Formosa, type description.

Parmelia nikkoensis Zahlbr., see *Hypogymnia nikkoensis* (Zahlbr.) Rass.

Parmelia nilgherrensis Nyl.   Alectoronic acid, atranorin, α-collatolic acid; rhodophyscin ($\pm$). Hale (1965*a*), from Ceylon, China, India, Nepal, Sikkim, and Thailand.
Alectoronic acid, atranorin. Asahina (1954*e*).
Atranorin, α-collatolic acid. Rangaswami and Rao (1955*b*), from India, discussion of α- and β-collatolic acids, β-collatolic acid also said to be present.
α-Collatolic acid. Asahina (1938*d*), description of microcrystal tests.
REVIEWS. Atranorin, salazinic acid: W. Brieger (1923); Hesse (1912); Thies (1932); Zopf (1907). — α-Collatolic acid: Shibata (1958, 1963).

Parmelia nimandairana Zahlbr.   Atranorin, salazinic acid. Asahina (1952*e*), from Formosa, Japan, and southern Manchuria.
Atranorin. Rangaswami and Rao (1955*b*), discussion.

OTHER REPORTS. Atranorin, lecanoric acid, gyrophoric acid, salazinic acid: Rangaswami and Rao (1954c), from India, by extraction. — Lecanoric acid: Rangaswami and Rao (1955a).
REVIEWS. Salazinic acid: Shibata (1958, 1963).

Parmelia nitens Müll. Arg.   Atranorin, lecanoric acid. Abbayes (1961), from Madagascar.

Parmelia nodakensis Asah.
Atranorin, protolichesterinic acid. Asahina (1953d), from Japan, no protocetraric acid, type description of *Parmelia koyaensis* f. *inactiva*; Asahina (1959b), type description.
Unidentified substance (KC+). Hale and Kurokawa (1964), mentioned in key.

Parmelia norstictica Hale, see *Parmelia microblasta* Vain.

Parmelia novella Vain.   Lichexanthone; unidentified substances. Hale (1960a), from Brazil.
Lichexanthone. Hale (1958c), mention UV+ brilliant orange-yellow.

Parmelia novomexicana Gyeln.   Fumarprotocetraric acid, usnic acid. Hale (1955b), from North America, as *Parmelia conspersa* Chem. strain 3, for plants black below see *P. hypomelaena*.

Parmelia nylanderi Lynge   Atranorin, salazinic acid, usnic acid. Hale (1960a), from Brazil.
Usnic acid. Hale and Kurokawa (1964), mentioned in key.

Parmelia obscurata Bitt., see *Hypogymnia obscurata* (Bitt.) Räs.

Parmelia obsessa Ach.   Atranorin, galbinic acid; zeorin; unidentified yellow pigment. Hale (1959b), galbinic acid as an unidentified colorless substance, yellow pigment in the medulla, as *Parmelia finkii*.
Atranorin, galbinic acid, a yellow pigment. Hale (1956a), from Alabama, Arkansas, Georgia, Kansas, Oklahoma, Texas, and West Virginia, galbinic acid as an unidentified colorless substance, yellow pigment thought to be entothein, as *Parmelia finkii*; Hale (1958b), from the U.S.A., galbinic acid as an unidentified colorless substance, pigment may be entothein, as *P. finkii*.
Atranorin, galbinic acid. Hale (1954a), from Arkansas, an exsiccati specimen, galbinic acid as an unidentified substance, as *Parmelia finkii*; Hale (1957a), from Arkansas, galbinic acid as an unknown substance, as *P. finkii*.
Galbinic acid. Asahina (1963c), from Virginia.

Parmelia ochroglauca Hale   Atranorin, usnic acid; protolichesterinic acid. Hale (1965a), from Transvaal, type description.

"*Parmelia ochroleuca* var. *sorediosa* Müll. Arg.," see *Parmelia rutidota* f. *sorediosa* (Müll. Arg.) Müll. Arg.

Parmelia olivacea (L.) Ach.   Unidentified substance (PD+); parmelia-brown. Asahina (1951b); Asahina (1952e), unidentified substance K+ orange-red.

No usnic acid. Gertig and Banasiewicz (1961), from Poland.

OTHER REPORTS. Lecanoric acid: Hale (1957b), but see *Parmelia glabra*. — Protocetraric acid and/or fumarprotocetraric acid: Krog (1951b), from Norway. — Protocetraric acid: Dahl (1952), from Norway, mention only.

REVIEWS. Olivaceaic acid, olivacein: Hesse (1912); Thies (1932), and parmelia-brown; Zopf (1907).

Parmelia olivacea var. *albopunctata* (Asah.) Ahti   Fumarprotocetraric acid. Ahti (1966), from Japan.

Parmelia olivacea var. *glabra* (Schaer.) Linds., see *Parmelia glabra* (Schaer.) Nyl.

Parmelia olivetorina (Zopf) Sandst., see *Pseudevernia olivetorina* (Zopf) Zopf

Parmelia olivetorum Nyl., see *Cetrelia olivetorum* (Nyl.) W. Culb. & C. Culb.

Parmelia omphalodes (L.) Ach.   Atranorin, lobaric acid (sometimes as usnetic acid), salazinic acid (sometimes as parmatic acid, saxatic acid, or saxatilic acid); arabitol, mannitol; enzymes; beryllium, chromium, cobalt, copper, germanium, iron, lead, manganese, molybdenum, nickel, tin, titanium, zinc.

Atranorin, lobaric acid, salazinic acid. Asahina (1951h); Dahl (1952), from Norway, mention only; Krog (1951b), from Norway; Ramaut (1960), from Belgium, as *Parmelia omphalodes* var. *discordans*; Ramaut (1961a), mention only, as *P. omphalodes* var. *discordans*; Solberg (1956), mention only.

Atranorin, salazinic acid. Hale (1959b), from North Carolina, Virginia, and West Virginia.

Lobaric acid. Asahina (1938d), discussion of microchemical tests; Asahina (1952e); Asahina and Nonomura (1935), mention of earlier work.

Salazinic acid. Schindler (1936).

Arabitol, mannitol. Lindberg, Misiorny, and Wachtmeister (1953), by paper chromatography, no volemitol.

Enzymes. Moissejeva (1961), enzymes tabulated.

Beryllium, chromium, cobalt, copper, germanium, lead, manganese, molybdenum, nickel, tin, titanium, zinc. Jenkins and Davies (1966), from Wales.

Iron, manganese, zinc. Lounamaa (1965), a quantitative study.

REVIEWS. Atranorin, lobaric acid, salazinic acid: W. Brieger (1923); Hesse (1912); both reports as *Parmelia saxatilis* var. *omphalodes*; Thies (1932), and as *P. saxatilis* var. *omphalodes*; Zopf (1907). — Lobaric acid: Asahina and Shibata (1954); Shibata (1958, 1963), and as *Lobaria adusta*.

Parmelia ontakensis Asah.   Atranorin, salazinic acid. Asahina (1954*d*), from Japan, type description.

Parmelia ornatula Hale   Alectoronic acid, atranorin. Hale (1965*a*), from Angola, type description.

Parmelia osseoalbida Lynge
No usnic acid, no lichexanthone. Hale and Kurokawa (1964), mention in key.

Parmelia osteoleuca Nyl.   Lichexanthone, olivetoric acid, unidentified pigment (K+ purple). Hale and Kurokawa (1964), no usnic acid, K+ pigment in medulla, mentioned in key.

Parmelia owariensis Asah.   Atranorin, divaricatic acid. Asahina (1953*b*), from Japan, type description.
Divaricatic acid. Hale and Kurokawa (1964), mentioned in key.

Parmelia pachyspora Hale   Atranorin, protocetraric acid. Hale (1965*a*), from southern Africa, type description.

Parmelia palmarum Lynge   Atranorin, unidentified substance (KC+ rose). Hale (1960*a*), from Brazil.

Parmelia pancheri Hue   Alectoronic acid, atranorin. Hale (1965*a*), from New Caledonia and Thailand.

Parmelia panniformis (Nyl.) Vain.   Atranorin, salazinic acid; unidentified aliphatic substance like protolichesterinic acid. Asahina (1951*h*).
Atranorin ($\pm$). Dahl (1952), from Norway, mention only; Krog (1951*b*), from Norway.

Parmelia papillosa Lynge *ex* Gyeln.   Fumarprotocetraric acid, usnic acid. Hale (1964), mention only.

Parmelia papyrina Fée   Gyrophoric acid. Hale and Kurokawa (1964), from tropical America, mentioned in key.

Parmelia paraguariensis Lynge, see *Parmelia andina* Müll. Arg.

Parmelia paulensis Zahlbr.   Atranorin, unidentified substance (KC+) (±); protolichesterinic acid. Hale (1965*a*), from Brazil, Madagascar, and Réunion, unidentified substance suggested by KC+ faint red reaction of medulla of some specimens.

Parmelia peralbida Hale   Atranorin, protocetraric acid. Hale (1965*a*), from Haiti, Honduras, Jamaica, Mexico, and Panama, type description.

Parmelia perforata (Jacq.) Ach.   Atranorin, norstictic acid; protolichesterinic acid. Moore (1966), from Florida, by microchemical tests, morphologically indistinguishable from *Parmelia rigida*.
    Atranorin, norstictic acid. Hale (1954*a*), from Arkansas, an exsiccati specimen, as *Parmelia erecta*; Hale (1956*a*), as *P. erecta*; Hale (1957*a*), from Missouri and Oklahoma, as *P. erecta*; Hale (1957*d*), from the U.S.A., as *P. hypotropoides*, two samples from Texas with stictic acid instead of norstictic acid; Hale (1965*a*), from Ireland, Jamaica, Madagascar, Connecticut, Delaware, Florida, Louisiana, Massachusetts, New York, North Carolina, Rhode Island, South Carolina, Tennessee, Texas, Virginia, and Washington.
    Norstictic acid. Hale (1960*a*), from southeastern U.S.A.
    OTHER REPORTS. Atranorin and salazinic acid: Hale (1957*a*), from Arkansas, Missouri, and Oklahoma; Hale (1957*d*). — Salazinic acid: Hale (1956*a*), mention only; Hale (1957*b*), mention only. — Atranorin, physodic acid: Klosa (1951*b*).
    REVIEWS. Atranorin, salazinic acid: W. Brieger (1923); Hesse (1912); Thies (1932); Zopf (1907).

Parmelia perlata (Huds.) Ach.   Atranorin, stictic acid. Asahina (1940*e*), from Japan, and a K+ violet substance in the cilia; Asahina (1952*e*), from Japan, and a K+ bluish substance in the cilia; Dahl (1952), from Norway, mention only; Hale (1965*a*), from Africa, Asia, Australia, Europe, Central and South America, North America, and the West Indies; Krog (1951*b*), from Portugal; Ramaut (1961*b*), from Belgium, mention only; Ramaut and Schumaker (1961), from Belgium.
    Stictic acid. Hale (1965*d*), from Japan; Hale and Kurokawa (1962), from temperate regions; Kurokawa (1964), mention only.
    OTHER REPORTS. Atranorin, chloroatranorin, salazinic acid, two unidentified phenolic substances: Huneck and Follmann (1965*d*), from Chile, 2.6% salazinic acid and 0.44% atranorin-chloroatranorin mixture by extraction of 25 g. of lichen. — Norstictic acid: Dahl (1953). —Vulpinic acid: Mittal and Seshadri (1957), mention only.

All reports are as *Parmelia trichotera.*
See also *Cetrelia cetrarioides* for reports on material misidentified as *Parmelia perlata.*

Parmelia permutata Stirt.　Atranorin, gyrophoric acid; unidentified yellow pigment. Hale (1965*a*), from Africa, Australia, Haiti, India, Sumatra, and Thailand, pigment in lower medulla also found in *Parmelia araucariarum, P. myelochroa,* and others.

Parmelia perreticulata (Räs.) Hale　Lecanoric acid. Hale (1959*b*), from Italy and Texas; Hale (1965*c*), from Italy and Texas.

Parmelia pertusa (Schrank) Schaer., see *Menegazzia terebrata* (Hoffm.) Mass.

Parmelia peruviana Nyl.　Unidentified pale yellow pigment (K−). Hale and Kurokawa (1964), no usnic acid, yellow pigment in medulla, mentioned in key.

Parmelia phlyctina Hale　Atranorin, norstictic acid. Hale (1959*a*), from Cuba and Jamaica, improperly dried samples with a pink color in the medulla and cortex from decomposing norstictic acid, type description; Ramaut (1964–1965), from the Dominican Republic, by thin-layer chromatography.
　Norstictic acid. Hale and Kurokawa (1964), mentioned in key.

Parmelia physcioides Nyl.　Barbatic acid. Hale and Kurokawa (1964), from tropical America, no usnic acid, mentioned in key.

Parmelia physodes (L.) Ach., see *Hypogymnia physodes* (L.) Nyl.

Parmelia physodes f. *labrosa* (Ach.) B. Stein, see *Hypogymnia physodes* (L.) Nyl.

Parmelia piedmontensis Hale　Fumarprotocetraric acid, usnic acid. Hale (1964), from North Carolina and Tennessee, type description, previously included with *Parmelia subramigera* as *P. isidiata* Chem. strain 3 in Hale (1955*b*).

Parmelia pilosella Hue, see *Parmelia crinita* Ach.

Parmelia placorodia Ach., see *Parmeliopsis placorodia* (Ach.) Nyl.

Parmelia planatilobata Hale　Atranorin, gyrophoric acid; rhodophyscin. Hale (1965*b*), from Java and Malaya, type description.

Parmelia planiuscula Kurok. Atranorin, usnic acid, unidentified substance (PD +), unidentified substance (C +).

Atranorin, usnic acid, unidentified substance (PD +). Hale and Kurokawa (1964), from Java, type description.

Unidentified substance (C +). Kurokawa (1965), substance also in *Parmelia echinocarpa* and *P. schizospatha*, by crystals in the GE solution.

Parmelia plittii Gyeln. Norstictic acid (±), stictic acid, usnic acid. Hale (1955*b*), included with *Parmelia conspersa* as *P. isidiata* Chem. strain 1; Hale (1964), North American distribution described.

See also *Parmelia conspersa*.

Parmelia pluriformis Nyl. Atranorin, gyrophoric acid. Hale (1960*a*), from Brazil.

Gyrophoric acid. Hale and Kurokawa (1964), mentioned in key, no usnic acid.

Parmelia portoalegrensis Lynge, see *Parmelia congensis* B. Stein

Parmelia praesignis Nyl. Lecanoric acid, usnic acid. W. L. Culberson (1962*a*), from Mexico; W. L. Culberson and Culberson (1956), from the U.S.A., by microchemical tests, as *Parmelia bolliana* Chem. strain 2.

Parmelia praesorediosa Nyl. Atranorin; caperatic acid. Hale (1965*a*), from Africa, Central and South America, India, Java, Mexico, New Caledonia, Trinidad, the West Indies, Alabama, Florida, Louisiana, South Carolina, and Texas.

Atranorin, unidentified fatty acid. Abbayes (1961), from Madagascar, as *Parmelia sanctae-crucis*; Abbayes (1962), from Africa, as *P. sanctae-crucis*; Asahina (1941*c*), from Formosa, Japan, and Manchuria, fatty acid near caperatic acid, type description of *P. neglecta*; Asahina (1952*e*), from Formosa, Indochina, Japan, and Manchuria, as *P. neglecta*; Hale (1958*a*), from Brazil, fatty acid near caperatic acid, by microchemical tests on the type specimen of *P. capitata*; Hale (1959*b*), from Africa, Honduras, Japan, Nicaragua, the West Indies, Alabama, Florida, Georgia, and Louisiana, as *P. sanctae-crucis*; Hale (1960*a*), from tropical regions, fatty acid near caperatic acid, as *P. capitata* and as *P. sanctae-crucis*.

OTHER REPORT. Atranorin only: Ramaut (1964–1965), from Africa, by thin-layer chromatography.

Parmelia proboscidea Tayl., see *Parmelia crinita* Ach.

Parmelia procera J. Stein. & Zahlbr. Alectoronic acid, atranorin. Hale (1965*a*), from Africa, New Caledonia, and Thailand.

Parmelia prolixa (Ach.) Carroll
REVIEWS. Olivaceaic acid, olivacein: W. Brieger (1923); Thies (1932), and parmelia-brown.

Parmelia prolongata Kurok.   Atranorin, unidentified substance (C+); unidentified pale yellow pigment. Hale and Kurokawa (1964), from Colombia, Haiti, Mexico, and Panama, unidentified substance "apparently related to olivetoric acid or anziaic acid," type description.

Parmelia pruinata Müll. Arg.   Lecanoric acid.   Hale and Kurokawa (1964), mentioned in key.

Parmelia pseudoborreri Asah., see *Parmelia borreri* (Sm.) Turn.

Parmelia pseudocrinita Abb.   Atranorin, gyrophoric acid; rhodophyscin (±). Hale (1965a), from Africa.
   Atranorin, gyrophoric acid. Abbayes (1961), from Africa.

Parmelia pseudoformosana Asah.   Atranorin, salazinic acid; zeorin. Asahina (1951e), from Formosa, type description.

Parmelia pseudohyporysalea Asah., see *Parmelia sancti-angelii* Lynge

Parmelia pseudohypotrypa Asah., see *"Hypogymnia pseudohypotrypa"*

Parmelia pseudolaevior Asah.   Atranorin, salazinic acid. Asahina (1951f), from Japan, type description; Asahina (1952e), from Japan.

Parmelia pseudolivetorum Asah., see *Cetrelia pseudolivetorum* (Asah.) W. Culb. & C. Culb.

Parmelia pseudonilgherrensis Asah.   Alectoronic acid, atranorin, α-collatolic acid. Asahina (1954e), from India, Korea, and Manchuria, type description; Hale (1965a), from Kenya, Korea, and Nepal.

Parmelia pseudophysodes Asah., see *Hypogymnia pseudophysodes* (Asah.) Rass.

Parmelia pseudophysodes f. *reagens* Asah., see *"Hypogymnia pseudophysodes* f. *reagens"*

Parmelia pseudoreticulata Tav.   Atranorin, chloroatranorin, salazinic acid. Huneck and Follmann (1965d), from Chile, 0.5% atranorin and chloroatranorin and 2.2% salazinic acid, by extraction of 32 g. of lichen.

Parmelia pseudorutidota Asah.   Atranorin, divaricatic acid. Asahina (1952*a*), type description; Asahina (1954*e*), from Japan, type description of *Parmelia tanakae*.
   Divaricatic acid. Asahina (1952*e*); Asahina (1957*c*).

Parmelia pseudosaxatilis Asah.   Atranorin, salazinic acid. Asahina (1951*g*), from Japan and Sakhalin, type description; Asahina (1952*e*), from Japan.

Parmelia pseudoshinanoana Asah.   Atranorin, salazinic acid. Asahina (1951*f*), from Japan, type description; Asahina (1952*e*), from Japan.

Parmelia pseudosinuosa Asah.   Atranorin, protocetraric acid. Asahina (1951*f*), from Japan, type description; Asahina (1952*e*).
   Protocetraric acid. Hale and Kurokawa (1964), no usnic acid, no lichexanthone, mentioned in key.

Parmelia pseudotinctorum Abb.   Atranorin, lecanoric acid. Hale (1965*a*), from Africa.

Parmelia pubescens (L.) Vain.   No substances found. Krog (1951*b*), from Norway.

Parmelia pulla (Schreb.) Ach.
   Divaricatic acid or glomelliferic acid. Krog (1951*b*), from Norway.
   Glomelliferic acid. Dahl (1952), from Norway, an unnamed variety with imbricaric acid.

Parmelia pulverulenta (Schreb.) Ach., see *Physcia pulverulenta* (Schreb.) Hampe

Parmelia pulvinata Fée
   No alectoronic acid, no usnic acid. Hale and Kurokawa (1964), mentioned in key.

Parmelia pustulata Hale   Atranorin, salazinic acid. Hale and Kurokawa (1964), from Africa, type description; Ramaut (1964–1965), from Africa, by thin-layer chromatography.

Parmelia pustulescens Kurok.   Atranorin; caperatic acid. Hale and Kurokawa (1964), from Angola, type description.

Parmelia quercina (Willd.) Vain.   Atranorin, lecanoric acid; isolichenin, lichenin; ascorbic acid, riboflavin; calcium, iron, nitrogen.
   Atranorin, lecanoric acid. Asahina (1951*d*), from Japan and Manchuria; Asahina (1952*e*), from Japan; W. L. Culberson (1961*c*), from California, a detailed microchemical and taxonomic study.

Lecanoric acid. Hale (1956*e*), from North America; Hale and Kurokawa (1964), mentioned in key.

Isolichenin, lichenin, ascorbic acid, riboflavin, calcium, iron, nitrogen. Lal and Rao (1959).

OTHER REPORTS. Atranorin, lecanoric acid, lichexanthone: Aghoramurthy and Seshadri (1953*a*); Aghoramurthy and Seshadri (1953*b*), from India, by extraction, 0.5% lichexanthone, species name unknown in this paper. — Lichexanthone: Grover, Shah, and Shah (1956), synthesis of lichexanthone; Neelakantan and Seshadri (1961), mention in discussion of biosynthesis.

REVIEWS. Lichexanthone: Neelakantan (1965); Shibata (1958, 1963).

Parmelia rahengensis Vain. Barbatic acid, usnic acid. Hale and Kurokawa (1964), mentioned in key.

Parmelia ramosissima Kurok. Usnic acid, unidentified substance (PD+).
Unidentified substance (PD+). Hale (1965*b*), no protocetraric acid as previously reported.

OTHER REPORT. Usnic acid, protocetraric acid: Hale and Kurokawa (1964), from the Tanimbar Islands and the Moluccas, type description.

Parmelia rampoddensis Nyl. Alectoronic acid, atranorin, α-collatolic acid; rhodophyscin (±).
Alectoronic acid, atranorin, α-collatolic acid. Ramaut (1964–1965), from Florida, by thin-layer chromatography, trace α-collatolic acid.

Alectoronic acid, atranorin, rhodophyscin (±). Hale (1965*a*), from Africa, Australia, India, Mexico, South America, the West Indies, Alabama, Florida, Louisiana, North Carolina, Texas, and South Carolina.

Atranorin, alectoronic acid. Hale (1959*a*), from the Dominican Republic, Honduras, Jamaica, Alabama, Florida, Georgia, Louisiana, Mississippi, North Carolina, and South Carolina, type description of *Parmelia subinvoluta*.

Alectoronic acid. Hale (1965*d*), from Japan.

Parmelia ramuscula Hale Atranorin, salazinic acid. Hale (1965*a*), from the Philippines, type description.

Parmelia raunkiaeri Vain. Atranorin, protocetraric acid. Hale (1958*a*), by microchemical tests on the type specimen of *Parmelia scabrosa* Vain.
Protocetraric acid. Hale and Kurokawa (1964), mentioned in key.

Parmelia recipienda Nyl., see *Parmelia subcaperata* Kremp.

Parmelia reddenda Stirt. Atranorin; two unidentified fatty acids. W. L. Culberson (1962*a*), from the British Isles and France.

Atranorin. Targé and Lambinon (1965), from the British Isles, France, and Holland, by thin-layer chromatography.
Unidentified fatty acids. Hale (1965c), from France, Great Britain, eastern North America, South Africa, South America, and Scandinavia, may contain protolichesterinic acid.

Parmelia reducens Nyl.   Norstictic acid, usnic acid. Hale and Kurokawa (1964), mentioned in key.

Parmelia regis Lynge   Lichexanthone; unidentified substances. Hale (1960a), from Brazil, medulla K + wine-red and KC + pale red, differs from *Parmelia silvatica* in lacking a medullary anthraquinone.

Parmelia regnellii Lynge   Barbatic acid, usnic acid; unidentified anthraquinone; unidentified substances. Hale (1960a), from Brazil, one sample PD + and containing stictic acid instead of barbatic acid, and also *Parmelia regnellii* f. *arida*.

Parmelia relicinella Nyl.   Fumarprotocetraric acid. Hale and Kurokawa (1964), mentioned in key.

Parmelia reparata Stirt.   Atranorin, salazinic acid. Hale (1965a), from Australia, Mexico, New Zealand, and Alabama.

Parmelia reticulata Tayl.   Atranorin, salazinic acid. Abbayes (1961), from Madagascar and Réunion; Abbayes (1962), from Africa; Asahina (1940e), from Formosa and Japan, and *Parmelia reticulata* f. *nuda* from Japan; Asahina (1951g), from China, Formosa, and Japan; Asahina (1952e), from China, Formosa, and Japan; Hale (1956b); Hale (1956g), from Oklahoma, an exsiccati specimen; Hale (1957a), from Arkansas, Missouri, and Oklahoma; Hale and Kurokawa (1962), from Africa, Asia, and North America; Targé (1964–1965), from Africa, by paper chromatography.
Salazinic acid. Asahina (1937e), by a microcrystal test, as *Parmelia cetrata* var. *sorediifera*; Asahina and Fuzikawa (1935c), by a microcrystal test, as *P. cetrata* var. *sorediifera*; Hale (1959a); Kurokawa (1964), mention only.
OTHER REPORTS. Atranorin, cetrataic acid, salazinic acid: Asahina and Asano (1933), from Japan, by extraction, as *Parmelia cetrata* var. *sorediifera*. — Cetrataic acid, salazinic acid: Asahina (1926), as *P. cetrata* var. *sorediifera*. — Norstictic acid: Asahina (1936d), as *P. cetrata* var. *sorediifera*.

Parmelia revoluta Flörke   Atranorin, gyrophoric acid. Abbayes (1961), from Madagascar; Asahina (1953e), from Japan; Dahl (1952), mention only; Krog (1951b), from Germany.

Gyrophoric acid. Hale and Kurokawa (1964), mentioned in key.

OTHER REPORTS. Atranorin, gyrophoric acid, lecanoric acid, orcinol, unidentified substance: Ramaut (1960), from Belgium by paper chromatography and extraction. — Atranorin, gyrophoric acid, lecanoric acid: Ramaut (1961a), mention only. — Lecanoric acid: Hale (1956e), from North America.

REVIEWS. Atranorin, gyrophoric acid: Zopf (1907). — Gyrophoric acid: W. Brieger (1923); Hesse (1912). — Parmelia-brown: Thies (1932).

Parmelia revolutella Nyl.   Gyrophoric acid. Hale and Kurokawa (1964), no usnic acid, mentioned in key.

Parmelia rhabdiformis Kurok.   Atranorin, norstictic acid. Hale and Kurokawa (1964), from Panama, no usnic acid, type description.

Parmelia rigida Lynge   Alectoronic acid, atranorin, norstictic acid (±). Moore (1966), from Florida, norstictic acid commonly present, by microchemical tests, morphologically indistinguishable from *Parmelia perforata*.

Alectoronic acid, atranorin. Hale (1958a), from Brazil, by microchemical tests on the type specimen, probably also traces of α-collatolic acid; Hale (1960a), from Brazil; Hale (1965a), from Argentina, Brazil, Uruguay, Alabama, Florida, Georgia, Mississippi, North Carolina, South Carolina, and Texas.

Parmelia rigidula Kurok.   Atranorin; unidentified pale yellow pigment. Hale and Kurokawa (1964), from northern India, type description, pale yellow pigment in medulla.

Parmelia rimulosa Dodge   Alectoronic acid, atranorin; rhodophyscin (±). Hale (1965a), from South Africa.

Parmelia riograndensis Lynge   Atranorin, gyrophoric acid. Hale (1960a), from Brazil.

Parmelia robusta Degel., see *Parmelia dilatata* Vain.

Parmelia rockii Zahlbr.   Atranorin, evernic acid, lecanoric acid.

Atranorin, lecanoric acid. Hale (1957b), mention as a chemical strain of *Parmelia laevigata*.

Evernic acid, lecanoric acid. Hale and Kurokawa (1964), medulla KC + red.

Lecanoric acid. Hale (1956e), from North America, as a chemical strain of *Parmelia laevigata*.

Parmelia rodriguesiana Hue   Divaricatic acid. Hale and Kurokawa (1964), mentioned in key.

Parmelia rubifaciens Hale   Atranorin, norstictic acid. Hale (1965*a*), from Brazil, Guatemala, Mexico, and Nicaragua, type description.

Parmelia rudecta Ach.   Atranorin, lecanoric acid. Abbayes (1961), from Madagascar; Asahina (1951*d*), from Japan and North America, small amount of atranorin, as *Parmelia ruderata*; Asahina (1952*e*), from Japan, as *P. ruderata*; W. L. Culberson (1957); W. L. Culberson (1962*a*), from Argentina, eastern Asia, and eastern North America; W. L. Culberson and Culberson (1956), from eastern U.S.A.; Hale (1954*a*), from Arkansas, an exsiccati specimen; Hale (1956*a*); Hale (1957*a*), from Arkansas, Missouri, and Oklahoma; Hale (1960*a*), two samples from South America with gyrophoric acid instead of lecanoric acid.

Lecanoric acid. Asahina (1926), as *Parmelia rudecta* var. *microphyllina*; Hale (1956*e*), from North America; Hale (1965*c*); Hale and Kurokawa (1964), mention only.

Parmelia ruderata Asah., see *Parmelia rudecta* Ach.

Parmelia rupicola Lynge   Atranorin, divaricatic acid. Hale (1960*a*), from Brazil.

Divaricatic acid. Hale and Kurokawa (1964), mentioned in key.

Parmelia rupta Lynge   Atranorin, salazinic acid. Hale (1960*a*), from Paraguay.

Parmelia rutidota Hook. f. & Tayl.   Protocetraric acid or fumarprotocetraric acid, usnic acid; unidentified pigment (K + purple-black).

Protocetraric acid or fumarprotocetraric acid, usnic acid. Hale (1960*a*), from Australia and South America.

Usnic acid, unidentified pigment (K + purple-black). Hale and Kurokawa (1964), mentioned in key, K + pigment in the medulla.

OTHER REPORTS. Atranorin, unidentified substance: Asahina (1952*a*), from Japan (but see also comments in Hale, 1960*a*). — Unidentified substance: Asahina (1957*c*).

Parmelia rutidota f. *filizans* Lynge   Fumarprotocetraric acid, usnic acid. Hale (1960*a*).

Parmelia rutidota f. *sorediosa* (Müll. Arg.) Müll. Arg.   Atranorin, protocetraric acid, usnic acid. Ramaut (1964–1965), from Katanga, by thin-layer chromatography, as *Parmelia ochroleuca* var. *sorediosa*.

Parmelia ryssolea (Ach.) Nyl.   Enzymes. Moissejeva (1961), enzymes tabulated.

Parmelia saccatiloba Tayl.   Atranorin, protocetraric acid. Hale (1958a), from Pitcairn Island, by microchemical tests on the type specimen; Hale (1960a), from Pitcairn Island; Hale (1965a), many South Pacific localities listed.

Parmelia salacinifera Hale   Atranorin, salazinic acid. Hale and Kurokawa (1964), from Brazil, Mexico, Florida, and Georgia, type description.

Parmelia samoensis Zahlbr.   Fumarprotocetraric acid. Hale and Kurokawa (1964).

Parmelia sanctae-crucis Vain., see *Parmelia praesorediosa* Nyl.

Parmelia sancti-angelii Lynge   Atranorin, gyrophoric acid; rhodophyscin (±). Hale (1965a), from Africa, Central and South America, China, India, Java, Mexico, Sikkim, Sumatra, Thailand, and the West Indies.
    Atranorin, gyrophoric acid. Abbayes (1961), from Madagascar, and an acetone-soluble yellow pigment; Asahina (1955a), from India, type description of *Parmelia pseudohyporysalea*; Dhar, Neelakantan, Ramanujam, and Seshadri (1959), from India, by extraction of 40 g. of lichen, as *P. pseudohyporysalea*; Hale (1960a), from Africa, China, Mexico, and the West Indies.

Parmelia saxatilis (L.) Ach.   Atranorin, lobaric acid (±) (sometimes as stereocaulic acid or usnetic acid), salazinic acid (sometimes as parmatic acid, saxatic acid, or saxatilic acid); unidentified fatty acid; arabitol, fructose, galactose, glucose, sucrose, volemitol; iron, manganese, zinc.
    Atranorin, lobaric acid (±), salazinic acid, unidentified fatty acid (±). Asahina (1951g), by microchemical tests on many samples, atranorin and salazinic acid (in specimens from Japan, Korea, Manchuria, and Sakhalin), atranorin and salazinic acid with lobaric acid (in specimens from Japan, atranorin, salazinic acid) and sometimes also with an unidentified fatty acid (in specimens from Japan, Manchuria, and Sakhalin); Asahina (1951h), unidentified fatty acid possibly identical to protolichesterinic acid, by microchemical tests on many samples; Asahina (1952e).
    Atranorin, salazinic acid. Asahina (1937e); Asahina (1938e), description of microcrystal tests; Asahina and Tanase (1934c), by extraction; Dahl (1952), from Norway, mention only; Hale (1957a), from Arkansas and Missouri; D. Hess (1958), from Germany, by paper chromatography; Krog (1951b), from Norway, by microcrystal tests; Ramaut (1960), from Belgium; Ramaut (1961a); Solberg (1956), mention only; Vartia (1950a).
    Atranorin. Hale (1957b), mention only; Vartia (1949), from Finland.
    Salazinic acid. Koller and Klein (1934a); Koller and Klein (1934b); Koller, Klein, and Pöpl (1933); Krog (1951a), mention only; Kurokawa (1964), mention only; J. Santesson (1965); Schindler (1936); Schindler

(1956/1957), from Tirol, mention only; Steiner (1955); Vartia (1950*b*), from Finland; Wachtmeister (1956*a*), used as an authentic source of this substance.

Arabitol, volemitol. Pueyo (1963*b*), from France, by paper chromatography.

Fructose, galactose, glucose, sucrose. Pueyo (1963*a*), from France, by paper chromatography.

No usnic acid. Gertig and Banasiewicz (1961), from Poland.

Iron, manganese, zinc. Lounamaa (1965), a quantitative study.

REVIEWS. Atranorin, lobaric acid, salazinic acid: W. Brieger (1923), and as *Parmelia saxatilis* var. *retiruga*; Hesse (1912), and as *P. saxatilis* var. *retiruga*; Thies (1932), and parmelia-brown, as *P. saxatilis* var. *retiruga*; Zopf (1907), and as *P. saxatilis* var. *retiruga*. — Atranorin, salazinic acid: Asahina and Shibata (1954); Shibata (1958, 1963). For a detailed review of early work on this species see Ryan and Riordan (1917).

Parmelia saxatilis var. *omphalodes* (L.) Fr., see *Parmelia omphalodes* (L.) Ach.

Parmelia saxatilis f. **panniformis** Cromb.
REVIEWS. Atranorin, lobaric acid (as usnetinic acid), salazinic acid (as parmatic acid and saxatilic acid): W. Brieger (1923); Hesse (1912). — Atranorin, lobaric acid, protocetraric acid: Zopf (1907). — Lobaric acid (as usnetinic acid): Thies (1932).

Parmelia saxatilis var. *sulcata* (Tayl.) Linds., see *Parmelia sulcata* Tayl.

Parmelia saximontana R. Anderson & W. Web. Gyrophoric acid. Anderson and Weber (1962), from Arizona, Colorado, Montana, New Mexico, South Dakota, and Utah, gyrophoric acid in medulla and upper cortex, type description.

Parmelia sbarbaronis B. de Lesd. Stictic acid. Hale (1959*b*), from Central America, Costa Rica, Italy, Honduras, and the West Indies; Hale (1959*c*).

Parmelia scabrosa Tayl. Norlobaridone (sometimes as loxodic acid), salazinic acid, (+)-usnic acid; D-mannitol. Briner, Gream, and Riggs (1960), from Australia, 2.2% norlobaridone, 0.64% salazinic acid, 0.50% (+)-usnic acid, by extraction, as *Parmelia conspersa*.

Norlobaridone, usnic acid. Asahina (1952*e*), from Japan, by microchemical tests, lichen apparently misidentified as *Parmelia loxodes*; Asahina (1959*c*), apparently misidentified as *P. loxodes*.

Norlobaridone. Gream and Riggs (1960), from Australia, as *Parmelia conspersa*, chemical structure determined for norlobaridone; Hale (1964), mention only.

OTHER REPORT. Alectoronic acid, usnic acid: Asahina (1951*b*), from Japan, as *Parmelia loxodes* (apparently misidentified substance, see Asahina, 1952*e*).

REVIEWS. Norlobaridone, salazinic acid: Shibata (1963), as *Parmelia conspersa*. — Norlobaridone: Neelakantan (1965), as *P. conspersa*.

Parmelia scabrosa Vain., see *Parmelia raunkiaeri* Vain.

Parmelia schimperi Müll. Arg.   Atranorin, cryptochlorophaeic acid; protolichesterinic acid. Hale (1965*a*), from Ethiopia.

Parmelia schizospatha Kurok.   Usnic acid, unidentified substance (PD+), unidentified substance (C+ yellow). Kurokawa (1965), from Taiwan, PD+ substance named "echinocarpic acid" also found in *Parmelia echinocarpa*.
Usnic acid, unidentified substance (PD+). Hale and Kurokawa (1964), from Java, type description.

Parmelia schmitzii Targé   Atranorin, lecanoric acid, norstictic acid, perlatolic acid, salazinic acid. Targé (1964–1965), from Katanga, by thin-layer chromatography, type description.

Parmelia scortea (Ach.) Ach., see *Parmelia tiliacea* (Hoffm.) Ach.

Parmelia scortella Nyl.   Atranorin, gyrophoric acid. Abbayes (1961), from Madagascar and Réunion; Asahina (1951*d*), from Japan; Asahina (1952*e*), from Japan.
Gyrophoric acid. Hale and Kurokawa (1964), mentioned in key.

Parmelia scrobicularis Kremp.   Atranorin, stictic acid. Hale (1960*a*), from Brazil and Paraguay.
Stictic acid. Hale and Kurokawa (1964), no usnic acid, mentioned in key.

Parmelia scytodes Kurok.   Atranorin, barbatic acid, unidentified substance (PD+). Hale and Kurokawa (1964), from India, medulla also KC+ orange, type description.

Parmelia scytophylla Kurok.   Atranorin, gyrophoric acid. Hale and Kurokawa (1964), from India, type description.

Parmelia segreganda Abb.   Atranorin, lecanoric acid. Abbayes (1961), from Madagascar, type description.

Parmelia sensibilis J. Stein. & Zahlbr.   Salazinic acid. Hale and Kurokawa (1964), from Africa, mentioned in key.

Parmelia separata Th. Fr.   Alectoronic acid, usnic acid. Hale (1957*b*), mention only.
   Alectoronic acid. Hale (1956*f*), mention only.
   Usnic acid. Hale (1954*b*), from Baffin Island.

Parmelia septentrionalis (Lynge) Ahti
   Probably fumarprotocetraric acid. Ahti (1966).

Parmelia setchellii Vain.   Atranorin, protocetraric acid. Hale (1965*a*), from Tahiti.

Parmelia seto-maritima Asah.   Norstictic acid, stictic acid, usnic acid. Asahina (1961*c*), from Japan, type description.

Parmelia setschwanensis Zahlbr.   Salazinic acid. Hale and Kurokawa (1964), mentioned in key.

Parmelia shinanoana Zahlbr.   Atranorin, salazinic acid. Asahina (1951*f*), from Japan; Asahina (1952*e*), from Japan and Korea.

Parmelia silvatica Lynge   Lichexanthone protocetraric acid, unidentified pigment (K + purple). Hale (1960*a*), from Brazil, K + pigment in medulla.
   Lichexanthone, unidentified pigment (K + purple). Hale and Kurokawa (1964), mentioned in key, K + pigment in the medulla.

Parmelia simodensis Asah., see *Parmelia grayana* Hue

Parmelia sinuosa (Sm.) Ach.   Salazinic acid (sometimes as usnaric acid), (+)-usnic acid.
   Salazinic acid, usnic acid. Abbayes (1961), from Réunion; Asahina (1951*b*), from Japan; Asahina (1952*e*), from Japan; Hale and Kurokawa (1962), from Europe; Hale and Kurokawa (1964), mentioned in key.
   Salazinic acid. Duvigneaud (1945), usnaric acid found to be identical to salazinic acid.
   OTHER REPORT. (Chloro)atranorin, salazinic acid: D. Hess (1958), from Germany.
   REVIEWS. Salazinic acid, (+)-usnic acid: W. Brieger (1923); Hesse (1912); Thies (1932); Zopf (1907).

Parmelia somaliensis Müll. Arg.   Protocetraric acid. Hale and Kurokawa (1964), from Africa, no usnic acid, mentioned in key.

Parmelia soredians Nyl.   Salazinic acid, usnic acid. Asahina (1951*h*); Hale and Kurokawa (1964), mentioned in key.

Parmelia sorediata (Ach.) Röhl., see *Parmelia sorediosa* Almb.

Parmelia soredica Nyl., see *Parmelia flaventior* Stirt.

Parmelia sorediosa Almb.
   Unidentified substance. Dahl (1952), from Norway, mention only; Krog (1951*b*), from Norway, description of microchemical test.
   Diffusin. Zopf (1900), compared with diffusin from *Platysma diffusum* (see *Parmeliopsis hyperopta*).
   REVIEWS. Diffusin, diffusinic acid: Zopf (1907). — Diffusinic acid, lecanoric acid: W. Brieger (1923); Hesse (1912); Thies (1932).
   All reports are as *Parmelia sorediata*.

Parmelia sorocheila Vain.   Atranorin, salazinic acid; unidentified fatty acid. Abbayes (1961), from Madagascar; Abbayes (1962), from Africa.

Parmelia soyauxii Müll. Arg.   Atranorin, lecanoric acid. Hale (1965*a*), from Angola, Madagascar, and Southern Rhodesia.

Parmelia spathulata Kurok.   Atranorin, gyrophoric acid. Hale and Kurokawa (1964), from South Africa, type description.

Parmelia spectabilis Asah.   Atranorin, protocetraric acid. Asahina (1951*e*), from Formosa and Japan, type description; Asahina (1952*e*), from Formosa and Japan.

Parmelia sphaerospora Nyl., see *Parmelia congruens* Ach.

Parmelia spumosa Asah.   Atranorin, gyrophoric acid. Abbayes (1961); Asahina (1951*d*), from Japan, type description; Asahina (1952*e*); Asahina (1953*e*).
   Gyrophoric acid. Hale and Kurokawa (1964), mention only.

Parmelia stenophylla (Ach.) Heug. Chem. strain 1, see *Parmelia taractica* Kremp. (pale below) and *Parmelia tasmanica* Tayl. (black below)

Parmelia stenophylla (Ach.) Heug. Chem. strain 2   Atranorin ($\pm$), norstictic acid ($\pm$), stictic acid, usnic acid. Hale (1955*b*), from the U.S.A.; Hale (1957*b*), mention only.
   For additional chemical data difficult to summarize see Hale (1956*d*).

Parmelia stictica (Duby) Nyl.   Atranorin, gyrophoric acid. Targé and Lambinon (1965), from France, by thin-layer chromatography.

Gyrophoric acid. Hale (1965c), from Africa, France, and South America.

Parmelia stuppea Tayl. Atranorin, salazinic acid. Abbayes (1963), from Vietnam, as *Parmelia maxima*; Hale (1965a), from Africa, Germany, Guatemala, India, Mexico, Portugal, California, Connecticut, Delaware, Maryland, Massachusetts, North Carolina, Pennsylvania, Tennessee, Vermont, Virginia, West Virginia, and Wisconsin; Ramaut (1961a), from Portugal, as *P. claudelii*; Ramaut (1961b), from Portugal, as *P. claudelii*.

Salazinic acid. Hale (1959a), from Europe, Mexico, and North America, as *Parmelia maxima*; Hale (1960a), as *P. maxima*; Hale (1960b), as *P. maxima*.

Parmelia stygia (L.) Ach. Atranorin ($\pm$), fumarprotocetraric acid and/or protocetraric acid; iron, manganese, zinc.

Atranorin ($\pm$), fumarprotocetraric acid and/or protocetraric acid. Krog (1951b), from Norway, by microcrystal and color tests; Solberg (1956), mention only.

Atranorin ($\pm$). Dahl (1952), from Norway, mention only.

Iron, manganese, zinc. Lounamaa (1965), a quantitative study.

REVIEW. Parmelia-brown: Thies (1932).

Parmelia subabstrusa Gyeln. Norstictic acid, usnic acid. Hale (1960a), from Brazil, as *Parmelia abstrusa* f. *laevigata*; Kurokawa (1965), from Taiwan.

Norstictic acid. Hale and Kurokawa (1964), mentioned in key.

Parmelia subaffinis Zahlbr. Lichexanthone, unidentified substance (PD +). Hale and Kurokawa (1964), mentioned in key.

Lichexanthone. Hale (1958c), as *Parmelia affinis*.

Parmelia subargentifera Nyl. Lecanoric acid. Dahl (1952), from Norway, mention only; Krog (1951b), from Norway; Dhar, Neelakantan, Ramanujam, and Seshadri (1959), from India, by extraction of 20 g. of lichen, no other products isolated.

Parmelia subarnoldii Abb. Atranorin, protocetraric acid; unidentified fatty acid. Abbayes (1961), from Madagascar, type description, fatty acid near protolichesterinic acid; Abbayes (1963), from Vietnam.

Atranorin, protocetraric acid. Hale (1965a), from Brazil, Madagascar, Mexico, and New Guinea.

Parmelia subaurifera Nyl. Lecanoric acid; subauriferin (sometimes reported as an unidentified yellow medullary pigment).

Lecanoric acid. Dahl (1952), from Norway, mention only; Hale (1956e), from North America; Krog (1951b), from Norway.

REVIEWS. Lecanoric acid, subauriferin: W. Brieger (1923); Hesse (1912); Thies (1932); Zopf (1907).

Parmelia subaurulenta Nyl. Atranorin; leucotylin, zeorin; entothein; unidentified acid ($C_{18}H_{30}O_3$, m.p. 182°). Asahina (1951c), from Japan. Atranorin, zeorin, entothein. Asahina (1952e), from Japan. Atranorin, zeorin. Kurokawa (1959a).

Parmelia subbalansae Gyeln. Atranorin. Hale (1960a), from Argentina, Brazil, and Uruguay.

Parmelia subcaperata Kremp. Atranorin, cryptochlorphaeic acid; protolichesterinic acid. Hale (1965a), from Argentina, Brazil, and Paraguay.
Atranorin, unidentified substance (KC+), unidentified fatty acid. Hale (1959b), from Brazil, chemically the same as *Parmelia haitiensis*, as *P. recipienda*, (unidentified KC+ substance identical to cryptochlorophaeic acid, unidentified fatty acid identical to protolichesterinic acid).
Atranorin, two colorless unidentified substances. Hale (1960a), from Brazil, medulla KC+ red, microcrystal tests described, as *Parmelia recipienda*.
OTHER REPORT. Atranorin, salazinic acid: Hale (1960a), from Brazil and Paraguay, an incorrect determination (M. E. Hale, Jr., private communication).

Parmelia subcolorata Hale Atranorin, unidentified pigment (K−). Hale (1965a), from Congo and Kenya, type description.

Parmelia subconspersa Nyl., see *Parmelia subramigera* Gyeln.

Parmelia subconspersa var. *hirosakiensis* (Gyeln.) Asah., see *Parmelia subramigera* Gyeln.

Parmelia subcorallina Hale Atranorin, protocetraric acid; protolichesterinic acid. Hale (1962a), from Formosa and Mauritius, type description; Hale (1965a), from Java, Mauritius, and São Tomé.

Parmelia subcoronata Müll. Arg. Norstictic acid. Hale and Kurokawa (1964), mentioned in key.

Parmelia subcrinita Nyl. Atranorin, salazinic acid, unidentified substance. Asahina (1961d), from Japan, microcrystal test in the GE solution described for the unidentified substance.

Atranorin, salazinic acid. Asahina (1952*e*), from Japan; Hale (1957*a*), from Arkansas, Missouri, and Oklahoma; Hale (1965*a*), from the Azores, Central and South America, Japan, Java, Mexico, New Caledonia, the West Indies, Alabama, Arkansas, Florida, Georgia, Mississippi, North Carolina, Oklahoma, South Carolina, Tennessee, Virginia, West Virginia, and Wisconsin.

Salazinic acid. Asahina (1952*b*); Hale (1965*d*), from Japan.

Parmelia subdissecta Nyl. Gyrophoric acid. Hale and Kurokawa (1964), from Asia, mentioned in key.

Parmelia subdivaricata Asah. Atranorin, salazinic acid. Asahina (1951*g*), from Formosa and Japan, type description; Asahina (1952*e*), from Formosa and Japan.

Parmelia suberadicata Abb. Stictic acid, usnic acid. Abbayes (1961), from Madagascar, type description.

Parmelia subfatiscens Kurok. Unidentified substance (KC+). Hale and Kurokawa (1964), from Jamaica and South Africa, type description.

Parmelia subflava Tayl. Atranorin, lecanoric acid. Hale (1960*a*).

Parmelia subglandulifera Hue Salazinic acid. Hale and Kurokawa (1964), mentioned in key.

Parmelia subinflata Hale Atranorin, protocetraric acid. Hale (1965*b*), from Malaya, Philippines, and Sabah, type description.

Parmelia subinvoluta Hale, see *Parmelia rampoddensis* Nyl.

Parmelia subisidiosa (Müll. Arg.) Dodge Atranorin, salazinic acid. Abbayes (1961), from Madagascar; Asahina (1940*e*), from Formosa, as *Parmelia cetrata* f. *granularis*.

Parmelia sublaevigata (Nyl.) Nyl. Atranorin, salazinic acid. Abbayes (1961), from Madagascar; Targé (1964–1965), from Africa, by paper chromatography.

Salazinic acid. Hale and Kurokawa (1964), no usnic acid.

OTHER REPORTS. Atranorin, zeorin, an unidentified substance (PD+): Asahina (1951*e*), from Japan, and as *Parmelia sublaevigata* f. *rugosa* and *P. sublaevigata* f. *subradiata*; Asahina (1952*e*), from Japan, (see *P. galbina*). — Atranorin, lecanoric acid: Aghoramurthy, Neelakantan, and Seshadri (1954), from India, by extraction.

Parmelia sublaevigata f. *rugosa* (Hue) Asah., see *Parmelia galbina* Ach.

Parmelia sublaevigata f. *subradiata* Asah., see *Parmelia galbina* Ach.

Parmelia sublanea Kurok. Protocetraric acid, usnic acid. Hale and Kurokawa (1964), from the Moluccas, type description.

Parmelia sublimbata Nyl. Unidentified substance (PD +), usnic acid. Hale and Kurokawa (1964), the unknown PD + substance also found in *Parmelia schizospatha*.
   Unidentified substance (PD +). Kurokawa (1965), not the same substance found in *Parmelia echinocarpa*.

Parmelia submarmariza Asah. Atranorin, salazinic acid. Asahina (1953a), from Japan, type description.

Parmelia submundata Oksn., see *Hypogymnia submundata* (Oksn.) Rass.

Parmelia submundata f. *colorans* Asah., see *Hypogymnia submundata* f. *colorans* (Asah.) Rass.

Parmelia subobscura Vain., see *Hypogymnia subobscura* (Vain.) Poelt

Parmelia subpraesignis Nyl. Atranorin, gyrophoric acid. W. L. Culberson (1962a), from Argentina, Mexico, and Texas; W. L. Culberson and Culberson (1956), from Mexico and Texas, as *Parmelia bolliana* Chem. strain 3.
   Gyrophoric acid. Hale (1956e), from North America, as a chemical strain of *Parmelia bolliana*.

Parmelia subproboscidea Lynge, see *Parmelia argentina* Kremp.

Parmelia subquercifolia Hue, see *Parmelia galbina* Ach.

Parmelia subramigera Gyeln. Fumarprotocetraric acid, usnic acid. Asahina (1949a), as *Parmelia subconspersa* (apparently misidentified according to a private communication from M. E. Hale, Jr.), Asahina (1951b), as *P. subconspersa* var. *hirosakiensis*; Asahina (1952e), from Japan, as *P. subconspersa*; Asahina (1959c), as *P. subconspersa*; Hale (1955b), as *P. isidiata* Chem. strain 3, except a few samples having black undersides (=*P. piedmontensis*); Hale (1957b), mention only, as *P. isidiata* Chem. strain 3; Hale (1964), from the U.S.A., taxonomy clarified in this paper.
   OTHER REPORT. Salazinic acid: Asahina (1951b), as *Parmelia conspersa* var. *hypoclysta* f. *isidiosa* (but see *P. mexicana*).

Parmelia subrudecta Nyl.   Atranorin, lecanoric acid. Asahina (1951*d*), from Europe, by microchemical tests, as *Parmelia dubia*; W. L. Culberson (1960*a*), mention only, as *P. borreri*; W. L. Culberson (1960*b*), mention only, as *P. dubia* and as *P. ulophylla*; W. L. Culberson (1962*a*), eastern Asia, Europe, and North America, as *P. borreri*; W. L. Culberson and Culberson (1956), from the U.S.A., as *P. dubia* Chem. strain 1, by microcrystal tests; Dahl (1952), from Norway, mention only, as *P. dubia*; Krog (1951*b*), from Ireland, as *P. dubia*; Ramaut (1961*a*), mention only, as *P. dubia*; Targé and Lambinon (1965), from Europe, by thin-layer chromatography, as *P. borreri* "chemovar." *borreri*; Zopf (1900), from Germany, by extraction, as *P. borreri*.

Lecanoric acid. Asahina (1936*d*), as *Parmelia borreri*, description of microcrystal tests; Hale (1956*e*), from North America, as *P. dubia*; Hale (1959*a*), from England, as *P. borreri*; Hale (1965*c*), from Australia, Europe, North America (map), South Africa, and southern Scandinavia; Hale and Kurokawa (1962), from England and North America, as *P. borreri*; Hale and Kurokawa (1964), mention only.

OTHER REPORT. Atranorin, lecanoric acid, orcinol, unidentified substance (R_f similar to orcellinic acid): Ramaut (1960), as *Parmelia dubia* (additional substances reported possibly artifacts of the chromatographic method).

REVIEWS. Atranorin, lecanoric acid: W. Brieger (1923); Hesse (1912); Thies (1932); Zopf (1907). — Lecanoric acid. Asahina and Shibata (1954); Shibata (1958, 1963). All review reports are as *Parmelia borreri*.

Parmelia subrugata Kremp.   Alectoronic acid, atranorin, α-collatolic acid; rhodophyscin (±).

Alectoronic acid, atranorin, α-collatolic acid. Abbayes (1962), from Africa; Abbayes (1963), from Vietnam; Targé (1964–1965), mention only.

Alectoronic acid, atranorin, rhodophyscin (±). Hale (1965*a*), from Africa, Central and South America, Australia, China, Haiti, and Mexico.

Alectoronic acid, atranorin. Hale (1960*a*), from Brazil.

OTHER REPORT. Alectoronic acid, atranorin, α-collatolic acid, divaricatic acid (±), fatty acid (±): Abbayes (1961).

Parmelia subsaxatilis B. de Lesd.   Salazinic acid. Hale and Kurokawa (1964), no usnic acid, mentioned in key.

Parmelia subscortea Asah.   Atranorin, salazinic acid. Asahina (1957*b*), from Formosa and India, type description.

Parmelia substygia Räs.   Unidentified substance. Krog (1951*b*), from Norway, unknown substance also found in *Parmelia sorediata*; Dahl (1952), from Norway, mention only, as *P. disjuncta*.

Parmelia subsulphurata Asah.   Atranorin; zeorin; entothein. Asahina (1951c), from Japan, type description; Asahina (1952e), from Japan.

Parmelia subsumpta Nyl.   Atranorin, salazinic acid (as Chem. strain 1, from Africa, Central and South America, Mexico, and Georgia). Atranorin, cryptochlorophaeic acid; protolichesterinic acid (as Chem. strain 2, from Brazil, Peru, and Nebraska). Hale (1965a).

Parmelia subtinctoria Zahlbr.   Atranorin, salazinic acid; protolichesterinic acid. Hale (1965a), from Africa, Asia, Central America, India, Mexico, Paraguay, the West Indies, Alabama, Arizona, Arkansas, Florida, Georgia, Illinois, Kansas, Kentucky, Mississippi, North Carolina, Oklahoma, Tennessee, Texas, Virginia, and West Virginia, as Chem. strain 2, see also *Parmelia haitiensis*.
    Atranorin, salazinic acid. Ramaut (1964–1965), by thin-layer chromatography.

Parmelia subturgida Kurok.   Norstictic acid, stictic acid, usnic acid. Kurokawa (1965), from Japan and Taiwan.

Parmelia suffixa Stirt.   Gyrophoric acid. Hale and Kurokawa (1964), mentioned in key.

Parmelia sulcata Tayl.   Atranorin, salazinic acid (sometimes as parmatic acid); enzymes.
    Atranorin, salazinic acid. Dahl (1952), from Norway, mention only; Hale (1956b); Hale (1956g), from Connecticut, an exsiccati specimen; Hale and Kurokawa (1962), from Asia, Europe, Ireland, and North America; Krog (1951b), from Norway; Ramaut (1960), from Belgium; Ramaut (1961a); Solberg (1956), mention only.
    Atranorin. Vartia (1949), from Finland.
    Enzymes. Moissejeva (1961), enzymes tabulated.
    No usnic acid. Gertig and Banasiewicz (1961), from Poland.
    OTHER REPORT. Atranorin, lecanoric acid, salazinic acid: Dhar, Neelakantan, Ramanujam, and Seshadri (1959), from India, by extraction of 50 g. of lichen.
    REVIEWS. Atranorin, lobaric acid (as usnetinic acid), salazinic acid: W. Brieger (1923); Hesse (1912); both reports as *Parmelia saxatilis* var. *sulcata*. — Lobaric acid (as usnetinic acid): Thies (1932), as *P. saxatilis* var. *sulcata*. — Atranorin, protocetraric acid: Zopf (1907).

Parmelia sulphurata Nees & Flot.   Atranorin; vulpinic acid. Hale (1960a), from the tropics; Hale (1965a), from Africa, Central and South America, Mexico, Sumatra, the West Indies, Florida, and Louisiana.
    Vulpinic acid. Hale (1959b), mention only.

Parmelia tabacina Mont. & Bosch Atranorin, salazinic acid. Ramaut (1964–1965), from Katanga, by thin-layer chromatography.

Parmelia tanakae Asah., see *Parmelia pseudorutidota* Asah.

Parmelia tananarivensis Gyeln. Norstictic acid ($\pm$), stictic acid, usnic acid. Abbayes (1961), from Madagascar.

Parmelia taractica Kremp. Atranorin ($\pm$), salazinic acid, ($+$)-usnic acid; mannitol; iron, manganese, zinc.
Atranorin ($\pm$), salazinic acid, usnic acid. Hale (1955*b*), from the U.S.A., as *Parmelia stenophylla* Chem. strain 1.
Atranorin, salazinic acid, usnic acid. Dahl (1952), from Norway, mention only, as *Parmelia stenophylla*; Krog (1951*b*), from Norway, as *P. stenophylla*; Solberg (1956), mention only, as *P. stenophylla*.
Atranorin. Vartia (1949), from Finland, as *Parmelia stenophylla*.
Salazinic acid, ($+$)-usnic acid. Asahina and Asano (1933), from Japan, by extraction, as *Parmelia conspersa* var. *hypoclysta*.
Salazinic acid, usnic acid. Hale (1956*b*), as *Parmelia stenophylla* Chem. strain 1; Hale (1956*g*), from Oklahoma, an exsiccati specimen, as *P. stenophylla* Chem. strain 1; Schatz (1963), from Kansas, as *P. stenophylla*.
Salazinic acid. Krog (1962), from Alaska.
($+$)-Usnic acid, mannitol. Dhar, Neelakantan, Ramanujam, and Seshadri (1959), from India, by extraction, as *Parmelia stenophylla*.
Iron, manganese, zinc. Lounamaa (1965), a quantitative study, as *Parmelia stenophylla*.
OTHER REPORT. Chemical data difficult to summarize: Hale (1956*d*), as *Parmelia stenophylla*.
Some reports, especially from the U.S.A., are based on samples which are black below and are more properly referred to *Parmelia tasmanica*. See also *P. stenophylla* Chem. strain 2.

Parmelia tasmanica Tayl., see *Parmelia taractica* Kremp.

Parmelia taylorensis Mitch. Atranorin, evernic acid, lecanoric acid (trace). Hale and Kurokawa (1962), from Europe and Ireland.
Evernic acid, lecanoric acid. Hale and Kurokawa (1964), no alectoronic acid, no usnic acid.

Parmelia tenuirima Hook. f. & Tayl. Atranorin, salazinic acid. W. L. Culberson (1966*c*), from New Zealand, by microcrystal tests and paper chromatography on the holotype specimen of *Aspidelia beckettii*.

Parmelia texana Tuck. Atranorin, divaricatic acid, unidentified substance (KC+ red) ($\pm$). Abbayes (1961), from Madagascar.
Atranorin, divaricatic acid. Hale (1957*a*).

Divaricatic acid. Hale (1960*a*), mention only; Hale and Kurokawa (1964), no usnic acid, mentioned in key; Kurokawa (1964), mention only.

Parmelia thomsonii (Stirt.) W. Culb. Alectoronic acid, atranorin, α-collatolic acid (rarely absent). W. L. Culberson (1962*b*), from India, Nepal, Réunion, and Thailand, and also *Parmelia thomsonii* var. *dissecta*.

Alectoronic acid, atranorin. Asahina (1955*a*), from India, as *Cetraria thomsonii*.

Parmelia thysanota Kurok. Atranorin, gyrophoric acid. Hale and Kurokawa (1964), from Mexico, type description.

Parmelia tiliacea (Hoffm.) Ach. Atranorin, lecanoric acid. Dahl (1952), from Norway, mention only, as *Parmelia scortea*; Hale (1957*b*), mention only, as *P. scortea*; Krog (1951*b*), from Norway, as *P. scortea*; Ramaut (1960), from Belgium, and an unidentified substance, by paper chromatography, as *P. scortea*; Ramaut (1961*a*), as *P. scortea*.

Lecanoric acid. Asahina (1936*d*), description of microcrystal tests, as *Parmelia scortea*; Hale (1956*e*), from North America, as *P. scortea*; Hale and Kurokawa (1964), mentioned in key; Klosa (1950).

No usnic acid. Gertig and Banasiewicz (1961), from Poland.

REVIEWS. Atranorin, lecanoric acid: W. Brieger (1923), and as *Parmelia tiliacea* f. *scortea* and *P. scortea*; Hesse (1912), and as *P. tiliacea* f. *scortea* and *P. scortea*; Thies (1932), and as *P. tiliacea* var. *scortea* and *P. scortea*; Zopf (1907), and as *P. scortea*. — Lecanoric acid: Asahina and Shibata (1954); Shibata (1958, 1963); all three reports as *P. scortea*.

Parmelia tinctina Mah. & Gillet Salazinic acid, usnic acid. Hale (1964), from Africa, Asia, Europe, and Minnesota.

OTHER REPORT. Atranorin, methylethersalazinic acid, stictic acid, usnic acid: Dahl (1952), from Norway, mention only.

See also *Parmelia conspersa*.

Parmelia tinctorum Nyl. Atranorin, chloroatranorin, lecanoric acid, methyl β-orcinolcarboxylate; isolichenin, lichenin; amino acids; ascorbic acid, riboflavin; calcium, iron, nitrogen, phosphorus.

Atranorin, chloroatranorin, lecanoric acid. Ahmann and Mathey (1967), from Mississippi, 25.6% lecanoric acid based on oven-dry lichen, 0.63% atranorin-chloroatranorin based on oven-dry lichen, by extraction and thin-layer chromatography, other PD + compounds reported by earlier workers not observed; Yamazaki and Shibata (1966), ^{3}H-labeling study on biosynthesis; Yamazaki, Matsuo, and Shibata (1965), biosynthesis study.

Atranorin, lecanoric acid. Abbayes (1961), from Madagascar; Aghoramurthy and Seshadri (1952), mention only; Asahina (1926); Asahina (1952e), from Japan, Korea, and Manchuria; Hale (1957a), from Arkansas; Hale (1958a), mention only; Hale (1960a), from tropical and subtropical regions; Hale (1965a), from tropical regions, many localities listed.

Atranorin. Aghoramurthy, Sarma, and Seshadri (1961b), used as an authentic source of this substance.

Lecanoric acid. Asahina (1936d), description of microcrystal tests; Hale (1956e), from North America; Kurokawa (1952); Kurokawa (1964), mention only; Mitsuno (1953), by paper chromatography; Rangaswami and Rao (1955a), 10% by extraction.

Isolichenin, lichenin; ascorbic acid, riboflavin, calcium, iron, nitrogen, phosphorus. Lal and Rao (1959).

Amino acids. Ramakrishnan and Subramanian (1964), from India, alanine, glycine, leucine, threonine, tyrosine, and valine as free amino acids and these with lysine, phenylalanine, and serine as combined amino acids; Ramakrishnan and Subramanian (1965), and also aspartic acid, glutamic acid, methionine, tryptophan, and an unidentified substance, by paper chromatography.

OTHER REPORTS. Atranorin, lecanoric acid, norstictic acid or salazinic acid: Neelakantan and Seshadri (1952a), from India; Neelakantan, Seshadri, and Subramanian (1951), from India, by extraction. — Atranorin, lecanoric acid, norstictic acid: Seshadri and Subramanian (1949a), from Java; Seshadri and Subramanian (1949c); Venkateswarlu and Rao (1962), from India, carotenes noted, by extraction. — Norstictic acid: Nair and Subramanian (1961a). — Atranorin, lecanoric acid, methyl β-orcinolcarboxylate, salazinic acid: Murty (1960), from India, by extraction.

REVIEWS. Atranorin, lecanoric acid: W. Brieger (1923); Hesse (1912); Thies (1932), and as *Parmelia coralloidea*; Zopf (1907), 23.5% lecanoric acid, and as *P. coralloidea*. — Atranorin, lecanoric acid, norstictic acid or salazinic acid: Neelakantan (1965), 20.3% lecanoric acid. — Lecanoric acid: Asahina and Shibata (1954); Shibata (1958, 1963).

Parmelia tortula Kurok. Atranorin, unidentified substance. Hale and Kurokawa (1964), from South Africa, KC+ faint rose reaction of the medulla not constant, type description.

Parmelia trabeculata Ahti Atranorin, norstictic acid. Ahti (1966), by a microcrystal test, type description.

Parmelia trichotera Hue, see *Parmelia perlata* (Huds.) Ach.

Parmelia tropica Vain. Atranorin; unidentified fatty acid. Hale (1958a), microchemical tests on type specimen.

Parmelia tubulosa (Schaer.) Bitt., see *Hypogymnia tubulosa* (Schaer.) Hav.

Parmelia tumescens Hale & Kurok.   Usnic acid, stictic acid. Hale and Kurokawa (1964), from Australia.
Stictic acid. Kurokawa (1965).

Parmelia ulophylla (Ach.) Merr., see *Parmelia subrudecta* Nyl.

Parmelia ulophyllodes (Vain.) Sav.   Lecanoric acid, usnic acid. Asahina (1941c), from Manchuria, type description of *Parmelia manchurica*; Asahina (1952e), from Japan and Manchuria, as *P. manchurica*; W. L. Culberson (1955), from Japan, Manchuria, and Wisconsin, as *P. manchurica*; W. L. Culberson (1960b), from Japan, Mexico, the U.S.S.R., Arizona, Michigan, Minnesota, and New Mexico.
    Lecanoric acid. Hale (1956e), from North America, as *Parmelia manchurica*.
    OTHER REPORT. Atranorin, lecanoric acid: Aghoramurthy, Neelakantan, and Seshadri (1954), from India, as *Parmelia manchurica*.

Parmelia uruguensis Kremp.   Atranorin, salazinic acid. Hale (1965a), from Argentina.
Salazinic acid. Hale (1960a), mentioned in key.

Parmelia usambarensis J. Stein. & Zahlbr.   Salazinic acid. Hale and Kurokawa (1964), mentioned in key.

Parmelia vagans (Nyl.) Nyl.   Salazinic acid, (+)-usnic acid; enzymes.
    Salazinic acid, (+)-usnic acid. Troshchenko (1957), from Uralak.
    (+)-Usnic acid, enzymes. Moissejeva (1961), 0.5% usnic acid, enzymes tabulated.
    (+)-Usnic acid. Moissejeva (1957), 0.5%; Savicz, Kuprevicz, Litvinov, Moissejeva, and Rassadina (1956), 0.5%; Zatulovskii (1956, 1957).
    REVIEW. (+)-Usnic acid: Savicz, Litvinov, and Moissejeva (1960), 0.5%.

Parmelia velloziae Vain.   Protocetraric acid, usnic acid; rhodophyscin. Hale and Kurokawa (1964), mentioned in key.

Parmelia ventricosa Hale & Kurok.   Atranorin, norstictic acid. Hale and Kurokawa (1964), from Mexico and South Africa.

Parmelia verruculifera Nyl.   Lecanoric acid.
    REVIEWS. Lecanoric acid: W. Brieger (1923); Hesse (1912); Thies (1932); Zopf (1907).

Parmelia vexans Zahlbr., see *Parmelia cirrhata* Fr.

Parmelia violacea Kurok.   Atranorin, unidentified hydroxyanthraquinone (K +). Hale and Kurokawa (1964), from South Africa, type description, hydroxyanthraquinone pigment in the medulla.

Parmelia virginica Hale   Atranorin, barbatic acid, unidentified substance (KC+ orange). Hale and Kurokawa (1964), from North Carolina and Virginia, type description.

Parmelia viridescens Lynge   No lichen substances found. Hale (1960*a*), from Brazil.

Parmelia viridiflava Hale   Atranorin (±), fumarprotocetraric acid, protocetraric acid, usnic acid. Hale (1965*a*), from the Dominican Republic, Haiti, and Mexico, type description.

Parmelia vittata (Ach.) Nyl., see *Hypogymnia vittata* (Ach.) Gas.

Parmelia vittata var. *hypotrypanea* Nyl., see *Hypogymnia vittata* (Ach.) Gas.

"*Parmelia vittata* f. *reagens*," see "*Hypogymnia vittata* f. *reagens*"

Parmelia vittata f. *stricta* (Hillm.) Asah., see *Hypogymnia vittata* (Ach.) Gas.

Parmelia wainii A. L. Sm.   Alectoronic acid, atranorin, α-collatolic acid; rhodophyscin (±).
  Alectoronic acid, atranorin, α-collatolic acid. Abbayes (1961), from Madagascar; Abbayes (1962), from Africa.
  Alectoronic acid, atranorin, rhodophyscin (±). Hale (1965*a*), from Africa and Brazil.
  Alectoronic acid, atranorin. Hale (1960*a*), from Africa and Brazil.

Parmelia wallichiana Tayl.   Atranorin, salazinic acid. Ramaut (1964–1965), by thin-layer chromatography.
  Salazinic acid. Hale and Kurokawa (1964), mentioned in key.

Parmelia xanthina (Müll. Arg.) Vain.   Atranorin, usnic acid, unidentified substance (KC+), (in samples from Madagascar, Mexico, Nicaragua, Southern Rhodesia, and Alabama). Atranorin, usnic acid; protolichesterinic acid, (in samples from India, Madagascar, Mexico, Alabama, Florida, Georgia, North Carolina, South Carolina, Tennessee, and Virginia). Hale (1965*a*).
  Usnic acid. Hale (1957*b*), as *Parmelia chrysantha*.

OTHER REPORTS. Atranorin, gyrophoric acid (±), protolichesterinic acid, usnic acid: Hale (1960*a*), from Brazil and Madagascar. — Atranorin, gyrophoric acid, usnic acid: Abbayes (1961), mention only.

Parmelia xantholepis Mont. & Bosch   Unidentified pigment. Hale and Kurokawa (1964), mentioned in key.

Parmelia xanthomelaena Müll. Arg.   Stictic acid. Hale and Kurokawa (1964), mentioned in key.

Parmelia yasudae Räs.   Atranorin, salazinic acid. Asahina (1951*f*), from Japan; Asahina (1952*e*), from Japan.

Parmelia zahlbruckneri Lynge   Barbatic acid. Hale and Kurokawa (1964), from tropical America, no usnic acid.
    OTHER REPORT. Atranorin, unidentified substances: Hale (1960*a*), from Brazil.

Parmelia zollingeri Hepp   Atranorin, protocetraric acid. Asahina (1952*a*), from Formosa and Japan; Asahina (1952*e*), from Formosa and Japan; Asahina and Tanase (1934*b*), from Formosa, 12% protocetraric acid by extraction, also a trace of lecanoric acid which was later ascribed to contamination with *Parmelia tinctorum* (see Asahina and Shibata, 1954); Hale (1960*a*), from tropical and subtropical regions, and as *P. latissima* var. *corniculata* and *P. latissima* var. *minima*; Hale (1965*a*), from tropical and subtropical regions.
    OTHER REPORT. Atranorin, norstictic acid (trace), protocetraric acid: Ramaut (1964–1965), from Africa.

Parmeliopsis aleurites (Ach.) Nyl.   Atranorin, thamnolic acid. W. L. Culberson (1956), by microchemical tests.
    Thamnolic acid. Asahina (1938*f*), original report of atranorin and lobaric acid (Zopf) apparently an error; Hale (1957*b*), mention only.
    REVIEWS. Atranorin, lobaric acid (sometimes as usnetinic acid): W. Brieger (1923); Hesse (1912); Zopf (1907). — Lobaric acid: Thies (1932). All review reports are as *Parmelia aleurites*.

Parmeliopsis ambigua (Wulf.) Nyl.   Divaricatic acid, usnic acid. Asahina (1936*e*), description of microcrystal tests; W. L. Culberson (1956); Hale (1957*c*), from West Virginia, an exsiccati specimen.
    Usnic acid. W. L. Culberson (1960*a*), mention only; Vartia (1949), from Finland.

Parmeliopsis americana Hillm., see *Parmeliopsis placorodia* (Ach.) Nyl.

Parmeliopsis hyperopta (Ach.) Vain.   Atranorin, divaricatic acid. Asahina (1936*e*), description of microcrystal tests; W. L. Culberson (1956).

Atranorin. W. L. Culberson (1960*a*), mention only; Thomson (1963), mention only, as *Parmelia hyperopta*.

OTHER REPORTS. Divaricatic acid, usnic acid: Asahina (1936*e*), description of microcrystal tests, as *Platysma diffusum*. — Diffusinic acid, usnic acid: Klosa (1951*b*), mention only, as *Cetraria diffusa*. — Thamnolic acid: Asahina (1938*f*), description of microcrystal tests, as *Platysma pallescens*.

REVIEWS. Atranorin: Thies (1932); Zopf (1907). — Diffusinic acid, (−)-usnic acid: W. Brieger (1923); Hesse (1912); Thies (1932); Zopf (1907); all four reports as *Cetraria diffusa* and *Platysma diffusum*. — Thamnolic acid: Asahina and Shibata (1954); Shibata (1963); both reports as *Platysma pallescens*.

Parmeliopsis pallescens (Neck.) Zahlbr., see *Parmeliopsis hyperopta* (Ach.) Vain.

Parmeliopsis placorodia (Ach.) Nyl. Atranorin, thamnolic acid. W. L. Culberson (1956), from Maryland, Massachusetts, New Jersey, North Carolina, and Wisconsin, by microchemical tests.

Thamnolic acid. Asahina (1938*f*), description of microcrystal tests, as *Parmelia americana* and *Parmelia placorodia*; W. L. Culberson (1961*b*), from Arizona, Colorado, and South Dakota; Hale (1957*c*), from West Virginia, an exsiccati specimen.

REVIEWS. Thamnolic acid: Asahina and Shibata (1954); Shibata (1963); both reports as *Parmeliopsis americana*.

Parmularia muralis (Schreb.) B. de Lesd., see *Lecanora muralis* (Schreb.) Rabenh.

Peltigera aphthosa (L.) Willd. Tenuiorin (sometimes as peltigerin); phlebin A, phlebin B, unidentified substances; D-arabitol, 3-O-β-D-glucopyranosyl-D-mannitol, *myo*-inositol, D-mannitol, sucrose; ascorbic acid; crude fat; enzymes, protein; calcium, phosphorus.

Tenuiorin, phlebin A, phlebin B, unidentified substances. Kurokawa, Jinzenji, Shibata, and Chiang (1966), from Europe, Japan, and Alaska, unidentified substances IV and V.

D-Arabitol, 3-O-β-D-glucopyranosyl-D-mannitol, *myo*-inositol, D-mannitol, sucrose. Lindberg, Silvander, and Wachtmeister (1963), a detailed chemical study of the last named glycoside.

Arabitol, mannitol. Lindberg, Misiorny, and Wachtmeister (1953), by paper chromatography, no volemitol.

3-O-β-D-Glucopyranosyl-D-mannitol. Lindberg, Silvander, and Wachtmeister (1964), mention only.

Ascorbic acid. Karev and Kochevykh (1962).

Crude fat, protein, calcium, phosphorus. Scotter (1965), from northern Canada.

Enzymes. Moissejeva (1961), enzymes tabulated.

OTHER REPORT. Tenuiorin, two unidentified acids (C+), mannitol: Zopf (1909), from Austria, no zeorin.

REVIEWS. 3-O-β-D-glucopyranosyl-D-mannitol: Neelakantan (1965). — Tenuiorin: W. Brieger (1923); Hesse (1912); Thies (1932). — No substances found: Zopf (1907).

Peltigera canina (L.) Willd.   Tenuiorin (sometimes as peltigerin); unidentified fatty acids; 3-O-β-D-glucopyranosyl-D-mannitol, D-mannitol; ergosterol; amino acids, chlorophyll *a*, enzymes, protein; ascorbic acid, riboflavin; calcium, iron, nitrogen, phosphorus.

Tenuiorin, unidentified fatty acids, D-mannitol, ergosterol. Zellner (1932), and "viskosin," by extraction.

Tenuiorin. Huneck and Tümmler (1965), by thin-layer chromatography; Steiner (1955).

3-O-β-D-Glucopyranosyl-D-mannitol. Lindberg, Silvander, and Wachtmeister (1964), by extraction.

Ergosterol. Aghoramurthy and Seshadri (1954), mention only; Murty and Subramanian (1959*d*), mention only.

Amino acids. Subramanian and Ramakrishnan (1964), from India, alanine, arginine, glycine, leucine, methionine, phenylalanine, threonine, tyrosine, and valine as free amino acids and these with isoleucine, lycine, serine, and tryptophan as combined amino acids, identifications and quantative estimations by chromatography; Ramakrishnan and Subramanian (1965), additional amino acids are aspartic acid, glutamic acid, histidine, and an unidentified substance.

Enzymes. Moissejeva (1961), enzymes tabulated.

Ascorbic acid, riboflavin, calcium, iron, nitrogen, phosphorus. Lal and Rao (1959), from India.

Chlorophyll *a*. Wilhelmsen (1959).

Protein, fat, calcium, phosphorus. Scotter (1965), from northern Canada.

OTHER REPORTS. Caninin, mannitol: Zopf (1909), from Tirol, no tenuiorin, no zeorin, no C+ red substances. — No substances found: Kurokawa, Jinzenji, Shibata, and Chiang (1966), from Japan, Mexico, and Taiwan, by thin-layer chromatography. — Polyoses of mannose, galactose, and glucose and a cellulose of glucose: Votoček and Burda (1926). — Unidentified fluorescent substance: D. N. Rao and Le Blanc (1965), by chromatography and microcrystal tests, no atranorin.

REVIEWS. Caninin, tenuiorin: W. Brieger (1923); Hesse (1912); Thies (1932). — Ergosterol, fatty acids: Steiner (1957). — Ergosterol: Neelakantan (1965). — D-Mannitol: Asahina and Shibata (1954).

Peltigera canina var. *rufescens* (Weiss) Mudd   Tenuiorin (±) (sometimes as peltigerin); 3-O-β-D-glucopyranosyl-D-mannitol, mannitol.

Tenuiorin. Huneck and Tümmler (1965), tenuiorin and peltigerin found to be identical, trace tenuiorin identified by thin-layer chromatography, as *Peltigera rufescens*.

3-O-β-D-Glucopyranosyl-D-mannitol. Lindberg, Silvander, and Wachtmeister (1964), by extraction.

Mannitol. Zopf (1909), from southern Tirol, also an unidentified substance forming fine colorless needles, no tenuiorin found, no zeorin, no caninin, unidentified C+ substance known in several other species of *Peltigera* not found, as *Peltigera rufescens*.

No substances found. Kurokawa, Jinzenji, Shibata, and Chiang (1966), from Canada, Japan, and Taiwan, by thin-layer chromatography, as "*Peltigera rubescens*."

Peltigera dilacerata (Gyeln.) Gyeln. No substances found. Kurokawa, Jinzenji, Shibata, and Chiang (1966), from Japan and Washington, by thin-layer chromatography.

Peltigera dolichorrhiza (Nyl.) Nyl. Tenuiorin; dolichorrhizin, zeorin; unidentified substances. Kurokawa, Jinzenji, Shibata, and Chiang (1966), from Japan and Taiwan, with unidentified substances I, II, IV, and V.

Zeorin. Shibata, Furuya, and Iizuka (1965), gas chromatographic study.

Peltigera erumpens (Gyeln.) Vain. No substances found. Kurokawa, Jinzenji, Shibata, and Chiang (1966), from Japan, by thin-layer chromatography.

Peltigera horizontalis (Huds.) Baumg. Tenuiorin (sometimes as peltigerin); dolichorrhizin, zeorin; scabrosin A ($\pm$), scabrosin B ($\pm$), unidentified substances; 3-O-β-D-galactofuranosyl-D-mannitol (sometimes as peltigeroside), 3-O-β-D-glucopyranosyl-D-mannitol, mannitol.

Tenuiorin, dolichorrhizin, zeorin, scabrosin A ($\pm$), scabrosin B ($\pm$), unidentified substances. Kurokawa, Jinzenji, Shibata, and Chiang (1966), from Japan with tenuiorin, dolichorrhizin, zeorin, scabrosin A, scabrosin B, and unidentified substances I and II, from Europe and Taiwan with no scabrosin A or B, but with unidentified substances IV and V.

Tenuiorin, unidentified substance (C+), mannitol. Zopf (1909), from Germany.

3-O-β-D-Galactofuranosyl-D-mannitol, 3-O-β-D-glucopyranosyl-D-mannitol. Lindberg, Silvander, and Wachtmeister (1964).

3-O-β-D-Galactofuranosyl-D-mannitol. Pueyo (1959); Pueyo (1960).

REVIEWS. Tenuiorin: W. Brieger (1923); Hesse (1912); Thies (1932). — 3-O-β-D-Galactofuranosyl-D-mannitol: Neelakantan (1965).

Peltigera lepidophora (Nyl.) Vain. Tenuiorin (as peltigerin); mannitol. Zopf (1909), from Norway, no zeorin, no C+ red substance, identification of mannitol not certain.

REVIEWS. Tenuiorin: W. Brieger (1923); Hesse (1912).

Peltigera leucophlebia (Nyl.) Gyeln. 3-O-β-D-Glucopyranosyl-D-mannitol. Lindberg, Silvander, and Wachtmeister (1964), by extraction.

Peltigera malacea (Ach.) Funck Tenuiorin (sometimes as peltigerin), unidentified substance (C+); mannitol; enzymes.

Tenuiorin, unidentified substance (C+), mannitol. Zopf (1909), from southern Tirol.

Enzymes. Moissejeva (1961), enzymes tabulated.

REVIEWS. Tenuiorin: W. Brieger (1923); Hesse (1912); Thies (1932).

Peltigera microphylla (And.) Gyeln. Tenuiorin; dolichorrhizin, zeorin; unidentified substance IV. Kurokawa, Jinzenji, Shibata, and Chiang (1966), by thin-layer chromatography.

Peltigera nigripunctata Bitt. Tenuiorin; zeorin; phlebin B, unidentified substances III, IV, and V. Kurokawa, Jinzenji, Shibata, and Chiang (1966), by thin-layer chromatography.

Peltigera polydactyla (Neck.) Hoffm. Tenuiorin (sometimes as peltigerin); dolichorrhizin, zeorin; unidentified substances; 3-O-β-D-glucopyranosyl-D-mannitol, mannitol; enzymes.

Tenuiorin, dolichorrhizin, zeorin, unidentified substances. Kurokawa, Jinzenji, Shibata, and Chiang (1966), from Finland, Germany, and Japan, unidentified substances I, II, IV, and V.

Tenuiorin, mannitol, unidentified substance (C+). Zopf (1909), from Germany, also peltidactylin and polydactylin, no zeorin.

Tenuiorin. Huneck and Tümmler (1965), 1% by extraction, peltigerin identical to tenuiorin.

3-O-β-D-Glucopyranosyl-D-mannitol. Lindberg, Silvander, and Wachtmeister (1964), by extraction.

Mannitol. Smith (1961).

Enzymes. Moissejeva (1961), enzymes tabulated.

OTHER REPORT. Polyoses of galactose, glucose, and mannose: Votoček and Burda (1926).

REVIEWS. Peltidactylin, tenuiorin: Thies (1932). — Tenuiorin: W. Brieger (1923); Hesse (1912).

Peltigera praetextata (Flörke *ex* Somm.) Vain. 3-O-β-D-Glucopyranosyl-D-mannitol, mannitol.

3-O-β-D-Glucopyranosyl-D-mannitol. Lindberg, Silvander, and Wachtmeister (1964), by extraction.

Mannitol. Zopf (1909), from Germany, no tenuiorin (as peltigerin), zeorin, or unidentified C+ substance.

OTHER REPORT. No substances found: Kurokawa, Jinzenji, Shibata, and Chiang (1966), from Japan, by thin-layer chromatography.

REVIEW. 3-O-β-D-Glucopyranosyl-D-mannitol: Neelakantan (1965).

Peltigera propagulifera B. Stein, see *Peltigera scutata* (Dicks.) Duby

Peltigera pruinosa (Gyeln.) Inum.   Tenuiorin; dolichorrhizin; unidentified substance IV. Kurokawa, Jinzenji, Shibata, and Chiang (1966), from Japan, by thin-layer chromatography.

"*Peltigera rubescens*," see *Peltigera canina* var. *rufescens* (Weiss) Mudd

Peltigera rufescens (Weiss) Humb., see *Peltigera canina* var. *rufescens* (Weiss) Mudd

Peltigera scabrosa Th. Fr.   Tenuiorin (sometimes as peltigerin); dolichorrhizin (±), zeorin; scabrosin A (±), scabrosin B (±); unidentified substances; 3-O-β-D-glucopyranosyl-D-mannitol.

Tenuiorin, dolichorrhizin, zeorin, scabrosin A (±), scabrosin B (±), unidentified substances. Kurokawa, Jinzenji, Shibata, and Chiang (1966), from Japan with tenuiorin, dolichorrhizin, zeorin, and unidentified substances I, II, IV, and V, and from Europe (Germany and Sweden) with tenuiorin, zeorin, scabrosin A, scabrosin B, and unidentified substances I and II.

Tenuiorin, unidentified acid (C+ red). Zopf (1909), possibly also mannitol, no zeorin.

3-O-β-D-Glucopyranosyl-D-mannitol. Lindberg, Silvander, and Wachtmeister (1964), by extraction.

REVIEWS. Tenuiorin: W. Brieger (1923); Hesse (1912); Thies (1932).

Peltigera scutata (Dicks.) Duby   Tenuiorin (sometimes as peltigerin); zeorin; unidentified substances; 3-O-β-D-glucopyranosyl-D-mannitol, mannitol.

Tenuiorin, zeorin, unidentified substances. Kurokawa, Jinzenji, Shibata, and Chiang (1966), from Germany and Sweden, unidentified substances I, II, IV, and V.

Tenuiorin, zeorin, unidentified acid. Zopf (1909), from northern Tirol, as *Peltigera propagulifera*, possibly also mannitol.

3-O-β-D-Glucopyranosyl-D-mannitol. Lindberg, Silvander, and Wachtmeister (1964), by extraction.

REVIEWS. Tenuiorin, zeorin: W. Brieger (1923); Hesse (1912); Thies (1932); all review reports as *Peltigera propagulifera*.

Peltigera spuria (Ach.) DC.   3-O-β-D-Glucopyranosyl-D-mannitol, mannitol.

3-O-β-D-Glucopyranosyl-D-mannitol. Lindberg, Silvander, and Wachtmeister (1964), by extraction.

Mannitol. Zopf (1909), from Germany, no caninin, no tenuiorin (as peltigerin), no zeorin, no unidentified C + substance.

No tenuiorin. Huneck and Tümmler (1965), by thin-layer chromatography.

OTHER REPORT. No substances found: Kurokawa, Jinzenji, Shibata, and Chiang (1966), from Japan, by thin-layer chromatography.

Peltigera subscutata Gyeln.   Tenuiorin; zeorin; unidentified substances I, II, and IV. Kurokawa, Jinzenji, Shibata, and Chiang (1966), from Japan, by thin-layer chromatography.

Peltigera variolosa (Mass.) Gyeln.   Tenuiorin; zeorin; phlebin A, phlebin B, unidentified substances IV and V. Kurokawa, Jinzenji, Shibata, and Chiang (1966), from Japan, by thin-layer chromatography.

Peltigera venosa (L.) Baumg.   Tenuiorin (sometimes as peltigerin), unidentified substance (C + red); zeorin; 3-O-β-D-glucopyranosyl-D-mannitol; phlebin A, phlebin B; unidentified substances.

Tenuiorin, zeorin, phlebin A, phlebin B, unidentified substances. Kurokawa, Jinzenji, Shibata, and Chiang (1966), from Austria, Finland, and Japan, unidentified substances III, IV, and V.

Tenuiorin, an unidentified substance (C + red). Zopf (1909), from Germany, no zeorin, possibly also mannitol present.

3-O-β-D-Glucopyranosyl-D-mannitol. Lindberg, Silvander, and Wachtmeister (1964), by paper chromatography.

REVIEWS. Tenuiorin: W. Brieger (1923); Hesse (1912); Thies (1932).

Perforaria cucurbitula (Mont.) Müll. Arg.   Norstictic acid (±), stictic acid. Asahina (1954*d*).

Pertusaria amara (Ach.) Nyl.   Picrolichenic acid; arabitol, mannitol.

Picrolichenic acid. T. A. Davidson and Scott (1960), a detailed chemical study on synthesis; T. A. Davidson and Scott (1961), mention only; Erdtman and Wachtmeister (1957*b*), by extraction; Wachtmeister (1958*a*), from southern Sweden, as *Variolaria amara*; Wachtmeister (1958*b*), mention in chemical study on this compound.

Arabitol, mannitol. Lindberg, Misiorny, and Wachtmeister (1953), by paper chromatography, mannitol isolated, no volemitol.

Mannitol. Pueyo (1960), mention only.

OTHER REPORTS. Salazinic acid: Klosa (1950). — Picrolichenic acid, physodalic acid: Klosa (1951*b*). — Picrolichenic acid, salazinic acid: Zopf (1900), early work reviewed.

REVIEWS. Picrolichenic acid, salazinic acid: W. Brieger (1923); Hesse (1912); Thies (1932). — Picrolichenic acid: Neelakantan (1965); Shibata (1958, 1963); Zopf (1907), no salazinic acid.

Pertusaria areolata (Ach.) Mass.
REVIEWS. Gyrophoric acid: W. Brieger (1923); Hesse (1912); Zopf (1907); all reports as *Pertusaria rupestris* var. *areolata*.

Pertusaria bryophaga Erichs. Lecanoric acid. Hale (1956e), from North America.

Pertusaria communis DC., see *Pertusaria pertusa* (L.) Tuck.

Pertusaria communis "var. *fraginea*," see *Pertusaria faginea* (L.) Leight.

Pertusaria communis var. *variolosa* Schaer., see *Pertusaria globulifera* (Turn.) Mass.

Pertusaria concreta Nyl., see *Pertusaria pseudocorallina* (Sw.) Arn.

Pertusaria concreta f. *westringii* (Ach.) Cromb., see *Pertusaria pseudo-corallina* (Sw.) Arn.

Pertusaria corallina (L.) Arn. Thamnolic acid (sometimes as ocellatic acid); arabitol, mannitol.
Thamnolic acid. Aghoramurthy, Sarma, and Seshadri (1961b), used as an authentic source of this substance; Wachtmeister (1955), thamnolic acid found to be identical to Hesse's ocellatic acid; Wachtmeister (1956a), used as an authentic source of this substance; Wachtmeister (1958a), mention only.
Arabitol, mannitol. Lindberg, Misiorny, and Wachtmeister (1953), by paper chromatography, no volemitol.
OTHER REPORT. Gyrophoric acid: Schindler (1956/1957).
REVIEWS. Thamnolic acid: W. Brieger (1923); Hesse (1912); Thies (1932); Zopf (1907).

Pertusaria coriacea (Th. Fr.) Th. Fr. Norstictic acid. Hale (1954b), from Baffin Island.

Pertusaria dealbata (Ach.) Nyl. Atranorin, thamnolic acid. Koller and Hamburg (1935b), from Germany, by extraction of 1 kg. of lichen.
Thamnolic acid. Asahina and Hiraiwa (1936), mention only.
REVIEWS. Thamnolic acid: Asahina and Shibata (1954); Shibata (1958, 1963). — No crystalline substances soluble in ethyl ether: Zopf (1907). — Variolarin: Thies (1932).

Pertusaria faginea; (L.) Leight.
    REVIEW. Orbiculatic acid: W. Brieger (1923); Hesse (1912); Thies (1932); all three reports as *Pertusaria communis* "var. *faginea.*" — Picrolichenic acid: Zopf (1907), as *Variolaria faginea.*

Pertusaria flavida (DC.) Laund.
    REVIEWS. Thiophanic acid: W. Brieger (1923); Hesse (1912); Thies (1932); Zopf (1907); all reports as *Pertusaria lutescens.*

Pertusaria globulifera (Turn.) Mass.
    REVIEWS. Orbiculatic acid, picrolichenic acid ($\pm$): Thies (1932), as *Pertusaria communis* var. *variolosa.* — Orbiculatic acid: W. Brieger (1923); Hesse (1912); both reports as *P. communis* var. *variolosa*; Zopf (1907), and as *P. communis* var. *variolosa.* — Pertusaric acid, pertusarin, salazinic acid: W. Brieger (1923), and picrolichenic acid ($\pm$); Hesse (1912), and picropertusarin; Thies (1932), and pertusaridin and picropertusaric acid; as *P. communis* var. *variolosa.*

Pertusaria glomerata (Ach.) Schaer., see *Pertusaria tuckermanii* Erichs.

Pertusaria inquinata (Ach.) Th. Fr.
    REVIEW. Thalloidima-green: Thies (1932).

Pertusaria lactea (L.) Arn.
    REVIEWS. Lecanoric acid, variolaric acid (sometimes as parellic acid): W. Brieger (1923); Hesse (1912); Thies (1932); all three reports also as *Variolaria lactea* (see discussions by Murphy, Keane, and Nolan, 1943, and by Asahina and Shibata, 1954); Zopf (1907). — Variolaric acid: Shibata (1958, 1963).

Pertusaria lutescens (Hoffm.) Lamy, see *Pertusaria flavida* (DC.) Laund.

Pertusaria multipuncta var. leptosporoides Erichs.   Salazinic acid. Galun and Lavee (1966), from Israel, by a microcrystal test (GAW).

Pertusaria ocellata (Wallr.) Körb.
    REVIEWS. Atranorin: W. Brieger (1923); Hesse (1912); Thies (1932); all reports as *Pertusaria ocellata* var. *variolosa.*

Pertusaria octomela (Norm.) Erichs.   Norstictic acid. Hale (1954b), from Baffin Island.

Pertusaria pertusa (L.) Tuck.
    Physodalic acid, unidentified substance. Klosa (1951b), as *Pertusaria communis.*

Pertusaria pseudocorallina (Sw.) Arn. Norstictic acid; concretin; mannitol. Breen, Keane, and Nolan (1937), from Ireland, by extraction, new substance (named concretin, $C_{14}H_7Cl_3O_5$) found to resemble Zopf's thiophanic acid, as *Pertusaria concreta* f. *westringii*.

REVIEWS. Norstictic acid: Asahina and Shibata (1954); Shibata (1958, 1963); all review reports as *Pertusaria concreta*.

Pertusaria rupestris (DC.) Schaer.
Gyrophoric acid. Klosa (1951*b*).
REVIEWS. Areolatin, areolin, gyrophoric acid: W. Brieger (1923); Hesse (1912); Thies (1932); Zopf (1907).

Pertusaria rupestris var. *areolata* (Ach.) Mudd, see *Pertusaria areolata* (Ach.) Mass.

Pertusaria subobducens Nyl.
REVIEW. Thalloidima-green: Thies (1932).

Pertusaria tuckermanii Erichs.
REVIEWS. Porin, porinic acid: W. Brieger (1923); Hesse (1912); Thies (1932); Zopf (1907); all reports as *Pertusaria glomerata*.

Pertusaria wulfenii DC.
REVIEWS. Thiophanic acid: W. Brieger (1923); Hesse (1912); Thies (1932); Zopf (1907).

Phaeographina elliptica Wirth & Hale   Unidentified substance. Wirth and Hale (1963), from Mexico, unidentified substance detected in the GA*o*-T solution.

Phaeographina isidiosa (Vain.) Zahlbr.   Norstictic acid. Wirth and Hale (1963).

Phaeographina lutescens (Fée) Zahlbr.   Norstictic acid. Wirth and Hale (1963).

Phaeographina oscitans (Tuck.) Zahlbr.   Stictic acid. Wirth and Hale (1963).

Phaeographina oxalifera Red.   Norstictic acid. Wirth and Hale (1963), from Brazil.

Phaeographina strigops Wirth & Hale   Unidentified substance. Wirth and Hale (1963), from Mexico, unidentified substance (K+ yellow, PD−) detected in the GA*o*-T solution, type description.

Phaeographis dendritica (Ach.) Müll. Arg.   Norstictic acid. Wirth and Hale (1963), from Mexico.

Phaeographis exaltata (Mont. & Bosch) Müll. Arg.   Unidentified substance. Wirth and Hale (1963), from Mexico, unidentified substance (K + yellow) detected in the GA*o*-T solution, also found in *Graphina triangularis, G. virginea,* and others.

Phaeographis inustoides (Fink) Red.   Norstictic acid; unidentified substance (±). Wirth and Hale (1963), from Puerto Rico, unidentified substance present in the holotype specimen.

Phaeographis pezizoidea var. *pruinosa* Red.   Norstictic acid. Wirth and Hale (1963).

Phaeographis radiatoramosa Red.   Norstictic acid. Wirth and Hale (1963).

Phialopsis rubra (Hoffm.) Körb., see *Gyalecta ulmi* (Sw.) Zahlbr.

Phlyctis agelaea (Ach.) Flot.   Norstictic acid. Schindler (1936).

Phlyctis argena (Ach.) Flot.   Norstictic acid. Schindler (1936).
    REVIEW. Salazinic acid: W. Brieger (1923); Hesse (1912); Thies (1932); Zopf (1907).

Physcia adscendens (Fr.) Oliv.   Atranorin. Galun and Lavee (1966), from Israel, by a microcrystal test; Hale (1957*b*), mention only; Vartia (1949), from Finland.
    REVIEWS. Atranorin: W. Brieger (1923); Hesse (1912); Thies (1932); all review reports as *Physcia stellaris* f. *adscendens.*

Physcia aegialita (Ach.) Nyl.   Atranorin, ramalinolic acid, sekikaic acid. Sasaki (1942).

Physcia aipolia (Ehrh.) Hampe   Atranorin. Hale (1957), from Arkansas, Missouri, and Oklahoma; Vartia (1949), from Finland.
    REVIEWS. Atranorin: W. Brieger (1923); Hesse (1912); Thies (1932), and parmelia-brown; Zopf (1907).

Physcia albata (Wils.) Hale   Atranorin. Hale (1963*a*), from Africa, Australia, Chile, and Hawaii, medulla and cortex K + yellow.

Physcia albicans (Pers.) Thoms.   Atranorin; zeorin. Asahina (1943*c*), as *Physcia crispa.*

Physcia caesia (Hoffm.) Hampe   Atranorin; zeorin. Vartia (1950*b*), from Finland; Zopf (1900), no haematommic acid.
   Atranorin. Klosa (1951*b*); Vartia (1949), from Finland.
   Zeorin. Neelakantan, Rangaswami, and Rao (1954), mention only.
   REVIEWS. Atranorin, zeorin: W. Brieger (1923); Hesse (1912); Thies (1932); Zopf (1907).

Physcia ciliaris (L.) DC., see *Anaptychia ciliaris* (L.) Körb.

Physcia ciliata (Hoffm.) Du Rietz   No substance found. Hale (1957*a*), from Arkansas, Missouri, and Oklahoma.
   OTHER REPORT. Atranorin: Vartia (1949), from Finland, as *Physcia obscura*.
   REVIEW. Parmelia-brown: Thies (1932), as *Physcia obscura*.

Physcia crispa (Pers.) Nyl., see *Physcia albicans* (Pers.) Thoms.

Physcia elaeina (Sm.) A. L. Sm.   No lichen substances found. Hale (1957*a*), from Arkansas.

Physcia endococcinea (Körb.) Th. Fr.
   Zeorin. Neelakantan, Rangaswami, and Rao (1954), mention only.
   REVIEWS. Atranorin, endococcin, rhodophyscin, zeorin: W. Brieger (1923); Hesse (1912); Thies (1932); Zopf (1907).

Physcia frostii (Tuck.) Zahlbr.   Divaricatic acid. Hale (1957*b*), mention only.

Physcia grisea (Lam.) Zahlbr.
   No lichen substances found. Galun and Lavee (1966), from Israel, and *Physcia grisea* f. *sorediosa*; Hale (1957*a*), from Arkansas, Missouri, and Oklahoma.
   Atranorin. Klosa (1951*b*), as *Physcia pityra*.
   REVIEWS. Atranorin: W. Brieger (1923); Hesse (1912); Thies (1932); Zopf (1907); all review reports as *Physcia pityra*.

Physcia integrata Nyl.   Atranorin; zeorin. Asahina (1943*c*).

Physcia millegrana Degel.   Atranorin. Hale (1954*a*), from Kansas, an exsiccati specimen; Hale (1956*a*); Hale (1957*a*), from Arkansas, Missouri, and Oklahoma.

Physcia muscigena (Ach.) Nyl.   Atranorin. Hale (1957*b*), mention only.

Physcia obscura (Ehrh.) Hampe, see *Physcia ciliata* (Hoffm.) Du Rietz

Physcia orbicularis f. *rubropulchra* Degel. Rhodophyscin. Hale (1957*b*), mention only.

Physcia parietina (L.) De Not., see *Xanthoria parietina* (L.) Th. Fr.

Physcia picta (Sw.) Nyl. Atranorin, divaricatic acid; zeorin (±). Sasaki (1942).

Physcia pityra Sandst., see *Physcia grisea* (Lam.) Zahlbr.

Physcia pulverulenta (Schreb.) Hampe   Atranorin. Vartia (1949), from Finland.
   REVIEW. Rhizoiden-green: Thies (1932), as *Parmelia pulverulenta*.

Physcia pulverulenta f. *venusta* (Ach.) Sandst.   No substances found. Galun and Lavee (1966), from Israel.

Physcia setosa (Ach.) Nyl.   Atranorin; volemitol. Dhar, Neelakantan, Ramanujam, and Seshadri (1959), from India, by extraction.

Physcia speciosa (Wulf.) Nyl., see *Anaptychia speciosa* (Wulf.) Mass.

Physcia stellaris (L.) Nyl.   Atranorin. Galun and Lavee (1966), from Israel, as *Physcia stellaris* f. *rosulata*; Hale (1957*a*), from Arkansas, Missouri, and Oklahoma; Vartia (1949), from Finland.
   REVIEW. Atranorin: Thies (1932).

Physcia stellaris var. *adscendens* Fr., see *Physcia adscendens* (Fr.) Oliv.

Physcia subtilis Degel.   Atranorin. Hale (1957*c*), from West Virginia, an exsiccati specimen.

Physcia syncolla Tuck.   No lichen substances found. Hale (1957*a*), from Arkansas and Missouri.

Physcia tenella (Scop.) DC.   Atranorin. Hale (1957*b*), mention only.
   REVIEWS. Atranorin: W. Brieger (1923); Hesse (1912); Thies (1932); Zopf (1907).

Physcia tribacoides Nyl.   Atranorin. Hale (1957*a*), from Arkansas, Missouri, and Oklahoma; Vartia (1949), from Finland.

Placodium alphoplacum (Wahlenb.) Link, see *Lecanora alphoplaca* (Wahlenb.) Ach.

Placodium chrysoleucum (Lam.) Link., see *Lecanora rubina* (Vill.) Ach.

Placodium circinatum (Pers.) S. Gray, see *Lecanora radiosa* (Hoffm.) Schaer.

Placodium circinatum var. *radiosum* (Hoffm.) Körb., see *Lecanora radiosa* (Hoffm.) Schaer.

Placodium circinatum var. *subcircinatum* (Nyl.) Zahlbr., see *Lecanora radiosa* var. *subcircinata* (Nyl.) Zahlbr.

Placodium cirrochroum (Ach.) Rabenh., see *Caloplaca cirrochroa* (Ach.) Th. Fr.

Placodium crassum (Huds.) Link, see *Squamarina crassa* (Huds.) Poelt

Placodium decipiens (Hedw.) Link, see *Lecidea decipiens* (Hedw.) Ach.

Placodium elegans (Link) DC., see *Xanthoria elegans* (Link) Th. Fr.

Placodium gypsaceum (Sm.) Trev., see *Squamarina gypsacea* (Sw.) Poelt

Placodium lagascae (Wahlenb.) Link, see *Squamarina lamarckii* (DC.) Poelt

Placodium lamarckii (DC.) Müll. Arg., see *Squamarina lamarckii* (DC.) Poelt

Placodium melanaspis (Ach.) Link, see *Lecanora melanaspis* (Ach.) Ach.

Placodium murorum (Hoffm.) DC., see *Caloplaca murorum* (Hoffm.) Th. Fr.

"*Placodium opacum*," see *Lecanora melanophthalma* (Ram.) Ram.

Placodium radiosum (Hoffm.) Ach., see *Lecanora radiosa* (Hoffm.) Schaer.

Placodium rubinum (Vill.) Müll. Arg., see *Lecanora rubina* (Vill.) Ach.

Placodium saxicolum (Hoffm.) Frege, see *Lecanora muralis* (Schreb.) Rabenh.

Placodium sympageum (Ach.) Bremme, see *Caloplaca aurantia* (Pers.) Hellb.

Placopsis gelida (L.) Linds.   Gyrophoric acid. Eigler and Poelt (1965), by paper chromatography; Hale (1956e), from North America, as *Lecanora gelida*.

Platismatia erosa W. Culb. & C. Culb.   Atranorin; caperatic acid. W. L. Culberson and Culberson (1968), from Formosa, Japan, Java, Nepal, the Philippines, Sikkim, and Vietnam, possibly also an unidentified yellow pigment, by microcrystal tests on 17 samples, type description.

Platismatia formosana (Zahlbr.) W. Culb. & C. Culb.   Atranorin; caperatic acid. W. L. Culberson and Culberson (1968), from Formosa, by microcrystal tests on two samples, possibly also an unidentified yellow pigment.

Platismatia glauca (L.) W. Culb. & C. Culb.   Atranorin; caperatic acid; arabitol, mannitol; iron, manganese, zinc.

Atranorin, caperatic acid. W. L. Culberson and Culberson (1968), world distribution mapped, by microcrystal tests on 11 samples; Dahl (1952), as *Platysma glaucum*; Tavares (1954), as *Platysma glaucum*, mention only; Vartia (1950b), from Finland, as *Cetraria glauca*.

Atranorin. Vartia (1949), from Finland; D. Hess (1958), from Germany, as *Cetraria glauca* and *C. fallax*.

Arabitol, mannitol. Lindberg, Misiorny, and Wachtmeister (1953), by paper chromatography, no volemitol, as *Cetraria glauca*.

Iron, manganese, zinc. Lounamaa (1965), as *Platysma glaucum*, a quantitative study.

OTHER REPORT. No usnic acid: Gertig and Banasiewicz (1961), from Poland, as *Cetraria glauca*.

REVIEWS. Atranorin, caperatic acid: W. Brieger (1923); Hesse (1912); Thies (1932); Zopf (1907); all review reports as *Cetraria glauca*.

Platismatia herrei (Imsh.) W. Culb. & C. Culb.   Atranorin; caperatic acid. W. L. Culberson and Culberson (1968), from North America, by microcrystal tests on five samples.

Platismatia interrupta W. Culb. & C. Culb.   Atranorin; caperatic acid. W. L. Culberson and Culberson (1968), from Japan and Sakhalin, by microcrystal tests on 20 samples, possibly also an unidentified yellow pigment, type description.

Platismatia lacunosa (Ach.) W. Culb. & C. Culb.   Atranorin, fumarprotocetraric acid; caperatic acid ($\pm$). W. L. Culberson and Culberson (1968), from extreme western North America, one sample without caperatic acid, by microchemical tests on 8 samples.

Fumarprotocetraric acid. Krog (1962), from Alaska, as *Cetraria lacunosa*, *C. lacunosa* var. *acharii*, and *C. lacunosa* var. *macounii*.

Platismatia norvegica (Lynge) W. Culb. & C. Culb.   Atranorin; caperatic acid. W. L. Culberson and Culberson (1968), from Canada, Miquelon Island, Norway, Scotland, Sweden, Alaska, Washington, and Oregon, by microcrystal tests on seven samples; Dahl (1952), mention only, as

Platysma norvegicum; Tavares (1954), mention only, as *Platysma norvegicum.*

Platismatia regenerans W. Culb. & C. Culb.   Atranorin; caperatic acid. W. L. Culberson and Culberson (1968), from Sabah, by microcrystal tests on four samples, type description.

Platismatia stenophylla (Tuck.) W. Culb. & C. Culb.   Atranorin; caperatic acid. W. L. Culberson and Culberson (1968), from western North America, by microcrystal tests on five samples.

Platismatia tuckermanii (Oakes) W. Culb. & C. Culb.   Atranorin; caperatic acid. W. L. Culberson and Culberson (1968), from Canada and the U.S.A., by microcrystal tests on seven samples.
   OTHER REPORT. Alectoronic acid, atranorin: Hale (1957b), mention only, as *Cetraria tuckermanii.*

Platysma chlorophyllum (Ach.) Vain., see *Cetraria scutata* (Wulf.) Poetsch

Platysma chrysanthum (Tuck.) Nyl., see *Asahinea chrysantha* (Tuck.) W. Culb. & C. Culb.

Platysma collatum Nyl., see *Cetrelia nuda* (Hue) W. Culb. & C. Culb.

Platysma complicatum (Laur.) Nyl., see *Cetraria laureri* Kremp.

Platysma cucullatum (Bell.) Hoffm., see *Cetraria cucullata* (Bell.) Ach.

Platysma diffusum (Web.) Nyl., see *Parmeliopsis hyperopta* (Ach.) Vain.

Platysma fahlunense (L.) Nyl., see *Cetraria commixta* (Nyl.) Th. Fr.

Platysma glaucum (L.) Frege, see *Platysmatia glauca* (L.) W. Culb. & C. Culb.

Platysma juniperinum (L.) Frege, see *Cetraria juniperina* (L.) Ach.

Platysma nivale (L.) Frege, see *Cetraria nivalis* (L.) Ach.

Platysma norvegicum (Lynge) Dahl, see *Platismatia norvegica* (Lynge) W. Culb. & C. Culb.

Platysma oakesianum (Tuck.) Nyl., see *Cetraria oakesiana* Tuck.

Platysma ochrocarpum Egg. (*nom. nud.*)   Atranorin, norstictic acid. Dahl (1952), mention only.
    Atranorin. Tavares (1954).

Platysma pinastri (Scop.) Frege, see *Cetraria pinastri* (Scop.) S. Gray

"*Platysma tubulosa* Schaer.," see *Cetraria alvarensis* (Wahlenb.) Vain.

Pleopsidium flavum f. *chlorophanum* (Wahlenb. *ex* Ach.) Korb., see *Acarospora chlorophana* (Wahlenb. *ex* Ach.) Mass.

Polycauliona regalis (Vain.) Hue   Parietin. Huneck and Follmann (1966*b*), from Chile, 0.12% by extraction of 26.0 g. of lichen.

Polychidium umhausense (Auersw.) Henss.   No substances found. C. F. Culberson (1967*b*), from France.

Porina lectissima (Fr.) Zahlbr.
    REVIEW. Lecanora-red: Thies (1932), as *Segestria lectissima*.

Porina mammillosa var. ***declivum*** (Bagl. & Carest.) Zahlbr.
    REVIEW. Lecanora-red, parmelia-brown: Thies (1932), as *Sagedia declivum*.

Protoblastenia ochracea (Hepp) Zahlbr.
    REVIEW. Parmelia-brown: Thies (1932), as *Biatora ochracea*.

Pseudevernia ceratea (Ach.) Zopf., see *Pseudevernia furfuracea* (L.) Zopf

Pseudevernia cladonia (Tuck.) Hale & W. Culb.   Atranorin, lecanoric acid. Hale (1955*a*), from North America, by microchemical tests on 55 samples; Hale (1957*b*), mention only.
    Lecanoric acid. Hale (1956*e*), from North America.
    All reports are as *Parmelia cladonia*.

Pseudevernia consocians (Vain.) Hale & W. Culb., see the note under *Pseudevernia intensa* (Nyl.) Hale & W. Culb.

Pseudevernia ericetorum (Fr.) Zopf, see *Pseudevernia furfuracea* (L.) Zopf

Pseudevernia furfuracea (L.) Zopf   Atranorin, chloroatranorin, physodic acid, unidentified substance ($FeCl_3$ + bluish); tetrahydroxy fatty acids; ergosterol, fungisterol; arabitol, erythritol, lichenin, mannitol; enzymes; folic acid-, folinic acid-, and vitamin B_{12}-group factors; iron, manganese, zinc. (For reports on samples with atranorin and olivetoric acid see

Pseudevernia olivetorina; for reports on samples with atranorin and lecanoric acid see *Pseudevernia intensa*.)

Atranorin, physodic acid, fatty acids, ergosterol, erythritol, lichenin. Zellner (1935), from Germany, as *Parmelia furfuracea*, no hypogymnol, no red pigment (=furfuracinic acid) previously reported by Zopf.

Atranorin, physodic acid, unidentified substance ($FeCl_3 +$ bluish). Nuno (1964), from Norway and Switzerland, by thin-layer chromatography, as *Parmelia furfuracea*.

Atranorin, physodic acid. C. F. Culberson (1965a), from Spain, a microchemical survey, some samples with olivetoric acid instead of physodic acid (see *Pseudevernia olivetorina*) and two samples with olivetoric acid and physodic acid, as *Parmelia furfuracea*; Dahl (1952), from Norway, mention only, as *Parmelia furfuracea*; D. Hess (1958), from Germany, by paper chromatography, as *Parmelia furfuracea*; Schatz (1963), mention only, as *Parmelia furfuracea*; Seshadri (1944), mention only, as *Parmelia furfuracea*; Shibata (1965), mention in discussion of biosynthesis, as *Parmelia furfuracea*; Solberg (1956), mention only, as *Parmelia furfuracea*.

Chloroatranorin. Koller and Pöpl (1934a), from Germany; Koller and Pöpl (1934b), by extraction; Spillane, Keane, and Nolan (1936), mention only; Yoshida (1951), mention only, as *Parmelia furfuracea*.

Physodic acid, unidentified substance (Barton-Evans-Gardner reagent + blue). Asahina and Nuno (1964), as *Parmelia furfuracea*.

Physodic acid. Asahina (1937e), as *Parmelia furfuracea*; Asahina (1938d), description of microcrystal tests, as *Parmelia furfuracea*; Erdtman and Wachtmeister (1957a), mention in discussion of biosynthesis, as *Parmelia furfuracea*; Krog (1951a), mention only, as *Parmelia furfuracea*; Lamb (1951b), as *Parmelia furfuracea* Chem. strain 1; Moissejeva (1961), enzymes tabulated, as *Parmelia furfuracea*; Stoll, Brack, and Renz (1950), as *Parmelia furfuracea*; Stoll, Renz, and Brack (1947), from Switzerland, by extraction, as *Parmelia furfuracea*; Vartia (1950b), from Finland, as *Parmelia furfuracea*; Wachtmeister (1958a), as *Parmelia furfuracea*.

Tetrahydroxy fatty acids. Solberg (1960), as *Parmelia furfuracea*.

Arabitol, mannitol. Lindberg, Misiorny, and Wachtmeister (1953), by paper chromatography, arabitol isolated, no volemitol, as *Parmelia furfuracea*.

Lichenin. Drake (1943), as *Parmelia furfuracea*; Karrer, Staub, and Staub (1924), as *Parmelia furfuracea*.

Folic acid-, folinic acid-, and vitamin B_{12}-group factors. Sjöström and Ericson (1953), as *Parmelia furfuracea*.

Iron, manganese, zinc. Lounamaa (1965), a quantitative study.

No usnic acid. Gertig and Banasiewcz (1961), from Poland.

OTHER REPORTS. Atranorin, physodalic acid, physodic acid: Vartia (1949), as *Parmelia furfuracea*. — Erythrin: Manaktala, Neelakantan, and Seshadri (1966), mention only, as *Evernia furfuracea*; V. S. Rao and

Seshadri (1942*b*), mention only, as *Parmelia furfuracea*. — A gelatinous polysaccharide: Votoček and Burda (1926), hydrolysis of various fractions described, as *Parmelia furfuracea*. — Usnic acid: Tomaselli (1957), from Italy, mention only, as *Parmelia furfuracea*.

REVIEWS. Apoolivoric acid, atranorin, evernuric acid, furevernic acid, fureverninic acid, furfuracinic acid, olivetoric acid, olivoric acid, physodylic acid (or as isidic acid): W. Brieger (1923), as *Evernia furfuracea*. — Atranorin, evernuric acid, furevernuric acid, fureverninic acid, furfurazinic acid, olivetoric acid, physodylic acid (as isidic acid): Hesse (1912), as *E. furfuracea*. — Atranorin, evernuric acid, furevernic acid, fureverninic acid: Thies (1932), as *Pseudevernia ceratea* (= *Pseudevernia furfuracea* var. *ceratea*). — Atranorin, furfurazinic acid, physodic acid: Thies (1932), and as *E. furfuracea* and *Pseudevernia ceratea*; Zopf (1907), as *Pseudevernia ceratea*. — Atranorin, physodic acid: Thies (1932); Zopf (1907), and a bitter substance; both reports as *Pseudevernia ericetorum*. —Atranorin, isidic acid, physodic acid: Zopf (1907). — Chloroatranorin, physodic acid, lichenin: Shibata (1958), as *Parmelia furfuracea*. — Chloroatranorin, physodic acid: Asahina and Shibata (1954); Shibata (1963); both reports as *Parmelia furfuracea*. — Fatty acids, ergosterol, fungisterol: Steiner (1957), as *Parmelia furfuracea*. — Isidic acid, physodic acid: Thies (1932), and as *E. isidiophora*. — Isidic acid: Hesse (1912), as *E. isidiophora*.

Pseudevernia intensa (Nyl.) Hale & W. Culb. Atranorin, chloroatranorin, lecanoric acid. Ahmann and Mathey (1967), from Arizona, 5.3% lecanoric acid, 1.1% atranorin-chloroatranorin mixture, by extraction and thin-layer chromatography.

Atranorin, lecanoric acid. C. F. Culberson (1965*a*), from Arizona, by microchemical tests, as *Parmelia furfuracea*; Hale (1955*a*), from North America, as *Parmelia furfuracea*; Hale (1956*a*), from Colorado, an exsiccati sample (see Hale, 1954*a*), reported to have atranorin and physodic acid, actually contains atranorin and lecanoric acid, as *Parmelia furfuracea*; Hale (1957*c*), from Virginia, an exsiccati sample, as *Parmelia furfuracea*.

Lecanoric acid. Hale (1956*e*), from North America, as *Parmelia furfuracea*.

Some reports as *Parmelia furfuracea* may be based on isidiate plants and are more properly referred to *Pseudevernia consocians*.

Pseudevernia isidiophora (Zopf) Zopf, see *Pseudevernia furfuracea* (L.) Zopf

Pseudevernia olivetorina (Zopf) Zopf Atranorin, olivetoric acid. Asahina (1937*c*), description of microcrystal tests, as *Parmelia olivetorina*; C. F. Culberson (1965*a*), from Spain, two samples also with physodic acid, as *Parmelia furfuracea*; Dahl (1952), from Norway, mention only, as

Hypogymnia furfuracea var. *olivetorina*; Hale (1956c), from Europe, as a chemical strain of *P. furfuracea*; Krog (1951b), as *H. furfuracea* var. *olivetorina*; Seshadri (1944), mention only, as *Evernia furfuracea*; Zopf (1900), as *E. furfuracea*, also an amorphous colorless substance and a yellow substance.

Olivetoric acid. Asahina (1937e), as *Parmelia olivetorina*; Lamb (1951b), as a chemical strain of *Parmelia furfuracea*; Wachtmeister (1958a), as *Parmelia furfuracea* var. *olivetorina*.

REVIEWS. Apoolivetoric acid, atranorin, olivetoric acid, olivoric acid: Thies (1932), as *Evernia furfuracea* var. *olivetorina*. — Atranorin, olivetoric acid: Zopf (1907). — Olivetoric acid: Shibata (1958, 1963), as *Parmelia olivetorina*.

Pseudevernia soralifera (Bitt.) Zopf
REVIEWS. Atranorin, physodic acid: Zopf (1907). — Physodic acid: Thies (1932).

Pseudocyphellaria aurata (Ach.) Vain.
Calycin, pulvinic acid, pulvinic dilactone (sometimes as stictaurin, the addition compound of pulvinic dilactone and calycin). Mitsuno (1955), by paper chromatography, as *Sticta aurata*; Hale (1957a), from Arkansas and Oklahoma.

Calycin, pulvinic dilactone. Asano and Kameda (1935a), as *Sticta aurata*; Mittal and Seshadri (1957), mention only.

Calycin. Bendz, Santesson, and Wachtmeister (1965d), used as an authentic source of this substance.

REVIEWS. Calycin, pulvinic acid, pulvinic dilactone: Shibata (1958, 1963). — Stictaurin: Asahina and Shibata (1954); W. Brieger (1923); Hesse (1912); Thies (1932); Zopf (1907). All review reports are as *Sticta aurata*.

Pseudocyphellaria crocata (L.) Vain.
Calycin, pulvinic dilactone (sometimes as pulvic anhydride or pulvinic anhydride). Maass, Towers, and Neish (1964), study on the biosynthesis of pulvinic dilactone; Maass and Neish (1967), from Canada, study on the biosynthesis of calycin.

OTHER REPORT. Epanorin: Mitsuno (1955), by paper chromatography, as *Sticta crocata*.

REVIEWS. Calycin: Asahina and Shibata (1954), as *Sticta crocata*. — Epanorin: Shibata (1958, 1963), as *Sticta crocata*. — Stictaurin (= calycin-pulvinic dilactone addition compound): W. Brieger (1923); Hesse (1912); Thies (1932); Zopf (1907); all four reports as *Stictina crocata*.

Psora limprichtii B. Stein, see *Lecidea limprichtii* (B. Stein) Zahlbr.

Psora lurida (Dill.) DC., see *Lecidea lurida* (Dill.) Ach.

Psora ostreata Hoffm., see *Lecidea scalaris* (Ach.) Ach.

Psoroma crassum (Huds.) S. Gray, see *Squamarina crassa* (Huds.) Poelt

Pycnothelia papillaria (Ehrh.) Duf.   Atranorin; (+)-protolichesterinic acid. Asahina (1942*a*), from Europe, not in Japan.
  Atranorin, protolichesterinic acid. Dahl (1952), mention only; Evans (1950).
  Atranorin. Evans (1944*b*), from Connecticut, and other lichen substances.
  Protolichesterinic acid. Evans (1955*a*).
  OTHER REPORTS. Atranorin, lichesterinic acid, protolichesterinic acid: Hale (1957*c*), from West Virginia, an exsiccati specimen. — Cladonin, protolichesterinic acid: Klosa (1951*b*).
  REVIEWS. Atranorin, cladonin, proto-α-lichesterinic acid: Thies (1932), as *Pycnothelia papillaria* var. *molariformis.* — Atranorin, proto-α-lichesterinic acid: W. Brieger (1923); Hesse (1912); both reports as *P. papillaria* var. *molariformis.* — (+)-Protolichesterinic acid: Asahina and Shibata (1954); Shibata (1958, 1963).
  Unless noted otherwise, all reports are as *Cladonia papillaria.*

Pyxine caesiopruinosa (Nyl.) Imsh.   Lichexanthone, unidentified phenolic acid; unidentified yellow pigment or pigments. W. L. Culberson and Hale (1965), unidentified yellow pigment or pigments in the medulla.

Pyxine coccifera (Fée) Nyl.   Pyxiferin.
  REVIEW. Pyxiferin. Neelakantan (1965).

Pyxine crysanthoides Vain.
  Probably lichexanthone. W. L. Culberson and Hale (1965), UV+, lectotype specimen examined.

Pyxine endochrysina Nyl.   Pyxinic acid. Yosioka, Matsuda, and Kitagawa (1966), a chemical study on this new triterpene.

Ramalina almquistii Vain.   Divaricatic acid, usnic acid. Asahina (1938*g*), from Japan.

Ramalina angustissima (Anzi) Vain., see *Ramalina subfarinacea* Nyl.

Ramalina armorica Nyl.   Stictic acid, usnic acid. Asahina (1938*g*).
  Stictic acid. Asahina (1939*f*), mention only; Asahina (1949*a*).
  REVIEWS. Armoric acid, armoricaic acid: W. Brieger (1923); Hesse (1912); Thies (1932).
  See also *Ramalina curnowii.*

Ramalina asahinana Zahlbr.   No depsides or depsidones. Asahina (1939*f*), from Formosa, Japan, and Korea.

Ramalina atlantica W. Culb.   Usnic acid only. C. F. Culberson (1965*c*), from France, Scotland, and Wales, as a chemical strain of *Ramalina siliquosa*, by microchemical tests.
   No medullary substances. W. L. Culberson (1967), from Scotland, type description.

Ramalina baltica Lett.   Usnic acid. Vartia (1950*b*), from Finland, as *Ramalina obtusata* f. *baltica*.

Ramalina boninensis Asah.   Boninic acid, (+)-usnic acid. Asahina and Kusaka (1937*a*), from the Bonin Islands, by extraction.
   Boninic acid. Asahina (1938*a*), type description; Asahina (1938*c*), description of microcrystal tests; Asahina (1939*b*); Asahina (1939*f*), and calcium oxalate, and *Ramalina boninensis* f. *subcalicariformis*.
   REVIEWS. Boninic acid: Asahina and Shibata (1954); Shibata (1958, 1963).

Ramalina boulhautiana Mah. & Gillet   Sekikaic acid, (+)-usnic acid; tetrahydroxy fatty acid mixture. Huneck and Trotet (1966), by extraction, aged form with 11.2% sekikaic acid, 0.02% (+)-usnic acid, and 0.5% tetrahydroxy fatty acid mixture (m.p. 185–186°), young form with 8% sekikaic acid, 0.01% (+)-usnic acid, 0.2% tetrahydroxy fatty acid mixture, and an unidentified red, medullary pigment.

Ramalina cactacearum Follm.   Methyl 3,5-dichlorolecanorate (as tumidulin), usnic acid. Huneck and Follmann (1967*a*), from Chile, 3.3% methyl 3,5-dichlorolecanorate by extraction of 18.0 g. of lichen, usnic acid by thin-layer chromatography.

Ramalina calicaris (L.) Fr.   Ramalinolic acid, sekikaic acid, (+)-usnic acid; D-arabitol, lichenin.
   Ramalinolic acid, sekikaic acid. Asahina (1938*c*), from Japan, discussion of microcrystal tests.
   Sekikaic acid, (+)-usnic acid, D-arabitol, lichenin. Mittal, Neelakantan, and Seshadri (1952), from India, by extraction, earlier literature reports discussed.
   OTHER REPORT. Evernic acid, obtusatic acid, (+)-usnic acid: Asahina and Fukuziro (1932*c*), from Japan, by extraction (but see *Ramalina commixta*).
   REVIEWS. Evernic acid, obtusatic acid, (+)-usnic acid: Thies (1932). — Ramalinolic acid, sekikaic acid, lichenin: Shibata (1958). — Ramalinolic acid, sekikaic acid: Asahina and Shibata (1954); Shibata (1963). — (+)-Usnic acid: W. Brieger (1923); Hesse (1912).

Ramalina calicaris var. *japonica* Hue   Salazinic acid (±), sekikaic acid (±), usnic acid; calcium oxalate. Asahina (1939*f*), from Japan, and also *Ramalina calicaris* var. *subfastigiata*.

Ramalina capitata (Ach.) Nyl.   (+)-Usnic acid. Stoll, Brack, and Renz (1947), 0.60% by extraction; Stoll, Renz, and Brack (1947), from Switzerland.
    Usnic acid. Tomaselli (1957), from Italy, mention only.

Ramalina ceruchis (Ach.) De Not.   Usnic acid; (−)-16α-hydroxykaurane (as ceruchdiol and as ceruchinol) (Chem. strain 1). Methyl 3,5-dichlorolecanorate (sometimes as tumidulin), usnic acid; (−)-16α-hydroxykaurane (Chem. strain 2). Bendz, Santesson, and Wachtmeister (1965*a*), from Chile, a detailed chemical study.
    Methyl 3,5-dichlorolecanorate, usnic acid, (−)-16α-hydroxykaurane. Follmann (1966), mention only.
    Methyl 3,5-dichlorolecanorate. Bendz, Santesson, and Wachtmeister (1965*b*), a detailed chemical study.
    REVIEWS. (+)-Usnic acid: W. Brieger (1923); Hesse (1912); Thies (1932). — Usnic acid: Zopf (1907).

Ramalina ceruchis var. *tumidula* (Tayl.) Nyl., see *Ramalina tumidula* (Tayl.) Hun. & Follm.

Ramalina ceruchoides Magn.   Usnic acid; (−)-16α-hydroxykaurane (as ceruchdiol and ceruchinol). Bendz, Santesson, and Wachtmeister (1965*a*), from Chile, a detailed chemical study; Follmann (1966), mention only.

Ramalina chilensis Bert.   Methyl 3,5-dichlorolecanorate (as tumidulin), norstictic acid (±), sekikaic acid, (+)-usnic acid, unidentified substance (±) (m.p. 294–295°, NaOCl+ yellow, FeCl₃+ wine-red). Huneck and Follmann (1966*a*), from Chile, 1.5% and 0.01% methyl 3,5-dichlorolecanorate, 0.18% and 0% norstictic acid, 3.6% and 4.0% sekikaic acid, 0.01% and 0.002% (+)-usnic acid, 0% and 0.01% unidentified substance, by extraction of two samples.
    Methyl 3,5-dichlorolecanorate. Huneck (1966*b*), mention only.

Ramalina combeoides Nyl.   Atranorin, stictic acid, usnic acid; (−)-16α-hydroxykaurane (as ceruchdiol and ceruchinol). Bendz, Santesson, and Wachtmeister (1965*a*), from California, a detailed chemical study; Follmann (1966), mention only.

Ramalina commixta Asah.   Evernic acid, obtusatic acid, (+)-usnic acid. Asahina and Fuzikawa (1932*c*), from Japan, reported as *Ramalina calicaris*, but see also Asahina (1939*f*); Mittal, Neelakantan, and Seshadri (1952), mention only.

Evernic acid, obtusatic acid. Asahina (1939*f*), from Japan, and calcium oxalate.

Evernic acid. Mitsuno (1953), by paper chromatography.

REVIEWS. Evernic acid, obtusatic acid: Asahina and Shibata (1954). — Evernic acid: Shibata (1958, 1963).

Ramalina crassa (Nyl.) Mot.   Atranorin ($\pm$), salazinic acid, usnic acid; D-arabitol.

Atranorin ($\pm$), salazinic acid, usnic acid. C. F. Culberson (1965*c*), from France, Germany, Sweden, and Wales, by microchemical tests, as a chemical strain of *Ramalina siliquosa*.

Salazinic acid, usnic acid, D-arabitol. Asahina (1949*a*), as *Ramalina scopulorum*; Asahina and Yanagita (1934), from Japan, by extraction, as *R. scopulorum*.

Salazinic acid. Asahina (1937*e*), from Japan, European samples with stictic acid (see *Ramalina curnowii*), as *R. scopulorum*; Asahina (1938*g*), from Europe, as *R. scopulorum*, suggestion that Curd and Robertson (1935) extracted *R. armorica* (see *R. curnowii*); Asahina (1939*f*), mention only, as *R. scopulorum* and *R. scopulorum* var. *minor*; W. L. Culberson (1967), distribution mapped, by microchemical tests, taxonomy clarified.

D-Arabitol. Asahina (1934*b*).

Ramalina crinalis (Ach.) Gyeln.

REVIEWS. ($-$)-Usnic acid: Thies (1932); Zopf (1907); both reports as *Alectoria crinalis*.

Ramalina curnowii Cromb. *ex* Nyl.   Atranorin ($\pm$), norstictic acid ($\pm$), stictic acid (sometimes as scopuloric acid), ($+$)-usnic acid; unidentified aliphatic substance; D-arabitol.

Atranorin ($\pm$), norstictic acid, stictic acid, usnic acid. C. F. Culberson (1965*c*), from England, France, Portugal, Scotland, and Sweden, by microchemical tests, as a chemical strain of *Ramalina siliquosa*.

Norstictic acid ($\pm$), stictic acid, unidentified substance ($\pm$). W. L. Culberson (1967), distribution mapped, by microchemical tests, taxonomy clarified.

Stictic acid, ($+$)-usnic acid, D-arabitol. Curd and Robertson (1935), from Ireland, by extraction, as *Ramalina scopulorum*; Ryan and O'Riordan (1917), from Ireland, as *R. scopulorum*.

Stictic acid, usnic acid. Asahina (1938*g*), from Europe, as *Ramalina cuspidata*; Solberg (1956), mention only, as *R. siliquosa*.

Stictic acid. Asahina, Yanagita, Hirakata, and Ida (1933), mention only, scopuloric acid identical to stictic acid, as *Ramalina scopulorum*.

($+$)-Usnic acid. Salkowski (1910), as *Ramalina scopulorum*.

D-Arabitol. Asahina (1934*b*), mention only, as *Ramalina scopulorum*; Asahina (1939*b*), mention only, as *R. scopulorum*; Mittal, Neelakantan, and Seshadri (1952), mention only, as *R. scopulorum*; Seshadri and Subramanian (1949*b*), mention only, as *R. scopulorum*.

REVIEWS. Stictic acid, (+)-usnic acid: W. Brieger (1923); Hesse (1912); Thies (1932); Zopf (1907). — Stictic acid, D-arabitol: Asahina and Shibata (1954); Shibata (1958, 1963). All review reports are as *Ramalina scopulorum*.

Ramalina cuspidata (Ach.) Nyl., see *Ramalina siliquosa* (Huds.) A. L. Sm.

Ramalina dilacerata (Hoffm.) Vain.   Divaricatic acid, (+)-usnic acid; calcium oxalate.

Divaricatic acid, usnic acid, calcium oxalate. Asahina (1938*g*), from Europe and Japan.

(+)-Usnic acid. Salkowski (1910), as *Ramalina minuscula*.

OTHER REPORT. Sekikaic acid: Seshadri and Subramanian (1949*b*), mention only.

REVIEWS. Obtusatic acid, sekikaic acid, (+)-usnic acid: Thies (1932), (see *Ramalina obtusata*). — (+)-Usnic acid: W. Brieger (1923); Hesse (1912); Thies (1932), as *R. minuscula*; Zopf (1907), and as *R. minuscula*.

Ramalina dilacerata var. *obtusata* (Arn.) Dalla Torre & Sarnth., see *Ramalina obtusata* (Arn.) Bitt. and *Ramalina sekika* Asah.

Ramalina druidarum W. Culb.   Atranorin, chloroatranorin, hypoprotocetraric acid, (+)-usnic acid; unidentified substance; D-arabitol. C. F. Culberson (1965*c*), from France, 0.0046% atranorin and chloroatranorin, 5.6% hypoprotocetraric acid, 0.12% (+)-usnic acid, 0.063% unidentified substance, 0.25% D-arabitol, by extraction, samples from England, France, Portugal, and Wales with atranorin (±), hypoprotocetraric acid, and usnic acid, by microchemical tests, as a chemical strain of *Ramalina siliquosa*.

Hypoprotocetraric acid. C. F. Culberson (1965*d*), from western Europe, as *Ramalina siliquosa*; W. L. Culberson (1967), distribution mapped, and an unidentified substance (±), type description; Huneck and Lehn (1966), mention only, coquimboic acid found to be identical to hypoprotocetraric acid.

Ramalina ecklonii (Spreng.) Meyen & Flot.

Usnic acid. Follmann (1964), mention only.

REVIEWS. Protocetraric acid (as ramalinic acid), (+)-usnic acid: W. Brieger (1923); Hesse (1912); Thies (1932). — Unidentified colorless acid, (+)-usnic acid: Zopf (1907).

All reports are as *Ramalina yemensis*.

Ramalina ecklonii var. *ambigua* Mont. Usnic acid; aspicilin. Huneck and Follmann (1967*a*), from Chile, 0.001% usnic acid, 0.002% aspicilin, by extraction of 60.0 g. of lichen.

Ramalina exilis Asah. Usnic acid. Asahina (1939*f*), from Japan, no depside present (FeCl₃−), and also *Ramalina exilis* f. *nana*, type descriptions.

Ramalina farinacea (L.) Ach. Protocetraric acid (sometimes as ramalinic acid), (+)-usnic acid; arabitol, mannitol; enzymes.
Protocetraric acid, (+)-usnic acid. Asahina (1937*e*).
Protocetraric acid, usnic acid. Asahina (1938*e*), from Europe, description of microcrystal tests; Galun and Lavee (1966), from Israel; D. Hess (1958), from Germany, by paper chromatography.
Protocetraric acid. C. F. Culberson (1965*d*), from France (see also *Ramalina hypoprotocetrarica*); W. L. Culberson (1966*b*), see also *R. hypoprotocetrarica*; Koller, Krakauer, and Pöpl (1934), from Germany, by extraction, ramalinic acid found to be identical to protocetraric acid; J. Santesson (1965).
(+)-Usnic acid. Moissejeva (1957), 0.75%; Moissejeva (1961), 0.75%, enzymes tabulated; Salkowski (1910).
Usnic acid. Tomaselli (1957), from Italy, mention only; Vartia (1949), from Finland; Vartia (1950*b*), from Finland.
Arabitol, mannitol. Lindberg, Misiorny, and Wachtmeister (1953), by paper chromatography, no volemitol.
OTHER REPORTS. Sekikaic acid, usnic acid: Asahina and Nonomura (1933), from Japan, by extraction (see *Ramalina nervulosa*). — Norstictic acid, sekikaic acid, usnic acid: Rangaswami and Rao (1954*a*), from India, by extraction of two samples. — Sekikaic acid: Seshadri and Subramanian (1949*b*), mention only. — Fumarprotocetraric acid: Asahina (1938*e*), description of microcrystal tests; Wachtmeister (1956*a*), used as an authentic source of this substance.
REVIEWS. Norstictic acid: Asahina and Shibata (1954). — Protocetraric acid, (+)-usnic acid: W. Brieger (1923); Hesse (1912); Thies (1932); Zopf (1907). — Protocetraric acid: Shibata (1958, 1963). — (+)-Usnic acid: Savicz, Litvinov, and Moissejeva (1960), 0.75%.

Ramalina farinacea var. *nervulosa* Müll. Arg., see *Ramalina nervulosa* (Müll. Arg.) Abb.

Ramalina fastigiata (Pers.) Ach. Usnic acid. Hale (1957*a*), from Arkansas, Missouri, and Oklahoma; Tomaselli (1957), from Italy, mention only; Vartia (1949), from Finland, as *Ramalina populina*; Vartia (1950*b*), from Finland, as *R. populina*.
OTHER REPORT. Usnic acid, unidentified substance: Klosa (1951*b*), as *Ramalina populina*.

REVIEWS. (+)-Usnic acid: W. Brieger (1923); Hesse (1912); Thies (1932), as *Ramalina populina*; Zopf (1907), as *R. populina*.

Ramalina flabelliformis Asah.   Usnic acid. Asahina (1955*a*), from India, no other substances found, type description.

Ramalina flaccescens Nyl.   Methyl 3,5-dichlorolecanorate (sometimes as tumidulin), usnic acid; (−)-16α-hydroxykaurane (as ceruchinol). Follmann (1966).

Methyl 3,5-dichlorolecanorate, usnic acid. Bendz, Santesson, and Wachtmeister (1965*a*), from Chile, a detailed chemical study.

Methyl 3,5-dichlorolecanorate. Bendz, Santesson, and Wachtmeister (1965*b*), a detailed chemical study.

Ramalina fraxinea (L.) Ach.   (+)-Usnic acid; D-arabitol. Briner, Gream, and Riggs (1960), from Australia, by extraction, 0.19% (+)-usnic acid, references to early work included.

Usnic acid. Dean (1952), mention only; Hale (1957*a*), from Arkansas and Oklahoma; Tomaselli (1947), from Italy, mention only; Vartia (1949), from Finland; Vartia (1950*b*), from Finland.

OTHER REPORT. Evernic acid, obtusatic acid (as ramalic acid), usnic acid: Klosa (1951*b*).

REVIEWS. (+)-Usnic acid: W. Brieger (1923); Hesse (1912); Thies (1932); Zopf (1907). — Usnic acid: Neelakantan (1965); Shibata (1958).

Ramalina geniculata Hook. f. & Tayl.   Ramalinolic acid, sekikaic acid, (+)-usnic acid; D-arabitol; calcium oxalate.

Ramalinolic acid, sekikaic acid, usnic acid, calcium oxalate. Asahina (1938*g*), from the Bonin Islands, Formosa, and Japan.

Ramalinolic acid, sekikaic acid. Asahina (1938*c*), from Japan, description of microcrystal tests.

Sekikaic acid, (+)-usnic acid. Asahina and Nonomura (1933), from Japan, by extraction.

Sekikaic acid, D-arabitol. Seshadri and Subramanian (1949*b*), mention only.

D-Arabitol. Asahina (1934*b*); Asahina (1939*b*), mention only; Asahina and Yanagita (1934); Mittal, Neelakantan, and Seshadri (1952), mention only.

REVIEWS. Ramalinolic acid, sekikaic acid, D-arabitol: Asahina and Shibata (1954); Shibata (1958, 1963).

Ramalina homalea Ach.   Divaricatic acid (±), usnic acid; (−)-16α-hydroxykaurane (as ceruchdiol and as ceruchinol). Bendz, Santesson, and Wachtmeister (1965*a*), from California, a detailed chemical study.

Divaricatic acid, usnic acid, (−)-16α-hydroxykaurane. Follmann (1966), mention only.

Divaricatic acid. Hale (1957*b*), mention only.

Ramalina hypoprotocetrarica W. Culb. Hypoprotocetraric acid. C. F. Culberson (1965*d*), from France, as *Ramalina farinacea* Chem. strain; W. L. Culberson (1966*a*), from France, type description.

Ramalina inanis Mont. Methyl 3,5-dichlorolecanorate, (+)-usnic acid. Huneck and Follmann (1966*c*), from Chile, 1.76% methyl 3,5-dichloro-lecanorate, 0.08% (+)-usnic acid, by extraction of 25 g. of lichen.

Ramalina intermediella Vain. Ramalinolic acid, sekikaic acid, usnic acid; calcium oxalate. Asahina (1939*f*), almost no usnic acid.
  REVIEWS. Ramalinolic acid, sekikaic acid: Asahina and Shibata (1954); Shibata (1958, 1963).

Ramalina kullensis Zopf, see *Ramalina siliquosa* (Huds.) A. L. Sm.

Ramalina landroënsis Zopf (+)-Usnic acid. Salkowski (1910).
  REVIEWS. (+)-Usnic acid, landroensin: Thies (1932); Zopf (1907). — (+)-Usnic acid: W. Brieger (1923); Hesse (1912).

Ramalina ligulata (Ach.) Brandt Evernic acid, obtusatic acid, usnic acid; calcium oxalate. Asahina (1939*f*), from Japan and Sakhalin.

Ramalina litoralis Asah. *non* Zahlbr. Divaricatic acid, usnic acid; calcium oxalate. Asahina (1939*f*), from the Bonin Islands and Japan, type description.

Ramalina minuscula Nyl., see *Ramalina dilacerata* (Hoffm.) Vain.

Ramalina nervosa (Nyl.) Räs. Usnic acid. Vartia (1950*b*), from Finland, by extraction.

Ramalina nervulosa (Müll. Arg.) Abb. Ramalinolic acid, sekikaic acid. Asahina and Kusaka (1936), by extraction; Asahina (1938*c*), from Japan, description of microcrystal tests.
  Sekikaic acid. Asahina and Nonomura (1933), by extraction, as an unknown variety of *Ramalina farinacea*.
  REVIEW. Ramalinolic acid, sekikaic acid: Asahina and Shibata (1954). Except where indicated, all reports are as *Ramalina farinacea* var. *nervulosa*.

Ramalina obtusata (Arn.) Bitt. Obtusatic acid, (+)-usnic acid.
  Obtusatic acid, usnic acid. Vartia (1949), from Finland.
  Obtusatic acid. Asahina (1937*d*), description of microcrystal tests.
  (+)-Usnic acid. Salkowski (1910).

Usnic acid. Vartia (1950*b*), from Finland.

OTHER REPORT. Obtusatic acid, sekikaic acid, (+)-usnic acid: Nakao (1923), as *Ramalina dilacerata* var. *obtusata*, the principal plant in the Chinese drug "Shi-hoa," misdetermined *R. sekika* according to Asahina and Shibata (1954).

REVIEWS. Obtusatic acid, ramalinellic acid: Thies (1932), ramalinellic acid possibly identical to evernic acid; Zopf (1907). — Obtusatic acid, usnic acid: Asahina and Shibata (1954). — (+)-Usnic acid: W. Brieger (1923); Hesse (1912).

Ramalina obtusata f. *baltica* (Lett.) Räs., see *Ramalina baltica* Lett.

Ramalina pacifica Asah.   Salazinic acid, usnic acid. Asahina (1939*f*), also a trace of sekikaic acid in one sample, type description.

Ramalina peruviana Ach.   Methyl 3,5-dichlorolecanorate (as tumidulin). Huneck (1966*b*), mention in structural study on this substance; Huneck and Follmann (1965*c*), from Chile, 4% by extraction of 5 g. of lichen, as *Ramalina peruviana* var. *pollinariaeformis*.

Ramalina pollinaria (Westr.) Ach.   Evernic acid, obtusatic acid (sometimes as ramalic acid), (+)-usnic acid; enzymes.

Evernic acid, obtusatic acid, (+)-usnic acid. Asahina and Nonomura (1933), from Europe, Japan, and Manchuria, mention only.

Evernic acid, obtusatic acid, usnic acid. Asahina (1939*f*), from Europe, mention only; Koller (1932*a*), by extraction.

Evernic acid, obtusatic acid. Asahina (1937*d*), description of microcrystal tests.

Evernic acid. Klosa (1950), mention only; W. L. Culberson (1966*b*), from Austria, Bulgaria, England, Finland, France, Germany, Hungary, Italy, Norway, Portugal, Spain, Switzerland, Sweden, and the U.S.S.R.

Obtusatic acid. Koller and Pfeiffer (1933*c*), ramalic acid found to be identical to obtusatic acid; Mitsuno (1953).

Usnic acid. Vartia (1950*b*), from Finland.

Enzymes. Moissejeva (1961), enzymes tabulated.

REVIEWS. Atranorin, evernic acid, obtusatic acid, (+)-usnic acid: W. Brieger (1923); Hesse (1912); Thies (1932). — Evernic acid, obtusatic acid, (+)-usnic acid: Zopf (1907). — Evernic acid, obtusatic acid: Asahina and Shibata (1954); Shibata (1958, 1963).

Ramalina pollinariella Nyl., see *Ramalina roesleri* (Hochst.) Nyl.

Ramalina polymorpha (Ach.) Ach.   Usnic acid; arabitol, mannitol.
Usnic acid. Vartia (1950*b*), Finland.
Arabitol, mannitol. Lindberg, Misiorny, and Wachtmeister (1953), by paper chromatography, mannitol isolated, no volemitol.
Mannitol. Pueyo (1960), mention only.

Ramalina populina (Hoffm.) Vain., see *Ramalina fastigiata* (Pers.) Ach.

Ramalina pseudosekika Asah. Evernic acid, obtusatic acid, usnic acid. Asahina (1941*d*), from Japan, type description.

Ramalina reagens (B. de Lesd.) W. Culb. Norstictic acid, salazinic acid. W. L. Culberson (1966*b*), from Canary Islands, Germany, Italy, Norway, Scotland, Sweden, and the U.S.S.R.

Ramalina reticulata (Noehd.) Kremp. (+)-Usnic acid; D-arabitol.
(+)-Usnic acid. Marshak, Barry, and Craig (1947); Vartia (1950*a*), mention only.
Usnic acid, D-arabitol. Stark, Walter, and Owens (1950), by extraction.
Usnic acid. Barry, O'Rourke, and Twomey (1947), mention only; F. T. Jones and Palmer (1950); Marshak (1947); Schindler (1957), mention only; Stoll, Brack, and Renz (1950), mention only.
REVIEW. Usnic acid: Asahina and Shibata (1954).

Ramalina roesleri (Hochst.) Nyl. Sekikaic acid, usnic acid. Asahina (1938*g*), from Japan, one sample with divaricatic acid instead of sekikaic acid.
OTHER REPORT. Obtusatic acid: Hale (1957*b*), as *Ramalina pollinariella*.

Ramalina scopulorum (Retz.) Ach., see *Ramalina crassa* (Nyl.) Mot. and *Ramalina curnowii* Cromb. *ex* Nyl.

Ramalina scopulorum "var. *minor* A. L. Sm.," see *Ramalina crassa* (Nyl.) Mot.

Ramalina sekika Asah. Sekikaic acid, usnic acid. Asahina (1941*d*), from Manchuria, type description.
REVIEW. Sekikaic acid: Asahina and Shibata (1954), see also *Ramalina obtusata*.

Ramalina siliquosa (Huds.) A. L. Sm. Atranorin (±), protocetraric acid (sometimes as kullensic acid), (+)-usnic acid.
Atranorin (±), protocetraric acid, usnic acid. C. F. Culberson (1965*c*), from France, as a chemical strain of *Ramalina siliquosa*.
Protocetraric acid, usnic acid. Asahina (1938*g*), from Norway, as *Ramalina cuspidata*; Asahina (1949*a*), as *R. cuspidata*.
Protocetraric acid. W. L. Culberson (1967), distribution mapped, taxonomy clarified.
(+)-Usnic acid. Salkowski (1910), as *Ramalina cuspidata*.
OTHER REPORTS. Unidentified substance, (+)-usnic acid: Ryan and O'Riordan (1917), from Ireland, by extraction, review of early literature included, as *Ramalina cuspidata*. — Cuspidatic acid: Solberg (1956), mention, as *R. cuspidata*.

REVIEWS. Cuspidatic acid: W. Brieger (1923); Hesse (1912); Thies (1932); Zopf (1907); all four reports as *Ramalina cuspidata*. — Protocetraric acid, (+)-usnic acid: W. Brieger (1923); Hesse (1912); Thies (1932); Zopf (1907); all four reports as *R. kullensis*.

See also *Ramalina atlantica, R. crassa, R. curnowii, R. druidarum*, and *R. stenoclada*.

Ramalina sinensis Jatta   (+)-Usnic acid; D-arabitol, isolichenin, lichenin; amino acids; ascorbic acid, riboflavin; calcium, iron, nitrogen, phosphorus.

(+)-Usnic acid, D-arabitol, lichenin. Mittal, Neelakantan, and Seshadri (1952), from India, by extraction.

Isolichenin, lichenin, ascorbic acid, riboflavin, calcium, iron, nitrogen, phosphorus. Lal and Rao (1959), from India.

Lichenin. Mittal and Seshadri (1954b), mention only.

Amino acids. Ramakrishnan and Subramanian (1966b), from India, alanine and tyrosine as free amino acids, these and arginine, aspartic acid, glutamic acid, glycine, isoleucine, leucine, lysine, methionine, serine, threonine, tryptophan, tyrosine, and an unidentified substance as combined amino acids.

Ramalina stenoclada W. Culb.   Atranorin ($\pm$), norstictic acid, usnic acid. C. F. Culberson (1965c), from England, France, and Wales, by microchemical tests, as a chemical strain of *Ramalina siliquosa*.

Norstictic acid. W. L. Culberson (1967), from England, by microchemical tests, type description.

Ramalina strepsilis (Ach.) Zahlbr.   (+)-Usnic acid.

REVIEWS. (+)-Usnic acid: W. Brieger (1923); Hesse (1912); Thies (1932); Zopf (1907).

Ramalina subbreviuscula Asah.   Divaricatic acid; calcium oxalate. Asahina (1939f), from Japan, usnic acid implied, type description.

Ramalina subfarinacea Nyl.   Norstictic acid. W. L. Culberson (1966b).

OTHER REPORTS. Salazinic acid, usnic acid: Galun and Lavee (1966), from Israel. — Salazinic acid: Salkowski (1910).

REVIEWS. Norstictic acid: Asahina and Shibata (1954); Shibata (1958, 1963); all three reports as *Ramalina angustissima*. — Salazinic acid, (+)-usnic acid: W. Brieger (1923); Hesse (1912); Thies (1932); Zopf (1907).

Ramalina subgeniculata Nyl.   Evernic acid, obtusatic acid; calcium oxalate. Asahina (1938g), from Formosa and Japan.

Ramalina sublitoralis Asah. Ramalinolic acid, sekikaic acid. Asahina (1939*f*), from Japan, usnic acid implied, type description.

Ramalina tayloriana Zahlbr. Sekikaic acid, (+)-usnic acid; D-arabitol, lichenin. Rangaswami and Rao (1954*a*), mention only; Seshadri and Subramanian (1949*b*).
Sekikaic acid, (+)-usnic acid, D-arabitol. Mittal, Neelakantan, and Seshadri (1952), mention only; Neelakantan and Seshadri (1952*a*), mention only.
(+)-Usnic acid. Murty and Subramanian (1959*b*), used as an authentic source of this substance.
D-Arabitol. Asahina (1951*i*), mention only.
REVIEWS. Sekikaic acid: Shibata (1958, 1963).

Ramalina terebrata Hook. f. & Tayl. (+)-Usnic acid. Huneck and Follmann (1964*a*), by extraction.
Usnic acid. Lamb (1964), from Graham Land.

Ramalina thrausta (Ach.) Nyl. (+)-Usnic acid. Salkowski (1910).
Usnic acid. Tomaselli (1957), from Italy, mention only; Vartia (1949), from Finland; Vartia (1950*b*), from Finland; Zopf (1900), from Tirol.
REVIEWS. (+)-Usnic acid: W. Brieger (1923); Hesse (1912); Thies (1932); Zopf (1907).

Ramalina tigrina Follm. Psoromic acid, (+)-usnic acid; (−)-16α-hydroxykaurane (sometimes as ceruchinol). Huneck and Follmann (1966*c*), from Chile, 0.082% psoromic acid, 0.014% (+)-usnic acid, 2.7% (−)-16α-hydroxykaurane, by extraction of 73.0 g. of lichen.
Psoromic acid, usnic acid, (−)-16α-hydroxykaurane. Follmann (1966), from Chile, type description.

Ramalina tumidula (Tayl.) Hun. & Follm. Hypoprotocetraric acid (sometimes as coquimboic acid), methyl 3,5-dichlorolecanorate (sometimes as tumidulin), (+)-usnic acid; (−)-16α-hydroxykaurane (as ceruchinol). Huneck and Follmann (1965*b*), from Chile, 1% hypoprotocetraric acid, 0.31% methyl 3,5-dichlorolecanorate, 0.08% (+)-usnic acid, 7.5% (−)-16α-hydroxykaurane, by extraction of 67 g. of lichen, as *Ramalina ceruchis* var. *tumidula*.
Hypoprotocetraric acid, methyl 3,5-dichlorolecanorate, usnic acid, (−)-16α-hydroxykaurane. Follmann (1966), mention only.
Hypoprotocetraric acid, methyl 3,5-dichlorolecanorate, (−)-16α-hydroxykaurane. Follmann and Villagrán (1965), mention only.
Hypoprotocetraric acid. Huneck and Lehn (1966), coquimboic acid found to be identical to hypoprotocetraric acid.

Methyl 3,5-dichlorolecanorate. Huneck (1966*b*), structural study on this substance.
Usnic acid. Follmann and Villagrán (1964), as *Ramalina ceruchis* var. *tumidula*.
(−)-16α-Hydroxykaurane. Lehn and Huneck (1965), a study on the identification of ceruchinol, found to be identical to (−)-16α-hydroxy-kaurane.

Ramalina usnea (L.) R. H. Howe   Ramalinolic acid, sekikaic acid. Asahina (1938*c*), description of microcrystal tests.
   REVIEWS. Ramalinolic acid, sekikaic acid: Asahina and Shibata (1954). — Ramalinolic acid: Shibata (1958, 1963).
   All reports are as *Ramalina usneoides*.

Ramalina usneoides (Ach.) Mont., see *Ramalina usnea* (L.) R. H. Howe

Ramalina yemensis (Ach.) Nyl., see *Ramalina ecklonii* (Spreng.) Meyen & Flot.

Raphiospora flavovirescens (Dicks.) Mass., see *Bacidia citrinella* (Ach.) Branth & Rostr.

Reinkella lirellina Darb.
   REVIEWS. Hydroxyroccellic acid or roccellic acid, roccellin: Thies (1932), as *Roccella lirellina*; Zopf (1907). — Roccellic acid, roccellin: W. Brieger (1923); Hesse (1912).

Rhizocarpon alpicola (Wahlenb.) Rabenh., see *Rhizocarpon oreites* (Vain.) Zahlbr.

Rhizocarpon atroalbescens (Nyl.) Zahlbr.   Norstictic acid or psoromic acid; rhizocarpic acid. Runemark (1956), circumpolar distribution described, psoromic acid in samples from Scandinavia and eastern Asia.

Rhizocarpon atroflavescens Lynge ssp. **atroflavescens**   Rhizocarpic acid. Runemark (1956), from Norway and Novaya Zemlya.

Rhizocarpon atroflavescens ssp. **pulverulentum** (Schaer.) Run.
   "Probably rhizocarpic acid and psoromic acid." Runemark (1956), from Scandinavia.

Rhizocarpon carpaticum Run.
   "Probably rhizocarpic acid and psoromic acid." Runemark (1956), from the Carpathian Mountains, type description.

Rhizocarpon cinereovirens (Müll. Arg.) Vain. Norstictic acid. Schindler (1936).

Rhizocarpon cookeanum Magn. Unidentified substance (K+ red). Runemark (1956), from Idaho and Washington, probably also rhizocarpic acid.

Rhizocarpon copelandii (Körb.) Th. Fr. Stictic acid. Hale (1957*b*), mention only.

Rhizocarpon dispersum Run. Norstictic acid; rhizocarpic acid. Runemark (1956), from Norway, type description.

Rhizocarpon distinctum Th. Fr.
REVIEW. Lecidea-green: Thies (1932).

Rhizocarpon effiguratum (Anzi) Th. Fr. Gyrophoric acid ($\pm$), psoromic acid; rhizocarpic acid. Runemark (1956), from southern Europe.

Rhizocarpon ferax Magn.
"Probably rhizocarpic acid and psoromic acid." Runemark (1956), from arctic North America, northeastern Siberia, and Scandinavia.

Rhizocarpon geographicum (L.) DC. Barbatic acid, psoromic acid; rhizocarpic acid.
Barbatic acid, psoromic acid. Wachtmeister (1956*a*), used as an authentic source of these substances.
Psoromic acid, rhizocarpic acid. Hale (1957*b*), mention only.
Psoromic acid. Hale (1956*f*); J. Santesson (1965).
Rhizocarpic acid. Asahina (1951*i*), mention only; Frank, Cohen, and Coker (1950); Mittal and Seshadri (1957), mention only.
REVIEWS. Barbatic acid, rhizocarpic acid: Asahina and Shibata (1954); Shibata (1958, 1963). — Psoromic acid: Thies (1932), and lecanora-red. — Rhizocarpic acid: Zopf (1907).

Rhizocarpon geographicum var. *contiguum* (Schaer.) Räs., see *Rhizocarpon tinei* ssp. *diabasicum* (Räs.) Run.

Rhizocarpon geographicum var. *geronticum* (Ach.) Räs., see *Rhizocarpon oreites* (Vain.) Zahlbr.

Rhizocarpon geographicum var. *lecanorinum* Körb., see *Rhizocarpon lecanorinum* (Körb.) And.

Rhizocarpon grande (Flörke *ex* Flot.) Arn. Gyrophoric acid, stictic acid. Hale (1957*b*), mention only.

Gyrophoric acid. Hale (1956e), from North America.
REVIEW. Parmelia-brown: Thies (1932), as *Rhizocarpon petraeum*.

Rhizocarpon inarense (Vain.) Vain.   Gyrophoric acid ($\pm$), norstictic acid ($\pm$), psoromic acid ($\pm$); rhizocarpic acid. Runemark (1956), plants from northeastern Asia sometimes lacking norstictic acid.

Rhizocarpon intermediellum Räs.
"Probably rhizocarpic acid." Runemark (1956), from the Alps, Greenland, and Scandinavia.

Rhizocarpon lecanorinum (Körb.) And.   Gyrophoric acid ($\pm$), stictic acid; rhizocarpic acid. Runemark (1956), from temperate Europe and North America.
REVIEWS. Psoromic acid (sometimes as parellic acid), rhizocarpic acid: W. Brieger (1923); Hesse (1912); Thies (1932), and lecanora-red; Zopf (1907); all four reports as *Rhizocarpon geographicum* var. *lecanorinum*. — Psoromic acid: Shibata (1958, 1963), as *R. geographicum* var. *lecanorinum*.

Rhizocarpon lindsayanum Räs., see *Rhizocarpon riparium* Räs.

Rhizocarpon lindsayanum ssp. *kittilense* (Räs.) Run., see *Rhizocarpon riparium* Räs.

Rhizocarpon lusitanicum (Nyl.) Arn.
"Probably rhizocarpic acid." Runemark (1956), from Portugal.

Rhizocarpon macrosporum Räs.   Stictic acid ($\pm$, rarely $+$); rhizocarpic acid. Runemark (1956), from Asia Minor, the Caucasus Mountains, Europe, and Scandinavia, stictic acid found in only three of 130 samples tested microchemically.

Rhizocarpon norvegicum Räs.
"Probably rhizocarpic acid and sometimes psoromic acid." Runemark (1956), from the Alps, Carpathian Mountains, Greenland, and Scandinavia.

Rhizocarpon obscuratum (Ach.) Mass.
REVIEW. Lecidia-green: Thies (1932).

Rhizocarpon oederi (G. Web.) Körb.   Copper, iron. Lange and Ziegler (1963), a quantitative study.

Rhizocarpon oportense (Vain.) Räs.   Stictic acid ($\pm$); rhizocarpic acid. Runemark (1956), from the Iberian Peninsula, by paper chromatography (one sample) and spot tests (nine samples).

Rhizocarpon oreites (Vain.) Zahlbr. Gyrophoric acid ($\pm$), psoromic acid (sometimes as parellic acid); rhizocarpic acid. Runemark (1956), from Europe, *Rhizocarpon alpicola*.

REVIEWS. Psoromic acid, rhizocarpic acid: W. Brieger (1923); Hesse (1912); Thies (1932); Zopf (1907), and rhizocarpinic acid; all review reports as *Catocarpus oreites* and as *Rhizocarpon geographicum* var. *geronticum*.

Rhizocarpon parvum Run.
"Probably rhizocarpic acid." Runemark (1956), from Canada, Greenland, and Norway.

Rhizocarpon petraeum (Wulf.) Mass., see *Rhizocarpon grande* (Flörke *ex* Flot.) Arn.

Rhizocarpon plicatile (Leight.) A. L. Sm. Norstictic acid. Schindler (1936), as *Rhizocarpon rubescens*.

Rhizocarpon pusillum Run.
"Probably rhizocarpic acid and psoromic acid." Runemark (1956), from southern Europe, type description.

Rhizocarpon ridescens (Nyl.) Zahlbr.
"Probably rhizocarpic acid and psoromic acid." Runemark (1956), from Romania.

Rhizocarpon riparium Räs. Barbatic acid or psoromic acid, gyrophoric acid ($\pm$); rhizocarpic acid. Runemark (1956), from Europe and North and South America, as *Rhizocarpon lindsayanum*, also *R. lindsayanum* ssp. *kittilense* with psoromic acid ($\pm$) and rhizocarpic acid, from Europe and Scandinavia.
Barbatic acid or psoromic acid. Wachtmeister (1958*a*), two chemical strains, as *Rhizocarpon lindsayanum*.

Rhizocarpon rubescens Th. Fr., see *Rhizocarpon plicatile* (Leight.) A. L. Sm.

Rhizocarpon saanaënse Räs.
Unidentified substance; rhizocarpic acid. Runemark (1956), from the Alps, the Arctic, Scandinavia, and Yunnan.

Rhizocarpon sorediosum Run. Barbatic acid or (probably) psoromic acid; rhizocarpic acid. Runemark (1956), from Germany, Italy, and Fenno-scandia.

Rhizocarpon sphaerosporum Räs.   Barbatic acid or stictic acid, psoromic acid ($\pm$, rarely +); rhizocarpic acid. Runemark (1956), from France, Germany, and Scandinavia.

Rhizocarpon sublucidum Räs.   Psoromic acid ($\pm$, usually +); rhizocarpic acid. Runemark (1956), from Ethiopia, Europe, India, Greenland, North Africa, and the U.S.A.

Rhizocarpon subtile Run.
"Probably rhizocarpic acid." Runemark (1956), from southwestern Greenland.

Rhizocarpon superficiale (Schaer.) Vain.   Stictic acid or norstictic acid, psoromic acid ($\pm$); rhizocarpic acid. Runemark (1956), from Europe, rarely with neither stictic nor norstictic acid.

Rhizocarpon superficiale ssp. *boreale* Run.   Stictic acid or norstictic acid; rhizocarpic acid. Runemark (1956), from Canada, Greenland, Iceland, Norway, Novaya Zemlya, Scandinavia, and Scotland, type description.

Rhizocarpon superficiale ssp. *splendidum* (Malme) Run.   Norstictic acid; rhizocarpic acid. Runemark (1956), from Norway and Sweden.

Rhizocarpon superficiale ssp. *superficiale*   Norstictic acid or stictic acid, psoromic acid ($\pm$, rarely +); rhizocarpic acid. Runemark (1956), from Africa and Europe.

Rhizocarpon tavaresii Räs.
"Probably rhizocarpic acid and stictic acid or psoromic acid." Runemark (1956), from Africa and Portugal.

Rhizocarpon tetrasporum Run.   Stictic acid; rhizocarpic acid. Runemark (1956), from southern Europe.

Rhizocarpon tinei (Torn.) Run.   Barbatic acid ($\pm$, rarely +), gyrophoric acid ($\pm$), psoromic acid; rhizocarpic acid. Runemark (1956), and also *Rhizocarpon tinei* ssp. *arcticum*, *R. t.* ssp. *diabasicum*, *R. t.* ssp. *frigidum*, *R. t.* ssp. *prospectans*, and *R. t.* ssp. *vulgare*.
Barbatic acid, psoromic acid. Galun and Lavee (1966), as *Rhizocarpon geographicum* var. *contiguum*.
REVIEWS. Barbatic acid, psoromic acid (sometimes as parellic acid), rhizocarpic acid: W. Brieger (1923); Hesse (1912); Thies (1932), and lecanora-red; Zopf (1907); all review reports as *Rhizocarpon geographicum* var. *contiguum*.

Rhizocarpon viridiatrum (Wulf.) Körb. Gyrophoric acid (±), perhaps physodic acid (±), stictic acid (±); rhizocarpic acid. Runemark (1956), from Europe.
Rhizocarpic acid. Zopf (1900).
REVIEWS. Rhizocarpic acid: Asahina and Shibata (1954); W. Brieger (1923); Hesse (1912); Shibata (1958, 1963); Thies (1932), and lecanora-red; Zopf (1907).

"*Rhizoplaca chrysoleuca*," see *Lecanora rubina* (Vill.) Ach.

"*Rhizoplaca opaca*," see *Lecanora melanophthalma* (Ram.) Ram.

Rinodina caluliformis W. Web. Atranorin, norstictic acid. Weber (1964), from Mexico, type description.

Rinodina ocellata (Hoffm.) Arn.
REVIEW. Urcellaria-red: Thies (1932), as *Urceolaria ocellata*.

Rinodina oreina (Ach.) Mass. Fumarprotocetraric acid, usnic acid (Chem. strain 1). Gyrophoric acid, usnic acid (Chem. strain 2). Usnic acid only (Chem. strain 3). (+)-Usnic acid; zeorin.
Gyrophoric acid, usnic acid (Chem. strain 2). Usnic acid only (Chem. strain 3). Hale (1952), from North America, a third chemical strain reported to contain protocetraric acid with usnic acid (Chem. strain 1) later shown to be fumarprotocetraric acid and usnic acid, mentions that Zopf found usnic acid and zeorin in European samples; Hale (1954b), from Baffin Island and North America.
Fumarprotocetraric acid (Chem. Strain 1). Gyrophoric acid (Chem. strain 2). Černohorský (1959), mention only; Hale (1957b), mention only.
Gyrophoric acid. Hale (1956e), from North America.
REVIEWS. (+)-Usnic acid, zeorin: Thies (1932); Zopf (1907). — Zeorin: W. Brieger (1923); Hesse (1912). All review reports are as *Dimelaena oreina*.

Roccella arboricola Follm. Lecanoric acid. Huneck and Follmann (1967b), from Chile, 3.7% by extraction of 0.4 g. of lichen.

Roccella babingtonii Mont. Erythrin; roccellic acid; erythritol.
Erythrin. Asahina (1936e), descriptions of microcrystal tests; Schindler (1956/1957), from Chile, Ecuador, Peru, and California.
Erythritol. Wehmer, Thies, and Hadders (1932).
REVIEWS. Erythrin, hydroxyroccellic acid, roccellic acid: W. Brieger (1923); Hesse (1912), and orcin; Thies (1932); Zopf (1907). — Roccellic acid: Asahina and Shibata (1954); Shibata (1958, 1963).
All reports are as *Roccella peruensis*.

Roccella canariensis Darb.
    REVIEWS. Erythrin, hydroxyroccellic acid, roccellic acid: Thies (1932).
— Lecanoric acid: W. Brieger (1923); Hesse (1912); Thies (1932); Zopf
(1907).

Roccella cervicornis Follm.   Norstictic acid.   Huneck and Follmann
(1967*b*), from Chile, 0.6% by extraction of 1.0 g. of lichen.

Roccella decipiens Darb.   Erythrin; roccellic acid. Hale (1957*b*), mention
only.
    Erythrin. Hale (1956*e*), from North America.
    REVIEWS. Roccellic acid: Thies (1932); Zopf (1907).

Roccella fuciformis (L.) DC.   Erythrin; roccellic acid; arabitol, erythritol,
galactose, mannitol, tagatose, unidentified di- and trisaccharides; picroroccellin; choline sulfate ester.
    Erythrin, picroroccellin. Asahina (1951*i*), mention only.
    Erythrin. Asahina (1936*e*), description of microcrystal tests; Schindler
(1956/1957), from western Africa, the Canary Islands, and western
Europe, mention only.
    Picroroccellin. Birkinshaw and Mohammed (1962), mention biosyn-
thesis; Forster, Onslow, and Saville (1922), detailed chemical study;
Jones, Keane, and Nolan (1944), mention only; Shibata (1965), mention
in discussion of biosynthesis; Subramanian and Swamy (1961).
    Arabitol, erythritol, galactose, mannitol, tagatose, unidentified di- and
trisaccharides, choline sulfate ester. Lindberg (1955), unidentified di-
and trisaccharides hydrolyzed to glucose and tagatose.
    Erythritol. Pueyo (1963*b*), from France, by paper chromatography;
Pueyo (1963*d*); Wehmer, Thies, and Hadders (1932).
    OTHER REPORT. No fructose, galactose, glucose, or sucrose: Pueyo
(1963*a*), from France, by paper chromatography.
    REVIEWS. Erythrin, hydroxyroccellic acid, picroroccellin, roccellic
acid: W. Brieger (1923); Hesse (1912), and β-erythrin; Thies (1932); Zopf
(1907). — Erythrin, picroroccellin, roccellic acid: Asahina and Shibata
(1954); Shibata (1958, 1963). — Picroroccellin: Neelakantan (1965).

Roccella fucoides (Dicks.) Vain.   Roccellic acid. Bendz, Santesson, and
Tibell (1966), used as an authentic source of this substance.

Roccella hypomecha (Ach.) Bory   Arabitol, erythritol, mannitol, unidentified di- and trisaccharides; choline sulfate ester. Lindberg (1955), unidentified di- and trisaccharides hydrolyzed to glucose and tagatose.

Roccella intricata Mont., see *Roccellaria mollis* (Hampe) Zahlbr.

Roccella linearis (Ach.) Vain. Arabitol, erythritol, galactose, mannitol, tagatose, unidentified di- and trisaccharides, choline sulfate ester. Lindberg (1955), unidentified di- and trisaccharides hydrolyzed to glucose and tagatose.

Roccella lirellina (Darb.) Choisy, see *Reinkella lirellina* Darb.

Roccella montagnei Bél. Erythrin, lecanoric acid, (+)-montagnetol (sometimes as picroerythrin), (±)-montagnetol, orcinol; roccellic acid; erythritol, isolichenin, lichenin; ergosterol; β-carotene, γ-carotene; free and combined amino acids; ascorbic acid, riboflavin; calcium, iron, nitrogen, phosphorus.

Erythrin, lecanoric acid, montagnetol, orcinol, roccellic acid, erythritol, isolichenin. V. S. Rao and Seshadri (1941), from India, first extraction of montagnetol.

Erythrin, lecanoric acid, montagnetol, orcinol, roccellic acid, erythritol. Manaktala, Neelakantan, and Seshadri (1966), mention in study on the synthesis of (±)-montagnetol and erythrin; V. S. Rao and Seshadri (1940), from India, references to early work included.

Erythrin, montagnetol, orcinol, roccellic acid, ergosterol, β-carotene, amino acids (especially leucine). Subramanian (1965), mention only.

Erythrin, orcinol, roccellic acid, ergosterol, carotene. Subramanian and Swamy (1961), from India, 3.0% erythritol, 1.4% erythrin, 2.3% roccellic acid, and 2.0–2.5 mg/g total carotene, by extraction.

Erythrin, lecanoric acid, (+)-montagnetol, (±)-montagnetol, roccellic acid, erythritol, β-carotene. Neelakantan and Seshadri (1952a), mention only.

Erythrin, lecanoric acid, montagnetol, orcinol, roccellic acid, erythritol, β-carotene, γ-carotene. Murty and Subramanian (1959c), separation of carotenes.

Erythrin, roccellic acid, β-carotene. Seshadri and Subramanian (1949b).

Erythrin, montagnetol. Asahina (1951i), mention only.

Erythrin. Asahina (1936e), description of microcrystal tests; Sakurai (1941), from Java, detailed chemical study; Schindler (1956/1957), from Africa, China, India, and Indonesia, mention only.

Montagnetol, orcinol. Seshadri (1944), mention only.

(+)-Montagnetol, (±)-montagnetol. V. S. Rao and Seshadri (1942a), from India.

Orcinol. Neelakantan and Seshadri (1959b), mention only.

Erythritol. Wehmer, Thies, and Hadders (1932).

Isolichenin, lichenin. Chanda, Hirst, and Manners (1957), mention only.

Isolichenin. Mittal and Seshadri (1954a), improved method of isolation; Mittal and Seshadri (1954b), mention only.

Ergosterol. Murty and Subramanian (1959d), mention β-carotene.

β-Carotene, γ-carotene. Murty and Subramanian (1959*a*), from India.
β-Carotene. Murty and Subramanian (1958), from India.

Free and combined amino acids. Ramakrishnan and Subramanian (1964), from India, alanine, arginine, glycine, leucine, threonine, tyrosine, and valine as free amino acids, and these with lysine, phenylalanine, and serine as combined amino acids; Ramakrishnan and Subramanian (1965), additional amino acids found by paper chromatography, aspartic acid, glutamic acid, methionine, tryptophan, and an unidentified substance.

Ascorbic acid, riboflavin, calcium, iron, nitrogen, phosphorus. Lal and Rao (1959), from India.

REVIEWS. Erythrin, hydroxyroccellic acid, roccellic acid: Thies (1932); Zopf (1907). — Erythrin, hydroxyroccellic acid: W. Brieger (1923); Hesse (1912). — Erythrin, roccellic acid, *meso*-erythritol, montagnetol: Asahina and Shibata (1954). — Erythrin, montagnetol, *meso*-erythritol, roccellic acid: Shibata (1958), and isolichenin; Shibata (1963). — Montagnetol, roccellic acid, ergosterol, β-carotene: Neelakantan (1965).

Roccella peruensis (Kremp.) Darb., see *Roccella babingtonii* Mont.

Roccella phycopsis (Ach.) Ach.   Erythrin; erythritol.

Erythrin. Asahina (1936*e*), description of microcrystal tests; Hale (1956*e*), from North America; Hale (1957*b*), mention only; Schindler (1956/1957), from the Canary Islands, Great Britain, Madagascar, and New Caledonia, mention only.

Erythritol. Wehmer, Thies, and Hadders (1932).

REVIEWS. Erythrin, hydroxyroccellic acid, erythritol: W. Brieger (1923); Hesse (1912); Thies (1932); Zopf (1907).

Roccella portentosa (Mont.) Darb.   Lecanoric acid; portentol.

Lecanoric acid. Follmann (1964), mention only; Follmann (1965); Follmann and Villagrán (1964); Follmann and Villagrán (1965); Schindler (1956/1957), from Chile and Peru, mention only.

Portentol. Huneck and Trotet (1967), from Galápagos, mention only.

REVIEWS. Erythrin, hydroxyroccellic acid, lecanoric acid, roccellic acid: Thies (1932), according to an old report. — Lecanoric acid, roccellic acid: Zopf (1907). — Lecanoric acid: W. Brieger (1923); Hesse (1912).

Roccella sinensis Nyl.   Lecanoric acid. Schindler (1956/1957), from China, mention only.

REVIEWS. Lecanoric acid, roccellic acid: Thies (1932); Zopf (1907). — Lecanoric acid: W. Brieger (1923); Hesse (1912).

Roccella tinctoria DC.

Erythrin. Manaktala, Neelakantan, and Seshadri (1966), mention in study on the synthesis of (±)-erythrin and (±)-montagnetol; V. S. Rao and Seshadri (1942*b*), mention only.

Lecanoric acid, orcinol. Seshadri (1944), mention only.

Lecanoric acid. Asahina (1936*e*), description of microcrystal tests; Schindler (1956/1957), from Africa and Europe, mention only.

Psoromic acid. Huneck and Follmann (1963), mention of an early report.

REVIEWS. Lecanoric acid, psoromic acid (sometimes as parellic acid), hydroxyroccellic acid, roccellic acid, roccellin: W. Brieger (1923); Hesse (1912), and orcin; Thies (1932). — Lecanoric acid, psoromic acid, hydroxyroccellic acid, roccellic acid: Zopf (1907). — Roccellic acid: Asahina and Shibata (1954); Shibata (1958, 1963).

Roccellaria intricata (Mont.) Darb., see *Roccellaria mollis* (Hampe) Zahlbr.

Roccellaria mollis (Hampe) Zahlbr.

Zeorin. Follmann (1964), as *Roccellaria intricata*, mention only; Neelakantan, Rangaswami, and Rao (1954), as *Roccella intricata*, mention only.

REVIEWS. Roccellaric acid, roccellic acid, zeorin: Thies (1932), report of zeorin doubtful; Zopf (1907), report of zeorin doubtful. — Roccellaric acid: W. Brieger (1923); Hesse (1912). All review reports are as *Roccella intricata*.

Sagedia declivum Bagl. & Carest., see *Porina mammillosa* var. *declivum* (Bagl. & Carest.) Zahlbr.

Sarcogyne cyclocarpa (Anzi) J. Stein.

REVIEW. Parmelia-brown: Thies (1932), as *Lithographa cyclocarpa*.

Sarcogyne pruinosa (Sm.) Körb., see *Sarcogyne regularis* Körb.

Sarcogyne regularis Körb.

REVIEW. Lecanora-red: Thies (1932), as *Sarcogyne pruinosa*.

Segestria lectissima Fr., see *Porina lectissima* (Fr.) Zahlbr.

Siphula ceratites (Wahlenb.) Fr.   Siphulin; siphulitol.

Siphulin. Bruun (1960), from Norway, by extraction, a detailed chemical study; Bruun (1965), chemical structure revised.

Siphulitol. Lindberg and Meier (1962), from Norway, a new acyclic polyol.

OTHER REPORT. Unidentified substance $(C_{19}H_{21}O_8)$: Solberg (1960).
REVIEWS. Siphulin, siphulitol: Neelakantan (1965). — Siphulin: Shibata (1958, 1963).

Siphula decumbens Nyl.   Decarboxythamnolic acid, thamnolic acid. Bendz, Santesson, and Wachtmeister (1965c), from New Zealand, decarboxythamnolic acid possibly only an artifact produced by thamnolic acid during its identification.

Siphula dissoluta (Nyl.) Zahlbr.   Hypothamnolic acid. Bendz, Santesson, and Wachtmeister (1965c), from New Zealand, a detailed chemical study.

Siphula fastigiata (Nyl.) Nyl.   Hypothamnolic acid. Bendz, Santesson, and Wachtmeister (1965c), from Argentina, as a new and unnamed variety.

Siphula moorei Zahlbr.   Baeomycesic acid, squamatic acid. Bendz, Santesson, and Wachtmeister (1965c), from New Zealand, a detailed chemical study.

Siphula roccellaeformis Nyl.   Hypothamnolic acid. Bendz, Santesson, and Wachtmeister (1965c), from New Zealand, a detailed chemical study.

Siphula torulosa (Thunb. *ex* Ach.) Nyl.   Baeomycesic acid, squamatic acid. Bendz, Santesson, and Wachtmeister (1965c), from South Africa, a detailed chemical study.

Siphula verrucigera (Gmel.) Sant.   Baeomycesic acid, squamatic acid. Bendz, Santesson, and Wachtmeister (1965c), from South Africa, a detailed chemical study.

Solorina crocea (L.) Ach.   Solorinic acid; D-mannitol.
    Solorinic acid. Koller and Russ (1937), from the Alps, by extraction; Neelakantan and Seshadri (1960a), mentioned in discussion of biosynthesis; Shibata (1965), mentioned in discussion of biosynthesis; Subramanian (1955).
    D-Mannitol. Wehmer, Thies, and Hadders (1962).
    Mannitol. Pueyo (1960), mention only.
    OTHER REPORT. Solorinic acid, solorinin, mannitol: Zopf (1909).
    REVIEWS. Hydrosolorinol, soloric acid, solorinic acid, solorinin, mannitol: Thies (1932). — Hydrosolorinol, soloric acid, solorinic acid: W. Brieger (1923). — Solorinic acid, D-mannitol: Asahina and Shibata (1954). — Solorinic acid: Hesse (1912); Shibata (1958, 1963); Zopf (1907).

Solorina saccata (L.) Ach.   Mannitol, unidentified substance (C+ red). Zopf (1909), from Tirol, unidentified substance not solorinin.

Sphaeromphale clopimoides (Bagl. & Carest.) Arn., see *Staurothele clopimoides* (Bagl. & Carest.) J. Stein.

Sphaerophorus compressus Ach., see *Sphaerophorus melanocarpus* (Sw.) DC.

Sphaerophorus coralloides Pers., see *Sphaerophorus globosus* (Huds.) Vain.

Sphaerophorus formosanus (Zahlbr.) Asah.   Sphaerophorin.   Mitsuno (1938), from Japan.

Sphaerophorus fragilis (L.) Pers.   Sphaerophorin; fragilin; arabitol, mannitol; ascorbic acid.
   Sphaerophorin. Asahina (1938*b*), description of microcrystal tests; Klosa (1951*b*); Mitsuno (1953), by paper chromatography; Ramaut (1961*c*), by extraction; Vartia (1949), from Finland; Vartia (1950*b*); Wachtmeister (1952).
   Arabitol, mannitol. Lindberg, Misiorny, and Wachtmeister (1953), by paper chromatography, no volemitol.
   Ascorbic acid. Karev and Kochevykh (1962).
   OTHER REPORT. Sphaerophorin, squamatic acid (as sphaerophoric acid): Mitsuno (1938), from Japan.
   REVIEWS. Fragilin, sphaerophorin, squamatic acid (as sphaerophoric acid): W. Brieger (1923); Hesse (1912); Thies (1932), sphaerophoric acid identical to ventosaric acid; Zopf (1907). — Sphaerophorin: Asahina and Shibata (1954); Shibata (1958, 1963).

Sphaerophorus globosus (Huds.) Vain.   Sphaerophorin; fragilin.
   Sphaerophorin. Asahina (1938*b*), description of microcrystal tests; Klosa (1951*b*), mention only, as *Sphaerophorus coralloides*.
   OTHER REPORT. Sphaerophorin, squamatic acid: Hale (1957*b*), mention only.
   REVIEWS. Fragilin, sphaerophorin, squamatic acid (as sphaerophoric acid): W. Brieger (1923); Hesse (1912); Thies (1932), sphaerophoric acid identical to ventosaric acid; Zopf (1907); all four reports as *Sphaerophorus coralloides*. — Sphaerophorin: Asahina and Shibata (1954), and as *S. coralloides*; Shibata (1958, 1963), as *S. coralloides*.

Sphaerophorus globosus f. *meiophorus* (Nyl.) Zahlbr., see *Sphaerophorus meiophorus* (Nyl.) Vain.

Sphaerophorus meiophorus (Nyl.) Vain.   Sphaerophorin, squamatic acid (sometimes as isosquamatic acid). Asahina and Hashimoto (1934), as *Sphaerophorus globosus* f. *meiophorus*; Mitsuno (1938), from Japan.

Sphaerophorus melanocarpus (Sw.) DC.   Sphaerophorin. Asahina (1938*b*), description of microcrystal tests; D. Hess (1958), from Germany; Mitsuno (1938), from Japan.

    OTHER REPORT. Fragilin, sphaerophorin: Asahina and Hashimoto (1934), from Japan, by extraction.

    REVIEWS. Sphaerophorin: Asahina and Shibata (1954), and as *Sphaerophorus compressus*; Shibata (1958, 1963).

Sphaerophorus turfaceus Asah.   Sphaerophorin, squamatic acid. Mitsuno (1938), from the Aleutian Islands, Kamchatka, Lapland, and North America, type description.

Sphyridium placophyllum (Ach.) Th. Fr., see *Baeomyces placophyllus* Ach.

Squamaria alphoplaca (Wahlenb.) Duby, see *Lecanora alphoplaca* (Wahlenb.) Ach.

Squamaria chrysoleuca (Lam.) Duby, see *Lecanora rubina* (Vill.) Ach.

Squamaria circinata (Ach.) Hook., see *Lecanora radiosa* (Hoffm.) Schaer.

Squamaria crassa (Huds.) DC., see *Squamarina crassa* (Huds.) Poelt

Squamaria gypsacea (Sm.) Nyl., see *Squamarina gypsacea* (Sm.) Poelt

Squamaria lagascae (Ach.) Balb., see *Squamarina lamarckii* (DC.) Poelt

Squamaria lamarckii (DC.) Nyl., see *Squamarina lamarckii* (DC.) Poelt

Squamaria melanaspis (Ach.) Elenk., see *Lecanora melanaspis* (Ach.) Ach.

"*Squamaria opaca*," see *Lecanora melanophthalma* (Ram.) Ram.

Squamaria radiosa (Hoffm.) Poetsch, see *Lecanora radiosa* (Hoffm.) Schaer.

Squamaria saxicola (Poll.) Howitt, see *Lecanora muralis* (Schreb.) Rabenh.

Squamarina concrescens (Müll. Arg.) Poelt   Psoromic acid, usnic acid. Eigler and Poelt (1965), from Liguria and Sicily, by paper chromatography.

Squamarina crassa (Huds.) Poelt   Psoromic acid (sometimes as parellic acid), (–)-usnic acid. Asahina and Hayashi (1933), mention only, as *Lecanora crassa*.

Psoromic acid, usnic acid. Eigler and Poelt (1965), from Germany, Gotland, and Italy, by paper chromatography.

Psoromic acid. Asahina (1943*d*), mention only, as *Lecanora crassa*; Murphy, Keane, and Nolan (1943), mention only, as *Squamaria crassa*.

REVIEWS. Psoromic acid, (–)-usnic acid: W. Brieger (1923); Hesse (1912); Thies (1932); all three reports as *Placodium crassum*; Zopf (1907), as *Squamaria crassa*. — Psoromic acid: Asahina and Shibata (1954); Shibata (1958, 1963); all three reports as *Lecanora crassa* and as *Psoroma crassum*.

Squamarina crassa var. *platyloba* (Matt.) Poelt   Psoromic acid, unidentified substance, usnic acid. Eigler and Poelt (1965), from Madeira, by paper chromatography, unidentified substance H.

Squamarina crassa f. *pseudocrassa* (Matt.) Poelt   Usnic acid, unidentified substance. Eigler and Poelt (1965), from Germany, by paper chromatography, unidentified substance G.

Squamarina gypsacea (Sm.) Poelt   Psoromic acid (sometimes as parellic acid), usnic acid. Eigler and Poelt (1965), from Czechoslovakia, by paper chromatography.

REVIEWS. Psoromic acid, (–)-usnic acid: W. Brieger (1923); Hesse (1912); Thies (1932); all three reports as *Placodium gypsaceum*; Zopf (1907), as *Squamaria gypsacea*. — Psoromic acid: Thies (1932), as *Lecanora gypsacea*.

Squamarina lamarckii (DC.) Poelt   Psoromic acid (sometimes as parellic acid), usnic acid. Eigler and Poelt (1965), from France, by paper chromatography.

REVIEWS. Psoromic acid, (–)-usnic acid: W. Brieger (1923), as *Placodium lamarckii* and *P. lagascae*; Hesse (1912), as *P. lamarckii* and *P. lagascae*; Thies (1932), as *P. lamarckii*, *P. lagascae*, and *Squamaria lamarckii*. — Psoromic acid, usnic acid: Zopf (1907), as *Squamaria lamarckii* and *Squamaria lagascae*.

Squamarina lentigera (G. Web.) Poelt   Usnic acid. Eigler and Poelt (1965), from Spain, by paper chromatography.

Squamarina nivalis Frey & Poelt   Usnic acid, unidentified substance. Eigler and Poelt (1965), from Germany, by paper chromatography, unidentified substance F.

Squamarina oleosa (Zahlbr.) Poelt   Psoromic acid, usnic acid. Eigler and Poelt (1965), from France, by paper chromatography, small concentration of usnic acid.

Squamarina periculosa (Duf.) Poelt   Psoromic acid, unidentified substance, usnic acid. Eigler and Poelt (1965), from France, by paper chromatography, unidentified substance E.

Staurothele clopimoides (Bagl. & Carest.) J. Stein.
REVIEW. Lecanora-red: Thies (1932), as *Sphaeromphale clopimoides*.

Stereocaulon albicans Th. Fr.
Fumarprotocetraric acid. Duvigneaud (1942*a*).
Unidentified substance (Chem. strain 1). Atranorin only (Chem. strain 2). Stictic acid (?) (Chem. strain 3). Lamb (1951*b*), with fatty acid accessory substances.

Stereocaulon alpinum Laur.   Atranorin, lobaric acid (sometimes as stereocaulic acid or usnetinic acid), norstictic acid. Ramaut (1962*a*), from Austria, France, and Switzerland, by paper chromatography.
Atranorin, lobaric acid. Duvigneaud (1942*a*); D. Hess (1958), from Germany, by paper chromatography.
Lobaric acid. Asahina and Nonomura (1935), mention only.
REVIEWS. Atranorin, lobaric acid: W. Brieger (1923); Hesse (1912); Thies (1932); Zopf (1907). — Lobaric acid: Asahina and Shibata (1954); Shibata (1958, 1963).

Stereocaulon antarcticum Vain.   Atranorin, lobaric acid; β-sitosterol. Huneck and Follmann (1966*d*), from Chile, 0.01% atranorin, 0.31% lobaric acid, 0.002% β-sitosterol, by extraction of 80 g. of lichen.

Stereocaulon aogasimense Asah.   Atranorin, norstictic acid, stictic acid. Asahina (1955*b*), from Aogashima, type description.

Stereocaulon arbuscula Nyl.   Squamatic acid (Chem. strain 1). Thamnolic acid (Chem. strain 3). Lamb (1951*b*).
Thamnolic acid. Duvigneaud (1942*a*).

Stereocaulon arbuscula var. *aberrans* Asah.   Fumarprotocetraric acid, unidentified substance. Asahina (1943*e*), from Japan, type description, unidentified substance in the GE solution.
Fumarprotocetraric acid. Lamb (1951*b*), as *Stereocaulon arbuscula* Chem. strain 2.

Stereocaulon argus Hook. f. & Tayl.   Stictic acid. Lamb (1951*b*).

Stereocaulon botryosum Ach.   Stictic acid. Lamb (1961), as *Stereocaulon spathuliferum*, mention only.
OTHER REPORT. Atranorin, lobaric acid: Duvigneaud (1942*a*).

Stereocaulon botryosum f. *congestum* (Magn.) Frey   Atranorin, unidentified substance. Ramaut (1962*a*), from Sweden, by paper chromatography.

Stereocaulon claviceps Th. Fr.   Atranorin ($\pm$), stictic acid. Lamb (1951*b*).

Stereocaulon condensatum Hoffm.   Atranorin. Lamb (1951*b*).
REVIEWS. Atranorin: Thies (1932); Zopf (1907).

Stereocaulon confluens Müll. Arg.   Atranorin, stictic acid. Duvigneaud (1942*a*).

Stereocaulon coniophyllum Lamb   Atranorin, lobaric acid. Lamb (1961), from Canada, India, Japan, Norway, Sweden, and the U.S.A., type description.

Stereocaulon coralligerum Meyer   Atranorin, divaricatic acid, unidentified substance. Duvigneaud (1942*a*), the unidentified substance also found in *Stereocaulon ramulosum*.

Stereocaulon coralloides Fr., see *Stereocaulon dactylophyllum* Flörke

Stereocaulon corticulatum var. *procerum* Lamb   Atranorin, lecanoric acid; parietin. Huneck and Follmann (1965*e*), from Chile, 0.66% atranorin, 1.7% lecanoric acid, and 0.02% parietin by extraction of 30 g. of lichen.

Stereocaulon curtatum Nyl.
Atranorin only. Lamb (1951*b*).
Atranorin, unidentified substance. Asahina (1961*b*), from Japan, unidentified substance found in the GE solution.

Stereocaulon dactylophyllum Flörke   Atranorin ($\pm$), stictic acid ($\pm$) (sometimes as pseudopsoromic acid), norstictic acid ($\pm$); copper, iron, zinc.
Atranorin, norstictic acid, stictic acid. Ramaut (1962*a*), from France, by paper chromatography; Ramaut and Schumacker (1962), from Belgium.
Atranorin, norstictic acid. Bachmann (1962), by paper chromatography, as *Stereocaulon coralloides*; D. Hess (1958), from Germany, by paper chromatography, as *S. coralloides*.
Atranorin ($\pm$), stictic acid. Lamb (1951*b*), as *Stereocaulon coralloides*.
Atranorin, stictic acid. Solberg (1956), mention only, as *Stereocaulon coralloides*.

Copper, iron. Lange and Ziegler (1963), a quantitative study.
Zinc. Lambinon, Maquinay, and Ramaut (1964), from Belgium, a quantitative study.
OTHER REPORT. Lobaric acid: Hale (1957b), mention only, as *Stereocaulon coralloides.*
REVIEWS. Atranorin, stictic acid: W. Brieger (1923); Hesse (1912); Thies (1932); Zopf (1907); all review reports as *Stereocaulon coralloides.*

Stereocaulon dendroides Asah., see *Stereocaulon octomerum* Müll. Arg.

Stereocaulon denudatum Flörke, see *Stereocaulon vesuvianum* Pers.

Stereocaulon denudatum var. *genuinum* Th. Fr., see *Stereocaulon vesuvianum* Pers.

Stereocaulon denudatum "var. *pulvinatum*," see *Stereocaulon vesuvianum* var. *pulvinatum* (Schaer.) Duncan

Stereocaulon depreaultii Del. Atranorin, lobaric acid. Asahina (1960c), from Japan.

Stereocaulon evolutoides (Magn.) Frey Atranorin, lobaric acid, norstictic acid. Ramaut (1962a), from Sweden, by paper chromatography.

Stereocaulon evolutum Graewe Atranorin, lobaric acid, norstictic acid. Ramaut (1962a), from France, by paper chromatography.

Stereocaulon exile Asah. Atranorin, norstictic acid, stictic acid. Asahina (1960c), from Japan, type description.

Stereocaulon exutum Nyl. Atranorin, lobaric acid. Asahina and Nonomura (1935), from Japan, by extraction; Asahina (1961b), from Japan, by microcrystal tests; Lamb (1951b); Kurokawa and Jinzenji (1965), by thin-layer chromatography.
REVIEWS. Lobaric acid: Asahina and Shibata (1954); Shibata (1958, 1963).

Stereocaulon flavireagens Gyeln.
Salazinic acid or α-methylethersalazinic acid or norstictic acid. Lamb (1951b).

Stereocaulon foliolosum var. ***strictum*** (Bab.) Lamb Stictic acid; D-arabitol; unidentified substance (m.p. 181–183°). Aghoramurthy, Sarma, and Seshadri (1961b), from India, by extraction.

Stereocaulon giltayi Duvign. Atranorin, stictic acid. Duvigneaud (1942*a*), type description.

Stereocaulon gracilius (Müll. Arg.) Duvign. Atranorin, stictic acid. Duvigneaud (1942*a*).

Stereocaulon hokkaidense Asah. & Lamb Atranorin, lobaric acid. Asahina (1961*a*), from Japan, type description.

Stereocaulon implexum Th. Fr. Atranorin only. Lamb (1951*b*).

Stereocaulon incrustatum Flörke Atranorin only. Lamb (1951*b*).
   REVIEWS. Atranorin, stictic acid (as pseudopsoromic acid and stereo-caulonic acid): W. Brieger (1923); Hesse (1912); Thies (1932); Zopf (1907).

Stereocaulon intermedium (Sav.) Magn. Lobaric acid. Lamb (1951*b*).

Stereocaulon japonicum Th. Fr. Atranorin, norstictic acid (trace), stictic acid. Asahina (1960*c*).
   Atranorin, stictic acid. Lamb (1951*b*), atranorin accessory, as Chem. strain 1 (see also below).
   Stictic acid. Asahina, Yanagita, and Yosioka (1936), mention only; Kurokawa (1964), mention only; Mitsuno (1953).
   OTHER REPORTS. Atranorin, lobaric acid, stictic acid: Asahina, Yanagita, Hirakata, and Ida (1933), mention only. — Lobaric acid: Asahina and Nonomura (1935), mention only; Lamb (1951*b*), as Chem. strain 2.
   REVIEWS. Atranorin, lobaric acid, stictic acid: Asahina aad Shibata (1954). — Stictic acid: Shibata (1958, 1963).

Stereocaulon japonicum var. *commixtum* Asah. Atranorin, norstictic acid (trace), stictic acid, unidentified substance. Asahina (1960*c*), from Japan, unidentified substance by microcrystal tests in the GE solution, type description.

Stereocaulon japonicum ssp. *etigoense* Asah. Atranorin, unidentified substance. Asahina (1960*c*), from Japan, type description.

Stereocaulon japonicum var. *subfastigiatum* Asah. Atranorin, norstictic acid (trace), stictic acid. Asahina (1960*c*), from Japan, type description.

Stereocaulon lavicola Magn.
   Salazinic acid or α-methylethersalazinic acid or norstictic acid. Lamb (1951*b*).

Stereocaulon massartianum Hue   Atranorin, stictic acid. Duvigneaud (1942*a*); Lamb (1965), from Borneo, Celebes, Java, Formosa, Malaya, New Guinea, the Philippines, and Sumatra, type specimen tested microchemically, other chemical strains with atranorin and norstictic acid (from Borneo, Celebes, Java, New Guinea, and the Philippines) or with atranorin and lobaric acid (from Borneo, the Himalayan region, Java, Malaya, and the Philippines).

Stereocaulon microscopicum (Vill.) Frey   Usnic acid; zeorin. Duvigneaud (1942*a*).
  Usnic acid. Asahina (1943*e*), from Europe, as *Stereocaulon nanum*; Lamb (1951*b*), as *S. quisquilare*.

Stereocaulon mixtum Nyl.   Atranorin, divaricatic acid, unidentified substance. Duvigneaud (1942*a*), the unidentified substance also in *Stereocaulon ramulosum*.

Stereocaulon montagneanum Lamb   Stictic acid. Lamb (1965), from Borneo, Celebes, Java, Malaya, and Sumatra, type description.

Stereocaulon myriocarpum var. *orizabae* Th. Fr., see *Stereocaulon tomentosum* ssp. *myriocarpum* var. *orizabae* (Th. Fr.) Lamb

Stereocaulon nabewariense Zahlbr.   Atranorin, stictic acid. Asahina, Yanagita, Hirakata, and Ida (1933), mention only.
  Stictic acid. Asahina, Yanagita, and Yosioka (1936), mention only.
  REVIEWS. Stictic acid: Asahina and Shibata (1954); Shibata (1958, 1963).

Stereocaulon nanodes Tuck.   Atranorin, lobaric acid, stictic acid; zinc.
  Atranorin, lobaric acid, stictic acid. Ramaut (1962*a*), from Germany and Italy, by paper chromatography; Ramaut and Schumacker (1962), from Belgium, as *Stereocaulon nanodes* f. *tyroliense*.
  Atranorin. Huneck (1966*c*), from Germany, 0.1% by extraction of 6.0 g. of lichen.
  Lobaric acid, stictic acid. Ramaut (1962*b*).
  Zinc. Lambinon, Maquinay, and Ramaut (1964), from Belgium, a quantitative study, and also *Stereocaulon nanodes* f. *tyroliense*; Maquinay, Lamb, Lambinon, and Ramaut (1961), from Belgium, as *S. nanodes* f. *tyroliense*.

Stereocaulon nanum (Ach.) Ach., see *Stereocaulon microscopicum* (Vill.) Frey

Stereocaulon nesaeum Nyl.   Atranorin, norstictic acid (Chem. strain 1). Atranorin, stictic acid (Chem. strain 2). Lamb (1951*b*).

Atranorin, norstictic acid. Duvigneaud (1942*a*), some samples with atranorin and another substance, possibly lobaric acid.

Stereocaulon nigrum Hue   Atranorin, lobaric acid. Asahina (1961*b*), from Japan.

Stereocaulon novo-arbuscula Asah.   Squamatic acid. Asahina (1943*e*), from Japan, type description.

Stereocaulon octomerellum Müll. Arg.   Lobaric acid. Lamb (1951*b*).

Stereocaulon octomerum Müll. Arg.   Atranorin, lobaric acid; dendroidin (±).
Atranorin, lobaric acid. Asahina (1961*b*), and also *Stereocaulon octomerum* f. *robustior*, from Japan.
Atranorin, dendroidin. Asahina (1961*b*), from Japan, by microcrystal tests, microcrystal test for dendroidin described, type description of *Stereocaulon dendroides*; Lamb (1967), *S. dendroides* considered a chemical strain of *S. octomerum*.
Lobaric acid. Lamb (1951*b*), as Chem. strain 1, Chem. strain 2 with atranorin and dendroidin.

Stereocaulon paschale (L.) Hoffm.   Atranorin, lobaric acid; friedelin, ergosterol; ascorbic acid; iron, manganese, zinc, ^{137}Cs, ^{40}K, ^{54}Mn, ^{106}Ru, ^{125}Sb.
Atranorin, lobaric acid. Asahina and Nonomura (1935), from South Sakhalin, by extraction; Lamb (1951*b*); Ramaut (1962*a*), from Sweden, by paper chromatography.
Atranorin. Vartia (1949), from Finland, by extraction; Vartia (1950*b*), from Finland.
Lobaric acid. Asahina (1938*d*), description of microcrystal tests; Mitsuno (1953), by paper chromatography.
Friedelin. Bruun (1954*a*), from Norway, by extraction, a detailed chemical study.
Ergosterol. Blix and Rydin (1932).
Ascorbic acid. Karev and Kochevykh (1962).
Iron, manganese, zinc. Lounamaa (1965), a quantitative study.
^{137}Cs, ^{40}K, ^{54}Mn, ^{106}Ru, ^{125}Sb. Häsänen and Miettinen (1966), from Lapland.
REVIEWS. Atranorin: Thies (1932); Zopf (1907). — Lobaric acid, ergosterol: Shibata (1958, 1963). — Lobaric acid: Asahina and Shibata (1954).

Stereocaulon pendulum Asah.   Atranorin; dendroidin. Lamb (1951*b*).
Dendroidin. Asahina (1961*b*), mention only.

Stereocaulon pileatum Ach.   Atranorin, lobaric acid (sometimes as stereocaulic acid or usnetinic acid), norstictic acid ($\pm$).
  Atranorin, lobaric acid, norstictic acid. Ramaut and Schumacker (1962), from Belgium, by paper chromatography.
  Lobaric acid, norstictic acid. Ramaut (1962*b*).
  Lobaric acid. Asahina and Nonomura (1935), mention of earlier work; Lamb (1951*b*).
  REVIEWS. Atranorin, lobaric acid: W. Brieger (1923); Hesse (1912); Thies (1932); Zopf (1907). — Lobaric acid: Asahina and Shibata (1954); Shibata (1958, 1963).

Stereocaulon piluliferum Th. Fr.   Atranorin, stictic acid. Duvigneaud (1942*a*).

Stereocaulon prostratum Zahlbr.   Lobaric acid. Lamb (1951*b*).

Stereocaulon proximum Nyl.   Atranorin, divaricatic acid, unidentified substance. Duvigneaud (1942*a*), unidentified substance also in *Stereocaulon ramulosum*.

Stereocaulon pseudoarbuscula Asah.   Thamnolic acid. Asahina (1943*e*), from Japan, type description.
  OTHER REPORT. Divaricatic acid and psoromic acid (Chem. strain 3), lecanoric acid (Chem. strain 7), probably fumarprotocetraric acid (Chem. strain 4): Lamb (1951*b*), and fatty acid accessories in several strains.

Stereocaulon pseudodepreaultii Asah.   Atranorin, lobaric acid. Asahina (1960*c*), from Japan, type description.

Stereocaulon quisquilare Nyl., see *Stereocaulon microscopicum* (Vill.) Frey

Stereocaulon ramulosum (Sw.) Räusch.
  Atranorin only (Chem. strain 1). Norstictic acid (Chem. strain 2). Lamb (1951*b*).
  Atranorin, divaricatic acid, unidentified substance. Duvigneaud (1942*a*), unidentified substance by a microcrystal test in the GE solution.
  REVIEWS. Atranorin: Thies (1932); Zopf (1907).

Stereocaulon rivulorum Magn.   Atranorin, lobaric acid. Duvigneaud (1942*a*).

Stereocaulon salazinum (Bory) Fée   Atranorin, norstictic acid. Duvigneaud (1942*a*).
  Norstictic acid. Asahina (1936*d*), description of microcrystal tests; Asahina and Fuzikawa (1935*c*).

OTHER REPORTS. Salazinic acid: Asahina and Asano (1933), mention early work of Zopf; Schindler (1936), mentions early work of Zopf; Mors (1952*a*), discusses extraction of "salazinic acid" by Zopf and the subsequent name change.

REVIEWS. Atranorin, salazinic acid: W. Brieger (1923); Hesse (1912); Thies (1932); Zopf (1907). — Norstictic acid: Asahina and Shibata (1954); Shibata (1958, 1963).

Stereocaulon scutelligerum Th. Fr.   Norstictic acid. Lamb (1951*b*).

Stereocaulon sorediiferum Hue   Atranorin, lobaric acid. Asahina and Nonomura (1935), from Japan, by extraction; Lamb (1965), from Formosa, Hong Kong, Japan, and the Philippines, one sample from northwestern Vietnam possibly containing stictic acid in addition.
   Lobaric acid. Lamb (1951*b*).
   REVIEWS. Lobaric acid: Asahina and Shibata (1954); Shibata (1958, 1963).

Stereocaulon spathuliferum Vain., see *Stereocaulon botryosum* Ach.

Stereocaulon sphaerophoroides Tuck.   Lobaric acid, stictic acid. Ramaut (1962*b*), mention only.

Stereocaulon strictum var. *lecanoreum* (Nyl.) Lamb   Atranorin; dendroidin. Lamb (1951*b*).
   Dendroidin. Asahina (1961*b*), mention only.

Stereocaulon subalbicans Lamb *in* Imsh.   Atranorin, divaricatic acid, psoromic acid (Chem. strain 1, from Chile). Atranorin, thamnolic acid; unidentified fatty acid (Chem. strain 2, from South America, Montana, Washington, and Colorado). Atranorin only (Chem. strain 4, from Peru). Imshaug (1957), type description, type specimen belonging to Chem. strain 1 tested by Y. Asahina, North American specimens of Chem. strain 2 tested by M. E. Hale, Jr.
   OTHER REPORT. Lecanoric acid. Imshaug (1957), from Peru, as Chem. strain 3.

Stereocaulon subcoralloides Nyl.   Atranorin, lobaric acid. Duvigneaud (1942*a*).

Stereocaulon symphycheilum Lamb   Atranorin, lobaric acid. Lamb (1961), from Finland, Sweden, and Alaska, morphologically similar to *Stereocaulon vesuvianum*, type description.

Stereocaulon tennesseense Magn.   Atranorin, lobaric acid, unidentified substance. Asahina (1961*a*), from Japan.

Stereocaulon tomentosum Fr.   Atranorin (sometimes as atranoric acid), stictic acid (Chem. strain 1). Atranorin, lobaric acid (Chem. strain 2). Atranorin, lobaric acid, stictic acid.

Atranorin, stictic acid (Chem. strain 1). Atranorin, lobaric acid (Chem. strain 2). Lamb (1951*b*); Ramaut and Schumacker (1962), Chem. strain 2 from Belgium, by paper chromatography.

Atranorin, lobaric acid (Chem. strain 2). Atranorin, lobaric acid, stictic acid. Ramaut (1962*b*), Chem. strain 2 from Belgium, stictic acid also present in a sample from Finland and in *Stereocaulon tomentosum* var. *alpestre* from Switzerland.

Atranorin, lobaric acid, stictic acid. Ramaut (1962*a*), from Finland, and also *Stereocaulon tomentosum* var. *alpestre* from Switzerland.

Stictic acid (Chem. strain 1). Lobaric acid (Chem. strain 2). Lamb (1951*a*), proposal for "chemical strain" nomenclature.

REVIEWS. Atranorin: W. Brieger (1923); Hesse (1912); Thies (1932); Zopf (1907).

Stereocaulon tomentosum ssp. *myriocarpum* var. *orizabae* (Th. Fr.) Lamb   Atranorin, unidentified substance (m.p. 255–268° d.); D-arabitol (from Homkund, India). Atranorin, stictic acid; ventosic acid; D-arabitol (from Pindari Snout, India). Aghoramurthy, Sarma, and Seshadri (1961*b*), by extraction, as *Stereocaulon myriocarpum* var. *orizabae*, unidentified substance named myriocarpic acid.

Stereocaulon uvuliferum Müll. Arg.   Atranorin, stictic acid. Asahina (1961*b*).

Stereocaulon verruculigerum Hue   Atranorin, lobaric acid. Asahina (1960*c*), from Java, and also *Stereocaulon verruculigerum* var. *formosanum* from Formosa.

Stereocaulon vesuvianum Pers.   Atranorin, stictic acid (sometimes as pseudopsoromic acid or stereocaulonic acid); copper, iron, zinc.

Atranorin, stictic acid. Duvigneaud (1942*a*), and as *Stereocaulon denudatum* (= *S. vesuvianum* var. *denudatum*); Lamb (1951*b*), and as *S. denudatum*; Ramaut (1962*a*), from Sweden, by paper chromatography, and also *S. vesuvianum* var. *denudatum*; Ramaut and Schumacker (1962), as *S. vesuvianum* var. *denudatum*; Vartia (1950*b*), from Finland, as *S. denudatum*.

Copper, iron. Lange and Ziegler (1963), a quantitative study.

Zinc. Lambinon, Maquinay, and Ramaut (1964), from Belgium, as *Stereocaulon vesuvianum* var. *denudatum*, a quantitative study.

REVIEWS. Atranorin, psoromic acid, stictic acid: W. Brieger (1923); Hesse (1912); Thies (1932); Zopf (1907); all four reports as *Stereocaulon denudatum* var. *genuinum*. — Atranorin, stictic acid: W. Brieger (1923); Hesse (1912); Thies (1932); Zopf (1907).

Stereocaulon vesuvianum var. *pulvinatum* (Schaer.) Duncan

REVIEWS. Atranorin, lobaric acid (as stereocaulic acid or usnetinic acid), stictic acid (as pseudopsoromic acid): Thies (1932); Zopf (1907). — Stictic acid (as pseudopsoromic acid and stereocaulonic acid): W. Brieger (1923); Hesse (1912). All reports are as *Stereocaulon denudatum* var. *pulvinatum*.

Stereocaulon vesuvianum var. *umbonatum* (Wallr.) Lamb

Atranorin, dendroidin. Lamb (1951*b*), but Asahina (1961*b*) finds no dendroidin.

Atranorin, stictic acid. Ramaut (1962*a*), from Sweden, by paper chromatography.

Stereocaulon virgatum Ach. Atranorin, norstictic acid. Duvigneaud (1942*a*).

OTHER REPORT. Salazinic acid or α-methylethersalazinic acid or norstictic acid: Lamb (1951*b*).

Stereocaulon virgatum f. *primaria* Vain.

REVIEWS. Atranorin, salazinic acid: Thies (1932); Zopf (1907). — Salazinic acid: W. Brieger (1923); Hesse (1912).

Stereocaulon vulcani (Bory) Ach. Atranorin, norstictic acid. Ahmadjian (1964), isolated fungus with a yellow pigment possibly an anthraquinone.

Stereocaulon wrightii Tuck. Atranorin, stictic acid (Chem. strain 1). Atranorin, lobaric acid (Chem. strain 2). Lamb (1951*b*). Atranorin, lobaric acid. Asahina (1936*c*).

Sticta aurata Ach., see *Pseudocyphellaria aurata* (Ach.) Vain.

Sticta colensoi Bab. Polyporic acid; calycin, pulvinic dilactone (as pulvinic lactone); two neutral substances. Murray (1952), from New Zealand, by extraction.

Polyporic acid. Neelakantan and Seshadri (1959*a*), mention in discussion of biosynthesis; Shibata (1965), mention in discussion of biosynthesis.

REVIEWS. Polyporic acid: Asahina and Shibata (1954); Shibata (1958, 1963).

Sticta coronata Müll. Arg. Two unidentified anthraquinones; calycin, polyporic acid, pulvinic dilactone (sometimes as stictaurin, a calycin-pulvinic dilactone addition compound); unidentified triterpene; unidentified substance. Murray (1952), from New Zealand, by extraction.

Polyphoric acid. Burton and Cain (1959); Cain (1961); Mittal and Seshadri (1957), mention only; Neelakantan and Seshadri (1959*a*), mention in discussion of biosynthesis; Shibata (1965), mention in discussion of biosynthesis.

REVIEWS. Two unidentified anthraquinones, calycin, polyporic acid, pulvinic dilactone, unidentified triterpene, unidentified substance: Asahina and Shibata (1954). — Polyporic acid: Shibata (1958, 1963).

Sticta crocata (L.) Ach., see *Pseudocyphellaria crocata* (L.) Vain.

Sticta desfontainei Del.
REVIEWS. Stictaurin: W. Brieger (1923); Hesse (1912); Thies (1932); Zopf (1907); all reports as *Sticta desfontainei* var. *munda*.

Sticta flavicans Hook. f. & Tayl.
REVIEWS. Stictaurin, one colorless substance: Zopf (1907). — Stictaurin: W. Brieger (1923); Hesse (1912); Thies (1932); Zopf (1907).

Sticta fuliginosa (Dicks.) Ach.   Lecanoric acid; ammonia, methylamine, trimethylamine.
Lecanoric acid. Dahl (1952), from Norway, as *Parmelia fuliginosa*; Hale (1956e), from North America, as *P. fuliginosa*; Koller (1932b), as *P. fuliginosa*; Koller and Pfeiffer (1933a), mention only, as *P. fuliginosa*; Krog (1951b), from Norway, as *P. fuliginosa*; Ramaut (1961c), by extraction, as *P. fuliginosa*; Steiner (1955), as *P. fuliginosa*; Wachtmeister (1952), as *P. fuliginosa*.
Ammonia, methylamine, trimethylamine. Stein von Kamienski (1958), from Tirol.
REVIEWS. Lecanoric acid: W. Brieger (1923); Hesse (1912); Thies (1932); Zopf (1907); all four reports as *Parmelia fuliginosa* and as *P. fuliginosa* var. *ferruginascens*. — Trimethylamine: Zopf (1907), no free lichen acids, as *Stictina fuliginosa*.

Sticta gilva (Ach.) Ach.
Pigments, neutral substances (stictinin and stictalbin). Murray (1952), mention only.
REVIEWS. Stictaurin, stictinin: W. Brieger (1923); Hesse (1912); Thies (1932); Zopf (1907); all review reports also as *Stictina gilva*.

Sticta glaucolurida Nyl.
Pigment, neutral substances (stictalbin and stictinin). Murray (1952), mention only.
REVIEWS. Stictaurin, stictalbin: W. Brieger (1923); Hesse (1912); Thies (1932); Zopf (1907).

Sticta impressa Hook. f. & Tayl.
REVIEWS. Stictaurin: W. Brieger (1923); Hesse (1912); Thies (1932); Zopf (1907).

Sticta origmaea Ach.
REVIEWS. Polyporic acid (as orygmaeaic acid), stictaurin: W. Brieger (1923); Hesse (1912); Thies (1932); Zopf (1907).

Sticta pulmonaria (L.) Bir., see *Lobaria pulmonaria* (L.) Hoffm.

Sticta "silvatica," see *Sticta sylvatica* (Huds.) Ach.

Sticta sylvatica (Huds.) Ach.   Norstictic acid, stictic acid; ammonia, methylamine, trimethylamine.
Norstictic acid, stictic acid. D. Hess (1958), from Germany, by paper chromatography.
Ammonia, methylamine, trimethylamine. Stein von Kamienski (1958), from Germany, study on fresh material, herbarium samples also with dimethylamine.
Both reports as *Sticta silvatica.*

Sticta weigelii (Ach.) Vain.   No lichen substances found. Hale (1957*a*), from Arkansas; Hale (1957*b*), mention only.
OTHER REPORT. Usnic acid: Follmann (1965).

Stictina crocata (L.) Nyl., see *Pseudocyphellaria crocata* (L.) Vain.

Stictina fuliginosa (Dicks.) Nyl., see *Sticta fuliginosa* (Dicks.) Ach.

Stictina gilva (Ach.) Nyl., see *Sticta gilva* (Ach.) Ach.

Stigmatidum venosum (S. Gray) Nyl., see *Enterographa crassa* (DC.) Fée

Teloschistes chrysophthalmus (L.) Th. Fr.
REVIEWS. Parietin (sometimes as physcion): W. Brieger (1923); Hesse (1912); Thies (1932); all reports as *Tornabenia chrysophthalma.*

Teloschistes exilis (Michx.) Vain.   Parietin. Hale (1956*b*); Hale (1956*g*), from Texas, an exsiccati specimen; Mors (1951), from Brazil, a detailed chemical study.
REVIEWS. Parietin: Shibata (1958, 1963).

Teloschistes flavicans (Sw.) Norm.   Vicanicin; fallacinal ($\pm$), parietin (sometimes as physcion), teloschistin (sometimes as fallacinol) ($\pm$).
Vicanicin, parietin, teloschistin. Neelakantan, Seshadri, and Subramanian (1962), from India, by column chromatography on magnesium carbonate.
Vicanicin. Neelakantan and Seshadri (1960*b*), mention only; Neelakantan, Seshadri, and Subramanian (1959), from India, separated from anthraquinone pigments by column chromatography, a detailed study of the chemical structure.

Fallacinal ($\pm$), parietin, teloschistin ($\pm$). Rajagapalan and Seshadri (1959), from India, extraction of a sample with fallacinal and parietin in the ratio 1:4, no teloschistin found.

Fallacinal, parietin, teloschistin. Neelakantan and Seshadri (1960*a*), mention in discussion of biosynthesis.

Parietin, teloschistin. Aghoramurthy and Seshadri (1954), mention only; Neelakantan, Rangaswami, and Subramanian (1951), and an unidentified colorless substance (vicanicin?), mention only; Neelakantan and Seshadri (1952*a*), and an unidentified colorless substance (vicanicin?), mention only; Neelakantan, Seshadri, and Subramanian (1956), mention only; Seshadri and Subramanian (1949*d*), and an unidentified colorless substance (vicanicin?), mention only; Subramanian (1955), from India, and an unidentified colorless substance (vicanicin?).

Parietin. Mors (1951), used as an authentic source of this substance; Shibata (1965), mention in discussion of biosynthesis.

Teloschistin. Neelakantan and Seshadri (1954), from India, mention in study on the synthesis of teloschistin.

REVIEWS. Acromelin: Thies (1932), as *Tornabenia flavicans*. — Fallacinal, parietin, teloschistin: Shibata (1963). — Fallacinal, parietin, teloschistin, vicanicin: Neelakantan (1965). — Parietin, teloschistin: Asahina and Shibata (1954); Shibata (1958). — Parietin, one neutral colorless substance: Zopf (1907). — Parietin: W. Brieger (1923); Hesse (1912); Thies (1932).

Teloschistes flavicans var. *acromelus* (Pers.) Müll. Arg.
REVIEWS. Acromelidin, acromelin, parietin (as physcion): W. Brieger (1923); Hesse (1912); Thies (1932); all reports as *Tornabenia flavicans* var. *acromela*

Teloschistes flavicans f. *cinerascens* (B. Stein) Müll. Arg.
REVIEWS. Acromelin: W. Brieger (1923); Hesse (1912); Thies (1932); as *Tornabenia flavicans* f. *cinerascens*.

Thalloidima candidum (Web.) Mass., see *Toninia candida* (Web.) Th. Fr.

Thalloidima coeruleonigricans (Lightf.) Poetsch, see *Toninia coeruleonigricans* (Lightf.) Th. Fr.

Thalloidima diffractum (Mass.) Mass., see *Toninia diffracta* (Mass.) Zahlbr.

Thalloidima mammillare (Sm.) Mass., see *Toninia tumidula* (Sm.) Zahlbr.

Thalloidima rosulatum Anzi, see *Toninia rosulata* (Anzi) Oliv.

Thamnolecania gerlachei (Vain.) Gyeln. Atranorin. Huneck and Follmann (1966*b*), 0.01% atranorin, by extraction of 9.5 g. of lichen.

Thamnolia subuliformis (Ehrh.) W. Culb. Baeomycesic acid, squamatic acid. Asahina (1937*b*), type description of *Thamnolia subvermicularis*; Asahina (1937*e*), as *T. subvermicularis*; Asahina (1938*e*), description of microcrystal tests, as *T. subvermicularis*; Asahina and Yasue (1937*b*), as *T. subvermicularis*; Bendz, Santesson, and Wachtmeister (1965*c*), mention only; W. L. Culberson (1963*b*), selection of neotype specimen; Satô (1963), world distribution mapped, as *T. subvermicularis*; Satô (1965), from New Zealand; Shibata (1965), mention in discussion of biosynthesis, as *T. subvermicularis*; Steiner (1960), as *T. subvermicularis*; Thomson (1963), mention only, as *T. subvermicularis*.

Baeomycesic acid. Mitsuno (1953), by paper chromatography, as *Thamnolia subvermicularis*; J. Santesson (1965).

OTHER REPORTS. Satô (1959), reports "mixture-ratios" of UV+ and UV− samples from different localities, as *Thamnolia subvermicularis* (see also *T. vermicularis*). — Thamnolic acid, unidentified substance, D-arabitol: Aghoramurthy, Sarma, and Seshadri (1961*b*), from India, as *T. subvermicularis*, by extraction and paper chromatography of UV+ lichen (see *T. vermicularis*).

REVIEWS. Baeomycesic acid, squamatic acid: Asahina and Shibata (1954); Shibata (1958, 1963); all review reports as *Thamnolia subvermicularis*.

Thamnolia subvermicularis Asah., see *Thamnolia subuliformis* (Ehrh.) W. Culb.

Thamnolia vermicularis (Sw.) Ach. *ex* Schaer. Decarboxythamnolic acid, thamnolic acid; arabitol, mannitol.

Decarboxythamnolic acid, thamnolic acid. Asahina and Nuno (1964), by thin-layer chromatography; J. Santesson (1965).

Thamnolic acid, D-arabitol. Agoramurthy, Sarma, and Seshadri (1961*b*), from India, by extraction.

Thamnolic acid. Asahina (1937*b*); Asahina (1937*e*); Asahina (1938*e*), description of microcrystal tests; Asahina and Ihara (1929), mention only; Asahina and Yasue (1937*b*); W. L. Culberson (1963*b*), selection of a neotype specimen; D. Hess (1958), from Germany; Koller and Hamburg (1935*b*); Kurokawa (1964), mention only; Mitsuno (1953), by paper chromatography; Satô (1963), world distribution mapped; Satô (1965), from New Zealand; Shibata (1965), mention in discussion of biosynthesis; Steiner (1960); Thomson (1963), mention only; Wachtmeister (1955), by extraction, used as an authentic source of this substance, thamnolic acid found to be identical to ocellatic acid; Wachtmeister (1956*a*); Wachtmeister (1958*a*), mention only.

Arabitol, mannitol. Lindberg, Misiorny, and Wachtmeister (1953), by paper chromatography, no volemitol.

OTHER REPORTS. Squamatic acid, thamnolic acid: Asahina and Hiraiwa (1935*b*), from Japan, by extraction (Asahina and Yasue (1937*b*) mention that this sample was probably mixed with *Thamnolia subuliformis*.); Satô (1959), reports "mixture-ratios" of UV + (squamatic acid and baeomycesic acid) samples of *T. subuliformis* and UV − (thamnolic acid) samples of *T. vermicularis*. — Baeomycesic acid, squamatic acid (but see *T. subuliformis*): Wachtmeister (1952), as *T. vermicularis* f. *taurica*; Wachtmeister (1956*a*), as *T. vermicularis* f. *taurica*. — Baeomycesic acid: Hale (1954*b*), from Baffin Island.

REVIEWS. Baeomycesic acid, thamnolic acid: Shibata (1958, 1963). — Squamatic acid, thamnolic acid: Shibata (1963), as *Thamnolia vermicularis* f. *taurica*. — Thamnolic acid: Asahina and Shibata (1954); W. Brieger (1923); Hesse (1912); Thies (1932); Zopf (1907).

Thelocarpon epibolum Nyl.   Pulvinic acid (traces), pulvinic dilactone, vulpinic acid. J. Santesson (1967), from Norway, by thin-layer chromatography.

Thelocarpon laureri (Flot.) Nyl.   Pulvinic acid, pulvinic dilactone, vulpinic Acid. J. Santesson (1967), from Ohio, by thin-layer chromatography.

Thysanothecium casuarinarum Groenh.   Divaricatic acid. Asahina (1957*a*), from Indonesia.

Thysanothecium casuarinarum ssp. *nipponicum* (Asah.) Asah.   Divaricatic Acid. Asahina (1956*a*), from Australia, Indochina, and Japan, type description of *Thysanothecium nipponicum*; Asahina (1957*a*).

Thysanothecium nipponicum Asah., see *Thysanothecium casuarinarum* ssp. *nipponicum* (Asah.) Asah.

Toninia candida (Web.) Th. Fr.
   REVIEW. Thalloidima-green: Thies (1932), as *Thalloidima candidum*.

Toninia cervina Lönnr.
   REVIEW. Parmelia-brown: Thies (1932), as *Toninia congesta*.

Toninia coeruleonigricans (Lightf.) Th. Fr.
   REVIEW. No lichen substances: Zopf (1907), as *Thalloidima coeruleonigricans*.

Toninia congesta Kremp., see *Toninia cervina* Lönnr.

Toninia diffracta (Mass.) Zahlbr.
REVIEW. Thalloidima-green: Thies (1932), as *Thalloidima diffractum*.

Toninia rosulata (Anzi) Oliv.
REVIEW. Thalloidima-green: Thies (1932), as *Thalloidima rosulatum*.

Toninia tumidula (Sm.) Zahlbr.
REVIEW. Parmelia-brown: Thies (1932), as *Thalloidima mammillare*.

Tornabenia atlantica (Ach.) Kurok.   No lichen substances found. Kurokawa (1962*b*), from Africa, the Canary Islands, Europe, and the
Himalayas.

Tornabenia chrysophthalma (L.) Mass., see *Teloschistes chrysophthalmus*
(L.) Th. Fr.

Tornabenia ephebea (Ach.) Kurok.   No lichen substances found. Huneck
and Follmann (1966*d*), from Chile, by extraction of 54.0 g. of lichen;
Kurokawa (1962*b*), from Peru.

Tornabenia flavicans (Sw.) Mass., see *Teloschistes flavicans* (Sw.) Norm.

"*Tornabenia flavicans* var. *acromela* Pers.," see *Teloschistes flavicans* var.
acromelus (Pers.) Müll. Arg.

Tornabenia flavicans f. *cinerascens* B. Stein, see *Teloschistes flavicans* f.
cinerascens (B. Stein) Müll. Arg.

Tornabenia intricata (Desf.) Kurok.
Zeorin. Follmann (1964), mention only.
No substances found. Galun and Lavee (1966), from Israel.
Both reports as *Anaptychia intricata*.

Trypetheliopsis boninensis Asah.   Unidentified yellow substance (K+
purple). Asahina (1937*b*), from the Bonin Islands, type description of
the genus and species.

Umbilicaria arctica (Ach.) Nyl.   Gyrophoric acid; arabitol, mannitol.
Gyrophoric acid. Hale (1956*e*), from North America; Schatz (1963),
mention only.
Arabitol, mannitol. Lindberg, Misiorny, and Wachtmeister (1953), by
paper chromatography, no volemitol.

Umbilicaria caroliniana Tuck.   Gyrophoric acid. Asahina (1936*d*), description of microcrystal tests.

Umbilicaria cinereorufescens (Schaer.) Frey   Gyrophoric acid. Hale (1956e), from North America.

Umbilicaria crustulosa (Ach.) Harm., see *Omphalodiscus crustulosus* (Ach.) Schol.

Umbilicaria cylindrica (L.) Del.   Arabitol, mannitol. Lindberg, Misiorny, and Wachtmeister (1953), by paper chromatography, no volemitol.
REVIEW. Parmelia-brown: Thies (1932), as *Gyrophora cylindrica*.

Umbilicaria decussata (Vill.) Zahlbr., see *Omphalodiscus decussatus* (Vill.) Schol.

Umbilicaria deusta (L.) Baumg.   Gyrophoric acid, umbilicaric acid; arabitol, mannitol.
Gyrophoric acid, umbilicaric acid. D. Hess (1958), from Germany, by paper chromatography; Klosa (1951b), mention only, as *Gyrophora deusta*; Koller and Pfeiffer (1933c), from Germany, by extraction.
Arabitol, mannitol. Lindberg, Misiorny, and Wachtmeister (1953), by paper chromatography, no volemitol.
REVIEWS. Gyrophoric acid, umbilicaric acid: Asahina and Shibata (1954); W. Brieger (1923); Hesse (1912); Thies (1932); Zopf (1907). — Umbilicaric acid: Shibata (1958, 1963). All review reports are as *Gyrophora deusta*.

Umbilicaria dillenii Tuck., see *Umbilicaria mammulata* (Ach.) Tuck.

Umbilicaria erosa (Web.) Ach., see *Umbilicaria torrefacta* (Lightf.) Schrad.

Umbilicaria esculenta (Miyoshi) Minks   Gyrophoric acid. Asahina and Kutani (1925); Asahina (1936d), description of microcrystal tests; Kurokawa (1952).
REVIEWS. Gyrophoric acid, lecanoric acid: Asahina and Shibata (1954). — Gyrophoric acid: Shibata (1958, 1963).
All reports are as *Gyrophora esculenta*.

Umbilicaria fuliginosa (Hav.) Zahlbr., see *Umbilicaria havaasii* Llano

Umbilicaria fuliginosa var. *wenckii* (Müll. Arg.) Zahlbr., see *Umbilicaria havaasii* Llano

Umbilicaria havaasii Llano
Probably gyrophoric acid. Dahl (1950), as *Umbilicaria fuliginosa* and *U. fuliginosa* var. *wenckii*.

Umbilicaria hirsuta (Sw. *ex* Westr.) Ach.   Gyrophoric acid; arabitol, mannitol, pustulin; enzymes; iron, manganese, zinc.

Gyrophoric acid. Huneck (1962*b*), used as an authentic source of this substance.

Arabitol, mannitol. Lindberg, Misiorny, and Wachtmeister (1953), by paper chromatography, no volemitol.

Pustulin. Drake (1943).

Enzymes. Moissejeva (1961), enzymes tabulated, as *Gyrophora hirsuta.*

Iron, manganese, zinc. Lounamaa (1965), a quantitative study.

REVIEWS. Gyrophoric acid: W. Brieger (1923); Hesse (1912); Thies (1932); Zopf (1907); all review reports as *Gyrophora hirsuta.*

Umbilicaria hyperborea (Ach.) Ach.   Gyrophoric acid, umbilicaric acid; arabitol, mannitol.

Gyrophoric acid. Hale (1956*e*), from North America.

Umbilicaric acid. Asahina (1937*d*), description of microcrystal tests, as *Gyrophora hyperborea*; Koller and Pfeiffer (1933*c*), mention only, as *G. hyperborea.*

Arabitol, mannitol. Lindberg, Misiorny, and Wachtmeister (1953), by paper chromatography, no volemitol.

REVIEWS. Gyrophoric acid, umbilicaric acid, parmelia-brown: Thies (1932). — Umbilicaric acid: W. Brieger (1923); Hesse (1912); Zopf (1907). All review reports are as *Gyrophora hyperborea.*

Umbilicaria lyngei Schol., see *Agyrophora lyngei* (Schol.) Llano

Umbilicaria mammulata (Ach.) Tuck.   Gyrophoric acid; unidentified fatty acid; ergosterol; glucose, mannitol, a polysaccharide similar to lichenin; a carotene-like substance; unidentified pigment; ^{40}K, ^{137}Cs.

Gyrophoric acid, unidentified fatty acid, ergosterol, glucose, mannitol, a polysaccharide similar to lichenin, a carotene-like substance, unidentified pigment. Zellner (1935), by extraction, also a substance like gyrophoric acid, possibly lecanoric acid, as *Gyrophora dillenii.*

Gyrophoric acid. Hale (1956*e*), from North America.

^{40}K, ^{137}Cs. Burley, Gilbert, and Clum (1962).

REVIEW. Ergosterol: Steiner (1957), as *Umbilicaria dillenii.*

Umbilicaria muehlenbergii (Ach.) Tuck., see *Actinogyra muehlenbergii* (Ach.) Schol.

Umbilicaria murina (Ach.) DC.   Galactose, mannitol, sucrose, volemitol.

Galactose, sucrose. Pueyo (1963*a*), from France.

Mannitol, volemitol. Pueyo (1963*b*), from France.

REVIEW. Parmelia-brown: Thies (1932).

All reports are as *Gyrophora murina.*

Umbilicaria papulosa (Ach.) Nyl., see *Lasallia papulosa* (Ach.) Llano

Umbilicaria pensylvanica Hoffm., see *Lasallia pensylvanica* (Hoffm.) Llano

Umbilicaria phaea Tuck.   Gyrophoric acid. Hale (1956*e*), from North America.

Umbilicaria polyphylla (L.) Baumg.   Gyrophoric acid, umbilicaric acid; arabitol, mannitol.
    Gyrophoric acid, umbilicaric acid. Asahina and Watanabe (1930), mention only, as *Gyrophora polyphylla*; D. Hess (1958), from Germany, by paper chromatography.
    Gyrophoric acid. Hale (1956*e*), from North America.
    Umbilicaric acid. Koller and Pfeiffer (1933*c*), mention only, as *Gyrophora polyphylla*.
    Arabitol, mannitol. Lindberg, Misiorny, and Wachtmeister (1953), by paper chromatography, no volemitol.
    OTHER REPORTS. Lecanoric acid, umbilicaric acid: Klosa (1951*b*), mention only, as *Gyrophora polyphylla*.
    REVIEWS. Umbilicaric acid, gyrophoric acid: W. Brieger (1923); Hesse (1912); Thies (1932), and parmelia-brown; Zopf (1907). — Umbilicaric acid: Asahina and Shibata (1954); Shibata (1958, 1963). All review reports are as *Gyrophora polyphylla*.

Umbilicaria polyrrhiza (L.) Ach., see *Actinogyra polyrrhiza* (L.) Schol.

Umbilicaria proboscidea (L.) Schrad.   Gyrophoric acid; arabitol, mannitol; ergosterol.
    Gyrophoric acid. Asahina (1936*d*), as *Gyrophora proboscidea*; Asahina and Kutani (1925), as *G. proboscidea*; Hale (1956*e*), from North America.
    Arabitol, mannitol. Lindberg, Misiorny, and Wachtmeister (1953), by paper chromatography, no volemitol.
    Ergosterol. Blix and Rydin (1932), as *Gyrophora proboscidea*.
    OTHER REPORT. Umbilicaric acid: Klosa (1951*b*), mention only, as *Gyrophora proboscidea*.
    REVIEWS. Gyrophoric acid: Asahina and Shibata (1954); W. Brieger (1923); Hesse (1912); Shibata (1958, 1963); Thies (1932); Zopf (1907); all review reports as *Gyrophora proboscidea*.

Umbilicaria pustulata (L.) Hoffm., see *Lasallia pustulata* (L.) Mér.

Umbilicaria rigida (Du Rietz) Frey, see *Agyrophora rigida* (Du Rietz) Llano

Umbilicaria spodochroa (Hoffm.) Schol., see *Omphalodiscus spodochrous* (Hoffm.) Schol.

Umbilicaria torrefacta (Lightf.) Schrad. Norstictic acid, stictic acid; arabitol, mannitol.

Norstictic acid, stictic acid. Bachmann (1962), from Germany and Sweden, by paper chromatography, and also α-methylethersalazinic acid, as *Umbilicaria erosa*.

Arabitol, mannitol. Lindberg, Misiorny, and Wachtmeister (1953), by paper chromatography, no volemitol.

Umbilicaria vellea (L.) Ach. Gyrophoric acid, umbilicaric acid; arabitol, mannitol.

Gyrophoric acid, umbilicaric acid. Asahina and Watanabe (1930), mention only, as *Gyrophora vellea*; Vartia (1950*b*), from Finland, as *G. vellea*.

Gyrophoric acid. Hale (1956*e*), from North America.

Umbilicaric acid. Koller and Pfeiffer (1933*c*), mention only.

Arabitol, mannitol. Lindberg, Misiorny, and Wachtmeister (1953), by paper chromatography, no volemitol.

OTHER REPORT. Gyrophoric acid, gyrophorin: Zopf (1900), as *Gyrophora vellea*.

REVIEWS. Gyrophoric acid, umbilicaric acid: Asahina and Shibata (1954); W. Brieger (1923); Hesse (1912); Thies (1932), and parmelia-brown; Zopf (1907). — Umbilicaric acid: Shibata (1958, 1963). All review reports are as *Gyrophora vellea*.

Umbilicaria virginis Schaer., see *Omphalodiscus virginis* (Schaer.) Schol.

Urceolaria albissima (Ach.) Fink, see *Diploschistes gypsaceus* (Ach.) Nyl.

Urceolaria ocellata (Hoffm.) Ach., see *Rinodina ocellata* (Hoffm.) Arn.

Urceolaria scruposa (Schreb.) Ach., see *Diploschistes scruposus* (Schreb.) Norm.

Urceolaria scruposa var. *arenaria* Schaer., see *Diploschistes scruposus* var. *arenarius* (Schaer.) Müll. Arg.

Urceolaria scruposa var. *bryophila* (Ehrh.) Ach., see *Diploschistes bryophilus* (Ehrh.) Zahlbr.

"*Urceolaria scruposa* var. *cretacea*," see *Diploschistes gypsaceus* (Ach.) Nyl.

Urceolaria scruposa var. *vulgaris* And., see *Diploschistes scruposus* (Schreb.) Norm.

Usnea aciculifera Vain. Stictic acid, usnic acid. Asahina (1956*c*), from Japan, and also *Usnea aciculifera* f. *abbreviata*, contrary to Motyka medulla reacts I −.

Usnea acromelana var. *decipiens* (Lamb) Lamb   Norstictic acid (±), usnic acid. Lamb (1964), from Antarctica and Tasmania, a sample from Chile with no norstictic acid (PD −).

Usnea angulata Ach.   Diffractaic acid, salazinic acid (as usnaric acid), usnic acid, unidentified substances. De Smet and Lambinon (1952), from the Belgian Congo, a specimen with more salazinic acid than the type specimen of the species with K + and PD + thallus reactions not found in type specimen named as a new variety, *Usnea goniodes* var. *reagens*.

Usnea antarctica Du Rietz   Fumarprotocetraric acid (±), usnic acid. Lamb (1964), from Graham Land, fumarprotocetraric acid in more northern localities.

Usnea arguta Mot.   Usnic acid; unidentified substance. De Smet and Lambinon (1952), from the Belgian Congo.

Usnea arizonica Mot.   Salazinic acid. Hale (1962*b*).

Usnea articulata (L.) Hoffm.
     Barbatic acid, potassium oxalate. Schulte (1905), barbatic acid by a microcrystal test, and also salazinic acid by color reactions, potassium oxalate by microchemical tests.
     REVIEWS. Salazinic acid (as usnaric acid), (+)-usnic acid: Thies (1932); Zopf (1907); both reports also as *Alectoria articulata*. — (+)-Usnic acid: W. Brieger (1923); Hesse (1912).

Usnea articulata var. *asperula* Müll. Arg.   Stictic acid, usnic acid.
     OTHER REPORTS. Atranorin, articulatic acid, (+)-usnic acid: Asahina (1926). — α-Methylethersalazinic acid (m.p. 210°), β-methylethersalazinic acid (m.p. 240°), (+)-usnic acid: Asahina and Tukamoto (1934), (but see *Usnea bismolliuscula*). — α-Methylethersalazinic acid, β-methylethersalazinic acid: Asahina (1937*e*), picture of crystals in the KK solution. (α-Methylethersalazinic acid and β-methylethersalazinic acid may have been mixtures of stictic acid and norstictic acid.)
     REVIEW. Stictic acid, usnic acid: Asahina and Shibata (1954).

Usnea articulata var. *intestiniformis* (Ach.) Cromb.
     REVIEWS. Articulatic acid, barbatic acid: Thies (1932). — Articulatic acid, (+)-usnic acid: W. Brieger (1923); Hesse (1912).

Usnea asahinai Mot.   Norstictic acid, usnic acid, unidentified substance. Asahina (1956*c*), from Japan.

Usnea aspera (Eschw.) Vain.
Norstictic acid, (+)-usnic acid. Mors (1952*a*), from Brazil, 2.8%
usnic acid, 4.4% norstictic acid, by extraction; Mors (1952*b*), from
Brazil.
Norstictic acid. Nair and Subramanian (1961*a*), from Brazil.
Psoromic acid, (+)-usnic acid. Dhar, Neelakantan, Ramanujam, and
Seshadri (1959), from India, by extraction.

Usnea aurantiaco-ater (Jacq.) Bory   Norstictic acid. Lamb (1964), from
the Falkland Islands.

Usnea aureola Mot.   Salazinic acid, (+)-usnic acid. Huneck and Follmann (1966*e*), from Chile, 3.6% and 3.6% salazinic acid, 0.43% and 0.5%
(+)-usnic acid, by extraction of two samples.

Usnea bakongoensis Duvign.   Salazinic acid (as usnaric acid), usnic acid,
unidentified pigment. De Smet and Lambinon (1952), from the Belgian
Congo.

Usnea barbata (L.) Wigg.   Barbatic acid, barbatolic acid, (+)-usnic acid;
unsaturated fatty acids; ergosterol; lichenin; ascorbic acid.
Barbatic acid, usnic acid. Robertson and Stephenson (1932), from
Scotland, by extraction of 1 kg. of lichen.
Barbatic acid. Asahina (1936*e*), description of microcrystal tests;
Schulte (1905), by a microcrystal test performed on sections of the
strands.
Barbatolic acid. Asahina (1940*d*), mention only.
(+)-Usnic acid. Klosa (1953*a*), mention only; Vartia (1950*a*), mention
only.
Usnic acid. Curd and Robertson (1933); Dean (1952), mention only;
J. C. Mitchell (1965), mention only; Schöpf and Heuck (1927), from
Germany.
Unsaturated fatty acids. Wagner and Friedrich (1965), C_{16}, C_{18}, and
C_{20} acids with one unsaturation, C_{18} acids with two and four unsatura-
tions and traces of unsaturated C_{20} and C_{22} acids, by thin-layer chromatography.
Ergosterol. Blix and Rydin (1932).
Lichenin. Karrer, Staub, and Staub (1924); Mittal and Seshadri
(1954*a*), mention only; Mittal and Seshadri (1954*b*), mention only.
OTHER REPORTS. Barbatolic acid, lobaric acid, salazinic acid (sometimes
as usnaric acid), (+)-usnic acid, ascorbic acid: Schindler (1957), mention
with references to the literature. — Norstictic acid, salazinic acid, D-
arabitol: Briner, Gream, and Riggs (1960), from Australia, a detailed
chemical study, good review with references to previous work.
REVIEWS. Barbatic acid, barbatolic acid, lobaric acid, (sometimes as
usnetinic acid), (+)-usnic acid: W. Brieger (1923), and salazinic acid;

Hesse (1912); Thies (1932), lobaric acid in South American material only. — Barbatic acid: Zopf (1907). — Barbatolic acid, usnic acid: Asahina and Shibata (1954); Shibata (1958). — Usnic acid: Neelakantan (1965).

Usnea barbata var. *ceratina* (Ach.) Schaer., see *Usnea ceratina* Ach.

Usnea barbata var. *dasypoga* (Ach.) Ach., see *Usnea dasypoga* (Ach.) Röhl.

Usnea barbata var. *hirta* (L.) Fr., see *Usnea hirta* (L.) Wigg.

Usnea bismolliuscula Zahlbr.
　　Salazinic acid, usnic acid. Asahina (1965*f*), from Maine.
　　Stictic acid, usnic acid. Asahina (1956*c*), from Formosa and Japan, and as *Usnea pygmea*, not α-methylethersalazinic acid and β-methylethersalazinic acid as previously reported (see *U. articulata* var. *asperula*).

Usnea bismolliuscula ssp. **pseudomolliuscula** Asah.　Thamnolic acid, usnic acid. Asahina (1956*c*), from Japan, and probably baeomycesic acid, type description.

Usnea californica Herre　Usnic acid. F. T. Jones and Palmer (1950).
　　REVIEW. Usnic acid: Asahina and Shibata (1954).

Usnea canariensis (Ach.) Du Rietz
　　Atranorin, chloroatranorin, unidentified pigment (m.p. 220°). Fernández and Pizarroso (1958), by extraction.
　　REVIEWS. Salazinic acid (as usnaric acid), (+)-usnic acid: W. Brieger (1923); Hesse (1912); Thies (1932); Zopf (1907); all review reports as *Alectoria canariensis*.

Usnea capilliformis Asah.　Protocetraric acid, usnic acid. Asahina (1956*c*), from Japan, type description.

Usnea ceratina Ach.　Diffractaic acid, (+)-usnic acid. Duvigneaud (1947), from western Europe.
　　Diffractaic acid. Asahina (1936*e*), description of microcrystal tests, no barbatic acid as reported by Zopf.
　　(+)-Usnic acid. Salkowski (1910).
　　Usnic acid. Gertig and Banasiewicz (1961), from Poland, 1.98%; D. Hess (1958), from Germany.
　　OTHER REPORTS. Norstictic acid, salazinic acid, D-arabitol: Briner, Gream, and Riggs (1960), from Australia, by extraction, a detailed chemical study, review of early literature. — Barbatic acid: St. Pfau (1928); Schulte (1905), by a microcrystal test performed on sections of the strands, no salazinic acid (as usnaric acid), by color reactions only, potassium oxalate by microchemical tests.

REVIEWS. Barbatic acid, barbatin, psoromic acid (as parellic acid), (+)-usnic acid: W. Brieger (1923); Hesse (1912); Thies (1932), psoromic acid from Java only. — Barbatic acid, (+)-usnic acid: Zopf (1907), barbatin mentioned from *Usnea barbata* var. *ceratina*. — Barbatic acid: Shibata (1958).

Usnea chilensis Mot.   Usnic acid. Follmann (1965), from Chile; Follmann and Villagrán (1965).

Usnea comosa (Ach.) Röhl.   Squamatic acid, usnic acid; mannitol; folic acid-, folinic acid-, and vitamin B_{12}-group factors.
  Squamatic acid, usnic acid. Asahina (1956c), from Japan and Manchuria, as *Usnea comosa* ssp. *eucomosa*.
  Usnic acid. Vartia (1949), from Finland, Vartia (1950b), from Finland.
  Mannitol. Pueyo (1964).
  Folic acid-, folinic acid-, and vitamin B_{12}-group factors. Sjöström and Ericson (1953).
  OTHER REPORTS. Thamnolic acid, usnic acid: Duvigneaud (1947), from the Belgian Congo. — Obtusatic acid (as ramalic acid), (+)-usnic acid: Asahina (1926), as *Usnea florida* var. *comosa*.

Usnea comosa ssp. *colorans* Asah.   Decarboxythamnolic acid, thamnolic acid, usnic acid. Asahina and Nuno (1964), by thin-layer chromatography.
  Thamnolic acid, usnic acid. Asahina (1956c), from Europe, Japan, and Korea, type description.

Usnea comosa ssp. *melanopoda* Asah.   Salazinic acid, usnic acid. Asahina (1956c), from Japan, Manchuria, and Switzerland, type description.

Usnea comosa ssp. *praetervisa* Asah.   Norstictic acid, usnic acid. Asahina (1956c), from Japan and Manchuria, type description.

Usnea confusa Asah.   Salazinic acid, usnic acid. Asahina (1956c), from Japan, type description.

Usnea confusa ssp. *rubroreagens* Asah.   Protocetraric acid, usnic acid. Asahina (1956c), type description.

Usnea contorta Jatta   Protocetraric acid, unidentified pigment (K+ purple). Asahina (1965d), from Madagascar, unidentified K+ pigment at base.

Usnea cornuta Körb.   Salazinic acid (sometimes as usnaric acid), usnic acid. Duvigneaud (1945), by microchemical tests on the type specimen, salazinic acid found to be identical to usnaric acid; Duvigneaud (1947), from western Europe.

538   *Chemical and Botanical Guide to Lichen Products*

OTHER REPORT. Barbatic acid, potassium oxalate: Schulte (1905), and salazinic acid by color reactions only, potassium oxalate by microchemical tests, barbatic acid by microcrystal tests.

REVIEWS. Salazinic acid, (+)-usnic acid: W. Brieger (1923); Hesse (1912); Thies (1932); Zopf (1907).

Usnea creberrima Vain.   Norstictic acid, usnic acid, unidentified red pigment.

Norstictic acid, usnic acid, unidentified red pigment. Asahina (1965*d*), by thin-layer chromatography.

Norstictic acid, usnic acid, unidentified substance. Asahina (1965*b*), unidentified substance by crystals in the GE solution, microcrystal tests on type specimen.

OTHER REPORTS. Diffractaic acid, usnic acid, unidentified red pigment: Asahina (1956*c*), from Japan, and as *Usnea creberrima* var. *fistulescens*.
— Diffractaic acid, usnic acid: Asahina (1950*b*).

Usnea cribosa Asah., see *Usnea merrillii* Mot.

Usnea croceorubescens Vain.   Unidentified substances, usnic acid. Asahina (1950*b*); Asahina (1956*c*), from Japan, an unidentified red pigment, and as *Usnea croceorubescens* f. *tenuiramea*, microcrystal tests described.

Unidentified red pigment. Asahina (1965*d*), by thin-layer chromatography.

Usnea dasypoga (Ach.) Röhl.   Salazinic acid (sometimes as usnaric acid), (+)-usnic acid; enzymes; ascorbic acid.

Salazinic acid, usnic acid. Asahina (1956*c*), from Japan; Duvigneaud (1945), from Belgium, salazinic acid found to be identical to usnaric acid; Duvigneaud (1947), from western Europe; D. Hess (1958), from Germany.

Salazinic acid. Schulte (1905), by color reactions only, also potassium oxalate by microchemical tests, no barbatic acid by a microcrystal test on sections, and as *Usnea schraderi*.

(+)-Usnic acid. Borkowski, Woźniak, Gertig, and Werakso (1964), by extraction; Moissejeva (1957), 1.12%; Moissejeva (1961), 1.12%, enzymes tabulated; Salkowski (1910), as *Usnea schraderi*; Savicz, Kuprevicz, Litvinov, Moissejeva, and Rassadina (1956), 1.12%; Stoll, Brack, and Renz (1947), 0.60% by extraction; Stoll, Renz, and Brack (1947), by extraction.

Usnic acid. Gertig (1961); Gertig and Banasiewicz (1961), from Poland, 3.47%; Tomaselli (1957), from Italy; Vartia (1950*b*), from Finland, and also *Usnea dasypoga* var. *spinosissima*, *U. dasypoga* var. *stramineola*, and *U. dasypoga* var. *tuberculata*.

Ascorbic acid. Karev and Kochevykh (1962).

OTHER REPORTS. Barbatolic acid, salazinic acid, thamnolic acid, (+)-usnic acid: Schindler (1957), mention only. — Thamnolic acid: Asahina (1938*f*), mentions Zopf found hirtellic acid, hirtellic acid found to be identical to thamnolic acid.

REVIEWS. Barbatolic acid (as alectoric acid), salazinic acid, thamnolic acid (as hirtellic acid), usnellin, (+)-usnic acid: Thies (1932); Zopf (1907); also as *Usnea schraderi* in both reports. — Barbatolic acid (as alectoric acid), salazinic acid, thamnolic acid, (as hirtellic acid), (+)-usnic acid: W. Brieger (1923); Hesse (1912); and as *U. barbata* var. *dasypoga* and *U. schraderi* in both reports. — (+)-Usnic acid: Savicz, Litvinov, and Moissejeva (1960), 1.12%.

Usnea dasypoga var. *plicata* f. *annulata* (Müll. Arg.) Hue   Diffractaic acid, usnic acid. Asahina (1956*c*), from China, Formosa, Japan, Korea, and Sakhalin.

Usnea deminuta Mot.   Diffractaic acid, usnic acid. Wall and Davis (unpublished communication), from Ethiopia, 0.6% diffractaic acid, 0.08% usnic acid, by extraction of 11 kg. of lichen, also 2% of an acidic fraction possibly containing more diffractaic and usnic acids.

Usnea diffracta Vain.   Barbatic acid (±), diffractaic acid, (+)-usnic acid.
Barbatic acid (±), diffractaic acid. Asahina (1958*d*), suggests barbatic acid may be a constant component; Nuno (1958), from China, Formosa, Japan, Korea, and Manchuria, by microcrystal tests, barbatic acid identified in about 20% of the samples.
Barbatic acid, (+)-usnic acid. Asahina (1926).
Diffractaic acid, (+)-usnic acid. Asahina and Fuzikawa (1932*b*), from Japan, 3.6% diffractaic acid, 0.8% (+)-usnic acid, by extraction.
Diffractaic acid, usnic acid. Asahina (1936*e*), from Japan, description of microcrystal tests; Asahina (1956*c*), from Japan.
Diffractaic acid. Asahina and Fuzikawa (1932*a*), diffractaic acid found to be identical to dirhizonic acid of Hesse; Mitsuno (1953); Stoll, Brack, and Renz (1950).
Usnic acid. Taguchi, Sankawa, and Shibata (1966), ^{14}C study on the biosynthesis of this pigment.
REVIEWS. Diffractaic acid: Asahina and Shibata (1954); Shibata (1958, 1963); Thies (1932).

Usnea diffracta ssp. *subdiffracta* Asah.   Unidentified substance, usnic acid. Asahina (1956*c*), from Japan, type description.

Usnea diplotypus Vain.   Salazinic acid, usnic acid. Hale (1958*d*), from Europe.

Usnea dorogawensis Asah.   Stictic acid, usnic acid; unidentified red pigment. Asahina (1953c), from Japan, pigment in "turgid hyphae of outer medulla," type description; Asahina (1956c), from Japan.

Usnea dusenii Du Rietz   Usnic acid. Follmann (1964), mention only.

Usnea eizanensis Asah.   Thamnolic acid, usnic acid; unidentified colorless substance; unidentified red pigment. Asahina (1956c), from Japan, type description.

Usnea elongata Mot.   Protocetraric acid, (+)-usnic acid; unidentified substances. Mors (1952a), from Brazil, one unidentified substance (m.p. near 100°) similar to hirtinic acid of Zopf, 0.07% unidentified substance (m.p. 124°), 0.4% (+)-usnic acid, 4.0% protocetraric acid, by extraction; Mors (1952b), from Brazil.

Usnea evansii Mot., see *Usnea strigosa* (Ach.) A. Eat.

Usnea fasciata Torrey   Fumarprotocetraric acid (±), usnic acid. Lamb (1964), from Antarctica and South America, and the type description of *Usnea fasciata* f. *strigulosa* from South Orkney Islands containing fumarprotocetraric acid.

Usnea flexilis Stirt.   Salazinic acid, usnic acid; amino acids.
    Salazinic acid, usnic acid. Asahina (1955a), from India; Asahina (1956c), from Formosa and Japan.
    Amino acids. Ramakrishnan and Subramanian (1964), from India, alanine, glycine, threonine, and tyrosine as free amino acids, but no arginine, leucine or valine, also find lysine, phenylalanine, and serine by alkaline hydrolysis of the residue.
    OTHER REPORT. Barbatic acid, norstictic acid, (+)-usnic acid: Nair and Subramanian (1961a), from India, by extraction.

Usnea florida (L.) Wigg.
    Barbatic acid, salazinic acid (as usnaric acid), thamnolic acid (±) (not in Japan), (+)-usnic acid. Schindler (1957), mention with literature references.
    Salazinic acid, stictic acid. Briner, Gream, and Riggs (1960), mention only.
    Salazinic acid, usnic acid. Rangaswami and Rao (1955c), from the Himalayas, 0.8% salazinic acid, 1.3% usnic acid, by extraction.
    Thamnolic acid, usnic acid. Duvigneaud (1947), from western Europe; Solberg (1956), mention only.
    Thamnolic acid. Hale (1962b), from Europe, mention only.
    (+)-Usnic acid. Dean, Halewood, Mongkolsuk, Robertson, and Whalley (1953), by extraction; Moissejeva (1957), 2.5%; Moissejeva

(1961), 2.5%, enzymes tabulated; Salkowski (1910); Savicz (1956), 2.5%; Stoll, Brack, and Renz (1947), 1.5% by extraction; Stoll, Renz, and Brack (1947), from Switzerland, by extraction.

Usnic acid. Burkholder and Evans (1945); Gertig and Banasiewicz (1961), from Poland, 1.31%; D. Hess (1958), from Germany; D. Hess (1960), mention only; Tomaselli (1957), from Italy, mention only.

No thamnolic acid. Asahina (1938*f*), from Japan.

Potassium oxalate. Schulte (1905), by microchemical tests, no barbatic acid by a microcrystal test on sections, no salazinic acid (as usnaric acid) by color reactions.

REVIEWS. Atranorin, barbatic acid, salazinic acid (as usnaric acid), thamnolic acid (as hirtellic acid), (+)-usnic acid: W. Brieger (1923); Hesse (1912). — Barbatic acid, salazinic acid (as usnaric acid), thamnolic acid (as hirtellic acid), (+)-usnic acid: Thies (1932); Zopf (1907), and an unidentified colorless substance. — (+)-Usnic acid: Savicz, Litvinov, and Moissejeva (1960), 2.5%; Schindler (1956/1957).

Usnea florida var. *comosa* (Ach.) Bir., see *Usnea comosa* (Ach.) Röhl.

Usnea florida var. *perplexans* (Stirt.) Vain., see *Usnea perplectans* Stirt.

Usnea fulvoreagens (Räs.) Räs.   Usnic acid. Vartia (1950*b*), from Finland.

Usnea cf. *fuscorubens* Mot.   Salazinic acid, usnic acid. Asahina (1956*c*), from Formosa.

Usnea galbinifera Asah.   Galbinic acid, norstictic acid, usnic acid. Asahina (1963*d*), from Formosa and India, by microcrystal tests and by thin-layer chromatography.

Usnea glabrata ssp. *pseudoglabrata* Asah.   Barbatic acid (±), salazinic acid, usnic acid. Asahina (1956*c*), from Japan, type description.

Usnea glabrescens (Nyl. *ex* Vain.) Vain.   Norstictic acid, usnic acid. Asahina (1959*b*), from Japan.
Usnic acid. Vartia (1949), from Finland; Vartia (1950*b*), from Finland.

Usnea glabrescens ssp. *asiatica* Asah.   Salazinic acid, usnic acid. Asahina (1959*b*), from Japan, type description.

Usnea glabrescens ssp. *pseudocolorans* Asah.   Thamnolic acid, usnic acid. Asahina (1959*c*), from Japan, type description.

Usnea goniodes var. *reagens* De Smet & Lambinon, see *Usnea angulata* Ach.

Usnea hakonensis Asah. Norstictic acid, salazinic acid, usnic acid. Asahina (1956*c*), from Japan, type description.

Usnea hakonensis f. *inactiva* Asah.
Medulla K−, PD− (apparently no salazinic or norstictic acids). Asahina (1956*c*), type description.

Usnea hirta (L.) Wigg. Salazinic acid (sometimes as usnaric acid), (+)-usnic acid; fat; enzymes, protein; calcium, phosphorus.
　　Salazinic acid, usnic acid. Rangaswami and Rao (1954*b*), from India; Rangaswami and Rao (1955*c*), mention only.
　　(+)-Usnic acid. Dean, Halewood, Mongkolsuk, Robertson, and Whalley (1953), by extraction; Moissejeva (1957), 3.0%; Moissejeva (1961), 3.0%, enzymes tabulated; Savicz, Kuprevicz, Litvinov, Moissejeva, and Rassadina (1956), 3.0%; Stoll, Brack, and Renz (1947), 1.15% by extraction; Stoll, Renz, and Brack (1947), from Switzerland, by extraction.
　　Usnic acid. Gertig and Banasiewicz (1961), from Poland, 4.28%; Schindler (1956/1957), mention only; Tichý (1955); Vartia (1949), from Finland; Vartia (1950*b*), from Finland.
　　Crude fat, protein, calcium, phosphorus. Scotter (1965), from northern Canada.
　　OTHER REPORTS. Thamnolic acid, usnic acid: Solberg (1956), mention only. — Hirtinic acid, usnic acid: Duvigneaud (1947). — Atranorin, thamnolic acid (as hirtellic acid), hirtinic acid: Klosa (1951*b*), mention only. — Potassium oxalate: Schulte (1905), by microchemical tests, no salazinic acid by color reactions, no barbatic acid by a microcrystal test on sections.
　　REVIEWS. Atranorin, barbatic acid, hirtinic acid, picrusunidic acid, salazinic acid, santhomic acid, thamnolic acid (as hirtellic acid), (+)-usnic acid, usnidic acid: Thies (1932), and as *Usnea barbata* var. *hirta.* — Atranorin, barbatic acid, hirtaic acid, salazinic acid, santhomic acid, thamnolic acid (as hirtellic acid), usnarinic acid, (+)-usnic acid: W. Brieger (1923); Hesse (1912); both reports also as *U. barbata* var. *hirta.* — Hirtaic acid, hirtinic acid, santhomic acid, thamnolic acid (as hirtellic acid), (+)-usnic acid: Zopf (1907), and as *U. barbata* var. *hirta.* — Salazinic acid: Shibata (1958, 1963). — (+)-Usnic acid: Savicz, Litvinov, and Moissejeva (1960), 3.0%. See also Schindler (1957).

Usnea hondoensis Asah. Barbatic acid (±), salazinic acid, usnic acid. Asahina (1956*c*), from Formosa and Japan, and also *Usnea hondoensis* ssp. *inflatula* f. *fujisanensis* from Formosa and Japan, and *U. hondoensis* ssp. *lacunosula* from Japan, type descriptions.

Usnea hossei Vain., see *Usnea pectinata* Tayl.

Usnea hossei f. *subtrichodea* Asah.   Diffractaic acid ($\pm$), stictic acid, usnic acid; unidentified yellow pigment (K+ purple). Asahina (1956c), with diffractaic acid in North America, without this substance in Japan, type description.

Usnea implicita (Stirt.) Zahlbr.   Norstictic acid, usnic acid; unidentified red pigment. Asahina (1956c), from Formosa and Japan, and also *Usnea implicata* var. *yokawensis* from Japan; Asahina (1965d).

Usnea indigena Mot.   Stictic acid, usnic acid; unidentified pigment (K+ purple). Asahina (1965d), from Madagascar, unidentified pigment at base of plant found to be acetone soluble.

Usnea intexta Stirt.   Norstictic acid, stictic acid, usnic acid. Duvigneaud (1947), from western Europe.

Usnea intumescens Asah.   Salazinic acid, usnic acid. Asahina (1956c), from Japan, type description; Asahina (1958e), mention only.

Usnea japonica Vain.   Norstictic acid, salazinic acid, usnic acid. Asahina (1938e), description of microcrystal tests; Asahina (1956c), from Japan, also mentioned in the type description of *Usnea japonica* var. *boninensis* from the Bonin Islands; Asahina and Tukamoto (1934), from the Bonin Islands, Japan, and Java.
   Norstictic acid, salazinic acid. Asahina (1958e).
   Norstictic acid. Mitsuno (1953), by paper chromatography.
   OTHER REPORTS. Barbatolic acid, stictic acid, (+)-usnic acid: Seshadri and Subramanian (1949c), from India; Neelakantan and Seshadri (1952a), from India.
   REVIEWS. Norstictic acid: Asahina and Shibata (1954); Shibata (1958, 1963).

Usnea jesoensis Asah., see *Usnea longissima* ssp. *jesoensis* Asah.

Usnea kinkiensis Asah.   Norstictic acid, salazinic acid, usnic acid. Asahina (1958e), from Japan, and also *Usnea kinkiensis* f. *gracilior*, type descriptions.

Usnea koyana Asah.   Decarboxythamnolic acid, thamnolic acid, usnic acid. Asahina and Nuno (1964), by thin-layer chromatography.
   Thamnolic acid. Asahina (1956c), from Japan, type description.

Usnea kurokawae Asah.   Norstictic acid, salazinic acid, usnic acid. Asahina (1956c), from Japan, type description.

Usnea kushiroensis Asah.   Norstictic acid, salazinic acid, usnic acid. Asahina (1956*c*), from Japan, type description.

Usnea kushiroensis ssp. *subasiriensis* Asah.   Norstictic acid, stictic acid, salazinic acid. Asahina (1956*c*), from Japan, type description.

Usnea kyotoensis Asah.   Norstictic acid, salazinic acid, usnic acid. Asahina (1958*e*), from Japan, type description.

Usnea lacerata Mot.   Protocetraric acid, (+)-usnic acid. Huneck and Follmann (1966*e*), from Chile, 0.6% protocetraric acid, 1.2% (+)-usnic acid, by extraction of 41.0 g. of lichen.

Usnea laricina Vain. *ex* Räs.   Salazinic acid, usnic acid. Duvigneaud (1947), from western Europe.

Usnea longissima Ach.   Barbatic acid, (+)-usnic acid; D-arabitol, isolichenin, lichenin; enzymes; ascorbic acid, riboflavin; calcium, iron, nitrogen, phosphorus.
    Barbatic acid, (+)-usnic acid, D-arabitol. Dhar, Neelakantan, Ramanujam, and Seshadri (1959), from India, by extraction; Rangaswami and Rao (1954*b*), mention only.
    Barbatic acid, (+)-usnic acid. Aghoramurthy, Sarma, and Seshadri (1961*b*), from India, used as an authentic source of these substances; Asahina (1937*e*); Asahina (1955*a*), from India; Asahina and Tukamoto (1933), from Europe and Japan, by extraction, also an unidentified substance (m.p. 211°); Neelakantan and Seshadri (1952*a*), from India, no substance of m.p. 211°, mention only.
    Barbatic acid, usnic acid. Asahina (1956*c*), from Japan, as *Usnea longissima* var. *robustior*; Seshadri (1953), from the Himalayas, 3–4% usnic acid, mention only.
    Barbatic acid. Fuzikawa, Shinamura, and Tarui (1941); Schulte (1905), by a microcrystal test on sections, potassium oxalate by microchemical tests, no salazinic acid (as usnaric acid) by color reactions.
    Usnic acid. Mikoshiba (1936); Moissejeva (1957), 2.5%; Moissejeva (1961), 2.5%, enzymes tabulated; Pereira, Sa, and Bhatnagar (1953), from the Himalayas, antibiotic study; Taguchi, Sankawa, and Shibata (1966), ^{14}C study on the biosynthesis of this pigment; Tomaselli (1957), from Italy, mention only.
    D-Arabitol. Mittal, Neelakantan, and Seshadri (1952), mention only.
    Isolichenin, lichenin, ascorbic acid, riboflavin, calcium, iron, nitrogen, phosphorus. Lal and Rao (1959).
    Isolichenin, lichenin. Chanda, Hirst, and Manners (1957), mention only.
    Lichenin. Mittal and Seshadri (1954*a*), improved method of isolation.

REVIEWS. Barbatic acid, diffractaic acid (as dirhizonic acid), proto-cetraric acid (as ramalinic acid), (+)-usnic acid: W. Brieger (1923); Hesse (1912); Thies (1932); Zopf (1907). — Barbatic acid, diffractaic acid: Asahina and Shibata (1954); Shibata (1958), and lichenin; Shibata (1963). — (+)-Usnic acid: Savicz, Litvinov, and Moissejeva (1960), 2.5%.

Usnea longissima ssp. *ambigua* Asah.   Diffractaic acid, (+)-usnic acid. Asahina (1937e), as *Usnea longissima*.
  Diffractaic acid, usnic acid. Asahina (1956c), from Japan.
  Diffractaic acid. Asahina (1936e), from Europe, description of micro-crystal tests, as *Usnea longissima*; Asahina and Fuzikawa (1932b), from Africa, mentions early work of Hesse, as *U. longissima*.

Usnea longissima ssp. *jesoensis* Asah.   Evernic acid, (+)-usnic acid. Asahina (1936e), from Japan and Sakhalin, description of microcrystal tests, as *Usnea longissima*; Asahina (1937e), as *U. longissima*; Asahina (1955a), from India, as *U. longissima*.
  Evernic acid, usnic acid. Asahina (1956c), from Japan.
  REVIEWS. Evernic acid: Asahina and Shibata (1954); Shibata (1958, 1963); all three reports as *Usnea jesoensis*.

Usnea longissima ssp. *persensibilis* Asah.   Atranorin (±), fumarproto-cetraric acid, usnic acid. Asahina (1956c), from Japan, type description.

Usnea longissima ssp. *persensibilis* f. *tingens* Asah.   Atranorin (±), fumarprotocetraric acid, usnic acid. Asahina (1956c), thallus K + yellow, type description.

Usnea longissima ssp. *sensibilis* Asah.   Salazinic acid, usnic acid. Asahina (1956c), from Japan, type description.

Usnea longissima var. *vulgata* Asah.   Barbatic acid, usnic acid. Asahina (1956c), from Japan, some samples with medulla K + probably due to barbatolic acid, type description.

Usnea ludicra Rizz.   Barbatic acid, norstictic acid, salazinic acid, (+)-usnic acid. Mors (1952a), from Brazil, 0.05% barbatic acid, trace norstictic acid, 3.6% salazinic acid, 1% (+)-usnic acid, by extraction of 110 g. of lichen; Mors (1952b), from Brazil.

Usnea lunaria Mot.   (+)-Usnic acid, unidentified substance (K + red). Mors (1952a), from Brazil, by extraction of 140 g. of lichen, actually a mixture of K + and K − types; Mors (1952b), from Brazil.

Usnea merrillii Mot.   Salazinic acid, usnic acid. Asahina (1956*c*), from Japan and Mexico, as *Usnea cribosa*; Asahina (1965*f*), from Maine and North Carolina.

Usnea microcarpa Pers.
  (+)-Usnic acid. Salkowski (1910).
  Barbatic acid, potassium oxalate. Schulte (1905), by microchemical tests, also salazinic acid by color tests, barbatic acid by a microcrystal test on sections.
  REVIEWS. Salazinic acid (as usnaric acid), (+)-usnic acid: W. Brieger (1923); Hesse (1912); Thies (1932); Zopf (1907).

Usnea misaminensis (Vain.) Mot.   Stictic acid; unidentified hydroxyanthraquinone (±). Asahina (1965*a*), from Japan, hydroxyanthraquinone pigment in the blackened portion of the base, and also *Usnea misaminensis* var. *subtrichodea.*
  Pigment at base. Asahina (1965*d*), K+ purple-red, also in *Usnea misaminensis* var. *subtrichodea.*

Usnea monstruosa Vain.   Usnic acid. Vartia (1950*b*), from Finland.

Usnea montis-fuji Mot.   Atranorin, protocetraric acid, (+)-usnic acid. Asahina and Tukamoto (1933), from Japan.
  OTHER REPORT. Salazinic acid, usnic acid: Asahina (1956*c*), from Japan.
  REVIEWS. Protocetraric acid: Asahina and Shibata (1954); Shibata (1958, 1963).

Usnea montis-fuji f. *cinerea* Asah.   Atranorin, salazinic acid, usnic acid. Asahina (1956*c*), from Japan, type description.

Usnea mutabilis Stirt.   Usnic acid, unidentified substances (soluble in benzene); unidentified red substance. Asahina (1965*d*), by thin-layer chromatography.

Usnea nipparensis Asah.   Stictic acid (±), usnic acid; unidentified substance (PD−). Asahina (1956*c*), from Japan, type description.

Usnea orientalis Mot.
  Barbatic acid, stictic acid, caperatic acid (from South India). Salazinic acid, (+)-usnic acid (from the Himalayas). Dhar, Neelakantan, Ramanujam, and Seshadri (1959), by extraction of 1.5 kg. and 500 g. of lichen respectively; Nair and Subramanian (1962), mention only.
  Psoromic acid, (+)-usnic acid, ergosterol. Nair and Subramanian (1962), from Nilgiris, 1.0% psoromic acid, 1.2% (+)-usnic acid, by extraction.

Stictic acid, usnic acid (from South India). Salazinic acid, usnic acid (from the Himalayas). Neelakantan and Seshadri (1952*a*), mention only; Rangaswami and Rao (1954*b*), mention only.

Stictic acid. Aghoramurthy, Sarma, and Seshadri (1961*b*), used as an authentic source of this substance.

Isolichenin, lichenin, ascorbic acid, riboflavin, calcium, iron, nitrogen, phosphorus. Lal and Rao (1959), from India.

Amino acids. Ramakrishnan and Subramanian (1964), from India, alanine, arginine, glycine, leucine, threonine, tyrosine, and valine as free amino acids, and also lysine, phenylalanine, and serine by alkaline hydrolysis of the residue.

Usnea orientalis f. *esorediosa* Asah. Salazinic acid, usnic acid. Asahina (1956*c*), from Formosa and Japan, type description.

Usnea pectinata Tayl. Stictic acid, usnic acid; unidentified yellow pigment (K+ purple). Asahina (1956*c*), from Formosa and Japan, as *Usnea hossei*.

Stictic acid. Asahina (1965*a*), from Formosa, the Himalayas, Indonesia, Java, and Thailand, also sometimes a hydroxyanthraquinone causing a blackened basal region.

OTHER REPORT. Barbatic acid, (+)-usnic acid, ventosic acid, D-arabitol: Aghoramurthy, Sarma, and Seshadri (1961*b*), from India, by extraction.

REVIEW. Ventosic acid: Neelakantan (1965).

Usnea perplectans Stirt. Barbatic acid, protocetraric acid, salazinic acid, (+)-usnic acid. Asahina and Tukamoto (1933), from Formosa, by extraction, as *Usnea florida* var. *perplexans*.

Usnea plicata (L.) Wigg. Salazinic acid (sometimes as usnaric acid), (+)-usnic acid. Solberg (1956), mention only.

Salazinic acid. Schulte (1905), by color reactions only, no barbatic acid by a microcrystal test, potassium oxalate by microchemical tests.

REVIEWS. Plicatic acid, salazinic acid, (+)-usnic acid: W. Brieger (1923); Hesse (1912); Thies (1932); Zopf (1907).

Usnea pseudintumescens Asah. Psoromic acid, usnic acid; unidentified substance. Asahina (1956*c*), from Japan, type description.

Psoromic acid, usnic acid. Asahina (1958*a*), from Japan, difficulties with microcrystal test described, confirmed psoromic acid by chromatography; Asahina (1958*e*).

Usnea pseudomontis-fuji Asah. Atranorin, fumarprotocetraric acid, usnic acid. Asahina (1956*c*), from Japan, type description.

Usnea pseudorubescens Asah.   Stictic acid. Asahina (1965*c*), from Formosa, Japan, Java, and Sabah, type description.

Usnea pygmea Mot., see *Usnea bismolliuscula* Zahlbr.

Usnea pygmea ssp. ***kitamiensis*** Asah.   Norstictic acid, salazinic acid (trace), usnic acid. Asahina (1956*c*), from Japan, type description.

Usnea roseola Vain.   Diffractaic acid, usnic acid; unidentified red pigment. Asahina (1956*c*), from Japan, type description of *Usnea subroseola*; Asahina (1965*e*), from Formosa, the Himalayas, Java, and Thailand, red pigment in medulla, and also *U. roseola* f. *subroseola* from Japan.

Usnea roseola ssp. ***pseudoroseola*** Asah.   Barbatic acid, usnic acid. Asahina (1956*c*), from Japan, type description.
    OTHER REPORT. Diffractaic acid, usnic acid: Asahina (1950*b*).

Usnea rubescens Stirt.   Norstictic acid, salazinic acid, usnic acid. Asahina (1953*c*), from Formosa and Japan; Asahina (1956*c*), from Formosa and Japan.
    Norstictic acid, salazinic acid. Asahina (1965*c*), from Formosa, Japan, Java, and eastern Nepal.

Usnea rubescens ssp. ***aberrans*** Asah.   Stictic acid. Asahina (1956*c*), from Japan, type description.

Usnea rubicunda Stirt.   Stictic acid, (+)-usnic acid. Huneck and Follmann (1966*e*), from Chile, 2.4% stictic acid, 0.67% (+)-usnic acid, by extraction of 37.0 g. of lichen.
    Stictic acid, usnic acid. Asahina (1953*c*), from Formosa and Japan; Asahina (1956*c*), from Formosa and Japan; Duvigneaud (1947), from western Europe.
    Stictic acid. Asahina (1965*c*), from Japan.
    Usnic acid. Hale (1957*a*), from Arkansas.
    OTHER REPORT. Barbatic acid, protocetraric acid, (+)-usnic acid, ergosterol: Nair and Subramanian (1961*b*), from Nilgiris.

Usnea rubicunda ssp. ***aberrans*** Asah.   Norstictic acid, salazinic acid. Asahina (1956*c*), from Japan, type description; Asahina (1965*c*).

Usnea rubiginea (Michx.) Mass.   Obtusatic acid (as ramalic acid), (+)-usnic acid. Asahina (1926).

Usnea rugulosa Vain.   Usnic acid. Vartia (1950*b*), from Finland.

Usnea scabrata Nyl.
Barbatic acid, potassium oxalate. Schulte (1905), barbatic acid by a microcrystal test on sections, potassium oxalate by microchemical tests and also salazinic acid by color reactions only.
REVIEWS. Salazinic acid (as usnaric acid), (+)-usnic acid: Thies (1932); Zopf (1907). — Salazinic acid (as usnaric acid): W. Brieger (1923); Hesse (1912).

Usnea "schraderi" (= *Usnea barbata* f. *schraderi* Dalla Torre & Sarnth.), see *Usnea dasypoga* (Ach.) Röhl.

Usnea shikokiana Asah. Stictic acid, usnic acid; unidentified yellow pigment (K + purple). Asahina (1965*d*), from Japan, unidentified yellow pigment in blackened basal portion found to be soluble in acetone.

Usnea similis (Mot.) Räs. Usnic acid. Vartia (1950*b*), from Finland.

Usnea sorediifera (Arn.) Lynge (+)-Usnic acid. Salkowski (1910).
REVIEWS. (+)-Usnic acid: W. Brieger (1923); Hesse (1912); Thies (1932). — Usnic acid: Zopf (1907).

Usnea spinigera Asah. Norstictic acid, salazinic acid, usnic acid. Asahina (1956*c*), from Japan, type description.

Usnea stirtoniana Zahlbr. Isolichenin, lichenin; ascorbic acid, riboflavin; calcium, iron, nitrogen. Lal and Rao (1959), from India.

Usnea strigosa (Ach.) A. Eat. Norstictic acid, usnic acid; unidentified red pigment (±) (Chem. strain 1). Psoromic acid, usnic acid; unidentified red pigment (±) (Chem. strain 2). Usnic acid, unidentified red pigment (±) (Chem. strain 3). Hale (1962*b*), Chem. strain 1 (from Alabama) identical to *Usnea evansii*, *U. tristis* possibly depauperate *U. strigosa* with norstictic acid.
Psoromic acid, usnic acid. Hale (1957*a*), from Arkansas, Missouri, and Oklahoma.

Usnea subfusca Stirt. Norstictic acid, usnic acid. Hale (1956*b*); Hale (1958*d*), from North America; Hale (1956*g*), from Arkansas, an exsiccati specimen.
OTHER REPORTS. Salazinic acid: Hale (1962*b*), from the Appalachian Mountains, morphologically identical to *Usnea florida*.

Usnea subroseola Asah., see *Usnea roseola* Vain.

Usnea sulphurea (Kön.) Th. Fr. Norstictic acid (±), usnic acid. Lamb (1964), from arctic and antarctic regions, usually no norstictic acid.

Usnea thomsonii Stirt.   Salazinic acid, usnic acid. Asahina (1955*a*), from India.

Usnea trichodea Ach.   Evernic acid, (+)-usnic acid. Asahina (1926). Usnic acid. Hale (1957*a*), from Arkansas.

Usnea trichodeoides Vain. *ex* Mot.   Usnic acid. Schindler (1956/1957), from Africa, mention only.

Usnea tristis Mot., see *Usnea strigosa* (Ach.) A. Eat.

Usnea venosa Mot.   Barbatic acid, salazinic acid, (+)-usnic acid; unidentified neutral substance (m.p. 230–232°), ergosterol; carotene; amino acids.

Barbatic acid, salazinic acid, (+)-usnic acid, unidentified neutral substance (m.p. 230–232°), carotene, ergosterol. Murty and Subramanian (1959*b*), from India, by extraction.

Barbatic acid. Nair and Subramanian (1961*a*); Nair and Subramanian (1961*b*), mention only.

Amino acids. Ramakrishnan and Subramanian (1964), from India, glycine, leucine, threonine, and valine as free amino acids, but no alanine, arginine, or tyrosine, and also lysine, phenylalanine and serine by alkaline hydrolysis of the residue.

REVIEW. Ergosterol: Neelakantan (1965).

Usnea wasmuthii Räs.   Salazinic acid, usnic acid. Duvigneaud (1947), from western Europe.

Usnea yakushimensis Asah.   Thamnolic acid, usnic acid. Asahina (1956*c*), from Japan, type description.

Variolaria amara Ach., see *Pertusaria amara* (Ach.) Nyl.

Variolaria faginea (L.) Pers., see *Pertusaria faginea* (L.) Pers.

Variolaria lactea (L.) Pers., see *Pertusaria lactea* (L.) Arn.

Xanthoria aureola (Ach.) Erichs.   Parietin, parietinic acid. Eschrich (1958), separation by chromatography.

REVIEWS. Parietinic acid: Neelakantan (1965); Shibata (1963).

All reports are as *Xanthoria parietina* var. *aureola*.

Xanthoria candelaria (L.) Th. Fr.   Parietin (sometimes as physcion). Hale (1957*a*), from Arkansas.

REVIEWS. Parietin: W. Brieger (1923); Hesse (1912); Thies (1932); Zopf (1907); all review reports as *Xanthoria lychnea* var. *pygmaea* (= *X. candelaria* var. *pygmaea*).

Xanthoria contortuplicata (Ach.) Boist. Parietin, parietinic acid. Eschrich (1958), as *Xanthoria parietina* var. *contortuplicata*.

Xanthoria elegans (Link) Th. Fr. Parietin (sometimes as physcion). D. Hess (1958), from Germany, as *Caloplaca elegans*; Neelakantan and Seshadri (1952*b*), from Kashmir, by extraction, as *C. elegans*; Neelakantan and Seshadri (1960*a*), mention only, as *C. elegans*; Schatz (1963), mention only; Thomas (1936), parietin produced by the lichen fungus in pure culture, as *C. elegans*.

REVIEWS. Parietin, rhizocarpic acid: W. Brieger (1923), as *Gasparrina elegans*; Hesse (1912), as *G. elegans*; Thies (1932), as *G. elegans* and *Placodium elegans*. — Parietin: Zopf (1907), as *P. elegans*.

Xanthoria fallax (Hepp) Arn. Atranorin; fallacinal, parietin, teloschistin. Atranorin. D. N. Rao and Le Blanc (1965), and quinone pigments.

Fallacinal, parietin, teloschistin. Murakami (1956), review of previous literature included, crude "fallacin" found to be a mixture of three pigments separated by chromatography; Rajagopalan and Seshadri (1959), mention only.

Fallacinal, teloschistin. Neelakantan and Seshadri (1960*a*), mention in discussion of biosynthesis.

OTHER REPORTS. Fallacin, parietin: Asano and Arata (1940), crude fallacin found to contain parietin. — Fallacin: Asano and Fuziwara (1936); Subramanian (1955), mention only.

REVIEWS. Fallacin, parietin: Asahina and Shibata (1954). — Fallacinal, parietin, teloschistin: Shibata (1958, 1963), teloschistin indicated as possibly equivalent to fallacinal rather than to fallacinol. — Teloschistin: Neelakantan (1965).

Xanthoria lychnea var. *polycarpa* (Ehrh.) Th. Fr., see *Xanthoria polycarpa* (Ehrh.) Oliv.

Xanthoria lychnea var. *pygmaea* (Bory) Nyl., see *Xanthoria candelaria* (L.) Th. Fr.

Xanthoria mandschurica (Zahlbr.) Asah. Parietin. Asahina (1954*c*), from China and Japan, type description.

Xanthoria parietina (L.) Th. Fr. Atranorin ($\pm$); parietin (sometimes as physcion), parietinic acid, unidentified anthraquinone; β-carotene, lutein, violoxanthin; D-mannitol, isolichenin, lichenin; chlorophyll *a*, chlorophyll *b*.

Atranorin, parietin. Hale (1955*a*), from North America.

Parietin, parietinic acid. Eschrich (1958).

Parietin, unidentified anthraquinone. Tomaselli (1963), water-soluble substances, parietin and an unidentified anthraquinone from the cultured fungus also found in the lichen.

Parietin, isolichenin, lichenin, mannitol. Dhar, Neelakantan, Ramanujam, and Seshadri (1959), from India, by extraction.

Parietin. Asahina and Fuzikawa (1935*d*), mention only; D. Hess (1958), from Germany; Krog (1951*a*), mention only; Ryan and Riordan (1917), from Ireland, as *Physcia parietina*, by extraction; Schratz and Vethake (1958); Steiner (1955); Tomaselli (1958), mention only.

β-Carotene, violoxanthin, xanthophyll. Nicola and Tomaselli (1961*a*), found in *Trebouxia decolorans* isolated from the lichen.

Carotene, chlorophyll *a*. Nicola and Tomaselli (1961*b*), chlorophyll *a* and total chlorophyll measured in *Trebouxia decolorans* isolated from the lichen.

Mannitol. Asahina (1934*b*); Pueyo (1960), mention only; Pueyo (1965), mention only.

Chlorophyll *a*, chlorophyll *b*. Wilhelmsen (1959).

OTHER REPORTS. Parietinic acid: Neelakantan and Seshadri (1960*a*), mention in discussion of biosynthesis. — Unidentified water-soluble substance with blue-green fluorescence: Cantone, Pignataro, and Tomaselli (1963), mass spectrographic study of water-soluble products of the lichen and the pure fungal culture.

REVIEWS. Atranorin, parietin: Asahina and Shibata (1954), and D-mannitol; W. Brieger (1923); Hesse (1912); Shibata (1958, 1963); Thies (1932), and mannitol, atranorin only in the shade form; Zopf (1907), and mannitol.

Xanthoria parietina var. *aureola* (Ach.) Th. Fr., see *Xanthoria aureola* (Ach.) Erichs.

Xanthoria parietina var. *contortuplicata* (Ach.) Oliv., see *Xanthoria contortuplicata* (Ach.) Boist.

Xanthoria polycarpa (Ehrh.) Oliv.   Parietin (sometimes as physcion).
REVIEWS. Parietin: W. Brieger (1923), as *Xanthoria lychnea* var. *polycarpa*; Hesse (1912), as *X. lychnea* var. *polycarpa*; Thies (1932), Zopf (1907).

Literature Cited in This Chapter

ABBAYES, H. DES. 1961. Lichens récoltés à Madagascar et à la Réunion. (Mission H. des Abbayes, 1956). I. — Introduction. II. — Parmeliacées. *Mém. Inst. Sci. Madagascar*, Sér. B, **10** (2), 81–121.

ABBAYES, H. DES. 1962. Lichens foliacés et fruticuleux d'Afrique Centrale récoltés par l'Expédition Suisse du Virunga en 1954–1955. *Rev. Bryol. Lichénol.* **31**, 239–250.

ABBAYES, H. DES. 1963. Lichens nouveaux ou intéressants du Vietnam. *Rev. Bryol. Lichénol.* **32**, 216–222.

ACKER, L., W. DIEMAIR, AND E. SAMHAMMER. 1955*a*. Über das Lichenin des Hafers. I. Mitteilung. Eigenschaften, Darstellung und Zusammensetzung des schleimbildenden Polysaccharids. *Z. Lebensm. Untersuch. Forsch.* **100**, 180–188.

ACKER, L., W. DIEMAIR, AND E. SAMHAMMER. 1955*b*. Über das Lichenin des Hafers. II. Mitteilung. Molekulargewichtsbestimmung und weitere Untersuchungen zur Konstitution. *Z. Lebensm. Untersuch. Forsch.* **102**, 225–231.

AGARWAL, S. C., K. AGHORAMURTHY, K. G. SARMA, AND T. R. SESHADRI. 1961. Variations in the chemical components of *Lobaria* lichen from Darjeeling. *J. Sci. Ind. Res.* (India) **20B**, 613–615.

AGARWAL, S. C. AND T. R. SESHADRI. 1965. Constitution of leprapinic acid. *Tetrahedron* **1965**, 3205–3208.

AGHORAMURTHY, K., S. NEELAKANTAN, AND T. R. SESHADRI. 1954. Chemical investigation of Indian lichens: Part XVII — Chemical components of some *Parmelia* lichens. *J. Sci. Ind. Res.* (India) **13B**, 326–328.

AGHORAMURTHY, K., K. G. SARMA, AND T. R. SESHADRI. 1959. The structure of thelephoric acid. *Tetrahedron Letters* **1959** (8), 20–24.

AGHORAMURTHY, K., K. G. SARMA, AND T. R. SESHADRI. 1961*a*. Chemical investigations of Indian lichens — Part XXIV. The chemical components of *Alectoria virens* Tayl. Constitution of a new depsidone, virensic acid. *Tetrahedron* **1961**, 173–177.

AGHORAMURTHY, K., K. G. SARMA, AND T. R. SESHADRI. 1961*b*. Chemical investigation of Indian lichens: Part XXV — Chemical components of some rare Himalayan lichens. *J. Sci. Ind. Res.* (India) **20B**, 166–168.

AGHORAMURTHY, K. AND T. R. SESHADRI. 1952. Nuclear oxidation in flavones and related compounds. Part XXXIX. Occurrence of nuclear oxidation in the biogenesis of lichen acids. *Proc. Indian Acad. Sci.* **35A**, 327–337.

AGHORAMURTHY, K. AND T. R. SESHADRI. 1953*a*. An improved synthesis of lichexanthone. *J. Sci. Ind. Res.* (India) **12B**, 350–352.

AGHORAMURTHY, K. AND T. R. SESHADRI. 1953*b*. Chemical investigation of Indian lichens: Part XV — A species of *Parmelia* containing lichexanthone. *J. Sci. Ind. Res.* (India) **12B**, 73–76.

AGHORAMURTHY, K. AND T. R. SESHADRI. 1954. A theory of biosynthesis of some mould products. *J. Sci. Ind. Res.* (India) **13A**, 114–124.

AHMADJIAN, V. 1963. The fungi of lichens. *Am. Scientist* **208** (2), 122–132.

AHMADJIAN, V. 1964. Further studies on lichenized fungi. *Bryologist* **67**, 87–98.

AHMANN, G. B. AND ANNICK MATHEY. 1967. Lecanoric acid and some constituents of *Parmelia tinctorum* and *Pseudevernia intensa*. *Bryologist* **70**, 93–97.

AHTI, T. 1964. Macrolichens and their zonal distribution in boreal and arctic Ontario, Canada. *Ann. Botan. Fennici* **1**, 1–35.

AHTI, T. 1966. *Parmelia olivacea* and the allied non-isidiate and non-sorediate corticolous lichens in the northern hemisphere. *Acta Botan. Fennica* **70**, 1–68.

ÅKERMARK, B. 1961. Studies on the chemistry of lichens. 14. The structure of calycin. *Acta Chem. Scand.* **15**, 1695–1700.

ÅKERMARK, B. 1962. Studies on the chemistry of lichens. 16. The absolute configuration of roccellic acid. *Acta Chem. Scand.* **16**, 599–606.

ÅKERMARK, B., H. ERDTMAN, AND C. A. WACHTMEISTER. 1959. Studies on the chemistry of lichens. XIII. The structure of pannaric acid. *Acta Chem. Scand.* **13**, 1855–1862.

ALERTSEN, A. R., T. BRUUN, AND ELLEN HEMMER. 1962. The occurrence of ergosterol in the lichen *Cornicularia normoerica* (Gunnerus) Lynge. *Acta Chem. Scand.* **16**, 541–542.

ALLISON, W. R. AND G. T. NEWBOLD. 1959. Lactones. Part VI. The preparation of 5,7-dihydroxyphthalide, its 5-methyl ether, and related compounds. *J. Chem. Soc.* **1959**, 3335–3340.

ANDERSON, R. A. 1962. The lichen flora of the Dakota sandstone in north-central Colorado. *Bryologist* **65**, 242–261.

ANDERSON, R. A. AND W. A. WEBER. 1962. Two species of *Parmelia* from western United States. *Bryologist* **65**, 234–241.

ARK, P., A. BOTTINI, AND J. P. THOMPSON. 1960. Sodium usnate as an antibiotic for plant diseases. *Plant Disease Reptr.* **44**, 200–203.

ASAHINA, Y. 1926. Untersuchungen über Flechtenstoffe. II. Bestandteile von einigen in Japan einheimischen Flechten. *Yakugaku Zasshi* **533**, 47–51. Reference from *Chem. Zentr.* **97** (2), 2728–2729 (1926). See also *Chem. Abstr.* **21**, 2262–2263 (1927).

ASAHINA, Y. 1933. *Lobaria*-Arten aus Japan. (1). *J. Japan. Botany* **9**, 333–339.

ASAHINA, Y. 1934a. Über die Reaktion von Flechten-Thallus. *Acta Phytochim.* **8**, 47–64.

ASAHINA, Y. 1934b. Zur Systematik der Flechtenstoffe. *Acta Phytochim.* **8**, 33–45.

ASAHINA, Y. 1935a. Flechten-Säuren. *Tabulae Biol. Periodicae* **4** (= *Tabulae Biologicae* **10**), 198–208.

ASAHINA, Y. 1935b. *Nephromopsis*-Arten aus Japan. *J. Japan. Botany* **11**, 10–27.

ASAHINA, Y. 1935c. Über den Nachweis der Usninsäure in den Flechten. *J. Japan. Botany* **11**, 692–695.

ASAHINA, Y. 1936a. *Alectoria*- und *Oropogon*-Arten aus Japan. *J. Japan. Botany* **12**, 690–698.

ASAHINA, Y. 1936b. Diagnose einiger *Alectoria*-Arten durch die Diamin-Probe. *J. Japan. Botany* **12**, 687–690.

ASAHINA, Y. 1936c. Lichenologische Notizen VIII. *J. Japan. Botany* **12**, 802–804.

ASAHINA, Y. 1936d. Mikrochemischer Nachweis der Flechtenstoffe (I.). *J. Japan. Botany* **12**, 516–525.

ASAHINA, Y. 1936*e*. Mikrochemischer Nachweis der Flechtenstoffe (II. Mitteil.). *J. Japan. Botany* 12, 859–872.

ASAHINA, Y. 1937*a*. *Anzia*-Arten aus Japan mit besonderer Berücksichtigung der chemischen Bestandteile. *J. Japan. Botany* 13, 219–226.

ASAHINA, Y. 1937*b*. Lichenologische Notizen. IX. *J. Japan. Botany* 13, 315–321.

ASAHINA, Y. 1937*c*. Mikrochemischer Nachweis der Flechtenstoffe (III. Mitteil.). *J. Japan. Botany* 13, 529–536.

ASAHINA, Y. 1937*d*. Mikrochemischer Nachweis der Flechtenstoffe IV. Mitteilung. *J. Japan. Botany* 13, 855–861.

ASAHINA, Y. 1937*e*. Über den taxonomischen Wert der Flechtenstoffe. *Botan. Mag.* (Tokyo) 51, 759–764.

ASAHINA, Y. 1938*a*. Lichenologische Notizen. X. *J. Japan. Botany* 14, 251–255.

ASAHINA, Y. 1938*b*. Mikrochemischer Nachweis der Flechtenstoffe. V. Mitteilung. *J. Japan. Botany* 14, 39–44.

ASAHINA, Y. 1938*c*. Mikrochemischer Nachweis der Flechtenstoffe. VI. Mitteilung. *J. Japan. Botany* 14, 244–250.

ASAHINA, Y. 1938*d*. Mikrochemischer Nachweis der Flechtenstoffe. VII. Mitteilung. *J. Japan. Botany* 14, 318–323.

ASAHINA, Y. 1938*e*. Mikrochemischer Nachweis der Flechtenstoffe. VIII. Mitteilung. *J. Japan. Botany* 14, 650–659.

ASAHINA, Y. 1938*f*. Mikrochemischer Nachweis der Flechtenstoffe. IX. Mitteilung. *J. Japan. Botany* 14, 767–773.

ASAHINA, Y. 1938*g*. *Ramalina*-Arten aus Japan (I). *J. Japan. Botany* 14, 721–730.

ASAHINA, Y. 1939*a*. *Cornicularia*-Arten aus Japan. *J. Japan. Botany* 15, 353–358.

ASAHINA, Y. 1939*b*. Flechtenstoffe. *Fortschr. Chem. Org. Naturstoffe* 2, 27–60.

ASAHINA, Y. 1939*c*. Japanische Arten der *Cocciferae* (*Cladonia-Coenomyce*). *J. Japan. Botany* 15, 602–620, 663–671.

ASAHINA, Y. 1939*d*. Lichenologische Notizen (XI). Ist *Chaudhuria* Zahlbr. wirklich eine selbständige Gattung? *J. Japan. Botany* 15, 277–280.

ASAHINA, Y. 1939*e*. Mikrochemischer Nachweis der Flechtenstoffe. X. Mitteilung. *J. Japan. Botany* 15, 465–472.

ASAHINA, Y. 1939*f*. *Ramalina*-Arten aus Japan (II). *J. Japan. Botany* 15, 205–223.

ASAHINA, Y. 1939*g*. Ueber den Chemismus der Flechten der *Cocciferae* (*Cladonia* subg. *Cenomyce*). *J. Japan. Botany* 15, 22–36.

ASAHINA, Y. 1940*a*. Chemismus der Cladonien unter besonderer Berücksichtigung der japanischen Arten. *J. Japan. Botany* 16, 709–727.

ASAHINA, Y. 1940*b*. *Cladonia verticillata* Hoffm. und *Cladonia calycantha* (Del.) Nyl. aus Japan. *J. Japan. Botany* 16, 462–470.

ASAHINA, Y. 1940*c*. Lichenologische Notizen XII. *J. Japan. Botany* 16, 401–404.

ASAHINA, Y. 1940*d*. Lichenologische Notizen (XIII). *J. Japan. Botany* **16**, 517–522.

ASAHINA, Y. 1940*e*. Lichenologische Notizen (XIV). *J. Japan. Botany* **16**, 592–603.

ASAHINA, Y. 1940*f*. Mikrochemischer Nachweis der Flechenstoffe. XI. Mitteilung. *J. Japan. Botany* **16**, 185–193.

ASAHINA, Y. 1941*a*. Chemismus der Cladonien unter besonderer Berücksichtigung der japanischen Arten. *J. Japan. Botany* **17**, 431–437.

ASAHINA, Y. 1941*b*. Chemismus der Cladonien unter besonderer Berücksichtigung der japanischen Arten. *J. Japan. Botany* **17**, 620–630.

ASAHINA, Y. 1941*c*. Lichenologische Notizen XV. *J. Japan. Botany* **17**, 71–76.

ASAHINA, Y. 1941*d*. Lichenologische Notizen XVI. *J. Japan. Botany* **17**, 136–143.

ASAHINA, Y. 1941*e*. Lichenologische Notizen XVII. *J. Japan. Botany* **17**, 485–489.

ASAHINA, Y. 1942*a*. Chemismus der Cladonien unter besonderer Berücksichtigung der japanischen Arten. *J. Japan. Botany* **18**, 489–502.

ASAHINA, Y. 1942*b*. Chemismus der Cladonien unter besonderer Berücksichtigung der japanischen Arten. *J. Japan. Botany* **18**, 663–683.

ASAHINA, Y. 1942*c*. Lichenologische Notizen (XVIII). *J. Japan. Botany* **18**, 549–552.

ASAHINA, Y. 1942*d*. Lichenologische Notizen (XIX). *J. Japan. Botany* **18**, 620–625.

ASAHINA, Y. 1943*a*. Chemismus der Cladonien unter besonderer Berücksichtigung der japanischen Arten. *J. Japan. Botany* **19**, 47–56.

ASAHINA, Y. 1943*b*. Chemismus der Cladonien unter besonderer Berücksichtigung der japanischen Arten. *J. Japan. Botany* **19**, 227–244.

ASAHINA, Y. 1943*c*. Lichenologische Notizen (XX). *J. Japan. Botany* **19**, 1–4.

ASAHINA, Y. 1943*d*. Lichenologische Notizen (XXII). *J. Japan. Botany* **19**, 189–196.

ASAHINA, Y. 1943*e*. Lichenologische Notizen (XXIII). *J. Japan. Botany* **19**, 279–283.

ASAHINA, Y. 1943*f*. Lichenologische Notizen (XXIV). *J. Japan Botany* **19**, 301–311.

ASAHINA, Y. 1944. Lichenologische Notizen XXV. *J. Japan. Botany* **20**, 129–134.

ASAHINA, Y. 1947*a*. Lichenologische Notizen (§61–64). *J. Japan. Botany* **21**, 3–7.

ASAHINA, Y. 1947*b*. Lichenologische Notizen (§65–67). *J. Japan. Botany* **21**, 83–86.

ASAHINA, Y. 1949*a*. Lichenologische Notizen (§70–71). *J. Japan. Botany* **23**, 1–4.

ASAHINA, Y. 1949*b*. Lichenologische Notizen (§72). Varieties, forms and related species of *Lobaria pulmonaria* from eastern Asia. *J. Japan. Botany* **23**, 65–68.

ASAHINA, Y. 1950*a*. Lichenes Japoniae novae vel minus cognitae. (1). *Acta Phytotaxon. Geobotan.* **14** (2), 33–35.

ASAHINA, Y. 1950*b*. Lichenologische Notizen (§73–74). *J. Japan. Botany* **25**, 65–68.

ASAHINA, Y. 1950*c*. Lichens of Japan. Vol. I. Genus *Cladonia*. Hirokawa Publishing Co., Tokyo.

ASAHINA, Y. 1951*a*. Lichenes Japoniae novae vel minus cognitae. (2). *J. Japan. Botany* **26**, 97–102.

ASAHINA, Y. 1951*b*. Lichenes Japoniae novae vel minus cognitae (3). *J. Japan. Botany* **26**, 193–198.

ASAHINA, Y. 1951*c*. Lichenes Japoniae novae vel minus cognitae. (4). *J. Japan. Botany* **26**, 225–228.

ASAHINA, Y. 1951*d*. Lichenes Japoniae novae vel minus cognitae. (5). *J. Japan. Botany* **26,** 257–261

ASAHINA, Y. 1951*e*. Lichenes Japoniae novae vel minus cognitae (6). *J. Japan. Botany* **26**, 289–293.

ASAHINA, Y. 1951*f*. Lichenes Japoniae novae vel minus cognitae. (7). *J. Japan. Botany* **26**, 329–334.

ASAHINA, Y. 1951*g*. Lichenes Japoniae novae vel minus cognitae. (8). *J. Japan. Botany* **26**, 353–357.

ASAHINA, Y. 1951*h*. Lichenologische Notizen. (§75–78). *J. Japan. Botany* **26**, 161–165.

ASAHINA, Y. 1951*i*. Neuere Entwicklungen auf dem Gebiete der Flechtenstoffe. *Fortschr. Chem. Org. Naturstoffe* **8**, 207–244.

ASAHINA, Y. 1952*a*. Lichenes Japoniae novae vel minus cognitae. (9). *J. Japan. Botany* **27**, 15–18.

ASAHINA, Y. 1952*b*. Lichenologische Notizen (§79–82). *J. Japan. Botany* **27**, 69–71.

ASAHINA, Y. 1952*c*. Lichenologische Notizen (§83–§84). *J. Japan. Botany* **27**, 239–242.

ASAHINA, Y. 1952*d*. Lichenologische Notizen (§85–§87). *J. Japan. Botany* **27**, 293–296.

ASAHINA, Y. 1952*e*. Lichens of Japan. Vol. II. Genus *Parmelia*. Research Institute for Natural Resources, Tokyo.

ASAHINA, Y. 1953*a*. Lichenes Japoniae novae vel minus cognitae (10). *J. Japan. Botany* **28**, 65–68.

ASAHINA, Y. 1953*b*. Lichenes Japoniae novae vel minus cognitae (11). *J. Japan. Botany* **28**, 134–140.

ASAHINA, Y. 1953*c*. Lichenes Japoniae novae vel minus cognitae (12). *J. Japan. Botany* **28**, 225–230.

ASAHINA, Y. 1953*d*. Lichenologische Notizen (§90–§94). *J. Japan. Botany* **28**, 114–122.

ASAHINA, Y. 1953*e*. Lichenologische Notizen (§95–§98). *J. Japan. Botany* **28**, 161–164.

Asahina, Y. 1954a. Lichenologische Notizen (§100–102). *J. Japan. Botany* **29**, 33–34.

Asahina, Y. 1954b. Lichenologische Notizen (§105–106). *J. Japan. Botany* **29**, 225–229.

Asahina, Y. 1954c. Lichenologische Notizen (§107–109). *J. Japan. Botany* **29**, 289–293.

Asahina, Y. 1954d. Lichenologische Notizen (§110–111). *J. Japan. Botany* **29**, 321–324.

Asahina, Y. 1954e. Lichenologische Notizen (§112–113). *J. Japan. Botany* **29**, 370–372.

Asahina, Y. 1955a. Lichens. Pp. 43–63 *in* H. Kihara [ed.], Fauna and Flora of Nepal Himalaya, Scientific Results of the Japanese Expeditions to Nepal Himalaya 1952–1953. Vol. I. Fauna and Flora Research Society, Kyoto University, Kyoto.

Asahina, Y. 1955b. Lichens collected in Aogashima and Mikurazima. *J. Japan. Botany* **30**, 222–224.

Asahina, Y. 1956a. Lichenologische Notizen (§117–119). *J. Japan. Botany* **31**, 65–70.

Asahina, Y. 1956b. Lichenologische Notizen (§120). A new arrangement of Japanese *Cladonia verticillata*-group. *J. Japan. Botany* **31**, 322–325.

Asahina, Y. 1956c. Lichens of Japan. Vol. III. Genus *Usnea*. Research Institute for Natural Resources, Tokyo.

Asahina, Y. 1957a. Lichenologische Notizen (§121–123). *J. Japan. Botany* **32**, 35–37.

Asahina, Y. 1957b. Lichenologische Notizen (§124–125). *J. Japan. Botany* **32**, 97–100.

Asahina, Y. 1957c. Lichenologische Notizen (§126–127). *J. Japan. Botany* **32**, 129–133.

Asahina, Y. 1957d. Lichenologische Notizen (§128–129). *J. Japan. Botany* **32**, 161–164.

Asahina, Y. 1957e. Lichenologische Notizen (§130). *J. Japan. Botany* **32**, 257–260.

Asahina, Y. 1957f. Lichenologische Notizen (§131–134). *J. Japan. Botany* **32**, 359–362.

Asahina, Y. 1958a. Lichenologische Notizen (§135–136). *J. Japan. Botany* **33**, 1–5.

Asahina, Y. 1958b. Lichenologische Notizen (§137–139). *J. Japan. Botany* **33**, 65–69.

Asahina, Y. 1958c. Lichenologische Notizen (§140–142). *J. Japan. Botany* **33**, 129–133.

Asahina, Y. 1958d. Lichenologische Notizen (§143). On the chemical ingredients of *Usnea diffracta* Vain. *J. Japan. Botany* **33**, 225–226.

Asahina, Y. 1958e. Lichenologische Notizen (§144–146). *J. Japan. Botany* **33**, 257–264.

Asahina, Y. 1958f. Lichenologische Notizen (§147–148). *J. Japan. Botany* **33**, 323–326.

ASAHINA, Y. 1959*a*. Lichenologische Notizen (§149). *J. Japan. Botany* **34**, 65–66.

ASAHINA, Y. 1959*b*. Lichenologische Notizen (§150–153). *J. Japan. Botany* **34**, 225–230.

ASAHINA, Y. 1959*c*. Lichenologische Notizen (§154–156). *J. Japan. Botany* **34**, 289–292.

ASAHINA, Y. 1959*d*. Lichenologische Notizen (§157–159). *J. Japan. Botany* **34**, 347–350.

ASAHINA, Y. 1960*a*. Lichenologische Notizen (§160–163). *J. Japan. Botany* **35**, 97–102.

ASAHINA, Y. 1960*b*. Lichenologische Notizen (§164). On *Cladonia corallifera* (Kunze) Nyl. collected in subarctic region. *J. Japan. Botany* **35**, 167–171.

ASAHINA, Y. 1960*c*. Lichenologische Notizen (§165–169). *J. Japan. Botany* **35**, 289–295.

ASAHINA, Y. 1961*a*. Lichenologische Notizen (§170–172). *J. Japan. Botany* **36**, 46–50.

ASAHINA, Y. 1961*b*. Lichenologische Notizen (§174–179). *J. Japan. Botany* **36**, (7), 225–232.

ASAHINA, Y. 1961*c*. Lichenologische Notizen (§180–181). *J. Japan. Botany* **36** (8), 225–230.

ASAHINA, Y. 1961*d*. Lichenologische Notizen (§182–184). *J. Japan. Botany* **36**, 289–291.

ASAHINA, Y. 1962. Lichenologische Notizen (§185). *J. Japan. Botany* **37**, 257–262.

ASAHINA, Y. 1963*a*. Lichenologische Notizen (§186–187). *J. Japan. Botany* **38**, 1–3.

ASAHINA, Y. 1963*b*. Lichenologische Notizen (§190). *J. Japan. Botany* **38**, 193–195.

ASAHINA, Y. 1963*c*. Lichenologische Notizen (§191). *J. Japan. Botany* **38**, 225–228.

ASAHINA, Y. 1963*d*. Lichenologische Notizen (§192). *J. Japan. Botany* **38**, 257–260.

ASAHINA, Y. 1964*a*. Lichenologische Notizen (§193). *J. Japan. Botany* **39**, 165–171.

ASAHINA, Y. 1964*b*. Lichenologische Notizen (§194). *J. Japan. Botany* **39**, 209–215.

ASAHINA, Y. 1965*a*. Lichenologische Notizen (§195). *J. Japan. Botany* **40**, 1–4.

ASAHINA, Y. 1965*b*. Lichenologische Notizen (§196). *J. Japan. Botany* **40**, 33–35.

ASAHINA, Y. 1965*c*. Lichenologische Notizen (§197). *J. Japan. Botany* **40**, 129–133.

ASAHINA, Y. 1965*d*. Lichenologische Notizen (§198–199). *J. Japan. Botany* **40**, 172–177.

ASAHINA, Y. 1965*e*. Lichenologische Notizen (§200). *J. Japan. Botany* **40**, 225–227.

ASAHINA, Y. 1965*f.* Lichenologische Notizen (§201). *J. Japan. Botany* **40**, 353–357.

ASAHINA, Y. AND H. AKAGI. 1938. Untersuchungen über Flechtenstoffe. LXXXVIII Mitteil.: Über die Zeorin-Gruppe (I). *Chem. Ber.* **71**, 980–985.

ASAHINA, Y., M. AOKI, AND F. FUZIKAWA. 1941. Untersuchungen über Flechtenstoffe, XCVI. Mitteil.: Über ein neues Depsid "Hypothamnolsäure." *Chem. Ber.* **74**, 824–831.

ASAHINA, Y. AND J. ASANO. 1932*a.* Untersuchungen über Flechtenstoffe, X. Mitteil.: Über Olivetorsäure (I.). *Chem. Ber.* **65**, 475–482.

ASAHINA, Y. AND J. ASANO. 1932*b.* Untersuchungen über Flechtenstoffe, XIII. Mitteil.: Über Olivetorsäure (II.). *Chem. Ber.* **65**, 584–586.

ASAHINA, Y. AND J. ASANO. 1933. Untersuchungen über Flechtenstoffe, XXI. Mitteil.: Über Salazinsäure (I.). *Chem. Ber.* **66**, 689–699.

ASAHINA, Y. AND F. FUZIKAWA. 1932*a.* Über die Identität der Diffractasäure mit der Hesseschen Dirhizoninsäure. *Chem. Ber.* **65**, 1668.

ASAHINA, Y. AND F. FUZIKAWA. 1932*b.* Untersuchungen über Flechtenstoffe, IX. Mitteil.: Über Diffractasäure, eine Monomethyläther-barbatinsäure. *Chem. Ber.* **65**, 175–178.

ASAHINA, Y. AND F. FUZIKAWA. 1932*c.* Untersuchungen über Flechtenstoffe, XI. Mitteil.: Über die Konstitution der Obtusatsäure. *Chem. Ber.* **65**, 580–583.

ASAHINA, Y. AND F. FUZIKAWA. 1934*a.* Untersuchungen über Flechtenstoffe, XXXV. Mitteil.: Über die Identität der α-Collatolsäure mit Lecanorolsäure. *Chem. Ber.* **67**, 169–170.

ASAHINA, Y. AND F. FUZIKAWA. 1934*b.* Untersuchungen über Flechtenstoffe, XLV. Mitteil.: Über die Identität der Coccellsäure mit der Barbatinsaüre. *Chem. Ber.* **67**, 1793–1795.

ASAHINA, Y. AND F. FUZIKAWA. 1935*a.* Untersuchungen über Flechtenstoffe, XLVIII. Mitteil.: Über Mikrophyllinsäure, ein neues Depsid aus *Cetraria collata* f. *micro-phyllina* A. Zahlbruckner. *Chem. Ber.* **68**, 80–82.

ASAHINA, Y. AND F. FUZIKAWA. 1935*b.* Untersuchungen über Flechtenstoffe, L. Mitteil.: Über die Bestandteile von *Parmelia perlata* Ach. *Chem. Ber.* **68**, 634–639.

ASAHINA, Y. AND F. FUZIKAWA. 1935*c.* Untersuchungen über Flechtenstoffe, LI. Mitteil.: Über das Vorkommen von Nor-stictinsäure in *Parmelia acetabulum* Ach. *Chem. Ber.* **68**, 946–947.

ASAHINA, Y. AND F. FUZIKAWA. 1935*d.* Untersuchungen über Flechtenstoffe, LV. Mitteil.: Über Endocrocin, ein neues Oxy-anthrachinon-Derivat. *Chem. Ber.* **68**, 1558–1565.

ASAHINA, Y. AND F. FUZIKAWA. 1935*e.* Untersuchungen über Flechtenstoffe, LX. Mitteil.: Über Mikrophyllinsäure und deren Spaltungsprodukte. *Chem. Ber.* **68**, 2022–2026.

ASAHINA, Y. AND F. FUZIKAWA. 1935*f.* Untersuchungen über Flechtenstoffe, LXI. Mitteil.: Über Olivetorsäure (III.). *Chem. Ber.* **68**, 2026–2028.

ASAHINA, Y. AND A. HASHIMOTO. 1933. Untersuchungen über Flechten-stoffe, XIX. Mitteil.: Über Alectoronsäure, einen neuen Bestandteil aus den hellfarbigen *Alectoria*-Arten. *Chem. Ber.* **66**, 641–649.

ASAHINA, Y. AND A. HASHIMOTO. 1934. Untersuchungen über Flechten-stoffe, XXXVII. Mitteil.: Über die Konstitution des Sphaerophorins. *Chem. Ber.* **67**, 416–420.

ASAHINA, Y. AND H. HAYASHI. 1928. Ueber die Bestandteile von *Alectoria sulcata* Nyl. (Untersuchungen ueber Flechtenstoffe. IV). *Yakugaku Zasshi* **48**, 1094–1098.

ASAHINA, Y. AND H. HAYASHI. 1933. Untersuchungen über Flechten-stoffe, XXVI. Mitteil.: Über Psoromsäure. *Chem. Ber.* **66**, 1023–1030.

ASAHINA, Y. AND M. HIRAIWA. 1935a. Untersuchungen über Flechten-stoffe, LVII. Mitteil.: Über ein neues Depsid (Anziasäure) und die Bestandteile einiger *Anzia*-Arten. *Chem. Ber.* **68**, 1705–1708.

ASAHINA, Y. AND M. HIRAIWA. 1935b. Untersuchungen über Flechten-stoffe, LVIII. Mitteil.: Über die Bestandteile von *Thamnolia vermicularis* Schaer. var. *taurica* Schaer. *Chem. Ber.* **68**, 1708–1710.

ASAHINA, Y. AND M. HIRAIWA. 1936. Untersuchungen über Flechten-stoffe, LXIV. Mitteil.: Über die Konstitution der Thamnolsäure (IV. Mitteil.). *Chem. Ber.* **69**, 330–333.

ASAHINA, Y. AND T. HIRAKATA. 1932. Untersuchungen über Flechten-stoffe, XV. Mitteil.: Über Divaricatsäure. *Chem. Ber.* **65**, 1665–1668.

ASAHINA, Y. AND S. IHARA. 1929. Untersuchungen über Flechtenstoffe, V.: Über die Konstitution der Thamnolsäure (I.). *Chem. Ber.* **62**, 1196–1207.

ASAHINA, Y. AND M. KAGITANI. 1934. Untersuchungen über Flechten-stoffe, XL. Mitteil.: Über das Vorkommen von Volemit in den Flechten. *Chem. Ber.* **67**, 804–805.

ASAHINA, Y., Y. KANAOKA, AND F. FUZIKAWA. 1933. Untersuchungen über Flechtenstoffe, XX. Mitteil.: Über Collatolsäure, eine Monomethyl-äther-alectoronsäure. *Chem. Ber.* **66**, 649–655.

ASAHINA, Y. AND T. KUSAKA. 1936. Untersuchungen über Flechten-stoffe, LXV. Mitteil.: Über ein neues Depsid "Ramalinolsäure." *Chem. Ber.* **69**, 450–455.

ASAHINA, Y. AND T. KUSAKA. 1937a. Untersuchungen über Flechten-stoffe, LXXXIII. Mitteil.: Über ein neues Depsid, die Boninsäure; Synthese der Boninsäure und der Homosekikasäure. *Chem. Ber.* **70**, 1815–1821.

ASAHINA, Y. AND T. KUSAKA. 1937b. Untersuchungen über Flechten-stoffe, LXXXIV. Mitteil.: Über das Vorkommen von Homosekikasäure in Cladonien. *Chem. Ber.* **70**, 1821–1823.

ASAHINA, Y. AND T. KUSAKA. 1942. Untersuchungen über Flechten-stoffe. XCVIII. Über ein neues Tridepsid "Hiascinsäure." *Bull. Chem. Soc. Japan* **17**, 152–159.

ASAHINA, Y. AND N. KUTANI. 1925. Untersuchungen ueber Flechten-stoffe. I. Ueber Gyrophorsäure. *Yakugaku Zasshi* **519**, 423–429.

ASAHINA, Y. AND H. NOGAMI. 1934. Untersuchungen über Flechtenstoffe, XLI. Mitteil.: Über die Konstitution der Physodsäure (I.). *Chem. Ber.* **67**, 805–811.

ASAHINA, Y. AND H. NOGAMI. 1935. Untersuchungen über Flechtenstoffe, XLVII. Mitteil.: Über die Konstitution der Physodsäure (II). *Chem. Ber.* **68**, 77–80.

ASAHINA, Y. AND H. NOGAMI. 1937. Untersuchungen über Flechtenstoffe, LXXXI. Mitteil.: Über die Glomellifersäure (I. Mitteil.). *Chem. Ber.* **70**, 1498–1499.

ASAHINA, Y. AND H. NOGAMI. 1942. Untersuchungen über Flechtenstoffe, XCVIII. Mitteil.: Über Lichexanthon, ein neues Stoffwechselprodukt der Flechte. *Bull. Chem. Soc. Japan* **17**, 202–207.

ASAHINA, Y. AND S. NONOMURA. 1933. Untersuchungen über Flechtenstoffe, XVI. Mitteil.: Bestandteile der *Ramalina*-Arten mit besonderer Berücksichtigung der Sekikasäure. *Chem. Ber.* **66**, 30–35.

ASAHINA, Y. AND S. NONOMURA. 1935. Untersuchungen über Flechtenstoffe, LVI. Mitteil.: Über die Konstitution der Lobarsäure (I. Mitteil.). *Chem. Ber.* **68**, 1698–1704.

ASAHINA, Y. AND M. NUNO. 1964. Note on the chromatogram of thamnolic acid and the use of Barton-Evans-Gardner's reagent for spot-detecting means. *J. Japan. Botany* **39**, 313–317.

ASAHINA, Y. AND Y. SAKURAI. 1951. Über die Bestandteile von *Cladonia submitis* Evans. *Yakugaku Zasshi* **71**, 1166.

ASAHINA, Y. AND T. SASAKI. 1942. Untersuchungen über Flechtenstoffe. C. Mitteilung: Über Rangiformsäure (I). *Bull. Chem. Soc. Japan* **17**, 495–498.

ASAHINA, Y. AND S. SHIBATA. 1939. Untersuchungen über Flechtenstoffe, XCIV. Mitteil.: Über das Vorkommen der Thelephorsäure in den Flechten. *Chem. Ber.* **72**, 1531–1533.

ASAHINA, Y. AND S. SHIBATA. 1954. Chemistry of Lichen Substances. Japan Society for the Promotion of Science, Tokyo.

ASAHINA, Y. AND Z. SIMOSATO. 1938. Untersuchungen über Flechtenstoffe, XC. Mitteil.: Über Monomethyläther-orcin-dicarbonsäuren und die Nicht-Existenz der sog. Iso-squamatsäure. *Chem. Ber.* **71**, 2561–2568.

ASAHINA, Y. AND Y. TANASE. 1933. Untersuchungen über Flechtenstoffe, XXII. Mitteil.: Über Cetrarsäure. *Chem. Ber.* **66**, 700–703.

ASAHINA, Y. AND Y. TANASE. 1934*a*. Untersuchungen über Flechtenstoffe, XXXVI. Mitteil.: Über Fumar-protocetrarsäure. *Chem. Ber.* **67**, 411–416.

ASAHINA, Y. AND Y. TANASE. 1934*b*. Untersuchungen über Flechtenstoffe, XXXVIII. Mitteil.: Über die Proto-cetrarsäure und ihre Alkyläther. *Chem. Ber.* **67**, 766–773.

ASAHINA, Y. AND Y. TANASE. 1934*c*. Untersuchungen über Flechtenstoffe, XLIII. Mitteil.: Über die Identität der Saxatilsäure mit Salazinsäure. *Chem. Ber.* **67**, 1434–1435.

ASAHINA, Y. AND Y. TANASE. 1937. Untersuchungen über Flechtenstoffe, LXXII. Mitteil.: Über die Konstitution der Squamatsäure. *Chem. Ber.* **70**, 62–63.

ASAHINA, Y., Y. TANASE, AND I. YOSIOKA. 1936. Untersuchungen über Flechtenstoffe, LXIII. Mitteil.: Über die Bestandteile der *Baeomyces*-Arten. *Chem. Ber.* **69**, 125–127.

ASAHINA, Y. AND T. TUKAMOTO. 1933. Untersuchungen über Flechtenstoffe, XXXI. Mitteil.: Bestandteile einiger *Usnea*-Arten unter besonderer Berücksichtigung der Verbindungen der Salazinsäure-Gruppe (I.). *Chem. Ber.* **66**, 1255–1263.

ASAHINA, Y. AND T. TUKAMOTO. 1934. Untersuchungen über Flechtenstoffe, XLII. Mitteil.: Bestandteile einiger *Usnea*-Arten unter besonderer Berücksichtigung der Verbindungen der Salazinsäure-Gruppe (II.). *Chem. Ber.* **67**, 963–971.

ASAHINA, Y. AND M. WATANABE. 1930. Untersuchungen über Flechtenstoffe, VI.: Über Gyrophorsäure. *Chem. Ber.* **63**, 3044–3048.

ASAHINA, Y. AND M. YANAGITA. 1933a. Untersuchungen über Flechtenstoffe, XVII. Mitteil.: Über Squamatsäure. *Chem. Ber.* **66**, 36–39.

ASAHINA, Y. AND M. YANAGITA. 1933b. Untersuchungen über Flechtenstoffe, XVIII. Mitteil.: Über Iso-squamatsäure, ein neues Depsid aus *Cladonia Boryi*, Tuck. *Chem. Ber.* **66**, 393–397.

ASAHINA, Y. AND M. YANAGITA. 1933c. Untersuchungen über Flechtenstoffe, XXX. Mitteil.: Über Caprarsäure. *Chem. Ber.* **66**, 1217–1220.

ASAHINA, Y. AND M. YANAGITA. 1933d. Untersuchungen über Flechtenstoffe, XXXII. Mitteil.: Über Tenuiorin, einen Monomethyläthergyrophorsäure-methylester. *Chem. Ber.* **66**, 1910–1912.

ASAHINA, Y. AND M. YANAGITA. 1934. Untersuchungen über Flechtenstoffe, XXXIX. Mitteil.: Über eine neue Flechten-Säure, die Norstictinsäure, und das Vorkommen von *d*-Arabit in den Flechten. *Chem. Ber.* **67**, 799–803.

ASAHINA, Y. AND M. YANAGITA. 1936. Untersuchungen über Flechtenstoffe, LXII. Mitteil.: Über die Bestandteile von *Cetraria islandica* Ach. *Chem. Ber.* **69**, 120–125.

ASAHINA, Y. AND M. YANAGITA. 1937. Untersuchungen über Flechtenstoffe, LXXVII. Mitteil.: Über die Flechten-Fettsäuren aus *Nephromopsis endocrocea*. *Chem. Ber.* **70**, 227–235.

ASAHINA, Y., M. YANAGITA, T. HIRAKATA, AND M. IDA. 1933. Untersuchungen über Flechtenstoffe, XXVIII. Mitteil.: Über das Vorkommen von Stictinsäure in verschiedenen Flechten. *Chem. Ber.* **66**, 1080–1086.

ASAHINA, Y., M. YANAGITA, AND T. OMAKI. 1933. Untersuchungen über Flechtenstoffe. XXV. Mitteil.: Über Stictinsäure. *Chem. Ber.* **66**, 943–947.

ASAHINA, Y., M. YANAGITA, AND I. YOSIOKA. 1936. Untersuchungen über Flechtenstoffe, LXVII. Mitteil.: Über Stictinsäure (III. Mitteil.). *Chem. Ber.* **69**, 1370–1375.

ASAHINA, Y. AND M. YASUE. 1936. Untersuchungen über Flechten-stoffe, LXXI. Mitteil.: Synthese der Diploschistes-Säure. *Chem. Ber.* **69**, 2327–2330.

ASAHINA, Y. AND M. YASUE. 1937a. Untersuchungen über Flechten-stoffe, LXXIX. Mitteil.: Über die Bestandteile von *Cetraria islandica* (L.) Ach. (II. Mitteil.). *Chem. Ber.* **70**, 1053–1059.

ASAHINA, Y. AND M. YASUE. 1937b. Untersuchungen über Flechten-stoffe, LXXX. Mitteil.: Über die Bestandteile der sog. *Thamnolia vermicularis* f. *taurica*. *Chem. Ber.* **70**, 1496–1497.

ASAHINA, Y. AND I. YOSIOKA. 1937. Untersuchungen über Flechten-stoffe, LXXXV. Mitteil.: Über die Synthese der Perlatolin- und Imbri-carsäure. *Chem. Ber.* **70**, 1823–1826.

ASAHINA, Y. AND I. YOSIOKA. 1940. Untersuchungen über Flechten-stoffe, XCV. Mitteil.: Über die Zeorin-Gruppe (II). *Chem. Ber.* **73**, 742–747.

ASANO, M. 1927. Ueber die Konstitution von Protolichesterinsäure. I. (III. Mitteilung der Untersuchungen über Flechtenstoffe von Y. Asahina.). *Yakugaku Zasshi* **539**, 1–6.

ASANO, M. AND H. ARATA. 1940. [Constituents of *Xanthoria fallax* (Hepp.) Arn.] *Yakugaku Zasshi* **60**, 521–525; German summary, 206–208.

ASANO, M. AND T. AZUMI. 1935. Über die Bestandteile von *Nephro-mopsis Stracheyi* f. *ectocarpisma* Hue (I. Mitteil.). *Chem. Ber.* **68**, 995–997.

ASANO, M. AND T. AZUMI. 1938. Über die Bestandteile von *Cetraria pseudocomplicata* Y. Asahina und *Nephromopsis cilialis* [sic] Hue. *Yakugaku Zasshi* **58**, 194.

ASANO, M. AND S. FUZIWARA. 1936. Ueber die Farbstoff von *Xanthoria fallax* (Hepp.) Arn. (vorläufige Mitteilung). *Yakugaku Zasshi* **56**, 1007–1010; German summary, 101.

ASANO, M. AND Y. KAMEDA. 1935a. Über die Konstitution des Calycins und dessen Synthese (IV. Mitteil. über Flechten-Farbstoffe der Pulvin-säure-Reihe). *Chem. Ber.* **68**, 1568–1571.

ASANO, M. AND Y. KAMEDA. 1935b. Über die Reduktion der Pinastrin-säure und der Vulpinsäure (III. Mitteil. über Flechten-Farbstoffe der Pulvinsäure-Reihe). *Chem. Ber.* **68**, 1565–1567.

ASANO, M. AND T. KANEMATSU. 1931. Über den Bestandteil von isländischen Moss. II. Mitteilung. *Yakugaku Zasshi* **51**, 390–395.

ASANO, M. AND T. KANEMATSU. 1932. Über die Konstitution der Proto-lichesterinsäure und Lichesterinsäure. *Chem. Ber.* **65**, 1175–1178.

ASANO, M. AND Z. OHTA. 1933. Über die Konstitution der Caperatsäure (I. Mitteil.). *Chem. Ber.* **66**, 1020–1023.

ASANO, M. AND M. TANIGUTI. 1939. Über die Bestandteile von *Nephro-mopsis Stracheyi* f. *ectocarpisma* Hue (III. Mitteil.). *Yakugaku Zasshi* **59**, 607–609; German summary, 216.

ASPINALL, G. O., E. L. HIRST, AND MARGARET WARBURTON. 1955. The alkali-soluble polysaccharides of the lichen *Cladonia alpestris* (reindeer moss). *J. Chem. Soc.* **1955**, 651–655.

BACHMANN, O. 1962. Über die Inhaltsstoffe von *Umbilicaria erosa* Ach. *Nova Hedwigia* **4**, 309–311.

BACHMANN, O. 1963. Dünnschichtchromatographische Trennung von Flechtensäuren der β-Orcin-Gruppe. *Österr. Botan. Z.* **110**, 103–107.

BARRY, V. 1946. The thyroid and tuberculosis. *Nature* **158**, 131–132.

BARRY, V. C. AND P. A. MCNALLY. 1945. Inhibitory action of dialkyl succinic acid derivatives on the growth *in vitro* of acid-fast bacteria. *Nature* **156**, 48–49.

BARRY, V. C., L. O'ROURKE, AND D. TWOMEY. 1947. Antitubercular activity of diphenylesters and related compounds. *Nature* **160**, 800–801.

BARTON, D. H. R. AND T. BRUUN. 1952. Triterpenoids. Part VI. Some observations on the constitution of zeorin. *J. Chem. Soc.* **1952**, 1683–1690.

BENDZ, GERD, J. SANTESSON, AND L. TIBELL. 1966. Chemical studies on lichens 2. Thin layer chromatography of aliphatic lichen acids. *Acta Chem. Scand.* **20**, 1181.

BENDZ, GERD, J. SANTESSON, AND C. A. WACHTMEISTER. 1965*a*. Studies on the chemistry of lichens. 20. The chemistry of the *Ramalina ceruchis* group. *Acta Chem. Scand.* **19**, 1185–1187.

BENDZ, GERD, J. SANTESSON, AND C. A. WACHTMEISTER. 1965*b*. Studies on the chemistry of lichens. 21. The isolation and synthesis of methyl 3,5-dichlorolecanorate, a new depside from *Ramalina* sp. *Acta Chem. Scand.* **19**, 1188–1190.

BENDZ, GERD, J. SANTESSON, AND C. A. WACHTMEISTER. 1965*c*. Studies on the chemistry of lichens. 22. The chemistry of the genus *Siphula*. I. *Acta Chem. Scand.* **19**, 1250–1252.

BENDZ, GERD, J. SANTESSON, AND C. A. WACHTMEISTER. 1965*d*. Studies on the chemistry of lichens. 23. Thin layer chromatography of pulvic acid derivatives. *Acta Chem. Scand.* **19**, 1776–1777.

BIRKINSHAW, J. H. AND Y. S. MOHAMMED. 1962. Studies in the biochemistry of microorganisms. 111. The production of *l*-phenylalanine anhydride (*cis*-L-3,6-dibenzyl-2,5-dioxopiperazine) by *Penicillium nigricans* (Bainer) Thom. *Biochem. J.* **85**, 523–527.

BLIX, G. AND H. RYDIN. 1932. Über das Vorkommen von Ergosterin und D-Vitamin in der Renntierflechte. *Acta Soc. Med. Upsalien.* **37**, 333–340.

BORKOWSKI, B., H. GERTIG, AND J. JELJASZEWICZ. 1958. [Yield and activity of antibiotics from some Polish lichens.] *Dissertationes Pharm.* **10**, 99–107. Reference from *Chem. Abstr.* **52**, 18700e (1958).

BORKOWSKI, B., WANDA WOŹNIAK, H. GERTIG, AND B. WERAKSO. 1964. Działanie bakteriostatyczne niektórych związków porostowych z plechy *Cetraria islandica* (L.) Ach. i kwasu usninowego. [The bacteriostatic action of some compounds from lichen *Cetraria islandica* (L.) Ach and usnic acid.] *Dissertationes Pharm.* **16**, 189–194.

BREADEN, T. W., J. KEANE, AND T. J. NOLAN. 1942. The chemical constituents of lichens found in Ireland. *Cladonia impexa* Harm. *Sci. Proc. Roy. Dublin Soc.* **23**, 6–9.

BREADEN, T. W., J. KEANE, AND T. J. NOLAN. 1944. The chemical constituents of lichens found in Ireland. *Cladonia sylvatica* (L.) Harm. emend. Sandst. *Sci. Proc. Roy. Dublin Soc.* **23**, 197–200.

BREEN, J., J. KEANE, AND T. J. NOLAN. 1937. The chemical constituents of lichens found in Ireland — *Pertusaria concreta* Nyl. form *Westringii* Nyl. *Sci. Proc. Roy. Dublin Soc.* **21**, 587–592.

BRIEGER, R. 1932. Flechenstoffe (Flechtensäuren). Pp. 413–429 *in* G. KLEIN [ed.], Handbuch der Pflanzenanalyse. 3. Band, Spezielle Analyse. 2. Teil, Organische Stoffe II. Julius Springer, Wein.

BRIEGER, W. 1923. Synthetische Versuche auf dem Gebiete der Flechtenstoffe und ihrer Bausteine. Pp. 205–438 *in* E. ABERHALDEN [ed.], Handbuch der biochemischen Arbeitsmethoden, Abteilung I: Chemische Methoden, Teil 10. Urban & Schwarzenberg, Berlin.

BRIGHTMAN, F. H. 1964. The distribution of the lichen *Lecanora conizaeoides* Cromb. in North Ireland. *Irish Naturalists' J.* **14**, 258–262.

BRINER, G. P., G. E. GREAM, AND N. V. RIGGS. 1960. Chemistry of Australian lichens. I. Some constituents of *Parmelia conspersa* (Ehrh.) Ach., *Ramalina fraxinea* (L.) Ach., *Usnea barbata* (L.) Wigg. and *U. ceratina* Ach., from the New England region. *Australian J. Chem.* **13**, 277–284.

BROWN, C. J., D. E. CLARK, W. D. OLLIS, AND P. L. VEAL. 1960. The synthesis of the depsidone, diploicin. *Proc. Chem. Soc.* **1960**, 393–394.

BRUUN, T. 1954a. Triterpenoids in lichens. I. The occurrence of friedelin and *epi*friedelinol. *Acta. Chem. Scand.* **8**, 71–75.

BRUUN, T. 1954b. Triterpenoids in lichens. II. Taxarene, a naturally occurring triterpene. *Acta. Chem. Scand.* **8**, 1291–1292.

BRUUN, T. 1960. Siphulin, a chromanone type lichen acid. *Tetrahedron Letters* **1960** (4), 1–4.

BRUUN, T. 1965. Siphulin, a chromenone lichen acid. *Acta Chem. Scand.* **19**, 1677–1693.

BRUUN, T. AND P. R. JEFFRIES. 1954. *epi*Friedelinol and derivatives, a re-investigation. *Acta Chem. Scand.* **8**, 1948–1949.

BRUUN, T. AND N. A. SÖRENSEN. 1954. A note on the occurrence of dimethyl sulphone in *Cladonia deformis* Hoffm. *Acta Chem. Scand.* **8**, 703.

BURKHOLDER, P. R. 1952. Coöperation and conflict among primitive organisms. *Am. Scientist* **40**, 601–631.

BURKHOLDER, P. R. AND A. W. EVANS. 1945. Further studies on the antibiotic activity of lichens. *Bull. Torrey Botan. Club* **72**, 157–164.

BURLEY, J. W. A., G. E. GILBERT, AND L. C. CLUM. 1962. Preliminary radiological investigations of the vegetation and soils of Neotoma. *Neotoma Ecol. Bioclim. Lab., Ohio State Univ. Ohio Agr. Expt. Sta.,* Spec. Rept. No. 10. 39 pp.

BURTON, J. F. AND B. F. CAIN. 1959. Antileukemic activity of polyporic acid. *Nature* **184**, 1326–1327.

BUSTINZA, F. 1951. A note on the antibacterial activity of *Cladonia rangiferina* Web., *Cladonia sylvatica* (L.) Hoffm. emend. Sandst, and *Cladonia impexa Harm. Antibiot. Chemother.* **1**, 443–446.

BUSTON, H. W. AND V. H. CHAMBERS. 1933. Some cell-wall constituents of *Cetraria islandica* ("Iceland Moss"). *Biochem. J.* **27**, 1691–1702.

CAIN, B. F. 1961. Potential anti-tumour agents. Part I. Polyporic acid series. *J. Chem. Soc.* **1961**, 936–940.

CANTONE, B., S. PIGNATARO, AND R. TOMASELLI. 1963. Ricerche su un prodotto del metabolismo della *Xanthoria parietina* con specttrometria di massa. *Boll. Sedute Accad. Gioenia Sci. Nat. Catania,* Ser. 4, 7 (9), 428–435.

CASTLE, H. AND FLORA KUBSCH. 1949. The production of usnic, didymic, and rhodocladonic acids by the fungal component in the lichen *Cladonia cristatella. Arch. Biochem.* **23**, 158–159.

CAVALTITO, C. J., DOROTHY MCKENICA FRUEHAUF, AND J. H. BAILEY. 1948. Lactone aliphatic acids as antibacterial agents. *J. Am. Chem. Soc.* **70**, 3724–3726.

ČERNOHORSKÝ, Z. 1959. Flechtensäuren als Indikatoren der verwandtschaftlichen Beziehungen bei den Flechten. Pp. 129–137 *in* Biologischen Gesellschaft in der DDR, Arbeitstagung zu Fragen der Evolution zum Gedenken an Lamarck — Darwin — Haeckel vom 20 bis 24 Oktober 1959 in Jena. Gustav Fisher Verlag, Jena.

CHANDA, N. B., E. L. HIRST, AND D. J. MANNERS. 1957. A comparison of *isolichenin* and lichenin from Iceland moss (*Cetraria islandica*). *J. Chem. Soc.* **1957**, 1951–1958.

CLAUZADE, G. AND J. POELT. 1961. *Parmelia koflerae* nova species. (Lichenes). *Nova Hedwigia* 3, 367–373.

COKER, J. N. 1950. The properties of some lichen and fungus pigments. University of Illinois Thesis Abstract. Urbana.

CULBERSON, CHICITA F. 1963. The lichen substances of the genus *Evernia. Phytochemistry* 2, 335–340.

CULBERSON, CHICITA F. 1964. Joint occurrence of a lichen depsidone and its probable depside precursor. *Science* **143**, 255–256.

CULBERSON, CHICITA F. 1965a. A note on the chemical strains of *Parmelia furfuracea. Bryologist* **68**, 435–439.

CULBERSON, CHICITA F. 1965b. Some constituents of the lichen *Parmelia cryptochlorophaea. J. Pharm. Sci.* **54**, 1815–1816.

CULBERSON, CHICITA F. 1965c. Some constituents of the lichen *Ramalina siliquosa. Phytochemistry* 4, 951–961.

CULBERSON, CHICITA F. 1965d. Some microchemical tests for the lichen depsidone hypoprotocetraric acid. *Bryologist* **68**, 301–304.

CULBERSON, CHICITA F. 1966a. Confluentinic acid, its microchemical identification, and its occurrence in *Herpothallon sanguineum*. *Bryologist* **69**, 312–317.

CULBERSON, CHICITA F. 1966b. The structure of a new depsidone from the lichen *Parmelia livida*. *Phytochemistry* **5**, 815–818.

CULBERSON, CHICITA F. 1967a. Some microchemical tests for two new lichen substances, scrobiculin and 4-O-methylphysodic acid. *Bryologist* **70**, 70–75.

CULBERSON, CHICITA F. 1967b. The structure of scrobiculin, a new lichen depside in *Lobaria scrobiculata* and *Lobaria amplissima*. *Phytochemistry* **6**, 719–725.

CULBERSON, CHICITA F. AND W. L. CULBERSON. 1958. Age and chemical constituents of individuals of the lichen *Lasallia papulosa*. *Lloydia* **21**, 189–192.

CULBERSON, CHICITA F. AND W. L. CULBERSON. 1966. The identification of imbricaric acid and a new imbricaric acid-containing lichen species. *Bryologist* **69**, 192–202.

CULBERSON, W. L. 1955. Notes on the *Parmelia caperata* group in Wisconsin. *Bryologist* **58**, 40–45.

CULBERSON, W. L. 1956. Note sur la nomenclature répartition et phytosociologie du *Parmeliopsis placorodia* (Ach.) Nyl. *Rev. Bryol. Lichénol.* **24**, 334–337.

CULBERSON, W. L. 1957. *Parmelia caroliniana* Nyl. and its distribution. *J. Elisha Mitchell Sci. Soc.* **73**, 443–446.

CULBERSON, W. L. 1958. The chemical strains of the lichen *Parmelia cetrarioides* Del. in North America. *Phyton* (Buenos Aires) **11**, 85–92.

CULBERSON, W. L. 1960a. *Parmelia pseudoborreri* Asahina, lichen nouveau pour la flore d'Europe et remarques sur les "espèces chimiques" en lichénologie. *Rev. Bryol. Lichénol.* **29**, 321–325.

CULBERSON, W. L. 1960b. *Parmelia ulophyllodes* (Vain.) Savicz in the North American lichen flora. *J. Elisha Mitchell Sci. Soc.* **76**, 141–142.

CULBERSON, W. L. 1961a. A second *Anzia* in North America. *Brittonia* **13**, 381–384.

CULBERSON, W. L. 1961b. The discovery of the lichen *Parmeliopsis placorodia* in western North America. *Madroño* **16**, 31.

CULBERSON, W. L. 1961c. The *Parmelia quercina* group in North America. *Am. J. Botany* **48**, 168–174.

CULBERSON, W. L. 1962a. Some pseudocyphellate *Parmeliae*. *Nova Hedwigia* **4**, 563–577.

CULBERSON, W. L. 1962b. The systematic position of *Platysma thomsonii* Stirton. *Bryologist* **65**, 304–307.

CULBERSON, W. L. 1963a. A summary of the lichen genus *Haematomma* in North America. *Bryologist* **66**, 224–236.

CULBERSON, W. L. 1963*b*. The lichen genus *Thamnolia*. *Brittonia* **15**, 140–144.

CULBERSON, W. L. 1965. *Cetraria chicitae*, a new and widely distributed lichen species. *Bryologist* **68**, 95–99.

CULBERSON, W. L. 1966*a*. Chemistry and taxonomy of the lichen genera *Heterodermia* and *Anaptychia* in the Carolinas. *Bryologist* **69**, 472–487.

CULBERSON, W. L. 1966*b*. Chimisme et taxonomie des lichens du groupe *Ramalina farinacea* en Europe. *Rev. Bryol. Lichénol.* **34**, 841–851.

CULBERSON, W. L. 1966*c*. The lichen genus *Aspidelia* Stirt. *Bryologist* **69**, 113–114.

CULBERSON, W. L. 1967. Analysis of variation in the *Ramalina siliquosa* species complex. *Brittonia* **19**, 333–352.

CULBERSON, W. L. AND CHICITA F. CULBERSON. 1956. The systematics of the *Parmelia dubia* group in North America. *Am. J. Botany* **43**, 678–687.

CULBERSON, W. L. AND CHICITA F. CULBERSON. 1965. *Asahinea*, a new genus in the Parmeliaceae. *Brittonia* **17**, 182–190.

CULBERSON, W. L. AND CHICITA F. CULBERSON. 1967*a*. A new taxonomy for the *Cetraria ciliaris* group. *Bryologist* **70**, 158–166.

CULBERSON, W. L. AND CHICITA F. CULBERSON. 1967*b*. Habitat selection by chemically differentiated races of lichens. *Science* **158**, 1195–1197.

CULBERSON, W. L. AND CHICITA F. CULBERSON. 1968. The lichen genera *Cetrelia* and *Platismatia* (Parmeliaceae). *Contrib. U.S. Natl. Herb.* **34**, i–vi + 449–558.

CULBERSON, W. L. AND M. E. HALE, JR. 1965. *Pyxine caesiopruinosa* in the United States. *Bryologist* **68**, 113–116.

CURD, F. H. AND A. ROBERTSON. 1933. Usnic acid. Part III. Usnetol, usnetic acid, and pyrousnic acid. *J. Chem. Soc.* **1933**, 1173–1179.

CURD, F. H. AND A. ROBERTSON. 1935. Lichen acids. Part VI. Constituents of *Ramalina scopulorum*. *J. Chem. Soc.* **1935**, 1379–1381.

CURD, F. H. AND A. ROBERTSON. 1937. Usnic acid. Part V. *J. Chem. Soc.* **1937**, 894–901.

CURD, F. H., A. ROBERTSON, AND R. J. STEPHENSON. 1933. Lichen acids. Part IV. Atranorin. *J. Chem. Soc.* **1933**, 130–133.

DAHL, E. 1950. Studies in the macrolichen flora of south west Greenland. *Medd. Grønland* **150** (2), 1–176.

DAHL, E. 1952. On the use of lichen chemistry in lichen systematics. *Rev. Bryol. Lichénol.* **21**, 119–134.

DAHL, E. 1953. Notes on some British macrolichens. *Ann. Mag. Nat. Hist.*, Ser. 12, **6**, 426–431.

DAVIDSON, T. A. AND A. I. SCOTT. 1960. Total synthesis of picrolichenic acid. *Proc. Chem. Soc.* **1960**, 390–391.

DAVIDSON, T. A. AND A. I. SCOTT. 1961. Oxidative pairing of phenolic radicals. Part II. The synthesis of picrolichenic acid. *J. Chem. Soc.* **1961**, 4075–4078.

DAVIDSON, V. E., J. KEANE, AND T. J. NOLAN. 1943. The chemical constituents of lichens found in Ireland — *Lecanora gangaleoides.* — Part 3. — The constitution of gangaleoidin. *Sci. Proc. Roy. Dublin Soc.* **23**, 143–163.

DEAN, F. M. 1952. Usnic acid. *Sci. Progr.* (London) **40**, 635–644.

DEAN, F. M., P. HALEWOOD, S. MONGKOLSUK, A. ROBERTSON, AND W. B. WHALLEY. 1953. Usnic acid. Part IX. A revised structure for usnolic acid and the resolution of (±)-usnic acid. *J. Chem. Soc.* **1953**, 1250–1261.

DE SMET, S. AND R. LAMBINON. 1952. Contribution à l'étude chimique de quelques *Usnea* du Bas Congo. *Bull. Soc. Roy. Botan. Belg.* **85**, 91–97.

DHAR, M. L., S. NEELAKANTAN, S. RAMANUJAM, AND T. R. SESHADRI. 1959. Chemical investigation of Indian lichens. Part XXII. *J. Sci. Ind. Res.* (India) **18B**, 111–113.

DINER, B., V. AHMADJIAN, AND H. ROSENKRANTZ. 1964. Preliminary fractionation of pigments from the lichen fungus *Acarospora fuscata.* *Bryologist* **67**, 363–368.

DRAKE, B. 1943. Untersuchungen über einige Polysaccharide der Flechten, vornehmlich das Lichenin und das neuentdeckte Pustulin. *Biochem. Z.* **313**, 388–399.

DUVIGNEAUD, P. 1939. Notes de michrochemie lichénique, I. Sur deux Cladoniacées nouvelles. *Bull. Soc. Roy. Botan. Belg.* **71**, 192–198.

DUVIGNEAUD, P. 1940. L'acide usnique et les espèces dites "chimiques" en lichénologie. Cas de *Evernia prunastri* (L.) Ach. et *Evernia Herinii* nov. spec. *Bull. Soc. Roy. Botan. Belg.* **72**, 148–154.

DUVIGNEAUD, P. 1942a. Contribution à l'étude systématique et chimique du genre *Stereocaulon.* *Biol. Jaarboek Konink. Natuurw. Genoot. Dodonaea Gent* **9**, 80–98.

DUVIGNEAUD, P. 1942b. *Hendricxia* Duvign. nouveau genre de Parméliacées des montagnes équatoriales. *Bull. Jardin Botan. État Bruxelles* **16**, 355–365.

DUVIGNEAUD, P. 1945. Notes de microchimie lichénique. III. Identité de l'acide usnarique et de l'acide salazinique. *Bull. Soc. Roy. Botan. Belg.* **77**, 68–69.

DUVIGNEAUD, P. 1947. Études sur le genre *Usnea.* I. Remarques sur le chimisme des *Usnea* de l'Europe occidentale. *Bull. Soc. Roy. Botan. Belg.* **79**, 141–147.

DUVIGNEAUD, P. AND L. BLERET. 1940. Notes de microchimie lichénique. II. Sur la valeur systématique de *Cladonia pycnoclada* (Pers.) Nyl. em. des Abb. *Bull. Soc. Roy. Botan. Belg.* **72**, 155–159.

EIGLER, G. AND J. POELT. 1965. Flechtenstoffe und Systematik der lobaten Arten der Flechtengattung *Lecanora* in der Holarktis. *Österr. Botan. Z.* **112**, 285–294.

ERDTMAN, H. AND C. A. WACHTMEISTER. 1953. Structure of porphyrilic acid. *Nature* **172**, 724–725.

ERDTMAN, H. AND C. A. WACHTMEISTER. 1956. The structure of porphyrilic acid. *Chem. Ind.* (London) **1956**, 960.

ERDTMAN, H. AND C. A. WACHTMEISTER. 1957*a*. Phenoldehydrogenation as a biosynthetic reaction. Pp. 144–165 *in* Festschrift Arthur Stoll. Sandoz Ltd., Basel.

ERDTMAN, H. AND C. A. WACHTMEISTER. 1957*b*. Picrolichenic acid, a new type of lichen acid. *Chem. Ind.* (London) **1957**, 1042.

ESCHRICH, W. 1958. Über Parietinsäure, einen neuen Inhaltsstoff der gelben Wandflechte *Xanthoria parietina* (L.). Th. Fr. *Biochem. Z.* **330**, 73–78.

EVANS, A. W. 1943*a*. Asahina's microchemical studies of the *Cladoniae*. *Bull. Torrey Botan. Club* **70**, 139–151.

EVANS, A. W. 1943*b*. Microchemical studies on the genus *Cladonia* subgenus *Cladina*. *Rhodora* **45**, 417–438.

EVANS, A. W. 1944*a*. On *Cladonia polycarpia* Merrill. *Bryologist* **47**, 49–56.

EVANS, A. W. 1944*b*. Supplementary report on the *Cladoniae* of Connecticut. *Trans. Conn. Acad. Arts Sci.* **35**, 519–626.

EVANS, A. W. 1947*a*. A study of certain North American *Cladoniae*. *Bryologist* **50**, 14–51.

EVANS, A. W. 1947*b*. The *Cladoniae* of Vermont. *Bryologist* **50**, 221–246.

EVANS, A. W. 1950. Notes on the *Cladoniae* of Connecticut — IV. *Rhodora* **52**, 77–123.

EVANS, A. W. 1951. On *Cladonia transcendens* Vainio. *Bryologist* **54**, 224–230.

EVANS, A. W. 1952*a*. *Cladonia ecmocyna* in North America. *Rhodora* **54**, 261–271.

EVANS, A. W. 1952*b*. The *Cladoniae* of Florida. *Trans. Conn. Acad. Arts Sci.* **38**, 249–336.

EVANS, A. W. 1955*a*. Notes on North American *Cladoniae*. *Bryologist* **58**, 93–112.

EVANS, A. W. 1955*b*. Three species of *Cladonia* from Patagonia. *Rev. Bryol. Lichénol.* **24**, 132–137.

FEDOSEEV, K. G. AND P. A. YAKIMOV. 1960. [Preparation of usnic acid from lichens, I. A. study of conditions of chemical extraction of usnic acid from *Cladonia* lichens.] *Tr. Leningr. Khim. Farmatsevt. Inst.* **9**, 139–149.

FERNÁNDEZ, O. AND A. PIZARROSO. 1958. Contribución a la química de los líquenes. Estudio de la *Usnea canariensis* (D.R.). *Rev. Real Acad. Cienc. Exact. Fís. Nat. Madrid* **52**, 557–563.

FISCHER, E. AND H. O. L. FISCHER. 1914. Synthese der *o*-Diorsellinsäure und Struktur der Evernsäure. *Chem. Ber.* **47**, 505–512.

FISCHER, R. AND D. TOTH. 1938. Über einige Wirkungen der Agaricinsäure, Abietinsäure und Lichesterinsäure. *Arch. Exptl. Pathol. Pharmakol.* **190**, 500–509.

FLEMING, M. AND D. J. MANNERS. 1966*a*. A comparison of the fine-structure of lichenin and barley glucan. *Biochem. J.* **100** (1), 4P–5P.

FLEMING, M. AND D. J. MANNERS. 1966*b*. The fine structure of iso-lichenin. *Biochem. J.* **100** (2), 24P.

FOLLMANN, G. 1964. Nebelflechten als Futterpflanzen des Küsten-guanacos. *Naturwissenschaften* **51**, 19–20.

FOLLMANN, G. 1965. Flechtenstoffe und Stecklingsbewurzelung. *Naturwissenschaften* **52**, 266.

FOLLMANN, G. 1966. Eine neue *Ramalina*-Art aus der *ceruchis*-Gruppe. *Willdenowia* **4**, 227–233.

FOLLMANN, G. AND VILMA VILLAGRÁN. 1964. Flechtenstoffe als Virus-inhibitoren. *Naturwissenschaften* **51**, 543.

FOLLMANN, G. AND VILMA VILLAGRÁN. 1965. Flechtenstoffe und Zellpermeabilität. *Z. Naturforsch.* **20b**, 723.

FORSTER, M. O. AND W. B. SAVILLE. 1922. Constitution of picrorocellin, a diketopiperazine derivative from *Roccella fuciformis*. *J. Chem. Soc.* **121**, 816–827.

FRANCK, B. AND T. RESCHKE. 1960. Mutterkorn-Farbstoffe, II. Isolierung der Hydroxy-anthrachinon-carbonsäuren Endocrocin und Clavorubin aus Roggenmutterkorn. *Chem. Ber.* **93**, 347–356.

FRANK, R. L., S. M. COHEN, AND J. N. COKER. 1950. The structures and syntheses of rhizocarpic acid and epanorin. *J. Am. Chem. Soc.* **72**, 4454–4457.

FUZIKAWA, F. 1939. Über die antiseptische Wirkung von verbindungen der Olivetonsäure Reihe auf Sojasauce. Über die antiseptische Wirkung von Phenolen und Phenolcarbonsäuren sowie von deren Estern, aus denen die Flechtenstoffe bestehen. (V. Mitteilung). *Yakugaku Zasshi* **59**, 245–247; German summary, 141–142.

FUZIKAWA, F. AND K. ISHIGURO. 1936. Ueber die Bestandteile der einigen in Japan einheimischen Flechten. *Yakugaku Zasshi* **56**, 992–997; German summary, 182–185.

FUZIKAWA, F., S. SHINAMURA, AND K. TARUI. 1941. [Antiseptics for foodstuff. X. β-Orcincarboxylic acid ethyl ester.] *Yakugaku Zasshi* **61**, 191–196. Reference from *Chem. Abstr.* **44**, 8888i (1950).

GALUN, MARGALITH AND HANNA LAVEE. 1966. Lichens from Har Meron (Jebel Jermak), Upper Galilee. *Bryologist* **69**, 324–333.

GERTIG, H. 1961. Oznaczanie zawartsci kwasu usniowego w porostach. [Determination of content of usninic acid in seaweeds.] *Acta Polon. Pharm.* **18**, 57–66.

GERTIG, H. 1963. Wyodrębnianie kwasu *d*-protolichesterynowego oraz fumaroprotocetrarowego z plech *Cetraria islandica* (L.) Ach.—pochod-zenia krajowego. [Isolation of *d*-protolichesterinic and fumarproto-cetraric acids from the protonema of *Cetraria islandica* (L.) Ach. growing domestically.] *Dissertationes Pharm.* **15**, 235–240.

GERTIG, H. AND Z. BANASIEWICZ. 1961. Rozpowszechnienie kwasu us-ninowego w niektórych porostach krajowych. [Distribution of usninic acid in various Polish seaweeds.] *Acta Polon. Pharm.* **18**, 67–71.

GIUDICI DE NICOLA, MARINA AND R. TOMASELLI. 1961*a*. Ricerche preliminari sui pigmenti nel ficosimbionte lichenico *Trebouxia decolorans* Ahm. I. — Carotenoidi. *Boll. Ist. Botan. Univ. Catania*, Ser. 3, **2**, 22–28.

GIUDICI DE NICOLA, MARINA AND R. TOMASELLI. 1961*b*. Ricerche preliminari sui pigmenti nel ficosimbionte lichenico *Trebouxia decolorans* Ahm. II. — Clorofille. *Boll. Ist. Botan. Univ. Catania*, Ser. 3, **2**, 29–34.

GODNEV, T. N., É. V. KHODASEVICH, AND A. I. ARNAUTOVA. 1966. Biosynthesis of pigments in lichens and overwintering plants at subzero temperatures. *Dokl., Biochem. Sect., Proc. Acad. Sci. USSR* (Engl. Transl.) **167**, 90–91.

GRANICHSTÄDTEN, H. AND E. G. V. PERCIVAL. 1943. The polysaccharides of Iceland moss (*Cetraria islandica*). Part I. Preliminary study of the hemicelluloses. *J. Chem. Soc.* **1943**, 54–58.

GREAM, G. E. AND N. V. RIGGS. 1960. Chemistry of Australian lichens II. A new depsidone from *Parmelia conspersa* (Ehrh.) Ach. *Australian J. Chem.* **13**, 285–295.

GROVER, P. K. AND T. R. SESHADRI. 1959. Chemical investigation of Indian lichens: Part XXIII — Imperfect lichens. *J. Sci. Ind. Res.* (India) **18B**, 238–240.

GROVER, P. K., G. D. SHAH, AND R. C. SHAH. 1956. Xanthones: Part V — A new synthesis of lichexanthone. *J. Sci. Ind. Res.* (India) **15B**, 629–630.

GUSTAFSON, F. G. 1954. A study of riboflavin, thiamine, niacin and ascorbic acid content of plants in northern Alaska. *Bull. Torrey Botan. Club* **81**, 313–322.

HALE, M. E., JR. 1950. The lichens of Aton Forest, Connecticut. *Bryologist* **53**, 181–213.

HALE, M. E., JR. 1952. Studies on the lichen *Rinodina oreina* in North America. *Bull. Torrey Botan. Club* **79**, 251–259.

HALE, M. E., JR. 1954*a*. Lichenes Americani Exsiccati. Fascicle I (No. 1–25). Wichita, Kansas.

HALE, M. E., JR. 1954*b*. Lichens from Baffin Island. *Am. Midland Naturalist* **51**, 232–264.

HALE, M. E., JR. 1955*a*. Studies on the chemistry and distribution of North American lichens (1–5). *Bryologist* **58**, 242–246.

HALE, M. E., JR. 1955*b*. Xanthoparmelia in North America I. The *Parmelia conspersa-stenophylla* group. *Bull. Torrey Botan. Club* **82**, 9–21.

HALE, M. E., JR. 1956*a*. A note on Lichenes Americani Exsiccati, Fascicle I. *Bryologist* **59**, 41–43.

HALE, M. E., JR. 1956*b*. A note on Lichenes Americani Exsiccati, Fascicle II. *Bryologist* **59**, 284–285.

HALE, M. E., JR. 1956*c*. Chemical strains of the lichen *Parmelia furfuracea*. *Am. J. Botany* **43**, 456–459.

HALE, M. E., JR. 1956*d*. Chemical strains of the *Parmelia conspersa-stenophylla* group in south central United States. *Bull. Torrey Botan. Club.* **83**, 218–220.

HALE, M. E., JR. 1956e. 2,4-Dihydroxy depsides in North American lichens. *Trans. Kansas Acad. Sci.* **59**, 229–232.

HALE, M. E., JR. 1956f. Fluorescence of lichen depsides and depsidones as a taxonomic criterion. *Castanea* **21**, 30–32.

HALE, M. E., JR. 1956g. Lichenes Americani Exsiccati. Fascicle II. (No. 26–50). Morgantown, West Virginia.

HALE, M. E., JR. 1956h. Studies on the chemistry and distribution of North American lichens (6–9). *Bryologist* **59**, 114–117.

HALE, M. E., JR. 1957a. Corticolous lichen flora of the Ozark Mountains. *Trans. Kansas Acad. Sci.* **60**, 155–160.

HALE, M. E., JR. 1957b. Lecture Notes. Lichenology. West Virginia University, Morgantown.

HALE, M. E., JR. 1957c. Lichenes Americani Exsiccati. Fascicle III. (No. 51–75). Morgantown, West Virginia.

HALE, M. E., JR. 1957d. The identity of *Parmelia hypotropoides*. *Bryologist* **60**, 344–347.

HALE, M. E., JR. 1957e. The *Lobaria amplissima-L. quercizans* complex in Europe and North America. *Bryologist* **60**, 35–39.

HALE, M. E., JR. 1958a. Chemical components of type specimens in *Parmelia* — I. *Brittonia* **10**, 177–180.

HALE, M. E., JR. 1958b. Studies on the chemistry and distribution of North American lichens (10–13). *Bryologist* **61**, 81–85.

HALE, M. E., JR. 1958c. The occurrence of *Parmelia formosana* in North America. *Castanea* **23**, 89–90.

HALE, M. E., JR. 1958d. The status of *Usnea diplotypus* in North America. *Bryologist* **61**, 247–248.

HALE, M. E., JR. 1958e. Vitamin requirements of three lichen fungi. *Bull. Torrey Botan. Club* **85**, 182–187.

HALE, M. E., JR. 1959a. New or interesting Parmelias from North and tropical America. *Bryologist* **62**, 123–132.

HALE, M. E., JR. 1959b. New or interesting species of *Parmelia* in North America. *Bryologist* **62**, 16–24.

HALE, M. E., JR. 1959c. The Mediterranean lichen *Parmelia perreticulata* in central Texas. *Southwestern Naturalist* **3**, 212.

HALE, M. E., JR. 1960a. A revision of the South American species of *Parmelia* determined by Lynge. *Contrib. U.S. Natl. Herb.* **36**, 1–41.

HALE, M. E., JR. 1960b. The typification of *Parmelia cetrarioides*. *Svensk Botan. Tidskr.* **54**, 269–272.

HALE, M. E., JR. 1961. The occurrence of *Lobaria amplissima* (Hoffm.) Schreb. in tropical America. *Lichenologist* **1**, 266–267.

HALE, M. E., JR. 1962a. A new species of *Parmelia* from Asia: *P. subcorallina*. *J. Japan. Botany* **37**, 345–347.

HALE, M. E., JR. 1962b. The chemical strains of *Usnea strigosa*. *Bryologist* **65**, 291–294.

HALE, M. E., JR. 1963a. Populations of chemical strains in the lichen *Cetraria ciliaris*. *Brittonia* **15**, 126–133.

HALE, M. E., JR. 1963*b*. The systematic position of *Parmelia albata* Wils. *Bryologist* **66**, 72–74.

HALE, M. E., JR. 1964. The *Parmelia conspersa* group in North America and Europe. *Bryologist* **67**, 462–473.

HALE, M. E., JR. 1965*a*. A monograph of *Parmelia* subgenus *Amphigymnia*. *Contrib. U.S. Natl. Herb.* **36**, 193–358.

HALE, M. E., JR. 1965*b*. Six new species of *Parmelia* from Southeast Asia. *J. Japan. Botany* **40**, 199–205.

HALE, M. E., JR. 1965*c*. Studies on the *Parmelia borreri* group. *Svensk Botan. Tidskr.* **59**, 37–48.

HALE, M. E., JR. 1965*d*. *Parmelia*, subgenus *Amphigymnia* of Japan. *Misc. Bryol. Lichenol.* (Japan) **3**, 161–162.

HALE, M. E., JR. AND S. KUROKAWA. 1962. *Parmelia* species first described from the British Isles. *Lichenologist* **2**, 1–5.

HALE, M. E., JR. AND S. KUROKAWA. 1964. Studies on *Parmelia* subgenus *Parmelia*. *Contrib. U.S. Natl. Herb.* **36**, 121–191.

HARDIMAN, JOSEPHINE, J. KEANE, AND T. J. NOLAN. 1935. Chemical constituents of lichens found in Ireland. *Lecanora gangaleoides* — Part I. *Sci. Proc. Roy. Dublin Soc.* **21**, 141–145.

HARPER, S. H. AND R. M. LETCHER. 1966. Chemistry of lichen constituents. II. Isolation of the internal salt of choline sulphuric acid from *Dermatiscum thunbergii*. *Chem. Ind.* (London) **1966**, 419–420.

HÄSÄNEN, E. AND J. K. MIETTINEN. 1966. Gamma emitting radionuclides in subarctic vegetation during 1962–64. *Nature* **212**, 379–382.

HENRIKSSON, ELISABET. 1963. The occurrence of carotenoids in some lichen species belonging to the Collemataceae. *Physiol. Plant.* **16**, 867–869.

HESS, D. 1958. Über die Papierchromatographie von Flechtenstoffen. *Planta* **52**, 65–76.

HESS, D. 1959. Untersuchungen über die Bildung von Phenolkörpern durch isolierte Flechtenpilze. *Z. Naturforsch.* **14b**, 345–347.

HESS, D. 1960. Untersuchungen über die hemmende Wirkung von Extrakten aus Flechtenpilzen auf das Wachstum von *Neurospora crassa*. *Z. Botan.* **48**, 136–142.

HESS, K. AND H. FRIES. 1927. Über Lichenin, Pringsheims Lichosan and Bergmanns Lichohexosan. (III. Mitteilung über Lichenin.). *Ann. Chem.* **455**, 180–205.

HESS, K. AND L. W. LAURIDSEN. 1940. Über die Konstitution des Lichenins (IV). *Chem. Ber.* **73**, 115–126.

HESSE, O. 1912. Die Flechtenstoffe. Pp. 32–144 *in* E. ABERHALDEN [ed.], Biochemisches Handlexikon. VII. Band. Julius Springer, Berlin.

HOLTZMANN, R. B. 1966. Natural levels of lead-210, polonium-210 and radium-226 in humans and biota of the Arctic. *Nature* **210**, 1094–1097.

HUNECK, S. 1961. Zur Struktur von Zeorin und Leucotylin. *Chem. Ber.* **94**, 614–622.

HUNECK, S. 1962a. Über das Vorkommen von Atranorin in *Crocynia neglecta* Hue. *Naturwissenschaften* 49, 608.

HUNECK, S. 1962b. Über das Vorkommen von Gyrophorsäure in *Acarospora fuscata* (Nyl.) Arnold. *Naturwissenschaften* 49, 396.

HUNECK, S. 1962c. Über die Inhaltsstoffe von *Lecidea tumida* Massalongo (*L. sorediza* Nyl.). *Naturwissenschaften* 49, 374–375.

HUNECK, S. 1962d. Über Flechteninhaltsstoffe, I. Konstitution der Confluentinsäure. *Chem. Ber.* 95, 328–332.

HUNECK, S. 1963a. Das Vorkommen von Divaricatsäure in *Lecidea kochiana* Hepp (*Biatora kochiana* Hepp). *Naturwissenschaften* 50, 645.

HUNECK, S. 1963b. Über ein weiteres Vorkommen von Norstictsäure in *Lecanora radiosa* (Hoffm.) Schaer. var *subcircinata* (Nyl.) Zahlbruckner. *Naturwissenschaften* 50, 646.

HUNECK, S. 1964. Über das Vorkommen von Norstictsäure in *Lecidea pantherina* (Ach.) Th. Fr. *Naturwissenschaften* 51, 536.

HUNECK, S. 1965a. XXI. Mitteilung über Flechteninhaltsstoffe. Über die Inhaltsstoffe von *Lecidea lithophila* (Ach.) Ach. emend. Th. Fr., *Lecidea macrocarpa* (D.C.) Steud. und *Lecidea fuscoatra* (L.) Ach. *Z. Naturforsch.* 20b, 1137–1138.

HUNECK, S. 1965b. Über das Vorkommen von Gyrophorsäure in *Lecidea tenebrosa* Flot. *Naturwiss̄ .ischaften* 52, 477.

HUNECK, S. 1965c. Über Flechteninhaltsstoffe, XVII. Konstitution der Planasäure, eines neuen Depsides aus *Lecidea plana* (Lahm ex Koerb). Nylander. *Z. Naturforsch.* 20b, 1119–1122.

HUNECK, S. 1966a. Über die Inhaltsstoffe von *Lecanora hercynica* Poelt et Ullrich, *Lecidea silacea* (Ach.) Ach. und *Acarospora montana* H. Magn. XXII. Mitteilung über Flechteninhaltsstoffe. *Z. Naturforsch.* 21b, 80–81.

HUNECK, S. 1966b. Flechteninhaltsstoffe, XXIV. Die Struktur von Tumidulin, einem neuen chlorhaltigen Depsid. *Chem. Ber.* 99, 1106–1110.

HUNECK, S. 1966c. Über Flechteninhaltsstoffe, XXVI. Die Inhaltsstoffe von *Lecanora handelii* Steiner und *Stereocaulon nanodes* Tuck. *Z. Naturforsch.* 21b, 199–200.

HUNECK, S. 1966d. Flechteninhaltsstoffe XXXII. Thiophansäure, ein neues chlorhaltiges Xanthon aus *Lecanora rupicola* (L.) Zahlbr. *Tetrahedron Letters* 1966, 3547–3549.

HUNECK, S. 1966e. XXXIII. Mitteilung über Flechteninhaltsstoffe. Die Inhaltsstoffe von *Lecanora viridula* (Flk.) Hillm., *Lecanora polytropa* (Ehrh.) Rabenh., *Lecanora badia* (Hoffm.) Ach. v. *milvina* Rabenh. und *Lecanora sulphurea* (Hoffm.) Ach. *Z. Naturforsch.* 21b, 888–890.

HUNECK, S. AND G. FOLLMANN. 1963. Zur Chemie chilenischer Flechten I. Das Vorkommen von Psoromsäure in *Ingaderia pulcherrima* Darbischire. VIII. Mitteilung über Flechteninhaltsstoffe. *Z. Naturforsch.* 18b, 991–992.

Huneck, S. and G. Follmann. 1964*a*. Das Vorkommen von Usninsäure in *Lecanora melanophthalma* Ram. und *Ramalina terebrata* Hook. et Taylor. *Naturwissenschaften* **51**, 291–292.

Huneck, S. and G. Follmann. 1964*b*. Zur Chemie chilenischer Flechten III. Das Vorkommen von Psoromsäure in *Chiodecton stalactinum* Nyl. und Roccellsäure in *Dirina lutosa* Zahlbr. *Z. Naturforsch.* **19b**, 658–659.

Huneck, S. and G. Follmann. 1965*a*. Zur Chemie chilenischer Flechten IV. Das Vorkommen von Gyrophorsäure in *Dolichocarpus chilensis* Sant. *Z. Naturforsch.* **20b**, 496.

Huneck, S. and G. Follmann. 1965*b*. Zur Chemie chilenischer Flechten V. Über die Inhaltsstoffe von *Ramalina ceruchis* (Ach.) De Not. var. *tumidula* (Tayl.) Nyl. *Z. Naturforsch.* **20b**, 611–612.

Huneck, S. and G. Follmann. 1965*c*. Acerca de la composición química de los liquenes chilenos. VI. La presencia de tumidulina en *Ramalina peruviana* Ach. *Bol. Univ. Chile* **7** (61), 56–57.

Huneck, S. and G. Follmann. 1965*d*. Zur Chemie chilenischer Flechten IX. Über die Inhaltsstoffe von *Lecanora dispersa* (Pers.) Roehl., *Parmelia perlata* (Huds.) Ach. und *Parmelia pseudoreticulata* Tav. *Z. Naturforsch.* **20b**, 1138–1139.

Huneck, S. and G. Follmann. 1965*e*. Zur Chemie chilenischer Flechten. VII. Über die Inhaltsstoffe von *Nephroma gyelnikii* (Raes.) Lamb, *Byssocaulon niveum* Mont. und *Stereocaulon corticulatum* Nyl. var. *procerum* Lamb. *Z. Naturforsch.* **20b**, 1012–1013.

Huneck, S. and G. Follmann. 1966*a*. Zur Chemie chilenischer Flechten VIII. Über die Inhaltsstoffe von *Ramalina chilensis* Bert. *Z. Naturforsch.* **21b**, 90–91.

Huneck, S. and G. Follmann. 1966*b*. Zur Chemie chilenischer Flechten X. Über die Inhaltsstoffe von *Himantormia lugubris* (Hue) Lamb, *Polycauliona regalis* (Wain.) Hue und *Thamnolecania gerlachei* (Wain.) Gyeln. *Z. Naturforsch.* **21b**, 91–92.

Huneck, S. and G. Follmann. 1966*c*. Zur Chemie chilenischer Flechten XI. Über die Inhaltsstoffe von *Ramalina tigrina* Follm. und *Ramalina inanis* Mont. *Z. Naturforsch.* **21b**, 713–714.

Huneck, S. and G. Follmann. 1966*d*. Zur Chemie chilenischer Flechten XII. Über die Inhaltsstoffe von *Stereocaulon antarcticum* Wain., *Anaptychia neoleucomelaena* Kur. und *Tornabenia ephebea* (Ach.) Kur. *Z. Naturforsch.* **21b**, 714–715.

Huneck, S. and G. Follmann. 1966*e*. Zur Chemie chilenischer Flechten XIII. Über die Inhaltsstoffe von *Usnea aureola* Mot., *Usnea lacerata* Mot. und *Usnea rubicunda* Stirt. var. *primaria* Mot. *Z. Naturforsch.* **21b**, 715–716.

Huneck, S. and G. Follmann. 1967*a*. Zur Chemie chilenischer Flechten. XV. Über die Inhaltsstoffe von *Ramalina cactacearum* Follm., *Ramalina ecklonii* (Spreng.) Mey. et Flot. var. *ambigua* Mont. und *Medusulina chilena* Dodge. *Z. Naturforsch.* **22b**, 110–111.

HUNECK, S. AND G. FOLLMANN. 1967*b*. Zur Chemie chilenischer Flechten. XVI. Über die Inhaltsstoffe einiger Roccellaceen. *Z. Naturforsch.* **22b**. 362–363.

HUNECK, S. AND J.-M. LEHN. 1966. 27. Mitteilung über Flechteninhaltsstoffe. Die Identität von Coquimbosäure und Hypoprotocetrarsäure. *Z. Naturforsch.* **21b**, 299.

HUNECK, S. AND M. SIEGEL. 1963. Über das Vorkommen von Norstictsäure in *Buellia sororioides* Erichsen. *Naturwissenschaften* **50**, 154–155.

HUNECK, S. AND G. TROTET. 1966. 34. Mitteilung über Flechteninhaltsstoffe. Über die Inhaltsstoffe von *Ramalina boulhautiana* Mah. et Gill. *Z. Naturforsch.* **21b**, 904.

HUNECK, S. AND G. TROTET. 1967. 40. Mitteilung über Flechteninhaltsstoffe. Über die Inhaltsstoffe von *Dirina repanda* (Nyl.) Fr. *Z. Naturforsch.* **22b**, 363.

HUNECK, S. AND R. TÜMMLER. 1965. Flechteninhaltsstoffe, XII. Die Struktur von Peltigerin. *Ann. Chem.* **685**, 128–133.

HUTCHINSON, W. A. AND J. T. REYNOLDS. 1960. Antibacterial activity of lichen extracts. *Bacteriol. Proc.* **60**, 100–101.

IMSHAUG, H. A. 1957. Alpine lichens of Western United States and adjacent Canada. I. The macrolichens. *Bryologist* **60**, 177–272.

IMSHAUG, H. A. AND I. M. BRODO. 1966. Biosystematic studies on *Lecanora pallida* and some related lichens in the Americas. *Nova Hedwigia* **12** (1-2), 1–59.

JENKINS, D. A. AND R. I. DAVIES. 1966. Trace element content of organic accumulations. *Nature* **210**, 129ʹ-1297.

JEWELL, M. E. AND H. B. LEWIS. 1918. The occurrence of lichenase in the digestive tract of invertebrates. *J. Biol. Chem.* **33**, 161–167.

JONES, F. T. AND K. J. PALMER. 1950. Optical, crystallographic and X-ray diffraction data for usnic acid. *J. Am. Chem. Soc.* **72**, 1820–1822.

JONES, M. P., J. KEANE, AND T. J. NOLAN. 1944. Lichen substances containing nitrogen. *Nature* **154**, 580.

KAREV, G. I. AND V. P. KOCHEVYKH. 1962. [On the ascorbic acid content in the fodder lichens of the tundra.] *Botan. Zh.* (Leningrad) **47**, 1686–1688.

KARRER, P., M. STAUB, AND J. STAUB. 1924. Polysaccharide XXIV. Über das Vorkommen von Lichenin (Reservecellulose) in Flechten und anderen Pflanzen. (5. Mitteilung über Lichenin). *Helv. Chim. Acta* **7**, 159–162.

KENNEDY, G., J. BREEN, J. KEANE, AND T. J. NOLAN. 1937. The chemical constituents of lichens found in Ireland — *Lecanora sordida* Th. Fr. *Sci. Proc. Roy. Dublin Soc.* **21**, 557–566.

KLIMA, J. 1933. Zur Chemie der Flechten II. *Alectoria ochroleuca* Ehrh. *Monatsh. Chem.* **62**, 209–213.

KLOSA, J. 1950. Antibiotika aus Flechten und ihre therapeutische Verwendungsmöglichkeit. *Med. Monatsschr.* (Stuttgart) **11**, 816–820.

KLOSA, J. 1951*a*. Antibiotika in Flechten. *Chem. Tech.* (Berlin) 3, 244–245.

KLOSA, J. 1951*b*. Über die antibiotische Wirkung der Flechtenstoffe. *Z. Physiol. Chem.* 287, 195–204.

KLOSA, J. 1952*a*. Beitrag zur Konstitution der Physodalsäure. (Einige Derivate der Proto-cetrarsäure). *Arch. Pharm.* 285, 432–438.

KLOSA, J. 1952*b*. Notiz zur Heilwirkungen der Parmeliaceae. *Pharmazie* 7, 755–756.

KLOSA, J. 1952*c*. Untersuchungen über die Inhaltsstoffe der Flechte *Lepraria flava. Pharmazie* 7, 687–688.

KLOSA, J. 1953*a*. Chemische Konstitution und antibiotische Wirkung der Flechtenstoffe. *Pharmazie* 8, 435–442.

KLOSA, J. 1953*b*. Über die Isolierung der Flechtensäuren von *Parmelia physodes. Pharm. Ind.* 15, 46–47.

KOLLER, G. 1932*a*. Über die Ramalsäure. *Monatsh. Chem.* 61, 286–292.

KOLLER, G. 1932*b*. Über eine Synthese des Diazetyl-evernsäure-methylesters und des Tetraazetylgyrophorsäure-methylesters. *Monatsh. Chem.* 61, 147–161.

KOLLER, G. AND H. HAMBURG. 1935*a*. Über die Konstitution der Diploschistessäure. *Monatsh. Chem.* 65, 367–374.

KOLLER, G. AND H. HAMBURG. 1935*b*. Über einen Inhaltsstoff der *Pertusaria dealbata* Ach., Nyl. *Monatsh. Chem.* 65, 375–379.

KOLLER, G. AND H. HAMBURG. 1936. Über die Rhodocladonsäure. *Monatsh. Chem.* 68, 202–206.

KOLLER, G. AND A. KLEIN. 1934*a*. Über die Saxatilsäure. *Monatsh. Chem.* 64, 80–86.

KOLLER, G. AND A. KLEIN. 1934*b*. Über die Saxatilsäure. *Monatsh. Chem.* 65, 91–92.

KOLLER, G. AND A. KLEIN. 1934*c*. Über eine Synthese der Pinastrinsäure. *Monatsh. Chem.* 63, 213–215.

KOLLER, G., A. KLEIN AND K. PÖPL. 1933. Über die Saxatilsäure und die Kaprarsäure. *Monatsh. Chem.* 63, 301–310.

KOLLER, G. AND E. KRAKAUER. 1929. Über die Konstitution der Cetrarsäure. *Monatsh. Chem.* 53/54, 931–951.

KOLLER, G., E. KRAKAUER AND E. PÖPL. 1934. Über die Ramalinsäure. *Monatsh. Chem.* 64, 3–5.

KOLLER, G. AND K. LOCKER. 1931. Über die Physodalsäure. *Monatsh. Chem.* 58, 209–212.

KOLLER, G. AND W. MAASS. 1935. Über einen Inhaltsstoff von *Baeomyces roseus* Pers. *Monatsh. Chem.* 66, 57–63.

KOLLER, G. AND G. PFEIFFER. 1933*a*. Über die Glabratsäure. *Monatsh. Chem.* 62, 169–171.

KOLLER, G. AND G. PFEIFFER. 1933*b*. Über die Konstitution der Pinastrinsäure. *Monatsh. Chem.* 62, 160–168.

KOLLER, G. AND G. PFEIFFER. 1933*c*. Über die Umbilikarsäure und die Ramalsäure. *Monatsh. Chem.* 62, 241–251.

KOLLER, G. AND G. PFEIFFER. 1933*d*. Über Enzyme der Flechten und über die Konstitution der Umbilikarsäure. *Monatsh. Chem.* **62**, 359–372.

KOLLER, G. AND K. PÖPL. 1934*a*. Über einen chlorhaltigen Flechtenstoff I. *Monatsh. Chem.* **64**, 106–113.

KOLLER, G. AND K. PÖPL. 1934*b*. Über einen chlorhaltigen Flechtenstoff II. *Monatsh. Chem.* **64**, 126–130.

KOLLER, G. AND H. RUSS. 1937. Über die Konstitution der Solorinsäure. *Monatsh. Chem.* **70**, 54–72.

KOLUMBE, E. 1927. Purpurbacterien und Flechten. *Mikrokosmos* **21**, 53–55.

KROG, HILDUR. 1951*a*. Litt om lavsyrer, noen saermerkte organiske stoffer i lav. *Blyttia* **8**, 91–98.

KROG, HILDUR. 1951*b*. Microchemical studies on *Parmelia. Nytt Mag. Naturvidenskapene* **88**, 57–85.

KROG, HILDUR. 1962. A contribution to the lichen flora of Alaska. *Arkiv Botan.*, Ser. 2, **4**, 489–513.

KROG, HILDUR. 1963. *Parmelia almquistii* Vain. and its distribution. *Bryologist* **66**, 28–31.

KUROKAWA, S. 1952. [What substances are contained in lichens?] *Kagaku no Jikken* **3**, 2–3, 24–27.

KUROKAWA, S. 1956. Notulae miscellaneae Lichenum japonicorum (2). *J. Japan. Botany* **31**, 193–196.

KUROKAWA, S. 1957. Lichens of Simokita Peninsula. *Shigen Kagaku Kenkyusho Iho* **1957** (43–44), 12–21.

KUROKAWA, S. 1959*a*. *Anaptychiae* (lichens) and their allies of Japan (1). *J. Japan. Botany* **34**, 117–124.

KUROKAWA, S. 1959*b*. *Anaptychiae* (lichens) and their allies of Japan (2). *J. Japan. Botany* **34**, 174–184.

KUROKAWA, S. 1959*c*. Notulae miscellaneae Lichenum japonicorum (6). *J. Japan. Botany* **34**, 23–24.

KUROKAWA, S. 1960*a*. *Anaptychiae* (lichens) and their allies of Japan (3). *J. Japan. Botany* **35**, 91–96.

KUROKAWA, S. 1960*b*. *Anaptychiae* (lichens) and their allies of Japan (4). *J. Japan. Botany* **35**, 240–243.

KUROKAWA, S. 1960*c*. *Anaptychiae* (lichens) and their allies of Japan (5). *J. Japan. Botany* **35**, 353–358.

KUROKAWA, S. 1961. *Anaptychiae* (lichens) and their allies of Japan (6). *J. Japan. Botany* **36**, 51–56.

KUROKAWA, S. 1962*a*. A monograph of the genus *Anaptychia. Nova Hedwigia* **6**, 1–115.

KUROKAWA, S. 1962*b*. A note on the lichen genus *Tornabenia* Trev. *J. Japan. Botany* **37**, 289–294.

KUROKAWA, S. 1964. [Lichens.] National Science Museum. Ueno Park, Tokyo.

KUROKAWA, S. 1965. Revision of series *Relicinae* of the genus *Parmelia* in Japan and Taiwan. *J. Japan. Botany* **40**, 264–269.

KUROKAWA, S. AND YOKO JINZENJI. 1965. Chemistry and nomenclature of Japanese *Anzia*. *Bull. Natl. Sci. Mus.* (Tokyo) **8**, 369–374.

KUROKAWA, S., YOKO JINZENJI, S. SHIBATA, AND HSÜCH-CHING CHIANG. 1966. Chemistry of Japanese *Peltigera* with some taxonomic notes. *Bull. Natl. Sci. Mus.* (Tokyo) **9**, 101–114.

LAAKSO, P. V. AND M. GUSTAFSSON. 1952. The colorimetric determination of usnic acid in the lichen *Cladonia alpestris*. *Suomen Kemistilehti* **25B**, 7–10.

LAL, B. M. AND K. R. RAO. 1956. The food value of some Indian lichens. *J. Sci. Ind. Res.* (India) **15C**, 71–73.

LAMB, I. M. 1951*a*. Biochemistry in the taxonomy of lichens. *Nature* **168**, 38.

LAMB, I. M. 1951*b*. On the morphology, phylogeny, and taxonomy of the lichen genus *Stereocaulon*. *Can. J. Botany* **29**, 522–584.

LAMB, I. M. 1961. Two new species of *Stereocaulon* occurring in Scandinavia. *Botan. Notiser* **114**, 265–275.

LAMB, I. M. 1964. Antarctic lichens. I. The genera *Usnea, Ramalina, Himantormia, Alectoria, Cornicularia*. British Antarctic Survey, Scientific Report No. 38. London.

LAMB, I. M. 1965. The *Stereocaulon massartianum* assemblage in East Asia. *J. Japan. Botany* **40**, 270–275.

LAMB, I. M. 1967. Private communication.

LAMBINON, J., A. MAQUINAY, AND J. L. RAMAUT. 1964. La teneur en zinc de quelques lichens des terrains calaminaires belges. *Bull. Jardin Botan. État Bruxelles* **34**, 273–282.

LANGE, O. L. AND H. ZIEGLER. 1963. Der Schwermetallgehalt von Flechten aus dem *Acarosporetum sinopicae* auf Erzschlackenhalden des Harzes. I. Eisen und Kupfer. *Mitteil. Florist-soziolog. Arbeitsgemeinschaft*, N.F. Heft. 10 (Festschr. für Prof. Dr. Otto Stocker), 156–183.

LEHN, J.-M. AND S. HUNECK. 1965. Über Flechteninhaltsstoffe, XVIII. Die erstmalige Isolierung des Diterpens $(-)$-16α-Hydroxykauran aus einer Flechte. *Z. Naturforsch.* **20b**, 1013.

LINDBERG, B. 1955. Studies on the chemistry of lichens VIII. Investigation of a *Dermatocarpon* and some *Roccella* species. *Acta Chem. Scand.* **9**, 917–919.

LINDBERG, B. AND J. MCPHERSON. 1954. Studies on the chemistry of lichens VI. The structure of pustulan. *Acta Chem. Scand.* **8**, 985–988.

LINDBERG, B. AND H. MEIER. 1962. Studies on the chemistry of lichens. 15. Siphulitol, a new polyol from *Siphula ceratites*. *Acta Chem. Scand.* **16**, 543–547.

LINDBERG, B., A. MISIORNY, AND C. A. WACHTMEISTER. 1953. Studies on the chemistry of lichens IV. Investigation of the low-molecular carbohydrate constituents of different lichens. *Acta Chem. Scand.* **7**, 591–595.

LINDBERG, B., B.-G. SILVANDER, AND C. A. WACHTMEISTER. 1963. Studies on the chemistry of lichens 18. 3-O-β-D-Glucopyranosyl-D-mannitol from *Peltigera aphthosa* (L.) Willd. *Acta Chem. Scand.* **17**, 1348–1350.

LINDBERG, B., B.-G. SILVANDER, AND C. A. WACHTMEISTER. 1964. Studies on the chemistry of lichens. 19. Mannitol glycosides in *Peltigera* species. *Acta Chem. Scand.* **18**, 213–216.

LINDBERG, B., C. A. WACHTMEISTER, AND B. WICKBERG. 1952. Studies on the chemistry of lichens. II. Umbilicin, an arabitol galactoside from *Umbilicaria pustulata* (L.) Hoffm. *Acta Chem. Scand.* **6**, 1052–1055.

LINDBERG, B. AND B. WICKBERG. 1953. Studies on the chemistry of lichens. III. Disaccharides from *Umbilicaria pustulata* (L.) Hoffm. *Acta Chem. Scand.* **7**, 140–142.

LINDBERG, B. AND B. WICKBERG. 1954. Studies on the chemistry of lichens V. The furanoside structure of umbilicin. *Acta Chem. Scand.* **8**, 821–824.

LINDBERG, B. AND B. WICKBERG. 1962. Studies on the chemistry of lichens. 17. The structure of umbilicin. *Acta Chem. Scand.* **16**, 2240–2244.

LINKO, P., M. ALFTHAN, J. K. MIETTINEN, AND A. I. VIRTANEN. 1953. Free sarcosine in reindeer-moss (*Cladonia sylvatica*). *Acta Chem. Scand.* **7**, 1310–1311.

LOUNAMAA, K. J. 1965. Studies on the content of iron, manganese and zinc in macrolichens. *Ann. Botan. Fennici* **2**, 127–137.

MAASS, W. S. G. AND A. C. NEISH. 1967. Lichen substances II. Biosynthesis of calycin and pulvinic dilactone by the lichen, *Pseudocyphellaria crocata*. *Can. J. Botany* **45**, 59–72.

MAASS, W. S. G., G. H. N. TOWERS, AND A. C. NEISH. 1964. Flechtenstoffe: I. Untersuchungen zur Biogenese des Pulvinsäureanhydrids. *Ber. Deut. Botan. Ges.* **77**, 157–161.

MANAKTALA, S. K., S. NEELAKANTAN, AND T. R. SESHADRI. 1966. Synthesis of (±) montagnetol and (±) erythrin. *Tetrahedron* **22**, 2373–2376.

MAQUINAY, A., I. M. LAMB, J. LAMBINON, AND J. L. RAMAUT. 1961. Dosage du zinc chez un lichen calaminaire belge: *Stereocaulon nanodes* Tuck. f. *tyroliense* (Nyl.) M. Lamb. *Physiol. Plant.* **14**, 284–289.

MARSHAK, A. 1947. A crystalline antibacterial substance from the lichen *Ramalina reticulata*. *Public Health Rept.* (U.S.) **62**, 3–19.

MARSHAK, A., G. T. BARRY, AND L. C. CRAIG. 1947. Antibiotic compound isolated from the lichen *Ramalina reticulata*. *Science* **106**, 394–395.

MARTIN, W. 1965. The lichen genus *Cladia*. *Trans. Roy. Soc. New Zealand, Botany* **3** (2), 7–12.

MĚRKA, V. 1951. Změny regulačních schopností a obsahu gyrophorové kyseliny v odumírajícím lišejníku *Umbilicaria pustulata* (L.) Hoffm. *Spisy Vydávané Přírodovědeckou Fak. Masary. Univ.* **327**, 97–119.

MEYER, K. H. AND P. GÜRTLER. 1947a. Recherches sur l'amidon. XXXI. La constitution de la lichénine. *Helv. Chim. Acta* **30**, 751–761.

MEYER, K. H. AND P. GÜRTLER. 1947*b*. Recherches sur l'amidon. XXXII. L'isolichénine. *Helv. Chim. Acta* **30**, 761–765.

MIKOSHIBA, K. 1936. Über die pharmakologischen Wirkungen der *d*-Usninsäure und ihrer Spaltungsprodukte. *Japan. J. Med. Sci., IV. Pharmacol.* **9**, 77–105.

MILLER, E. V., C. E. GRIFFIN, T. SCHAEFERS, AND MYRA GORDON. 1965. Two types of growth inhibitors in extracts of *Umbilicaria papulosa. Botan. Gaz.* **126**, 100–107.

MINAMI, K. 1944. [Synthesis of glomellin.] *Yakugaku Zasshi* **64**, 315–317. Reference from *Chem. Abstr.* **45**, 2939F (1951).

MITCHELL, J. C. 1965. Allergy to lichens. *Arch. Dermatol.* **92**, 142–146.

MITCHELL, M. E. 1960. Contribution à la lichénologie irlandaise. *Bull. Soc. Sci. Bretagne* **35**, 267–272.

MIT[S]UNO, M. 1938. *Sphaerophorus*-Arten aus Japan. *J. Japan. Botany* **14**, 659–669.

MITSUNO, M. 1953. Paper chromatography of lichen substances. I. *Pharm. Bull.* (Tokyo) **1**, 170–173.

MITSUNO, M. 1955. Paper chromatography of lichen substances. II. *Pharm. Bull.* (Tokyo) **3**, 60–62.

MITTAL, O. P., S. NEELAKANTAN, AND T. R. SESHADRI. 1952. Chemical investigation of Indian lichens: Part XIV — Chemical components of *Ramalina calicaris* and *Ramalina sinensis. J. Sci. Ind. Res.* (India) **11B**, 386–387.

MITTALL, O. P. AND T. R. SESHADRI. 1954*a*. Chemical investigation of Indian lichens: Part XVI — Purification and composition of lichenin and isolichenin from Indian lichens. *J. Sci. Ind. Res.* (India) **13B**, 244–245.

MITTALL, O. P. AND T. R. SESHADRI. 1954*b*. Chemistry of lichenin and isolichenin. *J. Sci. Ind. Res.* (India) **13A**, 174–177.

MITTALL, O. P. AND T. R. SESHADRI. 1955. Chemical investigation of Indian lichens. Part XIX. *Lepraria:* Constitution of leprapinic acid. *J. Chem. Soc.* **1955**, 3053–3055.

MITTALL, O. P. AND T. R. SESHADRI. 1956. Synthesis of leprapinic acid and constitution of pinastric acid. *J. Chem. Soc.* **1956**, 1734–1735.

MITTALL, O. P. AND T. R. SESHADRI. 1957. Occurrence of C₉ (forked) units in polyporic and pulvinic acid derivatives. *Current Sci.* (India) **26**, 4–6.

MOHAN, MARGARET, J. KEANE, AND T. J. NOLAN. 1937. The chemical constituents of lichens found in Ireland — *Parmelia conspersa* Ach. *Sci. Proc. Roy. Dublin Soc.* **21**, 593–594.

MOISSEJEVA, E. N. 1957. [The sodium salt of usnic acid (the substance binan), its production, physico-chemical properties, and methods of investigation.] Pp. 50–64 *in* N. V. LAZAREV AND V. P. SAVICZ [eds.], [The New Antibiotic Binan, or the Sodium Salt of Usnic Acid (Botanical and Medical Investigations).] Akademiia Nauk SSSR, Botanicheskii Institut im. V. L. Komarova, Moscow-Leningrad.

MOISSEJEVA, E. N. 1961. [Biochemical Properties of Lichens and Their Practical Importance.] Akademiia Nauk SSSR, Botanicheskii Institut im. V. L. Komarova, Moscow-Leningrad.

MONTIGNIE, E. 1935. La présence des stérols dans les cryptogames vasculaires. *Bull. Soc. Chim. France*, Sér. 5, **2**, 1219.

MOORE, BARBARA JO. 1966. The chemistry of the *Parmelia perforata* group in Florida. *Bryologist* **69**, 353–356.

MORS, W. B. 1951. Identificacāo microquímica de parietina no líquen "Theloschistes exilis." *Bol. Inst. Quím. Agr.* (Rio de Janeiro) **23**, 1–16.

MORS, W. B. 1952a. Chemical studies on Brazilian lichens I. The "Usneae" of the Organ Mountains. *Rev. Brasil. Biol.* **12**, 389–400.

MORS, W. B. 1952b. Investigações químicas sôbre líquens brasileiros: Estudo das *Usneae* da Serra dos Órgãos. *Bol. Inst. Quím. Agr.* (Rio de Janeiro) **29**, 1–23.

MOSBACH, K. 1964a. On the biosynthesis of lichen substances. Part 1. The depside gyrophoric acid. *Acta Chem. Scand.* **18**, 329–334.

MOSBACH, K. 1964b. On the biosynthesis of lichen substances. Part 2. The pulvic acid derivative vulpinic acid. *Biochem. Biophys. Res. Commun.* **17**, 363–367.

MOSBACH, K. 1964c. Studies on the Biosynthesis of Aromatic Compounds in Fungi and Lichens. Carol Bloms Boktryckeri A.-B., Lund.

MOZINGO, H. N. 1961. The genus *Cladonia* in eastern Tennessee and the Great Smoky Mountains. I. *Bryologist* **64**, 325–335.

MURAKAMI, T. 1956. The coloring matters of *Xanthoria fallax* (Hepp.) Arn. Fallacinal and fallacinol. *Pharm. Bull.* (Tokyo) **4**, 298–302.

MURPHY, D., J. KEANE, AND T. J. NOLAN. 1943. The chemical constituents of lichens found in Ireland. *Lecanora parella* Ach. The constitution of variolaric acid. *Sci. Proc. Roy. Dublin Soc.* **23**, 71–82.

MURRAY, J. 1952. Lichens and fungi. I. Polyporic acid in *Stictae*. *J. Chem. Soc.* **1952**, 1345–1350.

MURTY, T. K. 1960. Isolation of methyl β-orcinol carboxylate from *Parmelia tinctorum* Despr. (Awasthi). *J. Sci. Ind. Res.* (India) **19B**, 508–509.

MURTY, T. K. AND S. S. SUBRAMANIAN. 1958. Carotene content of *Roccella montagnei*. *J. Sci. Ind. Res.* (India) **17C**, 105–106.

MURTY, T. K. AND S. S. SUBRAMANIAN. 1959a. Carotene from lichens. *Res. Ind.* (New Delhi) **4**, 176.

MURTY, T. K. AND S. S. SUBRAMANIAN. 1959b. Chemical components of *Usnea venosa* Mot. *J. Sci. Ind. Res.* (India) **18B**, 394–395.

MURTY, T. K. AND S. S. SUBRAMANIAN. 1959c. Isolation of carotene from *Roccella montagnei*. *J. Sci. Ind. Res.* (India) **18B**, 162–163.

MURTY, T. K. AND S. S. SUBRAMANIAN. 1959d. Isolation of ergosterol from *Roccella montagnei*. *J. Sci. Ind. Res.* (India) **18B**, 91–92.

NAIR, A. G. R. AND S. S. SUBRAMANIAN. 1961a. Chemical components of *Usnea flexilis* Strtn. *J. Sci. Ind. Res.* (India) **20B**, 611–612.

NAIR, A. G. R. AND S. S. SUBRAMANIAN. 1961b. Chemical components of *Usnea rubicunda* Strtn. *J. Sci. Ind. Res.* (India) **20B**, 555–556.

NAIR, A. G. R. AND S. S. SUBRAMANIAN. 1962. Chemical strains of *Usnea orientalis*. *Current Sci.* (India) **31**, 60.

NAKAO, M. 1923. Chemical constituents of a Chinese drug "Shi-hoa" and their constituents. *Yakugaku Zasshi* No. 496, 423–497. Reference from *Chem. Abstr.* **17**, 3184⁹ (1923).

NEELKANTAN, S. 1965. Recent developments in the chemistry of lichen substances. Pp. 35–84 *in* "Advancing Frontiers in the Chemistry of Natural Products." Hindustan Publishing Corp., Delhi.

NEELKANTIAN, S., S. RANGASWAMI, AND V. S. RAO. 1954. Chemical components of some *Anaptychia* lichens of India. *Indian J. Pharm.* **16**, 173–175.

NEELAKANTAN, S., S. RANGASWAMI, T. R. SESHADRI, AND S. S. SUBRA-MANIAN. 1951. Chemical investigation of Indian lichens. Part XI. Constitution of teloschistin — the position of the methoxyl group. *Proc. Indian Acad. Sci.* **33A**, 142–147.

NEELAKANTAN, S. AND T. R. SESHADRI. 1952a. Chemical investigation of Indian lichens. *J. Sci. Ind. Res.* (India) **11A**, 338–340.

NEELAKANTAN, S. AND T. R. SESHADRI. 1952b. Chemical investigation of Indian lichens: Part XIII — A note on the chemistry of the lichen *Caloplaca elegans*. *J. Sci. Ind. Res.* (India) **11B**, 126–127.

NEELAKANTAN, S. AND T. R. SESHADRI. 1954. A new synthesis of telo-schistin. *J. Sci. Ind. Res.* (India) **13B**, 884–885.

NEELAKANTAN, S. AND T. R. SESHADRI. 1959a. Biogenesis of benzo-quinones and related substances. *Current Sci.* (India) **28**, 351–356.

NEELAKANTAN, S. AND T. R. SESHADRI. 1959b. Transformations of 6-methylsalicylic acid derivatives. *Current Sci.* (India) **28**, 394–397.

NEELAKANTAN, S. AND T. R. SESHADRI. 1960a. Biogenesis of naturally occurring anthraquinone derivatives. *J. Sci. Ind. Res.* (India) **19A**, 71–79.

NEELAKANTAN, S. AND T. R. SESHADRI. 1960b. Naturally occurring chlorodepsidones. *Proc. Natl. Inst. Sci. India*, Suppl. 1, **26A**, 84–98.

NEELAKANTAN, S. AND T. R. SESHADRI. 1961. Occurrence of C_8 units in xanthones. *Current Sci.* (India) **30**, 90–94.

NEELAKANTAN, S., T. R. SESHADRI, AND S. S. SUBRAMANIAN. 1951. Indian strains of the lichen *Parmelia tinctorum*. *J. Sci. Ind. Res.* (India) **10B**, 199–200.

NEELAKANTAN, S., T. R. SESHADRI, AND S. S. SUBRAMANIAN. 1956. Chemical investigation of Indian lichens. Part XX. A new synthesis of teloschistin. *Proc. Indian Acad. Sci.* **44A**, 42–45.

NEELAKANTAN, S., T. R. SESHADRI, AND S. S. SUBRAMANIAN. 1959. Constitution of vicanicin from the lichen *Teloschistes flavicans*. *Tetra-hedron Letters* **1959** (9), 1–4.

NEELAKANTAN, S., T. R. SESHADRI, AND S. S. SUBRAMANIAN. 1962. Chemical investigation of Indian lichens — XXVI. Constitution of vicanicin from *Teloschistes flavicans*. *Tetrahedron* **18**, 597–604.

NOGAMI, H. 1944. Didymic acid. *Yakugaku Zasshi* **64**, 47–50. Reference from *Chem. Abstr.* **45**, 2929g (1951).

Nolan, T. J. 1934*a*. A lichen substance containing chlorine. *Chem. Ind.* (London) **12**, 512–513.

Nolan, T. J. 1934*b*. Chemical constituents of lichens found in Ireland. *Buellia canescens.* — Part 1. *Sci. Proc. Roy. Dublin Soc.* **21**, 67–71.

Nolan, T. J., J. Algar, E. P. McCann, W. A. Manahan, and Niall Nolan. 1948. The chemical constituents of lichens found in Ireland. *Buellia canescens.* Part 3. The constitution of diploicin. *Sci. Proc. Roy. Dublin Soc.* **24**, 319–334.

Nolan, T. J. and J. Keane. 1933. Salazinic acid and the constituents of the lichen, *Lobaria pulmonaria. Nature* **132**, 281.

Nolan, T. J. and J. Keane. 1940. The chemical constituents of lichens found in Ireland. *Lecanora gangaleoides* — Part 2. *Sci. Proc. Roy. Dublin Soc.* **22**, 199–209.

Nolan, T. J., J. Keane, and V. E. Davidson. 1940. The chemical constituents of the lichen *Parmelia latissima* Fée. *Sci. Proc. Roy. Dublin Soc.* **22**, 237–239.

Nuno, Mariko. 1958. On the chemical ingredients of *Usnea diffracta* Vain. *J. Japan. Botany* **33**, 227–232.

Nuno, Mariko. 1962. Chemism of *Cladonia* subgenus *Clathrina* (Müll. Arg.) Vain. *J. Japan. Botany* **37**, 77–80.

Nuno, Mariko. 1963. On some brown *Nephromata* and chromatograms of their ingredients. *J. Japan. Botany* **38**, 196–202.

Nuno, Mariko. 1964. Chemism of *Parmelia* subgenus *Hypogymnia* Nyl. *J. Japan. Botany* **39**, 97–103.

Ollis, W. D. 1955–1957. Studies concerning the synthesis of diploicin. *Sci. Proc. Roy. Dublin Soc.* **27**, 161–164.

Paternò, E. and F. Crosa. 1894. Ueber eine neue aus Flechten erhaltene Verbindung. *Chem. Ber.* **27**, 399–400.

Pätiälä, Risto, J. Pätiälä, S. Siintola, and P. Heilala. 1948. *l*-Usniinihapon vaikutuksesta in vivo. *Suomen Kemistilehti* **21A**, 217.

Peat, S., W. J. Whelan, and J. G. Roberts. 1957. The structure of lichenin. *J. Chem. Soc.* **1957**, 3916–3924.

Peat, S., W. J. Whelan, J. R. Turvey, and K. Morgan. 1961. The structure of isolichenin. *J. Chem. Soc.* **1961**, 623–629.

Pereira, F. M., J. de Sa, and S. S. Bhatnagar. 1953. Antitubercular activity of a pure lichen chemical constituent. *Indian J. Pharm.* **15**, 287–289.

Perlin, A. S. and S. Suzuki. 1962. The structure of lichenin: selective enzymolysis studies. *Can. J. Chem.* **40**, 50–56.

Pišút, I. 1961. Bemerkungen über einige Arten der Flechtengattung *Cladonia* in der Slowakei. *Acta Fac. Rer. Nat. Univ. Comen.* **6**, 513–531.

Pueyo, G. 1959. Sur la présence d'un galactosido-mannitol dans le thalle de *Peltigera horizontalis. Compt. Rend.* **248**, 2788–2790.

Pueyo, G. 1960. Présence de mannitol et d'arabitol dans de nouvelles espèces de Lichens. Un hétéroside nouveau (peltigéroside) dans *Peltigera horizontalis* Hoffm. *Rev. Bryol. Lichénol.* **29**, 124–129.

PUEYO, G. 1963*a*. Identification par chromatographie sur papier des glucides solubles des Lichens. I. Sucres. *Rev. Bryol. Lichénol.* **32**, 279–284.

PUEYO, G. 1963*b*. Identification par chromatographie sur papier des glucides solubles des Lichens. II. — Polyalcools. *Rev. Bryol. Lichénol.* **32**, 285–289.

PUEYO, G. 1963*c*. Un polyalcool (mannitol) dans *Lichina pygmaea* Ag. *Bryologist* **66**, 74–76.

PUEYO, G. 1963*d*. Un polyalcool (erythritol) dans *Roccella fuciformis* D.C. *Bull. École Natl. Supér. Agronom. Nancy* **5**, 195–198.

PUEYO, G. 1964. Présence de mannitol chez *Usnea comosa* Röhl. *Bull. École Natl. Supér. Agronom. Nancy* **6**, 153–156.

PUEYO, G. 1964–1965*a*. Mannitol chez *Cladonia endiviaefolia* Mudd. *Rev. Bryol. Lichénol.* **33**, 595–596.

PUEYO, G. 1964–1965*b*. Polyalcools chez *Evernia prunastri* Ach. *Rev. Bryol. Lichénol.* **33**, 592–594.

PUEYO, G. 1965. Polyalcools chez *Cladonia rangiferina* (L.) Web. *Bryologist* **68**, 334–336.

RAJAGOPALAN, T. R. AND T. R. SESHADRI. 1959. Chemical investigation of Indian lichens. Part XXI. Occurrence of fallacinal in *Teloschistes flavicans*. *Proc. Indian Acad. Sci.* **49A** 1–5.

RAMAKRISHNAN, S. AND S. S. SUBRAMANIAN. 1964. Amino acids of *Roccella montagnei & Parmelia tinctorum*. *Indian J. Chem.* **2**, 467.

RAMAKRISHNAN, S. AND S. S. SUBRAMANIAN. 1965. Amino-acid composition of *Cladonia rangiferina, Cladonia gracilis* and *Lobaria isidiosa*. *Current Sci.* (India) **34**, 345–347.

RAMAKRISHNAN, S. AND S. S. SUBRAMANIAN. 1966*a*. Amino-acids of *Dermatocarpon moulinsii*. *Current Sci.* (India) **35**, 284–285.

RAMAKRISHNAN, S. AND S. S. SUBRAMANIAN. 1966*b*. Amino-acids of *Lobaria subisidiosa, Umbilicaria pustulata, Parmelia nepalensis,* and *Ramalina sinensis*. *Current Sci.* (India) **35**, 124–125.

RAMAUT, J. L. 1960. Étude par chromatographie de partage sur papier des acides lichéniques du genre *Parmelia* en Belgique. I. — *Parmelia* de la section "Hypotrachyna" Vain. *Rev. Bryol. Lichénol.* **29**, 307–320.

RAMAUT, J. L. 1961*a*. Contribution à l'étude chromatographique de quelques *Parmelia* de la section *Amphigymnia*, sous-section *Subglaucescentes* Vain. *Rev. Bryol. Lichénol.* **30**, 131–134.

RAMAUT, J. L. 1961*b*. La chromatographie de partage sur papier au service du lichénologue systématicien. *Mouvement Sci. Belg.* **5**, 347–352.

RAMAUT, J. L. 1961*c*. La chromatographie de partage sur papier des acides lichéniques. *Bull. Soc. Roy. Botan. Belg.* **93**, 27–40.

RAMAUT, J. L. 1962*a*. Contribution à l'étude chimique du genre *Stereocaulon* par chromatographie de partage sur papier. — II. *Stereocaulon* européens. *Rev. Bryol. Lichénol.* **31**, 251–255.

RAMAUT, J. L. 1962*b*. Réactions thallines, microcristallisations et chromatographie de partage sur papier en lichénologie. *Naturalistes Belges* **43**, 359–370.

RAMAUT, J. L. 1963*a*. Chromatographie en couche mince des depsidones du β-orcinol. *Bull. Soc. Chim. Belges* **72**, 97–101.

RAMAUT, J. L. 1963*b*. Chromatographie sur couche mince des despsides [*sic*] et des depsidones. *Bull. Soc. Chim. Belges* **72**, 316–321.

RAMAUT, J. L. 1964–1965. Étude chimique de quelques *Parmelia* tropicaux de la section *Amphigymnia*. *Rev. Bryol. Lichénol.* **33**, 587–591.

RAMAUT, J. L. 1965. Réflexions sur la valeur chimiotaxonomique des substances lichéniques à basses concentrations: le cas de l'acide usnique chez *Evernia prunastri* (L.) Ach. *Phytochemistry* **4**, 199–202.

RAMAUT, J. L., J. LAMBINON, AND A. TARGÉ. 1962. Le problème de l'acide usnique chez *Evernia prunastri* (L.) Ach. *Lejeunia*, N. Sér. No. 12, 11 pp.

RAMAUT, J. L. AND R. SCHUMACKER. 1961. Étude par chromatographie de partage sur papier des lichens du genre *Parmelia* en Belgique. II. — *Parmelia* de la section "Amphigymnia" Vain. *Rev. Bryol. Lichénol.* **30**, 125–130.

RAMAUT, J. L. AND R. SCHUMACKER. 1962. Étude par chromatographie de partage sur papier des acides lichéniques des espèces du genre *Stereocaulon*. I — *Stereocaulon* belges. *Lejeunia*, N. Sér. No. 4, 11 pp.

RANGASWAMI, S. AND V. S. RAO. 1954*a*. Chemical components of *Ramalina farinceae* [*sic*] Ach. *Indian J. Pharm.* **16**, 197.

RANGASWAMI, S. AND V. S. RAO. 1954*b*. Chemical components of *Usnea hirta* Hoffm. *Indian J. Pharm.* **16**, 151–152.

RANGASWAMI, S. AND V. S. RAO. 1954*c*. Chemical investigation of Indian lichens: Part XVIII — *Parmelia nimandairana* Zahlbr. *J. Sci. Ind. Res.* (India) **13B**, 403–405.

RANGASWAMI, S. AND V. S. RAO. 1955*a*. A note on the crystalline components of *Parmelia tinctorum* Despr. from Ceylon. *Indian J. Pharm.* **17**, 49–50.

RANGASWAMI, S. AND V. S. RAO. 1955*b*. Chemical components of some *Parmelia* lichens of India. *Indian J. Pharm.* **17**, 50–53.

RANGASWAMI, S. AND V. S. RAO. 1955*c*. Chemical components of *Usnea florida* Wiggs. *Indian J. Pharm.* **17**, 70.

RAO, D. N. AND F. LEBLANC. 1965. A possible role of atranorin in the lichen thallus. *Bryologist* **68**, 284–289.

RAO, P. S., K. G. SARMA, AND T. R. SESHADRI. 1965. Chemical components of the *Lobaria* lichens from the western Himalayas. *Current Sci.* (India) **34**, 9–11.

RAO, P. S., K. G. SARMA, AND T. R. SESHADRI. 1966. Chemical components of *Lobaria subisidiosa, L. retigera* and *L. subretigera* from the western Himalayas. *Current Sci.* (India) **35**, 147–148.

RAO, V. S. AND T. R. SESHADRI. 1940. Chemical investigation of Indian lichens. Part I. Chemical components of *Roccella montagnei*. *Proc. Indian Acad. Sci.* **12A**, 466–471.

RAO, V. S. AND T. R. SESHADRI. 1941. Chemical investigation of Indian lichens. Part III. The isolation of montagnetol, a new phenolic compound from *Roccella montagnei*. *Proc. Indian Acad. Sci.* **13A**, 199–202.

RAO, V. S. AND T. R. SESHADRI. 1942*a*. Chemical examination of Indian lichens. Part V. Occurrence of active montagnetol in *Roccella montagnei*. *Proc. Indian Acad. Sci.* **15A**, 429–431.

RAO, V. S. AND T. R. SESHADRI. 1942*b*. Chemical examination of Indian lichens. Part VI. Constitution of erythrin. *Proc. Indian Acad. Sci.* **16A**, 23–28.

RENNERT, ALDONA AND M. GUBANSKI. 1960. Ribonuclease in *Cetraria islandica*. *Naturwissenschaften* **47**, 18–19.

RIBEIRO, O. AND W. B. MORS. 1949. Ácido quiodectônico contribuição para o estudo de sua estrutura. *Bol. Inst. Quím. Agr.* (Rio de Janeiro) **15**, 1–14.

RIBEIRO, O. AND W. B. MORS. 1950. Caracterização do ácido quiodectônico. *Anais Assoc. Quím. Brasil* **9**, 182–189.

ROBERTSON, A. AND R. J. STEPHENSON. 1932. Lichen acids. Part III. The constitution of barbatic acid and the synthesis of *iso*rhizonic acid and methyl barbatate. *J. Chem. Soc.* **1932**, 1675–1681.

ROBINSON, H. 1959. Lichen succession in abandoned fields in the piedmont of North Carolina. *Bryologist* **62**, 254–259.

RUNEMARK, H. 1956. Studies in *Rhizocarpon* I. Taxonomy of the yellow species in Europe. *Opera Botanica* (Lund) **2** (1), 1–152.

RYABININ, A. A. AND L. G. MATYUKHINA. 1957. Study of the structure of the triterpene alcohol zeorin. *J. Gen. Chem. USSR* (Engl. Transl.) **27**, 311–315; *Zh. Obshch. Khim.* **27**, 277–281.

RYAN, H. AND W. M. O'RIORDAN. 1917. On the tinctorial constituents of some lichens which are used as dyes in Ireland. *Proc. Roy. Irish Acad.* **33B**, 91–104.

SAKURAI, Y. 1941. Constitution of erythrin. *Kakugaku Zasshi* **61**, 108–115; German summary, 45–46. Reference from *Chem. Abstr.* **36**, 1599[5] (1942).

SALKOWSKI, H. 1910. Zirkularpolarisation von Usninsäure und anderen Flechtenstoffen. III. *Ann. Chem.* **377**, 123–126.

SALO, MAIJA-LIISA. 1957. Lignin studies. II. The lignin content and properties of lignin in different materials. *J. Sci. Agr. Soc. Finland* **29**, 202–210.

SANTESSON, C. G. 1939. Einiges über die giftige Fuchs- oder Wolfsflechte (*Letharia vulpina* (L.) Vain.). *Acta Soc. Med. Upsalien.* **45**, 1–8.

SANTESSON, J. 1965. Studies on the chemistry of lichens. 24. Thin layer chromatography of aldehydic aromatic lichen substances. *Acta Chem. Scand.* **19**, 2254–2256.

SANTESSON, J. 1967. Chemical studies on lichens — III. The pigments of *Thelocarpon epibolum*, *T. laureri* and *Ahlesia lichenicola*. *Phytochemistry* **6**, 685–686.

SASAKI, I. 1942. Distinction between *Physica picta* and *Physica aegialita*. *J. Japan. Botany* **18**, 626–632.

SASTRY, V. V. K. AND V. S. RAO. 1941. A note on the production of colouring matters from certain Indian lichens. *Current Sci.* (India) **10**, 437–438.

SASTRY, V. V. K. AND T. R. SESHADRI. 1940. Chemical investigation of some Indian lichens. Part II. Synthetic uses of some lichen acids. *Proc. Indian Acad. Sci.* **12A**, 498–506.

SASTRY, V. V. K. AND T. R. SESHADRI. 1942. Chemical examination of Indian lichens. Part VII. Chemical components of *Parmelia abessinica* (Rathipuvvu). *Proc. Indian Acad. Sci.* **16A**, 137–140.

SATÔ, M. M. 1934. Studies on the lichens of Japan (I). *J. Japan. Botany* **10**, 17–22.

SATÔ, M. M. 1935. Studies on the lichens of Japan (V). *J. Japan. Botany* **11**, 238–244.

SATÔ, M. M. 1937. Enumeratio lichenum ins. Formosae (III). *J. Japan. Botany* **13**, 595–599.

SATÔ, M. M. 1940. East Asiatic lichens (IV). *J. Japan. Botany* **16**, 495–500.

SATÔ, M. M. 1954. Enumeration of lichens collected in Tohoku-district, Japan. (1) Anziaceae and Baeomycetaceae. *Yamagata Daigaku Kiyo, Shizen Kagaku* **3**, 113–126.

SATÔ, M. M. 1957. Range of the Japanese lichens (II). *Ibaraki Daigaku Bunrigakubu Kiyo (Shizen Kagaku)* **7**, 57–69.

SATÔ, M. M. 1959. Mixture ratio of the lichen genus *Thamnolia* collected in Japan and the adjacent regions. *Misc. Bryol. Lichenol.* (Japan) **22**, 1–6.

SATÔ, M. M. 1963. Mixture ratio of the lichen genus *Thamnolia*. *Nova Hedwigia* **5**, 149–155.

SATÔ, M. M. 1965. The mixture ratio of the lichen genus *Thamnolia* in New Zealand. *Bryologist* **68**, 320–324.

SAVICZ, V. P., V. F. KUPREVICZ, M. A. LITVINOV, E. N. MOISSEJEVA, AND K. A. RASSADINA. 1956. De antibiotico novo e lichenibus — natrium usninicum. *Tr. Botan. Inst., Akad. Nauk. SSSR*, Ser. 2, **11**, 5–37.

SAVICZ, V. P., M. A. LITVINOV, AND E. N. MOISSEJEVA. 1960. Ein Antibiotikum aus Flechten als Arzneimittel. *Planta Med.* **8**, 191–202.

SCHAFFNER, K., L. CAGLIOTI, D. ARIGONI, AND O. JEGER. 1958. Zur Kenntnis der Triterpene. 195. Mitteilung. Hydroxyhopanon, ein neuartiges pentacyclisches Triterpen. *Helv. Chim. Acta* **41**, 152–159.

SCHATZ, A. 1963. Soil microorganisms and soil chelation. The pedogenic action of lichens and lichen acids. *J. Agr. Food. Chem.* **11**, 112–118.

SCHINDLER, H. 1936. Über das Vorkommen der Norstictinsäure in der Lungenflechte *Lobaria pulmonaria* (L.) Hoffm. *Ber. Deut. Botan. Ges.* **54**, 240–246.

SCHINDLER, H. 1956/1957. Die Verwendung von Flechten und Flechtenstoffen in alter und neuer Zeit. *Botan. Beitr.* **2** (4/5), 1–15.

SCHINDLER, H. 1957. Die Inhaltsstoffe verschiedener *Usnea*-Arten unter besonderer Berücksichtigung des Usninsäure. *Arzneimittel-Forsch.* **7**, 69–72.

SCHÖPF, C. AND K. HEUCK. 1927. Die Konstitution der Usninsäure. *Ann. Chem.* **459**, 233–286.

SCHÖPF, C., K. HEUCK, AND R. DUNTZE. 1931. Die Konstitution der Barbatolsäure. *Ann. Chem.* **491**, 220–251.

SCHRATZ, E. AND H. J. VETHACKE. 1958. Papierchromatographische Wertbestimmung von *Radix rhei*. *Planta Med.* **6**, 44–69.

SCHULTE, F. 1905. Zur Anatomie der Flechtengattung *Usnea*. *Beih. Botan. Zentralblatt* **18**, 1–22.

SCHUMACKER, R. 1961. Étude par chromatographie sur papier des acides lichéniques de *Everniopsis pseudoreticulata* (Duvign.) Dodge. *Bull. Soc. Roy. Sci. Liège* **30**, 452–457.

SCOTT, G. D. 1956. Further investigation of some lichens for fixation of nitrogen. *New Phytol.* **55**, 111–116.

SCOTTER, G. W. 1965. Chemical composition of forage lichens from northern Saskatchewan as related to use by barren-ground caribou. *Can. J. Plant Sci.* **45**, 246–250.

SENFT, E. 1916. Beitrag zur Anatomie und zum Chemismus der Flechte *Chrysothrix nolitangere* Mont. *Ber. Deut. Botan. Ges.* **34**, 592–600.

SESHADRI, T. R. 1944. A theory of biogenesis of lichen depsides and depsidones. *Proc. Indian Acad. Sci.* **20A**, 1–4.

SESHADRI, T. R. 1953. Chemical investigation of Indian lichens. *Indian J. Pharm.* **15**, 286–287.

SESHADRI, T. R. AND G. B. V. SUBRAMANIAN. 1963. Synthesis of homodiploschistesic acid methyl ester. *J. Indian Chem. Soc.* **40**, 7–8.

SESHADRI, T. R. AND S. S. SUBRAMANIAN. 1949a. A lichen (*Parmelia tinctorum*) on a Java monument. *J. Sci. Ind. Res.* (India) **8B**, 170–171.

SESHADRI, T. R. AND S. S. SUBRAMANIAN. 1949b. Chemical investigation of Indian lichens. Part VIII. Some lichens growing on sandal trees (*Ramalina tayloriana* and *Roccella montagnei*). *Proc. Indian Acad. Sci.* **30A**, 15–22.

SESHADRI, T. R. AND S. S. SUBRAMANIAN. 1949c. Chemical investigations of Indian lichens. Part IX. Some lichens on sandal trees — *Parmelia tinctorum* and *Usnea japonica*. *Proc. Ind·an Acad. Sci.* **30A**, 62–66.

SESHADRI, T. R. AND S. S. SUBRAMANIAN. 1949d. Chemical investigation of Indian lichens. Part X. Chemical components of *Teloschistes flavicans*. *Proc. Indian Acad. Sci.* **30A**, 67–73.

SHAH, L. G. 1954. Chemical investigation of the lichens: *Parmelia kamtschadalis* and *Parmelia arnoldii*. *J. Indian Chem. Soc.* **31**, 253–256.

SHIBATA, S. 1944. Über Didymsäure, einen neuen Typus der Flechtenstoffe. *Acta Phytochim.* **14**, 9–38.

SHIBATA, S. 1958. Especial compounds of lichens. Pp. 560–623 *in* W. RUHLAND [ed.], Handbuch der Pflanzenphysiologie, Vol. X. Springer-Verlag, Berlin.

SHIBATA, S. 1963. Lichen substances. Pp. 155–193 *in* H. F. LINSKENS AND M. V. TRACEY [eds.], Modern Methods of Plant Analysis, Vol. VI. Springer-Verlag, Berlin.

SHIBATA, S. 1965. Biogenetical and chemotaxonomical aspects of lichen substances. Pp. 451–465 *in* Beitrage zur Biochemie und Physiologie von Naturstoffen, Festschrift Kurt Mothes zum 65. Geburtstag. Gustav Fischer Verlag, Jena.

SHIBATA, S. AND HSÜCH-CHING CHIANG. 1963. Grayanic acid, a new lichen depsidone. *Chem. Pharm. Bull.* (Tokyo) **11**, 926–930.

SHIBATA, S. AND HSÜCH-CHING CHIANG. 1965. The structures of cryptochlorophaeic acid and merochlorophaeic acid. *Phytochemistry* **4**, 133–139.

SHIBATA, S., T. FURUYA, AND H. IIZUKA. 1965. Gas-liquid chromatography of lichen substances. I. Studies on zeorin. *Chem. Pharm. Bull.* (Tokyo) **13**, 1254–1257.

SIINTOLA, SIPI, P. HEILALA, J. PÄTIÄLÄ, AND RISTO PÄTIÄLÄ. 1948. Usniinihaposta, acidum usnicum, ja sen mahdollisuuksista lääkkeelliseen käyttöön. *Suomen Kemistilehti* **21A**, 179–183.

SJÖSTRÖM, A. G. M. AND L.-E. ERICSON. 1953. The occurrence in lichens of the folic acid-, folinic acid-, and vitamin B_{12}-group of factors. *Acta Chem. Scand.* **7**, 870–872.

SMITH, D. C. 1961. The physiology of *Peltigera polydactyla* (Neck.) Hoffm. *Lichenologist* **1**, 209–226.

SOLBERG, Y. J. 1955. Studies on the chemistry of lichens I. D-Arabitol from *Alectoria jubata* Ach., var. *chalybeiformis* Th. Fr. *Acta Chem. Scand.* **9**, 1234–1235.

SOLBERG, Y. J. 1956. Dyeing of wool with lichens and lichen substances. *Acta Chem. Scand.* **10**, 1116–1123.

SOLBERG, Y. J. 1957. Studies on the chemistry of lichens II. Chemical components of *Haematomma ventosum* (L.) var. *lapponicum* (Räs.). *Acta Chem. Scand.* **11**, 1477–1484.

SOLBERG, Y. J. 1960. Studies on the chemistry of lichens III. Long-chain tetrahydroxy fatty acids from some Norwegian lichens. *Acta Chem. Scand.* **14**, 2152–2160.

SPILLANE, P. A., J. KEANE, AND T. J. NOLAN. 1936. The chemical constituents of lichens found in Ireland. *Buellia canescens* — Part 2. *Sci. Proc. Roy. Dublin Soc.* **21**, 333–343.

STARK, J. B., E. D. WALTER, AND H. S. OWENS. 1950. Method of isolation of usnic acid from *Ramalina reticulata*. *J. Am. Chem. Soc.* **72**, 1819–1820.

STEINER, M. 1955. Ein stabiles Diaminreagens für lichenologische Zwecke. *Ber. Deut. Botan. Ges.* **68**, 35–40.

STEINER, M. 1957. Die Fette der Pilze. Pp. 59–89 *in* W. RUHLAND [ed.], Handbuch der Pflanzenphysiologie, Vol. VII. Springer-Verlag, Berlin.

STEINER, M. 1959. *Maronella laricina* n. gen., n. spec. (Acarosporaceae), eine neue Flechte aus Tirol. *Österr. Botan. Z.* **106**, 440–455.

STEINER, M. 1960. Zur Unterscheidung von *Thamnolia vermicularis* und *subvermicularis* sensu Asahina. *Österr. Botan. Z.* **107**, 113–114.

STEIN VON KAMIENSKI, E. 1958. Untersuchungen über die flüchtigen Amine der Pflanzen. III. Mitteilung. Die Amine von Pilzen. *Planta* **50**, 331–352.

STICHER, O. 1965*a*. Über die antibakterielle Wirksamkeit von *Lichen islandicus* mit besonderer Berücksichtigung der Inhaltsstoffe. 1. Mitteilung. *Pharm. Acta Helv.* **40**, 385–394.

STICHER, O. 1965*b*. Über die antibakterielle Wirksamkeit von *Lichen islandicus* mit besonderer Berücksichtigung der Inhaltsstoffe. 2. Mitteilung. *Pharm. Acta Helv.* **40**, 483–495.

STOLL, A., A. BRACK, AND J. RENZ. 1947. Die antibakterielle Wirkung der Usninsäure auf Mykobakterien und andere Mikroorganismen. (Fünfte Mitteilung über antibakterielle Stoffe.). *Experimentia* **3**, 115–116.

STOLL, A., A. BRACK, AND J. RENZ. 1950. Die Wirkung von Flechtenstoffen auf Tuberkelbakterien und auf einige andere Mikroorganismen. (7. Mitteilung über antibakterielle Stoffe.). *Schweiz. Z. Allgem. Pathol. Bakteriol.* (Basel) 13, 729–751.

STOLL, A., J. RENZ, AND A. BRACK. 1947. Antibiotika aus Flechten. (Vierte Mitteilung über antibakterielle Stoffe.). *Experientia* 3, 111–113.

ST. PFAU, A. 1926. Zur Kenntnis der Flechtenbestandteile I. Die Konstitution des Atranorins. *Helv. Chim. Acta* **9**, 650–669.

ST. PFAU, A. 1928. Zur Kenntnis der Flechtenbestandteile II. Die Konstitution der Barbatinsäure. *Helv. Chim. Acta* **11**, 864–876.

ST. PFAU, A. 1934. Zur Kenntnis der Flechtenbestandteile IV. Über Chlor-atranorin. *Helv. Chim. Acta.* **17**, 1319–1328.

SUBRAMANIAN, S. S. 1955. A review of anthraquinone compounds in lichens. *J. Am. Pharm. Assoc., Sci. Ed.* **44**, 769–776.

SUBRAMANIAN, S. S. 1965. A note on the biogenesis of chemical components of *Roccella montagnei*. *Current Sci.* (India) **34**, 375.

SUBRAMANIAN, S. S. AND S. RAMAKRISHNAN. 1964. Amino-acids of *Peltigera canina*. *Current Sci.* (India) **33**, 522.

SUBRAMANIAN, S. S. AND M. N. SWAMY. 1961. A note on *Roccella* found in Pondicherry. *J. Sci. Ind. Res.* (India) **20C**, 275–276.

SUOMINEN, E. E. 1939. Die Konstitution der Barbatolsäure. *Suomen Kemistilehti* **12B**, 26–28. Reference from *Chem. Zentr.* **111**, 385–386 (1940).

TAGUCHI, H., U. SANKAWA, AND S. SHIBATA. 1966. Biosynthesis of usnic acid in lichens. *Tetrahedron Letters* **42**, 5211–5214.

594   *Chemical and Botanical Guide to Lichen Products*

TARGÉ, A. 1964–1965. Étude systématique et chimique d'une collection de *Parmelia katangais*. *Rev. Bryol. Lichénol.* **33**, 565–586.

TARGÉ, A. AND J. LAMBINON. 1965. Étude chimiotaxonomique du groupe de *Parmelia borreri* (Sm.) Turn. en Europe occidentale. *Bull. Soc. Roy. Botan. Belg.* **98**, 295–306.

TAVARES, C. N. 1954. Química e taxonomia nos líquenes. *Gaz. Fís.* (Lisbon) **3**, 17–20.

THIES, W. 1932. Systematische Verbreitung und Vorkommen der Flechtenstoffe (Flechtensäuren). Pp. 429–452 *in* G. KLEIN [ed.], Handbuch der Pflanzenanalyse. 3. Band, Spezielle Analyse. 2. Teil, Organische Stoffe II. Julius Springer, Wein.

THOMAS, E. A. 1936. Die Spezifizität des Parietins als Flechtenstoff. *Ber. Schweiz. Botan. Ges.* **45**, 191–197.

THOMAS, E. A. 1939. Über die Biologie von Flechtenbildnern. *Beitr. Kryptogamenflora Schweiz* **9** (1), 1–208.

THOMSON, J. W. 1963. I. Modern species concepts: Lichens. *Bryologist* **66**, 94–100.

TICHÝ, V. 1955. Podíl kyseliny usninové na fungistatických vlastnostech lišejníku *Usnea hirta* Hoffm. *Spisy Vydávané Přírodovědeckou Fak. Masary. Univ.* **364**, 185–196.

TICHÝ, V. AND V. RYPÁČEK. 1952. O fungistatickém účinku lišejníkových kyselin druhu *Parmelia physodes* (L.) Ach. [The fungistatic effect of the lichen acids of the species *P. p.*] *Spisy Vydávané Přírodovědeckou Fak. Masary. Univ.* **335**, 83–95.

TOMASELLI, R. 1957. Nuovo derivato dell'acido usnico usato in terapia. *Farmaco* (Pavia), *Ed. Prat.* **12**, 137–140.

TOMASELLI, R. 1958. La produzione di antrachinoni nella *Xanthoria parietina* in rapporto alle condizioni ecologiche. *Atti Soc. Ital. Sci. Nat. Mus. Civico Storia Nat. Milano* **97**, 357–361.

TOMASELLI, R. 1963. Modalità di crescita di vari ceppi italiani di *Xanthoriomyces* (fungo lichenizzante). II. Crescita su terreni completi. *Arch. Botan. Biogeogr. Ital.* **39** [=4a ser. 8 (1–2)], 1–20.

TROSHCHENKO, A. T. 1957. Investigation of the lichen *Parmelia vagans* Nyl. *Tr. Leningr. Tekhnol. Inst. im. Lensoveta* **42**, 39–42. Reference from *Chem. Abstr.* **53**, 13231 (1959).

VARTIA, K. O. 1949. Antibiotics in lichen I. *Ann. Med. Exptl. Biol. Fenniae* (Helsinki) **27**, 46–54.

VARTIA, K. O. 1950a. Antibiotics in lichens II. *Ann. Med. Exptl. Biol. Fenniae* (Helsinki) **28**, 7–19.

VARTIA, K. O. 1950b. On antibiotic effects of lichens and lichen substances. *Ann. Med. Exptl. Biol. Fenniae* (Helsinki) **28** (Suppl. 7), 1–82.

VENKATESWARLU, V. 1947. A note on the nuclear methylation of lecanoric acid. *Proc. Indian Acad. Sci.* **25A**, 331–332.

VENKATESWARLU, V. AND V. V. RAO. 1962. Chemical examination of lichens of the Araku Valley. *Current Sci.* (India) **31**, 192.

VERSEGHY, K. 1959. Studien über die Gattung *Ochrolechia*. III.

Angaben zur Chemie der *Ochrolechia*-Arten. *Ann. Hist.-Nat. Mus. Natl. Hung.* **51**, 145–159.

VERSEGHY, K. 1962. Die Gattung *Ochrolechia*. *Beih. Nova Hedwigia* **1**, 1–146.

VĚZDA, A. 1965. Lichenes Selecti Exsiccati. Fascicle XIV. (No. 326–350). Instituto Botanico Academiae Scientiarum Čechoslovacae, Průhonice prope Pragam.

VIRTANEN, O. E. AND N. KÄRKI. 1956. On the toxicity of an usnic acid preparation with the trade name USNO. *Suomen Kemistilehti* **29B**, 225–226.

VIRTANEN, O. E., H. VIITANEN, AND A. E. KORTEKANGAS. 1954. The antibiotic activity of some amino compound derivatives of L-usnic acid. I. *Suomen Kemistilehti* **27B**, 18–20.

VOTOČEK, E. AND J. BURDA. 1926. Sur les composants sucrés de quelques lichens (1). *Bull. Soc. Chim. France* **39**, 248–254.

WACHTMEISTER, C. A. 1952. Studies on the chemistry of lichens. I. Separation of depside components by paper chromatography. *Acta Chem. Scand.* **6**, 818–825.

WACHTMEISTER, C. A. 1954. Studies on the chemistry of lichens. VII. Structure of porphyrilic acid. *Acta Chem. Scand.* **8**, 1433–1441.

WACHTMEISTER, C. A. 1955. Studies on the chemistry of lichens. IX. On the identity of ocellatic acid and thamnolic acid. *Acta Chem. Scand.* **9**, 1395–1396.

WACHTMEISTER, C. A. 1956a. Identification of lichen acids by paper chromatography. *Botan. Notiser* **109**, 313–324.

WACHTMEISTER, C. A. 1956b. Studies on the chemistry of lichens. X. The structure of porphyrilic acid. *Acta Chem. Scand.* **10**, 1404–1413.

WACHTMEISTER, C. A. 1958a. Studies on the chemistry of lichens. *Svensk. Kem. Tidskr.* **70**, 117–133.

WACHTMEISTER, C. A. 1958b. Studies on the chemistry of lichens. XI. Structure of picrolichenic acid. *Acta Chem. Scand.* **12**, 147–164.

WAGNER, H. AND H. FRIEDRICH. 1965. Über die ungesättigten Fettsäuren von Moosen, Bärlappgewächsen und Flechten. *Naturwissenschaften* **52**, 305.

WALKER, ALMA T. 1965. Lichen acids of *Cornicularia*. *ASB Bull.* **12**, 53.

WALL, M. E. AND K. H. DAVIS (unpublished communication), Research Triangle Institute, Research Triangle Park, North Carolina, U.S.A.

WEBER, W. A. 1964. A bizarre new species of *Rinodina* (lichenized fungi) from Guadalupe Island, Mexico. *Bryologist* **67**, 473–476.

WEBER, W. A. 1965. *Hubbsia*, a new genus of Roccellaceae (lichenized fungi) from Mexico. *Svensk Botan. Tidskr.* **59**, 59–64.

WEHMER, C., W. THIES, AND M. HADDERS. 1932. Systematische Verbreitung und Vorkommen der Zuckeralkohole. Pp. 771–774 *in* G. KLEIN [ed.], Handbuch der Pflanzenanalyse, II. Band, Spezielle Analyse, Erster Teil. Julius Springer, Wien.

WETMORE, C. M. 1960. The lichen genus *Nephroma* in North and Middle America. *Publ. Mus., Michigan State Univ., Biol. Ser.* **1**, 369–452.

WILHELMSEN, J. B. 1959. Chlorophylls in the lichens *Peltigera, Parmelia*, and *Xanthoria. Botan. Tidsskr.* **55**, 30–36.

WIRTH, M. AND M. E. HALE, JR. 1963. The lichen family *Graphidaceae* in Mexico. *Contrib. U.S. Natl. Herb.* **36**, 63–119.

YAMAZAKI, M., M. MATSUO, AND S. SHIBATA. 1965. Biosynthesis of lichen depsides, lecanoric acid and atranorin. *Chem. Pharm. Bull.* (Tokyo) **13**, 1015–1017.

YAMAZAKI, M. AND S. SHIBATA. 1966. Biosynthesis of lichen substances. II. Participation of C_1-unit to the formation of β-orcinol type lichen depside. *Chem. Pharm. Bull.* (Tokyo) **14**, 96–97.

YANOVSKY, E. AND R. M. KINGSBURY. 1938. Analyses of some Indian food plants. *J. Assoc. Offic. Agr. Chemists* **21**, 648–665.

YOSHIDA, I. 1951. [Chlorine-containing metabolic products.] *Kagaku no Ryoiki* **5**, 406–409, 419.

YOSIOKA, I., A. MATSUDA, AND I. KITAGAWA. 1966. Pyxinic acid, a novel lichen triterpene with 3-β-hydroxyl function. *Tetrahedron Letters* **1966**, 613–616.

YOSIOKA, I. AND T. NAKANISHI. 1963. Structure of leucotylin. *Chem. Pharm. Bull.* (Tokyo) **11**, 1468–1470.

YOSIOKA, I., T. NAKANISHI, AND E. TSUDA. 1966. The structure of leucotylic acid, a new triterpenic acid from a lichen. *Tetrahedron Letters* **1966**, 607–612.

YOSIOKA, I., M. YAMAKI, AND I. KITAGAWA. 1966. On the triterpenic constituents of a lichen, *Parmelia entotheiochroa* Hue; zeorin, leucotylin, leucotylic acid, and five new related triterpenoids. *Chem. Pharm. Bull.* (Tokyo) **14**, 804–807.

ZAHLBRUCKNER, A. 1904. Schedae ad "Kryptogamas exsiccatas." Cent. X–XI. *Ann. K. K. Naturhist. Hofmus.* 379–427.

ZATULOVSKIĬ, B. G. 1956. [Antibiotic properties of *Parmelia*.] *Priroda* (Zagreb) **45**, 100–101.

ZATULOVSKIĬ, B. G. 1957. [The antibiotic activities of *Parmelia* and the materials isolated therefrom.] *Mikrobiol. Zh. Akad. Nauk Ukr. RSR* **19**, 34–37.

ZELLNER, J. 1932. Zur Chemie der Flechten (I. Mitteilung). Über *Peltigera canina* L. *Monatsh. Chem.* **59**, 300–304.

ZELLNER, J. 1934. Zur chemie der Flechten III. *Parmelia physodes* L. *Monatsh. Chem.* **64**, 6–11.

ZELLNER, J. 1935. Zur Chemie der Flechten IV. *Gyrophora dillenii* (Tuck.) Müll. Arg. und *Parmelia furfuracea* L. *Monatsh. Chem.* **66**, 81–86.

ZOPF, W. 1900. Zur Kenntniss der Flechtenstoffe. (Siebente Mitteilung.). *Ann. Chem.* **313**, 317–344.

ZOPF, W. 1907. Die Flechtenstoffe in chemischer, botanischer, pharmakologischer und technischer Beziehung. Gustav Fischer, Jena.

ZOPF, W. 1908. Beiträge zu einer chemischen Monographie der Cladoniaceen. *Ber. Deut. Botan. Ges.* **26**, 51–113.

ZOPF, W. 1909. Zur Kenntniss der Flechtenstoffe. (Siebzehnte Mittheilung.). Ueber die in den Lappenflechten (Peltigeraceen) vorkommenden Stoffe. *Ann. Chem.* **364**, 273–313.

Abbreviations of the Titles of the Periodicals Cited

Abhandl. Ber. Naturkundemus. Görlitz
  Abhandlungen und Berichte des Naturkundemuseums Görlitz
Accounts Chem. Res.
  Accounts of Chemical Research
Acta Botan. Fennica
  Acta Botanica Fennica
Acta Chem. Scand.
  Acta Chemica Scandinavica
Acta Fac. Rer. Nat. Univ. Comen.
  Acta Facultatis Rerum Naturalium Universitatis Comenianae
Acta Pharmacol. Toxicol.
  Acta Pharmacologica et Toxicologica
Acta Phytochim.
  Acta Phytochimica
Acta Phytotaxon. Geobotan.
  Acta Phytotaxonomica et Geobotanica
Acta Polon. Pharm.
  Acta Poloniae Pharmaceutica
Acta Soc. Med. Upsalien.
  Acta Societatis Medicorum Upsaliensis
Am. J. Botany
  American Journal of Botany
Am. Midland Naturalist
  American Midland Naturalist
Am. Scientist
  American Scientist
Anais Assoc. Quím. Brasil
  Anais da Associação Química do Brasil
Anal. Chem.
  Analytical Chemistry
Anal. Chim. Acta
  Analytica Chimica Acta

Analyst
  Analyst (Des Moines)
Ann. Botan. Fennici
  Annales Botanici Fennici
Ann. Chem.
  Justus Liebigs Annalen der Chemie and also the older *Annalen der Pharmacie* and *Annalen der Chemie und Pharmacie*
Ann. Hist.-Nat. Mus. Natl. Hung.
  Annales Historico-Naturales Musei Nationalis Hungarici
Ann. K. K. Naturhist. Hofmus.
  Annalen des K. K. Naturhistorischen Hofmuseums
Ann. Mag. Nat. Hist.
  Annals and Magazine of Natural History
Ann. Med. Exptl. Biol. Fenniae (Helsinki)
  Annales Medicinae Experimentalis et Biologiae Fenniae (Helsinki)
Antibiot. Chemother.
  Antibiotics and Chemotherapy
Apotheker-Ztg.
  Apotheker-Zeitung
Arch. Biochem.
  Archives of Biochemistry
Arch. Botan. Biogeogr. Ital.
  Archivio Botanico e Biogeografico Italiano
Arch. Dermatol.
  Archives of Dermatology
Arch. Exptl. Pathol. Pharmakol.
  Archiv für Experimentelle Pathologie und Pharmakologie
Arch. Pharm.
  Archiv der Pharmazie und Berichte der Deutschen Pharmazeutischen Gesellschaft
Arkiv Botan.
  Arkiv för Botanik
Arkiv Kemi
  Arkiv för Kemi
Arzneimittel-Forsch.
  Arzneimittel-Forschung
ASB Bull.
  A[ssociation of] *S*[outheastern] *B*[iologists] *Bulletin*
Atti Soc. Ital. Sci. Nat. Museo Civico Storia Nat. (Milano)
  Atti della Società Italiana di Scienze Naturali, e del Museo Civico di Storia Naturale (Milano)
Atti Ist. Botan. Lab. Crittogam. Univ. Pavia
  Atti dell'Istituto Botanico e Laboratorio Crittogamica dell'Università di Pavia
Australian J. Chem.
  Australian Journal of Chemistry

Australian J. Exptl. Biol. Med. Sci.
Australian Journal of Experimental Biology and Medical Science

Bacteriol. Proc.
Bacteriological Proceedings

Beih. Botan. Zentralblatt
Beihefte zum Botanischen Zentralblatt.

Beih. Nova Hedwigia
Beihefte zur Nova Hedwigia

Beitr. Kryptogamenflora Schweiz
Beiträge zur Kryptogamenflora der Schweiz

Ber. Deut. Botan. Ges.
Berichte der Deutschen Botanischen Gesellschaft

Ber. Schweiz. Botan. Ges.
Berichte der Schweizerischen Botanischen Gesellschaft

Biochem. Biophys. Res. Commun.
Biochemical and Biophysical Research Communications

Biochemistry
Biochemistry

Biochem. J.
Biochemical Journal

Biochem. Z.
Biochemische Zeitschrift

Biol. Jaarboek Konink. Natuurw. Genoot. Dodonaea Gent
*Biologisch Jaarboek, Uitgegeven door het Koninklijk Natuurweten-
schappelijk Genootschap Dodonaea te Gent*

Blyttia
Blyttia

Bol. Inst. Quím. Agr. (Rio de Janeiro)
Boletim do Instituto de Química Agrícola (Rio de Janeiro)

Boll. Ist. Botan. Univ. Catania
Bollettino dell'Istituto di Botanica dell'Università di Catania

Boll. Sedute Accad. Gioenia Sci. Nat. Catania
*Bollettino delle Sedute dell'Accademia Gioenia di Scienze Naturali
in Catania*

Bol. Univ. Chile
Boletín de la Universidad de Chile

Botan. Gaz.
Botanical Gazette

Botan. Mag. (Tokyo)
Botanical Magazine (Tokyo)

Botan. Notiser
Botaniska Notiser

Botan. Rev.
Botanical Review

Botan. Tidsskr.
Botanisk Tidsskrift
Botan. Zh. (Leningrad)
Botanicheskiĭ Zhurnal, Akademiya Nauk SSSR (Leningrad)
Botan. Ztg.
Botanische Zeitung
Brittonia
Brittonia
Bryologist
The Bryologist
Bull. Centre Études Rech. Sci. Biarritz
Bulletin du Centre d'Études et de Recherches Scientifiques, Biarritz
Bull. Chem. Soc. Japan
Bulletin of the Chemical Society of Japan
Bull. École Natl. Supér. Agronom. Nancy
Bulletin de l'École Nationale Supérieure Agronomique de Nancy
Bull. Jardin Botan. État Bruxelles
Bulletin du Jardin Botanique de l'État à Bruxelles
Bull. Natl. Inst. Sci. India
Bulletin of the National Institute of Sciences of India
Bull. Natl. Sci. Mus. (Tokyo)
Bulletin of the National Science Museum (Tokyo)
Bull. Soc. Chim. Belges
Bulletin des Sociétés Chimiques Belges
Bull. Soc. Chim. France
Bulletin de la Société Chimique de France
Bull. Soc. Roy. Botan. Belg.
Bulletin de la Société Royale de Botanique de Belgique
Bull. Soc. Roy. Sci. Liège
Bulletin de la Société Royale des Sciences de Liège
Bull. Soc. Sci. Bretagne
Bulletin de la Société Scientifique de Bretagne
Bull. Torrey Botan. Club
Bulletin of the Torrey Botanical Club

Can. J. Botany
Canadian Journal of Botany
Can. J. Chem.
Canadian Journal of Chemistry
Can. J. Plant Sci.
Canadian Journal of Plant Science
Castanea
Castanea
Chem. Abstr.
Chemical Abstracts

Chem. Ber.
  Chemische Berichte and also the older *Berichte der Deutschen Chemischen Gesellschaft*
Chem. Commun.
  Chemical Communications
Chem. Ind. (London)
  Chemistry and Industry (London)
Chem. Pharm. Bull. (Tokyo)
  Chemical and Pharmaceutical Bulletin (Tokyo)
Chem. Tech. (Berlin)
  Chemische Technik (Berlin)
Chem. Zentr.
  Chemische Zentralblatt
Compt. Rend.
  Comptes Rendus Hebdomadaires des Séances de l'Académie des Sciences (Paris)
Contrib. U.S. Natl. Herb.
  Contributions from the United States National Herbarium
Current Sci. (India)
  Current Science (India)

Dissertationes Pharm.
  Dissertationes Pharmaceuticae
Dokl., Biochem. Sect., Proc. Acad. Sci. USSR (Engl. Transl.)
  Doklady, Biochemistry Section, Proceedings of the Academy of Sciences of the USSR (English Translation)

Eesti NSV Teaduste Akad. Toimetised
  Eesti NSV Teaduste Akadeemia Toimetised
Experientia
  Experientia

Farmaco (Pavia), Ed. Prat.
  Il Farmaco (Pavia), *Edizione Pratica*
Fortschr. Chem. Org. Naturstoffe
  Fortschritte der Chemie Organischer Naturstoffe
Fuel
  Fuel

Gaz. Fís. (Lisbon)
  Gazeta de Física (Lisbon)
Gazz. Chim. Ital.
  Gazzetta Chimica Italiana
Giorn. Microbiol.
  Giornale Microbiologia

Helv. Chim. Acta
  Helvetica Chimica Acta

Ibaraki Daigaku Bunrigakubu Kiyo (Shizen Kagaku)
Ibaraki Daigaku Bunrigakubu Kiyo (Shizen Kagaku) = Bulletin of the Faculty of Liberal Arts, Ibaraki University (Natural Science)

Indian J. Chem.
Indian Journal of Chemistry

Indian J. Pharm.
Indian Journal of Pharmacy

Irish Naturalists' J.
Irish Naturalists' Journal

J. Agr. Food Chem.
Journal of Agricultural and Food Chemistry

J. Am. Chem. Soc.
Journal of the American Chemical Society

J. Am. Pharm. Assoc., Sci. Ed.
Journal of the American Pharmaceutical Association, Scientific Edition. Now: *Journal of Pharmaceutical Sciences*

Japan. J. Med. Sci. Biol.
Japanese Journal of Medical Science and Biology and also the older *Japanese Medical Journal*

Japan. J. Med. Sci., IV. Pharmacol.
Japanese Journal of Medical Sciences, IV. Pharmacology

J. Assoc. Offic. Agr. Chemists
Journal of the Association of Official Agricultural Chemists

J. Bacteriol.
Journal of Bacteriology

J. Biol. Chem.
Journal of Biological Chemistry

J. Chromatog.
Journal of Chromatography

J. Chem. Soc.
Journal of the Chemical Society (London)

J. Elisha Mitchell Sci. Soc.
Journal of the Elisha Mitchell Scientific Society

J. Gen. Chem. USSR (Engl. Transl.)
Journal of General Chemistry of the USSR (English Translation)

J. Indian Chem. Soc.
Journal of the Indian Chemical Society

J. Invest. Dermat.
Journal of Investigative Dermatology

J. Japan. Botany
Journal of Japanese Botany

J. Org. Chem.
Journal of Organic Chemistry

J. Pharm. Sci.
Journal of Pharmaceutical Sciences

J. Prakt. Chem.
  Journal für Praktische Chemie
J. Sci. Agr. Soc. Finland
  Journal of the Scientific Agricultural Society of Finland = Maata-loustieteelinen Aikakauskirja (Helsinki)
J. Sci. Ind. Res. (India)
  Journal of Scientific and Industrial Research (India)

Kagaku no Jikken
  Kagaku no Jikken
Kagaku no Ryoiki
  Kagaku no Ryoiki = Journal of Japanese Chemistry

Lejeunia
  Lejeunia, Revue de Botanique
Lichenologist
  The Lichenologist
Lloydia
  Lloydia

Madroño
  Madroño
Medd. Grønland
  Meddelelser om Grønland
Med. Monatsschr. (Stuttgart)
  Medizinische Monatsschrift (Stuttgart)
Mém Inst. Sci. Madagascar, Sér. B
  Mémoires de l'Institut Scientifique de Madagascar, Série B
Mikrobiol. Zh. Akad. Nauk Ukr. RSR
  Mikrobiologichniï Zhurnal, Akademiya Nauk Ukraïns'koï RSR, Institut Mikrobiologii imeni D. K. Zabolotonogo
Mikrokosmos
  Mikrokosmos
Misc. Bryol. Lichenol. (Japan)
  Miscellanea Bryologica et Lichenologica (Japan)
Mitteil. Florist-soziolog. Arbeitsgemeinschaft
  Mitteilungen der Florist-soziologischen Arbeitsgemeinschaft
Monatsh. Chem.
  Monatshefte für Chemie und Verwandte Teile Anderer Wissenschaften
Mouvement Sci. Belg.
  Le Mouvement Scientifique en Belgique

Naturalistes Belges
  Les Naturalistes Belges
Nature
  Nature

Naturwissenschaften
  Die Naturwissenschaften
Neotoma Ecol. Bioclim. Lab., Ohio State Univ. Ohio Agr. Expt. Sta.
  Neotoma Ecological and Bioclimatic Laboratory, Ohio State University and Ohio Agricultural Experiment Station
New Phytol.
  The New Phytologist
Nippon Kagaku Soran
  Nippon Kagaku Soran = Complete Chemical Abstracts of Japan
Nova Hedwigia
  Nova Hedwigia
Nytt Mag. Naturvidenskapene
  Nytt Magazin for Naturvidenskapene

Opera Botan. (Lund)
  Opera Botanica a Societate Botanica Lundensi
Österr. Botan. Z.
  Österreichische Botanische Zeitschrift

Pharm. Acta Helv.
  Pharmaceutica Acta Helvetiae
Pharmazie
  Die Pharmazie
Pharm. Bull. (Tokyo)
  Pharmaceutical Bulletin (Tokyo). Now: *Chemical and Pharmaceutical Bulletin* (Tokyo)
Pharm. Ind.
  Die Pharmazeutische Industrie
Pharm. Ztg., Ver. Apotheker-Ztg.
  Pharmazeutische Zeitung, Vereinigt mit Apotheker-Zeitung
Physiol. Plant.
  Physiologia Plantarum
Phytochemistry
  Phytochemistry
Phyton (Buenos Aires)
  Phyton (Buenos Aires)
Planta
  Planta
Planta Med.
  Planta Medica, Zeitschrift für Arzneipflanzenforschung
Plant Cell Physiol. (Tokyo)
  Plant and Cell Physiology (Tokyo)
Priroda (Zagreb)
  Priroda (Zagreb)
Proc. Chem. Soc.
  Proceedings of the Chemical Society (London)

Proc. Indian Acad. Sci.
  Proceedings of the Indian Academy of Sciences
Proc. Natl. Inst. Sci. India
  Proceedings of the National Institute of Sciences of India
Proc. Roy. Irish Acad., Sect. B.
  Proceedings of the Royal Irish Academy, Section B: Biological, Geological, and Chemical Science
Proc. Soc. Exptl. Biol. Med.
  Proceedings of the Society for Experimental Biology and Medicine
Proc. Trans. Rhodesian Sci. Assoc.
  Proceedings and Transactions of the Rhodesian Scientific Association
Public Health Rept. (U.S.)
  Public Health Reports (United States)
Publ. Mus., Michigan State Univ., Biol. Ser.
  Publications of the Museum—Michigan State University, Biological Series

Quart. Rev. (London)
  Quarterly Reviews (London)

Rend. Accad. Sci. Fis. Mat.
  Rendiconti dell'Accademia delle Scienze Fisische e Matematiche
Res. Ind. (New Delhi)
  Research and Industry (New Delhi)
Rev. Brasil. Biol.
  Revista Brasileira de Biologia
Rev. Bryol. Lichénol.
  Revue Bryologique et Lichénologique
Rev. Real Acad. Cienc. Exact. Fís. Nat. Madrid
  Revista de la Real Academia de Ciencias Exactas, Físicas y Naturales de Madrid
Rhodora
  Rhodora

Schriftenreihe, Schwabe: "Aus unserer Arbeit," Botan. Beitr.
  Schriftenreihe, Schwabe: "Aus unserer Arbeit," Botanische Beiträge
Schweiz. Z. Allgem. Pathol. Bakteriol. (Basel)
  Schweizerische Zeitschrift für Allgemeine Pathologie und Bakteriologie (Basel)
Science
  Science
Sci. Proc. Roy. Dublin Soc.
  Scientific Proceedings of the Royal Dublin Society
Sci. Progr. (London)
  Science Progress (London)
Shigen Kagaku Kenkyusho Iho

Shigen Kagaku Kenkyusho Iho = Miscellaneous Reports of the Research Institute for Natural Resources (Tokyo)

Southwestern Naturalist
  The Southwestern Naturalist

Spisy Vydávané Přírodovědeckou Fak. Masary. Univ.
  Spisy Vydávané Přírodovědeckou Fakultóu Masarykovy University.
  Now: *Spisy Přírodovědecké Fakulty University v Brně*

Suomen Kemistilehti
  Suomen Kemistilehti

Svensk Botan. Tidskr.
  Svensk Botanisk Tidskrift

Svensk Kem. Tidskr.
  Svensk Kemisk Tidskrift

Tabulae Biol. Periodicae
  Tabulae Biologicae Periodicae = Tabulae Biologicae

Tetrahedron
  Tetrahedron

Tetrahedron Letters
  Tetrahedron Letters

Trans. Conn. Acad. Arts Sci.
  Transactions of the Connecticut Academy of Arts and Sciences

Trans. Kansas Acad. Sci.
  Transactions of the Kansas Academy of Science

Trans. Roy. Soc. New Zealand, Botany
  Transactions of the Royal Society of New Zealand, Botany

Tr. Botan. Inst., Akad. Nauk SSSR
  Trudȳ Botanicheskogo Instituta im. V. L. Komarova, Akademiya Nauk SSSR

Tr. Leningr. Khim. Farmatsevt. Inst.
  Trudȳ Leningradskogo Khimiko-Farmatsevticheskogo Instituta

Tr. Leningr. Tekhnol. Inst. im. Lensoveta
  Trudȳ Leningradskogo Tekhnologicheskogo Instituta im. Lensoveta

Willdenowia
  Willdenowia

Yakugaku Zasshi
  Yakugaku Zasshi = Journal of the Pharmaceutical Society of Japan

Yamagata Daigaku Kiyo, Shizen Kagaku
  Yamagata Daigaku Kiyo, Shizen Kagaku = Bulletin of the Yamagata University, Natural Science

Z. Botan.
  Zeitschrift für Botanik

Zh. Obshch. Khim.
  Zhurnal Obshcheĭ Khimii

Z. Lebensm. Untersuch. Forsch.
   Zeitschrift für Lebensmittel-Untersuchung und -Forschung
Z. Naturforsch.
   Zeitschrift für Naturforschung
Z. Physiol. Chem.
   Hoppe-Seyler's Zeitschrift für Physiologische Chemie

Appendix 2

Abbreviations of the Authorities for the Names of the Species Cited

If the date of the birth or of the death of an author is unknown, the date of one of his more important publications is given in parentheses.

ABB.	Abbayes, Henry des. 1898–. France.
ACH.	Acharius, Erik. 1757–1819. Sweden.
AG.	Agardh, Karl Adolf. 1785–1859. Sweden.
AHTI	Ahti, Teuvo. 1934–. Finland.
ALMB.	Almborn, Ove. 1914–. Sweden.
AND.	Anders, Josef. 1863–1936. Austria.
R. ANDERSON	Anderson, Roger A. (1962). U.S.A.
ANZI	Anzi, Martino. 1812–1883. Italy.
ARN.	Arnold, Ferdinand Christian Gustav. 1828–1901. Germany.
ASAH.	Asahina, Yasuhiko. 1881–. Japan.
AUERSW.	Auerswald, Bernhard. 1818–1870. Germany.
AWAS.	Awasthi, Dharani Dhar. 1922–. India.
BAB.	Babington, Churchill. 1821–1889. New Zealand.
BAGL.	Baglietto, Francesco. 1826–1916. Italy.
BALB.	Balbis, Giovanni Ballista. 1765–1831. Italy.
BAUMG.	Baumgartner, Julius. b. 1870. Austria.
BAUSCH	Bausch, Wilhelm. 1804–1873. Germany.
BÉL.	Bélanger, Charles Paulus. 1805–1881. France.
BELL.	Bellardi, Carlo Antonio Lodovico. 1741–1826. Italy.
BELTR.	Beltramini de Casati, Francesco. b. 1828. Italy.
BERRY	Berry, Edward Cain. 1898–. U.S.A.
BERT.	Bertoloni, Antonio. 1775–1869. Italy.
BIR.	Biroli, Giovanni. 1772–1825. Italy.
BITT.	Bitter, Georg. 1873–1927. Germany.
BLOMB.	Blomberg, Olof Gotthard. 1838–1901. Sweden.

609

BORY — Bory de Saint-Vincent, Jean Baptiste Marcellin. 1780–1846. France.

BOSCH — Bosch, Roelof Benjamin van den. 1810–1862. Netherlands.

B. DE LESD. — Bouly de Lesdain, Maurice. 1869–1965. France.

BRANDT — Brandt, Th. 1877–1939. Germany.

BRANTH — Branth, Jacob Severin Deichmann. (1870). Denmark.

BREMME — Bremme. (1886). Germany.

BRODO — Brodo, Irwin. 1935–. U.S.A., Canada.

CAREST. — Carestia, Antonio (Fra.). 1825–1908. Italy.

CARROLL — Carroll, I. 1828–1880. Ireland.

CHAUB. — Chaubard, Louis Anastase. 1785–1854. France.

CHEV. — Chevallier, François Fulgis. 1796–1840. France.

CHOISY — Choisy, Maurice. 1897–1966. France.

COEM. — Coemans, Henri Eugène Lucien Gaëtan (Abbé). 1825–1871. Belgium.

COUT. — Coutinho, António Xavier Pereira. 1851–1939. Portugal.

CROMB. — Crombie, James Mascall Morrison (Rev.). 1830–1906. England.

C. CULB. — Culberson, Nan Chicita Frances. 1931–. U.S.A.

W. CULB. — Culberson, William Louis. 1929–. U.S.A.

CURT. — Curtis, Moses Ashley. 1808–1873. U.S.A.

DAHL — Dahl, Eilif. 1916–. Norway.

DALLA TORRE — Dalla Torre, Karl Wilhelm von. 1850–1928. Austria.

DARB. — Darbishire, Otto Vernon. 1870–1934. England.

DC. — De Candolle, Augustin Pyramus. 1778–1841. France, Switzerland.

DEGEL. — Degelius, Gunnar [*born* Nilsson, Gunnar]. 1903–. Sweden.

DEL. — Delise, Dominic François. 1780–1841. France.

DE NOT. — De Notaris, Giuseppe. 1805–1877. Italy.

DESF. — Desfontaines, Réné Louiche. 1750–1833. France.

DESM. — Desmazières, Jean Baptiste Henri Joseph. 1796–1862. France.

DE SMET — De Smet, S. (1952). Belgium.

DICKS. — Dickson, James. 1738–1822. Scotland, England.

DILL. — Dillenius, Johann Jacob. 1684–1747. Germany, England.

DODGE — Dodge, Carrol William. 1895–. U.S.A.

DUBY — Duby, Jean Étienne. 1798–1885. Switzerland.

DUF. — Dufour, Jean-Marie Léon. 1779–1865. France.

DUNCAN — Duncan, Ursula K. (1959). Scotland.

DU RIETZ — Du Rietz, Gustaf Einar. 1895–1967. Sweden.

DUVIGN. — Duvigneaud, Paul. 1913–. Belgium.

A. EAT.	Eaton, Amos. 1776–1842. U.S.A.
EGG.	Eggerth, C. 1860–1888. Austria.
EHRH.	Ehrhart, Friedrich. 1742–1795. Switzerland, Germany.
EITN.	Eitner, E. d. 1921. Germany.
ELENK.	Elenkin, Aleksander Aleksandrovitch. 1873–1942. U.S.S.R.
ERICHS.	Erichsen, Christian Friedo Eckhard. 1867–1945. Germany.
ESCHW.	Eschweiler, Franz Gerhard. 1796–1831. Germany.
EVANS	Evans, Alexander William. 1868–1959. U.S.A.
EVERSM.	Eversmann, Eduard Friedrich. 1794–1860. Germany, Russia.
FÉE	Fée, Antoine-Laurent-Apollinaire. 1789–1874. France.
FINK	Fink, Bruce. 1861–1927. U.S.A.
FLAG.	Flagey, Camille. 1837–1898. France.
FLÖRKE	Flörke, Heinrich Gustav. 1764–1835. Germany.
FLOT.	Flotow, Julius von. 1788–1856. Germany.
FOLLM.	Follmann, Gerhard. 1930–. Germany.
FORSS.	Forssell, Karl Bror Jakob. 1856–1898. Sweden.
FREGE	Frege, C. A. b. 1755. Germany.
FREY	Frey, Eduard. 1888–. Switzerland.
FR.	Fries, Elias Magnus. 1794–1878. Sweden.
TH. FR.	Fries, Theodor Magnus. 1832–1913. Sweden.
FUNCK	Funck, Heinrich Christian. 1771–1839. Germany.
GAROV.	Garovaglio, Santo. 1805–1882. Italy.
GÄRTN.	Gärtner, Joseph. 1732–1791. Germany.
GAS.	Gasilien, F.-G. (1898). France.
GAUD.	Gaudichaud-Beaupré, Charles. 1789–1864. France.
GILLET	Gillet, A. 1857–1927. France.
GMEL.	Gmelin, Johann Friedrich. 1748–1804. Germany.
GRAEWE	Graewe [*or* Graeve], Per Henrik Fredrik. 1819–1866. Sweden.
S. GRAY	Gray, Samuel Frederick. 1780–1828. England.
GROENH.	Groenhart, Pieter. 1894–. Netherlands.
GUNN.	Gunnerus, Johann Ernst. 1718–1773. Norway.
GYELN.	Gyelnik, Vilmos. 1906–1944/45. Hungary.
HAG.	Hagen, Karl Gottfried, 1749–1829. Germany.
HALE	Hale, Mason Ellsworth, Jr. 1928–. U.S.A.
HAMPE	Hampe, Georg Ernst Ludwig. 1795–1880. Germany.
HARM.	Harmand, Jean (Abbé). 1844–1915. France.
HAV.	Havås, Johan. 1864–1956. Norway.
HAZSL.	Hazslinsky, Frigyes Agost. 1818–1896. Hungary.
HEDL.	Hedlund, J. T. 1861–1953. Sweden.
HEDW.	Hedwig, Johann. 1730–1799. Germany.

HELLB.	Hellbom, Per Johan. 1827–1903. Sweden.
HEPP	Hepp, Philipp. 1797–1867. Germany.
HERRE	Herre, Albert William Christian Theodore. 1868–1962. U.S.A.
HEUG.	Heugel, Carl August. 1802–1876. Russia.
HILLM.	Hillmann, Johannes. 1881–1943. Germany.
HOCHST.	Hochstetter, Christian Friedrich. 1787–1860. Germany.
HOFFM.	Hoffmann, Georg Franz. 1761–1826. Germany, Russia.
HOOK. F.	Hooker, Joseph Dalton. 1817–1911. England.
HOOK.	Hooker, William Jackson. 1785–1865. England.
R. H. HOWE	Howe, Reginald Heber, Jr. 1875–1932. U.S.A.
HOWITT	Howitt, Godfrey. 1800–1873. Great Britain.
HUDS.	Hudson, William. 1730–1793. England.
HUE	Hue, Auguste Marie (Abbé). 1840–1917. France.
HUMB.	Humboldt, Friedrich Alexander von. 1796–1859. Germany.
HUN.	Huneck, Siegfried. 1928–. Germany.
IMSH.	Imshaug, Henry Andrew. 1925–. U.S.A.
INUM.	Inumaru, Sunao. 1899–. Japan.
JACQ.	Jacquin, Nicolaus Joseph von. 1727–1817. Netherlands, Austria.
JATTA	Jatta, Antonio. (1900). Italy.
KERNST.	Kernstock, Ernst. 1852–1900. Austria.
KICKX	Kickx, Jean Jacques. 1803–1864. Belgium.
KIEFF.	Kieffer, J. J. 1857–1925. Germany, France.
KÖN.	König, Johann Gerhard. 1728–1785. Denmark.
KÖRB.	Körber, Gustav Wilhelm. 1817–1885. Germany.
KREMP.	Krempelhuber, August von. 1813–1882. Germany.
KROG	Krog, Hildur. (1951). Norway.
KUNZE	Kunze, Gustav. 1793–1851. Germany.
KUROK.	Kurokawa, Syo. 1926–. Japan.
LA BILL.	La Billardière, Jacques Julien Houton de. 1755–1834. France.
LAHM	Lahm, G. 1811–1888. Germany.
LAM.	Lamarck, Jean Baptiste Antoine Pierre Monnet, Chevalier de. 1744–1829. France.
LAMB	Lamb, Ivan Mackenzie. 1911–. England, U.S.A.
LAMBINON	Lambinon, Jacques. (1964). Belgium.
LAMY	Lamy de la Chapelle, Pierre Marie Édouard. 1803–1886. France.
LAUND.	Laundon, J. R. 1934–. England.
LAUR.	Laurer, Johann Friedrich. 1798–1873. Germany.

LEIGHT.	Leighton, William Allport. 1805–1889. England.
LETT.	Lettau, Georg. 1878–1951. Germany.
LÉV.	Léveillé, Joseph Henri. 1796–1870. France.
LIGHTF.	Lightfoot, John (Rev.). 1735–1788. England.
LINDS.	Lindsay, William Lauder. 1829–1880. England.
LINK	Link, Heinrich Friedrich. 1769–1851. Germany.
L.	Linné, Carl von. 1707–1778. Sweden.
LLANO	Llano, George Albert. 1911–. U.S.A.
LÖNNR.	Lönnroth, K. J. 1826–1885. Sweden.
LYNGE	Lynge, Bernt. 1884–1942. Norway.
MAGN.	Magnusson, Adolf Hugo. 1885–1964. Sweden.
MAH.	Maheu, Jacques Marie Albert. 1873–1937. France.
MAK.	Makarevicz, M. F. 1906–. U.S.S.R.
MALME	Malme, Gustaf O. 1864–1937. Sweden.
MANN	Mann, Wenzel Blasius. 1799–1839. Austria (Bohemia).
MARTIN	Martin, William. (1956). New Zealand.
MASS.	Massalongo, Abramo Bartolommeo. 1824–1860. Italy.
MATS.	Matsumura, Jinzô. 1858–1928. Japan.
MATT.	Mattick, Fritz. 1901–. Germany.
MÉR.	Mérat de Vaumartoise, François Victor. 1780–1851. France.
MERESCH.	Mereschkovsky, Constantin de. 1854–1921. Russia, Switzerland.
MERR.	Merrill, George Knox. 1864–1927. U.S.A.
MEYEN	Meyen, Franz Julius Ferdinand. 1804–1840. Germany.
MEYER	Meyer, Georg Friedrich Wilhelm. 1782–1856. Germany.
MICHX.	Michaux, André. 1746–1802. France.
MIG.	Migula, Walter. 1863–1938. Germany.
MITCH.	Mitchell, Michael E. 1934–. Ireland.
MIYOSHI	Miyoshi, M. 1861–1939. Japan.
MONT.	Montagne, Jean François Camille. 1784–1866. France.
MOT.	Motyka, Józef. 1900–. Poland.
MUDD	Mudd, William. 1830–1879. England.
MÜLL. ARG.	Müller "Argoviensis," Jean. 1826–1896. Switzerland.
NÁDV.	Nádvorník, Josef. 1906–. Czechoslovakia.
NECK.	Necker, Noël Joseph de. 1729–1793. France, Germany.
NEES	Nees von Esenbeck, Christian Gottfried Daniel. 1776–1858. Germany, Poland.
NOEHD.	Noehden, Heinrich Adolph. 1775–1804. Germany.
NORM.	Norman, Johannes Musaeus. 1823–1903. Norway.
NORRL.	Norrlin, Johan Peter. 1842–1917. Finland.
NYL.	Nylander, William. 1822–1899. Finland, France.
OAKES	Oakes, William. 1799–1848. U.S.A.

OEDER	Oeder, Georg Christian. 1728–1791. Germany, Denmark.
OKSN.	Oksner, Alfred Nikolaevich. 1898–. U.S.S.R.
OLIV.	Olivier, Henri (Rév. Père). 1849–1923. France.
OPIZ	Opiz, Maximilian Philipp. 1787–1858. Austria (Prague).
PALL.	Pallas, Per Simon. 1741–1811. Germany.
PERS.	Persoon, Christian Hendrik. 1755–1837. Holland, Germany.
POELT	Poelt, Josef. 1924–. Germany.
POETSCH	Poetsch, Ignaz Sigmund. 1823–1884. Austria.
POLLINI	Pollini, Ciro. 1782–1833. Italy.
RABENH.	Rabenhorst, Gottlob Ludwig, Ritter von Albrechtsorders. 1806–1881. Germany.
RADDI	Raddi, Giuseppe. 1770–1829. Italy.
RAM.	Ramond de Carbonnières, Louis François Elisabeth. 1753–1827. France.
RÄS.	Räsänen, Veli. 1888–1953. Finland.
RASS.	Rassadina, Kseniya Aleksandrovna. 1903–. U.S.S.R.
RÄUSCH.	Räuschel, Ernest Adolf. (1797). Germany.
RAV.	Ravaud, Louis Célestin Mure (Abbé). 1822–1898. France.
REBENT.	Rebentisch, Johann Friedrich. 1772–1810. Germany.
RED.	Redinger, Karl. 1907–1940. Austria.
RETZ.	Retzius, Anders Johan. 1742–1821. Sweden.
RIZZ.	Rizzini, Carlos Toledo. (1952). Brazil.
ROBB.	Robbins, Charles Albert. 1874–1930. U.S.A.
RÖHL.	Röhling, Johann Christoph. 1757–1813. Germany.
ROSTR.	Rostrup, Frederik Emil Georg. 1831–1907. Denmark.
RUN.	Runemark, Hans. 1927–. Sweden.
SANDST.	Sandstede, Heinrich. 1859–1951. Germany.
SANT.	Santesson, Rolf. 1916–. Sweden.
SARNTH.	Sarnthein, Ludwig, Graf von. 1861–1914. Austria.
SATÔ	Satô, Masami. 1910–. Japan.
SAV.	Savicz, Vsevolod Pavlovich. 1885.– U.S.S.R.
SCHAER.	Schaerer, Ludwig Emanuel. 1785–1853. Switzerland.
SCHLEICH.	Schleicher, Johann Christoph. 1768–1834. Switzerland.
SCHNEID.	Schneider, Albert. 1863–1928. U.S.A.
SCHOL.	Scholander, Per Fredrik. 1905–. Norway, U.S.A.
SCHRAD.	Schrader, Heinrich Adolph. 1767–1836. Germany.
SCHRANK	Schrank, Franz von Paula. 1747–1836. Germany.
SCHREB.	Schreber, Johann Christian Daniel von. 1739–1810. Germany.
SCHWEIN.	Schweinitz, Lewis David von. 1780–1834. Germany, U.S.A.

Scop.	Scopoli, Johann Anton. 1722/23–1788. Austria, Italy.
Scriba	Scriba, Ludwig. 1847–1933. Germany.
Sheard	Sheard, John W. 1940–. England.
A. L. Sm.	Smith, Annie Lorrain. 1854–1937. England.
Sm.	Smith, James Edward. 1759–1828. England.
Somm.	Sommerfeldt, Sören Christian. 1794–1838. Norway.
Spreng.	Sprengel, Kurt Polycarp Joachim von. 1766–1833. Germany.
B. Stein	Stein, B. 1847–1899. Germany.
J. Stein.	Steiner, Julius. 1844–1918. Austria.
M. Stein.	Steiner, Maximilian. 1904–. Austria.
Steud.	Steudel, Ernst Gottlieb von. 1783–1856. Germany.
Stirt.	Stirton, James. 1833–1917. Scotland, Australia.
Stizenb.	Stizenberger, Ernst. 1827–1895. Germany.
Stuck.	Stuckenberg, E. K. 1883–. U.S.S.R.
Sw.	Swartz, Olof. 1760–1818. Sweden.
Syd.	Sydow, Paul. 1851–1925. Germany.
Szat.	Szatala, Ödön. 1889–1958. Hungary.
Targé	Targé, André. (1965). Belgium.
Tav.	Tavares, Carlos das Neves. 1914–. Portugal.
Tayl.	Taylor, Thomas. 1755–1848. Ireland.
Thoms.	Thomson, John Walter. 1913–. U.S.A.
Thunb.	Thunberg, Carl Pehr. 1743–1828. Sweden.
Tobl.	Tobler, Friedrich. 1879–1957. Switzerland.
Torn.	Tornabene, Francesco. 1813–1897. Italy.
Torrey	Torrey, John. 1796–1873. U.S.A.
Torss.	Torssell, Gustav. 1811–1849. Sweden.
Trev.	Trevisan de Saint-Léon, Vittore Benedetto Antonio, Conte. 1818–1897. Italy.
Tuck.	Tuckerman, Edward. 1817–1886. U.S.A.
Tul.	Tulasne, Louis René. 1815–1885. France.
Turn.	Turner, Dawson. 1775–1858. England.
Ullrich	Ullrich, Johannes. (1957). Germany.
Vahl	Vahl, Martin. 1749–1804. Norway, Denmark.
Vain.	Vainio [=Wainio (*born* Lang)], Edvard August. 1853–1929. Finland.
Van der Bly	Van der Bly. (1931).
Vers.	Verseghy, Klára. 1930–. Hungary.
Vill.	Villars, Dominique. 1745–1814. France.
Wahlenb.	Wahlenberg, Göran. 1780–1851. Sweden.
Wallr.	Wallroth, Karl Friedrich Wilhelm. 1792–1857. Germany.
W. Wats.	Watson, Walter. 1872–1960. England.
Web.	Weber, Friedrich. 1781–1823. Germany.

G. WEB.	Weber, Georg Heinrich. 1752–1828. Germany.
W. WEB.	Weber, William Alfred. 1918–. U.S.A.
WEIG.	Weigel, Christian Ehrenfried von. 1748–1831. Germany.
WEISS	Weiss, Friedrich Wilhelm. 1744–1826. Germany.
WESTR.	Westring, Johan Peter. 1753–1833. Sweden.
WETM.	Wetmore, Clifford M. 1934–. U.S.A.
WIGG.	Wiggers, Friedrich Heinrich. 1746–1811. Denmark.
WILLD.	Willdenow, Carl Ludwig. 1765–1812. Germany.
WILL.	Willey, Henry. 1824–1907. U.S.A.
WILS.	Wilson, William. 1799–1871. England.
WIRTH	Wirth, Michael. (1963). U.S.A.
WULF.	Wulfen, Franz Xavier von. 1728–1805. Austria.
YAS.	Yasuda, Atsushi. 1868–1924. Japan.
ZAHLBR.	Zahlbruckner, Alexander. 1860–1938. Austria.
ZOLL.	Zollinger, Heinrich. 1818–1859. Switzerland.
ZOPF	Zopf, Friedrich Wilhelm. 1846–1909. Germany.
ZSCH.	Zschacke, Hermann. 1867–1937. Germany.

Index

The index covers Chapters I and II and the names of the compounds in Chapter III. Chapters IV and V are not indexed because the material in them is presented alphabetically. Page references to the physical data, chemical structures, and annotated references for the lichen products are given in **boldface**.

A

Abbreviations, of microchemical reagents, 10; of titles of periodicals cited, 598–608; of authorities for names of species cited, 609–16
Acanthellin, 4
Acaranoic acid, **100**; relationship to other fatty acids, 57–58
Acarenoic acid, **101**; relationship to other fatty acids, 57–58
Acarospora oxytona, fatty acids in, 57
Accessory substances, 11
Acetate-polymalonate pathway, secondary products of, 25–58; **100–91**
Acetoacetyl coenzyme A in the mevalonic acid pathway, 59
7β-Acetoxy-22-hydroxyhopane, **196–97**
Acetyl coenzyme A, in the synthesis of evernic acid by a cell-free system, 21; in the mevalonic acid pathway, 59
Acetylenic compounds, 58
16α-O-Acetylleucotylic acid, **197**
6α-O-Acetylleucotylin, **198**
Acetylportentol, **111**
Acolic acid, 4
Acromelidin, 4
Adonitol, **73**; supplied to the fungus by the green algal symbiont, 63
Agaricic acid, relationship to caperatic acid, 55–56
Akromelin, 5
Alanine, **91, 93–94**

Alcohols, long-chain derivatives, 58
Alectoria, virensic acid in, 43
Alectorialic acid, **218**
Alectoric acid, 12
Alectoronic acid, **134**
Algae, as the source of certain lichen products, 20; control of nutrient balance to fungus, 24
Aliphatic acids and related substances, **100–11**
Allenes, 58
Allomyces arbuscula, calcium and strontium in, 66
Alternaria tenuis, 25
Alternariol, biosynthesis, 25
Amine derivatives, **90–91**
Amino acids, **91–94, 96–98**
α-Aminobutyric acid, **91, 93–94**
γ-Aminobutyric acid, **91, 93–94**
Ammonia, **90–91**
Amorphous pigmented products, 9
Amylase, **94, 95**
Anthraquinones, **181–91**; biosynthesis, 20–21, 25, 27; production of by cultured fungal component, 21; nonspecificity to lichens, 25; biogenesis in lichens, 54–55; chlorine-containing, 67
Anthrones, in the biogenesis of anthraquinones, 54
Antibiotics from lichens, 18
Antimony-125, **98, 99**
Anziaic acid, **114**

617

www.ingramcontent.com/pod-product-compliance
Lightning Source LLC
Chambersburg PA
CBHW021022210326
41598CB00016B/886